HUMAN FACTORS IN ENGINEERING AND DESIGN

HUMAN FACTORS IN ENGINEERING AND DESIGN

SEVENTH EDITION

Mark S. Sanders, Ph.D.

California State University, Northridge

Ernest J. McCormick, Ph.D.

Late Professor Emeritus of Psychological Sciences,
Purdue University

McGraw Hill Education (India) Private Limited

CHENNAI

McGraw Hill Education Offices

Chennai New York St Louis San Francisco Auckland Bogotá Caracas
Kuala Lumpur Lisbon London Madrid Mexico City Milan Montreal
San Juan Santiago Singapore Sydney Tokyo Toronto

McGraw Hill Education (India) Private Limited

HUMAN FACTORS IN ENGINEERING AND DESIGN

McGraw Hill Education (India) Edition 2013

Second reprint 2016
RXXZRDLQDXLAL

Reprinted in India by arrangement with McGraw-Hill Education (Asia)

Sales territories: India, Pakistan, Nepal, Bangladesh, Sri Lanka and Bhutan

Library of Congress Cataloging-in-Publication Data

Sanders, Mark S.
Human factors in engineering and design/Mark S. Sanders, Ernest McCormick.—7th ed.
p. cm.
Includes bibliographical references and index.
ISBN 0-07-054901-X
1. Human engineering I. McCormick, Ernest J. (Ernest James)
II. Title
TA166.S33 1933
620.8'2—dc20

ISBN-13: 978-1-25-906472-2
ISBN-10: 1-25-906472-7

Published by McGraw Hill Education (India) Private Limited, 444/1, Sri Ekambara Naicker Industrial Estate, Alapakkam, Porur, Chennai 600 116, Tamil Nadu, India, and printed at Adarsh Printers, New Delhi 110 032

Visit us at: www.mheducation.co.in

ABOUT THE AUTHORS

DR. MARK S. SANDERS received his M.S. and Ph.D. degrees in human factors from Purdue University. He is currently Professor and Chair of the Human Factors Applied Experimental Psychology Graduate Program at California State University, Northridge. Also Dr. Sanders consults with various organizations and serves as an expert witness in cases involving human factors issues. He has executed or directed over 100 research and development contracts, subcontracts, and consulting activities. In addition, he has authored or coauthored over 90 technical reports, journal articles, and professional presentations. He received the Human Factors Society's Jack A. Kraft Award for his research on human factors issues in the mining industry. Dr. Sanders is a member of the Ergonomics Society and Society of Automotive Engineers. He is a fellow in the Human Factors Society and has served that organization as president, secretary-treasurer, and chair of the Education Committee and Educators Professional Group. Dr. Sanders is also a fellow of Division 21 (Division of Applied Experimental and Engineering Psychologists) of the American Psychological Association.

DR. ERNEST J. McCORMICK (deceased) was Professor Emeritus, Purdue University. His academic career as an industrial psychologist covered a span of 30 years at Purdue. His first edition of this text (then titled *Human Engineering*) was published in 1957. Dr. McCormick's other major publications include *Industrial and Organizational Psychology* (now in its eighth edition) and *Job Analysis: Methods and Applications*. He was responsible for development of the position analysis questionnaire (PAQ), a structured, computerized job analysis procedure being used by numerous organizations; he was president of PAQ Services, Inc. He has served on various advisory panels and committees, including the Army Scientific Advisory Panel, the Navy Advisory Board for Personnel Research, and the Committee on Occupational Classification and Analysis of the National Academy of Sciences. His awards include the Paul M. Fitts award of the Human Factors Society, the Franklin V. Taylor award of the Society of Engineering Psychologists, and the James McKeen Cattell award of the Society of Industrial and Organizational Psychology.

**Dedicated to the Memory of
Ernest J. McCormick**

CONTENTS

PREFACE xi

PART 1 INTRODUCTION

1 **Human Factors and Systems** 3

2 **Human Factors Research Methodologies** 23

PART 2 INFORMATION INPUT

3 **Information Input and Processing** 47

4 **Text, Graphics, Symbols, and Codes** 91

5 **Visual Displays of Dynamic Information** 132

6 **Auditory, Tactual, and Olfactory Displays** 160

7 **Speech Communications** 197

PART 3 HUMAN OUTPUT AND CONTROL

8 **Physical Work and Manual Materials Handling** 225

9 **Motor Skills** 273

10 **Human Control of Systems** 301

11 Controls and Data Entry Devices 334

12 Hand Tools and Devices 383

PART 4 WORKPLACE DESIGN

13 Applied Anthropometry, Work-Space Design, and Seating 415

14 Arrangement of Components within a Physical Space 456

15 Interpersonal Aspects of Workplace Design 485

PART 5 ENVIRONMENTAL CONDITIONS

16 Illumination 511

17 Climate 551

18 Noise 589

19 Motion 622

PART 6 HUMAN FACTORS APPLICATIONS

20 Human Error, Accidents, and Safety 655

21 Human Factors and the Automobile 696

22 Human Factors in Systems Design 726

APPENDIXES

A List of Abbreviations 759

B Control Devices 764

C NIOSH Recommended Action Limit Formula for Lifting Tasks 769

INDEXES

Name Index 771

Subject Index 781

PREFACE

This book deals with the field of *human factors,* or *ergonomics,* as it is also called. In simple terms, the term *human factors* refers to *designing for human use*. Ten years ago, it would have been difficult to find very many people outside the human factors profession who could tell you what human factors or ergonomics was. Today, things are different. Human factors and ergonomics are in the news. Visual and somatic complaints of computer terminal users have been linked to poor human factors design. The incident at Three-Mile Island nuclear power station highlighted human factors deficiencies in the control room. The words *human factors* and especially *ergonomics* have also found their way into advertisements for automobiles, computer equipment, and even razors. The field is growing, as evidenced by the increase in the membership of human factors professional societies, in graduate programs in human factors, and in job opportunities.

We intended this book to be used as a textbook in upper-division and graduate-level human factors courses. We were also aware that this book has been an important resource for human factors professionals over the last six editions and 35 years. To balance these two purposes, we have emphasized the empirical research basis of human factors, we have stressed basic concepts and the human factors considerations involved in the topics covered, and we have supplied references for those who wish to delve into a particular area. We have tried to maintain a scholarly approach to the field. Unfortunately, there are times when our presentation may be a little technical or "dry," especially when we are presenting information that would be more appropriate for the practicing human factors specialist than for students. For this we apologize, but we hope the book will be one students will want to keep as a valuable reference.

For students, we have written a workbook to accompany this text (published by Kendall-Hunt Publishing Co., Dubuque, Iowa). Included in the workbook, for each chapter, are a list of key terms and self-contained projects that use concepts and information contained in this book.

There has been a virtual information explosion in the human factors field over the years. The first edition of this book, published in 1957, contained 16 chapters and 370 references. This edition contains 22 chapters and over 900 references. In 1972, the Human Factors Society (HFS) first published a proceedings of their annual meeting. It contained 106 papers and was 476 pages long. The proceedings for the 1991 HFS annual meeting contained over 350 papers and was 1600 pages long! In this book we have tried to cover both traditional and emerging areas of human factors, but it was impossible to include everything. The specific research material included in the text represents only a minute fraction of the vast amount that has been carried out in specific areas. It has been our interest to use as illustrative material examples of research that are relatively important or that adequately illustrate the central points in question. Although much of the specific material may not be forever remembered by the reader, we hope that the reader will at least develop a deep appreciation of the importance of considering human factors in the design of the features of the world in which we work and live. Appreciation is expressed to the many investigators whose research is cited. References to their work are included at the end of each chapter. To those investigators whose fine work we did not include, we apologize and trust they understand our predicament. We would also like to thank the following reviewers for their many helpful comments and suggestions: John G. Casali, Virginia Polytechnic Institute; Rick Gill, University of Idaho; Martin Helander, SUNY, Buffalo; John Lyman, University of California, Los Angeles; Joseph P. Meloy, Milwaukee School of Engineering; Edward J. Rinalducci, University of Central Florida; and William C. Howell, Rice University.

This edition represents some changes from the last edition. In addition to a general updating of the material (almost 30 percent of the figures are new), a new chapter on motor skills (Chapter 9) has been added. Several chapters have been extensively revised and renamed, including: Chapter 8, Physical Work and Manual Materials Handling; Chapter 20, Human Error, Accidents, and Safety; Chapter 4, Text, Graphics, Symbols, and Codes; and Chapter 13, Applied Anthropometry, Workspace Design, and Seating. We welcome comments and suggestions for making improvements in future editions.

It is with sadness that I report that Professor Ernest J. McCormick (Mac to his friends) died on February 9, 1990. Mac's passing came as we were starting to work on this edition of the book. I was deprived of a much admired colleague and a wonderful writing partner. His input and critiques are missing from this edition, but his contributions to the book over the last 35 years live on in every chapter. When I was a graduate student at Purdue University, twenty-odd years ago, I was asked to teach Mac's courses while he was in India. One of the perks was being allowed to use his office. I remember sitting in his chair for the first time. Although physically larger than Mac, I vividly recall feeling small sitting there. As I worked on this edition without Mac's help, I had a similar feeling. He has left an empty chair that will be hard to fill. Mac is survived by his wife Emily and two daughters Wynne and Jan. Mac was a model of professionalism and integrity. He was a person of quiet wit, keen analytic ability, and intellect. He will be missed.

Mark S. Sanders

HUMAN FACTORS IN ENGINEERING AND DESIGN

PART ONE

INTRODUCTION

1 Human Factors and Systems

2 Human Factors Research Methodologies

CHAPTER 1

HUMAN FACTORS AND SYSTEMS

In the bygone millennia our ancestors lived in an essentially ''natural'' environment in which their existence virtually depended on what they could do directly with their hands (as in obtaining food) and with their feet (as in chasing prey, getting to food sources, and escaping from predators). Over the centuries they developed simple tools and utensils, and they constructed shelter for themselves to aid in the process of keeping alive and making life more tolerable.

The human race has come a long way from the days of primitive life to the present with our tremendous array of products and facilities that have been made possible with current technology, including physical accoutrements and facilities that simply could not have been imagined by our ancestors in their wildest dreams. In many civilizations of our present world, the majority of the ''things'' people use are made by people. Even those engaged in activities close to nature—fishing, farming, camping—use many such devices.

The current interest in human factors arises from the fact that technological developments have focused attention (in some cases dramatically) on the need to consider human beings in such developments. Have you ever used a tool, device, appliance, or machine and said to yourself, ''What a dumb way to design this; it is so hard to use! If only they had done this or that, using it would be so much easier.'' If you have had such experiences, you have already begun to think in terms of human factors considerations in the design of things people use. Norman (1988), in an entertaining book, provides numerous examples of everyday things that were not designed from a human factors perspective, including single-control shower faucets, videocassette recorders, and stove-top controls. In a sense, the goal of human factors is to guide the applications of technology in the direction of benefiting humanity. This text offers an overview of the human factors field; its various sections and chapters deal with some of the more important aspects of the field as they apply to such objectives.

HUMAN FACTORS DEFINED

Before attempting to define human factors, we should say a word about terms. *Human factors* is the term used in the United States and a few other countries. The term *ergonomics*, although used in the United States, is more prevalent in Europe and the rest of the world. Some people have tried to distinguish between the two, but we believe that any distinctions are arbitrary and that, for all practical purposes, the terms are synonymous. Another term that is occasionally seen (especially within the U.S. military) is *human engineering*. However, this term is less favored by the profession, and its use is waning. Finally, the term *engineering psychology* is used by some psychologists in the United States. Some have distinguished engineering psychology, as involving basic research on human capabilities and limitations, from human factors, which is more concerned with the *application* of the information to the design of things. Suffice it to say, not everyone would agree with such a distinction.

We approach the definition of human factors in terms of its focus, objectives, and approach.

Focus of Human Factors

Human factors focuses on human beings and their interaction with products, equipment, facilities, procedures, and environments used in work and everyday living. The emphasis is on human beings (as opposed to engineering, where the emphasis is more on strictly technical engineering considerations) and how the design of things influences people. Human factors, then, seeks to change the things people use and the environments in which they use these things to better match the capabilities, limitations, and needs of people.

Objectives of Human Factors

Human factors has two major objectives. The first is to enhance the effectiveness and efficiency with which work and other activities are carried out. Included here would be such things as increased convenience of use, reduced errors, and increased productivity. The second objective is to enhance certain desirable human values, including improved safety, reduced fatigue and stress, increased comfort, greater user acceptance, increased job satisfaction, and improved quality of life.

It may seem like a tall order to enhance all these varied objectives, but as Chapanis (1983) points out, two things help us. First, only a subset of the objectives are generally of highest importance in a specific application. Second, the objectives are usually correlated. For example, a machine or product that is the result of human factors technology usually not only is safer, but also is easier to use, results in less fatigue, and is more satisfying to the user.

Approach of Human Factors

The approach of human factors is the systematic application of relevant information about human capabilities, limitations, characteristics, behavior, and motivation to the design of things and procedures people use and the environments in which they use them. This involves scientific investigations to discover relevant information about humans and their responses to things, environments, etc. This information serves as the basis for making design recommendations and for predicting the probable effects of various design alternatives. The human factors approach also involves the evaluation of the things we design to ensure that they satisfy their intended objectives.

Although no short catch phrase can adequately characterize the scope of the human factors field, such expressions as *designing for human use* and *optimizing working and living conditions* give a partial impression of what human factors is about. For those who would like a concise definition of human factors which combines the essential elements of focus, objectives, and approach discussed above, we present the following definition, modified slightly from Chapanis (1985): *Human factors discovers and applies information about human behavior, abilities, limitations, and other characteristics to the design of tools, machines, systems, tasks, jobs, and environments for productive, safe, comfortable, and effective human use.*

Discussion

There are several more or less established doctrines that characterize the human factors profession and that together distinguish it from other applied fields:

- Commitment to the idea that things, machines, etc. are built to serve humans and must be designed always with the user in mind
- Recognition of individual differences in human capabilities and limitations and an appreciation for their design implications
- Conviction that the design of things, procedures, etc. influences human behavior and well-being
- Emphasis on empirical data and evaluation in the design process
- Reliance on the scientific method and the use of objective data to test hypotheses and generate basic data about human behavior
- Commitment to a systems orientation and a recognition that things, procedures, environments, and people do not exist in isolation

We would be remiss if we did not at least mention what human factors is not. All too often, when people are asked what human factors is, they respond by saying what it is not. The following are three things human factors is not.

Human factors is not just applying checklists and guidelines. To be sure, human factors people develop and use checklists and guidelines; however, such aids are only part of the work of human factors. There is not a checklist or guideline in existence today that, if it were blindly applied, would ensure a good human factors product. Trade-offs, considerations of the specific application, and educated opinions are things that cannot be captured by a checklist or guideline but are all important in designing for human use.

Human factors is not using oneself as the model for designing things. Just because a set of instructions makes sense to an engineer, there is no guarantee others will understand them. Just because a designer can reach all the controls on a machine, that is no guarantee that everyone else will be able to do so. Human factors recognizes individual differences and the need to consider the unique characteristics of user populations in designing things for their use. Simply being a human being does not make a person a qualified human factors specialist.

Human factors is not just common sense. To some extent, use of common sense would improve a design, but human factors is more than just that. Knowing how large to make letters on a sign to be read at a specific distance or selecting an audible warning that can be heard and distinguished from other alarms is not determined by simple common sense. Knowing how long it will take pilots to respond to a warning light or buzzer is also not just common sense. Given the number of human factors deficiencies in the things we use, if human factors is based on just common sense, then we must conclude that common sense is not very common.

A HISTORY OF HUMAN FACTORS

To understand human factors, it is important to know from where the discipline came. It is not possible, however, to present more than just a brief overview of the major human factors developments. We have chosen to concentrate on developments in the United States, but several sources trace the history in other countries [see, for example, Edholm and Murrell (1973), Singleton (1982), and Welford (1976)].

Early History

It could be said that human factors started when early humans first fashioned simple tools and utensils. Such an assertion, however, might be a little presumptuous. The development of the human factors field has been inextricably intertwined with developments in technology and as such had its beginning in the industrial revolution of the late 1800s and early 1900s. It was during the early 1900s, for example, that Frank and Lillian Gilbreth began their work in motion study and shop management. The Gilbreths' work can be considered as one of the forerunners to what was later to be called human factors. Their work included the study of skilled performance and fatigue and the design of workstations and equipment for the handicapped. Their analysis of hospital surgical teams, for example, resulted in a procedure used today: a surgeon obtains an instrument by calling for it and extending his or her hand to a nurse who places the instrument in the proper orientation. Prior to the Gilbreths' work, surgeons picked up their own instruments from a tray. The Gilbreths found that with the old technique surgeons spent as much time looking for instruments as they did looking at the patient.

Despite the early contributions of people such as the Gilbreths, the idea of adapting equipment and procedures to people was not exploited. The major emphasis of behavioral scientists through World War II was on the use of tests for selecting the proper people for jobs and on the development of improved training procedures. The focus was clearly on fitting the person to the job. During World War II, however, it became clear that, even with the best selection and training, the operation of some of the complex equipment still exceeded the capabilities of the people who had to operate it. It was time to reconsider fitting the equipment to the person.

1945 to 1960: The Birth of a Profession

At the end of the war in 1945, engineering psychology laboratories were established by the U.S. Army Air Corps (later to become the U.S. Air Force) and U.S. Navy. At about the same time, the first civilian company was formed to do engineering psychology contract work (Dunlap & Associates). Parallel efforts were being undertaken in Britain, fostered by the Medical Research Council and the Department of Scientific and Industrial Research.

It was during the period after the war that the human factors profession was born. In 1949 the Ergonomics Research Society (now called simply the Ergonomics Society) was formed in Britain, and the first book on human factors was published, entitled *Applied Experimental Psychology: Human Factors in Engineering Design* (Chapanis, Garner, and Morgan, 1949). During the next few years conferences were held, human factors publications appeared, and additional human factors laboratories and consulting companies were established.

The year 1957 was an important year, especially for human factors in the United States. In that year the journal *Ergonomics* from the Ergonomics Research Society appeared, the Human Factors Society was formed, Division 21 (Society of Engineering Psychology) of the American Psychological Association was organized, the first edition of this book was published, and Russia launched *Sputnik* and the race for space was on. In 1959 the International Ergonomics Association was formed to link several human factors and ergonomics societies in various countries around the world.

1960 to 1980: A Period of Rapid Growth

The 20 years between 1960 and 1980 saw rapid growth and expansion of human factors. Until the 1960s, human factors in the United States was essentially concentrated in the military-industrial complex. With the race for space and staffed space flight, human factors quickly became an important part of the space program. As an indication of the growth of human factors during this period, consider that in 1960 the membership of the Human Factors Society was about 500; by 1980 it had grown to over 3000. More important, during this period, human factors in the United States expanded beyond military and space applications. Human factors groups could be found in many companies, including those dealing in pharmaceuticals, computers, automobiles, and other consumer products. Industry began to recognize the importance and contribution of human factors to the design of both workplaces and the products manufactured there. Despite all the rapid growth and recognition within industry, human factors was still relatively unknown to the average person in the street in 1980.

1980 to 1990: Computers, Disasters, and Litigation

Human factors continued to grow with membership in the Human Factors Society reaching almost 5000 in 1990. The computer revolution propelled human factors into the public limelight. Talk of ergonomically designed computer equipment, user-friendly software, and human factors in the office seems to be part and parcel of virtually any newspaper or magazine article dealing with computers and people. Computer technology has provided new challenges for the human factors profession. New control devices, information presentation via computer screen, and the impact of new technology on people are all areas where the human factors profession is making contributions.

The 1980s was, unfortunately, also a decade marred by tragic, large-scale, technological disasters. The incident at Three Mile Island nuclear power station in 1979 set the stage for the 1980s. Although no lives were lost and the property damage was confined to the reactor itself, the incident came very close to resulting in a nuclear meltdown. The 1980s would not be so fortunate. On December 4, 1984, a leak of methylisocyanate (MIC) at the Union Carbide pesticide plant in Bhopal, India, claimed the lives of nearly 4000 people and injured another 200,000. Two years later, in 1986, an explosion and fire at the Chernobyl nuclear power station in the Soviet Union resulted in more than 300 dead, widespread human exposure to harmful radiation, and millions of acres of radioactive contamination. Three years later, 1989, an explosion ripped through a Phillips Petroleum plastics plant in Texas. The blast was equivalent in force to 10 tons of TNT. It killed 23 people, injured another 100 workers, and resulted in the largest single U.S. business insurance loss in history ($1.5 billion). Meshkati (1989, 1991) analyzed several of these disasters and found that inadequate attention to human factors considerations played a significant role in contributing to each disaster he studied.

Another area that saw a dramatic increase in human factors involvement was forensics and particularly product liability and personal injury litigations. Courts have come to recognize the contribution of human factors expert witnesses for explaining human behavior and expectations, defining issues of defective design, and assessing the effectiveness of warnings and instructions. Approximately 15 percent of Human Factors Society members are involved in expert-witness work (Sanders, Bied, and Curran, 1986).

1990 and Beyond

What does the future hold for human factors? We, of course, have no crystal ball, but it seems safe to predict continued growth in the areas established during the short history of human factors. Plans for building a permanent space station will undoubtedly mean a heavy involvement of human factors. Computers and the application of computer technology to just about everything will keep a lot of human factors people busy for a long time. The National Research Council (Van Cott and Huey, 1991) estimates that the demand for human factors specialists will exceed supply well into the 1990s.

Other developments should also increase the demand for human factors. For example, the U.S. Occupational Safety and Health Administration (OSHA) will be formulating ergonomic regulations for general industry during the 1990s. A law passed by the U.S. Congress in 1988 ordered the Federal Aviation Administration (FAA) to expand its human factors research efforts to improve aviation safety. Two other area where human factors should expand are in the design of medical devices and in the design of products and facilities for the elderly. We hope that in the future human factors will become more involved and recognized for its contribution to the quality of life and work, contributions that go beyond issues of productivity and safety and embrace more intangible criteria such as satisfaction, happiness, and dignity. Human factors, for example, could play a greater role in improving the quality of life and work in underdeveloped countries.

HUMAN FACTORS PROFESSION

We offer here a thumbnail sketch of the human factors profession in the United States as depicted from surveys and from an analysis of human factors graduate education programs in the United States and Canada.

Graduate Education in Human Factors

The *Directory of Human Factors Graduate Programs in the United States and Canada* (Sanders and Smith, 1988) lists 65 programs. The programs (with but a few exceptions) are housed in engineering departments (46 percent) or psychology departments (42 percent) with an additional 6 percent being joint programs. From 1984 through 1987, engineering programs accounted for 55 percent of the master's degrees granted but only 42 percent of the doctoral degrees. In 1987 there were about 1000 students in graduate programs in the United States and Canada (Muckler and Seven, 1990).

Who Are Human Factors People?

It is not entirely fair to assume that membership in the Human Factors Society defines a person as a human factors specialist. Nonetheless, an analysis of the backgrounds of its membership should give some insight into the profession as a whole. Table 1-1 shows the composition of the Human Factors Society membership by academic specialty and highest degree held. Psychologists comprise almost half of the membership and account for 66 percent of all doctoral degrees but only 40 percent of the master's degrees. More than half of the members do not possess doctoral degrees. The number of different academic specialties in Table 1-1 attests to the multidisciplinary nature of the profession.

Where Do Human Factors Specialists Work?

The National Research Council (Van Cott and Huey, 1991) conducted an extensive survey of members and colleagues of members from 11 professional societies (including the Human Factors Society). These societies were thought to contain people doing human factors work even though they might not call themselves human factors specialists. The principle workplace of the respondents was in private business or industry (74 percent). In addition, 15 percent worked for government agencies and 10 percent worked in academia. (The percentage in academia is an underestimate because only those academicians engaged in outside consulting work were included.)

TABLE 1-1
ANALYSIS OF HUMAN FACTORS SOCIETY MEMBERSHIP BY ACADEMIC SPECIALTY AND HIGHEST DEGREE HELD

	Highest degree held (%)			
Academic specialty	**Bachelor's**	**Master's**	**Doctoral**	**Total (%)**
Psychology	6.8	12.0	26.3	45.1
Engineering	4.9	8.1	6.1	19.1
Human factors/ergonomics	1.3	3.8	2.6	7.7
Medicine, physiology, life sciences	0.8	1.1	1.1	3.0
Education	0.2	0.9	1.5	2.6
Industrial design	1.6	0.7	0.1	2.4
Business administration	0.5	1.3	0.1	1.9
Computer science	0.5	0.5	0.3	1.3
Other	2.1	4.1	2.1	8.3
Total degrees	18.7	32.5	40.2	91.4
Students	—	—	—	8.2
Not specified	—	—	—	0.4
Total				100.0

Source: Human Factors Society, 1991. Copyright by the Human Factors Society, Inc. and reproduced by permission.

The respondents were asked to identify the principle area in which they worked. Six areas accounted for 83 percent of the respondents as shown in Table 1-2.

In an earlier survey of Human Factors Society members, Sanders, Bied, and Curran (1986) found that a majority (57 percent) of the respondents reported working for a ''large'' organization, yet 49 percent indicated that their immediate work groups consisted of 10 or fewer people. For most of the respondents, the number of human factors people in their work group was small. Some 24 percent reported no other human factors people in their work group besides themselves, and 25 percent reported only one or two others besides themselves. Such a situation can get a little lonely, but it can also be quite challenging.

What Do Human Factors People Do?

As part of the survey of Human Factors Society members, Sanders, Bied, and Curran provided a list of 63 activities and asked respondents to indicate for each activity whether they performed it rarely (if ever), occasionally, moderately often, or frequently. Table 1-3 lists all the activities for which 30 percent or more of the respondents indicated they performed moderately often or frequently in their work. The activities group nicely into four major areas: communication, management, system development, and research and evaluation.

How Do Human Factors People Feel about Their Jobs?

Sanders (1982), in another survey of the Human Factors Society membership, found that respondents rated their jobs especially high (compared to other professional and technical people) on skill variety (the degree to which the job requires a variety of different activities, skills, and talents) and autonomy (the degree to which the job provides freedom, independence, and discretion to the

TABLE 1-2
PERCENTAGE OF HUMAN FACTORS SPECIALISTS WORKING IN VARIOUS AREAS OF WORK

Principal area of work	Percentage of respondents
Computers	22
Aerospace	22
Industrial processes	17
Health and safety	9
Communications	8
Transportation	5
Other	17

Source: Van Cott and Huey, 1991, Table 3-1.

TABLE 1-3
ACTIVITIES PERFORMED MODERATELY OFTEN OR FREQUENTLY BY OVER 30 PERCENT OF HUMAN FACTORS SOCIETY RESPONDENTS

Activity	Percentage responding moderately often or frequently
Communication	
Write reports	80
Conduct formal briefing and presentations	59
Edit reports written by others	54
Write proposals	52
Evaluate relevance, worth, and quality of reports written by others	47
Review and summarize the literature	35
Management	
Schedule project activities	53
Manage and supervise others	49
Prepare budgets and monitor fiscal matters	38
System development	
Determine system requirements	43
Verify system design meets human factors standards	43
Write system goals and objectives	40
Perform task analysis	37
Specify user requirements for hardware and software	31
Research and evaluation	
Develop experimental designs to test theories or evaluate systems	44
Design data collection instruments and procedures	39
Determine proper statistical test for particular data set	38
Plan and conduct user-machine evaluations	36
Collect data in controlled laboratory setting	32
Develop criterion measures of human-system performance	31

worker in scheduling and determining the procedures used to carry out the work). Sanders also assessed job satisfaction among his respondents and found them especially satisfied (compared to other professional and technical people) with their pay, security, and opportunities for personal growth. Their overall general level of satisfaction was also very high.

Human Factors beyond the Human Factors Society

We have emphasized the Human Factors Society and its membership because it is the largest human factors professional group in the world and because a lot of interesting data are available about its membership. There are, however, other human factors groups in other countries, including Great Britain, Germany, Japan, Indonesia, the Netherlands, Italy, France, Canada, Australia, Norway, Israel, Poland, Yugoslavia, Hungary, Belgium, China, Soviet Union,

Austria, Brazil, and Mexico. In addition, other professional organizations have divisions or technical groups dealing with human factors. Such organizations include the American Industrial Hygiene Association, American Psychological Association, Institute of Electrical and Electronics Engineers, Society of Automotive Engineers, American Society of Mechanical Engineers, Association for Computing Machinery, Aerospace Medical Association, American Institute of Industrial Engineers, American Nuclear Society, American Society of Safety Engineers, Environmental Design Research Association, and the Society for Information Displays.

THE CASE FOR HUMAN FACTORS

Since humanity has somehow survived for these many thousands of years without people specializing in human factors, one might wonder why—at the present stage of history—it has become desirable to have human factors experts who specialize in worrying about these matters. As indicated before, the objectives of human factors are not new; history is filled with evidence of efforts, both successful and unsuccessful, to create tools and equipment which satisfactorily serve human purposes and to control more adequately the environment within which people live and work. But during most of history, the development of tools and equipment depended in large part on the process of evolution, of trial and error. Through the use of a particular device—an ax, an oar, a bow and arrow—it was possible to identify its deficiencies and to modify it accordingly, so that the next "generation" of the device would better serve its purpose.

The increased rate of technological development of recent decades has created the need to consider human factors early in the design phase, and in a systematic manner. Because of the complexity of many new and modified systems it frequently is impractical (or excessively costly) to make changes after they are actually produced. The cost of retrofitting frequently is exorbitant. Thus, the initial designs of many items must be as satisfactory as possible in terms of human factors considerations.

In effect, then, the increased complexities of the things people use (as the consequence of technology) place a premium on having assurance that the item in question will fulfill the two objectives of functional effectiveness and human welfare. The need for such assurance requires that human factors be taken into account early in the (usually long) design and development process. In Chapter 22 we discuss the development process in more detail and the role of human factors in it.

SYSTEMS

A central and fundamental concept in human factors is the *system*. Various authors have proposed different definitions for the term; however, we adopt a very simple one here. A *system* is an entity that exists to carry out some purpose (Bailey, 1982). A system is composed of humans, machines, and other things that work together (interact) to accomplish some goal which these same components could not produce independently. Thinking in terms of systems serves to structure the approach to the development, analysis, and evaluation of complex collections of humans and machines. As Bailey (1982) states,

> The concept of a system implies that we recognize a purpose; we carefully analyze the purpose; we understand what is required to achieve the purpose; we design the system's parts to accomplish the requirements; and we fashion a well-coordinated system that effectively meets our purpose. (p. 192)

We discuss aspects of human-machine systems and then present a few characteristics of systems in general. Finally, we introduce the concept of system reliability.

Human-Machine Systems

We can consider a *human-machine system* as a combination of one or more human beings and one or more physical components interacting to bring about, from given inputs, some desired output. In this frame of reference, the common concept of machine is too restricted, and we should rather consider a "machine" to consist of virtually any type of physical object, device, equipment, facility, thing, or what have you that people use in carrying out some activity that is directed toward achieving some desired purpose or in performing some function. In a relatively simple form, a human-machine system (or what we sometimes refer to simply as a *system*) can be a person with a hoe, a hammer, or a hair curler. Going up the scale of complexity, we can regard as systems the family automobile, an office machine, a lawn mower, and a roulette wheel, each equipped with its operator. More complex systems include aircraft, bottling machines, telephone systems, and automated oil refineries, along with their personnel. Some systems are less delineated and more amorphous than these, such as the servicing systems of gasoline stations and hospitals and other health services, the operation of an amusement park or a highway and traffic system, and the rescue operations for locating an aircraft downed at sea.

The essential nature of people's involvement in a system is an active one, interacting with the system to fulfill the function for which the system is designed.

The typical type of interaction between a person and a machine is illustrated in Figure 1-1. This shows how the displays of a machine serve as stimuli for an operator, trigger some type of information processing on the part of the operator (including decision making), which in turn results in some action (as in the operation of a control mechanism) that controls the operation of the machine.

One way to characterize human-machine systems is by the degree of manual versus machine control. Although the distinctions between and among systems in terms of such control are far from clear-cut, we can generally consider systems in three broad classes: manual, mechanical, and automatic.

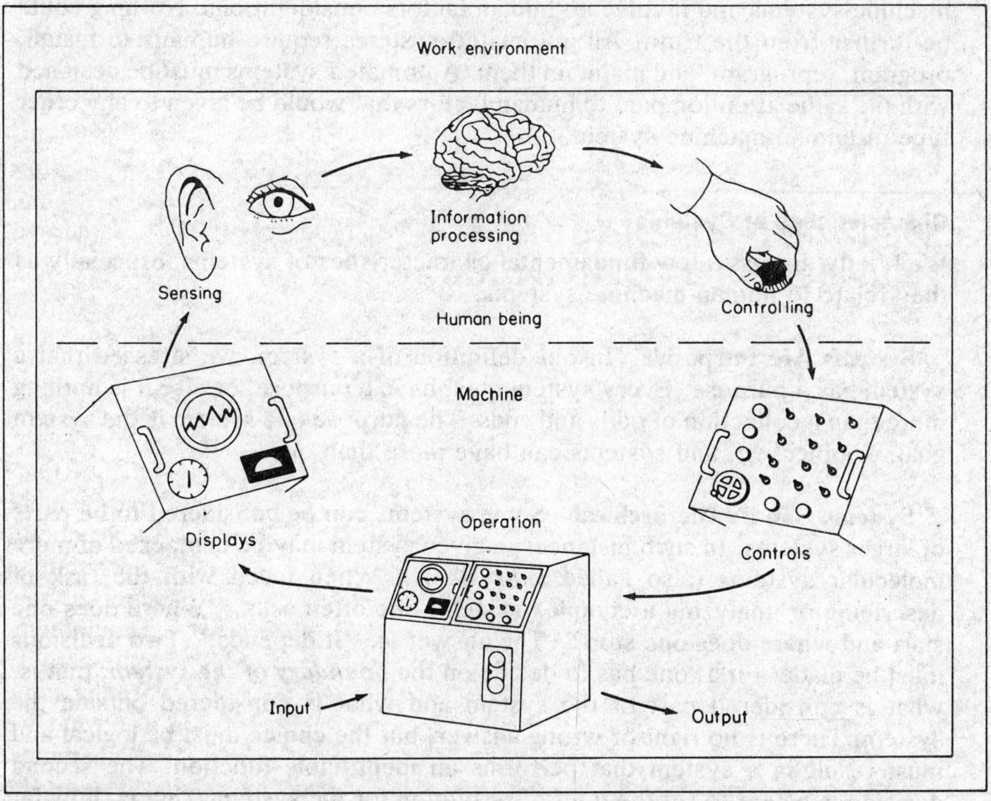

FIGURE 1-1
Schematic representation of a human-machine system. (*Source: Chapanis, 1976,* p. 701. *Used by permission of Houghton Mifflin Company.)*

Manual Systems A *manual system* consists of hand tools and other aids which are coupled by a human operator who controls the operation. Operators of such systems use their own physical energy as the power source.

Mechanical Systems These systems (also referred to as *semiautomatic* systems) consist of well-integrated physical parts, such as various types of powered machine tools. They are generally designed to perform their functions with little variation. The power typically is provided by the machine, and the operator's function is essentially one of control, usually by the use of control devices.

Automated Systems When a system is fully automated, it performs all operational functions with little or no human intervention. Robots are a good example of an automated system. Some people have the mistaken belief that since automated systems require no human intervention, they are not human-machine systems and involve no human factors considerations. Nothing could be further from the truth. All automated systems require humans to install, program, reprogram, and maintain them. Automated systems must be designed with the same attention paid to human factors that would be given to any other type of human-machine system.

Characteristics of Systems

We briefly discuss a few fundamental characteristics of systems, especially as they relate to human-machine systems.

Systems Are Purposive In our definition of a system, we stressed that a system has a purpose. Every system must have a purpose, or else it is nothing more than a collection of odds and ends. The purpose of a system is the system goal, or objective, and systems can have more than one.

Systems Can Be Hierarchical Some systems can be considered to be parts of larger systems. In such instances, a given system may be composed of more molecular systems (also called *subsystems*). When faced with the task of describing or analyzing a complex system, one often asks, "Where does one start and where does one stop?" The answer is, "It depends." Two decisions must be made. First, one has to decide on the *boundary of the system*, that is, what is considered part of the system and what is considered outside the system. There is no right or wrong answer, but the choice must be logical and must result in a system that performs an identifiable function. The second decision is where to set the *limit of resolution* for the system. That is, how far down into the system is one to go? At the lowest level of analysis one finds *components*. A component in one analysis may be a subsystem in another analysis that sets a lower limit of resolution. As with setting system boundaries, there is no right or wrong limit of resolution. The proper limit depends on why one is describing or analyzing the situation.

Systems Operate in an Environment The environment of a system is everything outside its boundaries. Depending on how the system's boundaries are drawn, the environment can range from the immediate environment (such as a workstation, a lounge chair, or a typing desk) through the intermediate (such as a home, an office, a factory, a school, or a football stadium) to the general (such as a neighborhood, a community, a city, or a highway system). Note that some aspects of the physical environment in which we live and work are part of the natural environment and may not be amenable to modification (although one can provide protection from certain undesirable environmental conditions such as heat or cold). Although the nature of people's involvement with their physical environment is essentially passive, the environment tends to impose certain constraints on their behavior (such as limiting the range of their movements or restricting their field of view) or to predetermine certain aspects of behavior (such as stooping down to look into a file cabinet, wandering through a labyrinth in a supermarket to find the bread, or trying to see the edge of the road on a rainy night).

Components Serve Functions Every component (the lowest level of analysis) in a system serves at least one function that is related to the fulfillment of one or more of the system's goals. One task of human factors specialists is to aid in making decisions as to whether humans or machines (including software) should carry out a particular system function. (We discuss this *allocation of function* process in more detail in Chapter 22.)

Components serve various functions in systems, but all typically involve a combination of four more basic functions: sensing (information receiving), information storage, information processing and decision, and action function; they are depicted graphically in Figure 1-2. Since information storage interacts with all the other functions, it is shown above the others. The other three functions occur in sequence.

1 *Sensing* (*information receiving*): One of these functions is sensing, or information receiving. Some of the information entering a system is from outside the system, for example, airplanes entering the area of control of a control-tower operator, an order for the production of a product, or the heat that sets off an automatic fire alarm. Some information, however, may originate inside the system itself. Such information can be feedback (such as the reading on the speedometer from an action of the accelerator), or it can be information that is stored in the system.

2 *Information storage:* For human beings, information storage is synonymous with memory of learned material. Information can be stored in physical components in many ways, as on magnetic tapes and disks, templates, records, and tables of data. Most of the information that is stored for later use is in coded or symbolic form.

3 *Information processing and decision:* Information processing embraces various types of operations performed with information that is received

FIGURE 1-2
Types of basic functions performed by human or machine components of human-machine systems.

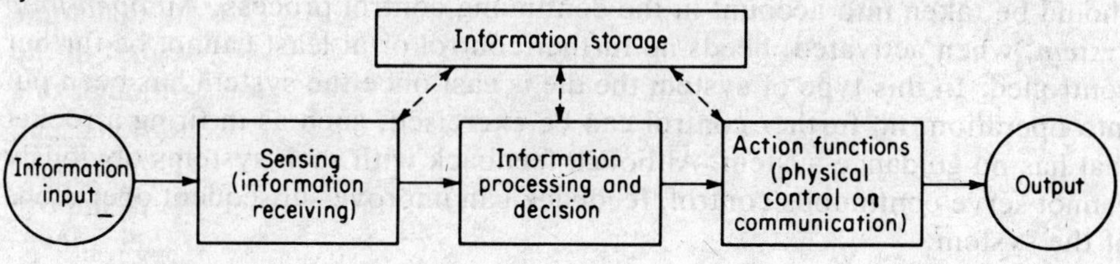

(sensed) and information that is stored. When human beings are involved in information processing, this process, simple or complex, typically results in a decision to act (or, in some instances, a decision *not* to act). When mechanized or automated machine components are used, their information processing must be programmed in some way. Such programming is, of course, readily understood if a computer is used. Other methods of programming involve the use of various types of schemes, such as gears, cams, electric and electronic circuits, and levers.

4 *Action functions:* What we call the *action* functions of a system generally are those operations which occur as a consequence of the decisions that are made. These functions fall roughly into two classes. The first is some type of *physical control action* or process, such as the activation of certain control mechanisms or the handling, movement, modification, or alteration of materials or objects. The other is essentially a *communication action*, be it by voice (in human beings), signals, records, or other methods. Such functions also involve some physical actions, but these are in a sense incidental to the communication function.

Components Interact To say that components interact simply means that the components work together to achieve system goals. Each component has an effect, however small, on other components. One outcome of a system's analysis is the description and understanding of these component and subsystem relationships.

Systems, Subsystems, and Components Have Inputs and Outputs At all levels of a complex system there are inputs and outputs. The outputs of one subsystem or component are the inputs to another. A system receives inputs from the environment and makes outputs to the environment. It is through inputs and outputs that all the pieces interact and communicate. Inputs can be physical entities (such as materials and products), electric impulses, mechanical forces, or information.

It might be valuable at this time to distinguish between open-loop and closed-loop systems. A *closed-loop system* performs some process which requires continuous control (such as in vehicular operation and the operation of certain chemical processes), and requires continuous feedback for its successful operation. The feedback provides information about any error that should be taken into account in the continuing control process. An *open-loop system*, when activated, needs no further control or at least cannot be further controlled. In this type of system the die is cast once the system has been put into operation; no further control can be exercised, such as in firing a rocket that has no guidance system. Although feedback with such systems obviously cannot serve continuous control, feedback can improve subsequent operations of the system.

In a system's analysis all the inputs and outputs required for each component and subsystem to perform its functions are specified. Human factors specialists are especially qualified to determine the inputs and outputs necessary for the human components of systems to successfully carry out their functions.

System Reliability

Unfortunately, nothing lasts forever. Things break or just fail to work, usually at the worst possible time. When we design systems, of course, we would like them to continue working. In this context, engineers speak of the *reliability* of a system or component to characterize its dependability of performance (including people) in carrying out an intended function. Reliability is usually expressed as the probability of successful performance (this is especially applicable when the performance consists of discrete events, such as starting a car). For example, if an automated teller machine gives out the correct amount of money 9999 times out of 10,000 withdrawal transactions, we say that the reliability of the machine, to perform the function, is .9999. (Reliabilities for electronic and mechanical devices are often carried out to four or more decimal places; reliabilities for human performance, on the other hand, usually are carried no further than three decimal places.)

Another measure of reliability is *mean time to failure* (abbreviated MTF). There are several possible variations, but they all relate to the amount of time a system or individual performs successfully, either until failure or between failures; this index is most applicable to continuous types of activities. Other variations could also be mentioned. For our present discussion, however, let us consider reliability in terms of the probability of successful performance.

If a system includes two or more components (machine or human or both), the reliability of the composite system will depend on the reliability of the individual components and how they are combined within the system. Components can be combined within a system in *series*, in *parallel*, or in a combination of both.

Components in Series In many systems the components are arranged in series (or sequence) in such a manner that successful performance of the total system depends on successful performance of each and every component, person or machine. By taking some semantic liberties, we could assume components to be *in series* that may, in fact, be functioning concurrently and interdependently, such as a human operator using some type of equipment. In analyzing reliability data in such cases, two conditions must be fulfilled: (1) failure of any given component results in system failure, and (2) the component failures are independent of each other. When these assumptions are fulfilled, the reliability of the system for error-free operation is the product of the reliabilities of the several components. As more components are added in series, the reliability of the system *decreases*. If a system consisted of 100 components in series, each with a reliability of .9900, the reliability of the entire system would be only .365 (that is, 365 times out of 1000 the system would properly perform its function). The maximum possible reliability in a series system is equal to the reliability of the *least* reliable component, which often turns out to be the human component. In practice, however, the overall reliability of a series system is often much *less* than the reliability of the least reliable component.

Components in Parallel The reliability of a system whose components are in parallel is entirely different from that whose components are in a series. With parallel components, two or more in some way are performing the same function. This is sometimes referred to as a *backup*, or *redundancy*, arrangement—one component backs up another so that if one fails, the other can successfully perform the function. In order for the entire system to fail, *all* the components in parallel must fail. Adding components in parallel *increases* the reliability of the system. For example, a system with four components in parallel, each with a reliability of .70, would have an overall system reliability of .992. Because humans are often the weak link in a system, it is common to see human-machine systems designed to provide parallel redundancy for some of the human functions.

Discussion We have been discussing system reliability as if it were static and unchanging. As our own experience illustrates, however, reliability changes as a function of time (usually it gets worse). The probability that a 10-year-old car will start is probably lower than it was when the car was 1 year old. The same sort of time dependency applies to the reliability of humans, only over shorter periods. The probability of successful human performance often deteriorates over just a few hours of activity. Human reliability is discussed further in Chapter 2.

COVERAGE OF THIS TEXT

Since a comprehensive treatment of the entire scope of human factors would fill a small library, this text must be restricted to a rather modest segment of the total human factors domain. The central theme is the illustration of how the achievement of the two primary human factors objectives (i.e., functional effectiveness and human welfare) can be influenced by the extent to which relevant human considerations have been taken into account during the design of the object, facility, or environment in question. Further, this theme is followed as it relates to some of the more commonly recognized human factors content areas (such as the design of displays for presenting information to people, human control processes, and physical environment). Pursuing this theme across the several subareas would offer an overview of the content areas of human factors.

The implications of various perceptual, mental, and physical characteristics as they might affect or influence the objectives of human factors probably can best be reflected by the result of relevant research investigations and of documented operational experience. Therefore the theme of the text will generally be carried out by presenting and discussing the results of illustrative research and by bringing in generalizations or guidelines supported by research or experiences that have relevance to the design process in terms of human factors considerations. Thus, in the various subject or content areas much of the material in this text will consist of summaries of research that reflect the relationships between design variables, on the one hand, and criteria of functional effectiveness or human welfare, on the other hand.

We recognize that the illustrative material brought in to carry out this theme is in no way comprehensive, but we hope it will represent in most content areas some of the more important facets.

Although the central theme will, then, deal with the human factors aspects of the design of the many things people use, there will be some modest treatment of certain related topics, such as how human factors fits in with the other phases of design and development processes.

REFERENCES

Bailey, R. (1982). *Human performance engineering: A guide for systems designers*. Englewood Cliffs, NJ: Prentice-Hall.

Chapanis, A. (1976). Engineering psychology. In M. D. Dunnette (ed.), *Handbook of industrial and organizational psychology*. Chicago: Rand McNally.

Chapanis, A. (1983). Introduction to human factors considerations in system design. In C. M. Mitchell, P. Van Balen, and K. Moe (eds.), *Human factors considerations in systems design* (NASA Conference Publ. 2246). Washington: National Aeronautic and Space Administration.

Chapanis, A. (1985). Some reflections on progress. *Proceedings of the Human Factors Society 29th Annual Meeting*. Santa Monica, CA: Human Factors Society, pp. 1–8.

Chapanis, A., Garner, W., and Morgan, C. (1949). *Applied experimental psychology: Human factors in engineering design*. New York: Wiley.

Edholm, O., and Murrell, K. F. H. (1973). *The Ergonomics Research Society: A History 1949–1970*. Winchester, Hampshire: Warren & Sons, Ltd.

Human Factors Society (1991). *The Human Factors Society 1991 directory and yearbook*. Santa Monica, CA: Human Factors Society.

Meshkati, N. (1989). An etiological investigation of micro- and macroergonomic factors in the Bhopal disaster: Lessons for industries of both industrialized and developing countries. *International Journal of Industrial Ergonomics*, 4, 161–175.

Meshkati, N. (1991). Human factors in large-scale technological systems' accidents: Three Mile Island, Bhopal, Chernobyl. *Industrial Crisis Quarterly*, 5(2).

Muckler, F., and Seven, S. (1990). National education and training. In H. Booher (ed.), *MANPRINT: An approach to systems integration*. New York: Van Nostrand Reinhold.

Norman, D. (1988). *The psychology of everyday things*. New York: Basic Books.

Sanders, M. (1982). HFS job description survey: Job dimensions, satisfaction, and motivation. *Proceedings of the Human Factors Society 25th Annual Meeting*. Santa Monica, CA: Human Factors Society, pp. 682–686.

Sanders, M., Bied, B., and Curran, P. (1986). HFS membership job survey. *Human Factors Society Bulletin*, 29(3), 1–3.

Sanders, M., and Smith, L. (eds.) (1988). *Directory of human factors graduate programs in the United States and Canada: 1988 edition*. Santa Monica, CA: Human Factors Society.

Singleton, W. T. (1982). *The body at work: Biological ergonomics*. London: Cambridge University Press.

Van Cott, H., and Huey, B. (eds.) (1991). *Human factors specialists education and utilization: Results of a survey*. Washington, D.C.: National Academy Press.

Welford, A. T. (1976). Ergonomics: Where have we been and where are we going: I. *Ergonomics*, 19(3), 275–286.

CHAPTER 2

HUMAN FACTORS RESEARCH METHODOLOGIES

Human factors is in large part an empirical science. The central approach of human factors is the application of relevant information about human capabilities and behavior to the design of objects, facilities, procedures, and environments that people use. This body of relevant information is largely based on experimentation and observation. Research plays a central role in this regard, and the research basis of human factors is emphasized throughout this book.

In addition to gathering empirically based information and applying it to the design of things, human factors specialists also gather empirical data to evaluate the "goodness" of their designs and the designs of others. Thus, empirical data, and hence research, play a dual role in the development of systems: at the front end as a basis for the design and at the back end as a means of evaluating and improving the design. For this reason, in this chapter we deal with some basic concepts of human research as it relates to human factors. Our purpose is not to present a handbook of research methods, but rather to introduce some of the purposes, considerations, and trade-offs involved in the research process. For a more complete discussion of research methodologies relevant to human factors, refer to Meister (1985), Meister and Rabideau (1965), Wilson and Corlett (1990), Keppel (1982), and Rosenthal and Rosnow (1984).

AN OVERVIEW

As one would expect, most *human* factors research involves the use of human beings as subjects, so we focus our attention there. Not all human factors research, however, involves human subjects. Sanders and Krohn (1983), for example, surveyed underground mining equipment to assess the field of view from the operator's compartment. Aside from the data collectors, no humans were involved. Human factors research can usually be classified into one of three types: descriptive studies, experimental research, or evaluation research. Actually, not all human factors research fits neatly into only one category; often a particular study will involve elements of more than one category. Although each category has different goals and may involve the use of slightly different methods, all involve the same basic set of decisions: choosing a research setting, selecting variables, choosing a sample of subjects, deciding how the data will be collected, and deciding how the data will be analyzed.

We describe each type of research briefly and then discuss the basic research decisions listed above. This will give us an opportunity to say a few things we want to say and introduce some concepts that will be popping up now and again in later chapters.

Descriptive Studies

Generally speaking, descriptive studies seek to characterize a population (usually of people) in terms of certain attributes. We present the results of many such studies throughout this book. Examples include surveys of the dimensions of people's bodies, hearing loss among people of different ages, people's expectations as to how a knob should be turned to increase the value on a display, and weights of boxes people are willing to lift.

Although descriptive studies are not very exciting, they are very important to the science of human factors. They represent the basic data upon which many design decisions are based. In addition, descriptive studies are often carried out to assess the magnitude and scope of a problem before solutions are suggested. A survey of operators to gather their opinions about design deficiencies and operational problems would be an example. In fact, the Nuclear Regulatory Commission (1981) required such a survey as part of its mandated human factors control room review process.

Experimental Research

The purpose of experimental research is to test the effects of some variable on behavior. The decisions as to what variables to investigate and what behaviors to measure are usually based on either a practical situation which presents a design problem, or a theory that makes a prediction about variables and behaviors. Examples of the former include comparing how well people can edit manuscripts with partial-line, partial-page, and full-page computer displays (Neal and Darnell, 1984) and assessing the effect of seat belts and shoulder harnesses on functional arm reach (Garg, Bakken, and Saxena, 1982). Experimental research of a more theoretical nature would include a study by Hull, Gill, and Roscoe (1982) in which they varied the lower half of the visual field to investigate why the moon looks so much larger when it is near the horizon than when it is overhead.

Usually in experimental research the concern is whether a variable has an effect on behavior and the direction of that effect. Although the level of performance is of interest, usually only the relative *difference* in performance between conditions is of concern. For example, one might say that subjects missed, on the average, 15 more signals under high noise than under low noise. In contrast, descriptive studies are usually interested in describing a population parameter, such as the mean, rather than assessing the effect of a variable. When descriptive studies compare groups that differ on some variable (such as sex or age), the means, standard deviations, and percentiles of each group are of prime interest. This difference in goals between experimental and descriptive studies, as we will see, has implications for subject selection.

Evaluation Research

Evaluation research is similar to experimental research in that its purpose is to assess the effect of "something." However, in evaluation research the something is usually a system or product. Evaluation research is also similar to descriptive research in that it seeks to describe the performance and behaviors of the people using the system or product.

Evaluation research is generally more global and comprehensive than experimental research. A system or product is evaluated by comparison with its goals; both intended consequences and unintended outcomes must be assessed. Often an evaluation research study will include a benefit-cost analysis. Examples of evaluation research include evaluating a new training program, a new software package for word processing, or an ergonomically designed life jacket. Evaluation research is the area where human factors specialists assess the "goodness" of designs, theirs and others, and make recommendations for improvement based on the information collected.

Evaluation is part of the overall systems design process, and we discuss it as such in Chapter 22. Suffice it to say, evaluation research is probably one of the most challenging and frustrating types of research endeavors to undertake. The rewards can be great because the results are often used to improve the design of an actual system or product, but conducting the research can be a nightmare. Murphy's law seems to rule: "If anything can go wrong, it will." Extra attention, therefore, must be paid to designing data collection procedures and devices to perform in the often unpredictable and unaccommodating field setting. One of your authors recalls evaluating the use of a helicopter patrol for fighting crime against railroads. Murphy's law prevailed; the trained observers never seemed to be available when they were needed, the railroad coordination chairman had to resign, the field radios did not work, the railroads were reluctant to report incidences of crime, and finally a gang of criminals threatened to shoot down the helicopter if they saw it in the air.

CHOOSING A RESEARCH SETTING

In choosing the research setting for a study, the fundamental decision is whether to conduct the study in the field, also referred to as the "real world," or in the laboratory.

Descriptive Studies

With descriptive studies the choice of research setting is somewhat moot. The primary goal of descriptive studies is to generate data that describe a particular population of people, be they coal miners, computer operators, or the general civilian population. To poll such people, we must go to the real world. As we will see, however, the actual data collection may be done in a laboratory—often a mobile laboratory—which is, in essence, like bringing the mountain to Mohammad.

Experimental Research

The choice of research setting for experimental research involves complex trade-offs. Research carried out in the field usually has the advantage of realism in terms of relevant task variables, environmental constraints, and subject characteristics including motivation. Thus, there is a better chance that the results obtained can be generalized to the real-world operational environment. The disadvantages, however, include cost (which can be prohibitive), safety hazards for subjects, and lack of experimental control. In field studies often there is no opportunity to replicate the experiment a sufficient number of times, many variables cannot be held constant, and often certain data cannot be collected because the process would be too disruptive.

The laboratory setting has the principal advantage of experimental control; extraneous variables can be controlled, the experiment can be replicated almost at will, and data collection can be made more precise. For this advantage, however, the research may sacrifice some realism and generalizability. Meister (1985) believes that this lack of realism makes laboratory research less than adequate as a source of applied human factors data. He believes that conclusions generated from laboratory research should be tested in the real world before they are used there.

For theoretical studies, the laboratory is the natural setting because of the need to isolate the subtle effects of one or more variables. Such precision probably could not be achieved in the uncontrolled real world. The real world, on the other hand, is the natural setting for answering practical research questions. Often, a variable that shows an effect in the highly controlled laboratory "washes out" when it is compared to all the other variables that are affecting performance in the real world.

In some cases, field research can be carried out with a good deal of control—although some people might say of the same situation that it is laboratory research that is being carried out with a good deal of realism. An example is a study in which Marras and Kroemer (1980) compared two distress signal designs (for flares). In one part of the study, subjects were taken by boat to an island in a large lake. They were told that their task was to sit in an inflatable rubber raft, offshore, and rate the visibility of a display which would appear on shore. They were given a distress signal to use in case of an emergency. While the subjects were observing the display, the raft deflated automatically, prompting the subjects to activate the distress signal. The time required for unpacking and successfully operating the device was recorded. The subjects were then pulled back to shore.

In an attempt to combine the benefits of both laboratory and field research, researchers often use *simulations* of the real world in which to conduct research. A distinction should be made between *physical* simulations and *computer* simulations. Physical simulations are usually constructed of hardware and represent (i.e., look like, feel like, or act like) some system, procedure, or environment. Physical simulations can range from very simple items (such as a picture of a control panel) to extremely complex configurations (such as a moving-base jumbo jet flight simulator with elaborate out-of-cockpit visual display capabilities). Some simulators are small enough to fit on a desktop; others can be quite large, such as a 400-ft^2 underground coal mine simulator built by one of your authors.

Computer simulation involves modeling a process or series of events in a computer. By changing the parameters the model can be run and predicted results can be obtained. For example, workforce needs, periods of overload, and equipment downtime can be predicted from computer simulations of work processes. To develop an accurate computer model requires a thorough understanding of the system being modeled and usually requires the modeler to make some simplifying assumptions about how the real-world system operates.

Evaluation Research

As with descriptive studies, choosing a research setting for evaluation research is also somewhat moot. For a true test of the "goodness" of a system or device, the test should be conducted under conditions representative of those under which the thing being tested will ultimately be used. A computerized map display for an automobile, as an example, should be tested in an automobile while driving over various types of roads and in different traffic conditions. The display may be very legible when viewed in a laboratory, but very hard to read in a moving automobile. Likewise, the controls may be easily activated when the device is sitting on a desk in a laboratory but may be very hard to operate while navigating a winding road or driving in heavy traffic.

SELECTING VARIABLES

The selection of variables to be measured in research studies is such a fundamental and important question that we have devoted two later sections of this chapter to it.

Descriptive Studies

In descriptive studies two basic classes of variables are measured: *criterion variables* and *stratification* (or *predictor*) *variables*.

Criterion Variables Criterion variables describe those characteristics and behaviors of interest in the study. These variables can be grouped into the following classes according to the type of data being collected: *physical characteristics*, such as arm reach, stomach girth, and body weight; *performance data*, such as reaction time, visual acuity, hand grip strength, and memory span; *subjective data*, such as preferences, opinions, and ratings; and *physiological indices*, such as heart rate, body temperature, and pupil dilation.

Stratification Variables In some descriptive studies (such as surveys), it is the practice to select *stratified* samples that are proportionately representative of the population in terms of such characteristics as age, sex, education, etc. Even if a stratified sample is not used, however, information is often obtained on certain relevant personal characteristics of those in the sample. Thus, the resulting data can be analyzed in terms of the characteristics assessed, such as age, sex, etc. These characteristics are sometimes called *predictors*.

Experimental Research

In experimental research, the experimenter manipulates one or more variables to assess their effects on behaviors that are measured, while other variables are controlled. The variables being manipulated by the experimenter are called *independent variables* (IVs). The behaviors being measured to assess the effects of the IVs are called *dependent variables* (DVs). The variables that are controlled are called *extraneous, secondary*, or *relevant variables*. These are variables that can influence the DV; they are controlled so that their effect is not confused (confounded) with the effect of the IV.

Independent Variables In human factors research, IVs usually can be classified into three types: (1) *task-related variables*, including equipment variables (such as length of control lever, size of boxes, and type of visual display) and procedural variables (such as work-rest cycles and instructions to stress accuracy or speed); (2) *environmental variables*, such as variations in illumination, noise, and vibration; and (3) *subject-related variables*, such as sex, height, age, and experience.

Most studies do not include more than a few IVs. Simon (1976), for example, reviewed 141 experimental papers published in *Human Factors* from 1958 to 1972 and found that 60 percent of the experiments investigated the effects of only one or two IVs and less than 3 percent investigated five or more IVs.

Dependent Variables DVs are the same as the criterion variables discussed in reference to descriptive studies, except that physical characteristics are used less often. Most DVs in experimental research are performance, subjective, or physiological variables. We discuss criterion variables later in this chapter.

Evaluation Research

Selecting variables for evaluation research requires the researcher to translate the goals and objectives of the system or device being evaluated into specific criterion variables that can be measured. Criterion variables must also be included to assess unintended consequences arising from the use of the system. The criterion variables are essentially the same as those used in descriptive studies and experimental research and are discussed further in later sections of this chapter and in Chapter 22.

CHOOSING SUBJECTS

Choosing subjects is a matter of deciding who to select, how to select them, and how many of them to select.

Descriptive Studies

Proper subject selection is critical to the validity of descriptive studies. Often, in such studies, the researcher expends more effort in developing a sampling plan and obtaining the subjects than in any other phase of the project.

Representative Sample The goal in descriptive studies is to collect data from a sample of people representative of the population of interest. A sample is said to be *representative* of a population if the sample contains all the *relevant* aspects of the population in the same proportion as found in the real population. For example, if in the population of coal miners 30 percent are under 21 years of age, 40 percent are between 21 and 40 years, and 30 percent are over 40 years of age, then the sample—to be representative—should also contain the same percentages of each age group.

The key word in the definition of representative is *relevant*. A sample may differ from a population with respect to an irrelevant variable and still be useful for descriptive purposes. For example, if we were to measure the reaction time of air traffic controllers to an auditory alarm, we would probably carry out the study in one or two cities rather than sampling controllers in every state or geographic region in the country. This is so because it is doubtful that the reaction time is different in different geographic regions; that is, geographic region is probably not a relevant variable in the study of reaction time. We would probably take great care, however, to include the proper proportions of controllers with respect to age and sex because these variables are likely to be relevant. A sample that is not representative is said to be *biased*.

Random Sampling To obtain a representative sample, the sample should be selected randomly from the population. *Random selection* occurs when each member of the population has an equal chance of being included in the sample. In the real world, it is almost impossible to obtain a truly random sample. Often the researcher must settle for those who are easily obtained even though they were not selected according to a strict random procedure.

Even though a particular study may not sample randomly or include all possible types of people in the proportions in which they exist in the population, the study may still be useful if the bias in the sample is not relevant to the criterion measures of interest. How does one know whether the bias is relevant? Prior research, experience, and theories form the basis for an educated guess.

Sample Size A key issue in carrying out a descriptive study is determining how many subjects will be used, i.e., the sample size. The larger the sample size, the more confidence one has in the results. Sampling costs money and takes time, so researchers do not want to collect more data than they need to make valid inferences about the population. Fortunately, there are formulas for determining the number of subjects required. [See, for example, Roebuck, Kroemer, and Thompson (1975).] Three main parameters influence the number of subjects required: *degree of accuracy desired* (the more accuracy desired, the larger the sample size required); *variance in the population* (the greater the degree of variability of the measure in the population, the larger the sample size needed to obtain the level of accuracy desired); and the *statistic being estimated*, e.g., mean, 5th percentile, etc. (some statistics require more subjects to estimate accurately than others; for example, more subjects are required to estimate the median than to estimate the mean with the same degree of accuracy).

Experimental Research

The issue in choosing subjects for experimental research is to select subjects representative of those people to whom the results will be generalized. Subjects do not have to be representative of the target population to the same degree as in descriptive studies. The question in experimental studies is whether the subjects will be affected by the IV in the same way as the target population. For example, consider an experiment to investigate the effects of room illumination (high versus low) on reading text displayed on a computer screen. If we wish to use the data to design office environments, do we need to use subjects who have extensive experience reading text from computer screens? Probably not. Although highly experienced computer screen readers can read faster than novice readers (hence it would be important to include them in a descriptive study), the effect of the IV (illumination) is likely to be the same for both groups. That is, it is probably easier to read computer-generated text under low room illumination (less glare) than under high—no matter how much experience the reader has.

Sample Size The issue in determining sample size is to collect enough data to reliably assess the effects of the IV with minimum cost in time and resources. There are techniques available to determine sample size requirements; however, they are beyond the scope of this book. The interested reader can consult Cohen (1977) for more details.

In Simon's (1976) review of research published in *Human Factors*, he found that 50 percent of the studies used fewer than 9 subjects per experimental condition, 25 percent used from 9 to 11, and 25 percent used more than 11. These values are much smaller than those typically used in descriptive studies. The danger of using too few subjects is that one will incorrectly conclude that an IV had no effect on a DV when, in fact, it did. The "danger," if you can call it that, of using too many subjects is that a tiny effect of an IV, which may have no practical importance whatsoever, will show up in the analysis.

Evaluation Research

Choosing subjects for evaluation research involves the same considerations as discussed for descriptive studies and experimental research. The subjects must be representative of the ultimate user population. The number of subjects must be adequate to allow predictions to be made about how the user population will perform when the system or device is placed in use. Unfortunately, most evaluation research must make do with fewer subjects than we would like, and often those that are made available for the evaluation are not really representative of the user population.

COLLECTING THE DATA

Data in *descriptive studies* can be collected in the field or in a laboratory setting. Bobo et al. (1983), for example, measured energy expenditure of underground coal miners performing their work underground. Sanders (1981) measured the strength of truck and bus drivers turning a steering wheel in a mobile laboratory at various truck depots around the country.

Often, surveys and interviews are used to collect data. Survey questionnaires may be administered in the field or mailed to subjects. A major problem with mail surveys is that not everyone returns the questionnaires, and the possibility of bias increases. Return rates of less than 50 percent should probably be considered to have a high probability of bias.

The collection of data for *experimental research* is the same as for descriptive studies. Because experimental studies are often carried out in a controlled laboratory, often more sophisticated, computer-based methods are employed. As pointed out by McFarling and Ellingstad (1977), such methods provide the potential for including more IVs and DVs in a study and permit greater precision and higher sampling rates in the collection of performance data. Vreuls et al. (1973), for example, generated over 800 different measures to evaluate the performance of helicopter pilots doing a few common maneuvers. One, of course, has to be careful not to drown in a sea of data.

Collecting data in an *evaluation research* study is often difficult. The equipment being evaluated may not have the capability of monitoring or measuring user performance, and engineers are often reluctant to modify the equipment just to please the evaluation researcher. All to often, the principle method of data collection is observation of users by the researcher and interviewing users regarding problems they encountered and their opinions of the equipment.

ANALYZING THE DATA

Once a study has been carried out and the data have been gathered, the experimenter must analyze the data. It is not our intention here to deal extensively with statistics or to discuss elaborate statistical methods.

Descriptive Studies

When analyzing data from descriptive studies, usually fairly basic statistics are compared. Probably most readers are already familiar with most statistical methods and concepts touched on in later chapters, such as frequency distributions and measures of central tendency (mean, median, mode). For those readers unfamiliar with the concepts of standard deviation, correlation, and percentiles, we describe them briefly.

Standard Deviation The standard deviation (S) is a measure of the variability of a set of numbers around the mean. When, say, the reaction time of a group of subjects is measured, not everyone has the same reaction time. If the scores varied greatly from one another, the standard deviation would be large. If the scores were all close together, the standard deviation would be small. In a normal distribution (bell-shaped curve), approximately 68 percent of the cases will be within $\pm$ 1S of the mean, 95 percent within $\pm 2S$ of the mean, and 99 percent within $\pm 3S$ of the mean.

Correlation A correlation coefficient is a measure of the degree of relationship between two variables. Typically, we compute a linear correlation coefficient (e.g., Pearson product-moment correlation r) which indicates the degree to which two variables are *linearly* related, that is, related in a straight-line fashion. Correlations can range from +1.00, indicating a perfect positive relationship, through 0 (which is the absence of any relationship), to −1.00, a perfect negative relationship. A positive relationship between two variables indicates that high values on one variable tend to be associated with high values on the other variable, and low values on one are associated with low values on the other. An example would be height and weight because tall people tend to be heavier than short people. A negative relationship between two variables indicates that high values on one variable are associated with low values on the other. An example would be age and strength because older people tend to have less strength than younger people.

If the correlation coefficient is squared (r^2), it represents the *proportion of variance* (standard deviation squared) in one variable accounted for by the other variable. This is somewhat esoteric, but it is important for understanding the strength of the relationship between two variables. For example, if the correlation coefficient between age and strength is .30 we would say that of the total variance in strength 9 percent ($.30^2$) was accounted for by the subjects' age. That means that 81 percent of the variance is due to factors other than age.

Percentiles Percentiles correspond to the value of a variable below which a specific percentage of the group fall. For example, the 5th percentile standing height for males is 63.6 in (162 cm). This means that only 5 percent of males are smaller than 63.6 in (162 cm). The 50th percentile male height is 68.3 in (173 cm), which is the same as the median since 50 percent of males are shorter than this value and 50 percent are taller. The 95th percentile is 72.8 in (185 cm), meaning that 95 percent of males are shorter than this height. Some investigators report the *interquartile range*. This is simply the range from the 25th to the 75th percentiles, and thus it encompasses the middle 50 percent of the distribution. (Interquartile range is really a measure of variability.) The concept of percentile is especially important in using anthropometric (body dimension) data for designing objects, workstations, and facilities.

Experimental Research

Data from experimental studies are usually analyzed through some type of inferential statistical technique such as analysis of variance (ANOVA) or multivariate analysis of variance (MANOVA). We do not discuss these techniques since several very good texts are available (Box, Hunter, and Hunter, 1978; Tabachnick and Fidell, 1989; Hays, 1988; Siegel and Castellan, 1988).

The outcome of virtually all such techniques is a statement concerning the statistical significance of the data. It is this concept that we discuss because it is central to all experimental research.

Statistical Significance Researchers make statements such as "the IV had a significant effect on the DV" or "the difference between the means was significant." The term *significance* is short for *statistical significance*. To say that something is statistically significant simply means that there is a low probability that the observed effect, or difference between the means, was due to chance. And because it is unlikely that the effect was due to chance, it is concluded that the effect was due to the IV.

How unlikely must it be that chance was causing the effect to make us conclude chance was not responsible? By tradition, we say .05 or .01 is a low probability (by the way, this tradition started in agricultural research, earthy stuff). The experimenter selects one of these values, which is called the *alpha level.* Actually, any value can be selected, and you as a consumer of the research may disagree with the level chosen. Thus, if the .05 alpha level is selected, the researcher is saying that if the results obtained could have occurred 5 times or less out of 100 by chance alone, then it is unlikely that they are due to chance. The researcher says that the results are significant at the .05 level. Usually the researcher will then conclude that the IV was causing the effect. If, on the other hand, the results could have occurred more than 5 times out of 100 by chance, then the researcher concludes that it is likely that chance was the cause, and not the IV.

The statistical analysis performed on the data yields the probability that the results could have occurred by chance. This is compared to the alpha level, and that determines whether the results are significant. Keep in mind that the statistical analysis does not actually tell the researcher whether the results did or did not occur by chance; only the probability is given. No one knows for sure.

Here are a couple of things to remember when you are reading about "significant results" or results that "failed to reach significance": (1) results that are significant may still be due only to chance, although the probability of this is low. (2) IVs that are not significant may still be influencing the DV, and this can be very likely, especially when small sample sizes are used. (3) Statistical significance has nothing whatever to do with importance—very small, trivial effects can be statistically significant. (4) There is no way of knowing from the statistical analysis procedure whether the experimental design was faulty or uncontrolled variables confounded the results.

The upshot is that one study usually does not make a fact. Only when several studies, each using different methods and subjects, find the same effects should we be willing to say with confidence that an IV does affect a DV. Unfortunately, in most areas, we usually find conflicting results, and it takes insight and creativity to unravel the findings.

CRITERION MEASURES IN RESEARCH

Criterion measures, as we discussed, are the characteristics and behaviors measured in descriptive studies, the dependent variables in experimental research, and the basis for judging the goodness of a design in an evaluation study. Any attempt to classify criterion measures inevitably leads to confusion and overlap. Since a little confusion is part and parcel of any technical book, we relate one simple scheme for classifying criterion measures to provide a little organization to our discussion.

In the human factors domain, three types of criteria can be distinguished (Meister, 1985): those describing the functioning of the *system*, those describing how the *task* is performed, and those describing how the *human* responds. The problem is that measures of how tasks are performed usually involve how the human responds. We briefly describe each type of criteria, keeping in mind the inevitable overlap among them.

System-Descriptive Criteria

System-descriptive criteria usually reflect essentially engineering aspects of the entire system. So they are often included in evaluation research but are used to a much lesser extent in descriptive and experimental studies. System-descriptive criteria include such aspects as equipment reliability (i.e., the probability that it will not break down), resistance to wear, cost of operation, maintainability, and other engineering specifications, such as maximum rpm, weight, radio interference, etc.

Task Performance Criteria

Task performance criteria usually reflect the outcome of a task in which a person may or may not be involved. Such criteria include (1) *quantity of output*, for example, number of messages decoded, tons of earth moved, or number of shots fired; (2) *quality of output*, for example, accidents, number of errors, accuracy of drilling holes, or deviations from a desired path; and (3) *performance time*, for example, time to isolate a fault in an electric circuit or amount of delay in beginning a task. Task performance criteria are more global than human performance criteria, although human performance is inextricably intertwined in task performance and in the engineering characteristics of the system or equipment being used.

Human Criteria

Human criteria deal with the behaviors and responses of humans during task performance. Human criteria are measured with performance measures, physiological indices, and subjective responses.

Performance Measures Human performance measures are usually *frequency measures* (e.g., number of targets detected, number of keystrokes made, or number of times the "help" screen was used), *intensity measures* (e.g., torque produced on a steering wheel), *latency measures* (e.g., reaction time or delay in switching from one activity to another), or *duration measures* (e.g., time to log on to a computer system or time on target in a tracking task).

Sometimes combinations of these basic types are used, such as number of missed targets per unit time. Another human performance measure is *reliability*, or the probability of errorless performance. There has been a good deal of work relating to human reliability, and we review some of it later in this chapter.

Physiological Indices Physiological indices are often used to measure strain in humans resulting from physical or mental work and from environmental influences, such as heat, vibration, noise, and acceleration. Physiological indices can be classified by the major biological systems of the body: *cardiovascular* (e.g., heart rate or blood pressure), *respiratory* (e.g., respiration rate or oxygen consumption), *nervous* (e.g., electric brain potentials or muscle activity), *sensory* (e.g., visual acuity, blink rate, or hearing acuity), and *blood chemistry* (e.g., catecholamines).

Subjective Responses Often we must rely on subjects' opinions, ratings, or judgments to measure criteria. Criteria such as comfort of a seat, ease of use of a computer system, or preferences for various lengths of tool handle are all examples of subjective measures. Subjective responses have also been used to measure perceived mental and physical workload. Extra care must be taken when subjective measures are designed because people have all sorts of built-in biases in the way they evaluate their likes, dislikes, and feelings. Subtle changes in the wording or order of questions, the format by which people make their response, or the instructions accompanying the measurement instrument can alter the responses of the subjects. Despite these shortcomings, subjective responses are a valuable data source and often represent the only reasonable method for measuring the criterion of interest.

Terminal versus Intermediate Criteria

Criterion measures may describe the terminal performance (the ultimate output of the action) or some intermediate performance that led up to the output. Generally, intermediate measures are more specific and detailed than terminal measures. Terminal measures are more valuable than intermediate ones because they describe the ultimate performance of interest. Intermediate measures, however, are useful for diagnosing and explaining performance inadequacies. For example, if we were interested in the effectiveness of a warning label that instructs people using a medicine to shake well (the bottle, that is) before using, the terminal performance would be whether the people shook the bottle. Intermediate criteria would include asking the people whether they recalled seeing the warning, whether they read the warning, and whether they could recall the warning message. The intermediate criterion data would be of value in explaining why people do not always follow warnings—is it because they do not see them, do not take the time to read them, or cannot recall them?

REQUIREMENTS FOR RESEARCH CRITERIA

Criterion measures used in research investigations generally should satisfy certain requirements. There are both practical and psychometric requirements. The psychometric requirements are those of reliability, validity, freedom from contamination, and sensitivity.

Practical Requirements

Meister (1985) lists six practical requirements for criterion measures, indicating that, when feasible, a criterion measure should (1) be objective, (2) be quantitative, (3) be unobtrusive, (4) be easy to collect, (5) require no special data collection techniques or instrumentation, and (6) cost as little as possible in terms of money and experimenter effort.

Reliability

In the context of measurement, reliability refers to the consistency or stability of the measures of a variable over time or across representative samples. This is different from the concept of human reliability, as we see later. Technically, reliability in the measurement sense is the degree to which a set of measurements is free from error (i.e., unsystematic or random influences).

Suppose that a human factors specialist in King Arthur's court were commanded to assess the combat skills of the Knights of the Roundtable. To do this, the specialist might have each knight shoot a single arrow at a target and record the distance off target as the measure of combat skill. If all the knights were measured one day and again the next, quite likely the scores would be quite different on the two days. The best archer on the first day could be the worst on the second day. We would say that the measure was unreliable. Much, however, could be done to improve the reliability of the measure, including having each knight shoot 10 arrows each day and using the average distance off target as the measure, being sure all the arrows were straight and the feathers set properly, and performing the archery inside the castle to reduce the variability in wind, lighting, and other conditions that could change a knight's performance from day to day.

Correlating the sets of scores from the two days would yield an estimate of the reliability of the measure. Generally speaking, test-retest reliability correlations around .80 or above are considered satisfactory, although with some measures we have to be satisfied with lower levels.

Validity

Several types of validity are relevant to human factors research. Although each is different, they all have in common the determination of the extent to which different variables actually measure what was intended. The types of validity relevant to our discussion are *face validity, content validity*, and *construct validity*. Another type of validity which is more relevant to determining the usefulness of tests as a basis for selecting people for a job is *criterion-related validity*. This refers to the extent to which a test predicts performance. We do not discuss this type of validity further; rather we focus on the other types.

Face Validity Face validity refers to the extent to which a measure *looks as though* it measures what is intended. At first glance, this may not seem important; however, in some measurement situations (especially in evaluation research and experimental research carried out in the field), face validity can influence the motivation of the subjects participating in the research. Where possible, researchers should choose measures or construct tasks that appear relevant to the users. To test the legibility of a computer screen for office use, it might be better to use material that is likely to be associated with offices (such as invoices) rather than nonsense syllables.

Content Validity Content validity refers to the extent to which a measure of some variable samples a domain, such as a field of knowledge or a set of job behaviors. In the field of testing, for example, content validity is typically used to evaluate achievement tests. In the human factors field, this type of validity would apply to such circumstances as measuring the performance of air traffic controllers. To have content validity, such a measure would have to include the various facets of the controllers' performance rather than just a single aspect.

Construct Validity Construct validity refers to the extent to which a measure is really tapping the underlying "construct" of interest (such as the basic type of behavior or ability in question). In our Knights of the Roundtable example, accuracy in shooting arrows at stationary targets would have only slight construct validity as a measure of actual combat skill. This is depicted in Figure 2-1, which shows the overlap between the construct (combat skill) and the measure (shooting accuracy with stationary targets). The small overlap denotes low construct validity because the measure taps only a few aspects of

FIGURE 2-1
Illustration of the concept of construct validity and contamination in the measurement of research criteria.

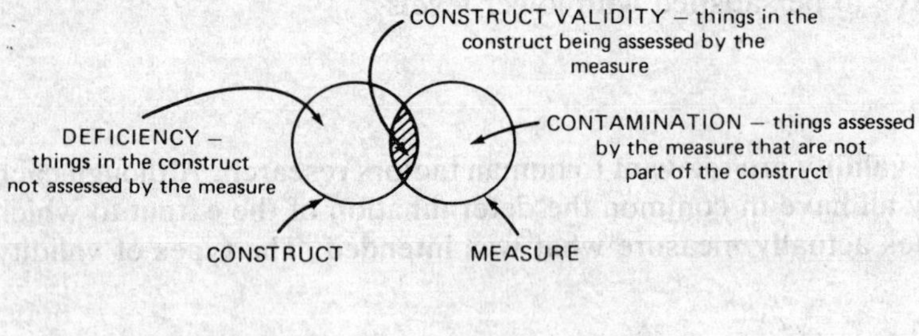

the construct. Construct validity is based on a judgmental assessment of an accumulation of empirical evidence regarding the measurement of the variable in question.

Freedom from Contamination

A criterion measure should not be influenced by variables that are extraneous to the construct being measured. This can be seen in Figure 2-1. In our example of the knights, wind conditions, illumination, and quality of the arrows could be sources of contamination because they could affect accuracy yet are unrelated to the concept being measured, namely combat skill.

Sensitivity

A criterion measure should be measured in units that are commensurate with the anticipated differences one expects to find among subjects. To continue with our example of the knights, if the distance off target were measured to the nearest yard, it is possible that few, if any, differences between the knights' performance would have been found. The scale (to the nearest yard) would have been too gross to detect the subtle differences in skill between the archers.

Another example would be using a 3-point rating scale (uncomfortable-neutral-comfortable) to measure various chairs, all of which are basically comfortable. People may be able to discriminate between comfortable chairs; that is, some are more comfortable than others. With the 3-point scale given, however, all chairs would be rated the same—comfortable. If a 7-point scale were used with various levels of comfort (a more sensitive scale), we would probably see differences between the ratings of the various chairs. An overly sensitive scale, however, can sometimes decrease reliability.

HUMAN RELIABILITY

Human reliability is inextricably linked to human error, a topic we discuss further in Chapter 20. As Meister (1984) points out, human reliability has been used to refer to a methodology, a theoretical concept, and a measure. As a methodology, human reliability is a procedure for conducting a quantitative analysis to predict the likelihood of human error. As a theoretical concept, human reliability implies an explanation of how errors are produced. As a measure, human reliability is simply the probability of successful performance of a task or an element of a task by a human. As such, it is the same as systems reliability, discussed in Chapter 1. Human reliability, the measure, is expressed as a probability. For example, if the reliability of reading a particular display is .992, then out of 1000 readings we would expect 992 to be correct and 8 to be in error. It is worth noting that human reliability is not the same as the reliability we discussed as a requirement for criterion measures.

Historical Perspective

Interest in human reliability began in the 1950s when there was a desire to quantify the human element in a system in the same way that reliability engineers were quantifying the hardware element. Back then, human factors activities, especially in the aerospace industry, were often part of the reliability, or quality assurance, divisions of the companies. Thus it is understandable that human factors folks would want to impress their hosts and contribute information to already existing analytical methods used by their cohorts.

In 1962, the first prototype human reliability data bank (now we call them data bases) was developed and called, appropriately, the *Data Store*. Work has continued in the area on a relatively small scale by a handful of people. The Three Mile Island nuclear power plant incident renewed interest in human reliability, and considerable work was done to develop data and methodologies for assessing human reliability in the nuclear industry.

Recently, a few good overviews of the field of human reliability have appeared (Meister, 1984; Dhillon, 1986; Miller and Swain, 1987) which can be referred to for a more detailed discussion of the topic.

Data Bases and Methodologies

Over the years, there have been several attempts to develop data bases of human reliability information. Topmiller, Eckel, and Kozinsky (1982) reviewed nine such attempts. Although these data bases represent formidable efforts, their usefulness is rather limited. For the most part, the data bases are small and the tasks for which data exist tend to be manual in nature, such as operation of a control or repair of a component. But as Rasmussen (1985) points out, modern technology has removed many repetitive manual tasks and has given humans more supervisory and trouble-shooting tasks to perform. Human errors are now more related to decision making and problem solving than in the past. The human reliability data bases currently available simply do not deal adequately with such situations.

In addition to human reliability data bases, there are also methodologies for determining human reliability. Probably the most well-developed methodology is the *technique for human error rate prediction (THERP)*. THERP has been developed in recent years mainly to assist in determining human reliability in nuclear power plants (Swain and Guttmann, 1983). THERP is a detailed procedure for analyzing a complex operation and breaking it down into smaller tasks and steps. Diagrams are used to depict the relationships between tasks and steps leading to successful and unsuccessful performance. Probabilities of successful performance, derived from empirical data and expert judgment, are assigned to the tasks and steps and are combined using rules of logic to determine the human reliability for performing the complex operation being analyzed.

A quite different approach for determining human reliability is used by stochastic simulation models (Siegel and Wolf, 1969). In essence, a computer is used to simulate the performance of a task or job. The computer performs the job over and over, sampling from probabilistic (i.e., stochastic) distributions of task element success and failure, and computes the number of times the entire task was successfully completed. Of course, the models and procedures are far more complex than depicted here. The advantage of using computer simulations is that one can vary aspects of the task and assess the effects on reliability. For example, Siegel, Leahy, and Wiesen (1977) in a study of a sonar system found that reliability for the simulated operators varied from .55 to .94 as time permitted the operators to do the task rose from 23 to 25 min. Here is an example in which a couple of minutes one way or the other could have profound effects on system performance. The results obtained from using such models, of course, are only as good as the models themselves and the data upon which the models are based.

Criticisms of Human Reliability

Despite all the effort at developing human reliability data banks and assessment methodologies, the majority of human factors specialists appear quite uninterested in the subject. Meister (1985) believes some of the problem is that the association of reliability with engineering is distasteful to some behavioral scientists and that some people feel behavior is so variable from day to day that it is ludicrous to assign a single point estimate of reliability to it.

Adams (1982) raises several practical problems with the concept of human reliability. Among them is that not all errors result in failure, and presently this problem is not handled well by the various techniques. Regulinski (1971) believes that point estimates of human reliability are inappropriate for continuous tasks, such as monitoring or tracking, and that tasks of a discrete nature may not be amenable to classical engineering reliability modeling. Williams (1985) points out that there has been little or no attempt made by human reliability experts to validate their various methodologies. Meister (1985), however, contends that although our estimates may not be totally accurate, they are good approximations and hence have utility.

Some scientists object to the illusion of precision implicit in reliability estimates down to the fourth decimal (for example, .9993) when the data upon which they were based come nowhere near such levels of precision. There is also concern about the subjective components that are part of all human reliability techniques. Most of the reliability data used were originally developed by expert judgment—subjectively. In addition, the analyst must interject personal judgment into the reliability estimate to correct it for a particular application. And, of course, the biggest criticism is that the data are too scanty and do not cover all the particular applications we would like to see covered.

Meister (1985), however, believes we cannot hibernate until some hypothetical time when there will be enough data—and as any human factors person will tell you, there are *never* enough data.

DISCUSSION

In this chapter we stressed the empirical data base of human factors. Research of all types plays a central role in human factors, and without it human factors would hardly be considered a science. A major continuing concern is the measurement of criteria to evaluate systems and the effects of independent variables. We outlined the major types of criteria used in human factors, with special reference to human reliability, and some of the requirements we would like to see in our measurement procedures. For some people, research methodology and statistics are not very exciting subjects. If these are not your favorite topics, think of them as a fine wine; reading about it cannot compare to trying the real thing.

REFERENCES

Adams, J. (1982). Issues in human reliability. *Human Factors*, 24, 1–10.

Bobo, M., Bethea, N., Ayoub, M., and Intaranont, K. (1983). Energy expenditure and aerobic fitness of male low seam coal miners. *Human Factors*, 25, 43–48.

Box, G., Hunter, W., and Hunter, J. (1978). *Statistics for experimenters: An introduction to design, data analysis, and model building*. New York: Wiley.

Cohen, J. (1977). *Statistical power analysis for the behavioral sciences*. New York: Academic.

Dhillon, B. (1986). *Human reliability with human factors*. New York: Pergamon Press.

Garg, A., Bakken, G., and Saxena, U. (1982). Effect of seat belts and shoulder harnesses on functional arm reach. *Human Factors*, 24, 367–372.

Hays, W. (1988). *Statistics* (4th ed.). New York: Holt, Rinehart & Winston.

Hull, J., Gill, R., and Roscoe, S. (1982). Locus of the stimulus to visual accommodation: Where in the world, or where in the eye? *Human Factors*, 24, 311–320.

Keppel, G. (1982). *Design and analysis: A researcher's handbook* (2d ed.). Englewood Cliffs, NJ: Prentice-Hall.

McFarling, L., and Ellingstad, V. (1977). Research design and analysis: Striving for relevance and efficiency. *Human Factors*, 19, 211–219.

Marras, W., and Kroemer, K. (1980). A method to evaluate human factors/ergonomics design variables of distress signals. *Human Factors*, 22, 389–400.

Meister, D. (1984). Human reliability. In F. Muckler (ed.) *Human Factors Review*. Santa Monica, CA: Human Factors Society.

Meister, D. (1985). *Behavioral analysis and measurement methods*. New York: Wiley.

Meister, D., and Rabideau, G. (1965). *Human factors evaluation in system development*. New York: Wiley.

Miller, D., and Swain, A. (1987). Human error and human reliability. In G. Salvendy (ed.) *Handbook of human factors*. New York: Wiley.

Neal, A., and Darnell, M. (1984). Text-editing performance with partial-line, partial-page, and full-page displays. *Human Factors*, 26, 431–442.

Nuclear Regulatory Commission (1981, September). *Guidelines for Control Room Design Process* (NUREG-0700). Washington, DC: Nuclear Regulatory Commission.

Rasmussen, J. (1985). Trends in human reliability analysis. *Ergonomics*, 28, 1185–1195.

Regulinski, T. (1971, February). Quantification of human performance reliability research method rationale. In J. Jenkins (ed.), *Proceedings of U.S. Navy Human Reliability Workshop 22–23 July 1970* (Rep. NAVSHIPS 0967-412-4010). Washington, DC: Naval Ship Systems Command.

Roebuck, J., Kroemer, K., and Thomson, W. (1975). *Engineering anthropometry methods*. New York: Wiley.

Rosenthal, R., and Rosnow, R. (1984). *Essentials of behavioral research: Methods and data analysis*. New York: McGraw-Hill.

Sanders, M. (1981). Peak and sustained isometric forces applied to a truck steering wheel. *Human Factors*, 23, 655–660.

Sanders, M., and Krohn, G. (1983, January). *Validation and extension of visibility requirements analysis for underground mining equipment*. Pittsburgh: Bureau of Mines.

Siegel, A., Leahy, W., and Wiesen, J. (1977, March). *Applications of human performance reliability—Evaluation concepts and demonstration guidelines*. Wayne, PA: Applied Psychological Sciences.

Siegel, A., and Wolf, J. (1969). *Man-machine simulation models: Performance and psychological interactions*. New York: Wiley.

Siegel, S., and Castellan, N. (1988). *Nonparametric statistics for the behavioral sciences* (2d ed.). New York: McGraw-Hill.

Simon, C. (1976, August). *Analysis of human factors engineering experiments: Characteristics, results, and applications*. Westlake Village, CA: Canyon Research Group.

Swain, A., and Guttmann, H. (1983). *Handbook of human reliability analysis with emphasis on nuclear power plant applications: Final report* (NUREG/CR-1278). Washington, DC: Nuclear Regulatory Commission.

Tabachnick, B., and Fidell, L. (1989). *Using multivariate statistics* (2d ed.). New York: Harper & Row.

Topmiller, D., Eckel, J., and Kozinsky, E. (1982). *Human reliability data bank for nuclear power plant operations:* vol. 1, *A review of existing human reliability data banks* (NUREG/CR-2744/1 of 2). Washington, DC: Nuclear Regulatory Commission.

Vreuls, D., Obermayer, R., Goldstein, I., and Lauber, J. (1973, July). *Measurement of trainee performance in a captive rotary-wing device* (Rep. NAVTRAEQUIPCEN 71-C-0194-1). Orlando, FL: Naval Training Equipment Center.

Williams, J. (1985). Validation of human reliability assessment techniques. *Reliability Engineering*, 11, 149–162.

Wilson, J., and Corlett, E. N. (1990). *Evaluation of human work*. London: Taylor & Francis.

PART TWO

INFORMATION INPUT

3 Information Input and Processing
4 Text, Graphics, Symbols, and Codes
5 Visual Displays of Dynamic Information
6 Auditory, Tactual, and Olfactory Displays
7 Speech Communications

CHAPTER **3**

INFORMATION INPUT AND PROCESSING

Today we hear a lot about information; in fact, we are reported to be in the throes of an information revolution. In industrialized countries the reception, processing, and storage of information are big business. Despite all the hoopla and advances in technology, still the most powerful, efficient, and flexible information processing and storage device on earth is located right between our ears—the human brain. On our jobs, in our homes, while driving in a car or just sitting on the beach, we are bombarded by all sorts of stimuli from our environment and from within our own bodies. These stimuli are received by our various sense organs (such as the eyes, ears, sensory receptors of the skin, and the proprioceptors in the muscles) and are processed by the brain.

The study of how humans process information about their environment and themselves has been an active area of research in psychology for over 100 years. Since about 1960, however, there has been an accelerated interest in the area, perhaps spurred by the information revolution and the associated advances in computer technology. At least one author (Anderson, 1980) gives some of the credit to the field of human factors and its early emphasis on practical problems of perception and attention. Various terms have been used to describe this growing area—*cognitive psychology, cognitive engineering*, and *engineering psychology*, to name a few. In this chapter we discuss the notion of information, including how it can be measured, coded, and displayed. A model of human information processing is presented and discussed along with the concept and methods of measuring mental workload. Finally, we close the chapter by looking at a few human factors implications of the information revolution itself.

INFORMATION THEORY

Our common notion of information is reflected in such everyday examples as what we read in the newspapers or hear on TV, the gossip over the backyard fence, or the instructions packed with every gadget we buy. This conception of information is fine for our everyday dealing; but when we investigate how people process information, it is important to have an operational definition of the term *information* and a method for measuring it. Information theory provides one such definition and a quantitative method for measuring it.

Historical Perspective

Before we discuss some of the rudiments of information within the context of information theory, it will be helpful to give a little historical perspective on the subject. During the late 1940s and early 1950s, information theory was developed within the context of communication engineering, not cognitive psychology. It quickly became, however, a source of ideas and experimentation within psychology.

Early investigators held out great promise that the theory would help unravel the mysteries of human information processing. Information theory was used to determine the information processing capacity of the different human sensory channels and was also employed extensively in problems of choice reaction time. Unfortunately, information theory has not lived up to its early promise in the field of human information processing, and its use has waned over the last 10 to 15 years until now its impact is quite negligible (Baird, 1984). It is interesting, however, that in newer areas of cognitive psychology, where detailed models and theories have yet to be formulated, information theory continues to be applied. The problems with information theory are (1) that most of its concepts are descriptive rather than explanatory and (2) that it offers only the most rudimentary clues about the underlying psychological mechanisms of information processing (Baird, 1984). Recognizing these limitations, however, information theory still provides a valuable definition of the concept of information and a methodology for measuring it. The concept and unit of measurement are often used to describe the stimuli and responses making up an information processing task.

Concept of Information

Information theory defines *information* as the reduction of uncertainty. The occurrences of highly certain events do not convey much information since they only confirm what was expected. The occurrences of highly unlikely events, however, convey more information. When the temperature warning light comes on in a car, for example, it conveys considerable information because it is an unlikely event. The "fasten seat belt" warning that comes on when the car is started conveys less information because it is expected. Note that the importance of the message is not directly considered in the definition of information; only the likelihood of its occurrence is considered. The "fasten seat belt" message is important; but because it comes on every time the car is started, it contains little (or no) information in the context of information theory.

Unit of Measure of Information

Information theory measures information in bits (symbolized by H). A *bit* is the amount of information required to decide between two equally likely alternatives. When the probabilities of the various alternatives are equal, the amount of information H in bits is equal to the logarithm, to the base 2, of the number N of such alternatives, or

$$H = \log_2 N$$

With only two alternatives, the information, in bits, is equal to 1.0 (because $\log_2 2 = 1$). The signal from the Old North Church to Paul Revere indicating whether the enemy was coming by land or by sea (two alternatives), for example, would convey 1.0 bit of information, assuming the two alternatives were equally probable. Four equally likely alternatives would convey 2.0 bits of information ($\log_2 4 = 2$). A randomly chosen digit from the set of numbers 0 to 9 would convey 3.322 bits ($\log_2 10 = 3.322$), and a randomly chosen letter from the set of letters A to Z would convey 4.7 bits ($\log_2 26 = 4.7$).

When the alternatives are not equally likely, the information conveyed by an event is determined by the following formula:

$$h_i = \log_2 \frac{1}{p_i}$$

where h_i is the information (in bits) associated with event i, and p_i is the probability of occurrence of that event. We are often more interested, however, in the average information conveyed by a series of events having different probabilities than in the information of just one such event. The average information H_{av} is computed as follows:

$$H_{av} = \sum_{i=1}^{N} p_i \left(log_2 \frac{1}{p_{i=1}} \right)$$

Let us go back to Paul Revere and the Old North Church. If the probability of a land attack were .9 and the probability of a sea attack were .1, then the average information would be .47 bit. The maximum possible information is always obtained when the alternatives are equally probable. In the case of Paul Revere, the maximum possible information in the Old North Church signal was 1.0 bit. The greater the departure from equal probability, the greater the reduction in information from the maximum. This leads to the concept of *redundancy,* which is the reduction in information from the maximum owing to unequal probabilities of occurrence. Percentage of redundancy is usually computed from the following formula:

$$\% \text{ Redundancy} = \left(1 - \frac{H_{av}}{H_{max}}\right) \times 100$$

For example, because not all letters of the English language occur equally often and certain letter combinations (e.g., th and qu) occur more frequently than others, the English language has a degree of redundancy of approximately 68 percent.

One other concept that may be encountered in the literature is bandwidth of a communication channel. The *bandwidth* is simply the rate of information transmission over the channel, measured in bits per second. For example, the bandwidth for the ear is estimated at from 8000 to 10,000 bits per second and for the eye (visual nerve fibers) 1000 bits per second. These are much higher than the amount of information that could possibly be absorbed and interpreted by the brain in that time. That is why much of what is received by our peripheral senses is filtered out before it even reaches our brain.

An Application of Information Theory

In choice reaction time experiments, the subject must make a discrete and separate response to different stimuli; for example, a person might be required to push one of four buttons depending on which of four lights comes on. Hick (1952) varied the number of stimuli in a choice reaction time task and found that reaction time to a stimulus increased as the number of equally likely alternatives increased. When he converted number of stimuli to bits and plotted the mean reaction time as a function of the information (bits) in the stimulus condition, the function was linear (that is, it was a straight line). (Scientists get very excited about linear relationships.) Reaction time appeared to be a linear function of stimulus information, but could it be just an artifact of the number of alternatives? That is, perhaps reaction time is just a logarithmic function of number of alternatives. Hyman (1953), however, realized that, according to information theory, he could alter stimulus information without changing the number of alternatives simply by changing the probability of occurrence of the alternatives. Lo and behold, using this technique, he found that reaction time was still a linear function of stimulus information when it was measured in bits. Thus, information theory and the definition of the bit made it possible to discover a basic principle: choice reaction time is a linear function of stimulus information. This is called the *Hick-Hyman law*.

Bishu and Drury (1988) used the amount of information, measured in bits, contained in a wiring task to predict task completion time. Amount of information was a function of both the number of wires to choose from and the number of terminals to be wired. They found that task completion time was linearly related to the amount of information contained in the task.

DISPLAYING INFORMATION

Human information input and processing operations depend, of course, on the sensory reception of relevant external stimuli. It is these stimuli that contain the information we process. Typically, the original source is some object, event, or environmental condition. Information from these original sources may come to us *directly* (such as by *direct* observation of an airplane), or it may come to us *indirectly* through some intervening mechanism or device (such as radar or a telescope). In the case of *indirect* sensing, the *new* stimuli may be of two types. First, they may be *coded* stimuli, such as visual or auditory displays. Second, they may be *reproduced* stimuli, such as those presented by TV, radio, or photographs or through such devices as microscopes, binoculars, and hearing aids. In such cases the reproduction may be intentionally or unintentionally modified in some way, as by enlargement, amplification, filtering, or enhancement.

The human factors aspect of design enters into this process in those circumstances in which *indirect* sensing applies, for it is in these circumstances that the designer can design displays for presenting information to people. *Display* is a term that applies to virtually any indirect method of presenting information, such as a highway traffic sign, a family radio, or a page of braille print. We discuss in later chapters the human factors considerations in the design of visual displays (Chapters 4 and 5); auditory, tactual, and olfactory displays (Chapter 6); and speech displays (Chapter 7). It is appropriate in our present discussion of information input, however, to take an overview of the various types of information that can be presented with displays.

Types of Information Presented by Displays

In a general sense, information presented by displays can be considered *dynamic* or *static*. We often speak of the display as being dynamic or static, although it is really the information that has that quality. Dynamic information continually changes or is subject to change through time. Examples include traffic lights that change from red to green, speedometers, radar displays, and temperature gauges. In turn, static information remains fixed over time (or at least for a time). Examples include printed, written, and other forms of alphanumeric data; traffic signs; charts; graphs; and labels. With the advent of computer displays or visual display terminals (VDTs), the distinction between static and dynamic information is becoming blurred. Thus, much of the "static" information previously presented only in "fixed" format (as on the printed page) is now being presented on VDTs. For our purposes, we still consider such presentation as static information (covered in Chapter 4). Note, however, that even if such material is shown on a VDT, most of it, *as such,* still is static. That is, the specific information does not itself change but rather can be replaced by other information. This is in contrast with most dynamic information (such as speed), which *itself* is changeable.

Although information can be grossly categorized as static or dynamic, a more detailed classification is given below:

- *Quantitative information:* Display presentations that reflect the quantitative value of some variable, such as temperature or speed. Although in most instances the variable is dynamic, some such information may be static (such as that presented in nomographs and tables).
- *Qualitative information:* Display presentations that reflect the approximate value, trend, rate of change, direction of change, or other aspect of some changeable variable. Such information usually is predicated on some quantitative parameter, but the displayed presentation is used more as an indication of the change in the parameter than for obtaining a quantitative value as such.
- *Status information:* Display presentations that reflect the condition or status of a system, such as on-off indications; indications of one of a limited number of conditions, such as stop-caution-go lights; and indications of independent conditions of some class, such as a TV channel. (What is called *status information* here is sometimes referred to in other sources as *qualitative information*. But the terminology used in this text is considered more descriptive.)
- *Warning and signal information:* Display presentation used to indicate emergency or unsafe conditions, or to indicate the presence or absence of some object or condition (such as aircraft or lighthouse beacons). Displayed information of this type can be static or dynamic.
- *Representational information:* Pictorial or graphic representations of objects, areas, or other configurations. Certain displays may present dynamic images (such as TV or movies) or symbolic representations (such as heartbeats shown on an oscilloscope or blips on a cathode-ray tube). Others may present static information (such as photographs, maps, charts, diagrams, and blueprints) and graphic representation (such as bar graphs and line graphs).
- *Identification information:* Display presentations used to identify some (usually) static condition, situation, or object, such as the identification of hazards, traffic lanes, and color-coded pipes. The identification usually is in coded form.
- *Alphanumeric and symbolic information:* Display presentations of verbal, numeric, and related coded information in many forms, such as signs, labels, placards, instructions, music notes, printed and typed material including braille, and computer printouts. Such information usually is static, but in certain circumstances it may be dynamic, as in news bulletins displayed by moving lights on a building.
- *Time-phased information:* Display presentations of pulsed or time-phased

signals, e.g., signals that are controlled in terms of duration of the signals and of intersignal intervals, and of their combinations, such as the Morse code and blinker lights.

Note that some displays present more than one type of information, and users "use" the displayed information for more than one purpose. The kinds of displays that would be preferable for presenting certain types of information are virtually specified by the nature of the information in question. For presenting most types of information, however, there are options available regarding the kinds of displays and certainly about the specific features of the display.

Selection of Display Modality

In selecting or designing displays for transmission of information in some situations, the selection of the sensory modality (and even the stimulus dimension) is virtually a foregone conclusion, as in the use of vision for road signs and the use of audition for many various purposes. Where there is some option, however, the intrinsic advantages of one over the other may depend on a number of considerations. For example, audition tends to have an advantage over vision in vigilance types of tasks, because of its attention-getting qualities. A more extensive comparison of audition and vision is given in Table 3-1, indicating the kinds of circumstances in which each of these modalities tends to be more useful. These comparisons are based on considerations of substantial amounts of research and experience relating to these two sensory modalities.

The tactual sense is not used very extensively as a means of transmission of information, but does have relevance in certain specific circumstances, such as with blind persons, and in other special circumstances in which the visual and auditory sensory modalities are overloaded. Further reference to such displays are made in Chapter 6.

TABLE 3-1
WHEN TO USE THE AUDITORY OR VISUAL FORM OF PRESENTATION

Use auditory presentation if:	Use visual presentation if:
1 The message is simple.	1 The message is complex.
2 The message is short.	2 The message is long.
3 The message will not be referred to later.	3 The message will be referred to later.
4 The message deals with events in time.	4 The message deals with location in space.
5 The message calls for immediate action.	5 The message does not call for immediate action.
6 The visual system of the person is overburdened.	6 The auditory system of the person is overburdened.
7 The receiving location is too bright or dark-adaptation integrity is necessary.	7 The receiving location is too noisy.
8 The person's job requires moving about continually.	8 The person's job allows him or her to remain in one position.

Source: Deatherage, 1972, p. 124, Table 4-1.

CODING OF INFORMATION

Many displays involve the coding of information (actually, of stimuli) as opposed to some form of direct representation or reproduction of the stimulus. Coding takes place when the original stimulus information is converted to a new form and displayed symbolically. Examples include radar screens that display aircraft as blips, maps that display population information about cities by using different-sized letters to spell the names of cities, and electronic resistors that use colored bands or stripes to indicate the resistance in ohms.

When information is coded, it is coded along various dimensions. For example, targets on a computer screen can be coded by varying size, brightness, color, shape, etc. An audio warning signal could be coded by varying frequency, intensity, or on-off pattern. Each of these variations constitutes a dimension of the displayed stimulus, or a *stimulus dimension*. The utility of any given stimulus dimension to convey information, however, depends on the ability of people (1) to identify a stimulus based on its position along the stimulus dimension (such as identifying a target as bright or dim, large or small) or (2) to distinguish between two or more stimuli which differ along a stimulus dimension (such as indicating which of two stimuli is brighter or larger).

Absolute versus Relative Judgments

The task of identifying a stimulus based on its position along a dimension and the task of distinguishing between two stimuli which differ along a dimension correspond to judgments made on an *absolute basis* and judgments made on a *relative basis*, respectively. A relative judgment is one which is made when there is an opportunity to compare two or more stimuli and to judge the relative position of one compared with the other along the dimension in question. One might compare two or more sounds and indicate which is the louder or compare two lights and indicate which is brighter. In some cases the judgment is merely whether the two stimuli are the same or different, as, for example, in judging the shape of two symbols.

In absolute judgments there is no opportunity to make comparisons; in essence, the comparison is held in memory. Examples of absolute judgments include identifying a given note on the piano (say, middle C) without being able to compare it with any others, grading the quality of wool into one of several quality categories without examples from each category to compare against, or identifying a symbol as belonging to one class of objects rather than another without other objects to compare it with.

Making Absolute Judgments along Single Dimensions

As one might expect, people are generally able to make fewer discriminations on an absolute basis than on a relative basis. For example, Mowbray and Gebhard (1961) found that people could discriminate on a relative basis about 1800 tone pairs differing only in pitch. On an absolute basis, however, people can identify only about 5 tones of different pitch.

In this connection Miller (1956) referred to the "magical number 7 ± 2" as the typical range of single-dimension identifications people can make on an absolute basis. This range of 7 ± 2 (that is, 5 to 9) seems to apply to many different dimensions, as shown in Table 3-2. With certain types of stimulus dimensions, however, people can identify more levels of stimuli. For example, most people can identify as many as 15 or more different geometric forms (some people might argue that geometric forms are multidimensional and hence not a valid test of the 7 ± 2 principle).

The limit to the number of absolute judgments that can be made appears to be a limitation of human memory rather than a sensory limitation. People have a hard time remembering more than about five to nine different standards against which stimuli must be judged. Note that the number of different stimuli that can be identified can be increased with practice.

Making Absolute Judgments along Multiple Dimensions

Given that we are not very good at making absolute judgments of stimuli, how can we identify such a wide variety of stimuli in our everyday experience? The answer is that most stimuli differ on more than one dimension. Two sounds may differ in terms of both pitch and loudness; two symbols may differ in terms of shape, size, and color.

When multiple dimensions are used for coding, the dimensions can be orthogonal or redundant. When dimensions of a stimulus are *orthogonal,* the value on one dimension is independent of the value on the other dimension. That is, each dimension carries unique information, and all combinations of the two dimensions are equally likely. If shape (circle-square) and color (red-green) were orthogonal dimensions, then there would be red circles, green circles, red squares, and green squares and each would signify something different in the coding scheme.

When dimensions are *redundant,* knowing the value on one dimension helps predict the value on the other dimension. There are various degrees of redundancy between stimulus dimensions depending on how they are used. If shape and color were completely redundant dimensions, then, for example, all circles would be green and all squares would be red; in this scheme, knowing the value of one dimension completely determines the value of the other. If 80 percent of

TABLE 3-2
AVERAGE NUMBER OF ABSOLUTE DISCRIMINATIONS AND CORRESPONDING NUMBER OF BITS TRANSMITTED BY CERTAIN STIMULUS DIMENSIONS

Stimulus dimension	Average number of discriminations	Number of bits
Pure tones	5	2.3
Loudness	4–5	2–2.3
Size of viewed objects	5–7	2.3–2.8
Brightness	3–5	1.7–2.3

the squares were red and 80 percent of the circles were green, color and shape would be considered partially redundant; knowing the value of one dimension would help predict the value of the other, but the prediction would not always be correct.

Combining Orthogonal Dimensions When dimensions are combined orthogonally, the number of stimuli that can be identified on an absolute basis increases. The total number, however, usually is less than the product of the number of stimuli that could be identified on each dimension separately. For example, if people could discriminate 7 different sizes of circles and 5 levels of brightness, we would expect that when size and brightness were combined, people would be able to identify more than 7 but fewer than 35 (7 × 5) different size-brightness combinations. As more orthogonal dimensions are added, the increase in the number of stimuli that can be identified becomes smaller and smaller.

Combining Redundant Dimensions When dimensions are combined in a redundant fashion, more levels can be identified than if only a single dimension is used. For example, if size and brightness were combined in a completely redundant fashion, people would be able to identify more than the 7 stimuli they could identify based on size alone. Thus, combining dimensions in either a redundant or orthogonal manner increases the number of stimuli that can be identified on an absolute basis. Orthogonal coding, however, usually results in a larger number of identifiable stimuli than does redundant coding.

Characteristics of a Good Coding System

In the next few chapters we discuss various aspects and uses of visual, auditory, tactual, and olfactory coding systems. The following characteristics, however, transcend the specific type of codes used and apply, across the board, to all types of coding systems.

Detectability of Codes To begin, any stimulus used in coding information must be detectable; specifically, it has to be able to be detected by the human sensory mechanisms under the environmental conditions anticipated. Color-coded control knobs on underground mining equipment, for example, would likely not be detectable in the low levels of illumination typically found in underground mines.

To measure detectability, a *threshold* is determined for the coding dimension such that stimuli at that value elicit a positive detection some fixed percentage of the time, usually 50 percent. That is, how bright must a stimulus be to be detected 50 percent of the time? How large must it be, or how loud must it be? Using the psychophysical *method of adjustment,* the subject manipulates a stimulus dimension that can be varied on a continuous scale (such as brightness or loudness) until it is just noticeable. Using the *method of serial exploration* (also called the *method of limits*), the stimulus dimension is presented by the experimenter in steps from either values below the threshold to values above that point, or vice versa. With the *method of constant stimuli*, the experimenter presents several values of the stimulus dimension in a random or near-random order. With both the method of serial exploration and method of constant stimuli, the subject merely indicates for each presentation whether the stimulus was detected. The threshold is then calculated from the data collected.

Discriminability of Codes Further, every code symbol, even though detectable, must be discriminable from other code symbols. Here the number of levels of coding becomes important, as we saw above. If 20 different levels of size are used to code information, it is highly likely that people will confuse one for another and hence fail to discriminate among them on an absolute basis. In this connection, the degree of difference between adjacent stimuli along a stimulus dimension may also influence the ease of discrimination, but not as much as one might expect. Pollack (1952), for example, found that people could identify between 4 and 5 tones when all were close together in terms of frequency (pitch). When the tones were spread out (20 times the original range), the number of stimuli that could be identified only increased to 8. Mudd (1961) reported evidence that in the case of auditory intensity, the greater the interstimulus distance, the faster the subjects could identify the signals. No such effect was found for frequency differences, however.

Discriminability is measured in much the same way as detectability, except the subject is presented an unchanging standard stimulus and then must judge whether the variable comparison stimulus was the same as or different from the standard. The methods of adjustment, serial exploration, and constant stimuli are used to determine the size of the difference that is just noticeably different from the standard some fixed percentage of the time, usually 50 percent. That difference is called the *just-noticeable-difference (JND)* and is sometimes referred to as the *difference threshold*.

Meaningfulness of Codes A coding system should use codes meaningful to the user. Meaning can be inherent in the code, such as a bent arrow on a traffic sign to denote a curved road ahead, or meaning can be learned, such as using red to denote danger. The notion of meaningfulness is bound up in the idea of conceptual compatibility, which we discuss in the next section.

Standardization of Codes When a coding system is to be used by different people in different situations, it is important that the codes be standardized and kept the same from situation to situation. If a new display is to be added to a factory already containing other displays, the coding system used should duplicate the existing coding schemes where the same information is being transmitted. Red, for example, should mean the same thing on all displays.

Use of Multidimensional Codes The use of multidimensional codes can increase the number and discriminability of coding stimuli used. Depending on the specific demands of the task and the number of different codes needed, redundant or orthogonal coding should be considered.

COMPATIBILITY

One cannot discuss human information processing without reference to the concept of compatibility. *Compatibility* refers to the relationship of stimuli and responses to human expectations, and as such is a central concept in human factors. A major goal in any design is to make the system compatible with human expectations. The term *compatibility* was originally used by A. M. Small and was later refined by Fitts and Seeger (1953). The concept of compatibility implies a process of information transformation, or recoding. It is postulated that the greater the degree of compatibility, the less recoding must be done to process the information. This in turn should result in faster learning, faster response times, fewer errors, and a reduced mental workload. In addition, people like things that work as they expect them to work.

Types of Compatibility

Traditionally, three types of compatibility have been identified: *conceptual, movement,* and *spatial*. A fourth type of compatibility has emerged from studies of human information processing—*modality compatibility*. We discuss each in turn.

Conceptual Compatibility Conceptual compatibility deals with the degree to which codes and symbols correspond to the conceptual associations people have. In essence, conceptual compatibility relates to how meaningful the codes and symbols are to the people who must use them. An aircraft symbol used to denote an airport on a map would have greater conceptual compatibility than would a green square. Another example of conceptual compatibility is the work being done on how to create meaningful abbreviations for use as command names in computer applications (Ehrenreich, 1985; Grudin and Barnard, 1985; Hirsh-Pasek, Nudelman and Schneider, 1982; Streeter, Ackroff, and Taylor, 1983). Some methods of generating abbreviations are easier to recall and use than others, but at present there is no conclusive evidence to recommend one technique over another when the goal is improved decoding performance (Ehrenreich, 1985).

Movement Compatibility Movement compatibility relates to the relationship between the movement of displays and controls and the response of the system being displayed or controlled. An example of movement compatibility is where clockwise rotation of a knob is associated with an increase in the system parameter being controlled. That is, to increase the volume on a radio, we expect to turn the knob clockwise. Another example is a vertical scale in which upward movement of the pointer corresponds to an increase in the parameter being displayed. We discuss movement compatibility in more detail in Chapter 10.

Spatial Compatibility Spatial compatibility refers to the physical arrangement in space of controls and their associated displays. For example, imagine five displays lined up in a horizontal row. Good spatial compatibility would be achieved by arranging the control knobs associated with those displays in a horizontal row directly below the corresponding displays. Poor spatial compatibility would result if the controls were arranged in a vertical column to one side of the displays. Spatial compatibility is also discussed further in Chapter 10.

Modality Compatibility Modality compatibility refers to the fact that certain stimulus-response modality combinations are more compatible with some tasks than with others (Wickens, 1984). This is illustrated by a portion of a study carried out by Wickens, Sandry, and Vidulich (1983). The experiment was rather involved, but we need to discuss only a portion of it here. Subjects performed one of two tasks, a verbal task (respond to commands such as "turn on radar-beacon tacan") or a spatial task (bring a cursor over a designated target). Each task was presented through either the auditory modality (speech) or the visual modality (displayed on a screen). The subject either spoke the response or manually performed the command (or used a controller to position the cursor). All combinations of presentation modality and response modality were used for the two types of tasks. The results are shown in Figure 3-1, which shows reaction times for the two tasks using the four presentation-response modality combinations. Note that in Figure 3-1 fast performance is shown at the top of the vertical axis. As can be seen, the most compatible combination

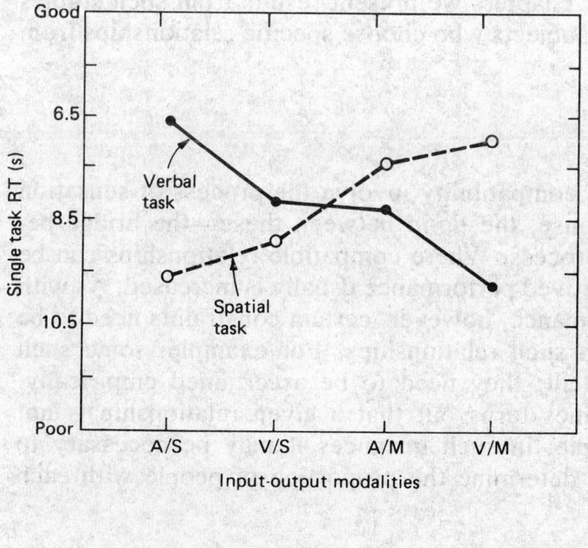

FIGURE 3-1
An example of modality compatibility. Input modality: A = auditory (speech) and V = visual (displayed on screen). Output modality: S = spoken response and M = manual response. For verbal tasks, the best input-output combination is A/S. For spatial tasks, the best combination is V/M. (*Source: Wickens, Sandry, and Vidulich, 1983, Fig. 6. Reprinted with permission of the Human Factors Society, Inc. All rights reserved.*)

for the verbal task is auditory presentation and spoken response. For the spatial task, the most compatible combination is visual presentation and manual response.

Origins of Compatibility Relationships

With the exception of modality compatibility, which probably has its origins in the way our brains are wired, compatibility relationships stem from two possible origins. First, certain compatible relationships are intrinsic in the situation, for example, turning a steering wheel to the right in order to turn to the right. In certain combinations of displays and controls, for example, the degree of compatibility is associated with the extent to which they are isomorphic or have similar spatial relationships. Second, other compatible relationships are culturally acquired, stemming from habits or associations characteristic of the culture in question. For example, in the United States a light switch is usually pushed up to turn it on, but in certain other countries it is pushed down. How such culturally acquired patterns develop is perhaps the consequence of fortuitous circumstances.

Identification of Compatibility Relationships

To take advantage of compatible relationships in designing equipment or other items, it is, of course, necessary to know what relationships are compatible. There generally are two ways in which these can be ascertained or inferred. First, certain relationships are obvious, or manifest; this is particularly true in the case of spatial compatibility. A word of caution is in order here. What is obvious to you may not be obvious to others, so it would be a good idea to check out the generality of the assumed relationship. Second, when compatible relationships are not obvious, it is necessary to identify them on the basis of empirical experiments. In later chapters we present results from such studies that indicate the proportion of subjects who choose specific relationships from several possibilities.

Discussion

Although different versions of compatibility involve the process of sensation and perception and also response, the tie-in between these—the bridge between them—is a mediation process. Where compatible relationships can be utilized, the probability of improved performance usually is increased. As with many aspects of human performance, however, certain constraints need to be considered in connection with such relationships. For example, some such relationships are not self-evident; they need to be ascertained empirically. When this is done, it sometimes turns out that a given relationship is not universally perceived by people; in such instances it may be necessary to "figure the odds," that is, to determine the proportion of people with each possible association or response tendency and make a design determination on this basis. In addition, in some circumstances, trade-off considerations may require that one forgo the use of a given compatible relationship for some other benefit.

A MODEL OF INFORMATION PROCESSING

A model is an abstract representation of a system or process. Models are not judged to be correct or incorrect; the criterion for evaluating a model is *utility*. Thus, a *good model* is one that can account for the behavior of the actual system or process and can be used to generate testable hypotheses that are ultimately supported by the behavior of the actual system or process.

Models can be mathematical, physical, structural, or verbal. Information theory (which we discussed previously) is a mathematical model of information transfer in communication systems. Signal detection theory (which we discuss later) is also a mathematical model. We present here a structural model of human information processing which will organize our discussion of the process.

Remember, in the human information processing literature, numerous models have been put forth to "explain" how people process information (e.g., Broadbent, 1958; Haber and Hershenson, 1980; Sanders, 1983; Welford, 1976). Most of these models consist of hypothetical black boxes or processing stages that are assumed to be involved in the system or process in question. Figure 3-2 presents one such model (Wickens, 1984). The model depicts the major components, or stages, of human information processing and the hypothesized relationships between them. Most information processing models would agree with this basic formulation; however, some might add a box or two or divide one box into several. As we will see, the notion of attention resources and how they are allocated is still very much up in the air.

In the next several sections we discuss *perception*, with special emphasis on detection and signal detection theory; *memory*, including short-term sensory storage, working memory, and long-term memory; *decision making*, with special emphasis on the biases found in human decision making; and *attention*, including some of the controversy in the area and the notion of limited capacity and time sharing multiple tasks. The discussion of *response execution* and *feedback* is deferred to Chapters 8, 9, and 10.

PERCEPTION

There are various levels of *perception* that depend on the stimulus and the task confronting a person. The most basic form of perception is simple *detection;* that is, determining whether a signal or target is present. This can be made more complicated by requiring the person to indicate in which of several different classes the target belongs. This gets into the realm of *identification* and *recognition*. The simplest form of multilevel classification is making absolute judgments along stimulus dimensions, which we discussed previously.

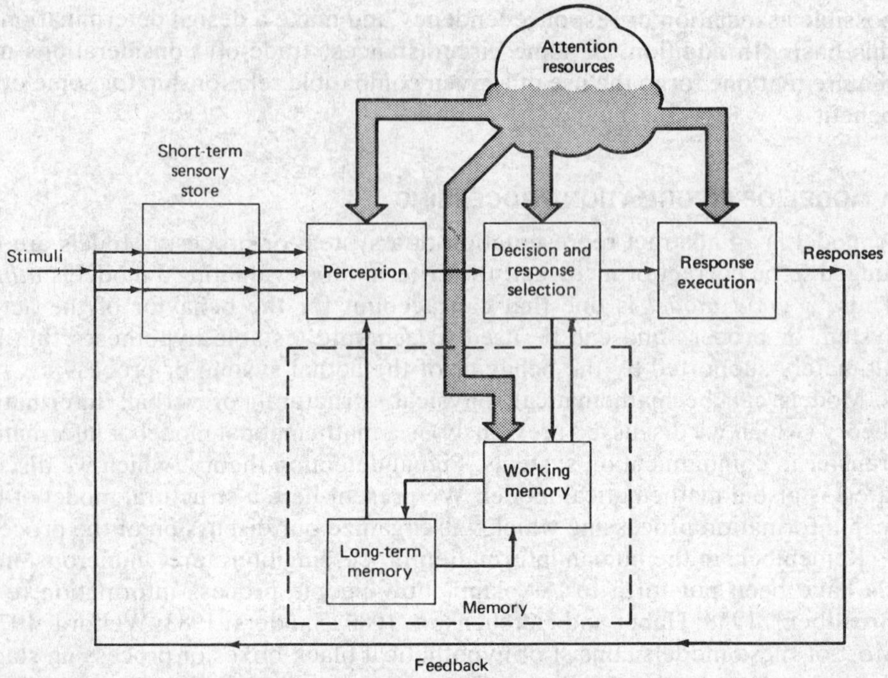

FIGURE 3-2
A model of human information processing showing the major processes, or stages, and the interrelationships. (*Source: Wickens, 1984, Fig. 1.1. Reprinted by permission of the publisher.*)

The act of perception involves our prior experiences and learned associations. This is depicted in Figure 3-2 by the line connecting the long-term memory box to the perception box. Even the act of simple detection involves some complex information processing and decision making. This complexity is embodied in signal detection theory (Swets, 1988; Green and Swets, 1988).

Signal Detection Theory

Signal detection theory (SDT), also referred to as the theory of signal detection (TSD), is applicable to situations in which there are two discrete states (signal and no signal) that cannot easily be discriminated (Wickens, 1984). Examples would include detecting a cavity on a tooth x-ray, detecting a blip on a radar screen, or detecting a warning buzzer in a noisy factory.

Concept of Noise A central concept in SDT is that in any situation there is *noise* that can interfere with the detection of a signal. This noise can be generated externally from the person (e.g., noises in a factory other than the warning buzzer, or electronic static and false radar returns on a radar screen) and internally within the person (e.g., miscellaneous neural activity). This noise varies over time, thus forming a distribution of intensity from low to high. The shape of this distribution is assumed to be normal (bell-shaped). When a "signal" occurs, its intensity is added to that of the background noise. At any given time a person needs to decide if the sensory input (what the person senses) consists of only noise or noise plus the signal.

Possible Outcomes Given that the person must say whether a signal occurred or did not occur, and given that there are only two possible states of reality (i.e., the signal did or did not occur), there are four possible outcomes:

- *Hit:* Saying there is a signal when there is a signal.
- *False alarm:* Saying there is a signal when there is no signal.
- *Miss:* Saying there is no signal when there is a signal.
- *Correct rejection:* Saying there is no signal when there is no signal.

Concept of Response Criterion One of the major contributions of SDT to the understanding of the detection process was the recognition that people set a criterion along the hypothetical continuum of sensory activity and that this criterion is the basis upon which a person makes a decision. The position of the criterion along the continuum determines the probabilities of the four outcomes listed above. The best way to conceptualize this is shown in Figure 3-3. Shown are the hypothetical distributions of sensory activity when only noise is present and when a signal is added to the noise. Notice that the two distributions overlap. That is, sometimes conditions are such that the level of noise alone will exceed the level of noise plus signal.

FIGURE 3-3
Illustration of the key concepts of signal detection theory. Shown are two hypothetical distributions of internal sensory activity, one generated by noise alone and the other generated by signal plus noise. The probabilities of four possible outcomes are depicted as the respective areas under the curves based on the setting of a criterion at X. Here *d′* is a measure of sensitivity, and beta is a measure of response bias. The letters *a* and *b* correspond to the height of the signal-plus-noise and noise-only distributions at the criterion.

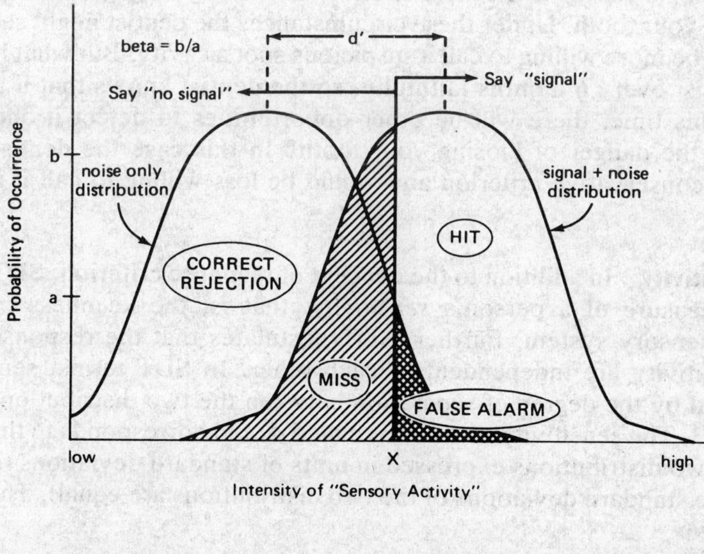

SDT postulates that a person sets a criterion level such that whenever the level of sensory activity exceeds that criterion, the person will say there is a signal present. When the activity level is below the criterion, the person will say there is no signal. Figure 3-3 also shows four areas corresponding to hits, false alarms, misses, and correct rejections based on the criterion shown in the figure.

Related to the position of the criterion is the quantity *beta*. Numerically, beta is the ratio (signal to noise) of the height of the two curves in Figure 3-3 at a given criterion point. If the criterion were placed where the two distributions cross, beta would equal 1.0. As the criterion is shifted to the right, beta increases and the person will say signal less often and hence will have fewer hits, but will also have fewer false alarms. We would say that such a person is being conservative. The criterion shown in Figure 3-3 is conservative and has a beta greater than 1.0. On the other hand, as the criterion is shifted to the left, beta decreases and the person will say signal more often and hence will have more hits, but also more false alarms. We would say such a person was being risky.

Influencing the Response Criterion SDT postulates two variables that influence the setting of the criterion: (1) the likelihood of observing a signal and (2) the costs and benefits associated with the four possible outcomes. Consider the first. If you told your dentist that one of your teeth was hurting, thus increasing the likelihood that you had a cavity, the dentist would be more likely to call a suspicious spot on the x-ray of that tooth a cavity. That is, as the probability of a signal increases, the criterion is lowered (reducing beta), so that anything remotely suggesting a signal (cavity in our example) might be called a signal.

With regard to the second factor, costs and benefits, again consider the example of the dentist. What is the cost of a false alarm (saying there is a cavity when there is not)? You get your tooth drilled needlessly. What is the cost of a miss (saying there is no cavity when there is one)? The cavity may get worse, and you might lose your tooth. Under these circumstances the dentist might set a low criterion and be more willing to call a suspicious spot a cavity. But what if you go to the dentist every 6 months faithfully, so the dentist knows that if a cavity is missed this time, there will be other opportunities to detect it and correct it without the danger of loosing your tooth? In this case the dentist would set a more conservative criterion and would be less willing to call it a cavity—this time.

Concept of Sensitivity In addition to the concept of response criterion, SDT also includes a measure of a person's *sensitivity,* that is, the keenness or resolution of the sensory system. Further, SDT postulates that the response criterion and sensitivity are independent of each other. In SDT terms, sensitivity is measured by the degree of separation between the two distributions shown in Figure 3-3. The sensitivity measure is called d' and corresponds to the separation of the two distributions expressed in units of standard deviations (it is assumed that the standard deviations of the two distributions are equal). The greater the separation, the greater the sensitivity and the greater the d'. In most applications of SDT, d' ranges between 0.5 and 2.0.

Some signal generation systems may create more noise than others, and some people may have more internal noise than others. The greater the amount of noise, the smaller d' will be. Also the weaker and less distinct the signal, the smaller d' will be. Another factor that influences d' is the ability of a person to remember the physical characteristics of a signal. When memory aids are supplied, sensitivity increases (Wickens, 1984). In the case of the dentist, better x-ray equipment and film, or x-rays of actual cavities to compare with the x-ray under consideration, would increase sensitivity.

Applications of SDT SDT has been applied to a wide range of practical situations, including sonar target detection (Colquhoun, 1967), industrial inspection tasks (Drury, 1975), medical diagnosis (Swets and Pickett, 1982), eyewitness testimony (Ellison and Buckhout, 1981), and air traffic control (Bisseret, 1981). Long and Waag (1981), however, caution that serious difficulties may result from the uncritical acceptance and application of SDT to some of these areas. SDT was developed from controlled laboratory experiments where subjects were given many, many trials with precisely controlled signals and background noise levels. Long and Waag believe that many applied situations do not match these conditions and that erroneous conclusions may be drawn if SDT is applied without due concern for the differences.

MEMORY

The human memory system has been conceptualized as three subsystems or processes: *sensory storage*, *working memory*, and *long-term memory*. Working memory is the gateway to long-term memory. Information in the sensory memory subsystem must pass through working memory in order to enter long-term memory. Human memory is vast, but imperfect. We possess incredible amounts of information in long-term storage, yet we often have trouble retrieving it when needed. Despite our vast storage capacity, we often forget a name or number just seconds after we have heard or read it. In this section we follow the lead of Wickens (1984) and stress the way information is coded in the three memory subsystems and some of the practical implications of these coding schemes. There are several models that attempt to explain various phenomena associated with memory. Some are models of only working memory (e.g., Baddeley, 1986), while others model working memory as part of a larger human cognitive model (e.g., Schneider and Detweiler, 1988). Unfortunately, none of the models can handle all the various phenomena associated with working memory (Card, 1989).

Sensory Storage

Each sensory channel appears to have a temporary storage mechanism that prolongs the stimulus representation for a short time after the stimulus has ceased. Although there is some evidence for tactual and olfactory sensory storage, the two sensory storage mechanisms we know most about are those associated with the visual system, called *iconic storage,* and the auditory system, called *echoic storage*.

If a visual stimulus is flashed very briefly on a screen, the iconic storage holds the image for a short time, allowing further processing of the image. The same sort of phenomenon occurs with auditory information in echoic storage. Iconic storage generally lasts less than about 1 s, while echoic storage can last a few seconds before the representation "fades" away. Information in the sensory stores is not coded but is held in its original sensory representation. Sensory storage does not require a person to attend to the information in order for it to be maintained there. In essence, sensory storage is relatively automatic, and there is little one can do to increase the length of the sensory representation. To retain information for a longer period, it must be encoded and transferred to working memory.

Working Memory*

To encode and transfer information from sensory storage to working memory and to hold information in working memory requires that the person direct attention to the process.

It is generally believed that information in working memory is coded with three types of codes: visual, phonetic, and semantic. Visual and phonetic codes are visual or auditory representations of stimuli. Each can be generated by stimuli of the opposite type or internally from long-term memory. For example, the visually presented word *DOG* is phonetically coded as the sound generated from reading the word. Further, if you hear the word *DOG,* you could generate a visual code (mental picture) of a dog. You can also form a visual image of an object from long-term memory without hearing or seeing it.

Semantic codes are abstract representations of the meaning of a stimulus rather than the sight or sound generated by the stimulus. Semantic codes are especially important in long-term memory. Although there is a natural progression of coding from the physical codes (visual and phonetic) to semantic codes, there is evidence that all three codes for a particular stimulus can exist, at some level of strength, at the same time in working memory.

One practical consequence of the various codes in working memory is the finding that letters read from a page are automatically converted to a phonetic code. When people are presented a list of letters and asked to recall them, errors of recall tend to reflect acoustic rather than visual confusion (Conrad, 1964). Thus the letter *E* is more likely to be recalled as *D* (which sounds like *E*) than as *F* (which looks like *E*). This has implications for selecting letters to code information on, say, a computer display. If a subset of letters is to be used, they should be phonetically dissimilar.

* Working memory is also called *short term memory,* but the term *working memory* is currently preferred.

Capacity of Working Memory The only way information can be maintained in working memory is by rehearsal (convert speech), that is, the direction of attention resources to the process. This can be illustrated by presenting a person with four letters (e.g., *J, T, N,* and *L*) and having that person count backward by 3s from 187. After 15 s of counting backward (which prevents rehearsal of the letters), the person probably will not be able to recall the letters. Even with rehearsal, information in working memory can decay over time. This decay occurs more rapidly with more items in working memory. This is probably due to the fact that rehearsing a long list of items increases the delay between successive rehearsals of any given item.

Given that rehearsal is delayed as the number of items in working memory is increased, what is the maximum number of items that can be held in working memory? The answer is the magical number 7 ± 2 (that is, 5 to 9) which we discussed in connection with absolute judgments along stimulus dimensions (Miller, 1956). The key to understanding the 7 ± 2 limit of working memory is to know how an item is defined. Miller (1956) in his formulation of the 7 ± 2 principle recognized that people can "chunk" information into familiar units, regardless of size, and these can be recalled as an entity. Thus the limit is approximately 7 ± 2 *chunks*. For example, consider the letters *C.A.T.D.O.G.R.A.T*. Rather than consider this string as composed of 9 items, Miller would argue that it contains 3 chunks, *CAT.DOG.RAT*, and hence is well within the 7 ± 2 limit. In a similar manner, words would be combined into meaningful sentences and chunked, thus effectively increasing the capacity of working memory.

The manner in which the information is presented can aid in chunking. Digits can be recalled better if they are grouped into chunks of three to four items (for example, 458 321 691 is easier to recall than 458321691). Also, the more meaningful the chunks, the easier they are to recall (*IB MJF KTV* is more difficult to recall than *IBM JFK TV*). The practical implications of all this are as follows: (1) Avoid presenting more than five to nine chunks of information for people to remember. (2) Present information in meaningful and distinct chunks. (3) Provide training on how to recall information by chunking. Schneiderman (1980) indicates that if a menu of options is presented on a computer screen and all alternatives must be compared simultaneously with one another, the number of alternatives should be kept within the capacity limits of working memory. If more than seven alternatives are displayed on the screen, a person may tend to forget the first alternative before the last ones are read.

Searching Working Memory The time to search a list of items in working memory, such as names or numbers, for a specific item increases linearly as the number of items in the memorized list increases. In addition to the time required to perceive and encode the to-be-searched-for item and the time required to respond, it takes about 38 ms per item in memory to search for the sought after item (Sternberg, 1966). Interestingly, this rate is the same whether the to-be-searched-for item is or is not one of the items actually in memory.

This implies that all items in memory are searched one at a time even if a match is found early in the search.

Long-Term Memory

Information in working memory is transferred to long-term memory by semantically coding it, that is, by supplying meaning to the information and relating it to information already stored in long-term memory. If you study for examinations by only repeatedly reading the textbook and course notes, you probably have difficulty recalling the information. One reason is that you have not taken the time to semantically encode the information. To recall more information, it must be analyzed, compared, and related to past knowledge. In addition, the more organized the information is initially, the easier it is to transfer to long-term memory; and the more organized the information is in long-term memory, the easier it is to retrieve it. Retrieval is often the weak link in utilizing information stored in long-term memory. The use of *mnemonics* to organize information makes retrieval easier. Mnemonics that can be used to learn a list of items include using the first letters of the items on the list to make a word or sentence and forming bizarre images connecting the items on the list.

DECISION MAKING

Decision making is really at the heart of information processing. For a good review see Slovic, Lichtenstein, and Fischhoff (1988). It is a complex process by which people evaluate alternatives and select a course of action. The process involves seeking information relevant to the decision at hand, estimating probabilities of various outcomes, and attaching values to the anticipated outcomes. Wickens (1984) distinguishes decision making from choice reaction time in that typical decision-making situations have a relatively long time frame in which to make a choice compared with the case of a choice reaction time.

People, unfortunately, are not optimal decision makers and often do not act "rationally," meaning that they do not act according to objective probabilities of gain and loss. A number of biases are inherent in the way people seek information, estimate probabilities, and attach values to outcomes that produce this irrational behavior. The following is a short list of some of these biases (Wickens, 1984):

1 People give an undue amount of weight to early evidence or information. Subsequent information is considered less important.

2 Humans are generally conservative and do not extract as much information from sources as they optimally should.

3 The subjective odds in favor of one alternative or the other are not assessed to be as extreme or given as much confidence as optimally they should.

4 As more information is gathered, people become more confident in their decisions, but not necessarily more accurate. For example, people engaged in troubleshooting a mechanical malfunction are often unjustly confident that they entertained all possible diagnostic hypotheses.

5 Humans have a tendency to seek far more information than they can absorb adequately.

6 People often treat all information as if it were equally reliable, even though it is not.

7 Humans appear to have a limited ability to entertain a maximum of more than a few (three or four) hypotheses at a time.

8 People tend to focus on only a few critical attributes at a time and consider only about two to four possible choices that are ranked highest on those few critical attributes.

9 People tend to seek information that confirms the chosen course of action and to avoid information or tests whose outcome could disconfirm the choice.

10 A potential loss is viewed as having greater consequence and therefore exerts a greater influence over decision-making behavior than does a gain of the same amount.

11 People believe that mildly positive outcomes are more likely than either mildly negative or highly positive outcomes.

12 People tend to believe that highly negative outcomes are less likely than mildly negative outcomes.

These biases explain, in part, why people do not always make the best decisions, given the information available. Maintenance workers troubleshooting a malfunctioning device, doctors diagnosing an illness, or you and I deciding which model of stereo or automobile to buy—all these are limited by the human capacity to process and evaluate information to arrive at an optimum decision. Understanding our capabilities and limitations can suggest methods for presenting and preprocessing information to improve the quality of our decisions. Computers can be used to aid the decision maker by aggregating the probabilities as new information is obtained; keeping track of several hypotheses, alternatives, or outcomes of sequential tests at one time; or selecting the best sources of information to test specific hypotheses. The area of computer-aided decision making will undoubtedly become more important in the future as new technologies combine with a better understanding of the capabilities and limitations of the human decision maker.

ATTENTION

We often talk about "paying attention." Children never seem to do it, and we occasionally do it to the wrong thing and miss something important. These sorts of everyday experiences suggest that attention can somehow be directed to objects or activities, and that things to which we are not paying attention are often not perceived (or at least recalled). Four types of situations or tasks involving the direction of attention are often encountered in work and at play. In the first, we are required to monitor several sources of information to determine whether a particular event has occurred. A pilot scanning the instruments, looking for a deviant reading, and a quarterback scanning the defensive line and backfield, looking for an opening, are examples of this type of task. Cognitive psychologists call this *selective attention.*

In the second type of task, a person must attend to one source of information and exclude other sources. This is called *focused attention,* and it is the sort of activity you engage in when you try to read a book while someone is talking on the telephone or you try to listen to a person talk in a crowded, noisy cocktail party.

In the third type of situation, called *divided attention,* two or more separate tasks must be performed simultaneously, and attention must be paid to both. Examples include driving a car while carrying on a conversation with a passenger and eating dinner while watching the evening news.

In the fourth type of task, a person sustains attention over prolonged periods of time, without rest, in order to detect infrequently occurring signals. This type of task is called *sustained attention, monitoring,* or *vigilance.* Examples include security guards viewing TV monitors for the infrequent intruder, air defense radar operators waiting for the missile we all hope will never appear, or an inspector on an assembly line looking for a defect in an endless line of products moving by. We discuss selective, focused, divided, and sustained attention in a little more detail and indicate some of the human capabilities and limitations associated with them. A few guidelines for designing each type of situation are also given.

Selective Attention

Selective attention requires the monitoring of several *channels* (or sources) of information to perform a single task. Consider a person monitoring a dial for a deviate reading (the signal). Let us say that 25 such deviate readings occur per minute. These 25 signals could all be presented on the one dial, or they could be presented on five dials, each presenting only 5 signals per minute. Which would be better in terms of minimizing missed signals? The answer is one dial presenting 25 signals per minute. As the number of channels of information (in this case, dials) increases, even when the overall signal rate remains constant, performance declines. Conrad (1951) referred to this as *load stress,* that is, the stress imposed by increasing the number of channels over which information is presented. Conrad also discussed *speed stress,* which is related to the rate of signal presentation. Goldstein and Dorfman (1978) found that load stress was more important than speed stress in degrading performance. Subjects, for example, could process 72 signals per minute from one dial better than 24 signals per minute presented over three dials (that is 8 signals per minute per dial).

When people have to sample multiple channels of information, they tend to sample channels in which the signals occur very frequently rather than those in which the signals occur infrequently. Because of limitations in human memory, people often forget to sample a source when many sources are present, and people tend to sample sources more often than would be necessary if they remembered the status of the source when it was last sampled (Moray, 1981). Under conditions of high stress, fewer sources are sampled and the sources that are sampled tend to be those perceived as being the most important and salient (Wickens, 1984).

Guidelines for Selective-Attention Tasks The following are a few guidelines gleaned from the literature that should improve selective-attention task performance:

1 Where multiple channels must be scanned for signals, use as few channels as possible, even if it means increasing the signal rate per channel.

2 Provide information to the person as to the relative importance of the various channels so that attention can be directed more effectively.

3 Reduce the overall level of stress on the person so that more channels will be sampled.

4 Provide the person with some preview information as to where signals will likely occur in the future.

5 Train the person to effectively scan information channels and develop optimal scan patterns.

6 If multiple visual channels are to be scanned, put them close together to reduce scanning requirements.

7 If multiple auditory channels are to be scanned, be sure they do not mask one another.

8 Where possible, stimuli that require individual responses should be separated temporally and presented at such a rate that they can be responded to individually. Extremely short intervals (say, less than 0.5 or 0.25 s) should be avoided. Where possible, the person should be permitted to control the rate of stimulus input.

Focused Attention

With focused attention, the problem is to maintain attention on one or a few channels of information and not to be distracted by other channels of information. One factor that influences the ability of people to focus attention is the proximity, in physical space, of the sources. For example, it is nearly impossible to focus attention on one visual source of information and completely ignore another if the two are within 1 degree of visual angle from each other

(Broadbent, 1982). Note that in a selective-attention task, close proximity would aid performance.

Anything that makes the information sources distinct from one another would also aid in focusing attention on one of them. For example, when more than one auditory message occurs at the same time, but only one must be attended to, performance can be improved if the messages are of different intensity (the louder one usually gets attended to), are spoken by different voices (e.g., male and female), or come from different directions.

Guidelines for Focused-Attention Tasks The following are some guidelines that should aid in the performance of focused-attention tasks:

1 Make the competing channels as distinct as possible from the channel to which the person is to attend.

2 Separate, in physical space, the competing channels from the channel of interest.

3 Reduce the number of competing channels.

4 Make the channel of interest larger, brighter, louder, or more centrally located than the competing channels.

Divided Attention

When people are required to do more than one task at the same time, performance on at least one of the tasks often declines. This type of situation is also referred to as *time-sharing*. It is generally agreed that humans have a limited capacity to process information, and when several tasks are performed at the same time, that capacity can be exceeded. Cognitive psychologists have put forth numerous models and theories to explain time-sharing or divided-attention performance [for a review see Lane (1982) or Wickens (1984)]. Most of the recent formulations are based on the notion of a limited pool of resources, or multiple pools of limited resources, that can be directed toward mental processes required to perform a task. We briefly discuss the two major classes of such theories: single-resource theories and multiple-resource theories.

Single-Resource Theories of Divided Attention Single-resource theories postulate one undifferentiated source of resources that are shared by all mental processes (e.g., Moray, 1967; Kahneman, 1973). These theories explain, very nicely, why performance in a time-sharing situation declines as the difficulty of one of the tasks increases. The increase in difficulty demands more resources from the limited supply, thus leaving fewer resources for performing the other task.

The problem with single-resource theories is they have difficulty explaining (1) why tasks that require the same memory codes or processing modalities interfere more than tasks not sharing common codes and modalities; (2) why with some combinations of tasks increasing the difficulty of one task has no effect on performance of the others; and (3) why some tasks can be time-shared perfectly (Wickens, 1984). To explain these phenomena, multiple-resource theories were developed.

Multiple-Resource Theories of Divided Attention Rather than postulating one central pool of resources, multiple-resource theories postulate several independent resource pools. The theories do not necessarily predict that tasks using separate resources will be perfectly time-shared, but only that time-sharing efficiency will be improved to the extent that the tasks use different resources. One multiple-resource theory proposed by Wickens (1984) suggests four dimensions along which resources can be allocated:

1 *Stages:* Perceptual and central processing versus response selection and execution. Wickens and Kessel (1980) found, for example, that a tracking task that demanded response selection and allocation resources was disrupted by another tracking task, but not by a mental arithmetic task that demanded central processing resources.

2 *Input modalities:* Auditory versus visual. Isreal (1980), for example, found that time-sharing was better when the two tasks did not share the same modality (i.e., one was presented visually and the other auditorially) than when both tasks were presented visually or both were presented auditorially.

3 *Processing codes:* Spatial versus verbal. Almost any task requiring moving, positioning, or orienting objects in space is predominantly spatial. Tasks for which words, language, or logical operations are used are verbal. Two spatial tasks or two verbal tasks will be time-shared more poorly than will one spatial and one verbal task.

4 *Responses:* Vocal versus manual response. Calling out numbers (vocal response) can be successfully time-shared with a tracking task (manual response), whereas entering numbers on a keypad will not be as successfully time-shared with the tracking task. This is because both keying and tracking compete for the same manual-response resources.

Consider the analysis of a time-sharing situation that is becoming more and more prevalent—driving a car while talking on the telephone. Little competition would be predicted for input modality or response resources. Driving essentially involves visual input and manual responses, while talking on the telephone involves auditory input and vocal responses. The processing code resources required for driving and talking on the phone are, for the most part, separate. Driving involves spatial coding (e.g., the location of other vehicles), while talking on the phone involves verbal codes. However, if the driver needs to read off-ramp signs, then verbal codes are being used and competition for the verbal-coding resources would be predicted. Telephone conversations require central processing resources. To the extent that driving requires the same resources (e.g., finding the way or computing miles before running out of gas), performance would be predicted to decline in the time-sharing situation. However, to the extent that driving involves well-practiced, almost automatic behavior, it would be placing more demands on the response selection and execution resources and would not be in as much competition for central processing resources being demanded by the telephone conversation.

Discussion Even with all the theory building and research in the area of divided attention, there is still much we do not understand. Even the notion of limited resources is not universally accepted by all (e.g., Navon, 1984, 1985). Some theories postulate competing influences, some of which should improve performance and others of which should degrade performance. How these factors come together to influence performance is still not entirely clear. Predictions on the outcome of time-sharing real-world tasks, therefore, are still relatively primitive.

Guidelines for Divided-Attention Tasks Despite the limitations referred to above, we can offer a few guidelines gleaned from the research evidence—some of which were cited above and some of which were not—that should improve performance in a time-sharing or divided-attention situation.

1 Where possible, the number of potential sources of information should be minimized.

2 Where time-sharing is likely to stress a person's capacity, the person should be provided with information about the relative priorities of the tasks so that an optimum strategy of dividing attention can be formulated.

3 Efforts should be made to keep the difficulty level of the tasks as low as possible.

4 The tasks should be made as dissimilar as possible in terms of demands on processing stages, input and output modalities, and memory codes.

5 Especially when manual tasks are time-shared with sensory or memory tasks, the greater the learning of the manual task, the less will be its effect on the sensory or memory tasks.

Sustained Attention

The study of sustained attention or vigilance concerns the ability of observers to maintain attention and remain alert to stimuli over prolonged periods of time (Parasuraman, Warm, and Dember, 1987). The importance of sustained attention will seemingly increase with the development of automated control equipment where the human operator assumes the role of system monitor, observing dials and computer screens for an occasional critical stimulus (signal) that demands action.

Vigilance has been studied for about 40 years and several good reviews and books have been published (Parasuraman, 1986; Craig, 1985; Davies and Parasuraman, 1982; Warm, 1984). The basic phenomenon under investigation is the *vigilance decrement*—a decline in both the speed of signal detection and the accuracy of detection with time on the task. Laboratory studies consistently find such a decrement occurring over the first 20 to 35 min of the vigil. Giambra and Quilter (1987) developed an exponential function that appears to describe rather well the time course of the vigilance decrement found in laboratory studies. This is shown in Figure 3-4 for probability of detection.

Although vigilance decrements are routinely found in laboratory studies, they are found much less frequently in real-life operational tasks (Mackie, 1984, 1987). Mackie postulates that this may be due to differences between laboratory studies and real-life situations in terms of the characteristics of the signals; nature of the task; and the training, experience, and motivation of the people doing the task.

Over the years, there have been many research studies dealing with vigilance. Davies and Parasuraman (1982), for example, cite over 700 references in their book on vigilance. And where there is research, there is theory! As pointed out by Loeb and Alluisi (1980), however, theories of vigilance are really not theories specific to vigilance, but rather are general theories applied to the vigilance situation. Examples include inhibition theory, expectancy theory, filter theory, arousal theory, habituation theory, and signal detection theory. Unfortunately, all describe and predict data almost equally well, and no one of them explains all vigilance phenomena.

Guidelines for Sustained-Attention Tasks Despite all the research and theory, there has been little documentation of successful countermeasures to the problems associated with vigilance in real-world tasks (Mackie, 1987). However, some suggestions can be gleaned from the literature for improving performance on tasks requiring sustained attention.

1 Provide appropriate work-rest schedules and task variation (e.g., interpolation of different activities).

2 Increase the conspicuity of the signal. Make it larger, more intense, of longer duration, and more distinctive.

3 Reduce uncertainty as to when or where the signal will likely occur.

4 Inject artificial signals and provide operators with feedback on their detection performance (Wylie and Mackie, 1987).

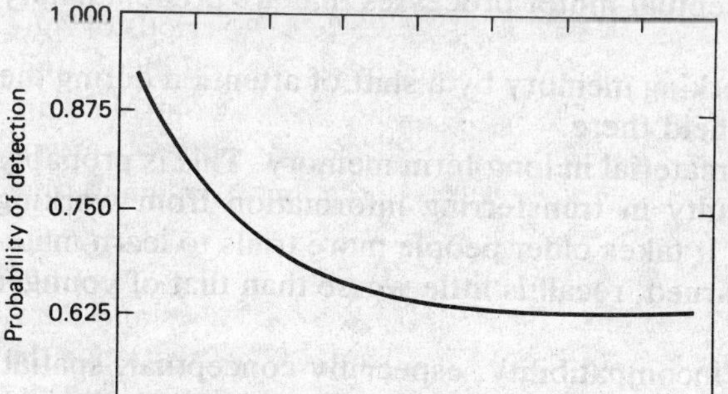

FIGURE 3-4
A typical vigilance decrement showing the probability of detecting a signal as a function of time into task. The equation for the function is: Predicted probability $= A \cdot e^{-T1 \cdot t} + A / (1 + e^{-T2 \cdot t})$. Where $A = 0.6419$; $T1 = 0.05319$; $T2 = 0.04633$; t = time into task in minutes. (*Source: Giambra and Quilter, 1987, Fig. 1. Reprinted with permission of the Human Factors Society, Inc. All rights reserved.*)

5 Provide adequate training of observers to make clear the nature of the signals to be identified.

6 Improve motivation by emphasizing the importance of the task being performed.

7 Reduce the rate at which stimuli (which may or may not be signals) are presented if the rate is high.

8 Maintain noise, temperature, illumination, and other environmental factors at optimal levels.

AGE AND INFORMATION PROCESSING

As medical knowledge and technology advance, the average life expectancy will tend to increase. The average age of people in the industrialized world is increasing. Despite the old adage that we grow older and wiser, there is evidence that age has some disruptive effects on our information processing capabilities.

Welford (1981) and Czaja (1988) summarized the effects of age on information processing. Although the effects are probably progressive throughout adulthood, they usually do not become noticeable until after the age of 65 or so. There are, of course, large individual differences in the aging process, so it is not really possible to pinpoint the exact age for an individual at which these effects will become apparent. Welford discusses the following changes in information processing capability that occur with age:

1 A slowing of performance. In general, older people take longer to retrieve information from long-term memory, to choose among response alternatives, and to execute responses. This slowing is most evident in tasks requiring continuous responses as opposed to tasks where an appreciable time lapses before the next response is required. There is not complete agreement as to the stages of information processing affected by age. Salthouse (1982) says that all processing stages, perceptual-motor and central-cognitive, are slowed with age. Cerella (1985) believes that central-cognitive stages are more affected than perceptual-motor stages. On the other hand, Strayer, Wickens, and Braune (1987) suggest that it is the perceptual-motor processes that are predominantly affected by age.

2 Increased disruption of working memory by a shift of attention during the time that the material is being held there.

3 Difficulty in searching for material in long-term memory. This is probably because of an increased difficulty in transferring information from working memory to long-term memory. It takes older people more trials to learn material; but once the material is learned, recall is little worse than that of younger subjects.

4 Difficulty in dealing with incompatibility, especially conceptual, spatial, and movement incompatibility.

5 Decrements in perceptual encoding of ambiguous stimuli. Older people

have increased difficulty identifying objects from incomplete representations or processing complex or confusing stimuli. They are more likely to experience interference from irrelevant or surplus information.

Welford interprets these effects within a signal detection framework. He attributes them to a reduction in sensitivity *d′* resulting from weaker signals coming from the sense organs to the brain and to increased random neural activity (noise) in the brain. The response criterion (beta) is adjusted to compensate for the reduced *d′*; sometimes it is made more conservative, sometimes more risky, depending on the task.

Guidelines for Designing Information Processing Tasks for the Elderly

The above findings suggest the following considerations in designing a task for the elderly that requires information processing:

1 Strengthen signals displayed to the elderly. Make them louder, brighter, etc.
2 Design controls and displays to reduce irrelevant details that would act as noise.
3 Maintain a high level of conceptual, spatial, and movement compatibility.
4 Reduce time-sharing demands.
5 Provide time between the execution of a response and the signal for the next response. Where possible, let the person set the pace of the task.
6 Allow more time and practice to initially learn material.

MENTAL WORKLOAD

In our everyday experiences we have encountered tasks that seem to tax our brains. In like manner, there are tasks that do not seem as demanding, such as balancing a checkbook (at least to some people it is not demanding). These experiences touch on the idea of mental workload as a measurable quantity of information processing demands placed on an individual by a task.

Since the middle to late 1970s, interest in defining and developing measures of mental workload has increased dramatically. A major area of focus of such work has been the assessment of the workload experienced by aircraft crew members, particularly pilots. As further indication of the growing concern, Wickens (1984) indicates that the Federal Aviation Administration will require certification of aircraft in terms of workload metrics and that the Air Force is imposing workload criteria on newly designed systems.

If a practical, valid method for measuring mental workload were available, it could be used for such purposes as (1) allocating functions and tasks between humans and machines based on predicted mental workload; (2) comparing alternative equipment and task designs in terms of the workloads imposed; (3) monitoring operators of complex equipment to adapt the task difficulty or allocation of function in response to increases and decreases in mental workload; and (4) choosing operators who have higher mental workload capacities for demanding tasks.

Concept of Mental Workload

The concept of mental workload is a logical extension of the resource models of divided attention discussed above. Resource models postulate a limited quantity of resources available to perform a task, and in order to perform a task, one must use some of or all these resources. Although there is no universally accepted definition of mental workload (McCloy, Derrick, and Wickens, 1983), the basic notion is related to the difference between the amount of resources available within a person and the amount of resources demanded by the task situation. This means that mental workload can be changed by altering eithe the amount of resources available within the person (e.g., keeping the person awake for 24 h) or the demands made by the task on the person (e.g., increasing the number of information channels). This concept of mental workload is similar to the concept of physical workload, which is discussed in Chapter 8.

Measurement of Mental Workload

Most of the research activity in the area of mental workload has been directed toward refining and exploring new measures of the concept. A useful measure of mental workload should meet the following criteria:

1 *Sensitivity:* The measure should distinguish task situations that intuitively appear to require different levels of mental workload.

2 *Selectivity:* The measure should not be affected by things generally considered not to be part of mental workload, such as physical load or emotional stress.

3 *Interference:* The measure should not interfere with, contaminate, or disrupt performance of the primary task whose workload is being assessed.

4 *Reliability:* The measure should be reliable; that is, the results should be repeatable over time (test-retest reliability).

5 *Acceptability:* The measuring technique should be acceptable to the person being measured.

The numerous measures of mental workload can be classified into four broad categories: primary task measures, secondary task measures, physiological measures, and subjective measures. Several reviews of mental workload measurement have appeared over the years (Eggemeier, 1988; Moray, 1988; O'Donnell and Eggemeier, 1986) and should be consulted for a more complete discussion of the various methods under study. We briefly review each major category and give an example or two to illustrate the sort of work being done.

Primary Task Measures Early attempts to quantify workload used a task analytic approach in which *workload* was defined simply as the time required to perform tasks divided by the time available to perform the tasks. Several computer-based modeling programs were developed to perform these calculations on a minute-by-minute basis throughout a scenario of tasks [one example is the statistical workload assessment model (SWAM) (Linton, 1975)]. The problem with this approach is that it does not take into account the fact that some tasks can be time-shared successfully while others cannot; nor do such methods take into account cognitive demands that are not reflected in increased task completion times. One computer-based tool, however, does take into account interference between time-shared tasks that compete for the same resources. The computer program is called Workload Index (W/INDEX) (North and Riley, 1988). The program incorporates a "conflict matrix," which specifies the degree of competition between time-shared tasks based on the multiple resource model proposed by Wickens (1984) and discussed previously. The technique has been shown to be superior to simply adding task demands based on a pure time-line analysis (Wickens, Harwood, Segel, Teakevic, and Sherman, 1988). W/INDEX has been applied to the analysis of a wide range of systems including the Army Apache helicopter and the National Aerospace Plane.

Task performance itself would seem, at first glance, an obvious choice for a measure of mental workload. It is known, however, that if two versions of a task are compared, both of which do not demand resources in excess of those available, performance may be identical, yet one task could still be more demanding than the other. To overcome this problem, the *primary task workload margin* (Wickens, 1984) can be computed for various versions of a task. Basically, this technique involves changing a parameter of the task known to affect the demand for resources until performance on the task can no longer be maintained at some preset criterion level. How much that parameter could be changed before performance broke down is the measure of primary task workload margin. Performance on a difficult version of a task would break down with less of an increase in resource demands than would an easy version of the task.

A problem faced by all primary task measures of mental workload is that they are task-specific, and it is difficult, if not impossible, to compare the workloads imposed by dissimilar tasks.

Secondary Task Measures The basic logic behind the use of secondary task measures is that spare capacity, not being directed to performance of the primary task, will be used by the secondary task. The greater the demand for resources made by the primary task, the less resources will be available for the secondary task and the poorer will be performance on the secondary task. In using this method, a person is usually instructed to maintain a level of performance on the primary task, so that differences in workload will be reflected in the performance of the secondary task. Sometimes this procedure is reversed; a person is instructed to devote all necessary resources to maintain performance on the secondary task, and performance on the primary task is taken as a measure of its difficulty. This is called a *loading task technique* (Ogden, Levine, and Eisner, 1979). The more demanding the primary task, the lower will be performance on it in the presence of a loading (secondary) task.

Some examples of secondary tasks that have been used include rhythmic tapping, reaction time, memory search, time estimation, production of random digits, and mental arithmetic. Casali and Wierwille (1984), in summarizing previous studies on the sensitivity of various measures of workload on perceptual, mediation, communication, and motor tasks, found that time estimation was the most sensitive secondary measure for all four types of tasks. The underlying philosophy of using time estimation as a secondary task is that timekeeping requires attention. Whenever attention is drawn away from the passage of time by a competing activity (the primary task), time continues to pass unnoticed. This leads to an underestimation of the passage of time (estimates that are too short or productions of time that are too long); the greater the attention demands of the primary task, the greater will be the underestimation of the passage of time.

There is a dilemma inherent in the secondary task methodology. In order to measure spare resource capacity, the secondary task should tap the same resources as those tapped by the primary task. If one accepts a multiple-resource model, then the tasks should share common modalities (visual, auditory, speech, manual) and common processing codes. Such tasks, however, by definition, interfere with the performance of the primary task. Thus, we cannot say whether we are measuring the workload imposed by just the primary task or that of the primary as interfered with by the secondary task. Further, a lack of secondary-task performance decrement may only indicate that the wrong secondary task was chosen.

Physiological Measures Most physiological measures of mental workload are predicated on a single-resource model of information processing rather than a multiple-resource model. The logic is that information processing involves central nervous system activity, and this general activity or its manifestations can be measured. Some physiological measures that have been investigated include variability in heart rate (also called *sinus arhythmia*), evoked brain potential, pupillary response, respiration rate, and body fluid chemistry.

An example of the use of pupillary response as a measure of mental workload can be seen in the work done by Ahern (as cited in Beatty, 1982). Figure 3-5 shows pupillary response (dilation) during the performance of mental multiplication tasks that differed in difficulty (multiplying pairs of one-digit numbers,

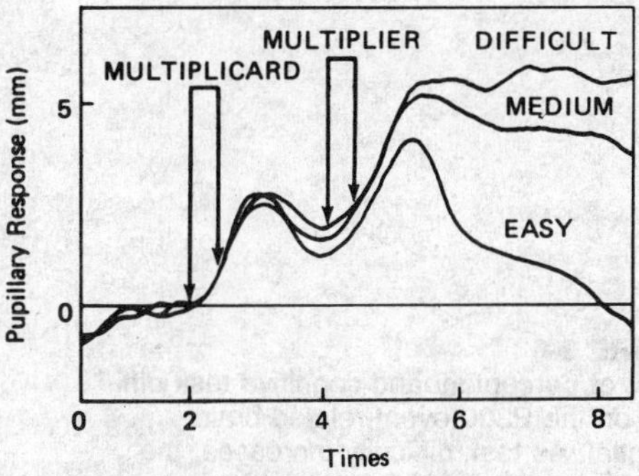

FIGURE 3-5
Pupillary responses during mental multiplication tasks that differ in cognitive difficulty. Both the amplitude and the duration of the dilation increase as a function of task difficulty. (*Source: Ahern as presented by Beatty, 1982, Fig. 5. Copyright 1982 by the American Psychological Association. Reprinted by permission of the author.*)

to multiplying pairs of two-digit numbers). As can be seen, both the peak and duration of the pupillary response were greater as the difficulty of the mental operations involved increased.

Kramer, Wickens, and Donchin (1983) report on a specific component of event-related brain potentials called *P300* as a measure of mental workload. Event-related brain potentials are transient voltage oscillations in the brain ("brain waves") that can be recorded on the scalp in response to a discrete stimulus event. There is evidence to suggest that P300—a positive polarity voltage oscillation occurring approximately 300 ms after the occurrence of a task-relevant, low-probability event—is a manifestation of neuronal activity that is involved whenever a person is required to update an internal "model" of the environment or task structure. Typically, P300 responses are elicited by requiring subjects to covertly count occurrences of one of two events that are occurring serially. For example, two tones, high and low, are randomly presented one after another, and a subject is required to count the high tones. Each time a high tone occurs and the subject updates the count (or internal model), a P300 wave occurs. To the degree to which a primary task demands perceptual and cognitive resources, fewer resources will be available to update the count and the magnitude of the P300 wave will be reduced. This can be seen in Figure 3-6. The P300 response is shown when no additional primary task was present (count only) and when the response was paired with a primary tracking task that varied in perceptual and cognitive difficulty. As the tracking task increased in difficulty, the P300 waveform became flatter and less pronounced.

Physiological measures of mental workload offer some distinct advantages: they permit continuous data collection during task performance, they do not often interfere with primary task performance, and they usually require no additional activity or performance from the subject. The disadvantages are that bulky equipment often must be attached to the subject to get the measurements, and, except for perhaps the P300 event-related brain potential, the measures do not isolate the specific stages of information processing being loaded by the primary task.

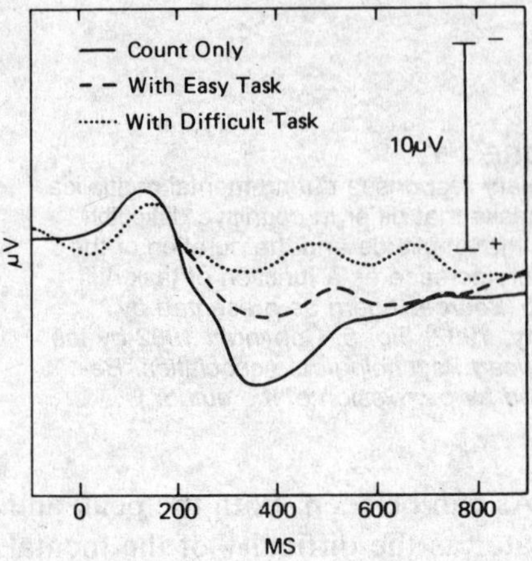

FIGURE 3-6
Effect of perceptual and cognitive task difficulty on the P300 event-related brain potential. As task difficulty increases, the magnitude of the P300 potential is decreased. (*Source: Adapted from Kramer, Wickens, and Donchin, 1983, Fig. 3. Copyright by the Human Factors Society, Inc., and reproduced by permission.*)

Subjective Measures Some investigators have expressed the opinion that subjective ratings of mental workload came closest to tapping the essence of the concept (Sheridan, 1980). Rating scales are easy to administer and are widely accepted by the people who are asked to complete them. Perhaps the oldest and most extensively validated subjective measure of workload (although it was originally developed to assess handling characteristics of aircraft) is the Cooper-Harper Scale (Cooper and Harper, 1969). Wierwille and Casali (1983) report that the scale can be used, with only minimal rewording, for a wide variety of motor or psychomotor tasks. In addition, Wierwille and Casali present a modified Cooper-Harper rating scale that can be used for perceptual, cognitive, and communications tasks. The scale combines a decision tree and a unidimensional 10-point rating scale. The rating scale goes from (1) very easy, highly desirable (operator mental effort is minimal and desired performance is easily attainable), through (5) moderately objectionable difficulty (high operator mental effort is required to attain adequate system performance), to (10) impossible (instructed task cannot be accomplished reliably).

Another approach to subjective measurement of mental workload treats the concept as a multidimensional construct. Sheridan and Simpson (1979) proposed three dimensions to define subjective mental workload: time load, mental effort load, and psychological stress. Reid, Shingledecker, and Eggemeier (1981) used these dimensions to develop the Subjective Workload Assessment Technique (SWAT). Subjects rate workload on each of the three dimensions (each with a 3-point scale), and the ratings are converted to a single interval scale of workload. Remember that the three underlying dimensions suggested by Sheridan and Simpson and used with SWAT were intuitively derived and have not been empirically validated. Further, Boyd (1983) found that when the dimensions were independently varied in a task, the ratings of the dimensions were not independent. That is, if only time load was increased in the task, subjects tended to increase their ratings on all three dimensions.

Another multidimensional approach to measuring workload subjectively has been taken by NASA-Ames Research Center (Hart and Staveland, 1988). The rating procedure, called NASA Task Load Index (NASA-TLX), provides an overall workload score based on a weighted average of ratings on six subscales (or dimensions): mental demands, physical demands, temporal (time) demands, own performance, effort, and frustration. The unique feature of the procedure is that the subscale ratings are weighted according to their subjective importance to the rater in a specific task situation, rather than their relevance to the rater's definition of workload in general as is done with SWAT. Results using SWAT and NASA-TLX have been similar, but NASA-TLX yields more consistent scores among people doing the same task (i.e., it reduces between-rater variability) (Vidulich and Tsang, 1985, 1986).

Discussion of Mental Workload Measures Despite all the work that has been done and all the measures that have been proposed, there is still no real agreement on an adequate measure of mental workload. Moray (1988) sums up the current stage of affairs:

> It is in the area of subjective measures that most progress has been made. Behavioral measures, particularly dual task measures, remain disorganized and largely arbitrary. Some progress in physiological measurement has been made, but it is of little practical use. Theory remains undeveloped. A major problem is the lack of reliability, and the lack of consistent correlation between estimates based on the different approaches.

It is probably safe to say that to measure mental workload, probably several measures will be needed to assess what is, in all likelihood, a multidimensional concept. In some cases where multiple measures of workload have been evaluated together, the different measures provide different results. When this occurs, the measures are said to *dissociate* (Yeh and Wickens, 1984; McCloy, Derrick, and Wickens, 1983). A common finding, for example, is that subjective measures dissociate from task performance. It appears that each may be measuring a different aspect of mental workload. Subjective measures appear to be more sensitive to the number of current tasks being performed, while task performance is more sensitive to the degree of competition for common resources among the various tasks being performed (Yeh and Wickens, 1984).

The ultimate success in developing a set of standardized mental workload assessment techniques will depend, in great part, on the development and refinement of our theories of human information processing.

HUMAN FACTORS IN THE INFORMATION REVOLUTION

Probably some of the most exciting developments in the information revolution are in the area of artificial intelligence (AI). We are not going to engage in philosophical discussion here as to whether computers will ever have the ability to think. The fact is that advances in computer technology and programming languages are endowing computers with abilities once considered strictly within the domain of humans. We have had to rethink the traditional allocation of functions between people and machines (see Chapter 22 for further discussion of allocation of function in system design). The new applications are presenting new and challenging human factors problems. We touch briefly on the two areas of expert systems and natural language interfaces, and we discuss a few general human factors areas of concern within each.

Expert Systems

An *expert system* is essentially a way to capture the knowledge and expertise of a subject-matter expert and transfer it to a computer program that, it is hoped, will emulate the problem-solving and decision-making performance of the expert (Hamill, 1984; Garg-Janardan et al., 1987). Expert systems have been successfully developed for attacking very specific, well-defined problem areas, such as diagnosing and treating infectious blood diseases, making inferences about the structures of unknown chemical compounds, freeing oil well drill bits stuck in a drill shaft, and troubleshooting locomotive engines.

Kidd (1983), Hamill (1984), Madni (1988), and Pew (1988) discuss a few of the human factors issues involved with the development and use of expert systems. Human factors specialists can contribute to the methodologies used to extract the knowledge and inference rules from the experts (Mitta, 1989). The interface between the system and user is an area also in need of human factors attention. Expert systems must communicate not only what the system is doing, but also why. This must be done in a manner that gives the user confidence in the system and permits intervention in the decision-making process by the user to redirect an obviously errant decision path (these systems are not perfect, and probably never will be). Cognitive psychologists should be able to contribute insights into human problem solving that could aid designers of future expert systems. Someday, maybe, an expert system will be created that emulates the knowledge and expertise of the human factors specialist—someday.

Natural Language Interfaces

A natural language interface allows a user of an interactive system to speak or write in "plain English" (or French, or Japanese, or whatever) and be "understood" by the machine. (Actually, no one really believes that the machine understands what is said, but rather only that it acts as if it does.) Natural language is incredibly complex; things we take for granted must be painstakingly "explained" to the computer. Consider the question "Which of our Washington staff lives in Virginia?" The computer has to figure out that Washington, in this question, refers to Washington, D.C., and not Washington state. Currently, natural language interfaces have been developed for very specific, well-defined domains, such as gathering facts from a specific data base of information. Human factors people have been involved in determining when and where natural language interfaces are likely to be more efficient than other types of interfaces (Small and Weldon, 1977; Thomas, 1978; Ehrenreich, 1981). Human factors people have also been helpful in determining the degree of structure that can be imposed in a natural language interface without destroying its appeal to the user (Gould, Lewis, and Becker, 1976; Scha, 1988).

Discussion

The prospects of artificial intelligence and its applications in expert systems and natural language interfaces have barely begun to crystallize. As these fields emerge, new problems and opportunities for integrating humans and machines into productive, efficient, and safe systems will emerge. Our understanding of the capabilities and limitations of human information processing will become more and more important in the design of future systems. Some people think that as computers take over more of the information processing and decision-making functions, human involvement in systems will become less and less. Do not believe it! The roles that humans play in systems of the future will undoubtedly change, but humans will be very much involved in such systems, doing tasks perhaps not even possible with today's technology. If we play our cards right, human factors will be a significant part of this exciting future.

REFERENCES

Anderson, J. (1980). *Cognitive psychology and its implications*. San Francisco: Freeman.

Baddeley, A. (1986). *Working memory*. Oxford: Clarendon Press.

Baird, J. (1984). Information theory and information processing. *Information Processing and Management*, 20, 373–381.

Beatty, J. (1982). Task-evoked pupillary response, processing load, and the structure of processing resources. *Psychological Bulletin*, 91, 276–292.

Bishu, R., and Drury, C. (1988). Information processing in assembly tasks—a case study. *Applied Ergonomics*, 19(2), 90–98.

Bisseret, A. (1981). Application of signal detection theory to decision making in supervisory control. *Ergonomics*, 24, 81–94.

Boyd, S. (1983). Assessing the validity of SWAT as a workload measurement instrument. *Proceedings of the Human Factors Society 27th Annual Meeting*. Santa Monica, CA: Human Factors Society, pp. 124–128.

Broadbent, D. (1958). *Perception and communication*. New York: Pergamon.

Broadbent, D. (1982). Task combination and selective intake of information. *Acta Psychologica*, 50, 253–290.

Card, S. (1989). Models of working memory. In J. Elkind, S. Card, J. Hochberg, and B. Huey (eds.) *Human performance models for computer-aided engineering*. Washington, DC: National Academy Press.

Casali, J., and Wierwille, W. (1984). On the measurement of pilot perceptual workload: A comparison of assessment techniques addressing sensitivity and intrusion issues. *Ergonomics*, 27(10), 1033–1050.

Cerella, J. (1985). Information processing rates in the elderly. *Psychological Bulletin,* 98, 67–83.

Colquhoun, W. (1967). Sonar target detection as a decision process. *Journal of Applied Psychology,* 51, 187–190.

Conrad, R. (1951). Speed and load stress in sensory-motor skill. *British Journal of Industrial Medicine,* 8, 1–7.

Conrad, R. (1964). Acoustic comparisons in immediate memory. *British Journal of Psychology,* 55, 75–84.

Cooper, G., and Harper, R. (1969, April). *The use of pilot ratings in the evaluation of aircraft handling qualities,* NASA Ames Tech. Rept. (NASA TN-D-5153). Moffett Field, CA: NASA Ames Research Center.

Craig, A. (1985). Vigilance: Theories and laboratory studies. In S. Folkard and T. Monk (eds.), *Hours of work: Temporal factors in work-scheduling*. Chichester: Wiley.

Czaja, S. (1988). Microcomputers and the elderly. In M. Helander (ed.), *Handbook of human-computer interaction*. Amsterdam: North-Holland.

Davies, D., and Parasuraman, R. (1982). *The psychology of vigilance*. London: Academic Press.

Deatherage, B. (1972). Auditory and other sensory forms of information presentation. In H. Van Cott and R. Kinkade (eds.), *Human engineering guide to equipment design* (rev. ed.). Washington, DC: Government Printing Office.

Drury, C. (1975). The inspection of sheet materials—Models and data. *Human Factors,* 17, 257–265.

Eggemeier, F. T. (1988). Properties of workload assessment techniques. In P. Hancock and N. Meshkati (eds.), *Human mental workload*. Amsterdam: North-Holland.

Ehrenreich, S. (1981). Query languages: Design recommendations derived from the human factors literature. *Human Factors,* 23, 709–725.

Ehrenreich, S. (1985). Computer abbreviations: Evidence and synthesis. *Human Factors,* 27(2), 143–155.

Ellison, K., and Buckhout, R. (1981). *Psychology and criminal justice*. New York: Harper & Row.

Fitts, P., and Seeger, C. (1953). S-R compatibility: Spatial characteristics of stimulus and response codes. *Journal of Experimental Psychology,* 46, 199–210.

Garg-Janardan, C., Eberts, R., Zimolong, B., Nof, S., and Salvendy, G. (1987). Expert systems. In G. Salvendy (ed.), *Handbook of human factors*. New York: Wiley.

Giambra, L., and Quilter, R. (1987). A two-term exponential description of the time course of sustained attention. *Human Factors,* 29(6), 635–644.

Goldstein, I., and Dorfman, P. (1978). Speed stress and load stress as determinants of performance in a time-sharing task. *Human Factors,* 20, 603–610.

Gould, J., Lewis, C., and Becker, C. (1976). *Writing and following procedural, descriptive and restrictive syntax language instructions*. IBM Research Rept. RC-5943. Yorktown Heights, NY: IBM Corp., Watson Research Center.

Green, D., and Swets, J. (1988). *Signal detection theory and psychophysics*. Los Altos, CA: Peninsula Publishing.

Grudin, J., and Barnard, P. (1985). When does an abbreviation become a word? and related questions. In *Proc CHI '85 Human Factors in Computing Systems* (San Francisco, April 14–18, 1985). New York: Association of Computing Machinery, pp. 121–125.

Haber, R., and Hershenson, M. (1980). *The psychology of visual perception* (2d ed.). New York: Holt.

Hamill, B. (1984). Psychological issues in the design of expert systems. *Proceedings of the Human Factors Society 28th Annual Meeting*. Santa Monica, CA: Human Factors Society, pp. 73–77.

Hart, S., and Staveland, L. (1988). Development of NASA-TLX: Results of empirical and theoretical research. In P. Hancock and N. Meshkati (eds.), *Human mental workload*. Amsterdam: North-Holland.

Hick, W. (1952). On the rate of gain of information. *Quarterly Journal of Experimental Psychology*, 4, 11–26.

Hirsh-Pasek, K., Nudelman, S., and Schneider, M. (1982). An experimental evaluation of abbreviation schemes in limited lexicons. *Behavior and Information Technology*, 1, 359–370.

Hyman, R. (1953). Stimulus information as a determinant of reaction time. *Journal of Experimental Psychology*, 45, 423–432.

Isreal, J. (1980). Structural interference in dual task performance: Behavioral and electrophysiological data. Unpublished Ph.D. dissertation. Champaign: University of Illinois.

Kahneman, D. (1973). *Attention and effort*. Englewood Cliffs, NJ: Prentice-Hall.

Kidd, A. (1983). Human factors in expert systems. In K. Coombes (ed.). *Proceedings of the Ergonomics Society Conference*. London: Taylor & Francis.

Kramer, A., Wickens, C., and Donchin, E. (1983). An analysis of the processing requirements of a complex perceptual-motor task. *Human Factors*, 25(6), 597–621.

Lane, D. (1982). Limited capacity, attention allocation, and productivity. In W. Howell and E. Fleishman (eds.), *Human performance and productivity: Information processing and decision making*, vol. 2. Hillsdale, NJ: Lawrence Erlbaum Associates.

Linton, P. (1975, May). *VFA-STOL crew loading analysis* (NADC-57209-40). Warminster, PA: U.S. Naval Air Development Center.

Loeb, M., and Alluisi, E. (1980). Theories of vigilance: A modern perspective. *Proceedings of the Human Factors Society 24th Annual Meeting*. Santa Monica, CA: Human Factors Society, pp. 600–603.

Long, G., and Waag, W. (1981). Limitations on the practical applicability of d' and β measures. *Human Factors*, 23(3), 285–290.

Mackie, R. (1984). Research relevance and the information glut. In F. Muckler (ed.), *Human factors review*. Santa Monica, CA: Human Factors Society.

Mackie, R. (1987). Vigilance research—are we ready for countermeasures? *Human Factors*, 29(6), 707–724.

Madni, A. (1988). The role of human factors in expert systems design and acceptance. *Human Factors*, 30(4), 395–414.

McCloy, T., Derrick, W., and Wickens, C. (1983, November). Workload assessment metrics—What happens when they dissociate? In *Second Aerospace Behavioral Engineering Technology Conference Proceedings* (P-132). Warrendale, PA: Society of Automotive Engineers, pp. 37–42.

Miller, G. (1956). The magical number seven, plus or minus two: Some limits on our capacity for processing information. *Psychological Review*, 63, 81–97.

Mitta, D. (1989). Knowledge acquisition: Human factors issues. *Proceedings of the Human Factors Society 33d Annual Meeting*. Santa Monica, CA: Human Factors Society, pp. 351–355.

Moray, N. (1967). Where is capacity limited? A survey and a model. *Acta Psychologica*, 27, 84–92.

Moray, N. (1981). The role of attention in the detection of errors and the diagnosis of errors in man-machine systems. In J. Rasmussen and W. Rouse (eds.), *Human detection and diagnosis of system failures*. New York: Plenum.

Moray, N. (1988). Mental workload since 1979. *International Reviews of Ergonomics*, 2, 123–150.

Mowbray, G., and Gebhard, J. (1961). Man's senses vs. information channels. In W. Sinaiko (ed.), *Selected papers on human factors in design and use of control systems*. New York: Dover.

Mudd, S. (1961, June). The scaling and experimental investigation of four dimensions of tone and their use in an audio-visual monitoring problem. Unpublished Ph.D. dissertation. Lafayette, IN: Purdue University.

Navon, D. (1984). Resources—A theoretical soup stone? *Psychological Review*, 91, 216–234.

Navon, D. (1985). *Attention division or attention sharing?* IPDM Rept. no. 29. Haifa, Israel: University of Haifa.

North, R., and Riley, V. (1988). W/INDEX: A predictive model of operator workload. In G. MacMillan (ed.), *Human Performance Models*. Orlando, FL: NATO AGARD Symposium.

O'Donnell, R., and Eggemeier, F. T. (1986). Workload assessment methodology. In K. Boff, L. Kaufman, and J. Thomas (eds.), *Handbook of perception and human performance*, vol. 2: *Cognitive processes and performance*. New York: Wiley.

Ogden, G., Levine, J., and Eisner, E. (1979). Measurement of workload by secondary tasks. *Human Factors*, 21(5), 529–548.

Parasuraman, R. (1986). Vigilance, monitoring, and search. In K. Boff, L. Kaufman, and J. Thomas (eds.), *Handbook of perception and human performance*, vol. 2: *Cognitive processes and performance*. New York: Wiley.

Parasuraman, R., Warm, J., and Dember, W. (1987). Overview paper: Vigilance: Taxonomy and utility. In L. Mark, J. Warm, and R. Huston (eds.), *Ergonomics and human factors: Recent research*. New York: Springer-Verlag.

Pew, R. (1988). Human factors issues in expert systems. In M. Helander (ed.), *Handbook of human-computer interaction*. Amsterdam: North-Holland.

Pollack, I. (1952). The information of elementary auditory displays. *Journal of the Acoustical Society of America*, 24, 745–749.

Reid, G., Shingledecker, C., and Eggemeier, F. (1981). Application of conjoint measurement to workload scale development. *Proceedings of the Human Factors Society 25th Annual Meeting*. Santa Monica, CA: Human Factors Society, pp. 522–526.

Salthouse, T. (1982). *Adult cognition: An experimental psychology of human aging*. New York: Springer-Verlag.

Sanders, A. F. (1983). Towards a model of stress and human performance. *Acta Psychologica*, 53, 61–97.

Scha, R. (1988). Natural language interface systems. In M. Helander (ed.), *Handbook of human-computer interaction*. Amsterdam: North-Holland.

Schneider, W., and Detweiler, M. (1988). A connectionist/control architecture for working memory. In G. Bower (ed.), *The psychology of learning and motivation: Advances in research and theory*. San Diego: Academic Press.

Schneiderman, B. (1980). *Software psychology*. Cambridge, MA: Winthrop.

Sheridan, T. (1980). Mental workload: What is it? Why bother with it? *Human Factors Society Bulletin*, 23, 1–2.

Sheridan, T., and Simpson, R. (1979, January). *Toward the definition and measurement of the mental workload of transport pilots* (FTL Rept. R 79-4). Cambridge, MA: Massachusetts Institute of Technology, Flight Transportation Laboratory.

Slovic, P., Lichtenstein, S., and Fischhoff, B. (1988). Decision making. In R. Atkinson, R. Herrnstein, G. Lindzey, and R. Luce (eds.), *Stevens' handbook of experimental psychology,* vol. 2: *Learning and cognition* (2d ed.). New York: Wiley.

Small, D., and Weldon, L. (1977). *The efficiency of retrieving information from computers using natural and structured query languages* (Tech. Rept. SAI-78-655-WA). Arlington, VA: Science Applications, Inc.

Sternberg, S. (1966). High-speed scanning in human memory, *Science,* 153, 652–654.

Strayer, D., Wickens, C., and Braune, R. (1987). Adult age differences in the speed and capacity of information processing: 2. An electrophysiological approach. *Psychology and Aging,* 2, 99–110.

Streeter, L., Ackroff, J., and Taylor, G. (1983). On abbreviating command names. *The Bell System Technical Journal,* 62, 1807–1826.

Swets, J. (ed) (1988). *Signal detection and recognition by human observers: Contemporary readings.* Los Altos, CA: Peninsula Publishing.

Swets, J., and Pickett, R. (1982). *The evaluation of diagnostic systems.* New York: Academic.

Thomas, J. (1978). A design-interpretation analysis of natural English with applications to man-computer interaction. *International Journal of Man-Machine Studies,* 10, 651–668.

Vidulich, M., and Tsang, P. (1985). Assessing subjective workload assessment: A comparison of SWAT and the NASA-Bipolar methods. *Proceedings of the Human Factors Society 29th Annual Meeting.* Santa Monica, CA: Human Factors Society, pp. 71–75.

Vidulich, M., and Tsang, P. (1986). Techniques of subjective workload assessment: A comparison of SWAT and the NASA-Bipolar methods. *Ergonomics,* 29(11), 1385–1398.

Warm, J. (ed.). (1984). *Sustained attention in human performance.* Chichester: Wiley.

Welford, A. T. (1981). Signal, noise, performance, and age. *Human Factors,* 23(1), 97–109.

Welford, H. (1976). *Skilled performance: Perceptual and motor skills.* Glenview, IL: Scott, Foresman.

Wickens, C. (1984). *Engineering psychology and human performance.* Columbus, OH: Merrill.

Wickens, C., Harwood, K., Segel, L., Teakevic, I., and Sherman, W. (1988). TASKILLAN: A simulation to predict the validity of multiple resource models of aviation workload. *Proceedings of the Human Factors Society 30th Annual Meeting.* Santa Monica, CA: Human Factors Society.

Wickens, C., and Kessel, C. (1980). The processing resource demands of failure detection in dynamic systems. *Journal of Experimental Psychology: Human Perception and Performance,* 6, 564–577.

Wickens, C., Sandry, D., and Vidulich, M. (1983). Compatibility and resource competition between modalities of input, central processing, and output. *Human Factors,* 25(2), 227–248.

Wierwille, W., and Casali, J. (1983). A validated rating scale for global mental workload measurement applications. *Proceedings of the Human Factors Society 27th Annual Meeting.* Santa Monica, CA: Human Factors Society, pp. 129–133.

Wylie, C. D., and Mackie, R. (1987). *Enhancing sonar detection performance through artificial signal injection* (Tech. Rept. 483-2). Goleta, CA: Essex Corporation.

Yeh, Y., and Wickens, C. (1984). Why do performance and subjective measures dissociate? *Proceedings of the Human Factors Society 28th Annual Meeting*. Santa Monica, CA: Human Factors Society, pp. 504–508.

CHAPTER 4

TEXT, GRAPHICS, SYMBOLS, AND CODES

We depend primarily on vision to gather information about the state of the world outside our own bodies. We read instructions, look up information in books, respond to traffic signs, use maps to navigate, and refer to dials and meters to understand speed, temperature, and the like. These are just some of the activities we do that involve the visual presentation of information. This chapter deals primarily with displays that present static information, that is, displays of visual stimuli that are unchanging or that remain in place for a reasonable time. Examples include written text, graphics, and symbols. With computers becoming such an integral part of our lives, this chapter also deals with the presentation of static information on computer screens. Chapter 5 deals primarily with displays that present stimuli representing dynamic or changing information, usually displayed on dials, meters, counters, etc. Before launching into a discussion of static visual displays, it will benefit us to summarize some basic information on the process of seeing and visual capabilities of people.

PROCESS OF SEEING

At a basic level, the eye is like a camera. Light rays pass through an opening and are focused with a lens on a photosensitive area. There are, however, important differences between the eye and a camera that make the analogy almost useless. Some of these will be discussed later.

Figure 4-1 illustrates the principal features of the eye in cross section. Light rays that are reflected from an object pass through the transparent *cornea* and

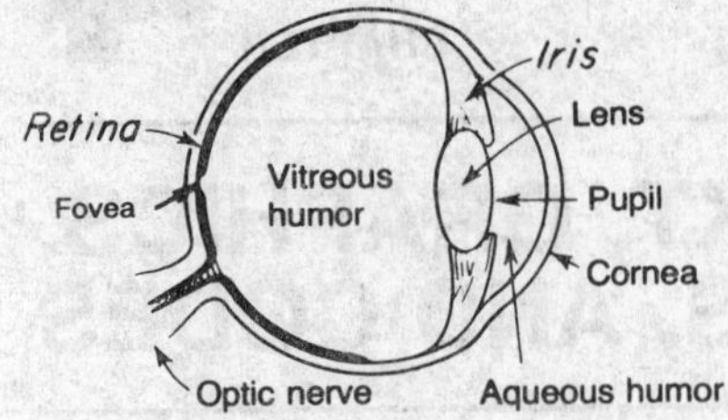

FIGURE 4-1
Principal features of the human eye in cross section. Light passes through the pupil, is refracted by the lens, and is brought to a focus on the retina. The retina receives the light stimulus and transmits an impulse to the brain through the optic nerve.

clear fluid (*aqueous humor*) that fills the space between the cornea and the pupil and lens. The *pupil* is a circular opening whose size is changed through action of the muscles of the *iris* (the colored part of the eye). The pupil becomes larger in the dark (up to about 8 mm in diameter) and smaller in bright surroundings (down to about 2 mm in diameter). The light rays that are transmitted through the pupil to the *lens* are refracted (bent) by the lens and then travel through the clear, jellylike *vitreous humor* that fills the eyeball behind the lens.

In persons with normal or corrected vision, the lens focuses the light rays on the *retina,* the photosensitive surface at the back of the eye. Muscles attached to the lens contract and relax to bring objects at different distances from the observer into focus. The muscles contract, causing the lens to bulge when focusing on objects close in, and relax, causing the lens to flatten when focusing on objects far away.

The retina consists of two types of photoreceptors: rods and cones. *Cones* function at high levels of illumination, such as during daylight, and can differentiate between colors. Without cones we could not see any color at all. *Rods* function at low levels of illumination, such as at night, and can only differentiate between shades of black and white. Rods and cones are not evenly distributed over the surface of the retina. The 6 or 7 million cones are concentrated near the center of the retina in an area called the *fovea*. The fovea is the area of greatest acuity. For an object to be seen clearly, there must be sufficient light to activate the cones and the eye must be directed so that the image is focused on the fovea. The 130 million or so rods predominate in the peripheral parts of the retina with maximum density at about 10 to 20 degrees from the fovea. Because rods function at low levels of illumination and are primarily in the periphery of the retina, we can see a dim object more effectively if we look slightly to one side of it rather than directly at it. Astronomers do this when looking for a dim star in the night sky.

Light rays when absorbed by the rods or cones cause a chemical reaction that, in turn, causes a neural impulse to be transmitted to the brain by way of the *optic nerve*. The brain integrates the many impulses to give us our visual impression of the outside world.

The important differences between the eye and a camera include the following. The eye's lens is in a fixed location and does not have to move back and forth as does a camera lens to focus objects at different distances. The eye, therefore, can focus faster than can a camera. A camera has a fixed type of film sensitivity, while the eye is differentially sensitive to light over its surface because of the distribution of rods and cones. The eye, therefore, can operate over a far greater range of illumination levels than can a camera with one type of film. Because we have two eyes, we can see in three dimensions, while a camera can record in only two dimensions. Finally, with the eye, our film, in essense, is put in backward! The photosensitive ends of the rods and cones actually face toward the back of the eye and are under blood vessels, nerves, and connective tissue. It is amazing that we can see anything at all.

VISUAL CAPABILITIES

The basic visual system gives rise to several visual capabilities that have important implications for the design of visual displays. A few of the more important capabilities are discussed below. More detailed treatment of these and other visual capabilities can be found in Boff, Kaufman, and Thomas (1986).

Accommodation

Accommodation refers to the ability of the lens of the eye to focus the light rays on the retina. This is what allows us to see details of an object, such as reading fine print on a contract or identifying a person across the street. There are limits on the degree to which the lens can accommodate objects. For example, if you try to focus the text on this page while bringing the page slowly toward your eyes, there will be a point at which the image becomes blurred and the eyes cannot clearly focus it. This is referred to as the *near point*, that is, the closest distance to which the eyes can focus. In a similar manner there is a *far point* beyond which the eyes can not clearly focus. In normal vision, the far point is usually near infinity.

Focal points and distances are usually expressed in diopters (D). One diopter is equal to 1/target distance in meters (m). The greater the value in diopters, the closer is the focal point. For example, 1D = 1 m; 2D = 0.5 m; 3D = 0.33 m; and 0D = infinity. We also speak of the focusing power of a lens in terms of diopters. The more powerful the lens, the higher the diopters and the more the lens is able to bend light.

In addition to the near and far points, there is a *dark focus*, which is the accommodative state to which the eyes tend to move spontaneously in darkness. Generally speaking, for a normal eye, the dark focus is about at arm's length (1 m or 1D). Accommodation is a compromise between the position of the stimulus to be focused and an individual's dark focus. Accommodation tends to undershoot distant objects, those beyond the dark focus, and overshoot near objects, those closer than the dark focus.

In some individuals the accommodation capacity of the eyes is inadequate. This causes the condition we sometimes call *nearsightedness* or *farsighted-*

ness. Nearsighted people can not see distant objects clearly; their far point is too close. Farsighted people can not see objects close up; their near point is too far away.

Looking for an excuse not to study? The National Research Council (1988) reports that as many as 40 percent of normal-sighted college freshmen will develop some degree of nearsightedness (*myopia*) before they graduate. This is a much higher rate than among their nonacademic peers. The cause is a mystery. It is correlated to reading ability and grade point average, but we do not know whether people become nearsighted because they read more or if they read more because they are myopic.

With age, people tend to become farsighted—their far point recedes slightly—but the major effect is a dramatic retreat of their near point. This is caused by changes in the lens and the muscles that control its shape. Corrective lenses can often correct the condition by refracting the light before it reaches the eye. Other effects of age on visual functioning are discussed in Chapter 16.

Visual Acuity

Visual acuity is the ability to discriminate fine detail and depends largely on the accommodation of the eyes. There are several different types of visual acuity depending on the type of target and detail one is asked to resolve. The most commonly used measure of acuity, *minimum separable acuity*, refers to the smallest feature, or the smallest space between the parts of a target that the eye can detect. Various targets are used in measuring minimum separable acuity, including letters and various geometric forms such as those illustrated in Figure 4-2. (As we will see, the acuity gratings of Figures 4-2*e* and 4-2*f* are also used in a special way to measure acuity). Acuity usually is measured in terms of the

FIGURE 4-2
Illustrations of various types of targets used in visual acuity tests and experiments. The features to be differentiated in targets *a, b, c,* and *d* are all the same size and would, therefore, subtend the same visual angle at the eye. With target *a* the subject is to identify each letter; with *c, e,* and *f* the subject is to identify the orientation (such as vertical or horizontal); and with *b* the subject is to identify any of four orientations. With target *d* the subject is to identify one checkerboard target from three others with smaller squares.

(*a*) Letter target (*b*) Landholt ring (*c*) Parallel bars (*d*) Checkerboard (*e*) Acuity grating (*f*) Acuity grating reciprocal of the visual angle subtended at the eye by the smallest detail that can be distinguished, such as the gap in a Landholt ring (Figure 4-2*b*). The visual angle (VA) is measured in minutes of arc or in seconds of arc. Each of the 360 degrees of a circle can be divided into 60 min of arc, and each minute can be divided into 60 s of arc. The concept of visual angle is illustrated in Figure 4-3. A formula for computing the visual angle for angles less than about 10 degrees is:

$$\text{VA (minutes)} = \frac{3438 \times H}{D}$$

Where H is the height of a stimulus or detail, and D is the distance from the eye. H and D must be in the same units, such as inches, feet, millimeters, etc.

"Normal" acuity is usually taken to be 1.0 (VA = 1 min), but it depends on the type of target used. For example, acuity for Landholt rings is better than for Snellen letters (the kind used on doctors' eye charts). If a person can distinguish a detail that subtends an arc of 1.5 min, the acuity for that person is 1/1.5 or 0.67. On the other hand, a person with better than average acuity who can distinguish a detail that subtends an arc of 0.8 min has an acuity score of 1/0.8 or 1.25. The higher the acuity score, the smaller the size of the detail that can be resolved.

In clinical testing of acuity with letter charts, an observer is usually 20 ft or 6 m from the eye chart. Acuity is expressed as a ratio, such as 20/30 (called *Snellen acuity*). This indicates that the person tested can barely read at 20 feet what a normal (20/20 vision) person can read at 30 feet. 20/10 indicates that the person can read at 20 ft what a normal person must bring to 10 ft before being able to read it. Normal 20/20 vision is assumed to be the ability to resolve a target detail of 1 min of arc at 20 ft (hence acuity equals 1.0). For example, 20/30 is equivalent to an acuity of 0.6.

There are other types of acuity besides minimum separable acuity. *Vernier acuity* refers to the ability to differentiate the lateral displacement, or slight offset, of one line from another that, if not so offset, would form a single continuous line (such as in lining up the "ends" of lines in certain optical devices). *Minimum perceptible acuity* is the ability to detect a spot (such as a round dot) from its background. In turn, *stereoscopic acuity* refers to the ability to differentiate the different images, or pictures, received by the retinas of the two eyes of a single object that has depth. (These two images differ most when the object is near the eyes and differ least when the object is far away.)

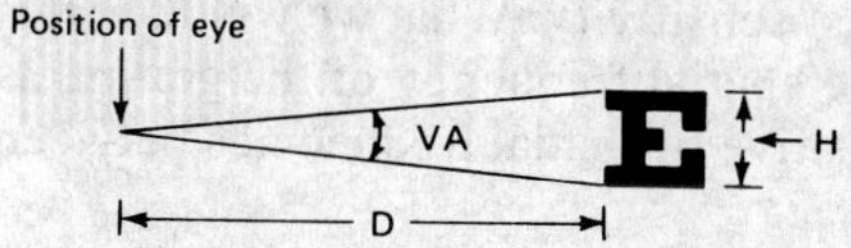

FIGURE 4-3
Illustration of the concept of visual angle; H = height of visual stimulus, and D = distance from the eye. In this illustration, the visual angle of the specific elements of *E* could be derived (the thickness of the elements would be the *H* value).

Contrast Sensitivity

Look at the two acuity gratings in Figure 4-2 and hold the page at arm's length. You will probably notice that the contrast of the wide grating (Figure 4-2*e*) appears greater, that is, the difference in lightness between the black-and-white bars appears more pronounced in the wide grating than in the narrow grating (Figure 4-2*f*). Of course, the difference in lightness between the black-and-white bars is the same in both gratings. The fact that we perceive a difference is very important for an understanding of how our visual system works.

The two gratings in Figure 4-2 differ in terms of *spatial frequency*, that is, the number of bars per unit distance. Spatial frequency is usually measured in terms of cycles per degree of visual angle, where one cycle would be a black bar plus a white bar (the unit of the grating that repeats itself). As you move a grating farther away, or make the bars narrower, the spatial frequency of the grating increases.

The waveform of a grating refers to the shape of the grating's luminance distribution (we discuss luminance in Chapter 16, but suffice to say it is the amount of light reflected by the parts of the grating). Although there are all sorts of possible distributions, Figure 4-4 illustrates the two most common types: squarewave and sinewave.

For stimuli such as gratings that deviate symmetrically above and below a mean luminance, *modulation contrast* (also called *Michelson contrast*) is generally computed as follows:

$$C = \frac{(L_{\max} - L_{\min})}{(L_{\max} + L_{\min})}$$

Where $L_{\max}$ and $L_{\min}$ are the maximum and minimum luminances in the pattern. Modulation contrast will take on a value between 0 and 1. It can be reduced, for example, by using gray instead of white in the grating.

Imagine an experiment in which we present a grating of low spatial frequency (wide bars) and contrast so low that it appears to be a homogeneous gray field. We then slowly increase the contrast until you report that you can just barely see its bars. This level of contrast is the *threshold contrast* for seeing the bars. Threshold can be converted into *contrast sensitivity* (contrast sensitivity = 1/threshold contrast). Thus, the lower the threshold contrast, the higher the contrast sensitivity. We would repeat this experiment with gratings of various spatial frequencies, each time determining the contrast required to just see the bars and converting it to contrast sensitivity. What we would find is that the contrast sensitivity depends on the spatial frequency of the grating as shown in Figure 4-5. Humans are most sensitive to spatial frequencies between 2 and 4 cycles per degree (cpd).

The more detailed the target, the higher is its spatial frequency. Typical visual acuity targets such as Snellen letters or Landholt rings are high-contrast,

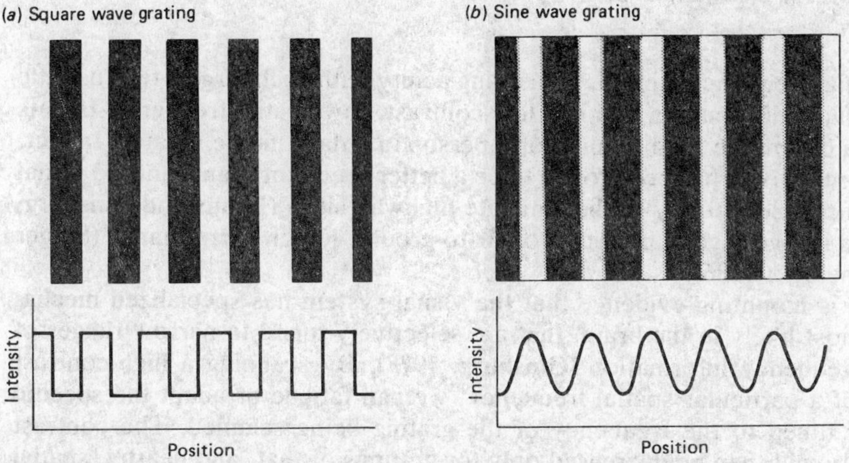

FIGURE 4-4
(*a*) A square wave grating and its intensity distribution. (*b*) A sine wave grating and its intensity distribution. The abrupt changes in intensity of the square wave grating are seen as sharp contours, whereas the more gradual changes in intensity of the sine wave grating are seen as fuzzy contours.

FIGURE 4-5
Contrast sensitivity curve using stationary gratings as the targets. The region between the two dashed lines encompasses contrast sensitivity limits for 75 percent of the sample, and the area between the outer solid lines encompasses 90 percent of the sample. (*Source: Adapted from Ginsburg, Evans, Cannon, Owsley, and Mulvanny, 1984.*)

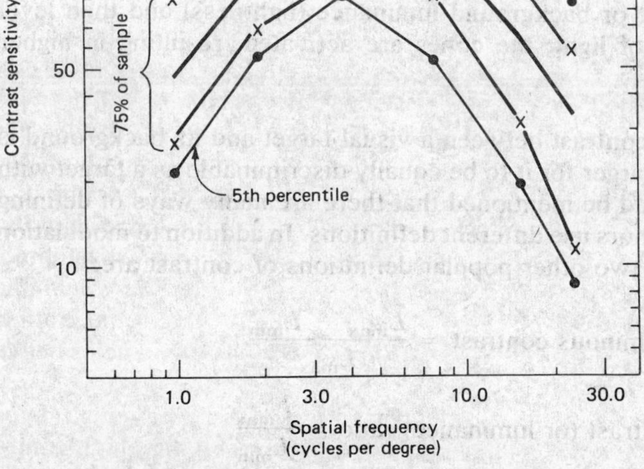

high-spatial-frequency stimuli. Measuring acuity with such targets tells us little about how well a person can see low-contrast, low-spatial-frequency targets such as a car on the road at night or a person in a dark movie theater. In fact, contrast sensitivity has been found to be a better predictor than standard visual acuity of people's ability to discriminate highway signs (Evans and Ginsburg, 1985) and is a very good predictor of air-to-ground search performance (Stager and Hameluck, 1986).

There is mounting evidence that the visual system has specialized mechanisms, most likely in the brain, that are selectively tuned to narrow ranges of spatial frequency information (Ginsburg, 1978). By scanning a high-contrast grating of a particular spatial frequency, we can fatigue or adapt the specific analyzer tuned to the frequency of the grating being scanned. The contrast sensitivity will then be decreased only for gratings of that, and nearby, spatial frequencies (Blakemore and Campbell, 1969). Lunn and Banks (1986) found this effect in subjects after they viewed lines of text on a computer screen for several minutes. Lunn and Banks found that the spatial frequency of the lines of text adapted or fatigued the specific spatial frequency analyzers in the 2 to 6 cpd range. Because frequencies within this range are a primary stimulus for the eye's natural reflexive accommodation, Lunn and Banks hypothesized that fatiguing the analyzers might reduce the reflexive accommodation of the eye and, thus, could be the cause of blurred vision and visual strain often experienced by computer terminal users.

Contrast sensitivity is fast replacing standard visual acuity tests as the measure of preference for assessing visual accommodative capabilities.

Factors Affecting Visual Acuity and Contrast Sensitivity

There are many variables that affect visual acuity and contrast sensitivity. Some of these variables are listed and briefly discussed below.

Luminance Level In general, acuity and contrast sensitivity increase with increasing levels of light or background luminance (lightness) and then level off. With higher levels of light, the cones are activated, resulting in higher acuity and sensitivity.

Contrast When the contrast between a visual target and its background is low, the target must be larger for it to be equally discriminable as a target with greater contrast. It should be mentioned that there are many ways of defining contrast and various authors use different definitions. In addition to modulation contrast defined above, two other popular definitions of contrast are:

$$\text{Luminous contrast} = \frac{L_{\text{max}} - L_{\text{min}}}{L_{\text{max}}}$$

$$\text{Contrast (or luminance) ratio} = \frac{L_{\text{max}}}{L_{\text{min}}}$$

These various contrast measures can be converted from one to another. For example, if given the contrast ratio (CR), luminous contrast equals (CR − 1)/CR, and modulation contrast equals (CR − 1)/(CR + 1).

Exposure Time In general, under high illumination conditions, acuity improves with increased exposure time up to 100 or 200 ms and then levels off.

Target Motion The movement of a target or the observer (or both) decreases visual acuity. The ability to make visual discriminations under such circumstances is called *dynamic visual acuity*. Dynamic visual acuity comes into play, for example, when driving in a car and looking at objects at the side of the road. Acuity deteriorates rapidly as the rate of motion exceeds about 60 degrees per s (Burg, 1966). The consensus has been that dynamic visual acuity and static visual acuity are not correlated. However, recent evidence suggests that they may indeed be moderately correlated ($r = 0.50$) (Scialfa et al., 1988).

Age As most of us are doomed to discover, visual acuity and contrast sensitivity declines with age. The decline generally begins after age 40 and is continuous throughout the rest of life. By age 75, acuity has declined to about 0.6 or 20/30 Snellen acuity (Pitts, 1982). After age 40 contrast sensitivity shows a decline at spatial frequencies above 2 cpd (Owsley, Sekuler, and Siemsen, 1983). Age-related vision changes are discussed further in Chapter 16. Suffice to say that if elderly people are to use visual displays, the displays should be designed accordingly, providing large targets and adequate illumination. Such displays will be easier for younger people to use as well.

Training One might think that short of wearing corrective lenses or having eye surgery, there is not much one can do to improve visual acuity. Evidence suggests that this is not true. Roscoe and Couchman (1987), for example, trained subjects to control their focus, and their Snellen acuity at 6 m improved by an average of 14 percent. Their contrast sensitivity at 12 cpd improved by an average of 32 percent. Long and Rourke (1989) reported improvements in dynamic visual acuity with practice. Although practice doesn't make perfect, it does appear to help.

Adaptation

When you first go into a dark movie theater you can hardly see anything except what is on the screen. But after a few minutes you can see empty seats and before long you can even recognize the face of a friend in the audience. These changes in our sensitivity to light are called *adaptation*. In the dark, the sensitivity of the visual system increases over time and we can see dimmer and dimmer objects. This process of dark adaptation takes place in two phases. The first phase takes about 5 min and represents the adaptation of the cones. The second phase takes about 30 to 35 min and represents the adaptation of the rods.

After leaving the movie theater and emerging outside on a sunny day, you will find that it is difficult to see until your eyes adapt to the new light level. The reason it is difficult to see at first is because the intense sun overwhelms the system that has become very sensitive to light having been in the dark for a period of time. The absolute sensitivity of the eye decreases when exposed to high levels of light. Light adaptation occurs more quickly than dark adaptation and is complete in a minute or so. If the eye becomes less sensitive when exposed to light, why is it that increasing the level of light increases our acuity? The reason is that with more light the cones are activated in the fovea, and when there is sufficient light for them to operate, they are very sensitive to differences in light levels (i.e., contrast).

Color Discrimination

Recall that the cones in the retina (fovea) are responsible for our perception of color. There are actually three types of cones, each sensitive to a range of wavelengths of light that center on those that correspond to one of the primary colors (red, green, and blue). Because cones must be activated to see color, color vision is reduced or nonexistent in dark conditions.

People with normal color vision (called *trichromats*) can distinguish hundreds of different colors. Complete lack of color vision is very, very rare (such people are called *monochromats*). More common are people with a color deficiency in one or another type of cone receptor, usually the red or green receptors (they are *dichromats* or *anomalous trichromats*). If the green cones are deficient, the person can distinguish only 5 to 25 color hues, all of which are perceived as shades of blue and yellow. A deficiency of the red receptors decreases ability to distinguish reds and oranges. Color-vision deficiency can be inherited or acquired as a result of injury, disease, or certain types of poisoning. Overall, about 8 percent of males and less than 0.5 percent of females are dichromats and anomalous trichromats. It is very difficult, however, to precisely determine the type and degree of a person's color deficiency, and wide variability exists within a class of color deficiency.

When comparing color defectives with normals on such practical tasks as judging signal lights at sea, sorting color-coded electronic components, or identifying and reacting to traffic signals, color defectives' performance is always poorer than normals' performance, but there can be considerable overlap between the groups. Among color defectives those with red deficiencies usually do worse than those with green deficiencies (Kinney, Paulson, and Beare, 1979).

Reading

Reading is more than just acuity and letter recognition. The process of reading involves complex eye movements that are affected by the physical characteristics and content of what is being read. As you read this sentence, your eyes are not moving smoothly across the page. In fact, a stable image of the text can only be formed when both the eyes and object are stationary. Therefore, eye movements during reading are characterized by a succession of fast movements (*saccades*) and stationary periods (*fixations*). It should be pointed out that even during the fixation periods, the eye makes small tremor movements so that individual receptors on the retina are not fatigued. The average length of a saccade is about 2 degrees of visual angle (6 to 8 character spaces) and the average fixation duration is about 200 to 250 ms for a skilled reader (Rayner, 1977). There are, needless to say, large differences between individuals, but there is also wide variation in saccade length and fixation times for a single person reading a particular passage.

You may have noticed that as you read the previous paragraph your eye movement (not counting the movements to return to the beginning of the next line) were not always from left to right, but occasionally would go in the opposite direction. These reverse movements are called *regressions* and occur about 10 to 20 percent of the time in skilled readers (Rayner, 1978). They seem to occur when the reader has difficulty understanding the text or overshoots the next fixation target.

Perception

People certainly need to be able to "see" the relevant features of visual displays they use (such as traffic signs, electrocardiograph tapes, or pointers on pressure gauges). But the ability to see such features usually is not enough to make appropriate decisions based on the information in question. People must also understand the meaningfulness of what they see. Here we bump into the concept of perception as discussed in Chapter 3. Although perception does involve the sensing of stimuli, it concerns primarily the interpretation of that which is sensed. This is essentially a cerebral rather than a sensory process. In many circumstances this interpretation process is, of course, fairly straightforward, but in the use of most visual displays it depends on previous learning (as by experience and training), such as learning shapes of road signs, color codes of electric wiring systems, abbreviations on a computer keyboard, implications of the zigs and zags of an electrocardiograph recording, or even the simple recognition of letters of the alphabet.

In considering visual displays, then, clearly the design should meet two objectives: the display must be able to be seen clearly and the design should help the viewer to correctly perceive the meaning of the display. Although displays are designed for specific work situations, of course, adequate training of workers to interpret the meaningfulness of the specific display indications is needed; but appropriate design should capitalize on those display features that help people to understand the meaning, that is, to perceive correctly what they sense. We should add that certain types of visual stimuli play tricks on our perception. For example, a straight line usually seems distorted when it is viewed against a background of curved or radiating lines. Vertical lines and spaces appear larger than equal-length horizontal lines or spaces. One of your authors used this *horizontal-vertical illusion* in a court case to explain why people often contact power lines when erecting antennas and get electrocuted. The vertical clearance to the power line appears larger than the horizontal antenna lying on the ground, even when it is not.

TEXT: HARDCOPY

The form and technology of presenting alphanumeric text has changed over the centuries, from stone carvings to handwritten scrolls, to printed books, magazines, and newspapers, and finally to electronic media such as computer screens. With the advent of computers it has become customary to speak about "hardcopy" when referring to information on paper. At one time, people talked about the computer creating a paperless office, but you don't hear much about that anymore. If anything, the computer has increased the hardcopy generated and used in offices. Office copiers (color, zoom, etc.), laser printers, and fax machines are the office equipment of the 1990s—and they all generate hardcopy.

In this section we discuss some human factors aspects of hardcopy text. In the next section we discuss issues involved in presenting text on a computer screen. Before focusing on specific issues, however, we should be familiar with the distinctions among certain human factors criteria related to text presented by any means:

- *Visibility:* The quality of a character or symbol that makes it separately visible from its surroundings. (This is essentially the same as the term *detectability*, as used in Chapter 3).
- *Legibility:* The attribute of alphanumeric characters that makes it possible for each one to be identifiable from others.* This depends on such features as stroke width, form of characters, contrast, and illumination. (This is essentially the same as the term *discriminability* as used in Chapter 3).
- *Readability:* A quality that makes possible the recognition of the information content of material when it is represented by alphanumeric characters in meaningful groupings, such as words, sentences, or continuous text. (This depends more on the spacing of characters and groups of characters, their combination into sentences or other forms, the spacing between lines, and margins than on the specific features of the individual characters.)

Typography

The term *typography* refers to the various features of alphanumeric characters, individually and collectively. For practical purposes, in everyday life most of

* For an excellent survey of legibility and related aspects of alphanumeric characters and related symbols, see Cornog and Rose (1967).

the variations in typography adequately fulfill the human factors criteria mentioned above (visibility, legibility, and readability). However, there are at least four types of circumstances in which it may be important to use "preferred" forms of typography: (1) when *viewing conditions are unfavorable* (as with poor illumination or limited viewing time), (2) when the *information is important or critical* (as when emergency labels or instructions are to be read), (3) when viewing occurs at a *distance*, and (4) when people with *poor vision* may use the displays. The following discussion of typography applies primarily to these conditions. (Under most other conditions, variations in typography can have esthetic values and still fulfill the criteria of visibility, legibility, and readability.)

Stroke Width The stroke width of an alphanumeric character usually is expressed as the ratio of the thickness of the stroke to the height of the letter or numeral, as illustrated in Figure 4-6. Some examples of letters varying in stroke width-to-height ratio are shown in Figure 4-7. The effects of stroke width are intertwined with the nature of the background (black on white versus white on black) and with illumination. A phenomenon called *irradiation* causes white features on a black background to appear to "spread" into adjacent dark areas, but the reverse is not true. Thus, in general, black-on-white letters should be thicker (have lower ratios) than white-on-black ones. Although these interactions are not yet fully understood, a few generalizations can be offered, adapted in part from Heglin (1973) (these are predicated on the assumption of good contrast):

- With reasonably good illumination, the following ratios are satisfactory for printed material: black on white, 1:6 to 1:8; and white on black, 1:8 to 1:10.
- As illumination is reduced, thick letters become relatively more readable than thin ones (this is true for both black-on-white and white-on-black letters).
- With low levels of illumination or low contrast with background, printed letters preferably should be boldface type with a low stroke width-to-height ratio (such as 1:5).

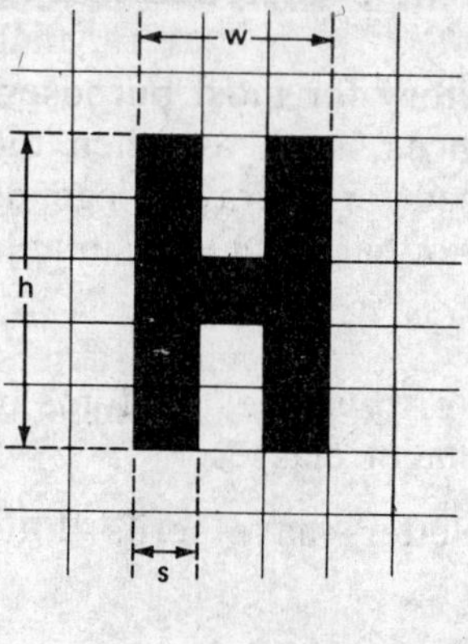

FIGURE 4-6
Dimensions used to compute stroke width-to-height and width-to-height ratios. Ratios can also be expressed as a proportion, e.g., 1:10 = 0.10. The letter shown has a stroke width-to-height ratio of 1:5 (0.20) and a width-to-height ratio of 3:5 (0.60).

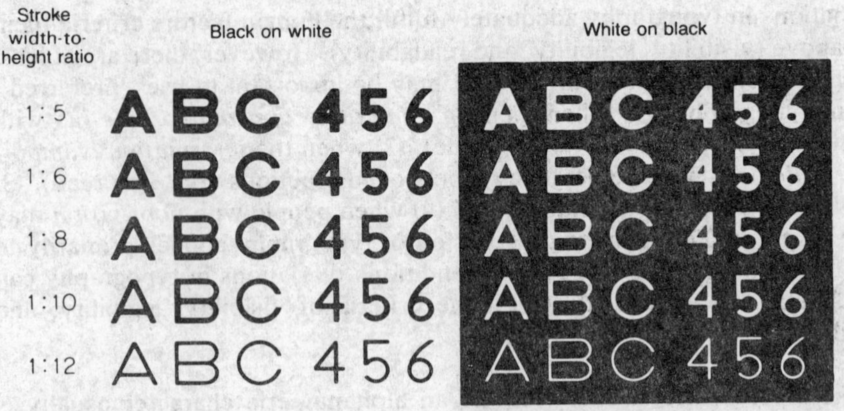

FIGURE 4-7
Illustrations of stroke width-to-height ratios of letters and numerals. With reasonably good illumination, the following ratios are satisfactory for printed material: black on white, 1:6 to 1:8; and white on black, 1:8 to 1:10.

- For highly luminous letters, ratios could be reduced to 1:12 to 1:20.
- For black letters on a very highly luminous background, very thick strokes are needed.

Width-to-Height Ratio The relationship between the width and height of a complete alphanumeric character is described as the *width-height ratio* and is expressed, as shown in Figure 4-6, as a ratio (such as 3:5) or as a proportion (such as 0.60). A 3:5 ratio has its roots in the fact that some letters have five elements in height (such as the three horizontal strokes of the letter B plus the two spaces between these) and three elements in width (such as the two vertical strokes of the letter B plus the space between them). Because of these geometric relationships, a 3:5 ratio has come into fairly common use and, in general, is reasonably well supported by research. However, Heglin (1973) argues against the use of a fixed ratio for all letters, pointing out that the legibility of certain letters would be enhanced if their width were adjusted to their basic geometric forms (such as O being a perfect circle and A and V being essentially equilateral triangles).

Although a 3:5 width-height ratio is quite satisfactory for most purposes, wider letters are appropriate for certain circumstances, such as when the characters are to be transilluminated or are to be used for engraved legends (Heglin, 1973). Figure 4-8 shows letters with a 1:1 ratio that would be suitable for such purposes.

Styles of Type There are over 30,000 type styles (or typefaces, or fonts of type) used in the printing trade. These fall into four major classes:

- *Roman:* This is the most common class. The letters have serifs (little flourishes and embellishments).

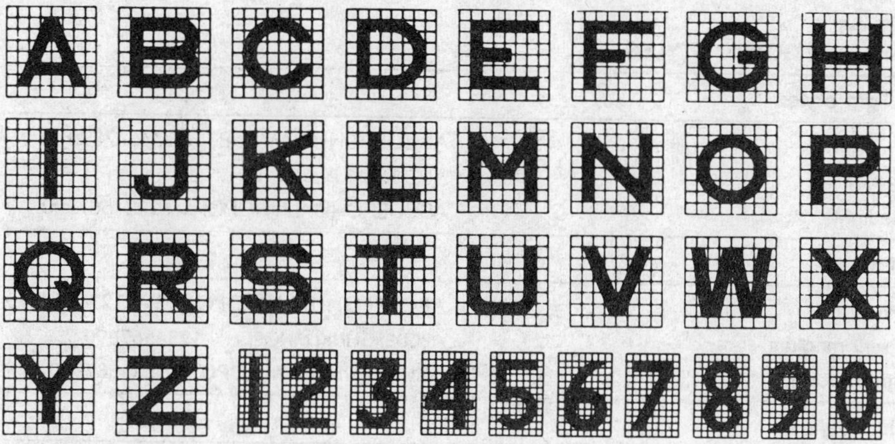

FIGURE 4-8
Letter and numeral font of United States Military Specification no. MIL-M-18012B (July 20, 1964); also referred to as NAMEL (Navy Aeronautical Medical Equipment Laboratory) or AMEL. The letters as shown have a width-height ratio of 1:1 (except for I, J, L, and W). The numerals have a width-height ratio of 3:5 (except 1 and 4).

- *Gothic:* The letters are of uniform stroke width and without serifs (these styles are also called *sans serif*).
- *Script:* These styles simulate modern handwriting and are used on such things as wedding invitations.
- *Block letter:* These styles resemble the German manuscript handwriting of the fifteenth century.

Roman types are used for conventional text materials (including this book) and are set in upper case (capitals) and lowercase (small) letters. Most Roman styles are about equally legible. The appearance of a type style can be changed by making it italic or boldface. *Italics, such as this, are used for emphasis, for titles, to indicate special usage of words, etc.* **Boldface, such as this, is used for headings and labels, in some instances for special emphasis, and to aid legibility under poor reading conditions.**

On the basis of research and experience in the military, certain type styles of uppercase letters and numerals have been developed and recommended for such things as words or abbreviations used in labels. One such set is shown in Figure 4-8. This type style is not specifically represented by standard type styles. Certain commercial types, however, have somewhat similar formats. A few Gothic styles are illustrated in Table 4-1.

Size

In the printing trade the size of type is measured in *points*. One point (pt) equals 1/72 in (0.35 mm). This height, however, is the height of the slug on which the type is set and includes space for *descenders* (e.g., the tail of the letter "q")

TABLE 4-1
EXAMPLE OF GOTHIC STYLES

Very light styles	
Futura light	ABCDEFGHIJKLMNOP 1234567890
Sans serif light	ABCDEFGHIJKLM 1234567890
Vogue light	ABCDEFGHIJKLMNOPQR 1234567890
Light styles	
Futura book	ABCDEFGHIJKLMNOPQRST 1234567890
Sans serif medium	ABCDEFGHIJKLMNOPQ 1234567890
Tempo medium	ABCDEFGHIJKLMNOPQ 1234567890
Medium styles	
Futura medium	ABCDEFGHIHKLMNOP 1234567890
Sans serif bold	**ABCDEFGHI 1234567890**

ascenders (e.g., the top of the letter "h"), capitals, and space between lines of text. Point size, therefore, is not a good representation of actual letter height. A close approximation of the height of capital letters in points is to consider 1 pt as equivalent to 1/100 in (0.25 mm) rather than 1/72 in. Some examples of different sizes are given below with the sizes of the slugs and the approximate heights of the capital letters in inches:

This line is set in 4-pt type (slug = 0.055; letters = 0.04).

This line is set in 6-pt type (slug = 0.084; letters = 0.06).

This line is set in 8-pt type (slug = 0.111; letters = 0.08).

This line is set in 9-pt type (slug = 0.125; letters = 0.09).

This line is set in 10-pt type (slug = 0.139; letters = 0.10).

This line is set in 11-pt type (slug = 0.153; letters = 0.11).

This line is set in 12-pt type (slug = 0.167; letters = 0.12).

For Close-Up Reading When reading a book or similar material, the normal reading distance is usually somewhere between 12 in and 16 in, with 14 in (35.5 cm) considered a nominal reading distance. In selecting the size of alphanumeric characters within such distances there are few absolute standards. If, however, we accept that the commonly used 9- to 11-pt print sizes of newspaper and magazines are suitable, then such sizes (0.09 to 0.11 in; 2.3 to 2.8 mm; 22 to 27 min of visual angle) would be acceptable as a basis for general use printed alphanumeric material.

When the reading is critical or is performed under poor illumination or when the characters are subject to change (as on some instruments), the character heights should be increased. Certain sets of recommendations have been made for the size of alphanumeric characters for visual displays, such as on instrument panels, that take such factors into account. One such set is given in Table 4-2.

For Distance Reading It is generally assumed that the legibility and readibility of alphanumeric characters are equal at various distances if the characters are increased in size for the viewing distance so that the visual angle subtended at the eye is the same. The National Bureau of Standards (Howett, 1983) developed a formula for determining the stroke width of letters to be read at various distances by people with various Snellen acuity scores. Knowing the stroke width-to-height ratio of the font to be used, it is a simple matter to determine the height of the letters. The formulas are as follows:

$$W_s = 1.45 \times 10^{-5} \times S \times d$$

$$H_L = W_s/R$$

Where W_s, d, and H_L are in same units (in or mm) and:

W_s = stroke width
S = denominator of Snellen acuity score (e.g., Snellen acuity = 20/20; S = 20; Snellen acuity = 20/40; S = 40)
d = reading distance
H_L = letter height
R = stroke width-to-height ratio of the font, expressed as a decimal proportion (e.g., R = 0.20 for a ratio of 1:5)

TABLE 4-2
ONE SET OF RECOMMENDED HEIGHTS OF ALPHANUMERIC CHARACTERS FOR CRITICAL AND NONCRITICAL USES UNDER LOW AND HIGH ILLUMINATION AT 28 IN VIEWING DISTANCE

	Height of numerals and letters*	
	Low luminance (down to 0.03 fL)	High luminance (1.0 fL and above)
Critical use, position variable	0.20–0.30 in (5.1–7.6 mm)	0.12–0.20 in (3.0–5.1 mm)
Critical use, position fixed	0.15–0.30 in (3.8–7.5 mm)	0.10–0.20 in (2.5–5.1 mm)
Noncritical use	0.05–0.20 (1.27–5.1 mm)	0.05–0.20 (1.27–5.1 mm)

* For other viewing distances (D), in inches, multiply tabled values by D/28.
Source: Adapted from Heglin (1973) and Woodson (1963).

Larger letters are required if the letters have low contrast against their background and/or are not adequately illuminated. Recognizing that Snellen acuity is not the best measure of people's ability to see letters at a distance (Evans and Ginsburg, 1985), it would probably be wise to design signs for Snellen acuities of, at best, 20/40. Using the formulas above, Table 4-3 presents recommended letter heights for various stroke width-to-height ratios at various distances using a Snellen acuity score of 20/40.

Case

Text can be presented in lowercase letters, such as this, OR IN UPPERCASE LETTERS, SUCH AS THIS. The general conclusion from the literature (Poulton, 1967) is that lowercase is easier to read than is all capitals. The reason suggested for the superiority of lowercase letters is that the shape of the envelope surrounding the whole word is more distinctive when lowercase letters are used than when all capitals are used. For example, the envelope around the word "pie" is somewhat different than that around the word "tie" because of the descender of *p* and the ascender of *t*. When the words are all uppercase, however, "PIE" and "TIE" are more similar; this probably means that the reader must examine the words more closely to distinguish them.

Labels, as on control panels, are often printed in all uppercase presumably to maximize the overall height of the letters within the space allocated. Phillips (1979) suggests that for search tasks, such as looking for a particular label on a control panel, the important aspect of the word is not the overall shape, but rather the initial letter. If the initial letter is larger than the other letters, the word should be found more quickly.

Layout

We have, for the most part, been discussing issues related to the design of the individual characters that make up text. Here we discuss just a few examples of how the layout of hardcopy text can influence reading performance.

TABLE 4-3
RECOMMENDED LETTER HEIGHTS (IN INCHES) FOR VARIOUS STROKE WIDTH-TO-HEIGHT RATIOS AT VARIOUS DISTANCES*

	Distance				
Stroke width-to-height ratio	**28 in**	**10 ft**	**20 ft**	**100 ft**	**1000 ft**
1:6	0.097	0.418	0.835	4.175	41.75
1:8	0.130	0.557	1.114	5.570	55.70
1:10	0.162	0.696	1.392	6.960	69.60

* Letter heights computed using formulas presented in the text and assuming a Snellen acuity score of 20/40.

Interletter Spacing The spacing between letters in text can be varied. Moriarty and Scheiner (1984) counted the words read by subjects reading copy from two sales brochures, one printed with regular interletter spacing and one printed with close-set (high-density) type. Portions of the brochures are reproduced in Figure 4-9. The authors found that the close-set type was read more rapidly than the regular-spaced type. Presumably, more characters are foveally visible at each fixation and fewer saccades are required to read a line of text.

Interline Spacing Wilkins and Nimmo-Smith (1987) suggest that the spatial frequency of lines of printed text on a page are similar to those that induce various visual effects and *may* provoke eye strain and headaches. They suggest that the clarity of printed text might be improved by increasing the spacing between lines of text.

Reading Ease

The readability of a passage of text depends on more than the physical characteristics of the letters or the particulars of how it is laid out. Readability is concerned with good writing style, sentence construction, and content. We cannot hope to present the many elements of good writing, some of which we have undoubtedly violated from time to time in this book. We relate a couple of aspects that have been researched, particularly within the context of writing instruction and manuals.

FIGURE 4-9
Portions of an advertising brochure used in a study of reading speed of regularly spaced types and of close-set (high-density) type. The close-set type was read more rapidly. (*Source: Adapted from Moriarty and Scheiner, 1984, Fig. 1. Copyright 1984 by the American Psychological Association. Adapted by permission of the author.*)

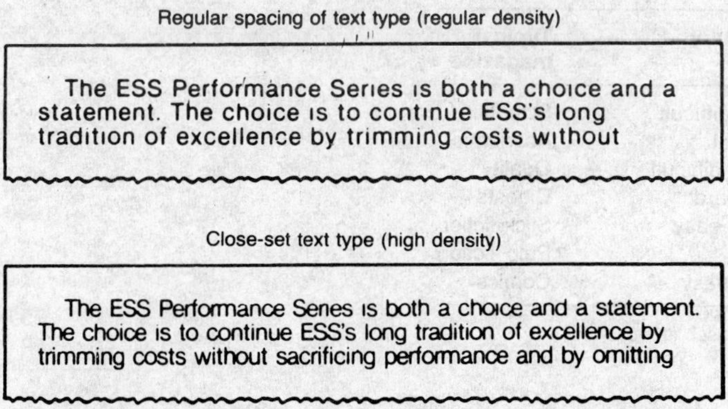

Type of Sentence A sentence can be active or passive, and it can be stated in the positive or negative. For example, the instruction "The green button starts the motor" is active. The passive form would be "The motor is started by the green button" and the negative form would be "The red button does not start the motor." Although Broadbent (1977) points out exceptions, the general rule is that simple, affirmative, active sentences are the easiest to understand.

Order of Words The instruction "Push the green button after turning off the gas" is likely to result in many green buttons being pushed with the gas on. The order of words in a sentence should match the order of actions to be taken. An improved instruction would be "Turn off the gas before pushing the green button."

Indices of Readability There are several different indices of readability (Klare, 1963). The most popular is probably the Flesch Reading Ease Score (Flesch, 1948). Based on a 100-word sample of the prose to be analyzed, the number of syllables (S) in the 100 words are counted in the way in which they would be read aloud and the average number of words per sentence (W) is computed. The following formula is then applied:

$$\text{Score} = 206.835 - (0.846 \times S) - (1.015 \times W)$$

Table 4-4 is then used to assess the reading ease of the material. Based on a 100-word sample from this chapter, the reading ease score was equal to 51, which would be rated "fairly difficult" to read.

It should be obvious that there is more to reading ease than just syllable and word counts. Randomly rearranging the words in sentences will not change the reading ease score, but it would surely affect ease of reading and understanding. Given that proper grammar is employed, however, the use of simple, short

TABLE 4-4
INTERPRETATION OF FLESCH READING EASE SCORES

Score	Difficulty level	Typical magazine
0 to 30	Very difficult	Scientific
30 to 50	Difficult	Academic
50 to 60	Fairly difficult	Quality
60 to 70	Standard	Digests
70 to 80	Fairly easy	Slick fiction
80 to 90	Easy	Pulp fiction
90 to 100	Very easy	Comics

Source: Adapted from Flesch (1948).

sentences made up of short words will certainly improve the reading ease of a passage of text and will be reflected in a higher reading ease score.

TEXT: VDT SCREENS

Reading text from a VDT (visual display terminal) or VDU (visual display unit, or computer screen) is not the same as reading text from hardcopy. Gould and Grischkowsky (1984) reported that people proofread about 20 to 30 percent slower when the material was presented on a VDT than when it was presented on hardcopy. These results were confirmed by Wilkinson and Robinshaw (1987). Harpster, Freivalds, Shulman, and Leibowitz (1989) found similar results for a visual search task.

The reason people read more slowly from VDTs than from hardcopy appears to be related to image quality (Gould et al., 1986, 1987; Harpster et al., 1989). The characters or images on a dot-matrix VDT are formed when combinations of many thousands of elements in a matrix are on or off (or lit to various levels of intensity). This matrix consists of many horizontal raster scan lines (or simply *scan lines*), each of which is made up of many separate elements called *pixels*. A scan line is an individual line formed by one horizontal sweep of the electron beam inside the VDT. Other display technologies, such as liquid crystal displays (LCD), use a different technology to illuminate points, but still can be thought of as consisting of a matrix of scan lines and pixels. Some typical VDT screens have resolutions of 320 × 200 pixels (320 pixels across by 200 lines down), 640 × 200, 640 × 480, and even 2000 × 2000.

The higher the resolution of the screen, the less the difference between reading from a VDT and reading from hardcopy. Harpster et al. (1989) report evidence that suggests that low-quality (resolution) characters stimulate different spatial frequency responders in the brain than do high-quality characters. This difference results in poorer accommodation to low-quality characters of VDTs and may be the cause of the poorer performance associated with reading from older VDTs. With newer, higher-resolution VDTs (monitors), however, reading from VDTs should be as easy as reading from hardcopy.

Typography

On dot-matrix displays, characters are made up from a matrix of pixels. The size of the matrix used for an individual character can range from 5 × 7 (5 wide by 7 high) to 15 × 24 and larger. An example of a letter formed from a 7 × 9 matrix is shown in Figure 4-10. Larger matrix characters are usually displayed on higher-resolution screens; otherwise the size of the letters would not permit the standard 24 lines of 80 characters each. With small matrices (5 × 7 or 7 × 9) the individual pixels making up the characters are visible and reading performance suffers. With larger matrices, the individual pixels are not as

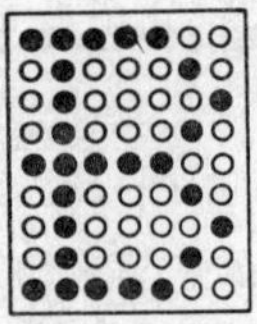

7 X 9 dot matrix

FIGURE 4-10
Example of a dot-matrix letter. All letters and numerals can be formed from combinations of the dots.

distinct and performance improves. A 7 × 9 matrix is usually considered to be the minimum size where reading of continuous text is required (Human Factors Society, 1988).

Reading Distance

Typically, a VDT screen is viewed at a somewhat greater distance than is hardcopy text. For example, Grandjean, Hunting, and Piderman (1983) reported eye-to-screen distances from 24 to 36 in (61 to 93 cm), with a mean of 30 in (76 cm). The American National Standards Institute (ANSI) (Human Factors Society, 1988) indicates that when the user is seated in an upright position, the typical eye-to-keyboard distance is about 18 to 20 in (45 to 50 cm). To reduce the need to accommodate the eyes to different distances, the screen should be placed at about that distance. The problem is that people often do not assume an upright posture when sitting at VDT terminals (see Chapter 13 for a discussion of VDT workstation layout). As we will see, recommendations for character height are often expressed in terms of visual angle. To translate visual angle to character height, a nominal viewing distance must be assumed. Taking the bull by the horns, and risking being gored by researchers in the field, we will take 20 in (45 cm) as a nominal VDT reading distance.

Size

Several investigators have recommended character sizes for VDTs that subtend a minimum visual angle of 11 or 12 min of arc (Shurtleff, 1967; Snyder and Taylor, 1979). These values would result in heights of 0.06 to 0.07 in (1.5 to 1.8 mm) at our nominal reading distance. This is somewhat smaller than recommended heights for hardcopy text.

ANSI (Human Factors Society, 1988) specifies that the minimum character height of a capital letter should be 16 min of visual angle for reading tasks in which legibility is important, with a preferred character height of 20 to 22 min of visual angle for such tasks. Translating these into heights at our nominal reading distance of 20 in gives a minimum height of 0.09 in (2.3 mm) and a preferred height of 0.116 to 0.128 in (2.9 to 3.3 mm). These, then, are more in line with height recommendations for hardcopy text.

ANSI also sets a maximum character height of 24 min of visual angle (0.14 in at 20 in reading distance). With larger characters, fewer of them can be foveally viewed during a fixation, thereby requiring more fixations to read a sentence.

The larger sizes of characters recommended by ANSI for VDTs reflects the threshold height for *comfortable* reading, rather than accurate legibility of a single character, which is 11 to 12 min of visual angle. The slightly larger character size recommended by ANSI is probably a good idea given the inherent differences in image quality between VDT screens and hardcopy.

Hardware Considerations

There are many hardware considerations of VDTs that affect legibility and readability, including such factors as *refresh rate* (how often the electron beam passes a particular pixel per unit time), *jitter* (variations in geometric location of a fixed object on the screen over time), *phosphor persistence* (how rapidly the intensity of the pixels decreases after being energized by the electron beam), and *flicker* (variations in intensity of the screen, which is dependent in part on the phosphor persistence and refresh rate of the screen). We have chosen to discuss two hardware-related issues—polarity and color.

Polarity Characters on a VDT screen can be dark against a light background (much like dark ink on white paper) or light against a dark background. A light background reduces glare and the visibility of reflections on the screen but increases the likelihood of perceiving flicker. To avoid perceptions of flicker, displays with light backgrounds must have a higher refresh rate than screens with dark backgrounds. This is true because sensitivity to flicker is greater the higher the background brightness. (Use of a light background may also decrease the life of the VDT tube.) Beyond these considerations, the research evidence is mixed. Some studies report no effect of polarity (Zwahlen and Kothari, 1986; Pawlak, 1986) while others find improved performance (speed and accuracy) with dark characters on a light background (Cushman, 1986; Snyder, Decker, Lloyd, and Dye, 1990).

Color Monochrome (single-color) monitors typically display green, amber, or white characters against a dark background; however, some lap-top computers are now even using red characters against a light background. The research comparing green, amber, and white has yielded mixed results with personal preference probably being the biggest factor to consider (National Research Council, 1983).

With full-color monitors, text and background can be almost any color from red on green to blue on black. There are no specific recommendations for the best color combination for reading text. The recommendations that are made include the following:

1 Use as few colors as possible; too many colors makes the display look like a Christmas tree and is distracting.

2 Avoid using the extremes of the color spectrum, the reds and blues (Matthews, 1987). There is, however, contradictory evidence on this point indicating that reds and blues need not be avoided (Matthews, Lovasik, and Mertins, 1989).

3 Avoid using the color pairs of saturated red and blue, and to a lesser extent, red and green, or blue and green on a dark background because of a phenomenon known as *chromostereopsis*, a false perception of depth in which one color appears closer to the observer than the other.

4 Maximize the color contrast between text and background; see Lippert (1986) for a discussion of, and methods for measuring, color contrast.

Screen Design Issues

People who use computers look at many screens of data: travel agents look up flight information; mechanics look up part numbers; and students look up reference information on library computers. Galitz (1980) reported the results of a study that estimated workers at a single insurance company using a single computer system would view 4.8 million screens of data per year. It is no wonder, then, that considerable thought has gone into producing guidelines for displaying information on computer screens (Galitz, 1985; Smith and Mosier, 1986; Brown, 1988). With whole books being written on the topic, we can only present a few global issues regarding screen design.

Density Overall density of information displayed on a screen is usually expressed in terms of the percentage of available character spaces being used. (On an 80-character by 24-line screen, there are 1920 available spaces.) Figure 4-11 presents two formats of the same information. The information, of course, would be meaningful only to persons familiar with the electrical test that yielded the data shown. The overall density of Figures 4-11*a* and 4-11*b* are 17.9 and 10.8 percent, respectively.

The results relating overall density to performance are generally consistent: as long as the necessary information is present, search time and errors increase with increasing density (Tullis, 1988). Tullis (1984) analyzed over 600 data screens and found that mean overall density was about 25 percent and that displays with densities higher than 40 or 50 percent were relatively rare.

Density can be minimized by appropriate use of abbreviations, avoiding unnecessary information, use of concise wording, and use of tabular formats with column headings. Compare Figures 14-11*a* and 14-11*b*. Note that the numbers are rounded off in Figure 14-11*b* because there was no need for levels of accuracy to hundredths of an ohm or volt. Units of measure (ohm, volt) were dropped because users of the system did not need them. Many words were also dropped from Figure 14-11*a* because they were unnecessary.

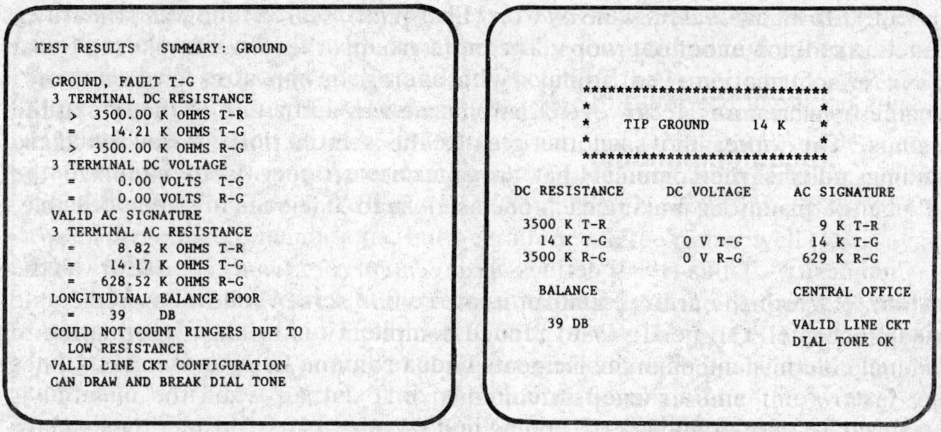

(*a*) (*b*)

FIGURE 4-11
Examples of a narrative format and a structured format in presenting information on a VDT screen. (These examples both present the same information about the condition of telephone lines, for use in diagnosing line problems.) (*Source: Tullis, 1981, Figs. 1 and 2. Copyright by the Human Factors Society, Inc., and reproduced by permission.*)

Overall density, although important, does not tell the whole story. Imagine displaying data that occupies 25 percent of the available character spaces and spacing it out over the entire screen versus squeezing it all into the upper right corner of the display. Both layouts would have the same overall density, but we would readily recognize that the displays are not equally easy to use. Tullis (1983) discusses the concept of *local density,* that is, the number of filled character spaces near other characters. Using blank lines and separating items on a line decrease local density. Tullis (1983) developed a mathematical procedure for determining local density which we will not go into here. Using the method to analyze Figure 14-11 yields local densities of 58 percent and 36 percent for *a* and *b,* respectively. This represents a greater difference than was the case for overall density. Tullis (1983) proposes that starting with low local densities and raising it increases performance, but at some point, increasing local density begins to degrade performance. That is to say, if a small amount of information is spread all over the screen, search time to find one item can be long. By packing the information a little more densely, search time can be improved, but eventually, the information becomes too crowded and performance deteriorates.

Grouping Grouping is the extent to which data items form well-defined perceptual groups. In Figure 4-11*a,* there appears to be one massive group of material, not counting the top title line. In Figure 4-11*b* the material forms several distinct perceptual groups. Tullis (1986*a*) found that for displays where the average size of data groups subtended less than 5 degrees of visual angle (about 12 to 14 characters wide by 6 or 7 lines high), search time was primarily a function of the number of groups. The more groups, the longer it takes to find a piece of information. For displays with average group sizes greater than 5 degrees of visual angle, however, search time was a function of the size of the groups. The larger the size, the greater the search time. Based on these findings, Tullis recommends that an optimum strategy is to minimize the number of groups by making each one as close to 5 degrees in size as feasible.

Complexity Tullis (1983) defines *arrangement* (or *layout*) *complexity* as the extent to which the arrangement of items on the screen follows a predictable visual scheme. The best way to reduce complexity is to align information in distinct columns and align the elements with a column. In general, search times are faster for items arranged in columns of text than when the material is arranged as a horizontal list of running text (Wolf, 1986). Consider the displays in Figure 4-12. The overall density of the displays are quite similar: 28.5 and 27.7 percent, respectively. Local density is considerably lower and the grouping sizes are near optimum (average about 5 degrees of visual angle) in Figure 4-12*b*. Beyond these factors is the alignment of the data in columns in Figure 14-12*b;* there is a predictable visual scheme: you know where to direct your eyes for the next piece of information. The complexity, therefore, is considerably lower in Figure 14-12*b* than in Figure 14-12*a*.

Highlighting One might think that using highlighting on a display (such as high-intensity letters, reverse video, color, or blinking) would reduce search time. The results, however, are not so neat (Fisher and Tan, 1989). Highlighted displays may be no better, and could even be worse, than unhighlighted screens. An important factor is the *validity* of highlighting, that is, the percentage of times the target being searched for is highlighted. If the highlighting has a

FIGURE 4-12
Illustrations of two formats for presenting information on VDT screens. Although the overall density level is nearly the same in the two formats, layout *b* has lower local density, near optimum grouping sizes, and reduced complexity because of the alignment of data in columns. (*Source: Adapted from Tullis, 1986*b, *Figs. 1 and 2. Reprinted with permission of the Human Factors Society, Inc. All rights reserved.*)

(*a*)

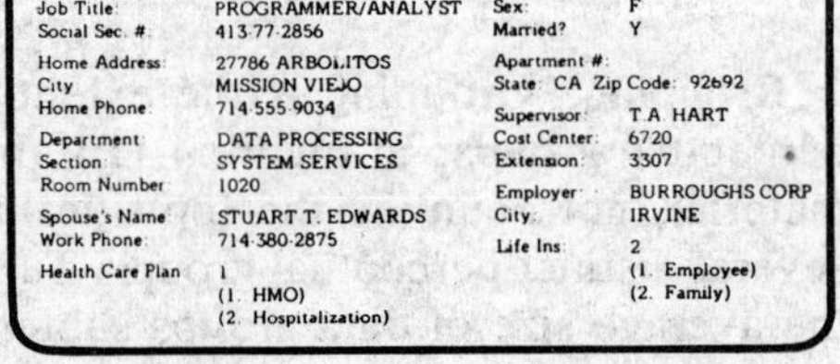

(*b*)

validity of over 50 percent, there will probably be some advantage of highlighting, and the higher the validity, the greater the advantage (Fisher and Tan, 1989).

Of all the various ways of highlighting material on a computer screen, one that should be avoided, except for critical-urgent information, is blinking. Blinking text is hard to read, is annoying, and can be distracting if other material must also be viewed. Blinking is best left for urgent messages and warnings, and even then, should be used sparingly.

GRAPHIC REPRESENTATIONS

Text or numeric data can be graphically represented. A description in a novel of an old Victorian house could be graphically represented by a drawing of the house itself. A table of data showing sales of various products over the years could be represented by a graph such as the ones shown in Figure 4-13.

Graphic Representations of Text

A picture is worth a thousand words—if it is the right picture. Often because of language differences, instructions and procedure manuals use pictures or drawings instead of words. For example, some airline passenger safety instruction cards (the ones located in the seatback pockets) use only pictures and drawings to explain such activities as how to put on and activate oxygen masks or how to evacuate the plane in an emergency. The general conclusion from the literature, however, is that pictorial information is important for speed, but text is

FIGURE 4-13
Illustrations of two graphic representations of data. Left, multiple-line graph; right, three-dimensional clustered bar chart. The data presented is the same in both representations.

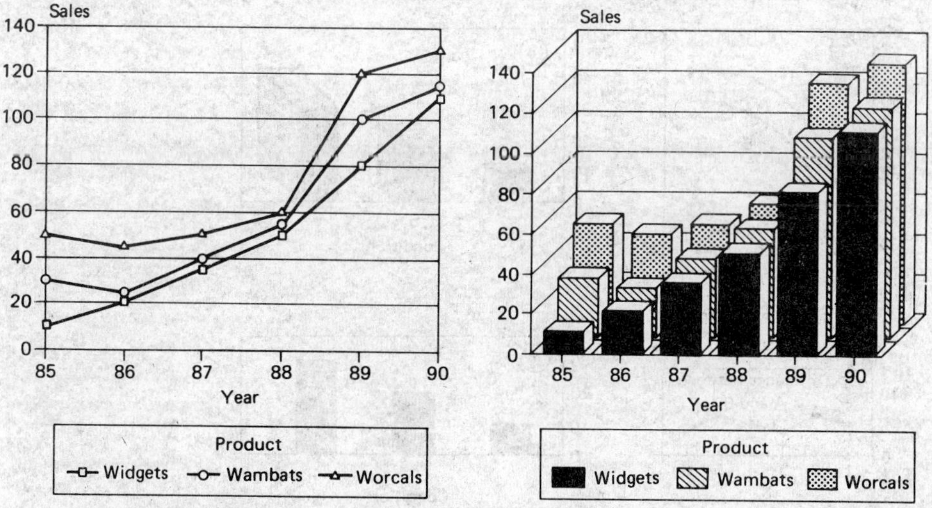

important for accuracy (Booher, 1975; Fisk, Scerbo, and Kobylak, 1986). The general recommendation for instructional materials is to combine pictures with text for speed, accuracy, and long-term retention.

Graphic Representation of Data

Many forms of graphic representations of data, that is, graphs, can be found in newspapers, on television, and in slide show sales presentations. Common types of graphs include pie charts, bar charts, and line graphs. Bar charts can be stacked or side by side. They can be presented in two dimensions or three dimensions. For example, Figure 4-13 shows a multiple-line graph and a three-dimensional bar chart of the same data. The research evidence suggests that there is no one best format for representing numeric data. Rather, different formats are best for conveying different types of information (Sparrow, 1989; Barfield and Robless, 1989).

Certain features of graphs can distort the perceptions of people and lead to inaccurate interpretations of the data. Examples of such features are shown in Figure 4-14. Think of part 1 as showing performance curves for two groups of subjects, A and B, over six trials (X) in an experiment. The form of these curves can create the perception (actually an illusion) that the *difference* between them increases over the trials. This, however, is not true, as shown in part 2. If one intended to focus on the differences between the two curves, the

FIGURE 4-14
Examples of possible distortions in perceptions of data presented in graphics. Part 1 can suggest that the difference between *A* and *B* increases; however, part 2 shows that this is not the case. Part 3 can suggest disproportionate increases from condition *a* to *b* to *c;* part 4 corrects for such an impression.

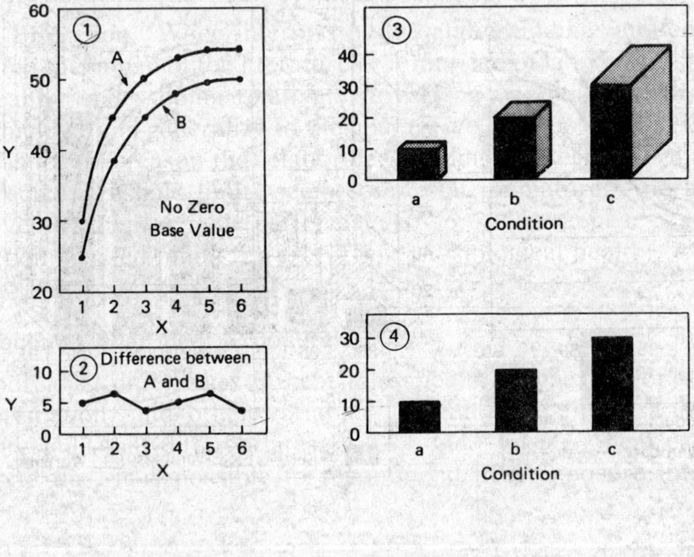

format of part 2 would be better. The common perception of part 3 probably is that the value for condition *b* is more than twice that for *a*, and that *c* is more than three times *a* because we "see" the volume of the block rather than their height. Part 4 corrects the distortion by showing the three heights only. Tufte (1983) provides many such examples of perceptual distortions and provides useful guidelines that would enhance the correct perception of the data displayed.

SYMBOLS

Civilization abounds with a wide assortment of visual symbols and signs that are intended to convey meaning to us. Examples include symbols for men's and women's rest rooms (which often are less distinguishable from one another than we would like them to be); symbols for windshield washers and wipers on some automobile dashboard controls; and little icons presented on computer screens of trash cans, file folders, and filing cabinets to symbolize various computer operations.

Comparison of Symbolic and Verbal Signs

Where there is some question as to whether to use a symbolic or verbal sign, a symbolic sign would probably be preferable *if* the symbol reliably depicted visually what it is intended to represent. One argument for this (supported by some research) is that symbols do not require the *recoding* that words or short statements do. For example, a road sign showing a deer conveys immediate meaning, whereas use of the words *Deer Crossing* requires the recoding from words to concept. Note, however, that some symbols do not visually resemble the intended concepts very well and thus may require learning and recoding.

A study that supports the possible advantage of symbols was conducted by Ellis and Dewar (1979) where subjects listened to a spoken traffic message (such as "two-way traffic") and were then shown a slide of a traffic sign (either a symbolic sign or a verbal sign) and asked to respond yes or no, depending on whether they perceived the sign as being the same as the spoken message. This was done under both visually "nondegraded" and "degraded" conditions. The mean reaction times, in Figure 4-15, show a systemic superiority for the symbolic signs, especially under the visually degraded conditions.

Objectives of Symbolic Coding Systems

When a system of symbols is developed, the objective is to use those symbols that best represent their referents (that is, the concepts or things the symbols are intended to represent). This basically depends on the strength of *association* of a *symbol* with its *referent*. This association, in turn, depends on either of two factors: any already established association [Cairney and Siess (1982) call this *recognizability*] and the ease of learning such an association.

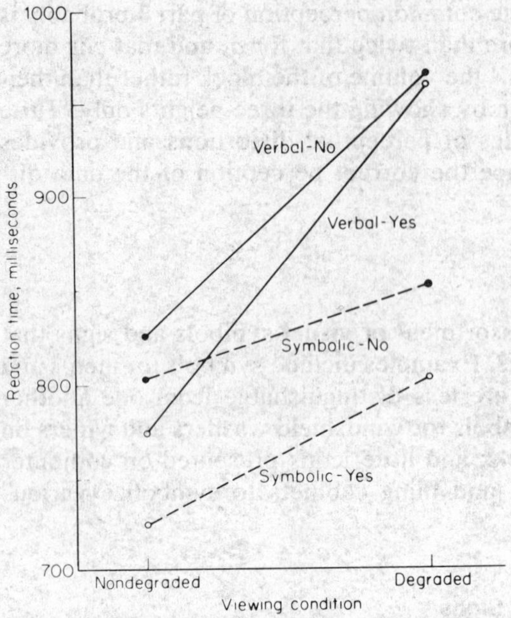

FIGURE 4-15
Mean reaction times of yes and no responses to symbolic and verbal traffic signs viewed under nondegraded and degraded viewing conditions. (*Source: Ellis and Deward, 1979, Fig. 1, p. 168.*)

At a more specific level, in Chapter 3 we discussed certain guidelines for using coding systems: detectability, discriminability, compatibility, meaningfulness, and standardization. These guidelines are equally applicable to the design of symbol systems. After all, symbols are really just another form of coding.

Criteria for Selecting Coding Symbols

Some coding symbols already exist (or could be developed) that could be used with reasonable confidence. If there is any question about their suitability, however, they should be tested by the use of some experimental procedure. Various criteria have been used in studies with such symbols. A few criteria are discussed below:

Recognition On this criterion, subjects usually are presented with experimental symbols and asked to write down or say what each represents.

Matching In some investigations, several symbols are presented to subjects along with a list of all the referents represented, and subjects are asked to match each symbol with its referent. A variation consists of presenting the subjects with a complete array of symbols of many types and asking them to identify the specific symbol that represents each referent. Sometimes reaction time is measured as well as the number of correct and incorrect matches.

Preferences and Opinions In other circumstances people are asked to express their preferences or opinions about experimental designs of symbols.

Examples of Code Symbol Studies

To illustrate the procedures used in code symbol studies, we give a few examples.

Mandatory-Action Symbols The first example deals with a set of so-called mandatory-action messages and their respective symbols, illustrated in Figure 4-16 (concerning use of protection devices for the ear, eye, foot, etc.) (Cairney and Siess, 1982). These symbols were shown to a group of newly arrived Vietnamese in Australia, who were first told where the signs were to be used and then asked to state what they thought each sign meant. A week later the subjects were given a recall test in the same manner. Figure 4-16 shows, for each symbol, the percentage of correct responses for both the original test (O) and for the second test (R). The results of this particular study show rather dramatically how well a group of people with a different cultural background could learn the intended meaning of those particular symbols. (Only the first symbol did not come through very well.)

This example illustrates the use of a recognition criterion and the desirable objective of using symbols that are easily learned, even if the symbols are not initially very recognizable.

Comparison of Exit Symbols for Visibility When a specific symbol has a poor association with its referent, there can be a problem in selecting or designing a better symbol. One approach is to create alternative designs for the same referent and to test them.

Such a procedure was followed by Collins and Lerner (1983) with 18 alternate designs of exit signs, particularly under difficult viewing conditions simu-

FIGURE 4-16
Symbols of mandatory-action messages used in a study of recognition and recall of such symbols. The percentages below the symbols are the percentages of correct recognition, as follows: O = original test; R = recall 1 week later. (*Source: Adapted from Cairney and Siess, 1982, Fig. 1.*)

	1: Must use ear protection	2: Must use eye protection	3: Must use foot protection	4: Must use hand protection	5: Must use head protection	6: Must use breathing protection
O →	10%	20%	30%	50%	37%	13%
R →	73	96	100	97	97	97

lating an emergency situation. (After all, you would like to be able to identify the exit if a building were on fire!) The subjects were presented with 18 designs of exit symbols under different levels of viewing difficulty (with very brief exposure time) and asked to indicate whether each was, or was not, an exit sign. Certain of the designs are shown in Figure 4-17, along with their percentages of error. (In reviewing the errors, it was found that certain symbols for "No exit" were confused with those for "Exit.") Certain generalizations about the features of the best signs can be made: (1) "filled" figures were clearly superior to "outline" figures, (2) circular figures were less reliably identified than those with square or rectangular backgrounds, and (3) simplified figures (as by reducing the number of symbol elements) seem beneficial.

Perceptual Principles of Symbolic Design

Much of the research regarding the design of symbols for various uses has to be empirical, involving, for example, experimentation with proposed designs. However, experience and research have led to the crystallization of certain principles that can serve as guidelines in the design of symbols. Easterby (1967, 1970), for example, postulates certain principles that are rooted in perceptual research and generally would enhance the use of such displays. Certain of these principles are summarized briefly and illustrated. The illustrations are in Figure 4-18. Although these particular examples are specifically applicable to machine displays (Easterby, 1970), the basic principles would be applicable in other contexts.

Figure to Ground Clear and stable figure-to-ground articulation is essential, as illustrated in Figure 4-18*a*. In the poor, unstable figure, the direction in which the arrow is pointing is ambiguous.

FIGURE 4-17
Examples of a few of the 18 exit signs used in a simulated emergency experiment, with percentages of errors in identifying them as exit signs. (*Source: Adapted from Collins and Lerner, 1983.*)

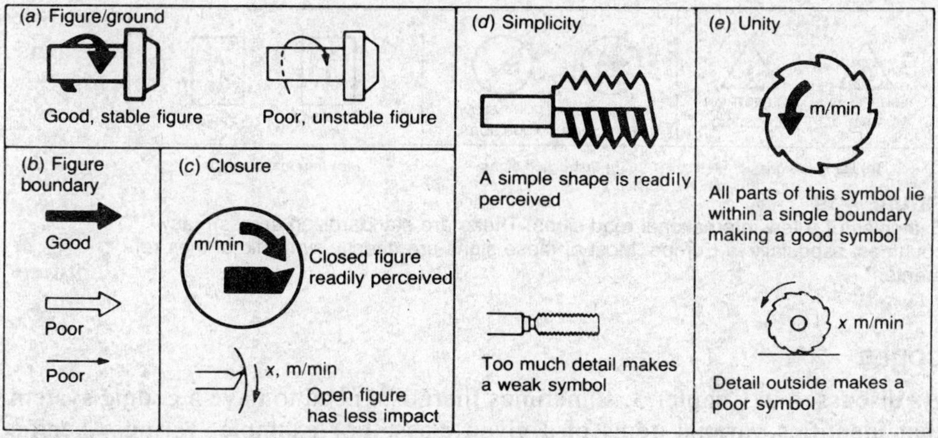

FIGURE 4-18
Examples of certain perceptual principles relevant to the design of visual code symbols. These particular examples relate to symbols used with machines. (*Source: Adapted from Easterby, 1970.*)

Figure Boundaries A contrast boundary (essentially a solid shape) is preferable to a line boundary, as shown in Figure 4-18*b*.

Closure A closed figure, as illustrated in Figure 4-18*c*, enhances the perceptual process and should be used unless there is reason for the outline to be discontinuous.

Simplicity The symbols should be as simple as possible, consistent with the inclusion of features that are necessary, as illustrated in Figure 4-18*d*.

Unity Symbols should be as unified as possible. For example, when solid and outline figures occur together, the solid figure should be within the line outline figure, as shown in Figure 4-18*e*.

Standardization of Symbolic Displays

When symbol displays might be used in various circumstances by the same people, the displays should be standardized with a given symbol *always* being associated with the same referent. One example of such standardization is the system of international road signs. A few examples are shown in Figure 4-19. The National Park Service also has a standardized set of symbols to represent various services and concepts such as picnic areas, bicycle trails, and playgrounds.

FIGURE 4-19
Examples of a few international road signs. These are standardized across many countries, especially in Europe. Most of these signs are directly symbolic of their referents.

CODES

As discussed in Chapter 3, sometimes there is a need to have a coding system that identifies various items of a given class. For example, various specific pipes in a plumbing system can be coded different colors, or various types of highway signs can be coded by different shapes. The items to be coded (such as different pipes or different types of highway signs) are called *referents*. Thus, a specific code such as an octagonal (eight-sided) road sign means "stop." The types of visual stimuli used in such circumstances are sometimes called *coding dimensions*. Some visual coding dimensions are color, geometric and other shapes, letters, numerals, flash rate (of a flashing light), visual angle (the positions of hands on a clock), size (variation in the size of circles), and even "chartjunk" [the term used by Tufte (1983) for the gaudy variations in shading and design of areas of graphs].

Single Coding Dimensions

When various coding dimensions can be used, an experiment can be carried out to determine what dimension would be best. One such study done by Smith and Thomas (1964) used the four sets of codes shown in Figure 4-20 in a task of counting the number of items of a specific class (such as red, gun, circle, or B-52) in a large display with many other items of the same class. (The density of items, that is, the total number of items in the display, was also varied.) The results, shown in Figure 4-21, show a clear superiority for the color codes for this task; this superiority was consistent for all levels of density.

Although these and other studies indicate that various visual coding dimensions differ in their relevance for various tasks and in various situations, definite guidelines regarding the use of such codes still cannot be laid down. Recognizing this, and realizing that good judgment must enter into the selection of visual codes for specific purposes, we see that a comparison such as that given in Table 4-5 can serve as at least partial guidance. In particular, Table 4-5 indicates the approximate number of levels of each of various visual codes that can be discriminated, along with some sideline comments about certain methods.

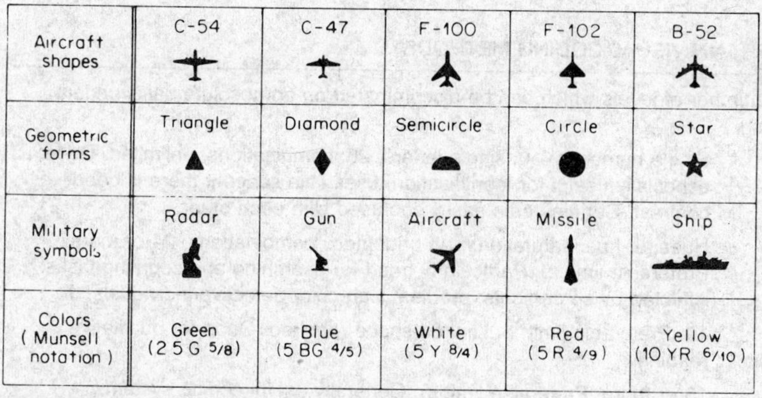

FIGURE 4-20
Four sets of codes used in a study by Smith and Thomas. The notations under the color labels are the Munsell color matches of the colors used. (*Source: Smith and Thomas, 1964. Copyright © 1964 by the American Psychological Association and reproduced by permission.*)

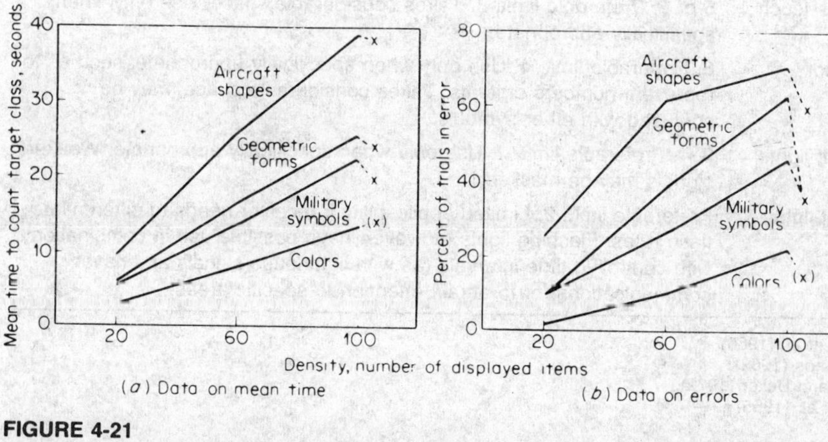

FIGURE 4-21
Mean time *a* and errors *b* in counting items of four classes of codes as a function of display density. The *X*s indicate comparison data for displays of 100 items with color (or shape) held constant. (*Source: Smith and Thomas, 1964. Copyright © 1964 by the American Psychological Association and reprinted by permission.*)

Color Coding

Since color is a fairly common visual code, we discuss it somewhat further. An important question relating to color deals with the number of distinct colors persons with normal color vision can differentiate on an absolute basis. It has generally been presumed that the number was relatively moderate; for example, Jones (1962) indicated that the normal observer could identify about nine

TABLE 4-5
SUMMARY OF CERTAIN VISUAL CODING METHODS

(Numbers refer to number of levels which can be discriminated on an absolute basis under optimum conditions.)

Alphanumeric	Single numerals, 10; single letters, 26; combinations, unlimited. Good; especially useful for identification; uses little space if there is good contrast. Certain items easily confused with each other.
Color (of surfaces)	Hues, 9; hue, saturation, and brightness combinations, 24 or more. Preferable limit, 9. Particularly good for searching and counting tasks. Affected by some lights; problem with color-defective individuals.*†
Color (of lights)	10. Preferable limit, 3. Limited space required. Good for qualitative reading.‡
Geometric shapes	15 or more. Preferable limit, 5. Generally useful coding system, particularly in symbolic representation; good for CRTs. Shapes used together need to be discriminable; some sets of shapes more difficult to discriminate than others.‡
Angle of inclination	24. Preferable limit, 12. Generally satisfactory for special purposes such as indicating direction, angle, or position on round instruments like clocks, CRTs, etc.§
Size of forms (such as squares)	5 or 6. Preferable limit, 3. Takes considerable space. Use only when specifically appropriate.
Visual number	6. Preferable limit, 4. Use only when specifically appropriate, such as to represent numbers of items. Takes considerable space; may be confused with other symbols.
Brightness of lights	3–4. Preferable limit, 2. Use only when specifically appropriate. Weaker signals may be masked.‡
Flash rate of lights	Preferable limit, 2. Limited applicability if receiver needs to differentiate flash rates. Flashing lights, however, have possible use in combination with controlled time intervals (as with lighthouse signals and naval communications) or to attract attention to specific areas.

*Feallock et al. (1966).
†M. R. Jones (1962).
‡Grether and Baker (1972).
§Muller et al. (1955).

surface colors, varying primarily in hue. However, it appears that, with training, people can learn to make upward of a couple of dozen such discriminations when combinations of hue, saturation, and lightness are prepared in a nonredundant manner (Feallock et al., 1966). When relatively untrained people are to use color codes, however, the better part of wisdom would argue for the use of only a small number of discriminable colors.

In general, color coding has been found to be particularly effective for "searching" tasks—tasks that involve scanning an array of many different stimuli to "search out" (or spot), or locate, or count those of a specific class. This advantage presumably is owing to the fact that colors "catch the eye" more readily than other visual codes. Examples of searching tasks include the use of maps and navigational charts, searching for items in a file, and searching for color-coded electric wires of some specific color. Christ (1975) analyzed the results of 42 studies in which color codes were compared with other types of codes and found that color codes were generally better for searching tasks than most other codes (geometric shapes, letters, and digits).

Christ also found that color codes were better for "identification" tasks than were certain other codes, but color codes were generally not as good for such tasks as letter and numerals were. In this sense, an identification task involves the conceptual recognition of the meaningfulness of the code—not simple visual recognition of items of one color from others (such as identifying red labels from all other labels). Thus, we can see the possible basis for the advantage of letters and numerals, since they could aid in tying down the meaning of codes (such as grade A or size 12).

In general, then, color is a very useful coding dimension, but it is obviously not a universally useful system.

Multidimension Codes

Chapter 3 included a discussion of multidimension codes, indicating some of the variations in combining two or more dimensions. Heglin (1973) recommends, however, that no more than two dimensions be used together if rapid interpretation is required. In combining two visual codes, certain combinations do not "go well" together. Some simply cannot be used in combination. The combinations that are potentially useful are shown as Xs in Figure 4-22. Although combinations of codes can be useful in many circumstances, they are not always more effective than single-dimension codes.

FIGURE 4-22
Potential combinations of coding systems for use in multidimension coding. (*Source: Adapted from Heglin, 1973, Tables VI-6, VI-22.*)

	Color	Numeral and letter	Shape	Size	Brightness	Location	Flash rate	Line length	Angular orientation
Color		X	X	X	X	X	X	X	X
Numeral and letter	X			X		X	X		
Shape	X			X	X		X		
Size	X	X	X		X		X		
Brightness	X		X	X					
Location	X	X						X	X
Flash rate	X	X	X	X					X
Line length	X					X			X
Angular orientation	X					X	X	X	

DISCUSSION

In our everyday lives, we are bombarded with scads of text, graphics, symbols, and codes. This chapter only scratched the surface of human factors issues involved with the display of such information. In the information age with increasing use of computers, more attention will be paid to human factors issues involved in the display of information in a form that can be quickly and accurately used without placing a burden on the receiver. Software packages such as word processors, data base programs, or file management utilities often fail or succeed in the marketplace based on how well information, both on the screen and in hardcopy manuals, is presented. Design of screen layouts and instruction manuals will continue to be areas in which human factors will make beneficial contributions.

REFERENCES

Barfield, W., and Robless, R. (1989). The effects of two- or three-dimensional graphics on the problem-solving performance of experienced and novice decision makers. *Behaviour and Information Technology,* 8, 369–385.

Blakemore, C., and Campbell, F. (1969). On the existence of neurones in the human visual system selectively sensitive to the orientation and size of retinal images. *Journal of Physiology,* 203, 237–260.

Boff, K., Kaufman, L., and Thomas, J, (eds.) (1986). *Handbook of perception and human performance,* vol. 1: *Sensory processes and perception.* New York: Wiley.

Booher, H. (1975). Relative comprehensibility of pictorial information and printed words in proceduralized instructions. *Human Factors,* 17, 266–277.

Broadbent, D. (1977). Language and ergonomics. *Applied Ergonomics,* 8, 15–18.

Brown, C. (1988). *Human-computer interface design guidelines.* Norwood, NJ: Ablex.

Burg, A. (1966). Visual acuity as measured by static and dynamic tests: A comparative evaluation. *Journal of Applied Psychology,* 50, 460–466.

Cairney, P., and Siess, D. (1982). Communication effectiveness of symbolic safety signs with different user groups. *Applied Ergonomics,* 13, 91–97.

Christ, R. (1975). Review and analysis of color coding research for visual displays. *Human Factors,* 17, 542–570.

Collins, B., and Lerner, N. (1983). *An evaluation of exit symbol visibility* (NBSIR 82-2685). Washington, DC: National Bureau of Standards.

Cornog, D., and Rose, F. (1967, February). *Legibility of alphanumeric characters and other symbols,* vol. 2: *A reference handbook* (NBS Misc. 262-2). Washington, DC: Government Printing Office.

Cushman, W. (1986). Reading from microfiche, a VDT, and printed page and subjective fatigue. *Human Factors,* 28(1), 63–73.

Easterby, R. (1967). Perceptual organization in static displays for man/machine systems. *Ergonomics,* 10, 195–205.

Easterby, R. (1970). The perception of symbols for machine displays. *Ergonomics,* 13, 149–158.

Ellis, J., and Dewar, R. (1979). Rapid comprehension of verbal and symbolic traffic sign messages. *Human Factors,* 21, 161–168.

Evans, D., and Ginsburg, A. (1985). Contrast sensitivity predicts age-related differences in highway-sign discriminability. *Human Factors*, 27, 637–642.

Feallock, J., Southard, J., Kobayashi, M., and Howell, W. (1966). Absolute judgments of colors in the Federal Standards System. *Journal of Applied Psychology*, 50, 266–272.

Fisher, D., and Tan, K. (1989). Visual displays: The highlighting paradox. *Human Factors*, 31, 17–30.

Fisk, A., Scerbo, M., and Kobylak, R. (1986). Relative value of pictures and text in conveying information: Performance and memory evaluations. *Proceedings of the Human Factors Society 30th Annual Meeting*. Santa Monica, CA: Human Factors Society, pp. 1269–1272.

Flesch, R. (1948). A new readability yardstick. *Journal of Applied Psychology*, 32, 221–233.

Galitz, W. (1980). *Human factors in office automation*. Atlanta, GA: Life Office Management Association.

Galitz, W. (1985). *Handbook of screen format design* (2d ed.). Wellesley Hills, MA: QED Information Sciences.

Ginsburg, A. (1978). *Visual information processing based on spatial filters constrained by biological data* (AFAMRL-TR-78-1129), vols. 1 and 2. Wright-Patterson Air Force Base, OH: Air Force Aerospace Medical Research Laboratory.

Ginsburg, A., Evans, D., Cannon, M., Owsley, C., and Mulvanny, P. (1984). Large-scale norms for contrast sensitivity. *American Journal of Optometry and Physiological Optics*, 61, 80–84.

Gould, J., Alfaro, L., Finn, R., Haupt, B., Minuto, A., and Salaun, J. (1986). Why is reading slower from CRT displays than from paper? *Proceedings of the Human Factors Society 30th Annual Meeting*. Santa Monica, CA: Human Factors Society, pp. 834–836.

Gould, J., Alfaro, L., Varnes, V., Finn, R., Grischkowsky, N., and Minuto, A. (1987). Reading is slower from CRT displays than from paper: Attempts to isolate a single-variable explanation. *Human Factors*, 29, 269–299.

Gould, J., and Grischkowsky, N. (1984). Doing the same work with hard copy and with cathode-ray tube (CRT) computer terminals. *Human Factors*, 26, 323–337.

Grandjean, E., Hunting, W., and Piderman, M. (1983). VDT workstation design: Preferred settings and their effects. *Human Factors*, 25, 161–175.

Grether, W., and Baker, C. (1972). Visual presentation of information. In H. Van Cott and R. Kinkade (eds.), *Human engineering guide to equipment design* (rev. ed.). Washington, DC: Government Printing Office.

Harpster, J., Freivalds, A., Shulman, G., and Leibowitz, H. (1989). Visual performance on CRT screens and hard-copy displays. *Human Factors*, 31, 247–257.

Heglin, H. (1973, July). *NAVSHIPS display illumination design guide*, vol. 2: *Human factors* (NELC-TD223). San Diego: Naval Electronics Laboratory Center.

Howett, G. (1983). *Size of letters required for visibility as a function of viewing distance and viewer acuity* (NBS Tech. Note 1180). Washington, DC: National Bureau of Standards.

Human Factors Society (1988). *American national standard for human factors engineering of visual display terminal workstations* (ANSI/HFS 100-1988). Santa Monica, CA: Human Factors Society.

Jones, M. (1962). Color coding. *Human Factors*, 4, 355–365.

Kinney, J., Paulson, H., and Beare, A. (1979). The ability of color defectives to judge signal lights at sea. *Journal of the Optical Society of America,* 69, 106–113.

Klare, G. (1963). *The measurement of readability.* Des Moines, IA: Iowa State University Press.

Lippert, T. (1986). Color difference prediction of legibility for raster CRT imagery. *Society of Information Displays Digest of Technical Papers,* 16, 86–89.

Long, G., and Rourke, D. (1989). Training effects on the resolution of moving targets—dynamic visual acuity. *Human Factors,* 31, 443–451.

Lunn, R., and Banks, W. (1986). Visual fatigue and spatial frequency adaptation to video displays of text. *Human Factors,* 28, 457–464.

Matthews, M. (1987). The influence of colour on CRT reading performance and subjective comfort under operational conditions. *Applied Ergonomics,* 18, 323–328.

Matthews, M., Lovasik, J., and Mertins, K. (1989). Visual performance and subjective discomfort in prolonged viewing of chromatic displays. *Human Factors,* 31, 259–271.

Moriarty, S., and Scheiner, E. (1984). A study of close-set type. *Journal of Applied Psychology,* 69, 700–702.

Muller, P., Sidorsky, R., Slivinske, A., Alluisi, E., and Fitts, P. (1955, October). *The symbolic coding of information on cathode ray tubes and similar displays* (TR-55-375). Wright-Patterson Air Force Base, OH.

National Research Council (1983). *Video displays, work, and vision.* Washington, DC: National Academy Press.

National Research Council (1988). *Myopia: Prevalence and progression.* Washington, DC: National Academy Press.

Owsley, C., Sekuler, R., and Siemsen, D. (1983). Contrast sensitivity throughout adulthood. *Vision Research,* 23, 689–699.

Pawlak, U. (1986). Ergonomic aspects of image polarity. *Behaviour and Information Technology,* 5, 335–348.

Phillips, R. (1979). Why is lower case better? *Applied Ergonomics,* 10, 211–214.

Pitts, D. (1982). The effects of aging on selected visual functions: Dark adaptation, visual acuity, stereopsis and brightness contrast. In R. Sekuler, D. Kline, and K. Dismukes (eds.), *Aging in human visual functions.* New York: Liss.

Poulton, E. (1967). Searching for newspaper headlines printed in capitals or lower-case letters. *Journal of Applied Psychology,* 51, 417–425.

Rayner, K. (1977). Visual attention in reading: Eye movements reflect cognitive processes. *Memory and Cognition,* 5, 443–448.

Rayner, K. (1978). Eye movements in reading and information processing. *Psychological Bulletin,* 85, 618–660.

Roscoe, S., and Couchman, D. (1987). Improving visual performance through volitional focus control. *Human Factors,* 29, 311–325.

Scialfa, C., Garvey, P., Gish, K., Deering, L., Leibowitz, H., and Goebel, C. (1988). Relationships among measures of static and dynamic visual sensitivity. *Human Factors,* 30, 677–687.

Shurtleff, D. (1967). Studies in television legibility: A review of the literature. *Information Display,* 4, 40–45.

Smith, S., and Mosier, J. (1986). *Guidelines for designing user interface software* (Tech. Rept. ESD-TR-86-278). Hanscom Air Force Base, MA: USAF Electronic Systems Division.

Smith, S., and Thomas, D. (1964). Color versus shape coding in information displays. *Journal of Applied Psychology,* 48, 137–146.

Snyder, H., Decker, J. Lloyd, C., and Dye, C. (1990). Effect of image polarity on VDT task performance. *Proceedings of the Human Factors Society 34th Annual Meeting*. Santa Monica, CA: Human Factors Society, pp. 1447–1451.

Snyder, H., and Taylor, G. (1979). The sensitivity of response measures of alphanumeric legibility to variations in dot matrix display parameters. *Human Factors*, 21, 457–471.

Sparrow, J. (1989). Graphical displays in information systems: Some data properties influencing the effectiveness of alternative forms. *Behaviour and Information Technology*, 8, 43–56.

Stager, P., and Hameluck, D. (1986, June). *Contrast sensitivity and visual detection in search and rescue*. Toronto: Defence and Civil Institute of Environmental Medicine.

Tufte, E. (1983). *The visual display of quantitative information*. Cheshire, CT: Graphics Press.

Tullis, T. (1981). An evaluation of alphanumeric, graphic, and color information displays. *Human Factors*, 25, 541–550.

Tullis, T. (1983). The formatting of alphanumeric displays: A review and analysis. *Human Factors*, 25, 657–682.

Tullis, T. (1984). *Predicting the usability of alphanumeric displays*. Ph.D. dissertation. Rice University. Lawrence, KS: The Report Store.

Tullis, T. (1986*a*). Optimizing the usability of computer-generated displays. *Proceedings of HCI '86 Conference on People and Computers: Designing for Usability*. London: British Computer Society, pp. 604–613.

Tullis, T. (1986*b*). A system for evaluating screen formats. *Proceedings of the Human Factors Society 30th Annual Meeting*. Santa Monica, CA: Human Factors Society, pp. 1216–1220.

Tullis, T. (1988). Screen design. In M. Helander (ed.), *Handbook of human-computer interaction*. Amsterdam: Elsevier Science, pp. 377–411.

Wilkins, A., and Nimmo-Smith, M. (1987). The clarity and comfort of printed text. *Ergonomics*, 30, 1705–1720.

Wilkinson, R., and Robinshaw, H. (1987). Proof-reading: VDU and paper text compared from speed, accuracy and fatigue. *Behaviour and Information Technology*, 6, 125–133.

Wolf, C. (1986). *BNA "HN" command display: Results of user evaluation*. Unpublished technical report. Irvine, CA: Unisys Corporation.

Woodson, W. (1963, Oct 21). *Human engineering design standards for spacecraft controls and displays* (General Dynamics Aeronautics Report GDS-63-0894-1). Orlando, FL: National Aeronautics and Space Administration.

Zwahlen, H., and Kothari, N. (1986). Effects of positive and negative image polarity VDT screens. *Proceedings of the Human Factors Society 30th Annual Meeting*. Santa Monica, CA: Human Factors Society, pp. 170–174.

CHAPTER 5

VISUAL DISPLAYS OF DYNAMIC INFORMATION

We live in a world in which things keep changing—or at least are subject to change. A few examples are natural phenomena (such as temperature and humidity), the speed of vehicles, the altitude of aircraft, our blood pressure and heart rate, traffic lights, the frequency and intensity of sounds, and the national debt. This chapter deals with some aspects of the design of displays used to present information about the parameters in our world that are subject to change, that is, that are dynamic. Such displays cover quite a spectrum, including those that present quantitative, qualitative, status, and representational information.

USES OF DYNAMIC INFORMATION

The information provided by dynamic displays can be used in a variety of ways. The following are four such uses:

1 *Quantitative readings:* The display is used to read a precise numeric value. For example, a quantitative reading would precede a response such as "The pressure is 125 psi."

2 *Qualitative readings:* The display is used to read an approximate value or to discern a trend, rate of change, or change in direction. For example, a qualitative reading would result in a response such as "The pressure is rising."

3 *Check readings:* The display is used to determine if parameters are within some "normal" bounds or that several parameters are equal. For example, check reading would elicit a response such as "All pressures are normal."

4 *Situation awareness:* The display (a representation of some physical space) is used to perceive and attach meaning to elements in a volume of time and space and to project the status of the elements into the near future (Endsley, 1988*a*). For example, situation awareness is involved when an air traffic controller, looking at a radar display, understands the meaning and positions of aircraft within the sector he or she is controlling and predicts where the aircraft will be in 5, 10, or 15 minutes.

When designing or selecting a display, it is critical that all the information needs of the user be fully understood. Failure to do so may result in providing incomplete information. For example, consider what drivers need to know about the temperature of their car's engine. At the most elementary level, only check reading is necessary; that is, is the engine overheated or not? A warning light (sometimes in this context called an ''idiot light'') would properly provide such information. Surely, drivers do not need to make a quantitative reading of the exact engine temperature. So do we need a engine-temperature gauge in a car? The answer is yes. The reason is that we do need to make qualitative readings such as whether the temperature is rising or whether it is getting near the overheating point. Such information would be valuable for taking corrective actions such as getting to a service station as quickly as possible. In a similar vein, why do we need a second hand on a wrist watch? We rarely need to know the time to the nearest second (quantitative reading). The answer is that we need to know whether the watch is running or not (check reading). In fact, that is why, on digital watches, the colon (:) between the hours and minutes blinks.

QUANTITATIVE VISUAL DISPLAYS

The objective of quantitative displays is to provide information about the quantitative value of some variable. In most cases, the variable changes or is subject to change (such as speed or temperature). But (with a bit of stretch) we also embrace the measurement of some variables that, in a strict sense, are more static (such as the length and weight of objects).

Basic Design of Quantitative Displays

Conventional quantitative displays are mechanical devices of one of the following types:

1 Fixed scale with moving pointer
2 Moving scale with fixed pointer
3 Digital display

The first two are analog indicators in that the position of the pointer is analogous to the value it represents. Examples of these are shown in Figure 5-1.

Although conventional quantitative displays have moving mechanical parts (the pointer, the scale itself, or the numerals of a digital counter), modern technology makes it possible to present electronically generated features, thus

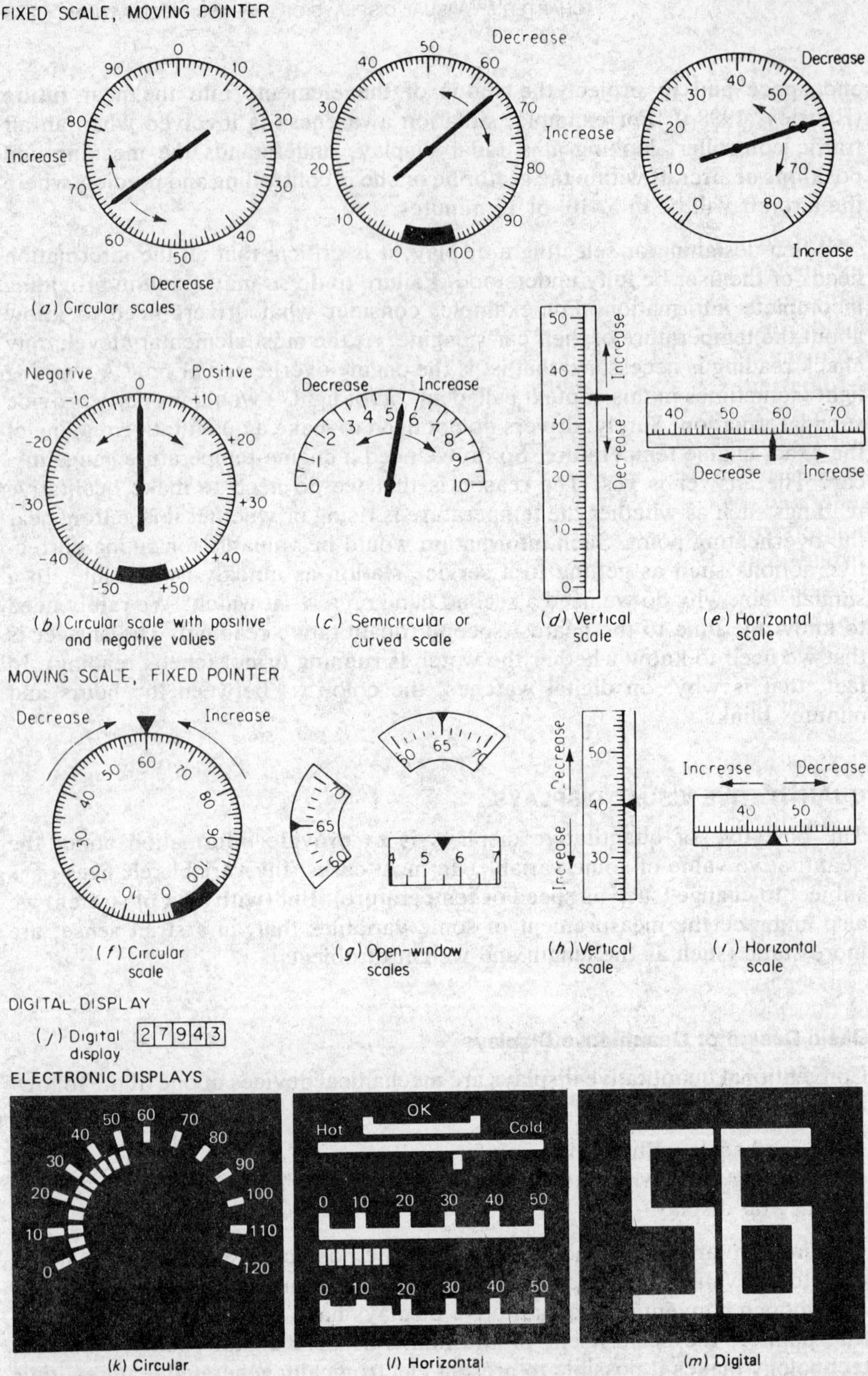

FIGURE 5-1

Examples of certain types of displays used in presenting quantitative information. The first three groups represent mechanical scales with moving pointers, scales, or counters (the digital display). The fourth group represents scales with electronically generated symbols or characters.

eliminating the need for moving mechanical components. Examples of a few such designs are given in Figure 5-1*k, l,* and *m*.

Comparison of Different Designs Over the years there have been a number of studies in which certain designs of conventional quantitative scales have been compared. Although the results of such studies are somewhat at odds with one another, certain general implications hold. For example, it is clear that for many uses digital displays (also called *counters*) are generally superior to analog displays (such as round, horizontal, and vertical scales) when the following conditions apply: (1) a precise numeric value is required (quantitative reading), and (2) the values presented remain visible long enough to be read (are not continually changing). In one study, for example, Simmonds, Galer, and Baines (1981) compared digital speedometers with three dial displays and an experimental curvilinear display. (All the designs used were electronically generated; two of the dials and the curvilinear display used bars rather than pointers to represent numeric values.) The experimenters found the digital display consistently better in terms of both accuracy of reading and preference.

Although digital displays have a definite advantage for obtaining specific numeric values that tend to remain fixed long enough to be read, analog displays have advantages in other circumstances. Fixed-scale moving-pointer displays, for example, are particularly useful when the values are subject to frequent or continual change that would preclude the use of a digital display (because of limited time for reading any given value). In addition, such analog displays have a positive advantage when it is important to observe the direction or rate of change of the values presented (qualitative reading). By and large, analog displays with fixed scales and moving pointers are superior to those with moving scales and fixed pointers. In this regard Heglin (1973) offers the following list of factors to consider in the selection of analog displays:

1 In general, a pointer moving against a fixed scale is preferred.

2 If numerical increase is typically related to some other natural interpretation, such as *more or less* or *up or down,* it is easier to interpret a straight-line or thermometer scale with a moving pointer (such as Figures 5-1*d* and 5-1*e*) because of the added cue of pointer position relative to the zero, or null, condition.

3 Normally, do not mix types of pointer-scale (moving-element) indicators when they are used for related functions—to avoid reversal errors in reading.

4 If manual control over the moving element is expected, there is less ambiguity between the direction of motion of the control and the display if the control moves the pointer rather than the scale.

5 If slight, variable movements or changes in quantity are important to the observer, these will be more apparent if a moving pointer is used.

Although fixed scales with moving pointers are generally preferred to moving scales with fixed pointers, the former do have their limitations, especially when the range of values is too great to be shown on the face of a relatively small scale. In such a case, certain moving-scale fixed-pointer designs (such as rectangular open-window and horizontal and vertical scales) have the practical advantage of occupying a small panel space, since the scale can be wound around spools behind the panel face, with only the relevant portion of the scale exposed.

Further, research and experience generally tend to favor circular and semicircular scales (Figure 5-1*a*, *b*, and *c*) over vertical and horizontal scales (Figures 5-1*d* and 5-1*e*). However, in some circumstances vertical and horizontal scales would have advantages, as discussed in item 2 above.

Basic Features of Quantitative Displays

Figure 5-2 illustrates a few important concepts in the design of quantitative displays. *Scale range* is the numerical difference between the highest and lowest values on the scale, whether numbered or not. *Numbered interval* is the numerical difference between adjacent numbers on the scale. *Graduation interval* is the numerical difference between the smallest scale markers. *Scale unit* is the smallest unit to which the scale is to be read. This may or may not correspond to the graduation interval. A medical thermometer, for example, usually is read to the nearest tenth of a degree, our conventional indoor and outdoor thermometer to the nearest whole degree, and some high-temperature industrial thermometers only to the nearest 10 or 100 degrees.

Specific Features of Conventional Quantitative Displays

The ability of people to make visual discriminations (such as those required in the use of quantitative scales) is influenced in part by the specific features to be discriminated. Some of the relevant features of quantitative scales (aside from

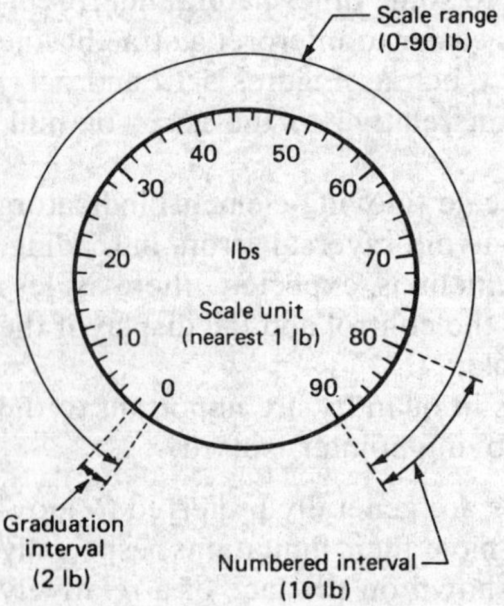

FIGURE 5-2
Illustration of several basic concepts in the design of quantitative displays. (*Source: Adapted from Sanders and Peay, 1988, Fig. 5-2.*)

the basic types) are the numeric progressions used, length of the scale units, "interpolation" required, use of scale markers, width (thickness) of markers, design of pointers, and location of scale numbers. On the basis of a modest experiment that involved variations in these and other features, Whitehurst (1982) found the numeric progression used and the length of scale units to be particularly important in terms of speed and accuracy of reading quantitative scales. A few features of such scales are discussed below. Although most of the research relating to such features has been carried out with conventional quantitative displays, it is probable that the implications would still be applicable to electronically generated displays.

Numeric Progressions of Scales Every quantitative scale has some intrinsic numeric progression system that is characterized by the graduation interval of the scale and by the numbering of the major scale markers. In general, the garden variety of progression by 1s (0, 1, 2, 3, etc.) is the easiest to use. This lends itself readily to a scale with major markers at 0, 10, 20, etc., with intermediate markers at 5, 15, 25, etc., and with minor markers at 1, 2, 3, etc. Progression by 5s is also satisfactory and by 25 is moderately so. Some examples of scales with progressions by 1s and 5s are shown in Figure 5-3. Where large numeric values are used in the scale, the relative readabilities of the scales are the same if they are all multiplied by 10, 100, 1000, etc. Decimals, however, make scales more difficult to use, although for scales with decimals the same relative advantages and disadvantages hold for the various numeric progressions. The zero in front of the decimal point should be omitted when such scales are used.

Unusual progression systems (such as by 3s, 8s, etc.) should be avoided except under very special circumstances.

FIGURE 5-3
Examples of certain generally acceptable quantitative scales with different numeric progression systems (1s/5s). The values to the left in each case are, respectively, the graduation interval *g* (the difference between the minor markers) and the numbered interval *n* (the difference between numbered markers). For each scale there are variations of the basic values of the system, these being decimal multiples or multiples of 1 or 5.

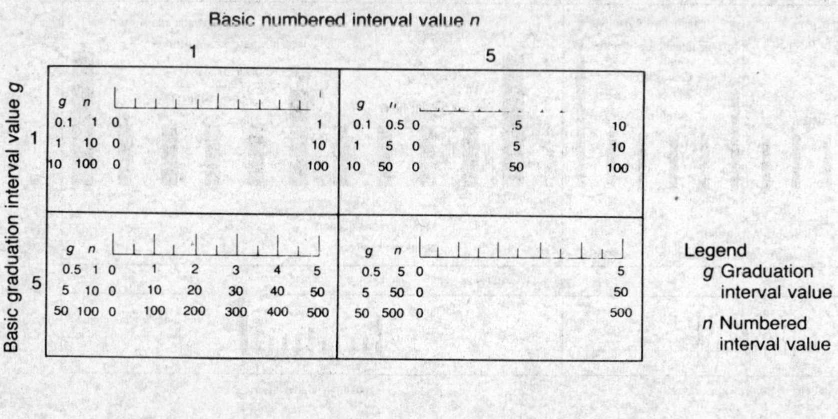

Length of Scale Unit The length of the scale unit is the length on the scale (in inches, millimeters, or degrees of arc) that represents the numeric value that is the smallest unit to which the scale is to be read. For example, if a pressure gauge is to be read to the nearest 10 lb (4.5 kg) and the scale is so constructed that a 0.05 in (0.13 cm) represents 10 lb (4.5 kg) of pressure, then the length of the scale unit is 0.05 in (0.13 cm).

The length of the scale unit should be such that the distinctions between the values can be made with optimum reliability in terms of human sensory and perceptual skills. Although certain investigators have reported acceptable accuracy in the reading of scales with scale units as low as 0.02 in (0.5 mm); most sets of recommendations provide for values ranging from about 0.05 to 0.07 in (1.3 to 1.8 mm), as shown in Figure 5-4. The larger values probably would be warranted when the use of instruments is under less than ideal conditions, such as when they are used by persons who have below-normal vision or when they are used under poor illumination or under pressure of time.

Design of Scale Markers It is good practice to include a scale marker for each scale unit to be read. Figure 5-4 shows the generally accepted design features of scale markers for normal viewing conditions and for low-illumination conditions. This illustration is for the conventional progression scheme with major markers representing intervals of 1, 10, 100, etc., and minor markers at the scale units (the smallest values to be read), such as 0.1, 1, 10.

Scale Markers and Interpolation The concept of interpolation (relative to quantitative scales) usually applies to the estimation of specific values between markers when not all scale units have markers. However, usually it is desirable to have a scale marker for each unit to be read. In such instances, pointers between markers are rounded to the nearest marker. (Although such "round-

FIGURE 5-4
Recommended format of quantitative scales, given length of scale unit and graduation markers. Format *a* is proposed for normal illumination conditions under normal viewing conditions and *b* for low illumination. (*Source: Adapted from Grether and Baker, 1972, p. 88.*)

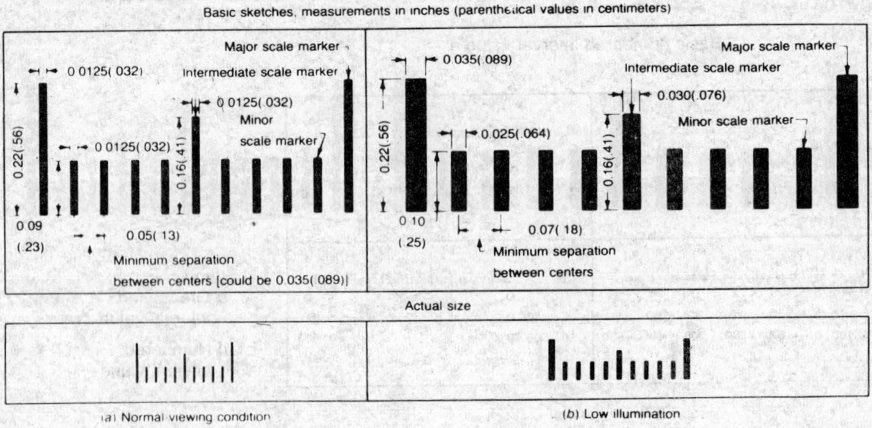

ing'' has also been called interpolation, we do not use the term in this sense.) If scales were to be much more compressed than those in Figure 5-4, the scale markers would be crowded together, and this could affect the reading accuracy (especially under time constraints and low illumination). In such circumstances, it is better to use a scale that does require interpolation. Actually, people are moderately accurate in interpolation; Cohen and Follert (1970) report that interpolation of fifths, and even of tenths, may yield satisfactory accuracy in many situations. Even so, where high accuracy is required (as with certain test instruments and fine measuring devices), a marker should be placed at every scale unit, even though this requires a larger scale or a closer viewing distance.

Scales are read most quickly when the pointers are precisely on the markers (Whitehurst, 1982). In other instances a person needs to round to the nearest scale unit, but such rounding must, of course, be tolerated.

Design of Pointers The few studies that have dealt with pointer design leave some unanswered questions, but some of the common recommendations follow: Use pointed pointers (with a tip angle of about 20°); have the tip of the pointer meet, but not overlap, the smallest scale markers; have the color of the pointer extend from the tip to the center of the scale (in the case of circular scales); and have the pointer close to the surface of the scale (to avoid parallax).

Combining Scale Features Several of the features of quantitative scales discussed above have been integrated into relatively standard formats for designing scales and their markers, as shown in Figure 5-4. Although these formats are shown in a horizontal scale, the features can, of course, be incorporated in circular or semicircular scales. Also, the design features shown in this figure should be considered as general guidelines rather than as rigid requirements, and the advantages of certain features may, in practical situations, have to be traded off for other advantages.

One does not have to look very hard to find examples of poor instrument design. The scale in Figure 5-5 (assuming that the amperes should be read to the nearest 5) suffers from shortness of scale units and inadequate intermediate markers. The original meter design in Figure 5-6*a* suffers from various ailments, which are corrected in the redesigned version (Figure 5-6*b*).

Scale Size and Viewing Distance The above discussion of the detailed features of scales is predicated on a normal viewing distance of 28 in (71 cm). If a display is to be viewed at a greater distance, the features have to be enlarged in order to maintain, at the eye, the same visual angle of the detailed features. To maintain that same visual angle, the following formula can be applied for any other viewing distance (x) in inches :

$$\text{Dimension at } x \text{ in} = \text{dimension at 28 in} \times \frac{x \text{ in}}{28}$$

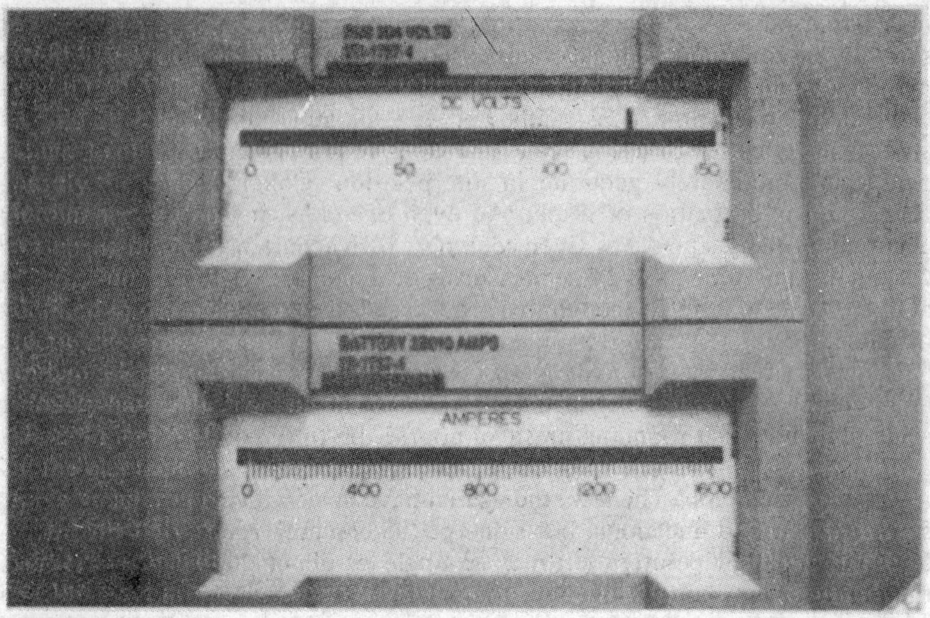

FIGURE 5-5
Example of a poorly designed ampere scale.

FIGURE 5-6
Illustration of two designs of a meter. The one at the right would be easier to read because it is bolder and less cluttered than the one at the left. It has fewer graduation markers, and the double arc line has been eliminated. The scale length is increased by placing the markers closer to the perimeter; although this requires that the numerals be placed inside the scale, the clear design and the fact that the numerals are upright probably would partially offset this disadvantage. (*Source: Adapted from* Applied ergonomics handbook, *1974, Fig. 3-1.*)

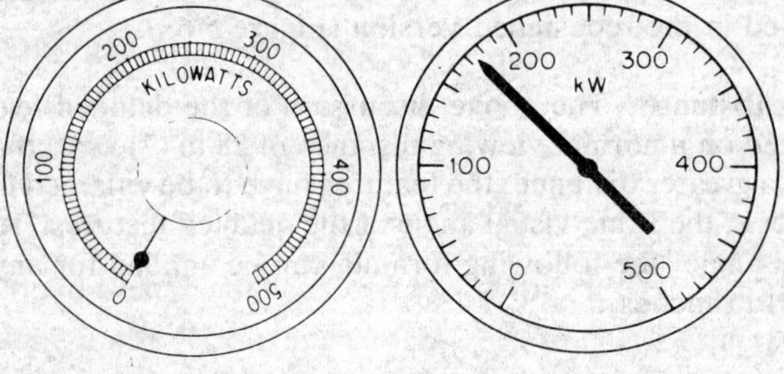

a. Original   b. Redesigned

Specific Features of Electronic Quantitative Displays

The previous section dealt largely with conventional, mechanical quantitative displays. Although much of that discussion presumably applies to corresponding features of electronic displays, certain features of electronic displays create special design problems.

One such problem, discussed by Green (1984), deals with the bars of bar-type displays (such as Figure 5-1*l*). In his study, various combinations of several automobile displays were used, including variations of bar-type designs for several purposes. He concluded that, for bar-type displays, only one of the segments of the bar should be shown. (An example of this feature is the top display of those in Figure 5-1*l*). The extension of the bar to the zero position (as shown in the middle example) was confusing to some people. However a bar-type design (whether with only a single segment or with an extended bar) would be particularly inappropriate for a fuel gauge because there would be *no* indication for an empty tank. The complete bar-type display, however, probably would be more appropriate for a variable such as speed, because it would grow longer or shorter with changing values.

Design of Altimeters

Aircraft altimeters represent a type of quantitative display of special interest. In fact, the initial interest in quantitative displays was triggered because of numerous instances in which aircraft accidents had been attributed to misreading of an earlier altimeter model. The model consisted of a dial with three pointers representing, respectively, 100, 1000, and 10,000 ft (30.5, 305, and 3050 m), like the second, minute, and hour hands of a clock. That model required the reader to combine the three pieces of information, and this combining was the basis for the errors made in reading the altimeter.

In connection with the design of altimeters, Roscoe (1968) reports a study that dealt directly with certain features of altimeters but that has implications for quantitative displays in certain other circumstances. In particular, the study compared three design variables: (1) vertical versus circular scales, (2) integrated presentations of three altitude values (present altitude, predicted altitude 1 minute away, and command altitude), and (3) circular (analog) versus counter (digital) presentation. The four designs that embody the combinations of these three variables are shown in Figure 5-7.

Without going into details, the results of the study (given in Figure 5-7) show that design *a* (the integrated vertical scale) was clearly the best in terms of time and errors. The explanation for this given by Roscoe is primarily its pictorial realism in representing relative positions in vertical space by a display in which *up* means *up* and *down* means *down*. Design *b*, which represents vertical space in a distorted manner (around a circle), did not fare as well as *a* but was generally superior to *c* and *d*, both of which consisted of *separate* displays of the three altitude values rather than an *integrated* display. The results can be interpreted as supporting the advantage of compatible designs (as illustrated by

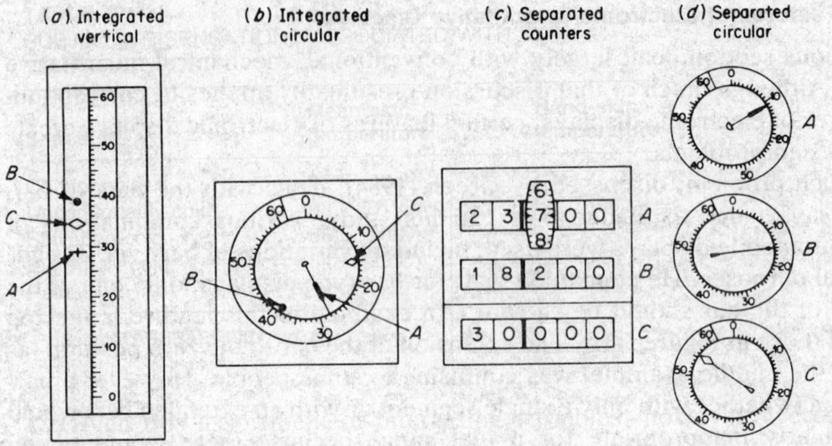

	(a)	(b)	(c)	(d)
Time, seconds	57	64	76	82
Number of errors	7	10	12	16
Percent making errors	12	33	29	50

Legend:
A. Present altitude
B. Predicted altitude (1 min)
C. Command altitude

FIGURE 5-7
Four display designs for presenting (*A*) present altitude, (*B*) predicted altitude (in 1 min), and (*C*) command altitude, and three criteria (mean time for 10 trials, number of errors, and percent of 24 subjects making errors). The displays are shown in overly simplified form. (*Source: Adapted from Roscoe, 1968*).

the vertical display representing altitude) and the advantage of integrated displays (where they are appropriate) as contrasted with separate indications for various related values. [Note that, although we have referred to the advantage of digital displays or counters, in this study they were not as good as a vertical (analog) display.]

Object Displays

In some situations operators must integrate information about several variables to understand the state of a system or process. Chemical plant operators integrate information regarding temperature, pressure, flow rates, and so forth to assess the state of the chemical process being controlled. Pilots integrate information about air speed, bank angle, flap setting, etc. to determine how close they are to stalling the aircraft. Wickens (1986, 1987) puts forward the *compatibility of proximity principle* for designing displays in which information about several variables is used for decision making. The principle states that tasks that require mental integration of information (i.e., "mental proximity") will benefit from close display proximity, whereas tasks that require focused attention on the individual sources (or variables) or independent processing of each source (or variable) will be harmed by close proximity. Proximity can be defined in a number of ways, including closeness in space, common color, etc. Wickens and Andre (1988), however, point out that proximity has a special status when defined as different variables representing different dimensions of a single object. Combining dimensions into a perceptual object can often produce what Pomerantz (1981) called *emergent features*. An emergent feature is a perceptual property of the whole that does not exist when the components are viewed separately. Combining two lines of different length into a rectangle, as shown in Figure 5-8, produces an emergent feature of area which does not exist when the line lengths are viewed separately.

Researchers have explored an application of the compatibility-of-proximity principle by assessing the advantages and disadvantages of what are called *object displays* or *configural displays*. Typically, such displays are polygons in which the shape is determined by the values of multiple variables. Figure 5-9, for example, shows a four-variable object display, as well as separate digital displays and a bar graph display of the same data. It is postulated that it is the emergent features of object displays that facilitate integration of information.

The results of the research have generally supported the compatibility-of-proximity principle. Object displays seem to have an advantage over separate digital displays when the task requires the operator to integrate the information, that is, to perceive a pattern in the data. However, object displays can be a hindrance when the operator must check individual values of the variables involved (Carswell and Wickens, 1987). Some controversy surrounds the relative merits of bar graph displays versus object displays. Coury, Boulette and Smith (1989), for example, found bar graphs to be superior to object displays when information integration was required, contrary to the findings of Carswell and Wickens (1987). Coury, Boulette, and Smith attributed the superiority of bar graph displays to the dual role that they may play. Bar graphs can easily be used as separate displays, but they also contain some configural properties in that a series of bar graphs can convey a perception of the emergent feature of contour (e.g., high, high, low, low). The relative contribution of information

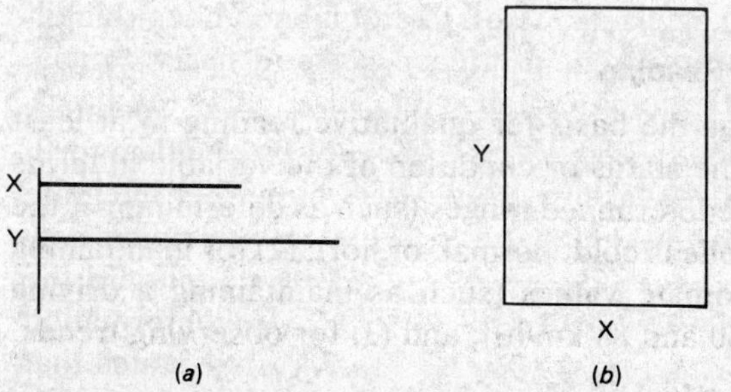

FIGURE 5-8
Demonstration of an emergent feature. The two separate dimensions in (*a*) when combined into a rectangle (*b*) show the emergent feature of area.

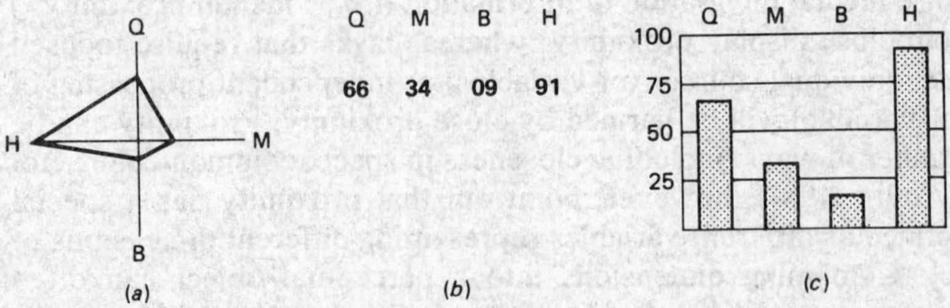

FIGURE 5-9
Three types of displays representing the values of four variables (Q, H, M, B): (*a*) an object or configural display; (*b*) separate digital displays; and (*c*) a bar graph display. (*Source: Coury and Purcell, 1988; Fig. 1. Reprinted with permission of the Human Factors Society, Inc. All rights reserved.*)

integration and focused attention may determine the relative superiority of bar graph displays and object displays.

From all this we can glean the following:

1 The choice of the type of display (object, separate digital, bar graph, etc.) must be predicated on a thorough understanding of the nature of the task confronting the user.

2 Object displays should be considered when integration of information is required, but separate displays must also be provided for situations in which the user needs to focus on specific variables.

3 The design of object displays may prove to be as much of a creative art as a systematic science (Wickens and Andre, 1988).

4 As with any display, but perhaps more important with object displays, the display configuration should be tested with subjects representative of the ultimate users under task conditions likely to be experienced in the real world.

QUALITATIVE VISUAL DISPLAYS

In using displays for obtaining qualitative information, the user is primarily interested in the approximate value of some continuously changeable variable (such as temperature, pressure, or speed) or in its trend, or rate of change. The basic underlying data used for such purposes usually are quantitative.

Quantitative Basis for Qualitative Reading

Quantitative data may be used as the basis for qualitative reading in at least three ways: (1) for determining the status or condition of the variable in terms of each of a limited number of predetermined ranges (such as determining if the temperature gauge of an automobile is cold, normal, or hot); (2) for maintaining some desirable range of approximate values [such as maintaining a driving speed between 50 and 55 mi/h (80 and 88 km/h)]; and (3) for observing trends, rates of change, etc. (such as noting the rate of change in altitude of an airplane). In the qualitative use of quantitative data, however, evidence suggests that a display that is best for a quantitative reading is not necessarily best for a qualitative reading task.

Some evidence for support of this contention comes from a study in which open-window, circular, and vertical designs are compared (Elkin, 1959). In one phase of this study, subjects made qualitative readings of high, OK, or low for three ranges of numerical values and also employed the same three scales to make strictly quantitative readings. The average times taken (shown in Table 5-1) show that, although the open-window (digital) design took the shortest time for quantitative reading, it took the longest time for qualitative reading. In turn, the vertical scale was best for qualitative reading but worst for quantitative reading. Thus different types of scales vary in effectiveness for qualitative versus quantitative reading.

Design of Qualitative Scales

As indicated above, many qualitative scales represent a continuum of values that are sliced into a limited number of ranges (such as cold, normal, and hot); in other instances, specific ranges have particular importance to the user (such as representing a danger zone). In such cases the perception of the correct reading is aided by some method of coding the separate ranges. One way to do this is to use color codes, as illustrated in Figure 5-10.

Another method is to use some form of *shape coding* to represent specific ranges of values. It is sometimes possible, in the design of such coded areas, to take advantage of natural "compatible" associations people have between coding features and the intended meanings. One study taking such an approach is reported by Sabeh, Jorve, and Vanderplas (1958) in which they were interested in identifying the shapes with the strongest associations with the intended meanings of certain military aircraft instrument readings. Having solicited a large number of designs initially, the investigators selected the seven shown in Figure 5-11. These were presented to 140 subjects, along with the seven

TABLE 5-1
TIMES FOR QUALITATIVE AND QUANTITATIVE READINGS WITH THREE TYPES OF SCALES

	Average reading time, s	
Type of scale	**Qualitative**	**Quantitative**
Open-window	115	102
Circular	107	113
Vertical	101	118

Source: Elkin, 1959.

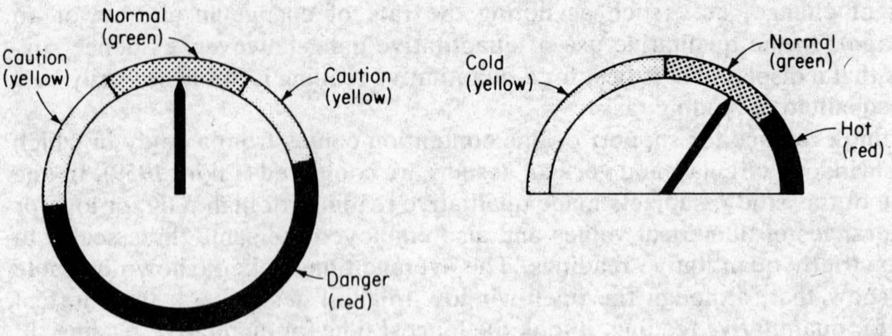

FIGURE 5-10
Illustration of color coding of sections of instruments that are to be read qualitatively.

following "meanings": caution, undesirable, mixture lean, mixture rich, danger—upper limit, danger—lower limit, and dangerous vibration. Figure 5-11 shows the percentage of subjects who selected the indicated meaning to a statistically significant level.

The argument for the use of precoded displays for qualitative reading (when this is feasible) is rooted in the nature of the human perceptual and cognitive processes. To use a strictly quantitative display to determine if a given value is within one specific range or another involves an additional cognitive process of

FIGURE 5-11
Association of coded zone markings with the intended meanings of specific scale values of military aircraft instruments. The numbers in parentheses are the percentages of individuals (out of 140) who reported significant associations with the meanings listed. (*Adapted from Sabeh, Jorve, and Vanderplas, 1958.*)

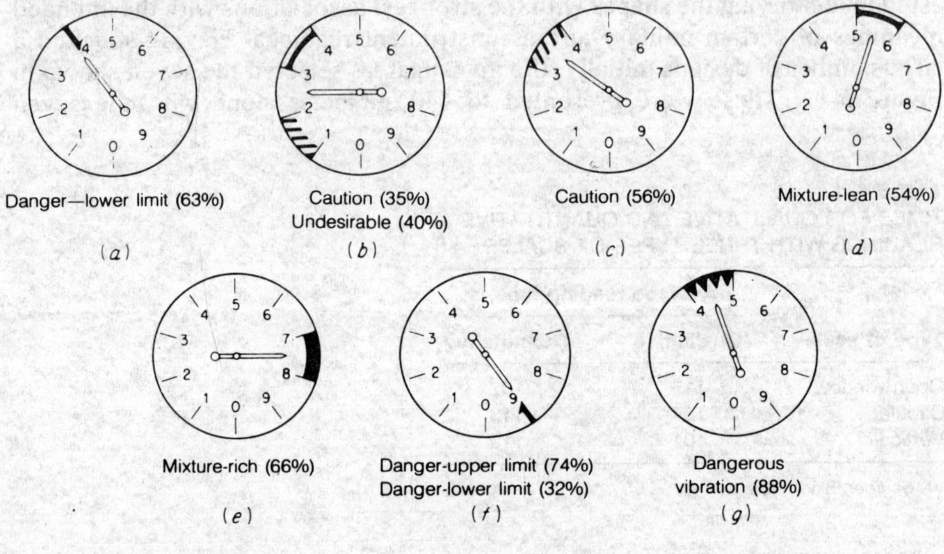

assigning the value read to one of the possible ranges of values that represent the categories. The initial perception of a precoded display immediately conveys the meaning of the display indicator. Note that quantitative displays with coded zones also can be used to reflect trends, directions, and rates of change. Further, they can be used for quantitative reading if the scale values are included.

In the use of electronically generated displays, segments of a total scale can be identified in various ways, as illustrated in Figure 5-1*l* as the top display for automobile temperature.

Check Reading

The term *check reading* refers to the use of an instrument to ascertain whether the reading is normal. Usually this is done with a quantitative scale, but the normal condition is represented by a specific value or narrow range of values. In effect, check reading is a special case of qualitative reading. As such, the normal reading value should be clearly coded, as discussed above.

If several or many instruments for check reading are used together in panels, their configuration should be such that any deviant reading stands out from the others. Research relating to various possible configurations has indicated that (with round instruments) the normal position preferably should be aligned at the 9 o'clock (or possibly 12 o'clock) positions. A 9 o'clock alignment is illustrated in Figure 5-12. The advantage of such a systematic alignment is based on human perceptual processes, in particular what is referred to as the *gestalt,* that is, the human tendency to perceive complex configurations as complete entities, with the result that any feature that is "at odds" with the configuration is immediately apparent. Thus, a dial that deviates from a systematic pattern stands out from the others. In the case of panels of such dials, the addition of extended lines *between* the dials can add to the gestalt, thus helping to make any deviant dial stand out more clearly (Dashevsky, 1964). This is illustrated in the top row of dials of Figure 5-12.

FIGURE 5-12
A panel of dials used for check reading. When all the "normal" readings are aligned at the 9 o'clock (or 12 o'clock) position, any deviant reading can be perceived at a glance. In some instances an extended line is shown between the dials, as illustrated in the top row; this can aid in making the deviant dial more distinct.

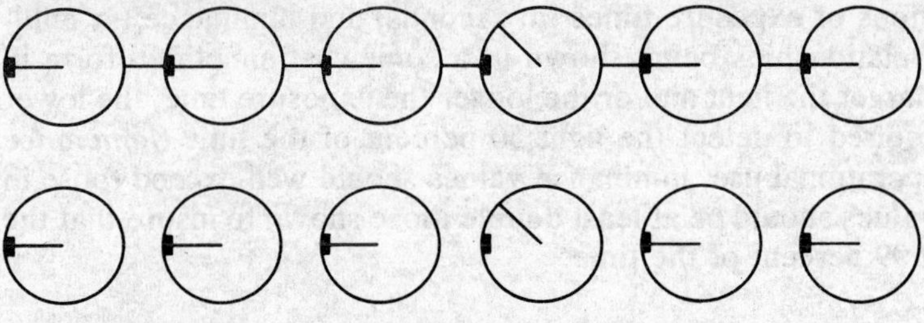

Status Indicators

Sometimes *qualitative* information indicates the *status* of a system or a component, such as the use of some displays for check reading to determine if a condition is normal or abnormal, or the qualitative reading of an automobile thermometer to determine if the condition is hot, normal, or cold. However, more strictly, status indications reflect separate, discrete conditions such as on and off or (in the case of traffic lights) stop, caution, and go. If a qualitative instrument is to be used *strictly* for check reading or for identifying a particular status (and *not* for some other purpose such as observing trends), the instrument could be converted to a status indicator.

The most straightforward, and commonly used, status indicators are lights, such as traffic lights. In some cases redundant codes are used, as with most traffic lights that are coded both by color (red, yellow, and green) and by location (top, middle, and bottom). Although many status indicators are lights, other coding systems can be used, such as the controls of some stoves that are marked to identify the off position.

SIGNAL AND WARNING LIGHTS

Flashing or steady-state lights are used for various purposes, such as indications of warning (as on highways); identification of aircraft at night; navigation aids and beacons; and to attract attention, such as to certain locations on an instrument panel. There apparently has been little research relating to such signals, but we can infer some general principles from our knowledge of human sensory and perceptual processes.

Detectability of Signal and Warning Lights

Of course, various factors influence the detectability of lights. Certain such factors are discussed below.

Size, Luminance, and Exposure Time The absolute threshold for the detection of a flash of light depends in part on a combination of size, luminance, and exposure time. Drawing in part on some previous research, Teichner and Krebs (1972) depicted the minimum sizes of lights (in terms of visual angle of the diameter in minutes of arc) that can be detected 50 percent of the time under various combinations of exposure times (in seconds) and luminance (in millilamberts), these relationships being shown in a somewhat simplified form in Figure 5-13. The larger the light and/or the longer the exposure time, the lower the luminance required to detect the light 50 percent of the time (*luminance threshold*). For operational use, luminance values should well exceed those in Figure 5-13; the values should be at least double those shown to insure that the target is detected 99 percent of the time.

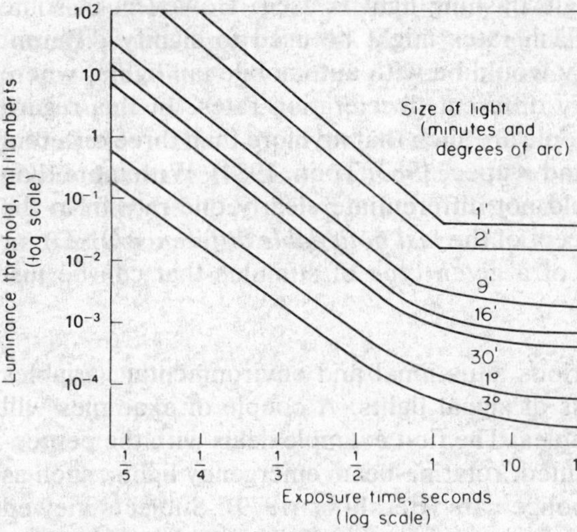

FIGURE 5-13
Minimum sizes of lights (in minutes and degrees of arc) that can be detected 50 percent of the time under varying combinations of exposure time and luminance. (*Source: Adapted from Teichner and Krebs, 1972.*)

Color of Lights Another factor related to the effectiveness of signal lights is color. Reynolds, White, and Hulgendorf (1972) used response time as an indication of the effectiveness of four different colors. The researchers found that the background color and ambient illumination can interact to influence the ability of people to detect and respond to lights of different colors. In general, if a signal has good brightness contrast against a dark background, and if the absolute level of brightness of the signal is high, the color of the signal is of minimal importance in attracting attention. But with low signal-to-background brightness contrast, a red signal has a marked advantage, followed by green, yellow, and white in that order.

Flash Rate of Lights In the case of flashing lights, the flash rate should be *well below* that at which a flashing light appears as a steady light (the flicker-fusion frequency), which is approximately 30 times per second. In this regard, flash rates of about 3 to 10 per second (with duration of at least 0.05 s) have been recommended for attracting attention (Woodson and Conover, 1964, pp. 2–26). Markowitz (1971) makes the point that the range of 60 to 120 flashes per minute (1 to 2 per second), as used on highways and in flyways, appears to be compatible with human discrimination capabilities and available hardware constraints.

In most circumstances, a single flashing light is used. However, in some situations lights with different flash rates might be used to signify different things. A very obvious possibility would be with automobile tail lights, where different flash rates could signify different deceleration rates. In this regard clearly the perceptual skills of people are such that no more than three different rates should be used (Mortimer and Kupec, 1983; Tolin, 1984). With more than three different rates, people could not differentiate clearly one rate from the others. [This is based on the concept of the *just noticeable difference* (JND), or the difference in the magnitude of a given type of stimulus that can be just noticeable to people.]

Context of Signal Lights Various situational and environmental variables can influence people's awareness of signal lights. A couple of examples will illustrate the effects of such variables. The first example deals with the perception of various types of roof-mounted, rotating-beam emergency lights, such as those used on ambulances and police cars (Berkhout, 1979). Subjects viewed the stimulus vehicle at night on an unused gravel road. The vehicle either was not moving or was moving toward or away from the subjects (who were in a parked car). The subjects viewed the stimulus vehicle for 7 s and were required to indicate if the vehicle was moving and, if so, to indicate the direction and whether it was moving fast or slow. The results were somewhat complex and are not discussed in detail here, but we do mention a couple of results.

Table 5-2 presents some of the results for six of the light systems viewed while the vehicle was standing still. First, in the majority of trials many of the subjects misperceived the situation and thought the stationary vehicle was moving either toward or away from them. Second, there was a tendency for the blue lights to be seen as moving toward the observer more often than the red

TABLE 5-2
SUBJECTS' RESPONSES TO VARIOUS EMERGENCY LIGHT SYSTEMS WHEN VEHICLE WAS STATIONARY

Light system		Subjects' response of movement, %		
		Toward	Still*	Away
Single dome	Red	17	47	36
	Blue	26	44	30
Twin sonic	Red	16	46	38
	Blue	31	42	27
Twin beacon	Red	9	36	55
	Blue	20	36	44
Blue lights (all three)		26	40	34
Red lights (all three)		14	43	43

* *Still* is the correct response.

Source: Adapted from Berkhout, 1979, Table 4. Copyright by the Human Factors Society, Inc. and reproduced by permission.

lights, while the red ones were seen more often as moving away. Berkhout points out that the red twin beacon (with a side-to-side flash pattern) is especially prone to produce a dangerous illusion of receding motion when actually the vehicle is standing still. An emergency vehicle parked on the shoulder of a road and displaying this light could, therefore, run an increased risk of a rear-end accident. (However, this same red twin beacon performed relatively well for indicating direction of motion when the vehicle was actually moving.) Given the responses of observers from other aspects of the study, it should be added that no single light system was consistently superior to the others. Rather, the results indicated the disturbing fact that a light that was most suitable for one particular dynamic situation did not necessarily perform well in another.

The second example deals with the problems of discriminating signal lights with other lights in the background. (Traffic lights in areas with neon and Christmas lights represent serious examples of such a problem.) In an interesting investigation Crawford (1963) used both steady and flashing signal lights against a background of (irrelevant) steady lights, flashing lights, and admixtures of both, using a criterion of time to identify the signal lights. His conclusions are summarized as follows:

Signal light	Background lights	Comment
Flashing	Steady	Best
Steady	Steady	OK
Steady	Flashing	OK
Flashing	Flashing	Worst

The presence of even one flashing background light (out of many) seriously affected the identification of a flashing signal light.

Recommendations Regarding Signal and Warning Lights

Here are some recommendations about signal and warning lights (based largely on Heglin, 1973):

- *When should they be used?* To warn of an actual or a potential dangerous condition.
- *How many warning lights?* Ordinarily only one. (If several warning lights are required, use a master warning or caution light and a word panel to indicate specific danger condition.)
- *Steady state or flashing?* If the light is to represent a continuous, ongoing condition, use a steady-state light unless the condition is especially hazardous; continuous flashing lights can be distracting. To represent occasional emergencies or new conditions, use a flashing light.
- *Flash rate:* If flashing lights are used, flash rates should be from about 3 to 10 per second (4 is best) with equal intervals of light and dark. If different flash

rates are to represent different levels of some variable, use no more than three different rates.

- *Warning-light intensity:* The light should be at least twice as bright as the immediate background.
- *Location:* The warning light should be within 30° of the operator's normal line of sight.
- *Color:* Warning lights are normally red because red means danger to most people. (Other signal lights in the area should be other colors.)
- *Size:* Warning lights should subtend at least 1° of visual angle.

REPRESENTATIONAL DISPLAYS

Most representational displays that depict changeable conditions consist of elements that tend to change positions or configuration superimposed on a background. An example would be an aircraft display that shows positions of other aircraft in the sky and terrain features on the ground. As we indicated at the beginning of this chapter, one purpose of such displays is to enhance operators' situation awareness. Fracker (1988) uses the concept of *zones of interest* to define situation awareness. An everyday example (at least for some of us) should clarify the concept—driving a car on a Southern California freeway. One zone of interest is the car you are driving. Being aware of that situation includes knowing the state of the car (is it running out of gas? overheating?), its speed, and the direction it is heading (straight, drifting to the right or left, etc.). Another zone of interest would be the vehicles around you (headways, side clearance, where they are, what they are likely to do, etc.). Expanding the zone of interest further would include the freeway system and city in which you are driving (where you are relative to landmarks and off ramps). At an even higher level is the zone of interest of goals (where you intend to go and the route you intend to take). Endsley (1988*b*) stresses that situation awareness involves matching information received from our senses and from displays to prelearned situational templates or *schema* that allow us to understand the situation and predict what is likely to happen. For example, a stream of red taillights in the distance might match a schema you have in long-term memory that tells you that you will be slowing down and that other cars are likely to be switching lanes to jockey for position (at least that's what happens in Southern California).

Although situation awareness is important in many contexts (including playing many video games), most of the research has concentrated on aircraft displays. We use such displays to illustrate the nature of representational displays and some of the human factors considerations involved in their design.

Aircraft Bank Angle Displays

The problem of representing the bank angle of an aircraft has haunted human factors specialists—and pilots—for years. One aspect of the problem relates to the basic movement relationships to be depicted by the display. The two basic relationships are shown in Figure 5-14 and are described as follows:

- *Moving aircraft:* The earth (specifically the horizon) is fixed and the aircraft moves in relation to it. Such displays are also called outside-in, bird's-eye, or ground-based displays. When the real plane banks to the left, the display indicator (the plane symbol) also rotates to the left. The problem is that the pilot sitting in the cockpit does not see the real horizon as level and his or her plane as tilted. What the pilot sees out the cockpit window is a tilted horizon and an aircraft that from the pilot's frame of reference is horizontal.
- *Moving horizon:* The aircraft symbol is fixed and the horizon moves in relation to it. Such displays are also called inside-out or pilot's-eye displays. Most aircraft bank angle displays are of this type. This type of display is congruent with the pilot's frame of reference, but when the pilot banks the real airplane to the left, generating a leftward rotation in the pilot's mind, the moving element of the display (the horizon) moves to the right.

Comparisons between moving-aircraft and moving-horizon displays have produced mixed results. Fogel (1959) proposed a bank angle display that combined the features of the moving-aircraft and moving-horizon displays. In this display, rapid movement of the controls and thus the aircraft induce the aircraft symbol to move in relation to the horizon (moving-aircraft display). However, when the pilot enters into a gradual turn and the bank angle is held constant for some period of time, the plane and horizon slowly rotate to a moving-horizon display. Sounds confusing, but Roscoe and Williges (1975) found such a display to be more effective than either the moving-aircraft or moving-horizon display.

FIGURE 5-14
The two basic movement relationships for depicting aircraft bank angle, namely, the moving-aircraft (outside-in) and the moving-horizon (inside-out) relationship.

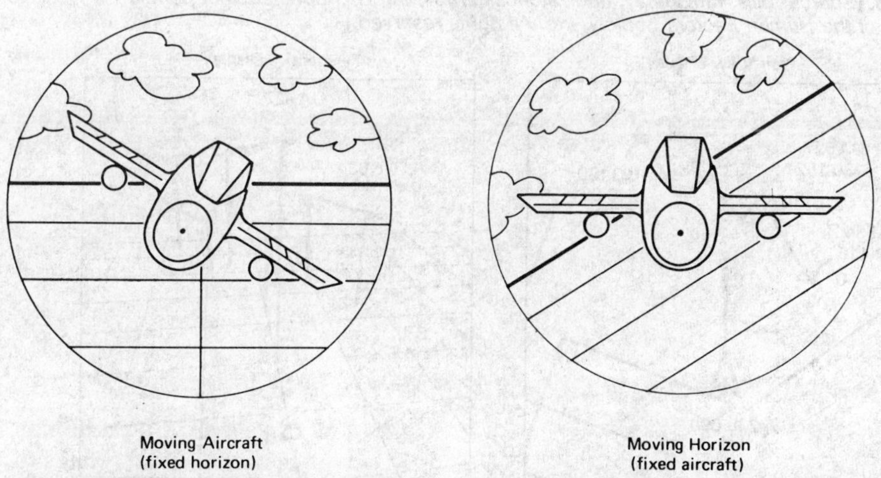

3-D Perspective Displays

Aircraft occupy a position in three-dimensional (3-D) space. Pilots must assimilate and integrate information from two-dimensional (2-D) displays into a coherent 3-D image or mental model of their environment. That is, pilots must develop situation awareness in 3-D. One method of presenting 3-D information on a 2-D display surface is to use artistic techniques to give the illusion of depth. Depth cues used by artists include linear perspective, interposition of objects, object size, texture gradients, and shadow patterns. Figure 5-15 presents a plan-view (2-D) display and an experimental perspective (3-D) display of the positions of three aircraft relative to the pilot's own plane. Ellis, McGreevy, and Hitchock (1987) compared these two types of displays on pilot's initiation of evasive maneuvers to avoid collisions. For all but head-on traffic where it was hard to tell on the 3-D display whether the other aircraft was coming or going, they found that pilot's decision time with the 3-D display was 3 to 6 s faster than with the 2-D display. If two planes are each coming toward one another at 600 mi/h (965 km/h), reducing decision time by 3 to 6 s means that evasive actions are initiated when the planes are 1 to 2 mi (1.6 to 3.2 km) farther apart. For head-on traffic, the 3-D display took about 5 s longer to interpret than did the 2-D display. Interestingly, with the 2-D display, 60 percent of the evasive maneuvers were level turns away or toward the intruder

FIGURE 5-15
An example of a plan-view (2-D) and a perspective (3-D) display of the positions of three aircraft relative to the pilot's own aircraft. Pilot's own plane was always presented so that it was in the center of the display. Each aircraft was presented with a 60-s ground-referenced predictor line and, to represent previous positions, a series of 10 trailing dots each 4 s apart. On the perspective display, the pilot's current altitude is represented by an X on the vertical reference lines of each plane. In the plan-view display, altitude is represented by the second three digits below the aircraft designation code. (*Source: Ellis, McGreevy, and Hitchock, 1987, Fig. 1. Reprinted with permission of the Human Factors Society, Inc. All rights reserved.*)

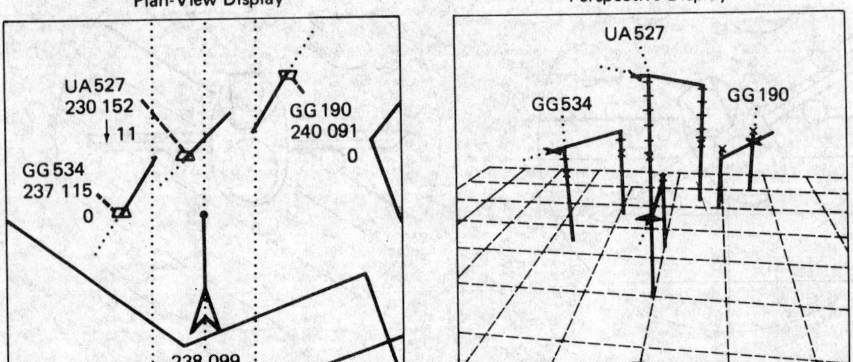

aircraft. (Turning toward an aircraft that is crossing one's flight path can avoid a collision by crossing behind the other aircraft.) With the 3-D display, only 33 percent of the evasive maneuvers were level turns, while 34 percent were changes in altitude (climbing or descending) with no turning. Apparently, with enhanced 3-D situation awareness provided by the perspective display, pilots were more likely to execute altitude maneuvers.

In addition to perspective 3-D displays, some experimental work is being carried out on true stereoscopic displays that produce real 3-D images (Zenyuh, Reising, Walchli, and Biers, 1988). Methods are used to present independent, horizontally disparate, images to the left and right eyes thus producing true 3-D images. The practicality of this line of research, however, has not yet been proven.

Principles of Aircraft-Position Displays

Although the evidence regarding aircraft-position displays is still not entirely definitive, Roscoe, Corl, and Jensen (1981) have crystallized a few principles that seem to have general validity for many types of representational displays. Four of these principles are discussed here. (Certain other principles are not discussed because they deal with some of the complex aspects of the control dynamics of aircraft.)

1 *Principle of pictorial realism:* This principle is an assertion that a display should present a spatial analog, or image, of the real world, in which the position of the object (the aircraft) is seen in depth as well as in up-down and left-right frames of reference.

2 *Principle of integration:* Following this principle, *related* information should be presented in integrated displays, as discussed before and illustrated in Figure 5-7*a* and Figure 5-9*a*.

3 *Principle of compatible motion:* In the use of aircraft displays, this principle is in conformity with the use of moving-aircraft (outside-in) displays with fixed horizons.

4 *Principle of pursuit presentation:* This refers to the preferable use of pursuit displays as contrasted with compensatory displays. (See Chapter 10 for a discussion of pursuit and compensation displays.)

Discussion of Representational Displays

Although the previous discussion dealt, for the most part, with aircraft-position displays, the same basic problems apply to most displays in which other moving objects (such as vehicles) change positions relative to a background. The same human factors principles usually would apply to other circumstances, albeit with situational variations (the displays in aircraft control towers, for example, have their own special features and problems).

Aside from displays of vehicles and their backgrounds, there are other displays that represent changing conditions and relationships, such as CRT displays used in medical and laboratory settings and in automobile repair shops. So far, relatively little research has been done on the use of such displays in these settings.

HEAD-UP DISPLAYS

A *head-up display* (HUD) is a display in which information is superimposed on the outside world via a windscreen or helmet visor (the latter is called a *helmet-mounted display,* or HMD). The information can be read with the head erect and with the outside world always in the field of view (Boff and Lincoln, 1988). HUDs are a technology which can be used to present any type of display, including quantitative, status, or representational displays. The principle application of HUD technology has been to military aircraft where it is important for the pilot to maintain visual contact with the outside world and where the time taken to look inside the cockpit and refocus on critical displays could adversely affect performance. For example, all tactical fighter aircraft and many military helicopters have HUDs. Soon you will not have to become a military pilot to use a HUD. Several automobile manufacturers are experimenting with HUD technology and soon HUDs should be available on selected models.

HUDs use collimated images that are optically focused at infinity. The idea is that one can view both distant objects through the windscreen and HUD information without changing the accommodation of the eye. Iavecchia, Iavecchia, and Roscoe (1988), however, provide evidence that seriously challenges the benign contribution of HUDs to visual accommodation. They found that the presence of a collimated HUD image causes the eye to accommodate to its dark focus, at about arm's length for most people, rather than at infinity. (See Chapter 4 for a discussion of accommodation and dark focus.) This causes misaccommodation of objects located beyond the dark focus point and in turn misperception of their size and distance. Objects in the real world appear smaller and more distant than they are. Needless to say, thinking you are farther away from an object than you are can contribute to plane crashes (Roscoe, 1987). The perceptual impact of HUDs in automobiles, however, is probably lessened by the richness of depth cues in the driving environment.

DISCUSSION

There are all sorts of visual displays that represent dynamic, changeable conditions. We covered the highlights of the major types in this chapter. A complete set of guidelines for all the variations that are possible is far beyond the scope of this text. However, certain guidelines offered by Heglin (1973) are given in Table 5-3. Some of these guidelines have been covered in this chapter, but a few have not.

It is fair to say that advances in technology will have a great effect on the future design of dynamic displays. The areas that are receiving attention are

TABLE 5-3
GENERAL GUIDE TO VISUAL DISPLAY SELECTION

To display	Select	Because	Example
Go, no go, start, stop, on, off	Light	Normally easy to tell if it is on or off.	
Identification	Light	Easy to see (may be coded by spacing, color, location, or flashing rate; may also have label for panel applications).	
Warning or caution	Light	Attracts attention and can be seen at great distance if bright enough (may flash intermittently to increase conspicuity).	
Verbal instruction (operating sequence)	Enunciator light	Simple "action instruction" reduces time required for decision making.	
Exact quantity	Digital counter	Only one number can be seen, thus reducing chance of reading error.	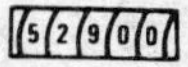
Approximate quantity	Moving pointer against fixed scale	General position of pointer gives rapid clue to the quantity plus relative rate of change.	
Set-in quantity	Moving pointer against fixed scale	Natural relationship between control and display motions.	
Tracking	Single pointer or cross pointers against fixed index	Provides error information for easy correction.	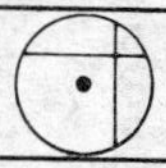
Vehicle attitude	Either mechanical or electronic display of position of vehicle against established reference (may be graphic or pictorial)	Provides direct comparison of own position against known reference or base line.	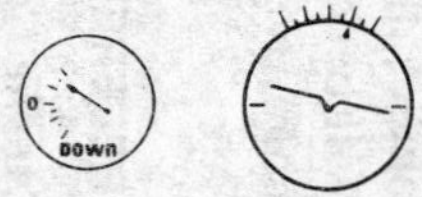

Source: Heglin, 1973, Table 11-4.

those that integrate information into meaningful forms for users, specifically, object displays and representational displays. Enhancing situation awareness will become a major goal as systems grow in complexity and place additional cognitive burdens on users.

REFERENCES

Applied ergonomics handbook. (1974). Guilford, Surrey, England: IPC Science and Technology Press.

Berkhout, J. (1979). Information transfer characteristics of moving light signals. *Human Factors,* 21(4), 445–455.

Boff, K., and Lincoln, J. (eds.). (1988). *Engineering data compendium: Human perception and performance,* vol. 3. Wright-Patterson Air Force Base, OH: Harry G. Armstrong Aerospace Medical Research Laboratory.

Carswell, C., and Wickens, C. (1987). Information integration and object display: An interaction of task demands and display superiority. *Ergonomics,* 30(3), 511–527.

Cohen, E., and Follert, R. L. (1970). Accuracy of interpolation between scale graduations. *Human Factors,* 12(5), 481–483.

Coury, B., Boulette, M., and Smith, R. (1989). Effect of uncertainty and diagnosticity on classification of multidimensional data with integral and separable displays of system status. *Human Factors,* 31(5), 551–569.

Coury, B., and Purcell, J. (1988). The bar graph as a configural and separable display. *Proceedings of the Human Factors Society 32d Annual Meeting.* Santa Monica, CA: Human Factors Society, pp. 1361–1365.

Crawford, A. (1963). The perception of light signals: The effect of mixing flashing and steady irrelevant lights. *Ergonomics,* 6, 287–294.

Dashevsky, S. G. (1964). Check-reading accuracy as a function of pointer alignment, patterning, and viewing angle. *Journal of Applied Psychology,* 48, 344–347.

Elkin, E. H. (1959, February). *Effect of scale shape, exposure time and display complexity on scale reading efficiency* (TR 58-472). Wright-Patterson Air Force Base, OH: USAF, WADC.

Ellis, S., McGreevy, M., and Hitchock, R. (1987). Perspective traffic display format and airline pilot traffic avoidance. *Human Factors,* 29(4), 371–382.

Endsley, M. (1988*a*, May). *Situation awareness global assessment technique (SAGAT).* Paper presented at the National Aerospace and Electronic Conference (NAECON), Dayton, OH.

Endsley, M. (1988*b*). Design and evaluation for situation awareness enhancement. *Proceedings of the Human Factors Society 32d Annual Meeting.* Santa Monica, CA: Human Factors Society, pp. 97–101.

Fogel, L. (1959). A new concept: The kinalog display system. *Human Factors,* 1(1), 30–37.

Fracker, M. (1988). A theory of situation assessment: Implications for measuring situation awareness. *Proceedings of the Human Factors Society 32d Annual Meeting.* Santa Monica, CA: Human Factors Society, pp. 102–106.

Green, P. (1984). *Driver understanding of fuel and engine gauges.* Tech. Paper Series 840314. Warrendale, PA: Society of Automotive Engineers.

Grether, W. F., and Baker, C. A. (1972). Visual presentation of information. In H. A. Van Cott and R. G. Kinkade (eds.), *Human engineering guide to equipment design* (rev. ed.). Washington, DC: Government Printing Office.

Heglin, H. J. (1973, July). *NAVSHIPS display illumination design guide: II. Human factors* (NELC/TD223). San Diego: Naval Electronics Laboratory Center.

Iavecchia, J., Iavecchia, H., and Roscoe, S. (1988). Eye accommodation to head-up virtual images. *Human Factors,* 30(6), 689–702.

Markowitz, J. (1971). Optimal flash rate and duty cycle for flashing visual indicators. *Human Factors,* 13(5), 427–433.

Mortimer, R. G., and Kupec, J. D. (1983). Scaling of flash rate for a deceleration signal. *Human Factors,* 25(3), 313–318.

Pomerantz, J. (1981). Perceptual organization in information processing. In R. Kubovy and J. Pomerantz (eds.), *Perceptual organization*. Hillsdale, NJ: Erlbaum.

Reynolds, R. F., White, R. M., Jr., and Hilgendorf, R. I. (1972). Detection and recognition of color signal lights. *Human Factors,* 14(3), 227–236.

Roscoe, S. N. (1968). Airborne displays for flight and navigation. *Human Factors,* 10(4), 321–332.

Roscoe, S. N. (1987, July). The trouble with HUDs and HMDs. *Human Factors Society Bulletin,* 30(7), 1–3.

Roscoe, S. N., Corl, L., and Jensen, R. S. (1981). Flight display dynamics revisited. *Human Factors,* 23(3), 341–353.

Roscoe, S., and Williges, R. (1975). Motion relationships and aircraft attitude guidance displays: A flight experiment. *Human Factors,* 17, 374–387.

Sabeh, R., Jorve, W. R., and Vanderplas, J. M. (1958, March). *Shape coding of aircraft instrument zone markings* (Tech. Note 57–260). Wright-Patterson Air Force Base, OH: USAF, WADC.

Sanders, M., and Peay, J. (1988). *Human factors in mining* (IC 9182). Washington, DC: U.S. Department of the Interior, Bureau of Mines.

Simmonds, G. R. W., Galer, M., and Baines, A. (1981). *Ergonomics of electronic displays,* Tech. Paper Series 810826. Warrendale, PA: Society of Automotive Engineers.

Teichner, W. H., and Krebs, M. J. (1972). Estimating the detectability of target luminances. *Human Factors,* 14(6), 511–519.

Tolin, P. (1984). An information transmission of signal flash rate discriminability. *Human Factors,* 26(4), 489–493.

Whitehurst, H. O. (1982). Screening designs used to estimate the relative effects of display factors on dial reading. *Human Factors,* 24(3), 301–310.

Wickens, C. (1986). *The object display: Principles and a review of experimental findings* (Tech. Rept. CPL 86-6). Champaign, IL: University of Illinois Cognitive Psychophysiology Laboratory, Department of Psychology.

Wickens, C. (1987). Ergonomic-driven displays: A set of principles of the perceptual-cognitive interface. *Proceedings of the Fourth Mid-Central Ergonomics/Human Factors Conference*. Champaign, IL: University of Illinois.

Wickens, C., and Andre, A. (1988). Proximity compatibility and the object display. *Proceedings of the Human Factors Society 32d Annual Meeting*. Santa Monica, CA: Human Factors Society, pp. 1335–1339.

Woodson, W. E., and Conover, D. W. (1964). *Human engineering guide for equipment designers* (2d ed.). Berkeley: University of California Press.

Zenyuh, J., Reising, J., Walchli, S., and Biers, D. (1988). A comparison of a stereographic 3-D display versus a 2-D display using an advanced air-to-air format. *Proceedings of the Human Factors Society 32d Annual Meeting*. Santa Monica, CA: Human Factors Society, pp. 53–57.

CHAPTER **6**

AUDITORY, TACTUAL, AND OLFACTORY DISPLAYS

We all depend on our auditory, tactual, and olfactory senses in many aspects of our lives, including hearing our children cry or the doorbell ring, feeling the smooth finish on fine furniture, or smelling a cantaloupe to determine if it is ripe. As discussed in Chapter 3, information can come to us directly or indirectly. A child's natural cry is an example of direct information, while a doorbell's ring would be indirect information that someone is at the door. It is becoming increasingly possible to convert stimuli that are intrinsically, or directly, associated with one sensory modality into stimuli associated (indirectly) with another modality. Such technological developments have resulted in increased use of the auditory, tactual, and—to a lesser extent—olfactory senses. Using buzzers to warn of fire or to warn blind people of physical objects in their path are examples. Often the indirect stimulus can be more effective than the actual direct stimulus. For example, Kahn (1983) reported that sleeping subjects responded faster and more often to an auditory fire alarm, of sufficient intensity, than to the presence of either heat or the smell of smoke. Heat or smoke awoke the subjects only 75 percent of the time, while the alarm was effective virtually 100 percent of the time.

In this chapter we discuss the use of auditory, tactual, and olfactory senses as means of communication. Our focus is on indirect information sources rather than direct, and the topic of speech is deferred until Chapter 7.

HEARING

In discussing the hearing process, first we describe the physical stimuli to which the ear is sensitive, namely, sound vibrations, and then we discuss the anatomy of the ear.

Nature and Measurement of Sound

Sound is created by vibrations from some source. Although such vibrations can be transmitted through various media, our primary concern is with those transmitted through the atmosphere to the ear. Two primary attributes of sound are *frequency* and *intensity* (or amplitude).

Frequency of Sound Waves The frequency of sound waves can be visualized if we think of a simple sound-generating source such as a tuning fork. When it is struck, the tuning fork is made to vibrate at its *natural* frequency. In so doing, the fork causes the air molecules to be moved back and forth. This alternation creates corresponding increases and decreases in the air pressure.

The vibrations of a simple sound-generating source, such as a tuning fork, form *sinusoidal* (or *sine*) *waves,* as shown in Figure 6-1. The height of the wave above the midline, at any given time, represents the amount of above-normal air pressure at that point. Positions below the midline in turn, represent the reduction in air pressure below normal.

One feature of simple sine waves is that the waveform above the midline is the mirror image of the waveform below the midline. Another feature is that the waveform pattern repeats itself again and again. One complete cycle is shown in Figure 6-1. The number of cycles per second is called the *frequency* of the sound. Frequency is expressed in hertz (abbreviated Hz), which is equivalent to cycles per second. On the musical scale, middle C has a frequency of 256 Hz. Any given octave has double the frequency of the one below it, so one octave above middle C has a frequency of 512 Hz. In general terms, the human ear is sensitive to frequencies in the range of 20 to 20,000 Hz, but it is not equally sensitive to all frequencies. In addition, there are marked differences among people in their relative sensitivities to various frequencies.

The frequency of a physical sound is associated with the human sensation of pitch. *Pitch* is the name given to the highness or lowness of a tone. Since high frequencies yield high-pitched tones and low frequencies yield low-pitched tones, we tend to think of pitch and frequency as synonymous. Actually there are other factors that influence our perception of pitch besides frequency. For example, the intensity of a tone can influence our perception of pitch. When

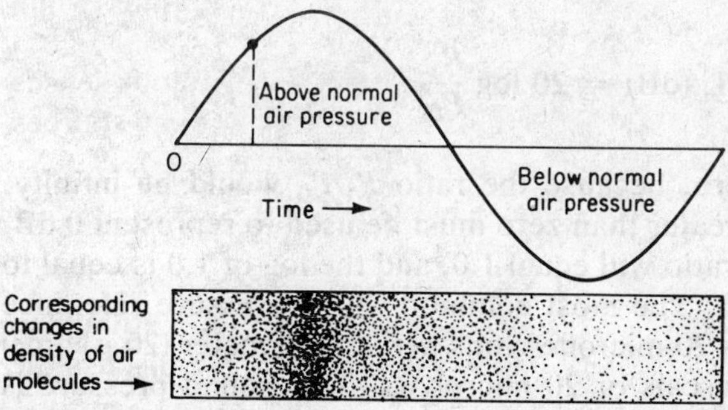

FIGURE 6-1
Sinusoidal wave created by a simple sound-generating source. The magnitude of the alternating changes in air pressure can be represented by a sine wave, shown in the upper part of the figure. The lower part of the figure depicts the changes in the density of the air molecules caused by the vibrating source and corresponding to the sine wave changes in pressure.

they are increased in intensity, low-frequency tones (less than about 1000 Hz) become lower in pitch, while high-frequency tones (greater than about 3000 Hz) become higher in pitch. Tones between 1000 and 3000 Hz, on the other hand, are relatively insensitive to such intensity-induced pitch changes. Incidentally, complex sounds such as those produced by musical instruments are very stable in perceived pitch regardless of whether the instruments are played loudly or softly—happily for music lovers!

Intensity of Sound Sound intensity is associated with the human sensation of *loudness*. As in the case of frequency and pitch, several factors besides intensity can influence our perception of loudness. We, however, have deferred the discussion of loudness to Chapter 18.

Sound intensity is defined in terms of power per unit area, for example, watts per square meter (W/m^2). Because the range of power values for common sounds is so tremendous it is convenient to use a logarithmic scale to characterize sound intensity. The *bel* (B), named after Alexander Graham Bell, is the basic unit of measurement used. The number of bels is the logarithm (to the base 10) of the ratio of two sound intensities. Actually, a more convenient and commonly used measure of sound intensity is the *decibel* (dB), where 1 dB = 0.1 B.

Most sound measuring instruments do not directly measure the sound power of a source. However, since sounds are pressure waves that vary above and below normal air pressure, these variations in air pressure can be directly measured. Luckily, sound power is directly proportional to the *square* of the sound pressure. The sound-pressure level (SPL), in decibels, is therefore defined as

$$\text{SPL (dB)} = 10 \log \frac{P_1^2}{P_0^2}$$

where P_1^2 is the sound pressure squared of the sound one wishes to measure and P_0^2 is the reference sound pressure squared that we choose to represent 0 dB. With a little algebraic magic, we can simplify the above equation so that we deal with sound pressure rather than sound pressure squared.

The equation for SPL becomes

$$\text{SPL (dB)} = 20 \log \frac{P_1}{P_0}$$

Note that P_0 cannot equal zero, because the ratio P_1/P_0 would be infinity. Therefore, a sound pressure greater than zero must be used to represent 0 dB. Then, when P_1 equals P_0, the ratio will equal 1.0, and the log of 1.0 is equal to zero; i.e., SPL will equal 0 dB. The most common sound-pressure reference value used to represent 0 dB is 20 micronewtons per square meter (20 $\mu N/m^2$) or the equivalents, 20 micropascals or 20 microbars. This sound pressure is roughly equivalent to the lowest intensity, 1000 Hz pure tone, which a healthy adult can just barely hear under ideal conditions. It is possible, therefore, to have sound-pressure levels that are actually less than 0 dB; that is, they have sound pressures less than 20 microbars.

The decibel scale is a logarithmic scale, so an increase of 10 dB represents a 10-fold increase in sound power and a 100-fold increase in sound pressure (remember that power is related to the square of pressure). In like manner, it can be shown that doubling the sound power will raise the SPL by 3 dB. Another consequence of using a logarithmic scale is that the ratio of two sounds is computed by subtracting (rather than dividing) one decibel level from the other. So when we speak of the *signal-to-noise ratio,* we are actually referring to the difference between a meaningful signal and the background noise in decibels. For example, if a signal is 90 dB and the noise is 70 dB, the signal-to-noise ratio is +20 dB.

Sound pressure is measured by sound-level meters. The American National Standards Institute (ANSI) and the American Standards Association have established a standard to which sound-level meters should conform. This standard requires that three different *weighting* networks (designated *A, B,* and *C*) be built into such instruments. Each network responds differently to low or high frequencies according to standard *frequency-response* curves. We have more to say about this in Chapter 18 when we discuss noise. For now, we need to be aware that when a sound intensity is reported, the specific weighting network used for the measure must be specified, for example, "the *A*-weighted sound level is 45 dB," "sound level (*A*) = 45 dB," "SL*A* = 45 dB." or "45 dB*A*." Figure 6-2 shows the decibel scale with examples of several sounds that fall at varying positions.

Complex Sounds Very few sounds are pure tones. Even tones from musical instruments are not pure, but rather consist of a fundamental frequency in combination with others (especially harmonic frequencies that are multiples of the fundamental). Most complex sounds, however, are nonharmonic. Complex sounds can be depicted in two ways. One is a waveform which is the composite of the waveforms of its individual component sounds. Figure 6-3 shows three component waveforms and the resultant composite waveform.

The other method of depicting complex sounds is by use of a *sound spectrum* which divides the sound into frequency bands and measures the intensity of the sound in each band. A frequency-band analyzer is used for this purpose. The four curves of Figure 6-4 illustrate spectral analyses of the noise from a rope-closing machine. Each curve was generated with a different-sized frequency band (*bandwidth*), namely, an octave, a half octave, one-third of an octave, and a thirty-fifth of an octave. The narrower the bandwidth, the greater the detail of the spectrum and the lower the sound level within each bandwidth. There have been various practices in the division of the sound spectrum into octaves, but the current preferred practice as set forth by ANSI (1966) is to divide the audible range into 10 bands, with the *center* frequencies being 31.5,

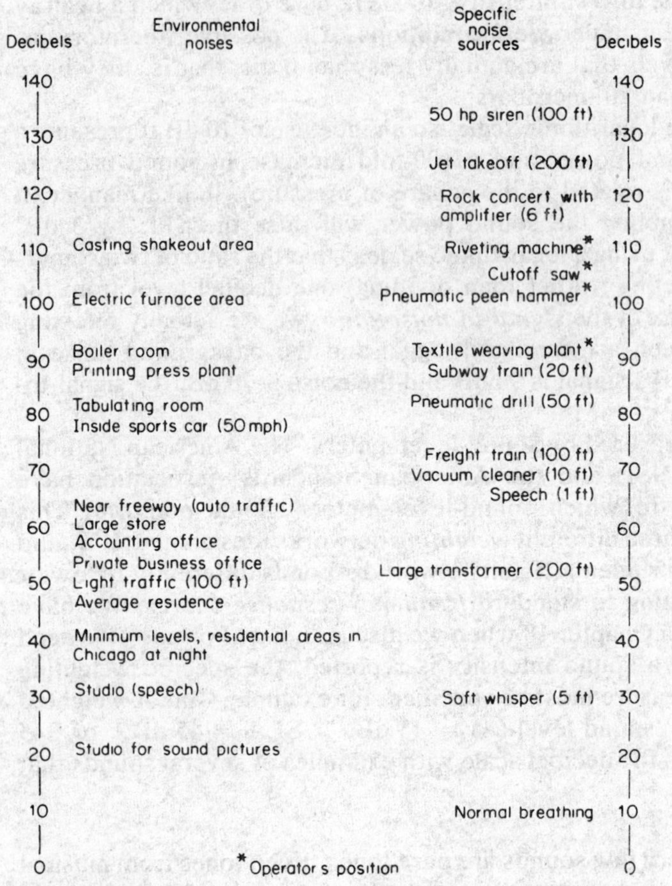

FIGURE 6-2
Decibel levels for various sounds. Decibel levels are *A*-weighted sound levels measured with a sound-level meter. (*Source: Peterson and Gross, 1972, Fig. 2-1, p. 4.*)

63, 125, 250, 500, 1000, 2000, 4000, 8000, and 16,000 Hz. (Many existing sets of sound and hearing data are presented that use the previous practice of defining octaves in terms of the *ends* of the class intervals instead of their *centers*.)

Anatomy of the Ear

The ear has three primary anatomical divisions: the outer ear, the middle ear, and the inner ear. These are shown schematically in Figure 6-5.

Outer Ear The outer ear, which collects sound energy, consists of the external part (called the *pinna* or *choncha*); the auditory canal (the *meatus*), which is a bayonet-shaped tube about 1 in long that leads inward from the

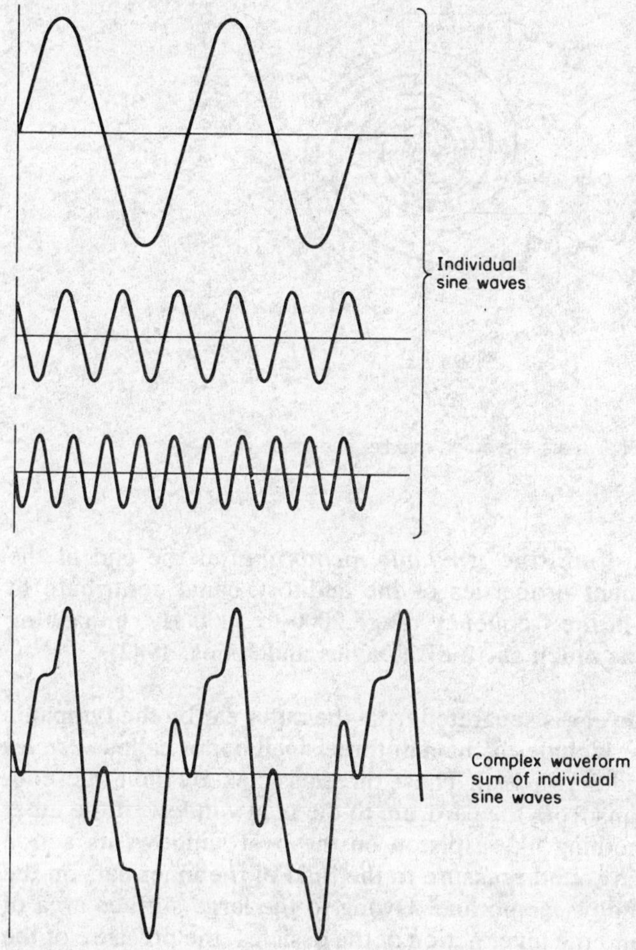

FIGURE 6-3
Waveform of a complex sound formed from three individual sine waves. (*Source: Minnix, 1978, Fig. 1-11.*)

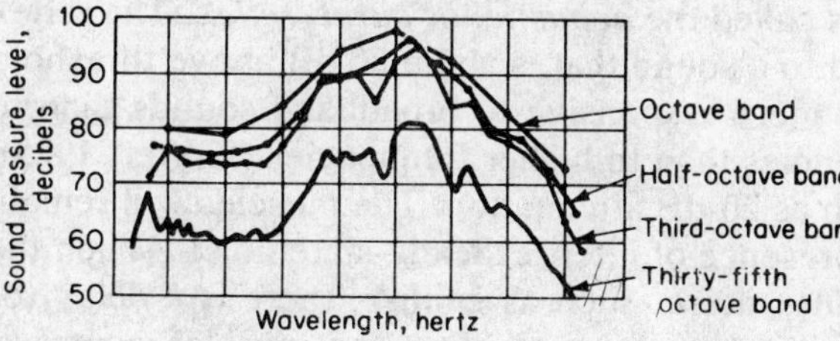

FIGURE 6-4
Spectral analyses of noise from a rope-closing machine. Analyzers used varying bandwidths. The narrower the bandwidth, the greater the detail and the lower the sound-pressure level within any single bandwidth. (*Source: Adapted from* Industrial noise manual, *1966, p. 25.*)

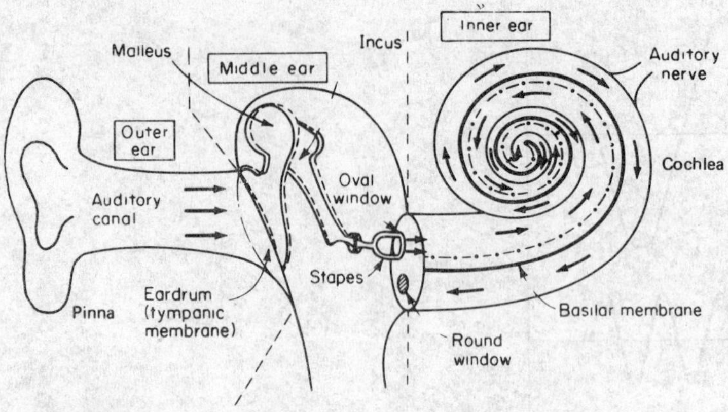

FIGURE 6-5
Schematic drawing of the ear. Shown are the outer, middle, and inner ear structures.

external part; and the eardrum (the *tympanic membrane*) at the end of the auditory canal. The resonant properties of the auditory canal contribute to the sensitivity of the ear in the frequency range 2000 to 5000 Hz, enhancing sound-pressure levels by as much as 12 dB (Davies and Jones, 1982).

Middle Ear The middle ear is separated from the outer ear by the tympanic membrane. The middle ear includes a chain of three small bones called *ossicles* (the *malleus*, the *incus*, and the *stapes*). These three ossicles, by their interconnections, transmit vibrations from the eardrum to the oval window of the inner ear. The stapes acts something like a piston on the oval window, its action transmitting the changes in sound pressure to the fluid of the inner ear, on the other side of the oval-window membrane. Owing to the large surface area of the tympanic membrane and the lever action of the ossicles, the pressure of the foot of the stapes against the oval window is amplified to about 22 times that which could be effected by applying sound waves directly to the oval window (Guyton, 1969, p. 300).

The middle ear also contains two muscles attached to the ossicles. The *tensor tympani muscle* attaches to the malleus, and the *stapedius muscle* attaches to the stapes. In response to loud noises, the stapedius muscle tightens and reduces sound transmission to the inner ear, thus providing protection against intense sounds. This reaction is called the *acoustic*, or *aural, reflex*. The reflex occurs when the ear is exposed to a sound that is about 80 dB above threshold level; the reflex appears to be more responsive to broadband sounds than to pure tones and to lower frequencies than to higher frequencies (Kryter, 1985). The reflex can provide as much as 20-dB attenuation. The muscles will remain flexed for up to 15 min in the presence of intense steady-state noise. When the ear is exposed to intense impulse noise, such as gunfire, there is a delay (or latency) of about 35 to 150 milliseconds (ms) before the muscles contract.

Because of this delay there is not much protection provided against the initial impulse; however, the relaxation time following an impulse can be as long as 2 to 3 s, with most of it occurring within about 0.5 s. Protection, therefore, is provided against subsequent impulses when the time between impulses is less than about 1 s.

Inner Ear The inner ear, or *cochlea,* is a spiral-shaped affair that resembles a snail. If uncoiled, it would be about 30 mm long, with its widest section (near the oval window) being about 5 or 6 mm. The inner ear is filled with a fluid. The stapes of the middle ear acts on this fluid like a piston, driving it back and forth in response to changes in the sound pressure. These movements of the fluid force into vibration a thin membrane called the *basilar membrane,* which in turn transmits the vibrations to the *organ of Corti.* The organ of Corti contains hair cells and nerve endings that are sensitive to very slight changes in pressure. The neural impulses picked up by these nerve endings are transmitted to the brain via the *auditory nerve.*

Conversion of Sound Waves to Sensations

Although the mechanical processes involved in the ear have been known for some time, the procedures by which sound vibrations are ''heard'' and differentiated are still not entirely known. In this regard, the numerous theories of hearing fall generally into two classes. The *place* (or resonance) theories are postulated on the notion that the fibers at various positions (or places) along the basilar membrane act, as Geldard (1972) puts it, as harp or piano strings. Since these fibers vary in length, they are differentially sensitive to different frequencies and thus give rise to sensations of pitch. On the other hand, the *temporal* theories are based on the tenet that pitch is related to the time pattern, or interval, of the neural impulses emitted by the fibers. For more indepth discussions of the physiology and psychophysics of hearing, see Kiang and Peake (1988) and Green (1988).

As Geldard (1972) comments, it is not now in the cards to be able to accept one basic theory to the exclusion of any others as explaining all auditory phenomena. Actually, Geldard indicates that the current state of knowledge would prejudice one toward putting reliance in a place theory as related to high tones and in a temporal theory as related to low tones, one principle thus giving way to the other in the middle range of frequencies. As with theories in various areas of life, this notion of the two theories being complementary still needs to be regarded as tentative pending further support or rejection from additional research. Moore (1982), for example, offers a theory or model based on both place and temporal information.

One consequence of all this is that the ear is not equally sensitive to all frequencies of sound. Although we discuss this more fully in Chapter 18, it is important to know that, in general, the ear is less sensitive to low-frequency sounds (20 to approximately 500 Hz) and more sensitive to higher frequencies

(1000 to approximately 5000 Hz). That is, a 4000-Hz tone, at a given sound-pressure level, will seem to be louder than, say, a 200-Hz tone at the same sound-pressure level.

Masking

Masking is a condition in which one component of the sound environment reduces the sensitivity of the ear to another component. Operationally defined, *masking* is the amount that the threshold of audibility of a sound (the masked sound) is raised by the presence of another (masking) sound. In studying the effects of masking, an experimenter typically measures the absolute threshold (the minimum audible level) of a sound (the sound to be masked) when presented by itself and then measures its threshold in the presence of the masking sound. The difference is attributed to the masking effect. The concept of masking is central to discussions of auditory displays. In selecting a particular auditory signal for use in a particular environment, we must consider the masking effect of any noise on the reception of that signal.

The effects of masking vary with the type of masking sound and with the masked sound itself—whether pure tones, complex sound, white noise, speech, etc. Figure 6-6 illustrates the masking of a pure tone by another pure tone. A few general principles can be gleaned from this. The greatest masking effect occurs near the frequency of the masking tone and its harmonic overtones. As we would expect, the higher the intensity of the masking tone, the greater the masking effect. Notice, however, that with low-intensity masking tones (20 to 40 dB) the masking effect is somewhat confined to the frequencies around that of the masking tone; but with higher-intensity masking tones (60 to 100 dB), the masking effect spreads to higher frequencies. The masking of pure tones by narrowband noise (i.e., noise concentrated in a narrow band of frequencies) is very similar to that shown in Figure 6-6.

In the masking of pure tones by wideband noise, the primary concern is with the intensity of the masking noise in a "critical band" around the frequency of

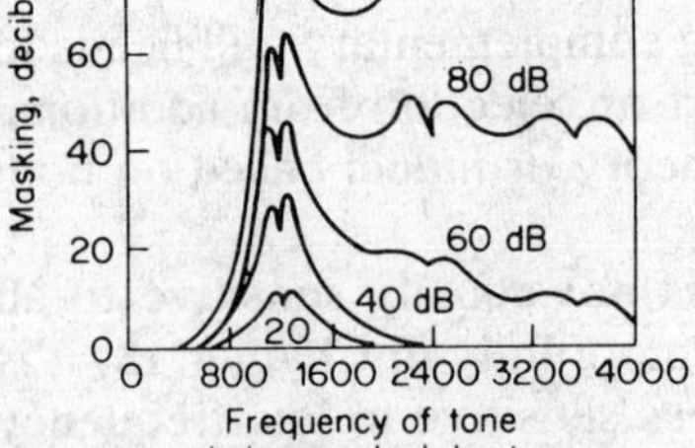

FIGURE 6-6
The effects of masking pure tones by a pure tone of 1200 Hz (100, 80, 60, 40, and 20 dB). (*Source: Wegel and Lane, 1924.*)

the masked tone. The size of the critical band is a function of the center frequency, with larger critical bands being associated with higher frequencies.

The nature of masking effects depends very much upon the nature of the two sounds in question, as discussed by Geldard (1972, pp. 215–220) and by Deatherage (1972). Although these complex effects are not discussed here, the effects of masking in our everyday lives are of considerable consequence, as, for example, the noise of a hair dryer drowning out the sound of the telephone ringing.

AUDITORY DISPLAYS

The nature of the auditory sensory modality offers certain unique advantages for presenting information as contrasted with the visual modality. One set of comparisons of these two senses was given in Table 3-1 (Chapter 3). On the basis of such comparisons and of other cues, it is possible to identify certain types of circumstances in which auditory displays would be preferable to visual displays:

- When the origin of the signal is itself a sound
- When the message is simple and short
- When the message will not be referred to later
- When the message deals with events in time
- When warnings are sent or when the message calls for immediate action
- When continuously changing information of some type is presented, such as aircraft, radio range, or flight-path information
- When the visual system is overburdened
- When speech channels are fully employed (in which case auditory signals such as tones should be clearly detectable from thc spccch)
- When illumination limits use of vision
- When the receiver moves from one place to another
- When a verbal response is required

Obviously the application of these guidelines should be tempered with judgment rather than being followed rigidly. There are, of course, other circumstances in which auditory displays would be preferable. In the above guidelines particular mention should be made of the desirability of restricting auditory messages to those that are short and simple (except in the case of speech), since people do not recall complex messages from short-term memory very well.

We can consider four types of human functions, or tasks, involved in the reception of auditory signals: (1) *detection* (determining whether a given signal is present, such as a warning signal), (2) *relative discrimination* (differentiating between two or more signals presented close together), (3) *absolute identification* (identifying a particular signal of some class when only one is presented), and (4) *localization* (determining the direction from which the signal is coming). Relative discrimination and absolute identification can be made on the basis of any of several stimulus dimensions, such as intensity, frequency, and duration.

Detection of Signals

The detection of auditory signals can be viewed within the framework of signal detection theory (as discussed in Chapter 3). Signals can occur in "peaceful" surroundings or in environments that are permeated by ambient noise. As indicated in the discussion of signal-detection theory, if at all possible, the signal plus noise (SN) should be distinct from the noise (N) itself. When this is not the case, the signal cannot always be detected in the presence of the noise; this confusion is, in part, a function of masking. When the signal (the masked sound) occurs in the presence of noise (i.e., the masking sound), the threshold of detectability of the signal is elevated—and it is this elevated threshold that should be exceeded by the signal if it is to be detected accurately. Failure to consider this forced New York City in 1974 to abandon legislation aimed at regulating the decibel output of newly manufactured automobile horns—the quieter horns could not be heard over the noise of traffic (Garfield, 1983).

In quiet surroundings (when one does not have to worry about clacking typewriters, whining machines, or screeching tires), a tonal signal about 40 to 50 dB above absolute threshold normally would be sufficient to be detected. However, such detectability would vary somewhat with the frequency of the signal (Pollack, 1952) and its duration. With respect to duration, the ear does not respond instantaneously to sound. For pure tones it takes about 200 to 300 ms to "build up" (Munson, 1947) and about 140 ms to decay (Stevens and Davis, 1938); wideband sounds build up and decay more rapidly. Because of these lags, signals less than about 200 to 500 ms in duration do not sound as loud as those of longer duration. Thus, auditory signals (especially pure tones) should be at least 500 ms in duration; if they have to be shorter than that, their intensity should be increased to compensate for reduced audibility. Although detectability increases with signal duration, there is not much additional effect beyond a few seconds in duration.

In rather noisy conditions, the signal intensity has to be set at a level far enough above that of the noise to ensure detectability but not so loud as to become annoying or to disrupt thought or communications. In the context of civil aircraft, Patterson (1982) recommends a minimum level of 15 dB above the masked threshold to ensure detectability and a maximum level of 25 dB above the masked threshold to guard against annoyance and disruption. (Using a gradual signal onset will also reduce annoyance.) Deatherage (1972) has suggested a rule of thumb for specifying the optimum signal level: The signal intensity (at the entrance of the ear) should be about midway between the masked threshold of the signal in the presence of noise and 110 dB.

Fidell, Pearsons, and Bennett (1974) developed a method for predicting the detectability of complex signals (complex in terms of frequency) in noisy backgrounds. Although we do not present the details of the procedure, it is based on the observation that human performance is a function of the signal-to-noise ratio in the single one-third octave band to which human sensitivity is highest. The basic data employed in the proposed method are therefore the one-third octave band spectra of both the signal and background noise.

Use of Filters Under some circumstances it is possible to enhance the detectability of signals by filtering out some of the noise. This is most feasible when the predominant frequencies of the noise are different from those of the signal. In such a case, some of the noise can be filtered out and the intensity of the remaining sound raised (signal plus the nonfiltered noise). This has the effect of increasing the signal-to-noise ratio, thus making the signal more audible.

The central nervous system is also capable of filtering out noise if the circumstances are right. Presenting a signal to one ear and noise to both ears will increase detectability compared to the situation where both the noise and signal are presented to both ears. This appears to be due to filtering of the noise somewhere in the central nervous system. The exact mechanism is not known, but the brain has the capacity to compare the signals received by both ears and, in effect, cancel some of the noise.

Increasing Detectability of Signals Mulligan, McBride, and Goodman (1984) list several procedures to increase the detectability of a signal in noise, several of which we have already mentioned: (1) reduce the noise intensity in the region of frequencies (i.e., critical bandwidth) around the signal's frequency; (2) increase the intensity of the signal; (3) present the signal for at least 0.5 to 1 s; (4) change the frequency of the signal to correspond to a region where the noise intensity is low; (5) phase-shift the signal and present the unshifted signal to one ear and the shifted signal to the other; and (6) present the noise to both ears and the signal to only one ear. Patterson (1982) adds another recommendation: the signal should be composed of four or more prominent frequency components in the range from 1000 to 4000 Hz because complex sounds are more difficult to mask than simpler sounds.

Relative Discrimination of Auditory Signals

The relative discrimination of signals on the basis of intensity and frequency (which are the most commonly used dimensions) depends in part on interactions between these two dimensions. In many real-life situations, however, discriminations depend on much more than just frequency and intensity differences. Talamo (1982), for example, discusses the difficulties faced by equipment operators monitoring machine sounds for indications of impending mechanical failure. Discriminations in such cases may involve relative strengths of harmonics or even the relative phases of sounds.

A common measure of discriminability is the *just-noticeable difference* (JND). The JND is the smallest change or difference along a stimulus dimension (e.g., intensity or frequency) that can just be detected (noticed) 50 percent of the time by people. The smaller the JND found, the easier it is for people to detect differences on the dimension being changed. A small JND indicates that subjects could detect small changes; a large JND indicates that a large change was necessary before people could detect it. As we will see, the JND often varies as a function of where along the stimulus dimension the change occurs.

Discriminations of Intensity Differences Figure 6-7 (Deatherage, 1972) reflects the JNDs for certain pure tones and for wideband noise of various sound-pressure levels. It is clear that the smallest differences can be detected with signals of higher intensities, at least 60 dB above the absolute threshold. The JNDs of signals above 60 dB are smallest for the intermediate frequencies (1000 and 4000 Hz). These differences in JNDs, of course, have implications for selecting signals if the signals are to be discriminated by intensity.

Discrimination of Frequency Differences Some indication of the ability of people to tell the difference between pure tones of different frequencies is shown by the results of an early study by Shower and Biddulph (1931) as depicted in Figure 6-8. This figure shows the JNDs for pure tones of various frequencies at various levels above threshold. The JNDs are smaller for frequencies below about 1000 Hz (especially for high intensities) but increase rather sharply for frequencies above that. Thus, if signals are to be discriminated on the basis of frequency, it usually would be desirable to use signals of lower frequencies. This practice may run into a snag, however, if there is much ambient noise (which usually tends to consist of lower frequencies) that might mask the signals. A possible compromise is to use signals in the 500- to 1000-Hz range (Deatherage, 1972). It is also obvious from Figure 6-8 that the JNDs are smaller for signals of high intensity than for those of low intensity, thus suggesting that signals probably should be at least 30 dB above absolute threshold if frequency discrimination is required. Duration of the signal is also important in discrimination of frequency differences. Gales (1979) reports that discrimination is best when stimulus duration is in excess of 0.1 s.

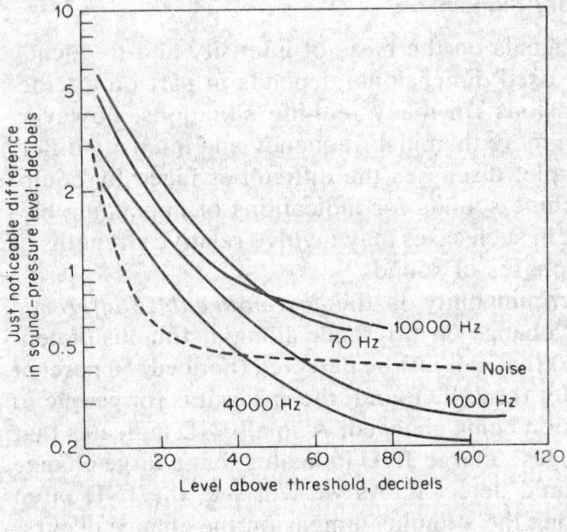

FIGURE 6-7
The just-noticeable differences (JNDs) in sound intensity for pure tones of selected frequencies and for wideband noise. (*Source: Deatherage, 1972, p. 147, as based on data from Riesz, 1928, and Miller, 1947.*)

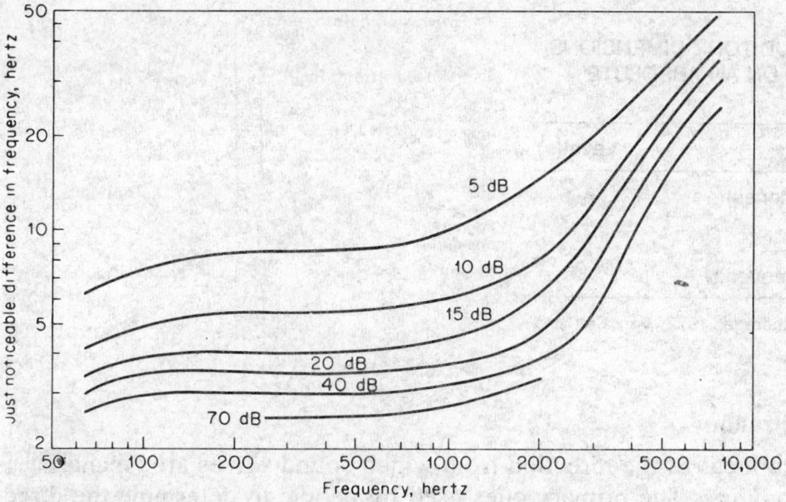

FIGURE 6-8
Just-noticeable differences in frequency for pure tones at various levels above threshold. (*Source: Shower and Biddulph, 1931.*)

Absolute Identification of Auditory Signals

The JND for a given stimulus dimension reflects the minimum difference that can be discriminated on a relative basis. But in many circumstances it is necessary to make an absolute identification of an individual stimulus (such as the frequency of a single tone) presented by itself. The number of "levels" along a continuum that can be so identified usually is quite small, as mentioned in Chapter 3. Table 6-1 gives the number of levels that can be identified for certain common auditory dimensions.

Multidimensional Coding If the amount of information to be transmitted by auditory codes is substantial (meaning that absolute identification of numerous signals needs to be made), it is possible to use a multidimensional code system. For example, Pollack and Ficks (1954) used various combinations of several dimensions, such as direction (right ear versus left ear), frequency, intensity, repetition rate, on-and-off time fraction, and duration. Such a system obviously imposes the requirement for training the receiver regarding the "meaning" of each unique combination. In using such multidimensional codes, however, it is generally better to use more dimensions with fewer steps or levels of each (such as eight dimensions with two steps of each dimension) than to use fewer dimensions and more levels of each (such as four dimensions with four steps of each). It is on the basis of the many different facets (dimensions) of sounds that we can identify the voices of individuals, a dripping water faucet, or a squeaking hinge.

TABLE 6-1
LEVELS OF AUDITORY DIMENSIONS IDENTIFIABLE ON AN ABSOLUTE BASIS

Dimension	Level(s)
Intensity (pure tones)	4–5
Frequency	4–7
Duration	2–3
Intensity and frequency	9

Source: Deatherage, 1972; Van Cott and Warrick, 1972.

Sound Localization

The ability to localize the direction from which sound waves are emanating is called *stereophony*. The primary cues used by people to determine the direction of a sound source are differences in both intensity and phase of the sounds. If a sound source is directly to one side of the head, the sound reaches the nearer ear approximately 0.8 ms before it reaches the other ear. This results in the sound in one ear being out of phase with the sound in the other ear, and this is an effective cue for localizing a sound source when the frequency of the sound is below 1500 Hz. The problem with using phase differences as the localization cue is that one cannot tell whether the sound is forward of or behind the head without turning the head from side to side. At low frequencies, sound waves go around the head easily, and even if the sound is on one side, there is little or no intensity difference at the two ears. At high frequencies (above 3000 Hz), however, the head effectively shadows the ear and intensity differences can be marked. The effectiveness of intensity differences as a cue to sound localization can be demonstrated with a home stereo recording system. If speakers are placed at each end of a room and the volume of one speaker is increased while the volume of the other is decreased, the sound will appear to move across the room. Listeners will be able to "follow" the sound by pointing to it as it "moves."

Localizing sound in the midrange frequencies (1500 to 3000 Hz) is relatively difficult because such frequencies produce neither effective phase nor intensity difference cues. The ability to turn the head from side to side greatly increases accuracy of localization under almost all circumstances. Without moving (tilting) the head, localization in the vertical (up-down) direction is rather poor.

Caelli and Porter (1980) provide a practical example of research on sound localization, i.e., determining the direction of an ambulance siren. Subjects sat in a car and heard a 2-s burst of a siren (120 dB*A* at the ambulance) 100 m from the car. They were required to estimate its direction and distance. Subjects tended to overestimate the distance of the siren, often by a factor of 2 or more. That is, the subjects thought the ambulance was farther away than it really was, a dangerous type of error in a real-life situation. With respect to localization, subjects were often 180° off in their perceptions of direction. One-third of all the test trials were classified as reversals of perceived direction. Particularly poor were perceptions when the siren was behind the driver. One-half to two-thirds of the time the subjects thought the siren was in front of them! It was also found that localization was poorer when the driverside window was rolled down. With the window down, the sound intensity to the window-side ear was higher than to the passenger-side ear, even when the sound was coming from the passenger side. This gave false localization cues to the subjects and resulted in poor localization performance.

A futuristic application of sound localization is being explored for military cockpits—*head-coupled auditory displays* (Furness, 1986). Specifically, threats, targets, radio communications, etc. are heard as if they originated from their specific locations in three-dimensional space by manipulating the signal's intensity and phase to each ear. A computer senses and compensates for the pilot's head position so that the sounds are directionally accurate and stabilized in space regardless of the position of the pilot's head. Such a system increases situation awareness (see Chapter 5), enhances audio communications by giving each source a different apparent direction, and provides a natural method of cuing where to look (Calhoun, Janson, and Valencia, 1988). For cuing where to look, initial experimental results with head-coupled auditory displays (Calhoun et al., 1988; Sorkin, Wrightman, Kistler, and Elvers, 1989) have been encouraging, at least with respect to localizing sounds left-right (azimuth). Localizing sounds up-down (elevation) have been less successful.

Carrying sound localization a step further, Strybel (1988) explored the ability of people to detect the motion of a sound. The measurement used is the *minimum audible movement angle* (MAMA). This is defined as the minimum angle traveled by a sound source which enables a listener to detect motion. The MAMA is dependent on the speed of movement and the initial position of the sound in the auditory field. For example, Grantham (1986) *simulated movement* of 500 Hz tone and found a MAMA of 5° when the target's initial position was straight ahead (0°). When the target was at 90°, the MAMA increased to 33°. By comparison, Manligas (1988), using *real motion* of a broadband noise source, found a MAMA of 1° for sources straight ahead and MAMAs of 3 to 4° for sources at 80° left or right of straight ahead.

Strybel provides a few conclusions gleaned from prior research: (1) the auditory system is relatively insensitive to motion displacement when compared to the visual system where displacement thresholds are measured in seconds of arc; (2) detection of simulated sound movement is worse than detection of real movement, especially in the periphery; and (3) detection of movement of broadband noise is easier (smaller MAMAs) than detection of movement of a pure tone. Whether head-coupled auditory displays will become commonplace in cockpits of the future is an open question. Such displays, however, will present interesting human factors challenges.

Principles of Auditory Display

As in other areas of human factors engineering, most guidelines and principles have to be accepted with a few grains of salt because specific circumstances, trade-off values, etc., may argue for their violation. With such reservations in mind, a few guidelines for the use of auditory displays are given. These generally stem from research and experience. Some are drawn in part from Mudd (1961) and Licklider (1961).

1 General principles

- **A** *Compatibility:* Where feasible, the selection of signal dimensions and their encoding should exploit learned or natural relationships of the users, such as high frequencies associated with up or high and wailing signals with emergency.
- **B** *Approximation:* Two-stage signals should be considered when complex information is to be presented. The signal stages would consist of:
  - **(1)** Attention-demanding signal: to attract attention and identify a general category of information.
  - **(2)** Designation signal: to follow the attention-demanding signal and designate the precise information within the general class indicated by the first signal.
- **C** *Dissociability:* Auditory signals should be easily discernible from any ongoing audio input (be it meaningful input or noise). (See Figure 6-9.) For example, if a person is to listen concurrently to two or more channels, the frequencies of the channels should be different if it is possible to make them so.
- **D** *Parsimony:* Input signals to the operator should not provide more information than is necessary.
- **E** *Invariance:* The same signal should designate the same information at all times.

2 Principles of presentation

- **A** *Avoid extremes of auditory dimensions:* High-intensity signals, for example, can cause a startle response and actually disrupt performance.
- **B** *Establish intensity relative to ambient noise level:* The intensity level should be set so that it is not masked by the ambient noise level.
- **C** *Use interrupted or variable signals:* Where feasible, avoid steady-state signals and, rather, use interrupted or variable signals. This will tend to minimize perceptual adaptation.
- **D** *Do not overload the auditory channel:* Only a few signals should be used in any given situation. Too many signals can be confusing and will overload the operator. (For example, during the Three Mile Island nuclear crisis, over 60 different auditory warning signals were activated.)

3 Principles of installation of auditory displays

- **A** *Test signals to be used:* Such tests should be made with a representative sample of the potential user population to be sure the signals can be detected and discriminated by them.

FIGURE 6-9
An example of violating the principle of dissociability. Auditory signals should be easily discernible from any ongoing audio input—including birds. (*By permission of Johnny Hart and NAS, Inc.*)

B *Avoid conflict with previously used signals:* Any newly installed signals should not be contradictory in meaning to any somewhat similar signals used in existing or earlier systems.

C *Facilitate changeover from previous display:* Where auditory signals replace some other mode of presentation (e.g., visual), preferably continue both modes for a while to help people become accustomed to the new auditory signals.

Auditory Displays for Specific Purposes

For illustration, a few examples of special-purpose auditory displays are discussed.

Warning and Alarm Signals The unique features of the auditory system make auditory displays especially useful for signaling warnings and alarms. Various types of available devices each have their individual characteristics and corresponding advantages and limitations. A summary of such characteristics and features is given in Table 6-2.

As one example of research in this area, Adams and Trucks (1976) tested reaction time to eight different warning signals (''wail,'' ''yelp,'' ''whoop,''

TABLE 6-2
CHARACTERISTICS AND FEATURES OF CERTAIN TYPES OF AUDIO ALARMS

Alarm	Intensity	Frequency	Attention-getting ability	Noise-penetration ability
Diaphone (foghorn)	Very high	Very low	Good	Poor in low-frequency noise
Horn	High	Low to high	Good	Good
Whistle	High	Low to high	Good if intermittent	Good if frequency is properly chosen
Siren	High	Low to high	Very good if pitch rises and falls	Very good with rising and falling frequency
Bell	Medium	Medium to high	Good	Good in low-frequency noise
Buzzer	Low to medium	Low to medium	Good	Fair if spectrum is suited to background noise
Chimes and gong	Low to medium	Low to medium	Fair	Fair if spectrum is suited to background noise
Oscillator	Low to high	Medium to high	Good if intermittent	Good if frequency is properly chosen

Source: Deatherage (1972, Table 4-2).

"yeow," intermittent horns, etc.) in five different ambient noise environments. Across all five environments, the most effective signals were the "yeow" (descending change in frequency from 800 to 100 Hz every 1.4 s) and the "beep" (intermittent horn, 425 Hz, on for 0.7 s, off for 0.6 s). The least effective signal was the "wail" (slow increase in frequency from 400 to 925 Hz over 3.8 s and then a slow decrease back to 400 Hz). Figure 6-10 shows the mean reaction time to these three signals as a function of overall signal intensity in one of the ambient-noise environments tested. As can be seen, reaction time decreased with increased signal intensity. The difference in reaction time between the most effective and the least effective signals also decreased somewhat as intensity increased.

This relationship between reaction time and signal intensity may only hold for simple reactions, that is, where the same response is to be made for all signals. Where a different response is to be made to different signals (choice reaction time), the relationship between reaction time and signal intensity is not so simple. Van der Molen and Keuss (1979) used two signals (1000 and 3000 Hz) and required the subjects to make either the same response to both (simple reaction time) or a different response to each (choice reaction time). Figure 6-11 illustrates their results. For simple reaction time, increases in signal

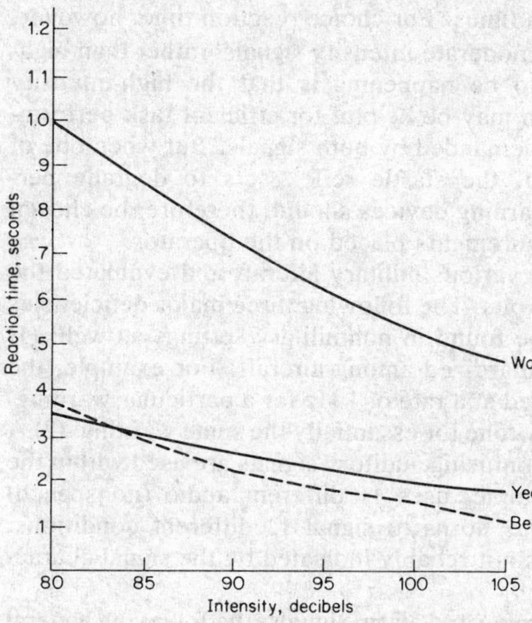

FIGURE 6-10
Mean reaction time to three auditory warning signals in an ambient noise environment. Reaction time decreases with increased signal intensity. See text for description of signals. (*Source: Adapted from Adams and Trucks, 1976, Fig. 4.*)

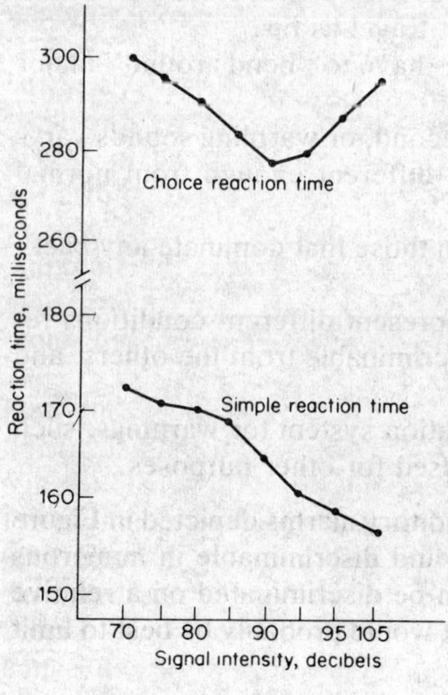

FIGURE 6-11
Relationship between reaction time and signal intensity for choice reaction time and simple reaction time tasks. (*Source: Adapted from Van der Molen and Keuss, 1979, Fig. 1.*)

intensity resulted in faster reaction times. For choice reaction time, however, the fastest reaction times were for moderate-intensity signals, rather than high-intensity signals. What appears to be happening is that the high-intensity signals elicit a startle reflex, which may be helpful for efficient task performance only if the same response is demanded by both signals. But when one of two responses must be provided, the startle reflex acts to degrade performance. The intensity level of warning devices should, therefore, be chosen with concern for the response requirements placed on the operator.

Doll and Folds (1986) surveyed various military aircraft and evaluated the auditory warnings used in the cockpits. The following three major deficiencies were noted, many of which can be found in nonmilitary settings as well: (1) Auditory signals are not well standardized among aircraft. For example, the F-15 uses a 1600 Hz tone interrupted at a rate of 1 Hz for a particular warning, while the F-16 uses a 250 Hz steady tone for essentially the same warning. (2) A fairly large number of potentially confusing auditory signals are used within the same cockpit. The F-15, for example, uses 11 different audio (nonspeech) signals, and the C-5A uses various horns to signal 12 different conditions. (3) The urgency of the condition is not reliably indicated by the signal characteristics.

In the selection or design of warning and alarm signals, the following general design recommendations have been proposed by Deatherage (1972, pp. 125, 126) and by Mudd (1961), these being in slightly modified form.

- Use frequencies between 200 and 5000 Hz, and preferably between 500 and 3000 Hz, because the ear is most sensitive to this middle range.
- Use frequencies below 1000 Hz when signals have to travel long distances (over 1000 ft), because high frequencies do not travel as far.
- Use frequencies below 500 Hz when signals have to "bend around" major obstacles or pass through partitions.
- Use a modulated signal (1 to 8 beeps per second, or warbling sounds varying from 1 to 3 times per second), since it is different enough from normal sounds to demand attention.
- Use signals with frequencies different from those that dominate any background noise, to minimize masking.
- If different warning signals are used to represent different conditions requiring different responses, each should be discriminable from the others, and moderate-intensity signals should be used.
- Where feasible, use a separate communication system for warnings, such as loudspeakers, horns, or other devices not used for other purposes.

Weiss and Kershner (1984) report that the auditory alarms depicted in Figure 6-12 have been found to be easily discernible and discriminable in numerous tests at nuclear power plants. Although 12 can be discriminated on a relative basis, if an absolute identification is required, it would probably be best to limit the number to 5.

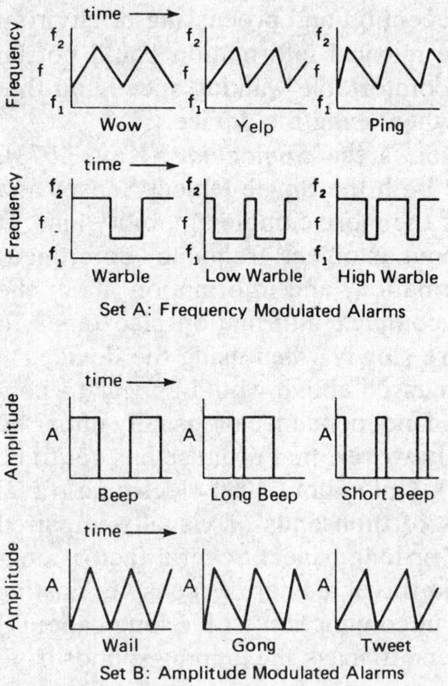

FIGURE 6-12
Twelve auditory alarms that are easily discernible and discriminable on a relative basis as determined in numerous tests at nuclear power plants. (*Source: Weiss and Kershner, 1984, Figs. 6 and 7.*)

Aids for the Blind With advances in technology, several devices have been built which use auditory information to aid the blind in moving about in their environment. Kay et al. (1977) distinguish two types of mobility aids—clear-path indicators (where the aim is to provide essentially a go–no-go signal indicating whether it is safe to proceed along the line of travel) and environmental sensors (which attempt to allow users to build up a picture of their environment). Examples of clear-path indicators would include *Pathsounder* (Russell, 1969), which emits an audible buzzing sound when there is something 6 ft ahead and a distinctively different high-pitched beeping sound when an object is within 3 ft of the user; *Single-Object Sensor* (Kay et al., 1977), which emits repetitive clicks, coded in terms of repetition rate to the distance of objects; and *Sonic Pathfinder* (Heyes, 1984), which emits eight musical notes to indicate the range of an object located within a 120° arc in eight 30-cm zones in front of the user. Dodds, Clark-Carter, and Howarth (1984) evaluated the Sonic Pathfinder with six totally blind subjects. All subjects used a long cane, with and without the device, to navigate a course. The Sonic Pathfinder was designed to be used with a long cane or guide dog because it cannot detect holes and curbs. The results clearly showed that fewer contacts occurred and that the subjects were more aware of obstacles when they used the device. Not surprisingly, from our discussion of resource competition in Chapter 3, the auditory display did not interfere with the subjects' reception and processing of tactile changes in the ground's surface, but it did reduce their reception and processing of environmental sounds. It was also noted that the musical information could not be processed at a rate commensurate with a comfortable walking speed and that the subjects tended to walk more slowly when using the device.

An example of an environmental sensor is the *Sonicguide* (Kay, 1973), which is built into eyeglass frames (as are both the Single-Object Sensor and the Sonic Pathfinder). Sonicguide is one of the more complex mobility aids. It gives the user three types of information about an object: frequency differences for determining distance, directional information, and information about the surface characteristics of an object. The complex information also tends to overload users, and they tend to walk more slowly when using the device.

Electronic travel aids such as those discussed above would appear to hold great promise for increasing the mobility and independence of visually impaired travelers. Even after 20 years of research, however, that promise has yet to be fulfilled (National Research Council, 1987). Only about 3000 electronic travel aids have ever been sold, while hundreds of thousands of visually impaired Americans continue to rely on guide dogs or long canes. Several factors contribute to this state of affairs: (1) electronic travel aids are expensive, costing from $300 to $4000; (2) they must be used in conjunction with a long cane; (3) the auditory signals from them, however, tend to mask the tapping sounds from the long cane; (4) they lack comfort and grace, important criteria for the user; and (5) travelers need more than just obstacle detectors to find their way around—they need cues to orient themselves in their environment and to direct their progress toward their intended destination (National Research Council, 1987). More research is needed to understand what information is required by the visually impaired to navigate and the best way of providing it to them.

In another application of using sound to aid the blind, Mansur (1984) reported on the use of a continuously varying pitch to represent two-dimensional graphical data. The pitch corresponded to the *Y* axis (ordinate) of the graph, and time was used to represent the *X* axis (abscissa). The method was compared with tactual graphs and was found to convey concepts (such as symmetry, monotonicity, and slopes) more quickly and with the same level of accuracy as the tactual presentations.

CUTANEOUS SENSES

In everyday life people depend on their cutaneous (somesthetic, or skin) senses much more than they realize. There is, however, an ongoing question as to how many cutaneous senses there are; this confusion exists in part because of the bases on which the senses can be classified. As Geldard (1972, pp. 258, 259) points out, we can classify them *qualitatively* (on the basis of their observed similarity—that is, the sensations generated), in terms of the *stimulus* (i.e., the form of energy that triggers the sensation, as thermal, mechanical, chemical, or electrical), or *anatomically* (in accordance with the nature of the sense organs or tissues involved). With respect to the anatomic structures involved, it is still not clear how many distinct types of nerve endings there are, but for convenience Geldard considers the skin as housing three more or less separate systems of sensitivity, one for pressure reception, one for pain, and one responsive to temperature changes. Although it has been suggested that any particular sensation is triggered by the stimulation of a specific corresponding type of nerve ending or receptor, this notion seems not to be warranted. Rather, it is now generally believed that the various receptors are specialized in their functions through the operation of what Geldard refers to as the principle of *patterned response*. Some of the cutaneous receptors are responsive to more than one form of energy (such as mechanical pressure and thermal changes) or to certain ranges of energy. Through complex interactions among the various types of nerve endings, being stimulated by various forms and amounts of energy, we experience a wide ‘‘variety’’ of specific sensations that we endow with such labels as *touch, contact, tickle,* and *pressure* (Geldard, 1972).

For the most part, tactual displays have utilized the hand and fingers as the principal receptors of information. Craft workers, for centuries, have used their sense of touch to detect irregularities and surface roughness in their work. Interestingly, Lederman (1978) reports that surface irregularities can be detected more accurately when the person moves an intermediate piece of paper or thin cloth over the surface than when the bare fingers are used alone. Inspection of the coachwork of Aston Martin sports cars, for example, is done by rubbing over the surface with a cotton glove on the hand.

Not all parts of the hand are equally sensitive to touch. One common measure of touch sensitivity is the *two-point threshold,* the smallest distance between two pressure points at which the points are perceived as separate. Figure 6-13 shows the median two-point threshold for fingertip, finger, and

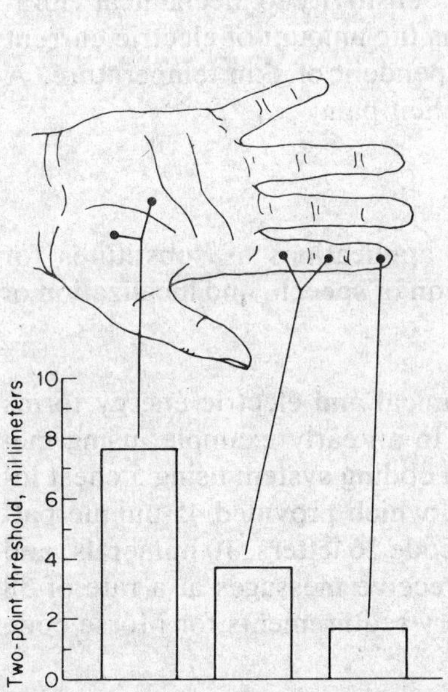

FIGURE 6-13
Median two-point threshold for fingertip, finger, and palm. (*Source: Vallbo and Johansson, 1978, Fig. 6.*)

palm (Vallbo and Johansson, 1978). The sensitivity increases (i.e., the two-point threshold becomes smaller) from the palm to the fingertips. Thus, tactual displays that require fine discriminations are best designed for fingertip reception. Tactual sensitivity is also degraded by low skin temperatures; therefore, tactual displays should be used with extreme caution in low-temperature environments. Stevens (1982), however, found that cooling or warming the object being felt caused a marked improvement (averaging 41 percent) in the skin's acuity relative to thermally neutral stimulation. With this bit of background information, let us look at or, more appropriately, feel our way through some specific types of tactual displays.

TACTUAL DISPLAYS

Although we depend very much upon our cutaneous senses in everyday living, these senses have been used only to a limited degree as the basis for the intentional transmission of information to people by the use of tactual displays. The primary uses of tactual displays to date have been as substitutes for hearing, especially as aids to the deaf, and as substitutes for seeing, especially as aids to the blind.

The most frequent types of stimuli used for tactual displays have been mechanical vibration or electric impulses. Information can be transmitted by mechanical vibration based on such physical parameters as location of vibrators, frequency, intensity, or duration. Electric impulses can be coded in terms of electrode location, pulse rate, duration, intensity, or phase. Electric energy has some advantages over mechanical vibrators, as noted by Hawkes (1962). Electrodes are easily mounted on the body surface and have less bulk and a lower power requirement than mechanical vibrators. Sensitivity to mechanical vibration is dependent on skin temperature, whereas the amount of electric current required to reach threshold apparently is independent of skin temperature. A disadvantage of electric energy is that it can elicit pain.

Substitutes for Hearing

Tactual displays have generally found three applications as substitutes for hearing: reception of coded messages, perception of speech, and localization of sound.

Reception of Coded Messages Both mechanical and electric energy forms have been used to transmit coded messages. In an early example, using mechanical vibrators, Geldard (1957) developed a coding system using 5 chest locations, 3 levels of intensity, and 3 durations, which provided 45 unique patterns (5 × 3 × 3), which in turn were used to code 26 letters, 10 numerals, and 4 frequently used words. One subject could receive messages at a rate of 38 words per minute, somewhat above the military requirements for Morse code reception of 24 words per minute.

Perception of Speech Kirman (1973) reviews numerous attempts to transmit speech to the skin through mechanical and electrical stimulation. All attempts to build successful tactile displays of speech have been disappointing at best. One reason given for the poor performance of such devices is that the resolving power of the skin, both temporally and spatially, is too limited to deal with the complexities of speech (Loomis, 1981).

Localization of Sounds Richardson and his colleagues (Frost and Richardson, 1976; Richardson and Frost, 1979; Richardson, Wuillemin, and Saunders, 1978) report that subjects are able to localize a sound source by use of a simple tactual display with an accuracy level comparable to normal audition. The sound intensity is picked up by microphones placed over each ear. The outputs of these microphones are amplified and fed to two vibrators, upon which the subjects rest their index fingers. The intensity of vibration on each finger is then proportional to the intensity of the sound reaching the ear. It is this difference in intensity, as discussed earlier, that allows localization of sound with normal audition.

Substitutes for Seeing

Tactual displays have been most extensively used as substitutes for seeing—as an extension of our eyes. The most exciting possibilities in the use of tactual displays lie in this realm, especially with the advances being made in low-cost microelectronics technology. As an example of possible future applications, Dobelle (1977) has developed and tested an experimental unit that uses direct electrical stimulation of the brain. It produces genuine sensations of vision—indeed, volunteers who have been blind for years complain of eyestrain. A television camera picks up a black-and-white picture which is converted to a matrix of 60 or so electrodes implanted in the primary visual cortex of the brain. Each electrode, when stimulated, produces a single point of light in the visual field. Several subjects have had the implant and have been able to recognize white horizontal and vertical lines on black backgrounds. There are, of course, problems with such a technique, and the best that can be expected in the foreseeable future, according to Dobelle, is low-resolution black-and-white slow-scan images.

We now review some applications of tactual displays as substitutes for seeing, ranging from the relatively mundane to the more sophisticated.

Identification of Controls A very practical use of the tactual sense is with respect to the design of control knobs and related devices. Although these are not *displays* in the conventional sense of the term, the need to correctly identify such devices may be viewed within the general framework of displays. The coding of such devices for tactual identification includes their shape, texture, and size. Figure 6-14, for example, shows two sets of shape-coded knobs such

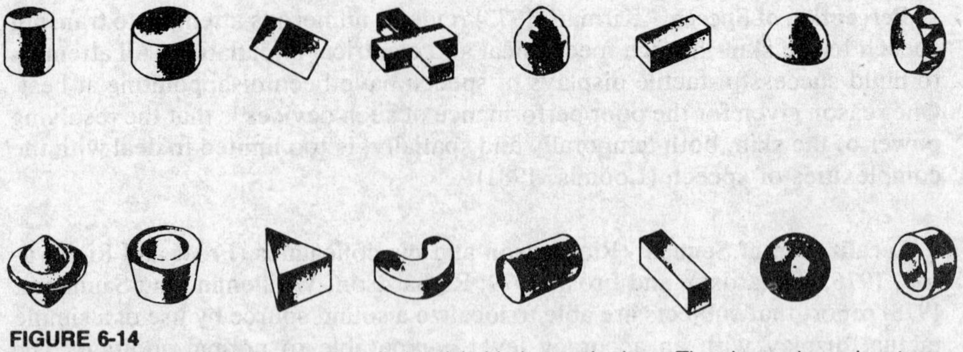

FIGURE 6-14
Two sets of knobs for levers that are distinguishable by touch alone. The shapes in each set are rarely confused with each other. (*Source: Jenkins, 1947.*)

that the knobs within each group are rarely confused with each other (Jenkins, 1947). In Chapter 11, we discuss in more detail methods of tactually coding controls.

Reading Printed Material Probably one of the most widely known tactual displays for reading printed material is braille printing. Braille print for the blind consists of raised dots formed by the use of all the possible combinations of six dots numbered and arranged thus:

1 . . 4
2 . . 5
3 . . 6

A particular combination of these represents each letter, numeral, or common word. The critical features of these dots are position, distance between dots, and dimension (diameter and height), all of which have to be discriminable to the touch.

Because of the cost of braille printing there are only limited amounts of braille reading material available to blind persons. The combination of research relating to the sensitivity of the skin and new technological developments, however, now permit us to convert optical images of letters and numbers to tactile vibrations that can be interpreted by blind persons. The Optacon (for *op*tical-to-*ta*ctile *con*verter), one such device developed at the Stanford University Electronics Laboratories, consists of essentially two components. One is an optical pickup "camera," about the size of a pocketknife, and a "slave" vibrating tactile stimulator that reproduces images that are scanned by the camera (Bliss et al., 1970; Linvill, 1973). A photograph of the Optacon is shown in Figure 6-15. The 144 phototransistors of the camera, in a 24 × 6 array, control the activation of 144 corresponding vibrating pins that "reproduce" the

FIGURE 6-15
Photograph of Optacon for use in converting visual images to tactile vibrations. (*Source: Stanford Electronics Laboratories. The Optacon is produced by Telesensory Systems, Inc.*)

visual image. The vibrating pins can be sensed by one fingertip. The particular features of the vibrating pins and of their vibrations were based on extensive psychophysical research relating to the sensitivity of the skin to vibrations (Bliss et al., 1970). The letter O, for example, produces a circular pattern of vibration that moves from right to left across the user's fingertip as the camera is moved across the letter.

Learning to "read" with the Optacon takes time. For example, one subject took about 160 h of practice and training to achieve a reading rate of about 50 words per minute. Most users, however, only achieve rates of about half that. This rate is well below that of most readers of braille, some of whom achieve a rate of 200 words per minute. At the same time, such a device does offer the possibility to blind persons of being able to read conventional printing at least at a moderate rate.

Navigation Aids To navigate in an environment, we need a spatial map so that we can "get around." Tactual maps for the blind are very similar to conventional printed maps, except the density of information must be reduced because symbols must be large and well-spaced to be identified by the sense of touch. Figure 6-16 shows an example of a portion of a university campus tactual map. If Figure 6-16 is hard to read, remember it was not made to be looked at, but rather to be felt. The map is located inside a building on campus.

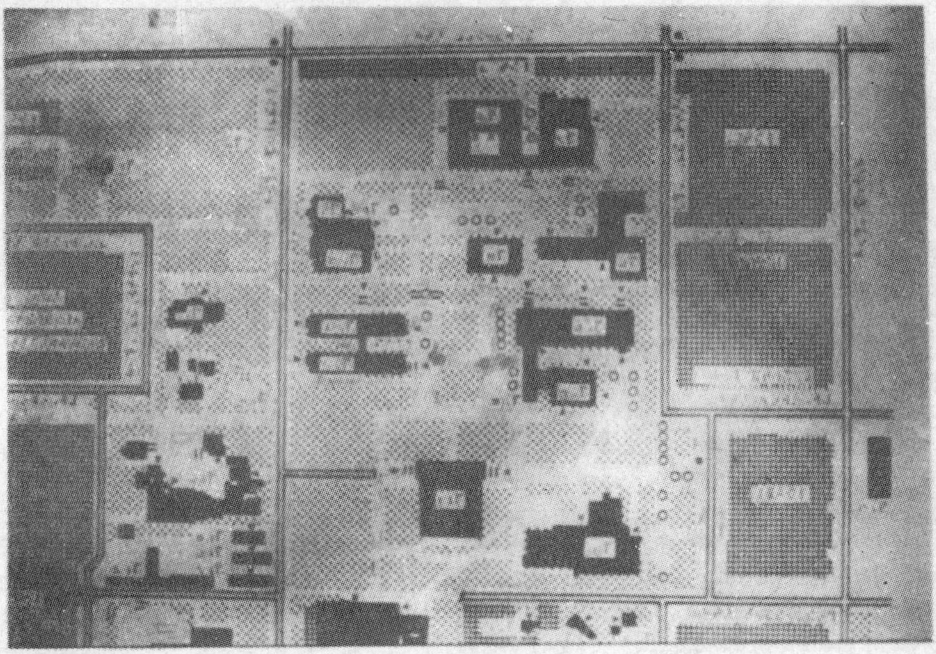

FIGURE 6-16
An example of a tactual map of a university campus. (*Source: Courtesy of California State University, Northridge.*)

How a blind person would know that such a map exists and where it is located raises questions that need answers in order to increase the utility of such displays.

Environmental features can be classified as *point* features (e.g., a bus stop), *linear* features (e.g., a road) and *areal* features (e.g., a park). Much research effort has been expended to develop sets of discriminable symbols for maps. For example, James and Armstrong (1976) have suggested a standard set of point and line symbols, which have been used in the United Kingdom and other parts of the world (Armstrong, 1978). James and Gill (1975) investigated areal symbols and identified five discriminable patterns.

Although a fair amount of research has been done on symbol discriminability, there has been relatively little research studying the ability of blind persons to use spatial maps. What few studies there have been, however, seem to indicate that such maps can be used effectively to aid navigation in unfamiliar environments (James and Armstrong, 1975).

Tactual Graphs One can hardly read a newspaper or magazine today without encountering a graphic display of data. Be they about the decline in college entrance examination scores or the increase in the federal deficit, graphs are being used with increasing frequency. Communicating such information to the blind requires careful design to optimize the information transfer and reduce confusion. Apparently it is fairly easy for blind people to learn to read two-dimensional tactual graphs such as the one shown in Figure 6-17. Research has concentrated on assessing whether a raised-background grid would cause confusion for blind subjects. The results indicate that such a background does not cause confusion and can even be helpful for certain tasks (Lederman and Campbell, 1982; Aldrich and Parkin, 1987). Barth (1984) took the question a little deeper (pun intended) by using incised lines (grooves) for the background grid and raised lines for the data. Performance with incised-grid graphs turned out to be even better than performance with raised-grid graphs on several tasks.

Tracking-Task Displays Although the task of tracking is discussed in Chapter 10, we want to illustrate here the use of tactual displays for such purposes. Jagacinski, Gilson, and their colleagues have conducted a series of tracking studies, using the tactual display shown in Figure 6-18 (Jagacinski, Miller, and Gilson, 1979; Burke, Gilson, and Jagacinski, 1980; Jagacinski, Flach, and Gilson, 1983). Both single- and dual-task compensatory tracking situations were studied. In the single-task situation, the addition of quickening (see Chapter 10 for further discussion of quickening) to the tactual display produced tracking performance comparable to that obtained with traditional visual displays. In the dual-task situation subjects performed two tracking tasks simultaneously. The primary task was to control, by using one hand, the vertical movement of a target which was always shown on a visual display. The secondary task was to control, by using the other hand, the horizontal move-

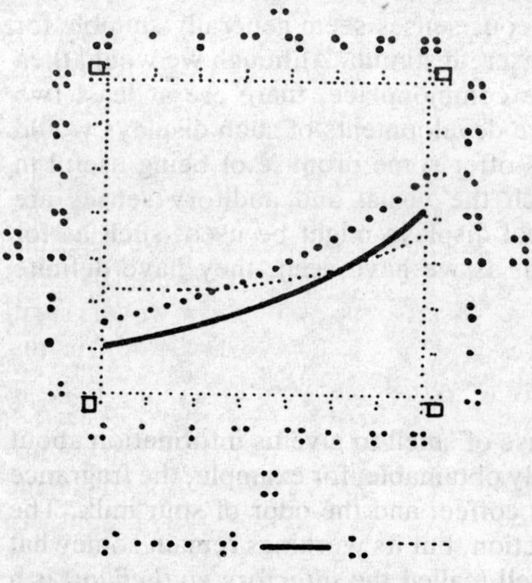

FIGURE 6-17
An example of a tactual graph for the blind. (*Source: Lederman and Campbell, 1982; Fig. 1. VI.*)

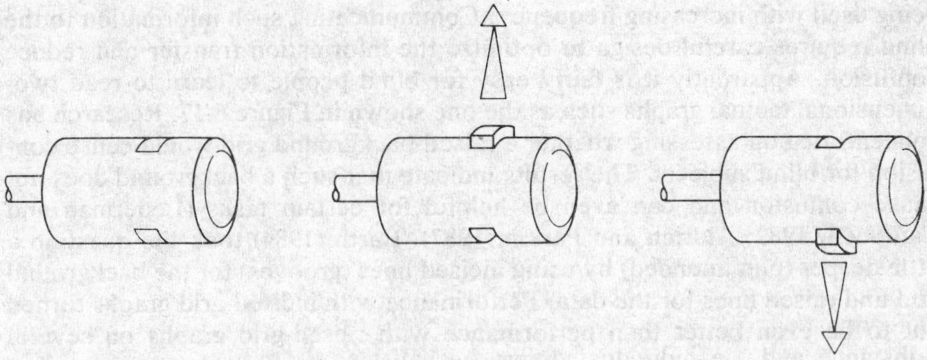

FIGURE 6-18
The tactual display used to present directional information to subjects in a tracking task. The correct response to condition *b* was to move the control stick up; to condition *c*, to move the stick down. (*Source: Jagacinski, Miller, and Gilson, 1979, Fig. 1.*)

ment of the target which was displayed on either the visual display or the quickened tactual display. Performance was better with the visual-tactual primary-secondary task combination than with the visual-visual combination. Further, the superiority of the tactual display was not due to the addition of quickening but was attributed to its utilization of a different sensory modality, which, it is believed, resulted in less competition for information processing resources.

Discussion

Relative to the other senses, the cutaneous senses seem generally suitable for transmitting only a limited number of discrete stimuli. Although we would then not expect tactual displays to become commonplace, there are at least two types of circumstances for which future developments of such displays would seem appropriate. First, such displays offer some promise of being useful in certain special circumstances in which the visual and auditory senses are overburdened; in such instances tactual displays might be used, such as for warning or tracking purposes. Second, as we have seen, they have definite potential as aids to blind persons.

OLFACTORY SENSE

In everyday life we depend on our sense of smell to give us information about things that otherwise would not be easily obtainable, for example, the fragrance of a flower, the smell of fresh-brewed coffee, and the odor of sour milk. The olfactory sense is simple in its construction, but its workings remain somewhat of a mystery. The sense organ for smell (called the *olfactory epithelium*) is a small, 4- to 6-cm^2, patch of cells located in the upper part of each nostril. These *olfactory cells* contain *olfactory hairs* which actually do the detecting of different odors. The cells are connected directly to the olfactory areas of the brain.

There is little doubt that the sensation of odor is caused by airborne molecules of a volatile substance entering the nose. Research workers, however, are still not in agreement on just what property of these molecules is responsible for the kind of odor perceived, or whether the sense of smell is composed of distinct primary odors, and if so, what they are (Engen, 1982).

Many people think that the sense of smell is very good—in one way it is, and in other ways it is not. The nose is apparently a very sensitive instrument for detecting the presence of odors; this sensitivity depends on the particular substance and the individual doing the sniffing. Ethyl mercaptan (a skunk odor), for example, can be detected when as little as 0.5 ml of the substance is dispersed in 10,000 liters of air (Engen, 1982). Although we are good at detecting the presence of an odor, we also tend to have a high false-alarm rate, that is, to report the existence of an odor when none is present (Richardson and Zucco, 1989).

Surprisingly, however, our sense of smell is not outstanding when it comes to making absolute identifications of specific odors (we do far better comparing odors on a relative basis). The number of different odors we can identify depends on a number of factors including the types of odors to be identified and the amount of training. Desor and Beauchamp (1974), for example, found that untrained subjects could identify approximately 15 to 32 common stimuli (coffee, paint, banana, tuna, etc.) by their odors alone. With training, however, some subjects could identify up to 60 stimuli without error. This may seem high compared with the visual or auditory sense, but it is not. These odor stimuli were complex "whole odors" as opposed to simple chemical odors. The analogy with the visual sense is the difference between simple color patches and complex pictures or faces. We can identify literally hundreds of different faces visually. Apparently, one major reason for our relatively poor ability to identify odors is that we have an impoverished vocabulary for describing odors (Cain, 1988; Richardson and Zucco, 1989).

When it comes to identifying odors that differ only in intensity, we can identify about three or four different intensities (Engen, 1982). Thus, overall, the sense of smell is probably not to be counted on for identifying a specific stimulus from among many, but it can be effective in detecting the presence of an odor.

OLFACTORY DISPLAYS

Olfactory displays, needless to say, have not found widespread application. Part of the reason for this is that they cannot be depended on as a reliable source of information because people differ greatly with respect to their sensitivity to various odors; a stuffy nose can reduce sensitivity; people adapt quickly to odors so that the presence of an odor is not sensed after a little while of exposure; the dispersion of an odor is hard to control; and some odors make people feel sick.

Despite all these problems, olfactory displays do have some useful applications—primarily as warning devices. The gas company, for example, adds an odorant to natural gas so that we can detect gas leaks in our homes. This is no less an information display than if a blinking light were used to warn us of the leak.

The author has seen an unusual sign in a computer room which reads, "If the red light is blinking or you smell wintergreen, evacuate the building." The fire prevention system in the building releases carbon dioxide when a fire is detected. When the system was initially activated, the people in the building did not detect the presence of the carbon dioxide, and several people lost consciousness from the lack of oxygen. A wintergreen odor was added to the gas as a warning.

In the prior examples odor was added to a gas to act as a warning that the gas was being released, intentionally in the case of carbon dioxide and unintentionally in the case of natural gas. Another example of an olfactory display uses odor to signal an emergency not associated with the gas itself. Several underground metal mines in the United States use a "stench" system to signal workers to evacuate the mine in an emergency. The odor is released into the mine's ventilation system and is quickly carried throughout the entire mine. This illustrates one general advantage of olfactory displays; they can penetrate vast areas which might not be economically reached by visual or auditory displays. Note, however, this is somewhat of a unique situation, consisting of vast areas (often hundreds of miles of tunnels) with no visual or auditory access and a closed ventilation system for controlling the dispersion of the odorant.

Olfactory displays will probably never become widespread in application, but they do represent a unique form of information display which could be creatively integrated into very special situations to supplement more traditional forms of displays.

DISCUSSION

In this chapter we explored some alternative modes for displaying information other than the traditional visual mode of presentation. No one has suggested that auditory, tactual, or olfactory displays will ever become the dominant mode for information display, except for the visually impaired, but they are important and their importance will probably grow in the future. Where the visual sense is overloaded these alternative displays have a role to play. They can also be useful for reinforcing or directing attention to visually presented information. The important point is that in the design of systems, creative use of auditory, tactual, and even olfactory displays can enhance human performance; we do not have to be entirely dependent on visual displays to present information.

REFERENCES

Adams, S., and Trucks, L. (1976). A procedure for evaluating auditory warning signals. In *Proceedings of the 6th Congress of the International Ergonomics Association and Technical Program for the 20th Annual Meeting of the Human Factors Society*. Santa Monica, CA: Human Factors Society, pp. 166–172.

Aldrich, F., and Parkin, A. (1987). Tangible line graphs: An experimental investigation of three formats using capsule paper. *Human Factors,* 29(3), 301–309.

American National Standards Institute (ANSI) (1966). *Standard specification for octave, half-octave, and third-octave band filter sets* (ANSI S1.11-1966), New York.

Armstrong, J. (1978). The development of tactual maps for the visually handicapped. In G. Gordon (ed.), *Active touch*. Elmsford, NY: Pergamon.

Barth, J. (1984). Incised grids: Enhancing the readability of tangible graphs for the blind. *Human Factors,* 26, 61–70.

Bliss, J. C., Katcher, M. H., Rogers, C. H., and Shepard, R. P. (1970). Optical-to-tactile image conversion for the blind. *IEEE Transactions on Man-Machine Systems,* 11(1), 58–65.

Burke, M., Gilson, R., and Jagacinski, R. (1980). Multi-modal information processing for visual workload relief. *Ergonomics,* 23, 961–975.

Caelli, T., and Porter, D. (1980). On difficulties in localizing ambulance sirens. *Human Factors,* 22, 719–724.

Calhoun, G., Janson, W., and Valencia, G. (1988). Effectiveness of three-dimensional auditory directional cues. *Proceedings of the Human Factors Society 32d Annual Meeting*. Santa Monica, CA: Human Factors Society, pp. 68–72.

Cain, W. (1988). Olfaction. In R. Atkinson, R. Herrnstein, G. Lindzey, and R. Luce (eds.), *Stevens' handbook of experimental psychology,* vol. 1 (2d ed.). New York: Wiley, pp. 409–460.

Davies, D., and Jones, D. (1982). Hearing and noise. In W. T. Singleton (ed.), *The body at work*. New York: Cambridge University Press.

Deatherage, B. H. (1972). Auditory and other sensory forms of information presentation. In H. P. Van Cott and R. G. Kinkade (eds.), *Human engineering guide to equipment design*. Washington, DC: Government Printing Office.

Desor, J., and Beauchamp, G. (1974). The human capacity to transmit olfactory information. *Perception and Psychophysics,* 16, 551–556.

Dobelle, W. (1977). Current status of research on providing sight to the blind by electrical stimulation of the brain. *Visual Impairment and Blindness,* 71, 290–297.

Doll, T., and Folds, D. (1986). Auditory signals in military aircraft: Ergonomics principles versus practice. *Applied Ergonomics,* 17(4), 257–264.

Dodds, A., Clark-Carter, D., and Howarth, C. (1984). The Sonic Pathfinder: An evaluation. *Journal of Visual Impairment and Blindness,* 78, 203–206.

Engen, T. (1982). *The perception of odors*. New York: Academic.

Fidell, S., Pearsons, K., and Bennett, R. (1974). Prediction of aural detectability of noise signals. *Human Factors,* 16, 373–383.

Frost, B., and Richardson, B. (1976). Tactile localization of sounds: Acuity, tracking moving sources, and selective attention. *Journal of the Acoustical Society of America,* 59, 907–914.

Furness, T. (1986). The Super Cockpit and its human factors challenges. *Proceedings of the Human Factors Society 30th Annual Meeting,* Santa Monica, CA: Human Factors Society, pp. 48–52.

Gales, R. (1979). Hearing characteristics. In C. M. Harris (ed.), *Handbook of noise control* (2d ed.). New York: McGraw-Hill.

Garfield, E. (1983, July 11). The tyranny of the horn—Automobile, that is. *Current Contents,* 28, 5–11.

Geldard, F. A. (1957). Adventures in tactile literacy. *American Psychologist,* 12, 115–124.

Geldard, F. A. (1972). *The human senses* (2d ed.). New York: Wiley.

Grantham, C. (1986). Detection and discrimination of simulated motion of auditory targets in the horizontal plane. *Journal of the Acoustical Society of America,* 79, 1939–1949.

Green, D. (1988). Audition: Psychophysics and perception. In R. Atkinson, R. Herrnstein, G. Lindzey, and R. Luce (eds.), *Stevens' handbook of experimental psychology,* vol. 1 (2d ed.). New York: Wiley, pp. 327–376.

Guyton, A. (1969). *Function of the human body* (3d ed.). Philadelphia: Saunders.

Hawkes, G. (1962). *Tactile communication* (Rept. 61–11). Oklahoma City: Federal Aviation Agency, Aviation Medical Service Research Division, Civil Aeromedical Research Institute.

Heyes, A. (1984). The Sonic Pathfinder: A new electronic travel aid. *Journal of Visual Impairment and Blindness,* 78, 200–202.

Industrial noise manual (2d ed.). (1966). Detroit: American Industrial Hygiene Association.

Jagacinski, R., Flach, J., and Gilson, R. (1983). A comparison of visual and kinesthetic-tactual displays for compensatory tracking. *IEEE Transactions on Systems, Man, and Cybernetics,* 13, 1103–1112.

Jagacinski, R., Miller, D., and Gilson, R. (1979). A comparison of kinesthetic-tactual and visual displays via a critical tracking task. *Human Factors,* 21, 79–86.

James, G., and Armstrong, J. (1975). An evaluation of a shopping center map for the visually handicapped. *Journal of Occupational Psychology,* 48, 125–128.

James, G., and Armstrong, J. (1976). *Handbook on mobility maps,* Mobility Monograph no. 2. University of Nottingham.

James, G., and Gill, J. (1975). A pilot study on the discriminability of tactile areal and line symbols. *American Foundation for the Blind Research Bulletin,* 29, 23–33.

Jenkins, W. O. (1947). The tactual discrimination of shapes for coding aircraft-type controls. In P. M. Fitts (ed.), *Psychological research on equipment design,* Research Rept. 19. Army Air Force, Aviation Psychology Program.

Kahn, M. J. (1983) Human awakening and subsequent identification of fire-related cues. *Proceedings of the Human Factors Society 27th Annual Meeting.* Santa Monica, CA: Human Factors Society, pp. 806–810.

Kay, L. (1973). Sonic glasses for the blind: A progress report. *AFB Research Bulletin,* 25, 25–28.

Kay, L., Bui, S., Brabyn, J., and Strelow, E. (1977). A single object sensor: A simplified binaural mobility aid. *Visual Impairment and Blindness,* 71, 210–213.

Kiang, N., and Peake, W. (1988). Physics and physiology of hearing. In R. Atkinson, R. Herrnstein, G. Lindzey, and R. Luce (eds.), *Stevens' handbook of experimental psychology,* vol. 1 (2d ed.). New York: Wiley, pp. 377–408.

Kirman, J. (1973). Tactile communication of speech: A review and an analysis. *Psychological Bulletin,* 80, 54–74.

Kryter, K. (1985). *The effects of noise on man* (2d ed.). Orlando, FL: Academic.

Lederman, S. (1978). Heightening tactile impressions on surface texture. In G. Gordon (ed.), *Active touch*. Elmsford, NY: Pergamon.

Lederman, S., and Campbell, J. (1982). Tangible graphs for the blind. *Human Factors*, 24, 85–100.

Licklider, J. C. R. (1961). *Audio warning signals for Air Force weapon systems* (TR 60-814). USAF, Wright Air Development Division, Wright-Patterson Air Force Base.

Linvill, J. G. (1973). *Research and development of tactile facsimile reading aid for the blind (the Optacon)*. Stanford, CA: Stanford Electronics Laboratories, Stanford University.

Loomis, J. (1981). Tactile pattern perception. *Perception*, 10, 5–28.

Manligas, C. (1988). Minimum audible movement angle as a function of the spatial location of a moving sound source in the lateral dimension. Unpublished master's thesis. Los Angeles: California State University.

Mansur, D. (1984). *Graphs in sound: A numerical data analysis method for the blind* (UCRL-53548). Berkeley, CA: Lawrence Livermore National Laboratory.

Miller, G. A. (1947). Sensitivity to changes in the intensity of white noise and its relation to masking and loudness. *Journal of the Acoustical Society of America*, 19, 609–619.

Minnix, R. (1978). The nature of sound. In D. M. Lipscomb and A. C. Taylor, Jr. (eds.), *Noise control: Handbook of principles and practices*. New York: Van Nostrand Reinhold.

Moore, B. (1982). *An introduction to the psychology of hearing* (2d ed.). New York: Academic.

Mudd, S. A. (1961). The scaling and experimental investigation of four dimensions of pure tone and their use in an audio-visual monitoring problem. Unpublished Ph.D. thesis. Lafayette, IN: Purdue University.

Mulligan, B., McBride, D., and Goodman, L. (1984). *A design guide for nonspeech auditory displays* (SR-84-1). Pensacola, FL: Naval Aerospace Medical Research Laboratory.

Munson, W. A. (1947). The growth of auditory sensitivity. *Journal of the Acoustical Society of America*, 19, 584.

National Research Council (1987). *Electronic travel aids: New directions for research*. Washington, DC: National Research Council.

Patterson, R. (1982). *Guidelines for auditory warning systems on civil aircraft* (CAA Paper 82017). London: Civil Aviation Authority.

Peterson, A. P. G., and Gross, E. E., Jr. (1972). *Handbook of noise measurement* (7th ed.). New Concord, MA: General Radio Co.

Pollack, I. (1952). Comfortable listening levels for pure tones in quiet and noise. *Journal of the Acoustical Society of America*, 24, 158.

Pollack, I., and Ficks, L. (1954). Information of elementary multidimensional auditory displays. *Journal of the Acoustical Society of America*, 26, 155–158.

Richardson, B., and Frost, B. (1979). Tactile localization of the direction and distance of sounds. *Perception and Psychophysics*, 25, 336–344.

Richardson, B., Wuillemin, D., and Saunders, F. (1978). Tactile discrimination of competing sounds. *Perception and Psychophysics*, 24, 546–550.

Richardson, J., and Zucco, G. (1989). Cognition and olfaction: A review. *Psychological Bulletin*, 105(3), 352–360.

Riesz, R. R. (1928). Differential intensity sensitivity of the ear for pure tones. *Physiological Review*, 31, 867–875.

Russell, L. (1969). *Pathsounder instructor's handbook.* Cambridge: Massachusetts Institute of Technology, Sensory Aids Evaluation and Development Center.

Shower, E. G., and Biddulph, R. (1931). Differential pitch sensitivity of the ear. *Journal of the Acoustical Society of America,* 3, 275–287.

Sorkin, R., Wrightman, F., Kistler, D., and Elvers, G. (1989). An exploratory study of the use of movement-correlated cues in an auditory head-up display. *Human Factors,* 31(2), 161–166.

Stevens, J. (1982). Temperature can sharpen tactile acuity. *Perception and Psychophysics,* 31, 577–580.

Stevens, S. S., and Davis, H. (1938). *Hearing, its psychology and physiology.* New York: Wiley.

Strybel, T. (1988). Perception of real and simulated motion in the auditory modality. *Proceedings of the Human Factors Society 32d Annual Meeting,* Santa Monica, CA: Human Factors Society, pp. 76–80.

Talamo, J. D. (1982). The perception of machinery indicator sounds. *Ergonomics,* 25, 41–51.

Vallbo, A., and Johansson, R. (1978). The tactile sensory innervation of the glabrous skill of the human hand. In G. Gordon (ed.), *Active touch.* Elmsford, NY: Pergamon.

Van Cott, H. P., and Warrick, M. J. (1972). Man as a system component. In H. P. Van Cott and R. G. Kinkade (eds.), *Human engineering guide to equipment design* (rev. ed.). Washington, DC: Superintendent of Documents.

Van der Molen, M., and Keuss, P. (1979). The relationship between reaction time and intensity in discrete auditory tasks. *Quarterly Journal of Experimental Psychology,* 31, 95–102.

Wegel, R. L., and Lane, C. E. (1924). The auditory masking of one pure tone by another and its probable relation to the dynamics of the inner ear. *Physiological Review,* 23, 266–285.

Weiss, C., and Kershner, R. (1984). A case study: Nuclear power plant alarm system human factors review. In D. Attwood and C. McCann (eds.), *Proceedings of the 1985 International Conference on Occupational Ergonomics,* Rexdale, Ontario, Canada: Human Factors Association of Canada, pp. 81–85.

CHAPTER 7

SPEECH COMMUNICATIONS

From a human factors perspective, speech is an information "display," a form of auditory information. The source of this information is usually a human, but it can also be a machine. With our current level of technology, we can synthesize speech very inexpensively. Many of us have experienced synthesized speech in such consumer products as cameras and automobiles, and telephone companies are making greater use of synthesized speech to reduce the burden on operators.

Usually the receiver of speech (be it generated by a human or a microchip) is also a human, but it can be a machine. Although not as advanced as speech synthesis, speech recognition by computer systems is a reality. In this context, speech serves a control function, telling the computer system what to do, and so we defer our discussion of speech recognition to Chapter 11, which deals with control devices.

It is not feasible in this chapter to deal extensively with the many facets and intracacies of speech perception. For an introductory overview, consult Goldstein (1984), Jusczyk (1986), or Cole (1980). For our purposes, we first review briefly the nature of speech and some of the criteria used to evaluate it. Most of the chapter deals with how the components of a speech information system affect speech intelligibility. Finally, we close the chapter with a discussion of some human factors aspects of synthetic speech.

THE NATURE OF SPEECH

The ability of people to speak is, of course, integrally tied to and dependent on the respiratory process—if you do not breathe, you cannot speak. Speech sounds are generated by the sound waves that are created by the breath stream as it is exhaled from the lungs and modified by the speech organs. The various organs involved in speech include the lungs, the larynx (with its vocal cords), the pharynx (the channel between the larynx and the mouth), the mouth or oral cavity (with tongue, teeth, lips, and velum), and the nasal cavity. The vocal cords consist of folds that can vibrate very rapidly, 80 to 400 times per second. (The opening between the folds is called the *glottis*.) In speech, the rate of vibration of the vocal folds controls the basic frequency of the resulting speech sounds. As the sound wave is transmitted from the vocal cords, it is further modified in three resonators: the pharynx, the oral cavity, and the nasal cavity. By manipulating the mandible (in the back of the oral cavity) and the articulators (the tongue, lips, and velum) we are able to modify the sound wave to utter various speech sounds. In addition, by controlling the velum and the pharyngeal musculature we can influence the nasal quality of the articulation.

Types of Speech Sounds

The basic unit of speech is the phoneme. A *phoneme* is defined as the shortest segment of speech which, if changed, would change the meaning of a word. The English language has about 13 phonemes with vowel sounds (e.g., the *u* sound in *put* or the *u* sound in *but*) and about 25 phonemes with consonant sounds (e.g., the *t* sound in *tie*, the *g* sound in *gyp*, or the *g* sound in *gale*), and a couple of other phonemes called *diphthongs* (e.g., the *oy* sound in *boy* or the *ou* sound in *about*). Other languages have phonemes that do not exist in English and vice versa.

Phonemes are built up into syllables which form words that compose sentences. As such, phonemes are analogous to letters in printed text—although there are more phonemes than there are letters. One characteristic of phonemes is that their acoustical sound properties can change when they are paired with other phonemes, yet we usually do not perceive the change. For example, can you detect the subtle difference in the *d* sound and of *di* and *du?* Acoustically the sounds are quite different, although we perceive the *d* sounds as virtually the same.

Depicting Speech

The variations in air pressure with any given sound, including speech, can be represented graphically in various ways, as shown in Figure 7-1. One such representation is a *waveform* (Figure 7-1*a*), which shows the variation in air pressure (intensity) over time. Another form of representation is a *spectrum* (Figure 7-1*b*), which shows, for a given period, phoneme, or word, the intensity of the various frequencies occurring during that sample. The third form of representation is called a *sound spectrogram* (Figure 7-1*c*). Here frequency is plotted on the vertical axis, time is on the horizontal axis, and intensity is depicted by the degree of darkness of the plot.

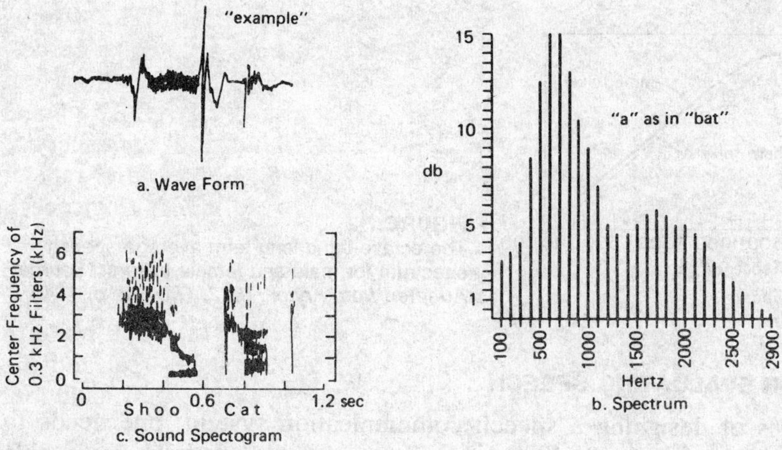

FIGURE 7-1
Three methods of graphically representing speech. (*Source: c. Kiang, 1975.*)

Intensity of Speech

The average intensity, or speech power, of individual speech sounds varies tremendously, with the vowels generally having much greater speech power than the consonants. For example, the *a* as pronounced in *talk* has roughly 680 times the speech power of *th* as pronounced in *then*. This is a difference of about 28 dB. Unfortunately, the consonants, with their lower speech power, are the important phonemes for speech intelligibility.

The overall intensity of speech, of course, varies from one person to another and also for the same individual. For example, the speech intensity of males tends to be higher than for females by about 3 to 5 dB. Speech intensity usually varies from about 45 dB*A* for weak conversational speech; to 55 dB*A* for conversational speech; to 65 dB*A* for telephone-lecture speech; to 75 dB*A* for loud, near-shouting speech; to 85 dB*A* for shouting (Kryter, 1985).

Frequency Composition of Speech

As indicated above, each speech sound has its own unique spectrum of sound frequencies, although there are, of course, differences in these among people and individuals can and do shift their spectra up and down the register depending in part on the circumstances, such as in talking in a quiet conversation (when frequency and intensity are relatively low) or screeching at the children to stay out of trouble (when frequency and intensity typically are near their peak). In general, the spectrum for men has more dominant components in the lower frequencies than for women. These differences are revealed in Figure 7-2, which shows long-time average speech spectra for men and women.

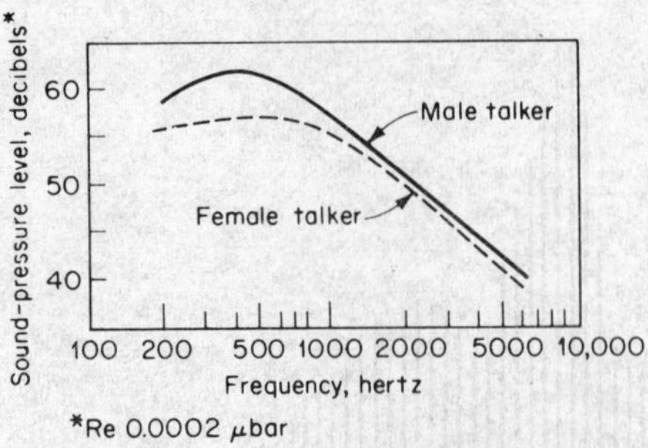

FIGURE 7-2
The octave-band long-term average speech spectrum for male and female talkers. (*Source: Adapted from Kryter, 1972, Fig. 5-2, p. 164.*)

CRITERIA FOR EVALUATING SPEECH

In the process of designing a speech communication system, one needs to establish the criteria or standards that the system should meet to be acceptable for the intended uses. Criteria are also required to evaluate the effects of noise and ear protectors on speech communications and to measure the effectiveness of hearing aids. The major criterion for evaluating a speech communication system is intelligibility, but there are others as well, including quality or naturalness of the speech.

Speech Intelligibility

Intelligibility is simply the degree to which a speech message (e.g., a list of words) is correctly recognized. Intelligibility is assessed by transmitting speech material to individuals who are asked to repeat what they hear or to answer questions about what they heard. Various speech intelligibility tests exist which differ in terms of the types of materials presented [e.g., nonsense syllables, phonetically balanced (PB) word lists that contain all the various speech sounds, and sentences]. In general, in a given situation, intelligibility is highest for sentences, less for isolated words, and lowest for nonsense syllables. In fact, in everyday conversational speech about half the words are unintelligible when they are taken out of their fluent context and presented alone—even when one listens to one's own voice (Pollack and Pickett, 1964). In a normal rate of speaking, try saying, "Did you go to the store?" "Dijoo"? or "Did you"? Intelligibility, therefore is very dependent on context and expectations; without the context of the sentence, the individual words may not be recognized.

Speech Quality

Speech quality, or naturalness, goes beyond intelligibility. It is important in situations where it is desirable to recognize the identity of a speaker, such as on the telephone. Speech quality is also important as one determinant of user satisfaction with a communication system. For example, mobile telephones for automobiles have become relatively inexpensive and available, and telephones for the home that allow a person to talk and hear from anywhere in the room (called *speaker phones*) are also popular. Automatic telephone answering machines have also become common. In all these cases the quality of the speech heard over the system is often the critical factor that accounts for the purchase of one system over another.

Speech quality is defined in terms of preference. Usually samples of speech are presented over the system, and people are asked to either rate the quality (excellent, fair, poor, unacceptable) or to compare it to some standard and indicate which would be preferred. This type of methodology is often used to evaluate synthetic speech as well.

COMPONENTS OF SPEECH COMMUNICATION SYSTEMS

A speech communication system can be thought to consist of the speaker, the message, the transmission system, the noise environment, and the hearer. We discuss each in terms of its effects on intelligibility of speech communications, and we provide some guidance for improving intelligibility within the system.

Speaker

As we all know, the intelligibility of speech depends in part on the character of the speaker's voice. Although, in common parlance, we refer to such features of speech as *enunciation,* research has made it possible to trace certain specific features of speech that affect its intelligibility. Bilger, Hanley, and Steer (1955), for example, found that the speech of "superior" speakers (as contrasted with less intelligible speakers) (1) had a longer "syllable duration," (2) spoke with greater intensity, (3) utilized more of the total time with speech sounds (and less with pauses), and (4) varied their speech more in terms of fundamental vocal frequencies.

The differences in intelligibility of speakers generally are due to the structure of their articulators (the organs that are used in generating speech) and the speech habits people have learned. Although neither factor can be modified very much, it has been found that appropriate speech training can usually bring about moderate improvements in the intelligibility of speakers.

Message

Several characteristics of the message affect its intelligibility, including the phonemes used, the words, and the context.

Phoneme Confusions Certain speech sounds are more easily confused than others. Hull (1976), for example, found letters within each of the following groups to be frequently confused with one another: DVPBGCET, FXSH, KJA, MN. Kryter (1972) specified two groups of consonants which in noise tend to be confused within sets but not between sets: MNDGBVZ and TKPFS. As you can see, the groupings are not consistent. In general, avoid using single letters as codes where noise is present.

Word Characteristics Intelligibility is greater for familiar words than for unfamiliar words. Therefore, in designing a speech communication system, the receiver should be familiar with the words that the sender will use. In general, intelligibility is higher for long words than for short words. Part of the reason is that if only parts of a long word are intelligible, then often a person can figure out the word. With short words, there is less opportunity to pick up pieces of the word. An application of this principle is the use of a word-spelling alphabet (alpha, bravo, charlie, delta, . . .) in place of single letters (A, B, C, D, . . .) in the transmission of alphanumeric information.

Context Features As we already discussed, intelligibility is higher for sentences than for isolated words because the context supplies information if a particular word is not understood. Further, intelligibility is higher when words are arranged in meaningful sentences (*the boat is carrying passengers*) than when the same words are arranged in nonsense sentences (*boat the carrying passengers is*). In addition, the more words in the vocabulary being used, the lower will be the intelligibility under conditions of noise. For example, Miller, Heise, and Lichten (1951) presented subjects with words from vocabularies of various sizes (2, 4, 8, 16, 32, and 256 words plus unselected monosyllables). These were presented under noise conditions with signal-to-noise ratios ranging from −18 to +9 dB. (The signal-to-noise ratio is simply the difference in sound-pressure level between the signal and the noise; negative numbers indicate that the noise is more intense than the signal.) The results are shown in Figure 7-3; the percentage of words correctly recognized was very strongly correlated with the size of the vocabulary that was used.

To improve intelligibility, especially under noisy conditions, the following guidelines should be observed when messages are designed for a speech communication system: (1) keep the vocabulary as small as possible; (2) use

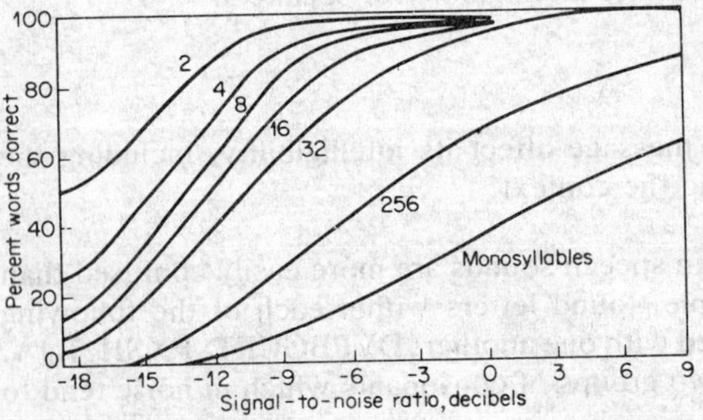

FIGURE 7-3
Intelligibility of words from vocabularies of different sizes under varying noise conditions. The numbers 2, 4, 8, etc., refer to the number of words in the vocabulary used. (*Source: Miller, Heise, and Lichten, 1951. Copyright © 1951 by the American Psychological Association and reproduced by permission.*)

standard sentence constructions with information always transmitted in the same order (e.g., bin location, part number, part name, quantity); (3) avoid short words, and rather than use letters, use a word-spelling alphabet; and (4) familiarize the receiver with the words and sentence structure to be used.

Transmission System

Speech transmission systems (such as telephones and radios) can produce various forms of distortion, such as frequency distortion, filtering, amplitude distortion, and modifications of the time scale. If intelligibility (and not fidelity) is important in a system, then certain types of distortion (especially amplitude distortion) still can result in acceptable levels of intelligibility. Since high-fidelity systems are very expensive, it is useful to know what effects various forms of distortion have on intelligibility to be able (it is hoped) to make better decisions about the design or selection of communication equipment. For illustrative purposes, we discuss the effects of filtering and amplitude distortion.

Effects of Filtering on Speech The filtering of speech consists basically in blocking out certain frequencies and permitting only the remaining frequencies to be transmitted. Filtering may be an unintentional consequence, or in some instances an intentional consequence, of the design of a component. Most filters eliminate frequencies *above* some level (a *low-pass* filter) or frequencies *below* some level (a *high-pass* filter). Typically, however, the cutoff, even if intentional, is not precisely at a specific frequency, but rather tapers off over a range of adjacent frequencies. Filters that eliminate all frequencies above, or all frequencies below, 2000 Hz *in the quiet* will still transmit speech quite intelligibly, although the speech does not sound natural. The effects on speech of filtering out certain frequencies are summarized in Figure 7-4. This shows, for high-pass filters and for low-pass filters with various cutoffs, the intelligibility of the speech. The filtering out of frequencies above 4000 Hz or below about 600 Hz has relatively little effect on intelligibility. But look at the effect of filtering out frequencies above 1000 Hz or below 3000 Hz! Such data as those given in Figure 7-4 can provide the designer of communications equipment with some guidelines about how much filtering can be tolerated in the system.

Effects of Amplitude Distortion on Speech *Amplitude distortion* is the deformation which results when a signal passes through a nonlinear circuit. One form of such distortion is *peak clipping*, in which the peaks of the sound waves are clipped and only the center part of the waves are left. Although peak clipping is produced by electronic circuits for experiments, some communication equipment has insufficient amplitude-handling capability to pass the peaks of the speech waves and at the same time provide adequate intensity for the lower-intensity speech components. Since peak clipping impairs the quality of speech and music, it is not used in regular broadcasting; but it is sometimes

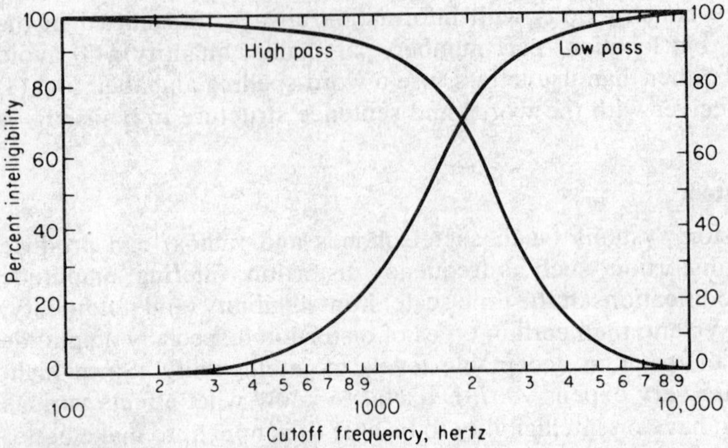

FIGURE 7-4
Effect on intelligibility of elimination of frequencies by the use of filters. A low-pass filter permits frequencies below a given cutoff to pass through the communication system and eliminates frequencies above that cutoff. A high-pass filter, in turn, permits frequencies above the cutoff to pass and eliminates those below. (*Source: Adapted from French and Steinberg, 1947.*)

used in military and commercial communication equipment. *Center clipping*, on the other hand, eliminates the amplitudes below a given value and leaves the peaks of the waves. Center clipping is more of an experimental procedure than one used extensively in practice. The amount of clipping can be controlled through electronic circuits. Figure 7-5 illustrates the speech waves that would result from both forms of clipping.

The effects of these two forms of clipping on speech intelligibility are very different, as shown in Figure 7-6. Peak clipping does not cause major degradation of intelligibility even when the amount of clipping (in decibels) is reasonably high. On the other hand, even a small amount of center clipping results in rather thorough garbling of the message. The reason for this difference in the effects of peak and center clipping is that when the peaks are lopped off, we

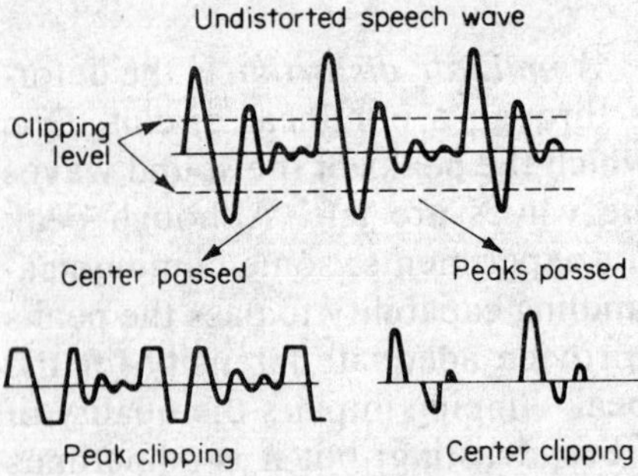

FIGURE 7-5
Undistorted speech wave and the speech waves that would result from peak clipping and center clipping. (*Source: Miller, 1963, p. 72.*)

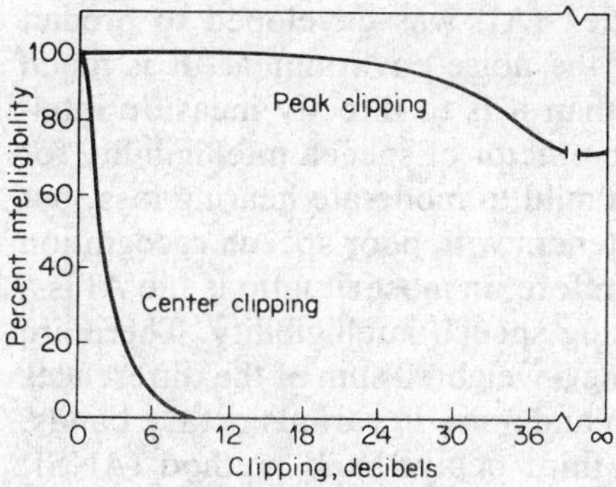

FIGURE 7-6
The effects on speech intelligibility of various amounts of peak clipping and of center clipping. (*Source: Licklider and Miller, 1951.*)

reduce the power of the vowels, which are less critical in intelligibility, and leave the consonants essentially unscathed. But when we cut out the center amplitudes, the consonants fall by the wayside, thus leaving essentially the high peaks of the vowels. Since intelligibility is relatively insensitive to peak clipping, one can clip off the peaks and then amplify the remaining signal to the original level. This increases the intensity level of the consonants relative to the vowels and reduces the detrimental effects of noise. Intelligibility is higher if the speech signal is clipped before being added to noise then if it is clipped after being added to the noise. Speech peak clipped 6 dB or less sounds essentially normal and the effect of clipping is barely detectable. Higher levels of clipping (18 dB or more), however, make the speech sound "grainy."

Noise Environment

Noise, be it external in the environment or internal to the transmission system, is the bane of speech intelligibility. Various approaches have been developed to evaluate the effects of noise on speech communications. The simplest approach is to compute the signal-to-noise (S/N) ratio (the algebraic difference between the signal and noise in decibels). When the sound pressure level of noise is between 35 and 110 dB, the S/N ratio required to reach the threshold of intelligibility for prose passages is constant at about +12 dB (Hawkins and Stevens, 1952). To maintain such an S/N ratio at high noise levels would require either a public address system or excessive vocal effort (shouting); this may not be possible for some people to achieve and, even if achievable, would be less intelligible than speech of normal intensity because of articulation distortions at vocal extremes. The major problem with S/N ratio as a measure of speech interference is that the frequency of the noise is not taken into consideration. Various indices for evaluating the effects of noise on speech intelligibility have been developed that do take noise frequency into account. A few of these are described briefly, namely, the articulation index (AI), the preferred-octave speech interference level (PSIL), and the preferred noise criteria (PNC) curves.

Articulation Index The articulation index (AI) was developed to predict speech intelligibility given a knowledge of the noise environment. It is much easier to measure the noise environment than it is to directly measure intelligibility in the environment. AI is a good predictor of speech intelligibility for normal and hearing-impaired listeners with mild to moderate hearing loss, but is a poor predictor for hearing-impaired listeners with poor speech recognition ability (Kamm, Dirks, and Bell, 1985). Therefore, in most situations the AI is a cost-effective method for indirectly assessing speech intelligibility. There are actually several articulation indices, all being a weighted sum of the differences between the sound-pressure level of speech and noise in various octave bands. One such technique, which uses the one-third octave-band method (ANSI, 1969; Kryter, 1972), is described below and illustrated in Figure 7-7:

1 For each one-third octave band, plot on a work sheet, such as in Figure 7-7, the band level of the speech peak reaching the listener's ear. (The specific procedures for deriving such band levels are given by Kryter.) An idealized spectrum for males is presented in Figure 7-7 as the basis for illustrating the derivation of the AI.

2 Plot on the work sheet the band levels of steady-state noise reaching the ear of the listener. An example of such a spectrum is presented in Figure 7-7.

3 Determine, at the center frequency of each of the bands on the work sheet, the difference in decibels between the level of speech and that of the noise. When the noise level exceeds that of speech, assign a zero value. When the speech level exceeds that of noise by more than 30 dB, assign a difference of 30 dB. Record this value in column 2 of the table given as part of Figure 7- 7.

4 Multiply the value for each band derived by step 3 by the weight for that band as given in column 3 of the table given with Figure 7-7, and enter that value in column 4.

5 Add the values in column 4. This sum is the AI.

The relationships between AI and intelligibility (percentage of material correctly understood) for various types of materials are shown in Figure 7-8. Using the AI of 0.47 computed in Figure 7-7, we would predict that about 75 percent of a 1000-word vocabulary would be correctly understood, while almost 90 percent of a 256-word vocabulary would be understood. If, on the other hand, complete sentences known to the receiver were used, we would expect near 100 percent intelligibility.

In general, expected satisfaction with a speech communication system is related to the AI of the system, as shown in Table 7-1.

Preferred-Octave Speech Interference Level The PSIL index, reported by Peterson and Gross (1978), can be used as a rough estimate of the effects of noise on speech reception. PSIL is simply the numeric average of the decibel levels of noise in three octave bands centered at 500, 1000, and 2000 Hz. Thus, if the decibel levels of the noise were 70, 80, and 75 dB, respectively, the PSIL would be 75 dB. The PSIL is useful when the noise spectrum is relatively flat.

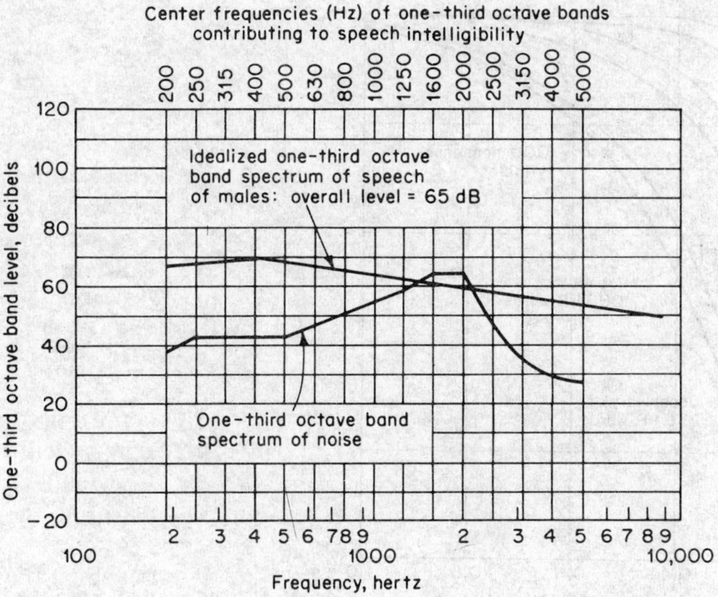

1. Band	2. Speech peaks minus noise, dB	3. Weight	4. Column 2 x 3
200	30	0.0004	0.0120
250	26	0.0010	0.0260
315	27	0.0010	0.0270
400	28	0.0014	0.0392
500	26	0.0014	0.0364
630	22	0.0020	0.0440
800	16	0.0020	0.0320
1000	8	0.0024	0.0192
1250	3	0.0030	0.0090
1600	0	0.0037	0.0000
2000	0	0.0038	0.0000
2500	12	0.0034	0.0408
3150	22	0.0034	0.0758
4000	26	0.0024	0.0624
5000	25	0.0020	0.0500
		AI =	0.4738

FIGURE 7-7
Example of the calculation of an articulation index (AI) by the one-third octave-band method. In any given situation the difference (in decibels) between the level of speech and the level of noise is determined for each band. These differences are multiplied by their weights. The sum of these is the AI. (*Source: Adapted from Kryter, 1972.*)

The PSIL loses some of its utility if the noise has intense components outside the range included in the PSIL, has an irregular spectrum, or consists primarily of pure tones. An older index called the *speech interference level* (SIL) is an average of the decibel levels of different octave bands, namely, 600 to 1200, 1200 to 2400, and 2400 to 4800 Hz.

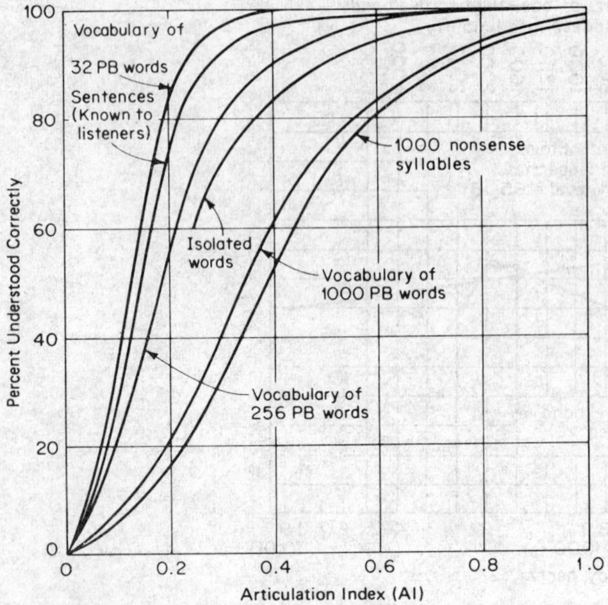

FIGURE 7-8
Relationship between the AI and the intelligibility of various types of speech test materials. (*Source: Adapted from Kryter, 1972, and French and Steinberg, 1947.*)

Some indication of the relationship between PSIL (and SIL) values and speech intelligibility under different circumstances is shown in Figure 7-9. In particular, this shows the effects of noise in face-to-face communications when the speaker and the hearer are at various distances from each other. For the distance and noise level combinations in the white area of the figure, a normal voice would be satisfactory; but for those combinations above the "maximum vocal effort line," it would probably be best to send up smoke signals.

TABLE 7-1
RELATIONSHIP BETWEEN ARTICULATION INDEX AND EXPECTED USER SATISFACTION WITH A COMMUNICATION SYSTEM

AI	Rating of satisfaction
<0.3	Unsatisfactory to marginal
0.3–0.499	Acceptable
0.5–0.7	Good
>0.7	Very good to excellent

Source: Beranek (1947).

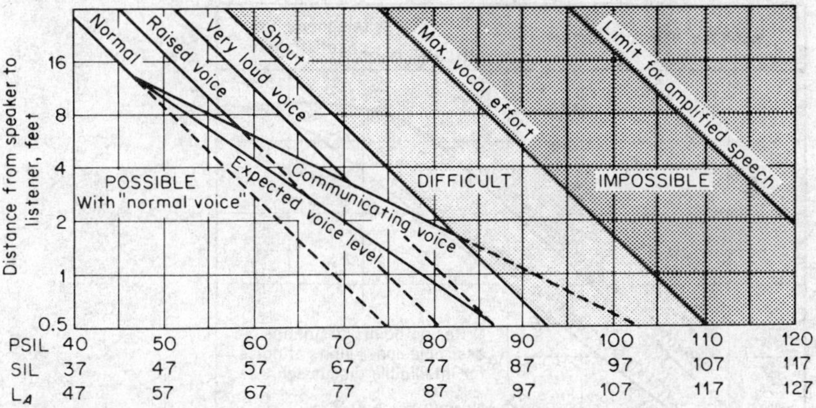

FIGURE 7-9
Voice level and distance between talker and listener for satisfactory face-to-face communication as limited by ambient noise level (expressed in PSIL and SIL). For any given noise level (such as PSIL of 70) and distance [such as 8 ft (2.4 m)] it is possible to determine the speech level that would be required (in this case a "shout"). (*Source: Adapted from Webster, 1969, Fig. 19, p. 69.*)

The subjective reactions of people to noise levels in private and large offices (secretarial, drafting, business machine offices, etc.) were elicited by Beranek and Newman (1950) by questionnaires. The results of this survey were used to develop a rating chart for office noises, as shown in Figure 7-10. The line for each group represents PSILs (baselines) that were judged to exist at certain subjective ratings (vertical scale). The dot on each curve represents the judged upper limit for intelligibility. The judged PSIL limit for private offices [normal voice at 9 ft (2.7 m)] was slightly above 45 dB, and for larger offices [slightly raised voice at 3 ft (0.9 m)] was 60 dB.

The judgments with regard to telephone use are also given in Figure 7-10 at the top of the figure. These judgments were made for long-distance or suburban calls. For calls within a single exchange, about 5 dB can be added to each of the listed levels, since there is usually better transmission within a local exchange.

Criteria for background noise in various communication situations have been set forth by Peterson and Gross, based on earlier standards by Beranek and Newman. These, expressed as PSILs, are given in Table 7-2.

Preferred Noise Criteria Curves Beranek, Blazier, and Figwer (1971) developed a set of curves to evaluate the noise environment inside office buildings and in rooms and halls in which speech communications are important. The curves are based on the older noise criteria (NC) curves originally developed by Beranek in 1957. A set of PNC curves is shown in Figure 7-11. To use these curves, the noise spectrum of the area being evaluated is plotted on the chart of curves, and each octave-band level is compared with the PNC curves to find

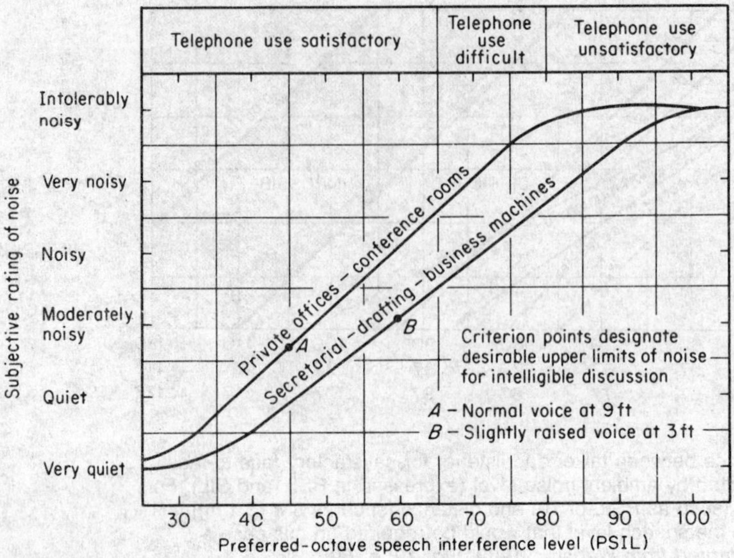

FIGURE 7-10
Rating chart for office noises. (*Source: Beranek and Newman, 1950, as modified by Peterson and Gross, 1978, to reflect the current practice of using octave bands with centers at 500, 1000, and 2000 Hz.*)

TABLE 7-2
MAXIMUM PERMISSIBLE PSIL FOR CERTAIN TYPES OF ROOMS AND SPACES

Type of room	Maximum permissible PSIL (measured when room is not in use)
Secretarial offices, typing	60
Coliseum for sports only (amplification)	55
Small private office	45
Conference room for 20	35
Movie theater	35
Conference room for 50	30
Theaters for drama, 500 seats (no amplification)	30
Homes, sleeping areas	30
Assembly halls (no amplification)	30
Schoolrooms	30
Concert halls (no amplification)	25

Source: Peterson and Gross, 1978, Table 3-5, p. 39.

the one that penetrates the highest PNC level. The highest PNC curve penetrated is the PNC rating for the environment. As an example, the noise spectrum of an office is shown in Figure 7-11 by several encircled crosses. This noise would have a PNC rating of 37 since that is the value of the highest PNC curve penetrated by the noise spectrum. Table 7-3 lists a few recommended PNC levels for various purposes and spaces.

Reverberation *Reverberation* is the effect of noise bouncing back and forth from the walls, ceiling, and floor of an enclosed room. As we know from experience, in some rooms or auditoriums this reverberation seems to obliterate speech or important segments of it. Figure 7-12 shows approximately the reduction in intelligibility that is caused by varying degrees of reverberation (specifically, the time in seconds that it takes the noise to decay 60 dB after it is shut off). This relationship essentially is a straight-line one.

Hearer

The hearer, or listener, is the last link in the communication chain. For receiving speech messages under noise conditions, the hearer should have normal hearing, should be trained in the types of communications to be re-

FIGURE 7-11
Preferred noise criterion (PNC) curves. (*Source: Beranek, Blazier, and Figwer, 1971.*)

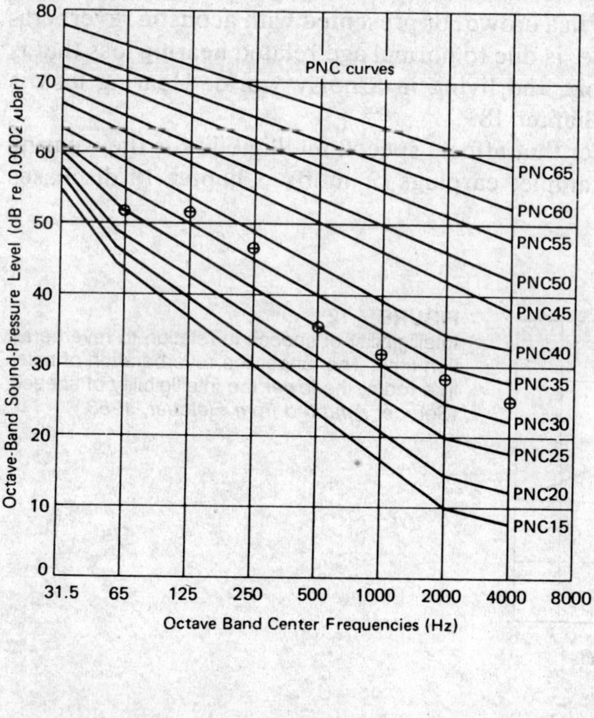

TABLE 7-3
RECOMMENDED PNC LEVELS FROM STEADY BACKGROUND NOISE ACCORDING TO TYPE OF SPACE FOR INDOOR ACTIVITIES INVOLVING SPEECH COMMUNICATIONS

Type of space	Recommended PNC level
Concert halls, opera houses	10–20
Large auditoriums, theaters, churches	Not to exceed 20
Small auditoriums, theaters, churches	Not to exceed 35
Bedrooms	25–40
Private or semiprivate offices, classrooms	30–40
Large offices, retail stores, restaurants	35–45
Shops, garages, power plant control rooms	50–60

Source: Beranek, Blazier, and Figwer, 1971.

ceived, should be reasonably able to withstand the stresses of the situation, and should be able to concentrate on one of several conflicting stimuli.

The age of the listener also affects speech reception (intelligibility) as shown in Figure 7-13. Depicted is the percent decrement in intelligibility, using 20 to 29-year-olds as the base, for various speech conditions. The decrement in intelligibility increases markedly after 60 years of age, especially when speech is speeded up (300 words/min versus 120 words/min), heard with other simultaneously presented voices as in a crowd, or presented with acoustic reverberation. The decrement, of course, is due to normal age-related hearing loss that is part and parcel of growing old and living in a noisy world. Hearing loss is discussed in more detail in Chapter 18.

Another aspect of the hearer that affects speech intelligibility is the wearing of hearing protectors, for example, earplugs or muffs. Chapter 18 discusses

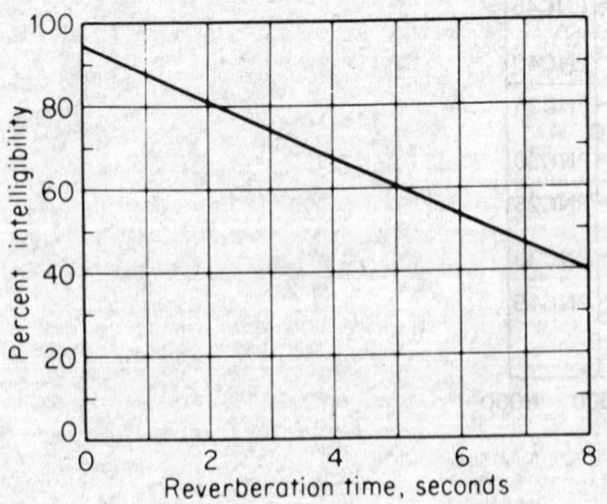

FIGURE 7-12
Intelligibility of speech in relation to reverberation time. The longer the reverberation of noise in a room, the lower the intelligibility of speech. (*Source: Adapted from Fletcher, 1953.*)

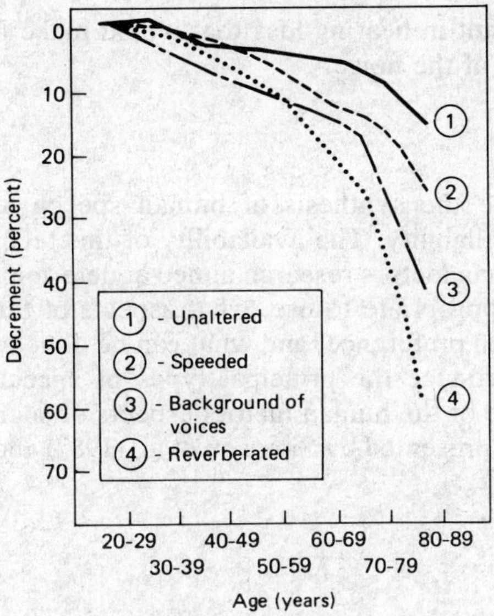

FIGURE 7-13
Speech intelligibility as a function of the age of the listener. Performance of 20- to 29-year-olds is taken as the base level, and decrement in performance relative to that level is presented for each age group. Speeded speech is presented at 300 wpm versus 120 wpm for unaltered speech. (*Source: Adapted from Bergman et al., 1976.*)

characteristics of hearing protection in more depth; here we deal with its effects on speech intelligibility. The effect of hearing protection on speech intelligibility is complex because one must consider the effect of wearing hearing protection on both the hearer and the speaker. In a noisy environment, people wearing hearing protection speak 2 to 4 dB more softly, 20 percent faster, and take 25 percent shorter pauses than when they are not wearing hearing protection (Hormann, Lazarus-Mainka, Schubeius, and Lazarus, 1984). Interestingly in quiet surroundings, wearing hearing protection has an opposite effect; people speak 3 to 4 dB louder than when they are not wearing them (Casali, 1989).

Although contradictions exist among studies, a few generalizations regarding the effects of hearing protection on speech intelligibility are possible (Casali, 1989). When the speech loudness can be controlled, as, for example, when it is broadcast over a public address system, hearing protection worn by the listener has little or no degrading effect on intelligibility when the ambient noise level is above about 80 dB*A* and may even enhance intelligibility at higher noise levels. However, at low noise levels below about 80 dB*A*, hearing protection will degrade intelligibility of broadcast speech. It should be pointed out that hearing protection is usually not worn when noise levels are below 80 dB*A*, unless there is intermittent noise at higher levels. On the other hand, when both the speaker and listener are in a noisy environment communication is generally poorer (by 15 to 40 percent) if both are wearing hearing protection than if they are not. This is because of the difficulty of maintaining adequate voice intensity in noise while wearing hearing protection. Impaired communications, however, should not be a reason for forgoing the use of hearing protection in a noisy environment. Not using them could result in hearing loss that would make it difficult to understand speech in or out of the noise.

SYNTHESIZED SPEECH

Advances in technology made possible the synthesis of human speech at relatively low cost and with excellent reliability. The availability of this technology has generated a great deal of human factors research aimed at determining when synthesized speech is most appropriate to use, what aspects of the system influence human performance and preference, and what can be done to improve such systems. We briefly introduce the principal types of speech synthesis systems, and we discuss some of the human factors aspects of such systems. A good overview of the area is presented by Simpson et al. (1987) and McCauley (1984).

Types of Speech Synthesis Systems

The traditional method of reproducing speech (as opposed to synthesizing it) is to record it as an analog signal on tape for later playback. The major problem with this technology is that moving parts, inherent in a record or tape playback system, wear out and break down. Further, only messages recorded in advance can be played back, and it takes time to access specific messages on the tape or record. A more advanced technology, but one similar in concept, involves digitizing speech and storing it in computer memory; this eliminates the moving-parts problem. Digitizing involves sampling the speech signal very rapidly and storing information about each sample (such as its intensity) for later decoding back to speech. The problem with this technique is that massive amounts of storage are required to store the digitized information. It is not uncommon to sample a speech signal 8000 or more times *per second*. Depending on how much information is encoded in each of the samples, to store a dozen or so English words could require 1 million bits of memory or more. Although the quality of analog and digital recordings is very good, the reliability and storage problems have made them somewhat impractical for most applications. To get around these problems, two methods have been developed that produce speech: synthesis by analysis and synthesis by rule. We discuss each briefly; but for a more in-depth introduction, see Michaelis and Wiggins (1982).

Synthesis by Analysis In this technique, digitized human speech is transformed into a more compressed data format. Various techniques are employed to reduce the amount of information needed to store a speech segment (e.g., linear predictive coding or waveform parameter encoding). The principal advantage is that considerably less computer memory is required to store the speech information than with straight digitization. As with straight digitized speech, however, digitized speech messages generated by synthesis by analysis can be formed from only those words or phrases that were previously encoded and stored. If new vocabulary words need to be added, the human, whose voice was originally encoded, must be used to maintain consistency within the system. Further when individual words are linked, they can sound very awkward and unnatural because of a lack of *coarticulation,* or the natural blending and modification of speech sounds caused by words and phonemes that precede and follow a particular sound. Consider the word *bookcase*. A synthesis-by-analysis system linking the two words *book* and *case* would pronounce the *k* sound from *book* and the *k* sound from *case*. When a human says *bookcase,* only one *k* sound is made. Technically, one could say that synthesis by analysis is really not synthesized speech per se; rather it is merely reconstructed digitized speech.

Synthesis by Rule Synthesis by rule is true synthesized speech. Such systems generate speech entirely based on sets of rules for generating elementary speech sounds, rules for combining elementary sounds into words and sentences (including coarticulation rules), and rules for stressing particular sounds or words that produce the prosody of speech. *Prosody* is the rhythm or singsong quality of natural speech. Synthesis-by-rule systems do not depend on digitized recordings of human speech at all. The main advantage of such systems is that very large vocabularies are possible with relatively small amounts of computer memory. New vocabulary can be generated without need of a human voice. A principal application of synthesis-by-rule systems is direct translation of text into speech. The problem is that the voice quality is usually not as good as that achieved by digitized speech-based systems. Most synthesis-by-rule systems produce male-sounding speech, although a few systems are available that sound like a female voice.

Uses of Synthesized Speech

Synthesized speech has taken on a "whiz-bang" high-tech connotation. Many consumer products can be purchased with built-in speech synthesizers, including automobiles ("Your door is ajar."), cameras ("Out of film."), microwave ovens ("What power level?"), scales "You weigh 150 pounds."), and wristwatches ("It is 2 o'clock and 15 minutes."). Several children's toys use the technology for educational purposes. In the field of aviation, synthesized speech has been used for cockpit warnings (Werkowitz, 1980) and interactive training systems (McCauley and Semple, 1980). Telephone companies have made wide use of synthesized speech to present telephone numbers to callers. The possibility of using synthesized speech to aid the handicapped has also not gone unnoticed. Devices include machines for the visually handicapped that convert printed materials (typewritten or typeset) directly into speech at rates of up to 200 words per minute (Kurzweil Computer Products, Cambridge, MA), signs that talk when scanned by a special infrared receiver, and systems for providing a voice for the vocally handicapped. The widespread use of synthesized speech has introduced some interesting human factors considerations in terms of performance with, preference for, and appropriate use of synthesized speech.

Performance with Synthesized Speech

Human factors specialists are concerned about how well people can perform when they are presented messages in synthesized speech. This often boils down to questions of how intelligible synthesized speech is compared to natural speech and how well people can remember synthesized speech messages.

Intelligibility of Synthesized Speech As we intimated earlier, the intelligibility of synthesized speech can be worse than that of natural speech depending on the type of system used to produce the sound and the sophistication of the speech production model built into the system. Logan, Greene, and Pisoni (1989) tested ten synthesis-by-rule systems and found a wide range of intelligibility across the systems. The researchers used a modified rhyme test (MRT) that assesses intelligibility of isolated monosyllabic words (actually the MRT assesses the intelligibility of individual phonemes that make up such words, and what it measures is referred to as *segmental intelligibility*). This is a very sensitive test of intelligibility. Using the MRT, they found an error rate for the best system of 3 percent, while the worst system had an error rate of 35 percent. Natural speech produced an error rate of less than 1 percent. Nusbaum and Pisoni (1985) present results from four synthesis-by-rule systems for correct word identification within meaningful sentences. This is not as sensitive a test of intelligibility as is the MRT test, but it is closer to how real-world synthesis systems are used. Natural speech was 99.2 percent intelligible. The best synthesized speech was 95.3 percent intelligible, while the worst was 83.7 percent intelligible. Often for familiarized subjects and limited vocabularies used under high positive signal-to-noise ratios, a good system can be 99 to 100 percent intelligible (Simpson et al., 1985). Brabyn and Brabyn (1982), for example, report intelligibility scores of 91 to 99 percent for talking signs ("drinking fountain") used to guide the visually handicapped. People apparently adapt rather quickly to synthesized speech; and with just a little exposure or training to a system's voice, intelligibility increases significantly. (Thomas, Rosson, and Chodorow, 1984; Schwab, Nusbaum, and Pisoni, 1985).

Memory for Synthesized Speech A fairly common finding is that people have more trouble remembering messages spoken in synthesized speech than in natural speech (Luce, Feustel, and Pisoni, 1983; Waterworth and Thomas, 1985). It is generally agreed that listening to synthesized speech requires more processing capacity than does listening to natural speech. Apparently this is due to encoding difficulties that disrupt working memory and the transfer of information to long-term memory. (For a discussion of processing capacity and memory see Chapter 3.) It appears, however, that once encoded, synthesized speech is stored just as efficiently as is natural speech. Luce, Feustel, and Pisoni (1983) found that increasing the interword interval from 1 to 5 s during the presentation of a list of words almost doubled the number of words recalled. In a similar vein, Waterworth (1983) reported that increasing the pause between groups of digits improved their recall.

Preference for Synthesized Speech

A broad consideration is whether people really want inanimate objects such as cars and cameras to talk to them. Creitz (1982), for example, reports that sales of a microwave oven that told the user what to do were so bad that the manufacturer stopped producing them. Similarly, Stern (1984) found that people using automated teller machines (ATMs) did not like voice messages (for example, "Wrong keys used"), especially when other people were around.

The quality of synthesized speech can be quite poor even though the messages are intelligible. Conversely, a system can be rated as pleasant and yet be totally incomprehensible. Synthesized speech is often described as machinelike, choppy, harsh and grainy, flat, and noisy (Edman and Metz, 1983). Usually what is lacking is coarticulation (where words are blended) and natural intonation. New techniques are constantly being developed to improve the quality of synthesized speech, but preference and rated quality are very subjective and differ from person to person for the same system. In some applications, such as aircraft cockpit warnings, many pilots do not want a natural-sounding synthesized voice; they want a voice that sounds like a machine, to distinguish it from human voices in the cockpit (Simpson et al., 1985).

As an illustration of the subtle effect of intonation on preference, Waterworth (1983) compared various ways of saying digit strings on people's preferences. Various combinations of neutral, terminator (dropping pitch), and continuant (rising-pitch) intonations were tested. Figure 7-14 shows both the original intonation pattern used by British Telecom to deliver telephone num-

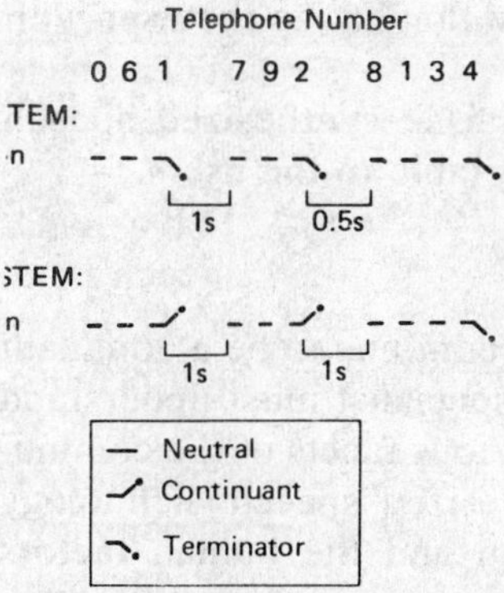

FIGURE 7-14
Two intonation patterns tested by Waterworth for use in delivering telephone numbers by synthesized speech. (*Source: Waterworth, 1983.*)

bers on their system and the new, improved intonation pattern developed as a result of Waterworth's research.

Guidelines for Use of Synthesized Speech

It is probably a little premature to offer specific guidelines for using synthesized speech because the technology is evolving so rapidly. In Chapter 3 we present some general guidelines for deciding when auditory or visual displays are appropriate. Synthesized speech is one form of an auditory display, and as such, its use should conform to the guidelines in Chapter 3. The design of the messages for synthesized speech systems should also follow those mentioned earlier in this chapter for natural speech communication systems. In addition, the following guidelines were gleaned from a number of sources (Simpson and Williams, 1980; Thomas, Rosson, and Chodorow, 1984; Wheale, 1980) and deal with various applications of synthesized speech:

1 Voice warnings should be presented in a voice that is qualitatively different from other voices that will be heard in the situation.

2 If synthesized speech is used exclusively for warnings, there should be no alerting tones before the voice warning.

3 If synthesized speech is used for other types of information in addition to warnings, some means of directing attention to the voice warning might be required.

4 Maximize intelligibility of the messages.

5 For general-purpose use, maximize user acceptance by making the voice as natural as possible.

6 Consider providing a replay mode in the system so users can replay the message if they desire. Roelofs (1987), however, found that although appreciated, replay modes are hardly used.

7 If a spelling mode is provided, its quality may need to be better than that used for the rest of the system.

8 Give the user the ability to interrupt the message; this is especially important for experienced users who do not need to listen to the entire message each time the system is used.

9 Provide an introductory or training message to familiarize the user with the system's voice.

10 Do not get caught up in "high-tech fever." Use synthesized speech sparingly and only where it is appropriate and acceptable to the users.

DISCUSSION

Speech communication by human or microchip will continue to be a dominant form of information display, so the human factors specialist must understand the basics of speech perception and the effects of various facets of the communication system on speech intelligibility. As synthesized speech technology evolves, whole new areas of application will open and the human factors problems and challenges will be there in force.

REFERENCES

American National Standards Institute (ANSI) (1969). *Methods for the calculation of the articulation index* (Rept. 535). New York.

Beranek, L. (1947). The design of speech communication systems. *Proceedings of the Institute of Radio Engineers,* 35, 880–890.

Beranek, L. (1957). Revised criteria for noise in buildings. *Noise Control,* 3(1), 19–27.

Beranek, L., Blazier, W., and Figwer, J. (1971). Preferred noise criteria (PNC) curves and their application to rooms. *Journal of the Acoustical Society of America,* 50, 1223–1228.

Beranek, L., and Newman, R. (1950). Speech interference levels as criteria for rating background noise in offices. *Journal of the Acoustical Society of America,* 22, 671.

Bergman, M., Blumenfeld, V., Cascardo, D., Dash, B., Levitt, H., and Margulies, M. (1976). Age-related decrement in hearing for speech. Sampling and logitudinal studies. *Journal of Gerontology,* 31, 533–538.

Bilger, R., Hanley, T., and Steer, M. (1955). *A further investigation of the relationships between voice variables and speech intelligibility in high level noise* (TR for SDC, 104-2-26, Project 20-F-8, Contract N6ori-104). Lafayette, IN: Purdue University (mimeographed).

Brabyn, J., and Brabyn, L. (1982). Speech intelligibility of the talking signs. *Visual Impairment and Blindness,* 76, 77–78.

Casali, J. (1989, July). Multiple factors affect speech communication in the work place. *Occupational Health & Safety,* 32–42.

Cole, R. (ed.) (1980). *Perception and production of fluent speech.* Hillsdale, NJ: Erlbaum.

Creitz, W. (ed.) (1982). *Voice news.* Rockville, MD: Stoneridge Technical Services.

Edman, T., and Metz, S. (1983). A methodology for the evaluation of real-time speech digitization. *Proceedings of the Human Factors Society 27th Annual Meeting.* Santa Monica, CA: Human Factors Society, pp. 104–107.

Fletcher, H. (1953). *Speech and hearing in communication.* Princeton, NJ: Van Nostrand Reinhold.

French, N., and Steinberg, J. (1947). Factors governing the intelligibility of speech sounds. *Journal of the Acoustical Society of America,* 19, 90–119.

Goldstein, E. (1984). *Sensation and perception* (2d ed.). Belmont, CA: Wadsworth.

Hawkins, J. Jr., and Stevens, S. (1952). The masking of pure tones and of speech by white noise. *Journal of the Acoustical Society of America,* 22, 6–13.

Hormann, H., Lazarus-Mainka, G., Schubeius, M., and Lazarus, H. (1984). The effect of noise and the wearing of ear protectors on verbal communication. *Noise Control Engineering Journal,* 23(2), 69–77.

Hull, A. (1976). Reducing sources of human error in transmission of alphanumeric codes. *Applied Ergonomics,* 7(2), 75–78.

Jusczyk, P. (1986). Speech perception. In K. Boff, L. Kaufman, and J. Thomas (eds.), *Handbook of perception and human performance,* vol. 2: *Cognitive processes and performance.* New York: Wiley, pp. 27-1–27-57.

Kamm, C., Dirks, D., and Bell, T. (1985). Speech recognition and the Articulation Index for normal and hearing-impaired listeners. *Journal of the Acoustical Society of America,* 77, 281–288.

Kiang, N. (1975). Stimulus representation in the discharge patterns of auditory neurons. In E. L. Eagles (ed.), *The nervous system,* vol. 3. New York: Raven.

Kryter, K. (1972). Speech communication. In H. Van Cott and R. Kinkade (eds.),

Human engineering guide to equipment design. Washington, DC: Government Printing Office.

Kryter, K. (1985). *The effects of noise on man* (2d ed.). Orlando, FL: Academic.

Licklider, J., and Miller, G. (1951). The perception of speech. In S. S. Stevens (ed.), *Handbook of experimental psychology*. New York: Wiley.

Logan, J., Greene, B., and Pisoni, D. (1989). Segmental intelligibility of synthetic speech produced by rule. *Journal of the Acoustical Society of America*, 86(2), 566–581.

Luce, P., Feustel, T., and Pisoni, D. (1983). Capacity demands in short-term memory for synthetic and natural speech. *Human Factors*, 25, 17–32.

McCauley, M. (1984). Human factors in voice technology. In F. Muckler (ed.), *Human factors review—1984*. Santa Monica, CA: Human Factors Society.

McCauley, M., and Semple, C. (1980). *Precision approach radar training system (PARTS)* (NAVTRAEQUIPCEN 79-C-0042-1). Orlando, FL: Naval Training Equipment Center.

Michaelis, P., and Wiggins, R. (1982). A human factors engineer's introduction to speech synthesizers. In A. Badre and B. Shneiderman (eds.), *Directions in human computer interaction*. Norwood, NJ: Ablex Publishing.

Miller, G. (1963). *Language and communication*. New York: McGraw-Hill.

Miller, G., Heise, G., and Lichten, W. (1951). The intelligibility of speech as a function of the context of the test materials. *Journal of Experimental Psychology*, 41, 329–335.

Nusbaum, H., and Pisoni, D. (1985). Constraints on the perception of synthetic speech generated by rule. *Behavior Research Methods, Instruments, & Computers*, 17(2), 235–242.

Peterson, A., and Gross, E., Jr. (1978). *Handbook of noise measurement* (8th ed.). New Concord, MA: General Radio Co.

Pollack, I., and Pickett, J. (1964). The intelligibility of excerpts from conversational speech. *Language and Speech*, 6, 165–171.

Roelofs, J. (1987). Synthetic speech in practice: Acceptance and efficiency. *Behavior and Information Technology*, 6(4), 403–410.

Schwab, E., Nusbaum, H., and Pisoni, D. (1985). Some effects of training on the perception of synthetic speech. *Human Factors*, 27(4), 395–408.

Simpson, C., McCauley, M., Roland, E., Ruth, J., and Williges, B. (1985). System design for speech recognition and generation. *Human Factors*, 27(2), 115–141.

Simpson, C., McCauley, M., Roland, E., Ruth, J., and Williges, B. (1987). Speech controls and displays. In G. Salvendy (ed.), *Handbook of human factors*. New York: Wiley, pp. 1490–1525.

Simpson, C., and Williams, D. (1980). Response time effects of alerting tone and semantic context for synthesized voice cockpit warnings. *Human Factors*, 22, 319–320.

Stern, K. (1984). An evaluation of written, graphics, and voice messages in proceduralized instructions. *Proceedings of the Human Factors Society 28th Annual Meeting*. Santa Monica, CA: Human Factors Society, pp. 314–318.

Thomas, J., Rosson, M., and Chodorow, M. (1984). Human factors and synthetic speech. *Proceedings of the Human Factors Society 28th Annual Meeting*. Santa Monica, CA: Human Factors Society, pp. 763–767.

Waterworth, J. (1983). Effect of intonation form and pause duration of automatic telephone number announcements on subjective preference and memory performance. *Applied Ergonomics,* 14(1), 39–42.

Waterworth, J., and Thomas, C. (1985). Why is synthetic speech harder to remember than natural speech? In *CHI-85 Proceedings*. New York: Association for Computing Machinery.

Webster, J. (1969). *Effects of noise on speech intelligibility* (ASHA Rept. 4). Washington, DC: American Speech and Hearing Association.

Werkowitz, E. (1980, May). Ergonomic considerations for the cockpit application of speech generation technology. In S. Harris (ed.), *Voice-interactive systems: Applications and pay-offs*. Warminster, PA: Naval Air Development Center.

Wheale, J. (1980). Pilot opinion of flight deck voice warning systems. In D. Oborne and J. Levis (eds.), *Human factors in transport research,* vol. 1. New York: Academic.

PART **THREE**

HUMAN OUTPUT AND CONTROL

8 Physical Work and Manual Materials Handling

9 Motor Skills

10 Human Control of Systems

11 Controls and Data Entry Devices

12 Hand Tools and Devices

CHAPTER **8**

PHYSICAL WORK AND MANUAL MATERIALS HANDLING

With the emphasis on the computer age, automation, and the cognitive aspects of work we tend to forget that a great deal of our work activities require physical effort and the manual handling of materials, supplies, and tools. Occupations in mining, construction, and manufacturing often require workers to expend moderate to high levels of physical energy to perform jobs. Back injuries resulting from overexertion are common in most occupations, accounting for about 25 percent of all occupational injuries (Bureau of Labor Statistics, 1982). Even occupations not typically associated with heavy lifting are equally affected, including professional-technical workers, managers and administrators, clerical workers, and service workers. Interestingly, nurses experience more back injuries than most other occupational groups (Jensen, 1985, 1986). This chapter deals with the capabilities of humans to expend energy to perform useful work and their strength and endurance characteristics. Special attention is given to manual materials handling.

Before getting into such matters, however, we introduce the concepts of stress and strain because they are relevant to various aspects of human work, including physical work. In general terms, *stress* refers to some undesirable condition, circumstance, task, or other factor that impinges upon the individual. Some possible sources of stress include: heavy work, immobilization, heat and cold, noise, sleep loss, danger, information overload, boredom, loneliness, financial insecurity, etc. *Strain* refers to the effects of the stress on the individual. Strain can be measured by observing changes in such things as: blood chemistry, oxygen consumption, electrical activity of the muscles or the brain, heart rate, body temperature, work rate, errors, and attitudes.

Certain work-related factors would be sources of stress to virtually anyone. It must be recognized, however, that some factors might be sources of stress for some people but not for others. In addition, people vary widely in the manner and level of strain produced by a given level of stress. These differences can arise from various individual characteristics including physical, mental, attitudinal, or emotional factors. We often forget the range of individual differences in response to stress when we confront mounds of statistics, neat curves, and precise formulas. This chapter should test your resolve in this regard.

MUSCLE PHYSIOLOGY

There are three types of muscles in the human body: striated (skeletal), cardiac (heart), and smooth (walls of blood vessels and internal organs). Our focus is on skeletal, striated muscles.

Nature of Muscles

There are over 600 muscles in the human body, among which are over 400 skeletal muscles that appear in pairs on both sides of the body. Most vigorous activities, however, are carried out by fewer than 80 pairs of such muscles (Jensen, Schultz, and Bangerter, 1983). Skeletal muscles are attached to bones by way of tendons. When muscles contract, they cause movement around articulated joints such as the elbow, shoulder, hip, or knee.

Each muscle is composed of many muscle fibers. These typically range in length from 0.2 to 5.5 in (0.5 to 14 cm), but some as long as 12 in (30 cm) have been traced. A muscle has upwards of several hundred thousand fibers. The fibers are connected by tissue through which nerve impulses enter and are carried throughout the fiber. Tiny blood vessels carrying oxygen and nutrients to the muscle form a complex capillary network permeating the muscle tissue.

Contractibility of Muscles

The only active action a muscle can perform is to contract and shorten its length. The contractile units of a muscle fiber are the *myofibrils,* which are composed of two basic types of protein filaments: *myosin* and *actin*. Each muscle fiber has many myofibrils. The myosin and actin filaments are arranged in bands as shown in Figure 8-1. When the muscle contracts, the actin filaments slide together between the myosin filaments. This combined action along a muscle fiber can cause the muscle fiber to contract to about half its original length. Each fiber contracts with a given force and the force applied by the complete muscle is a function of the number of muscle fibers that were activated. When a muscle contracts, electrical impulses can be detected and recorded by placing electrodes on the skin over the muscle group. The recording process is called electromyography (EMG). We discuss EMG recordings as a measure of local muscle activity later in the chapter.

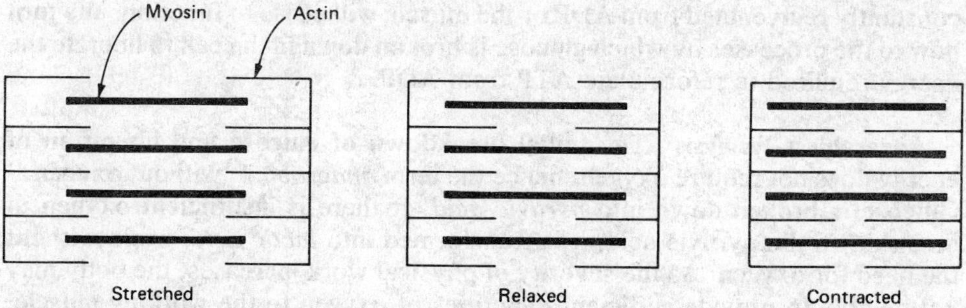

FIGURE 8-1
Schematic drawing of the protein filaments of a myofibril in a stretched, relaxed, and contracted condition. The myosin and actin filaments slide together to cause the muscle to shorten when contracted.

Control of Muscle Action

Muscle fibers are enervated by motor nerves emanating from the spinal cord. Muscles can be enervated at a conscious level or on a reflex level. A single motor-nerve fiber enervates a group of muscle fibers called a *motor unit*. The fibers within a motor unit are dispersed somewhat throughout a muscle so that stimulating one motor unit does not cause a strong contraction at only one specific location in the muscle, but rather causes a weak contraction throughout the muscle. We discuss further the neural control of muscles at a more cognitive level in the next chapter.

Muscle Metabolism

Energy is required for a muscle to contract. Energy is needed to slide the actin filaments over the myosin filaments. Energy for muscle action is provided by the foods we eat and digest, primarily the carbohydrates and fats. (Protein can also be used to provide energy, but it plays a small role except during severe malnutrition.) We follow the path of carbohydrates, the primary source of muscle energy. Carbohydrates are converted into glucose (or glycogen, a storage form of glucose). This occurs primarily in the liver.

In the muscle, glucose or glycogen is broken down to liberate energy for use in contracting the muscle. The muscle, however, can not directly use the energy liberated by the breakdown of glucose or glycogen. The energy must be utilized through an intermediate process. Energy is stored in molecules of *adenosine triphosphate* (ATP). ATP breaks down into *adenosine diphosphate* (ADP) plus a phosphate radical and, in so doing, releases the energy for the cell to use. The ADP then captures more energy and recombines into ATP. ATP is somewhat like a spoon, feeding the cell. It takes a load of energy (ADP to ATP) and then unloads it (ATP to ADP) for the cell to use.

Another source of immediate energy for a muscle is *creatinine phosphate* (CP), which serves as an energy store for ATP generation but is useful for only a short period of time, on the order of a few seconds to a minute. ATP must be constantly rejuvenated from ADP or the muscle will cease to function. We turn now to the processes by which glucose is broken down in the cell to liberate the energy required to reformulate ATP from ADP.

Anaerobic Glycolysis The initial breakdown of glucose and liberation of energy does not require oxygen, hence the term *anaerobic*, "without oxygen." Glucose is broken down into *pyruvic acid*. If there is insufficient oxygen to break down the pyruvic acid, it is transformed into *lactic acid*, again without the need for oxygen. As the severity of physical work increases, the body may not be able to provide sufficient quantities of oxygen to the working muscle, thus causing a buildup of lactic acid in the muscle and an eventual spillage into the blood stream and ultimately through the kidneys into the urine. Buildup of lactic acid in the muscle is a primary cause of muscle fatigue.

Aerobic Glycolysis If oxygen is available, pyruvic acid is oxidized into carbon dioxide and water rather than being converted into lactic acid. This breakdown of pyruvic acid is called *aerobic* because it requires oxygen to be accomplished. Aerobic glycolysis liberates about 20 times the energy liberated by anaerobic glycolysis. After a strenuous bout of physical activity, the lactic acid can be converted back into pyruvic acid. And, if oxygen is available, the pyruvic acid can then undergo aerobic glycolysis. The energy liberated during aerobic glycolysis is used to build up ATP and CP stores for future muscle activity.

Figure 8-2 shows the time course of how the body uses the various forms of energy during the first few minutes of moderately heavy physical work. The ATP stores are exhausted in a few seconds, while the CP stores last about 45 s.

FIGURE 8-2
Sources of energy during the first few minutes of moderately heavy work. High-energy phosphate stores (ATP and CP) provide most of the energy during the first seconds of work. Anaerobic glycolysis supplies less and less of the energy required as the duration of work increases and aerobic metabolism takes over. (*Source: Jones, Moran-Campbell, Edwards, and Robertson, 1975.*)

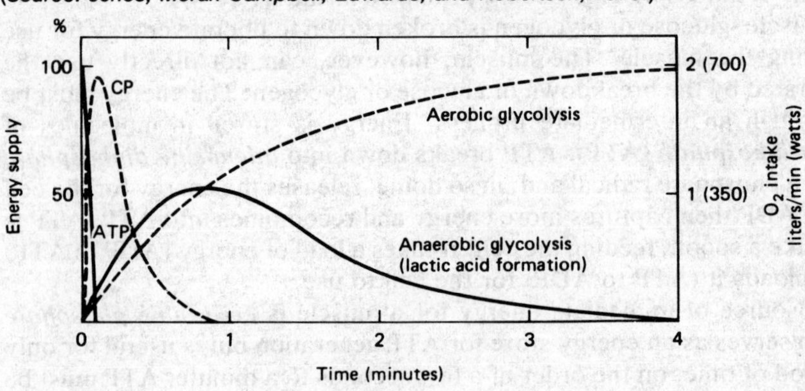

Anaerobic glycolysis supplies less and less of the energy required as the duration of exercise increases and the more efficient aerobic metabolism takes over.

Basal Metabolism Metabolism is the chemical process by which foodstuffs are converted into heat or mechanical energy. The large muscles of the body convert approximately 30 percent of the energy liberated into mechanical work, while the other 70 percent is given off as heat. The body requires a certain amount of energy just to stay alive even if no physical work is being performed, that is, lying down without moving a muscle. The amount of energy needed per unit time to sustain life is called the *basal metabolic rate (BMR)*. There are large differences between people's BMR. BMR depends in part on the following: body size (large people tend to have higher BMRs); age (younger people tend to have higher BMRs); and sex (men tend to have higher BMRs than females). Grandjean (1988) reports the following BMRs: for a 155-lb (70-kg) male, 1.2 kcal/min; and for a 133-lb (60-kg) female, 1.0 kcal/min. (We define kcal, a measure of energy, later in this chapter.)

WORK PHYSIOLOGY

The metabolic processes being carried out within the muscle must be supported by the cardiovascular and respiratory system of the body. An understanding of some of these more macrolevel, physiological responses will aid in understanding how physiological strain from physical work is measured.

Respiratory Response

Additional oxygen is needed by working muscles for aerobic glycolysis. At rest, oxygen consumption is usually less than 0.5 L/min. Under extremely heavy work, it can increase to about 5 L/min. The respiratory response is to increase the rate of breathing and the volume of air inspired with each breath.

Unfortunately, the body cannot instantaneously increase the amount of oxygen to working muscles when they begin working. There is a lag, or more properly, a sluggish response during the first couple of minutes of work. It is during this time that anaerobic glycolysis and the depletion of ATP and CP stores provide energy to the muscles. This is shown in Figure 8-2. During these first few minutes, the body experiences *oxygen debt*. Oxygen debt is the amount of oxygen required by the muscles after the beginning of work, over and above that which is supplied to them during the work activity. This debt needs to be "repaid" after the cessation of work. The repayment of the oxygen debt is reflected in a rate of oxygen consumption during the recovery period that is above the resting level. This situation is illustrated in Figure 8-3, which represents the theoretical pattern as it would be expected to occur under a steady-state condition of work. The additional oxygen that is consumed after cessation of work is used to break down pyruvic and lactic acid that has accumulated and to replenish the ATP and CP energy stores in the muscle.

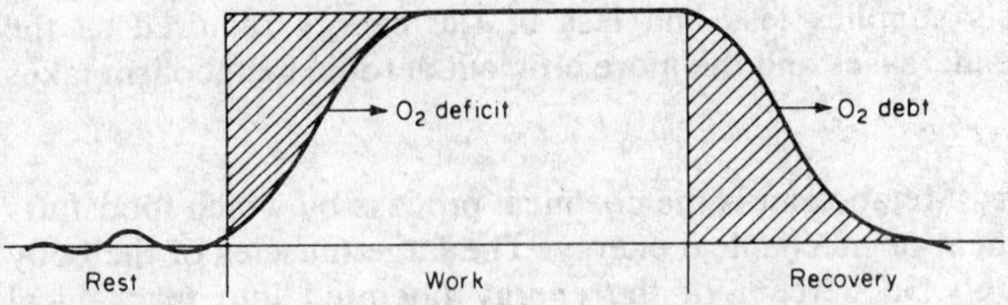

FIGURE 8-3
Illustration of the oxygen debt under conditions of steady state work.

Cardiovascular Response

Working muscles need oxygen and glucose, and these products are delivered in the blood through the cardiovascular system. Under normal resting conditions, blood leaves the lungs essentially saturated with oxygen (about 97 percent saturated). Therefore, just breathing harder would not increase the amount of oxygen reaching the muscles.

Increased Cardiac Output In order to provide more oxygen (and more glucose), more blood must flow to the muscles. To do this, the heart must pump more blood per unit time, that is, increase *cardiac output*. Cardiac output can be increased in two ways: by increasing the number of pumping actions per minute (i.e., *heart rate*) and/or by increasing the volume of blood pumped during each beat (i.e., *stroke volume*). Cardiac output at rest is about 5 L/min. During heavy work, cardiac output can increase to 25 L/min or more.

Heart rate and stroke volume begin to increase at the outset of physical work. At approximately 40 percent of a person's maximum capacity for physical work, the stroke volume stabilizes while heart rate continues to increase.

Increased Blood Pressure One consequence of the increased cardiac output is an increase in blood pressure during physical work. The adrenal glands release hormones (called *catecholamines*) that strengthen the heart beat and increase blood pressure. Because the heart must pump against the increased pressure, the strain on the heart is greater. The increased blood pressure is necessary so that sufficient quantities of blood return to the heart to fill it between beats. At 120 beats per min, which is not excessive, the heart only has about a half second to fill between beats.

Redistribution of Blood Flow Accompanying the increased cardiac output and blood pressure is a redistribution of blood flowing throughout the body. Table 8-1 shows the differences in blood distributions between rest and heavy work. Because there is more blood flowing per unit time during heavy work, the heart muscle receives more blood, although the percentage is the same. In the case of the brain, the percentage is lower during work, but the same volume of blood is being distributed as during rest. The big winners, of course, are the

TABLE 8-1
DISTRIBUTION OF BLOOD DURING REST AND WORK SHOWN AS A PERCENTAGE OF CARDIAC OUTPUT

	Blood flow distribution (%)	
Part of body	**Resting**	**Heavy work**
Muscles	15–20	70–75
Skin	5	10
Brain	15	3–4
Bones	3–5	0.5–1
Kidneys	20	2–4
Digestive system	20–25	3–5
Heart muscle	4–5	4–5

Source: Adapted from Astrand and Rodahl, 1986, Fig. 4-9. Reproduced with permission of McGraw-Hill.

muscles, which receive a greater share of a larger volume of blood. The volume of blood going to the skin increases during work to help dissipate heat that is generated by the metabolic process.

Discussion

The full range of physiological responses that accompany physical work are intricate and complex. We have touched on only a few gross responses. For more detail, see Astrand and Rodahl (1986). The responses we discussed serve as bases for measuring the stress level experienced by a person during physical work. Actually, what is measured is the strain placed on the individual, from which we infer the level of stress. We now turn our attention to measures of physiological strain.

MEASURES OF PHYSIOLOGICAL STRAIN

Before discussing measures of physiological strain, let us define a few basic physics concepts and measurement units:

- *Force:* A unit of force is the newton (N), which is equal to 1 kg-m/s^2 or the force which gives a 1 kg mass an acceleration of 1 m/s^2. [1 N = 0.225 lbf (pounds-force).]
- *Work or energy:* Work is done or energy is consumed when a force is applied over a distance. One N acting over a distance of 1 m requires 1 joule (J) of energy or does 1 J of work. A commonly used measure of energy is the kilocalorie (kcal), which is the amount of heat (a form of energy) required to raise the temperature of 1 kg of water from 15°C to 16°C. The Calorie (with a capital *C*) used for measuring the energy content of food is actually a kilocalorie

(with a lowercase *c*). Although the joule is becoming the preferred unit of measure, we will continue to use the kilocalorie because of its direct link to the familiar dietary Calorie. The following conversions allow one to move from one unit of measure to another:

1 kcal = 1000 cal = 1 Cal
1 kJ = 1000 J
1 kcal = 4.1868 kJ
1 kcal = 3087.4 ft·lbs

- *Power:* Work per unit time is power. A measure is the watt (W), equal to 1 J/s. Another unit of power is horsepower (hp), where 1 hp = 736 W.

Within the field of work physiology we are especially interested in measuring the energy expenditure or energy consumption (in kcal or kJ) of individuals engaged in various activities. As we will discuss, there are relationships between energy consumption (expenditure) and oxygen consumption, and between oxygen consumption and heart rate. We explore these relationships and other measures used to assess physiological strain.

Oxygen Consumption (Uptake)

As discussed, oxygen is used to metabolize food stuffs and release energy. The amount of energy released depends, in part, upon what food stuffs are being metabolized. For a typical diet, one liter of oxygen liberates approximately 5 kcal of energy. Therefore, by measuring the oxygen consumption (also called uptake) during work, we can directly estimate the energy consumption.

Measurement of Oxygen Consumption (Uptake) The measurement of oxygen consumption of people at work is rather cumbersome, but it can be done as shown in Figure 8-4. The volume of inspired air per unit time is measured. Air contains approximately 21 percent oxygen, so the amount of oxygen inspired per unit time can be easily determined. The volume of expired air per unit time is also measured and samples of expired air are taken. The samples are analyzed to determine the percentage of oxygen in them. From these measures, the volume of oxygen consumed per unit time (or *oxygen uptake*) can be determined.

Maximum Aerobic Power (MAP)

Increasing the rate of work causes a linear increase in oxygen uptake, to a point, after which oxygen uptake levels off. A further increase in work rate will no longer be accompanied by a further increase in oxygen uptake. The level at which oxygen uptake levels off is called the *maximal oxygen uptake* or *maximum aerobic power (MAP)*. Maximum aerobic power is defined as the highest oxygen uptake an individual can attain during exercise while breathing air at sea level. At the level of maximum aerobic power, energy metabolism is largely

FIGURE 8-4
Apparatus for monitoring heart rate and obtaining oxygen consumption levels of a person while working on the job. (*Photograph courtesy of Human Factors Department of Eli Lilly and Company.*)

anaerobic and lactic acid level in the muscle tissue and blood rises sharply. The higher a person's MAP, the greater the efficiency of the cardiovascular system.

Age, Sex, and MAP Figure 8-5 shows the general relationship between age and MAP for males and females. Before puberty, girls and boys show no significant differences in MAP. After puberty, however, women's MAP is, on average, 65 to 75 percent of that of men. In both sexes, there is a peak at 18 to 21 years of age, followed by a gradual decline. At the age of 65, the mean MAP is about 70 percent of what it was at 25 years of age. An average 65-year-old man has about the same maximum aerobic power as a 25-year-old woman (Astrand, Astrand, Hallback, and Kilbom, 1973).

Genetics seems to play a major role in determining maximum aerobic power. Physical training consisting of three 30-min training sessions per week commonly results in an average increase in maximum aerobic power of only 10 to 20 percent (Astrand and Rodahl, 1986).

Measurement of Maximum Aerobic Power The most straightforward method for determining a person's maximum aerobic power is to have the individual exercise on a treadmill or bicycle ergometer (exercise involving large muscle groups). The workload (incline, resistance, or rate of work) is increased until oxygen uptake levels off. The nature of the exercise, however, affects the maximum aerobic power measured. For example, higher values will be achieved with a treadmill at an incline greater than or equal to 3° than if the treadmill is level (Taylor, Buskirk, and Henschel, 1955). Methods for estimat-

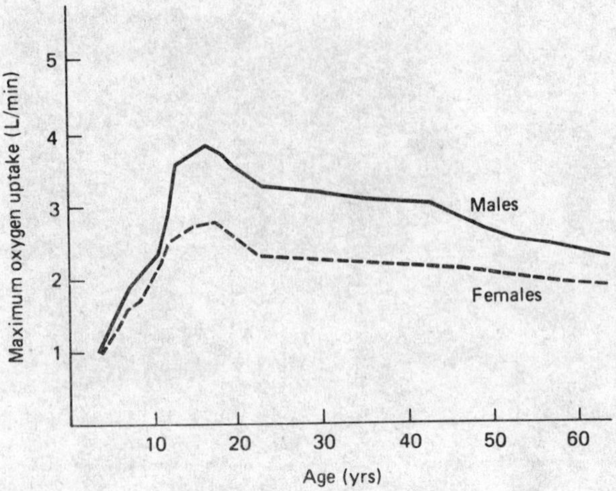

FIGURE 8-5
Mean maximum oxygen uptake (also called maximum aerobic power) in males and females as a function of age. (*Source: Astrand, Astrand, Hallback, and Kilbom, 1973.*)

ing maximum aerobic power using less than maximum effort have been proposed; however, such methods are generally not as accurate as methods that require maximum exertions, although they are safer (Astrand and Rodahl, 1986).

Heart Rate

Oxygen consumption is directly related to energy expenditure but is cumbersome to assess in a real-world work situation. As it turns out, there is a linear relationship between oxygen consumption and heart rate, and heart rate is very easy to measure. A few wires attached to the person and a telemetry or recording device about the size of a cigarette package is all that is needed. Heart rate, then, can be used to estimate oxygen consumption, which in turn can be converted into energy expenditure.

There are a few complications, however. First, the linear relationship between heart rate and oxygen consumption is different for different people. The better one's physical condition and the greater one's maximum aerobic power, the higher will be the oxygen consumption for a given heart rate. That is, the heart does not have to work as hard in order to provide a given level of oxygen to the working muscles. This requires that each person be "calibrated" to determine what their particular heart rate–oxygen consumption relationship is. To do this, heart rate and oxygen consumption are measured at several levels of work. A treadmill or bicycle ergometer is commonly used. Maas, Kok, Westra, and Kemper (1989), however, show that the heart rate–oxygen consumption relationship determined using dynamic work (e.g., a treadmill) may not be good for predicting oxygen consumption when static work is performed. Static work (e.g., holding a weight) results in a higher heart rate at any given level of oxygen consumption than does dynamic work.

The second problem with using heart rate to predict oxygen consumption is that the oxygen consumption is more predictable after the stroke volume of the heart has stabilized, which, as we indicated, occurs at about 40 percent of maximum aerobic power. Heart rate, then, is best used as a predictor of oxygen consumption when moderate to heavy work is performed.

Finally, there are a host of other factors, such as emotional stress, fatigue, and heat stress that affect heart rate but not oxygen consumption. The upshot is that the conditions under which the heart rate–oxygen consumption relationship is determined should, as closely as possible, match the conditions under which the relationship will be used to predict oxygen consumption.

Heart rate, continuously sampled over a workday or task, is useful as a general indicator of physiological stress without reference to oxygen consumption or energy expenditure. For example, Figure 8-6 presents the heart rate of an airline flight attendant while carrying out various activities during a flight.

Another measure based on heart rate is the *heart rate recovery curve* as used by Brouha (1967). This is a curve of the heart rate measured at certain intervals after work (such as 1, 2, and 3 min). In general terms, the more strenuous the work activity, the longer it takes the heart rate to settle down to its prework level.

Other Measures of General Physiological Strain

It is well known that the adrenal glands secrete hormones when the body is under stress (both physical and mental). The principal hormones are epinephrine and norepinephrine (catecholamines). The level of these catecholamines in the blood or urine can be used as a measure of occupational stress. Given the reluctance of people to give repeated samples of blood, urinary catecholamines are usually assessed. But even here it is difficult to obtain samples more frequently than every 2 hours or so. Astrand, Fugelli, Karlsson, Rodahl, and Vokac (1973) reported 10-fold increases in epinephrine excretion among coastal fishermen during the workday. Astrand and Rodahl (1986), however, report that extensive studies by the Norwegian Institute of

FIGURE 8-6
Continuous record of an airline flight attendant's heart rate over a 4-h flight. Such records can be useful as general indicators of physiological stress. (*Source: Kilbom, 1990, Fig. 21.8.*)

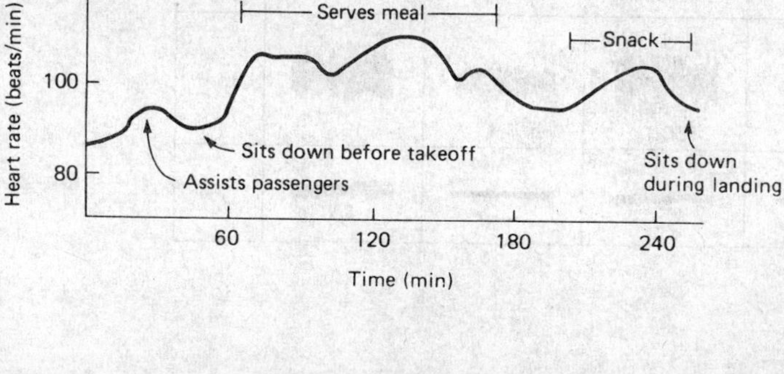

Work Physiology have failed to show any significant elevations of urinary catecholamines in most industries.

Measures of Local Muscle Activity

The measures discussed above relate to the overall level of physical work and stress placed on the body. It is often desirable, however, to measure the physiological strain of individual muscles or muscle groups. This can be done by placing recording electrodes on the skin surface over the muscles of interest and recording the electrical activity resulting from the contraction of the muscles. This procedure is called *electromyography (EMG)*, and the resulting recordings indicate EMG activity.

The EMG signal can be electrically analyzed, and a pen tracing of the electrical activity can be made. Figure 8-7 shows, for example, EMG recordings for four muscles of one subject when two levels of torque were applied to a steel socket. The signal can also be analyzed with respect to its frequency or amplitude. The amplitude is usually expressed as a percentage of the amplitude produced by a maximum voluntary contraction of the muscle taken prior to the study.

It has been shown that a high correlation exists between EMG activity and muscular force; this can also be seen in Figure 8-7. The relationship was once thought to be linear, but it is now believed to be exponential (Lind and Petrofsky, 1979). When a muscle begins to fatigue, there is an increase in activity in the low-frequency range of the EMG signal and a reduction in activity in the high-frequency range (Petrofsky, Glaser, and Phillips, 1982).

Subjective Measures of Exertion

The measures discussed so far require expensive apparatus and intrusive methodologies. The idea of substituting subjective ratings for physiological

FIGURE 8-7
Electromyograms recorded for four muscles of a subject maintaining a constant torque of 60 and 15 ft·lb (8.3 and 2.1 kg·m). The sum of the four values is an index of the total amount of energy expended. (*Source: Adapted from Khalil, 1973, Fig. 3.*)

Foot-pounds	Deltoid	Biceps	Triceps	Brachioradialis	Total
60	4.7	30.1	7.1	19.7	63.6
15	1.9	9.7	1.2	5.1	17.9

measurements is attractive because of the ease of administering and scoring rating scales. Borg (1985) has developed a scale for assessing perceived exertion, the so-called *Borg-RPE* (*rating of perceived exertion*) Scale. The scale is constructed so that the ratings, 6 to 20, are linearly related to the heart rate expected for that level of exertion (expected heart rate is 10 times the rating given). The Borg-RPE Scale is as follows:

6 - No exertion at all
7
 - Extremely light
8
9 - Very light
10
11 - Light
12
13 - Somewhat hard
14
15 - Hard (heavy)
16
17 - Very hard
18
19 - Extremely hard
20 - Maximal exertion

The scale is intended to rate exertion during dynamic work. The results obtained, however, are affected by the previous experience and motivation of the person making the ratings. For example, highly motivated people tend to underestimate their level of exertion (Kilbom, 1990).

Borg's RPE scale is not the only measure of perceived exertion. Promising results have also been obtained with a perceived effort scale developed by Fleishman, Gebhardt, and Hogan (1984).

PHYSICAL WORKLOAD

There are many factors that contribute to the workload experienced by people while engaged in physical work. Astrand and Rodahl (1986) present the major factors that influence the body's level of energy output as shown in Figure 8-8. The factors include the nature of the work, somatic factors, training, motivation, and environmental factors. These influence energy output through the physiological service function of supplying fuel and oxygen for muscle metabolism.

Work Efficiency

As noted, all energy expended by a person does not end up in useful work. Much of it (about 70 percent) ends up as heat, and some of it is expended in unproductive static efforts (such as holding or supporting things). We can think

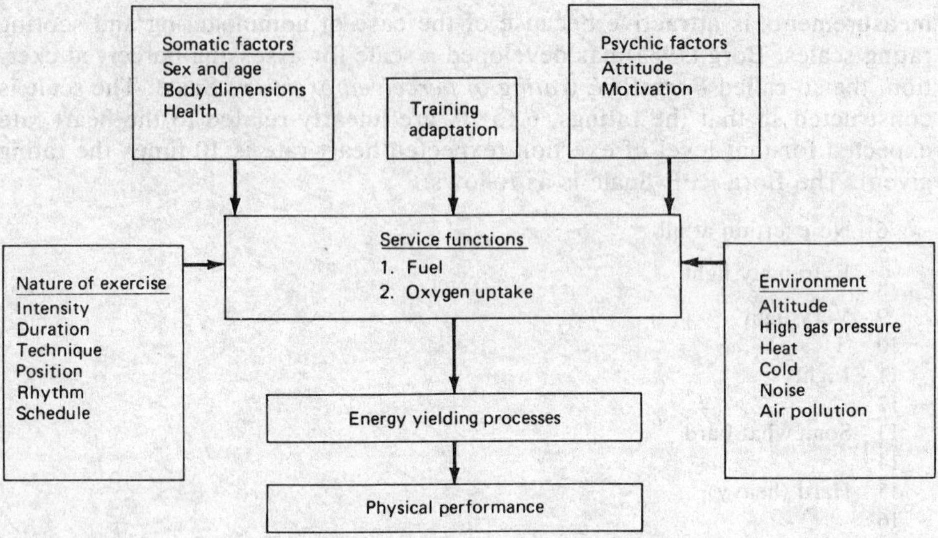

FIGURE 8-8
Factors influencing the power and capacity for physical activity. (*Adapted from Astrand and Rodahl, 1986, Fig. 7-1. Reproduced with permission of McGraw-Hill.*)

of a person as a machine and compute the work efficiency using the following basic equation:

$$\text{Efficiency (\%)} = \frac{\text{work output}}{\text{energy consumption}} \times 100$$

The estimated efficiencies of a few activities are shown in Table 8-2.

An example of the effects of the equipment used on efficiency is illustrated in Figure 8-9. This shows the work calories expended and time required when

TABLE 8-2
ESTIMATED WORK EFFICIENCY FOR VARIOUS ACTIVITIES

Activity	Efficiency (%)
Shoveling (stooped posture)	3
Shoveling (normal posture)	6
Using heavy hammer	15
Going up and down stairs (no load)	23
Pulling a cart	24
Pushing a cart	27
Cycling	25
Walking on level (no load)	27

Source: Grandjean, 1988, Table 14.

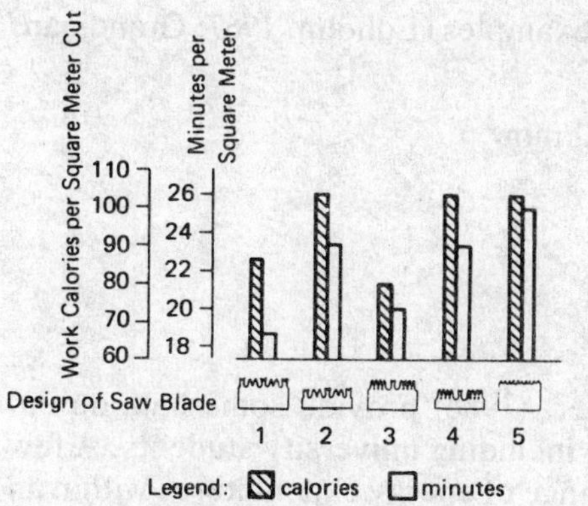

FIGURE 8-9
Calories expended and time spent in cutting a standard unit of work with five timber saws (unit of work: 1 m² of area cut). Saws 1 and 3 were clearly most efficient. (*From Grandjean, 1988, Fig. 73, p. 89.*)

five types of timber saws are used in cutting slices or disks of 1 m² area. It is clear that saws 1 and 3 resulted in the lowest calories consumed per unit of useful work (per cut). You may be able to relate this study to your own experiences. If you use a dull saw or the wrong type of saw, you can expend a lot of extra energy that is, in effect, wasted. Thus, in considering the above equation, the work output component should be restricted to useful work only.

To illustrate the concept of efficiency further, let us consider the matter of walking, as shown in Figure 8-10. This example brings us back to rate of work discussed before. Here we can see that the most efficient walking speeds for people with shoes and with bare feet are about 65 to 80 m/min (2.4 to 3.0 mi/h).

Energy Consumption

To get some feel for the numerical values of energy consumption for different kinds of physical activities, it is useful to present the physiological costs of

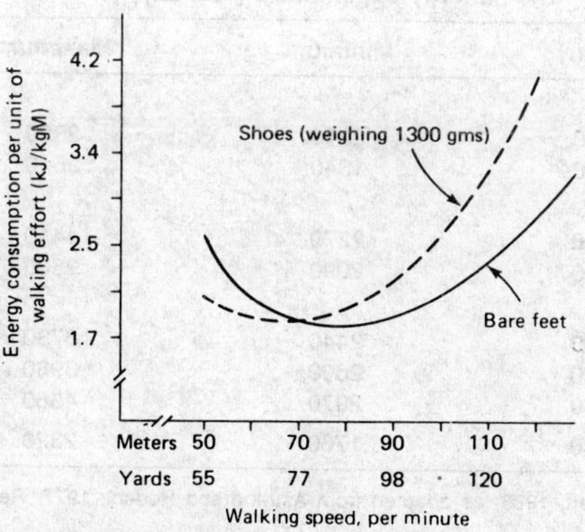

FIGURE 8-10
Efficiency, expressed as energy consumption (kJ) per unit of walking effort (kg-m), of walking with shoes and bare foot as a function of walking speed. (*Adapted from Grandjean, 1988, Fig. 75.*)

certain everyday activities. The following examples (Edholm, 1967; Grandjean, 1988) are given:

Sleeping	1.3 kcal/min
Sitting	1.6
Standing	2.25
Walking (level)	2.1
Cleaning/ironing	2.0–3.0
Cycling [10 mi/h (16 km/h)]	5.2

Kroemer, Kroemer, and Kroemer-Elbert (1986) provide some data on energy expenditure for various occupations, including university students. A few examples are provided in Table 8-3. The range of energy expenditures within an occupational group is quite large, with considerable overlap between occupations.

The energy consumption of several types of work is illustrated in Figure 8-11. The energy costs for these activities range from 1.6 to 16.2 kcal/min.

Grades of Work

Table 8-4 provides some definitions of various grades of work in terms of energy expenditure (kcal/min). Shown also in Table 8-4 are typical heart rates and oxygen consumption levels associated with the various grades. Because of the vagaries of the relationship between heart rate and oxygen consumption, the heart rate values given in Table 8-4 should be considered only rough approximations.

TABLE 8-3
ENERGY COST PER DAY FOR VARIOUS OCCUPATIONS

	Energy expenditure (kcal/day)		
Occupation	**Mean**	**Minimum**	**Maximum**
Laboratory technicians			
Males	2840	2240	3820
Females	2130	1340	2540
University students			
Males	2930	2270	4410
Females	2290	2090	2500
Males only			
Construction workers	3000	2440	3730
Steel workers	3280	2600	3960
Coal miners	3660	2970	4560
Housewives (middle age)	2090	1760	2320

Source: Kroemer, Kroemer, and Kroemer-Elbert, 1986; as adapted from Astrand and Rodahl, 1977. Reproduced with permission of McGraw-Hill.

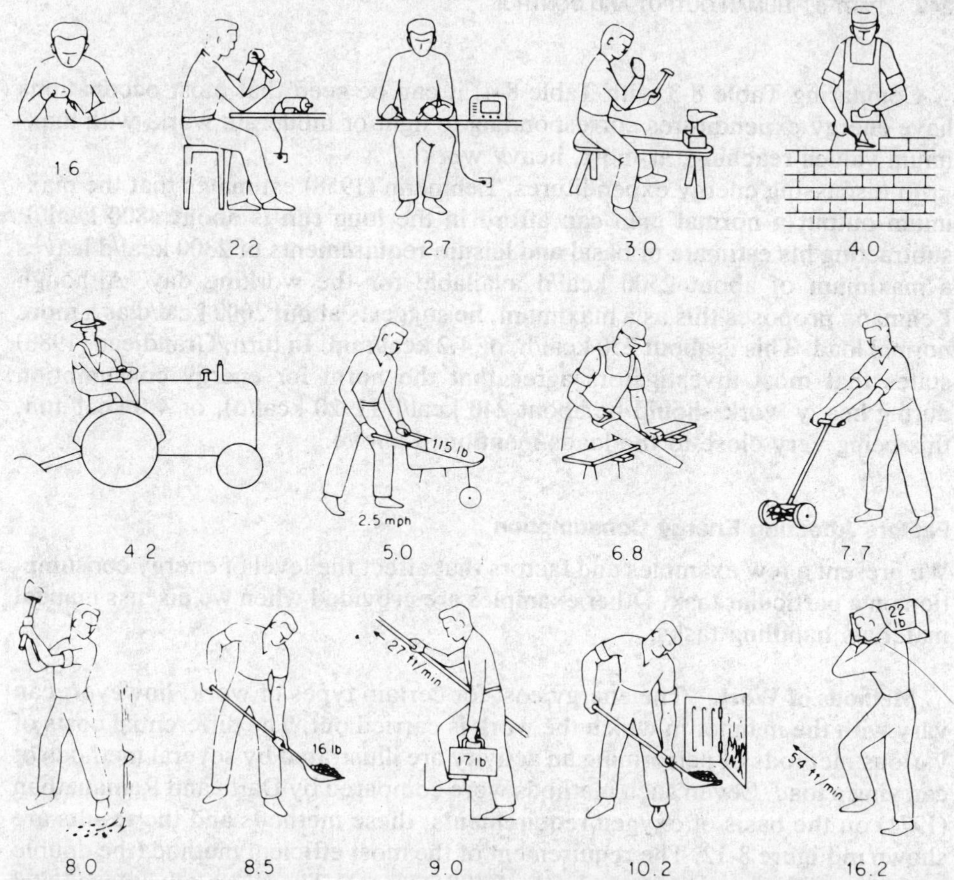

FIGURE 8-11
Examples of energy costs of various types of human activity. Energy costs are given in kilocalories per minute. (*Source: Passmore and Durnin, 1955, as adapted and presented by Gordon, 1957.*)

TABLE 8-4
GRADE OF PHYSICAL WORK BASED ON ENERGY EXPENDITURE LEVEL (ASSUMING A REASONABLY FIT ADULT MALE)

Grade of work	Energy expenditure, kcal/min	Energy expenditure, 8 h (kcal/d)	Heart rate, beats per minute	Oxygen consumption, L/min
Rest (sitting)	1.5	<720	60–70	0.3
Very light work	1.6–2.5	768–1200	65–75	0.3–0.5
Light work	2.5–5.0	1200–2400	75–100	0.5–1.0
Moderate work	5.0–7.5	2400–3600	100–125	1.0–1.5
Heavy work	7.5–10.0	3600–4800	125–150	1.5–2.0
Very heavy work	10.0–12.5	4800–6000	150–180	2.0–2.5
Unduly heavy work	>12.5	>6000	>180	>2.5

Source: Adapted from American Industrial Hygiene Association, 1971. Reprinted with permission by American Industrial Hygiene Association.

Comparing Table 8-3 with Table 8-4, it can be seen that most occupations have energy expenditures corresponding to light or moderate work, with maximum values reaching, at most, heavy work.

In discussing energy expenditures, Lehmann (1958) estimates that the maximum output a normal man can afford in the long run is about 4800 kcal/d; subtracting his estimate of basal and leisure requirements of 2300 kcal/d leaves a maximum of about 2500 kcal/d available for the working day. Although Lehmann proposes this as a maximum, he suggests about 2000 kcal/d as a more normal load. This is about 250 kcal/h, or 4.2 kcal/min. In turn, Grandjean (1988) states that most investigators agree that the norm for energy consumption during heavy work should be about 240 kcal/h (1920 kcal/d), or 4.0 kcal/min, this being very close to the loads mentioned above.

Factors Affecting Energy Consumption

We present a few examples and factors that affect the level of energy consumption on a particular task. Other examples are provided when we discuss manual materials handling tasks.

Methods of Work The energy cost for certain types of work, however, can vary with the manner in which the work is carried out. The differential costs of various methods of performing an activity are illustrated by several methods of carrying a load. Seven such methods were compared by Datta and Ramanathan (1971) on the basis of oxygen requirements; these methods and the results are shown in Figure 8-12. The requirement of the most efficient method (the double pack) is used as an arbitrary base of 100 percent. There are advantages and disadvantages to the various methods over and above their oxygen requirements, but the common denominator of the most efficient methods reported in this and other studies is maintenance of good postural balance, one that affects the body's center of gravity the least.

Work Posture The posture of workers while performing some tasks is another factor that can influence energy requirements. In this regard certain agricultural tasks in particular have to be carried out at or near ground level, as in picking strawberries. When such work is performed, however, any of several postures can be assumed. The energy costs of certain such postures were measured in a study by Vos (1973) in which he used a task of picking up metal tags placed in a standard pattern on the floor. A comparison of the energy expenditures of five different postures is given in Figure 8-13. This figure shows that a kneeling posture with hand support and a squatting posture required less energy than the other postures. (The kneeling posture, however, precludes the use of one hand and might cause knee discomfort over time.) On the basis of another phase of the study, it was shown that a sitting posture (with a low stool) is a bit better than squatting, but a sitting posture is not feasible if the task requires moving from place to place. Although this particular analysis dealt

FIGURE 8-12
Relative oxygen consumption of seven methods of carrying a load, with the double-pack method used as a base of 100 percent. This illustrates that the manner in which an activity is carried out can influence the energy requirements. (*Source: Adapted from Datta and Ramanathan, 1971.*)

with postures used on ground-level tasks, differences in postures used in certain other tasks also can have differential energy costs.

Work Rate Still another factor that affects the workload is the work rate, or pace. Given any particular task, it is generally true that—at some specific pace—the task can be carried out over an extended time without any appreciable physiological cost. For such reasonable paces the heart rate typically increases somewhat and maintains that level for some time. However, if the rate of work is increased appreciably, the workload can cause a continued rise in heart rate and in other physiological costs. This is illustrated in Figure 8-14, which shows the heart rate for two work rates. Aside from the differences in the increases in the heart rates, this figure shows marked differences in the heart rate recovery curves.

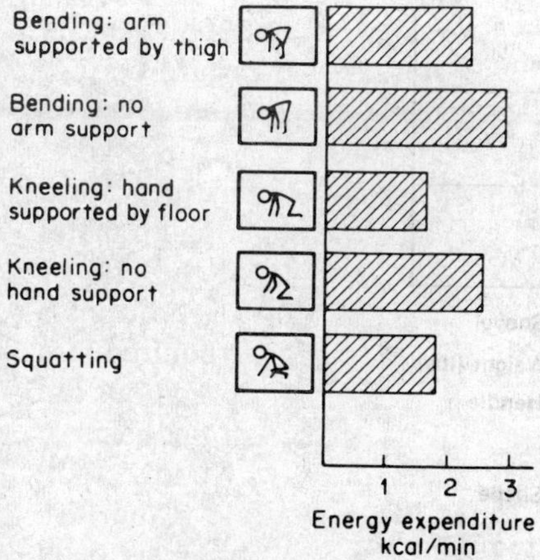

FIGURE 8-13
Human energy expenditures (kilocalories per minute) for five postures used in the task of picking up light objects from ground level. (*Source: Adapted from Vos, 1973, Fig. 5.*)

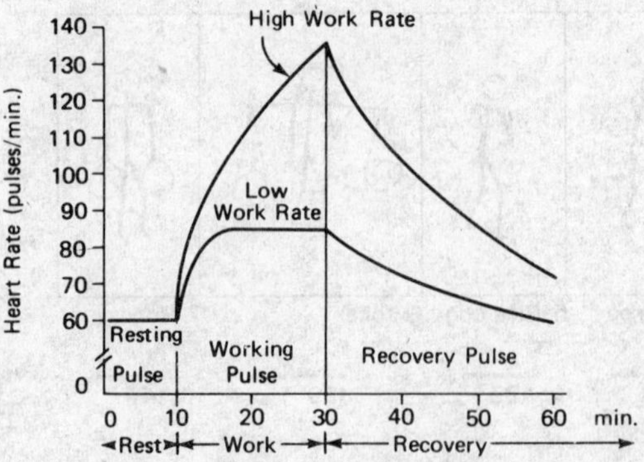

FIGURE 8-14
Heart rate for two different work rates. At the higher work rate the heart rate continues to increase, whereas at the lower work rate the heart rate levels off at a plateau, or steady state. (*From Grandjean, 1988, Fig. 78, p. 95.*)

Tool Design The design of a tool can also influence energy expenditure as well as the amount of work accomplished. We presented the effect of the design of a saw blade on work efficiency in Figure 8-9. Another example is shown in Figure 8-15 illustrating the effect of shovel design on energy expenditure.

KEEPING ENERGY EXPENDITURE WITHIN BOUNDS

If those who are concerned with the nature of human work activities want to keep energy costs within reasonable bounds, it is necessary for them to know both what those bounds should be and what the costs are (or would be) for specific activities (such as those shown in Figure 8-11).

FIGURE 8-15
Efficiency of shoveling (energy expenditure per shovelful) as a function of shovel design. (*Source: Freivalds, 1986, Fig. 4.*)

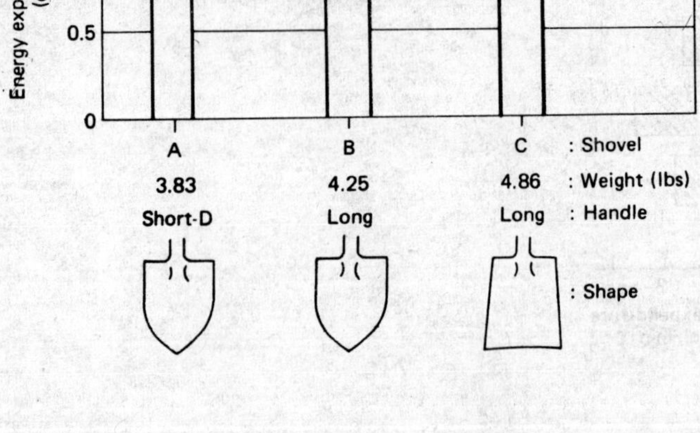

Recommended Limits

Various upper limits of energy expenditure for daily work have been proposed. Michael, Hutton, and Horvath (1961) and Blink (1962), for example, suggest that workloads should not exceed about 35 percent of a person's maximum aerobic power over an 8-h day. Ayoub and Mital (1989) recommend that over an 8-h day, energy expenditure should average less than 5.0 kcal/min for men and 3.35 kcal/min for women. These values correspond to 33 percent of maximum aerobic power for men and women, respectively. For 4 h of work, Ayoub and Mital recommend a limit of 6.25 and 4.20 kcal/min as the upper limit for men and women, respectively.

Some authors have recommended workload limits based on heart rate. For example, Brouha (1967) and Suggs and Splinter (1961) recommend that the mean heart rate should not exceed 115 beats/min. Snook and Irvine (1969) recommend a maximum average heart rate of 112 beats/min for leg work and 99 beats/min for arm work.

Work-Rest Cycles

If the overall level of workload cannot be maintained within recommended limits, then there must be rest to compensate for the excess requirements demanded. There have been several suggestions for determining the amount of rest required for various levels and durations of work [see Ayoub and Mital (1989) for a review]. Figure 8-16 presents one such methodology. For a given grade of work and duration, the required rest allowance (expressed as a percentage of work time) can be determined. As an example, a person working for 10 min at a very heavy task (say 8 kcal/min) would require a rest allowance of 80 percent (80 percent of 10 min), or 8 min.

Another approach is suggested by Murrell (1965). He provides a simple formula for estimating the total amount of rest required for any given work activity. The formula is:

$$R = \frac{T(W - S)}{K - 1.5}$$

Where R = rest required in minutes
T = total work time in minutes
W = average energy consumption of work in kcal/min
S = recommended average energy expenditure in kcal/min (usually taken as 4 or 5 kcal/min)

The value of 1.5 in the denominator of the formula is an approximation of the energy expenditure (in kcal/min) at rest, which is slightly higher than the basal metabolic rate. If we apply Murrell's formula to our example above using a conservative standard (S) of 4 kcal/min, we find that after 10 min of work at 8 kcal/min, the person would need only 6 min of rest, this being somewhat less than predicted using Figure 8-16. Ayoub and Mital (1989) present other procedures for determining rest allowances, some more complicated than the two presented here, that yield still different estimates of the rest required.

The total amount of rest required for a given period of work (or the work/rest ratio) is important; however, the actual duration of the work before a rest period is given may be of even greater importance for adequate recovery. Consider the example shown in Figure 8-17. A constant ratio of 1:1 between work and rest is maintained (i.e., the rest allowance is always 100 percent), but the actual duration of the work period before rest is taken is varied. Shown are the heart rates and blood lactic acid levels during the work periods. Shorter work periods with shorter rest periods result in better physiological recovery and lower stress levels than longer work periods with longer rest periods. This is probably a good general principle to keep in mind when scheduling work and rest to maximize recovery and minimize stress.

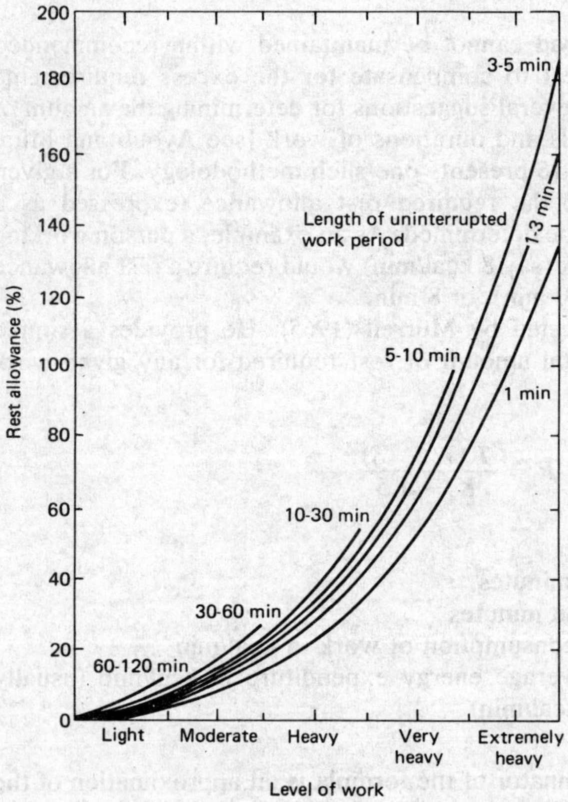

FIGURE 8-16
Rest allowance (as percent of work time) as a function of level of work and length of uninterrupted work period. Levels of work correspond to the following levels of energy expenditure (kcal/min): light = 1–2.5; moderate = 2.6–3.75; heavy = 3.8–6.0; very heavy = 6.1–10.0; extremely heavy > 10.0. (*Source: Kodak, 1986, Fig. 11-2; based on data from Rohmert, 1973*a,b.)

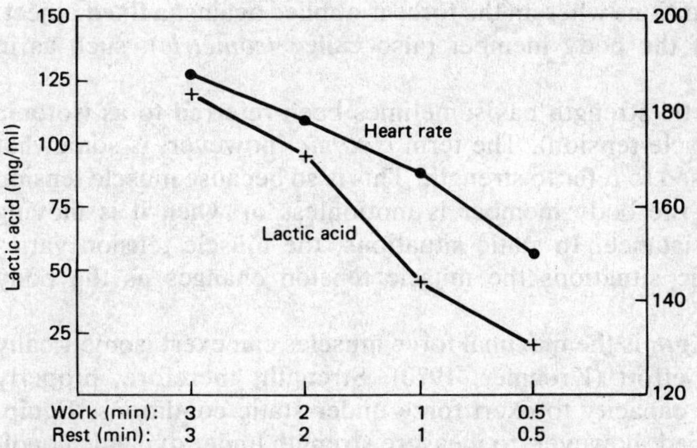

FIGURE 8-17
Effect of length of work and rest period on recovery and physiological strain. In all cases, equal periods of work and rest were performed. The shorter the periods, the greater the recovery and the lower the physiological strain. (*Adapted from Simonson, 1971, Chapter 18. Courtesy of Charles C. Thomas, Publisher, Springfield, IL.*)

Exercise Training

We know that exercise training makes us more physically fit and better able to perform physical work without fatigue. Astrand and Rodahl (1986) discuss in depth the effects of exercise on the body and features of efficient exercise programs. In general, exercise training can increase maximum aerobic power, reduce heart rate, reduce blood pressure, and increase muscle strength and endurance. All of these changes reduce the level of strain experienced for a given level of workload stress.

STRENGTH AND ENDURANCE

As we discussed in previous sections, work capacity is limited by the cardiovascular system's ability to transport fuel and oxygen to muscles. In addition, strength and muscular endurance are also limiting factors in many tasks and activities. (Endurance can be considered, in great part, as just a manifestation of cardiovascular system capability.)

Definition of Strength

Strength is associated with specific muscles or groups of muscles, such as arm, leg, or back. There are two conditions in which strength is important: *dynamic* conditions, wherein the body member (arm, leg, etc.) is actually being moved (also called *isokinetic*), such as in lifting a box or doing an arm curl with a barbell; and *static* conditions wherein the force is applied against a fixed object, with no movement of the body member (also called *isometric*), such as in holding an object.

Dynamic or isokinetic strength has sometimes been referred to as isotonic strength (constant muscle tension). The term *isotonic,* however, is somewhat of a misnomer when used to refer to strength. This is so because muscle tension is not constant when the body member is motionless or when it is moving against a constant resistance. In static situations, the muscle tension varies over time. In dynamic situations the muscle tension changes as the body member moves.

A definition of *strength* is the maximal force muscles can exert isometrically in a single voluntary effort (Kroemer, 1970). Strength, therefore, properly refers to the muscles' capacity to exert force under static conditions. Equipment has been developed, however, to measure strength under dynamic conditions.

Measurement of Strength

Strength is assessed by measuring the force exerted on an external object by a muscle group or groups. It is not possible to actually measure the force within the muscle, and the relationship between measured strength and muscle force is not known.

Static Strength Static strength is assessed by having a subject exert a maximum force against an immovable object. The duration is usually less than 10 s to avoid fatiguing the muscle. The American Industrial Hygiene Association (Chaffin, 1975) recommends that static exertions be maintained for 4 to 6 s, during which the instantaneous maximum and 3-s mean force values be recorded.

The measured strength depends on such factors as posture, angle of the joints, motivation, and the manner in which the force is exerted (e.g., slow buildup versus maximum impulse). Multiple trials are required to increase the reliability of the data collected. Strength measures can vary ± 10 percent within a single day on repeated measurements from the same person.

Dynamic Strength Dynamic strength is difficult to measure because of the effects that acceleration and changes in joint angles have on the force exerted and measured. Dynamic, or isokinetic, strength testing devices have been developed that control the speed of movement to a preset value, regardless of the effort exerted. By holding the velocity of movement constant, such devices eliminate most acceleration effects during the measurement. A record of the instantaneous forces at various positions throughout the movement can be obtained. Figure 8-18, for example, shows the results of an analysis of a dynamic lift at a velocity of 1 m/s (39 in/s) presented as percentages of shoulder height and maximum force.

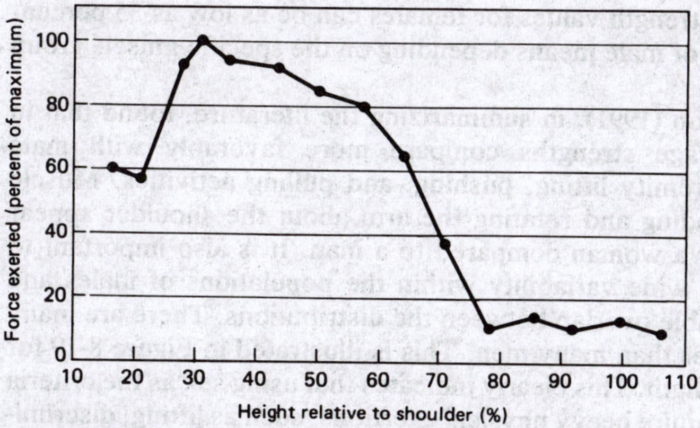

FIGURE 8-18
Dynamic lifting profile for one subject lifting at a velocity of 1 m/s (39 in/s). The values are given as a percentage of shoulder height (55 in; 140 cm) and maximum force exerted (68 lbf; 305 N). (*Source: Adapted from Kodak, 1986, Table 27-1. Reprinted courtesy of Eastman Kodak Co.*)

The speed of movement is an important factor in dynamic strength testing. Slower movements usually result in higher levels of measured strength. For example, Pytel and Kamon (1981) found dynamic strength values were 50 percent higher when lifting at 28.7 in/s (0.73 m/s) than when lifting slightly faster at 38.2 in/s (0.93 m/s).

In most work activities, it is a combination of muscles acting around various joints that produce the force on an external object. The interactions are complex. This, in part, explains the difficulty of using static strength to predict dynamic strength and even using dynamic strength testing to predict force productions during real-world activities.

Personal Factors Affecting Strength

There is wide variability in strength between people, from weight lifters to couch potatoes. Even within an industrial population, the strength of the strongest person can be 6 to 8 times that of the weakest person (Chaffin and Andersson, 1991). Numerous personal factors affect strength including genetics, body dimensions, physical training, and motivation. We, however, briefly discuss two factors, sex and age, that are relevant to job placement decisions on jobs requiring strength.

Sex and Strength Of the many factors that affect strength, sex accounts for the largest differences in mean values of any easily definable population parameter. The rule of thumb, typically cited, is that females' mean strength is approximately two-thirds (67 percent) of males' mean strength (Roebuck, Kroemer, and Thomson, 1975). This is an average value from various muscle groups. Specific mean strength values for females can be as low as 35 percent or as high as 85 percent of male means depending on the specific muscle group measured.

Chaffin and Andersson (1991), in summarizing the literature, found that in general, females' average strengths compare more favorably with male strengths for lower-extremity lifting, pushing, and pulling activities. Muscle exertions involving bending and rotating the arm about the shoulder appear particularly difficult for a woman compared to a man. It is also important to remember that there is wide variability within the populations of males and females, with considerable overlap between the distributions. There are many women who are stronger than many men. This is illustrated in Figure 8-19 for leg and torso lifting strength. This clearly indicates that using sex as the criteria for hiring in jobs that require heavy physical exertions, such as lifting, discriminates against the many women who are capable and endangers many men who are not. Strength testing to identify those capable of performing the job, whether male or female, is a far more equitable basis for screening new hires.

Age and Strength In general, maximum strength reaches its peak between the ages of 25 to 35 years, shows a slow or imperceptible decrease in the forties, and then an accelerated decline thereafter. Figure 8-20 presents strength data

FIGURE 8-19
Distribution of static strength for males and females illustrating the degree of overlap between the two distributions. (*Source: Chaffin, Herrin, and Keyserling, 1978; as redrawn by Chaffin and Andersson, 1991, Fig. 4.12. © American College of Occupational Medicine.*)

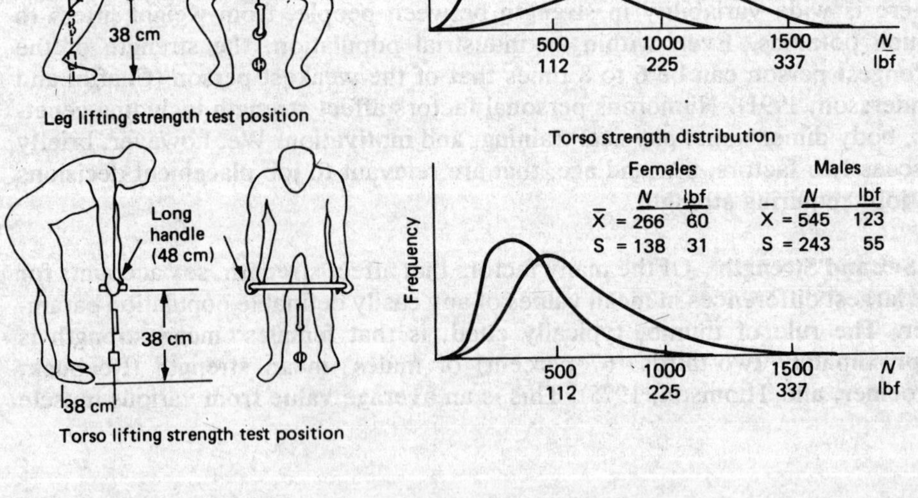

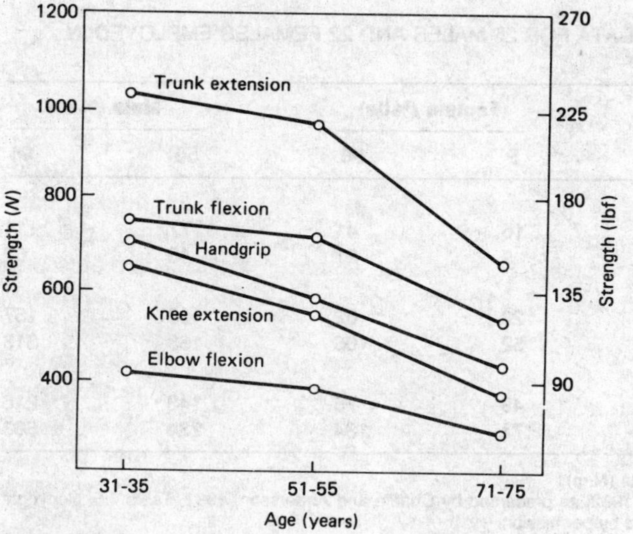

FIGURE 8-20
Maximum isometric strength for five muscle groups as a function of age (cross-sectional study). (*Source: Viitasalo, Era, Leskinen, and Heikkinen, 1985.*)

from three groups of men ages 31–35, 51–55, and 71–75 years. The general decline with age is evident; however, different muscle groups are affected differently. Knee extension and hand grip show the largest percentage declines, while trunk flexion (bending forward) and elbow flexion show the least change. On average, strength at 51–55 years of age is about 80 percent, and at 71–75 years of age it is about 60 percent of what it was at 31–35 years of age. The effects of age are not completely preordained. Physical exercise can increase muscle strength by 30 to 40 percent in adults. However, age inevitably takes its toll.

Examples of Strength Data

Table 8-5 presents various static strength values from male and female workers. Shown are the medians, 5th percentile female, and 95th percentile male values. Another example of strength data is illustrated by Hunsicker (1955): arm strength of males in each of several directions, as depicted in Figure 8-21*a*. Figure 8-21*b* shows the maximum strength of the 5th percentile male. Clearly, pull and push movements are the strongest, but these are noticeably influenced by the angle of the elbow, with the strongest being at 150° and 180°. The difference among the other movements are not great, but what patterns do emerge are the consequence of the mechanical advantages of such movements, considering the angles involved and the effectiveness of the muscle contractions in applying leverage to the body members. Although left-hand data are not shown, the strength of left-hand movements is roughly 10 percent less than those of the right hand.

TABLE 8-5
STATIC MUSCLE STRENGTH DATA FOR 25 MALES AND 22 FEMALES EMPLOYED IN MANUAL JOBS IN INDUSTRY*

	Female (%ile)		Male (%ile)	
Body joint/description	**5**	**50**	**50**	**95**
Elbow				
Flexion (90° arm at side)	16	41	77	111
Extension (70° arm at side)	9	27	46	67
Knee				
Flexion (135° seated)	22	62	100	157
Extension (120° seated)	52	106	168	318
Torso				
Flexion (100° seated)	49	75	143	216
Extension (100° seated)	71	184	234	503

* Measured as strength moments (N·m).
Source: Adapted from Strobbe, 1982; as presented by Chaffin and Andersson, 1991, Table 4.5. Copyright John Wiley & Sons, Inc. Reproduced by permission.

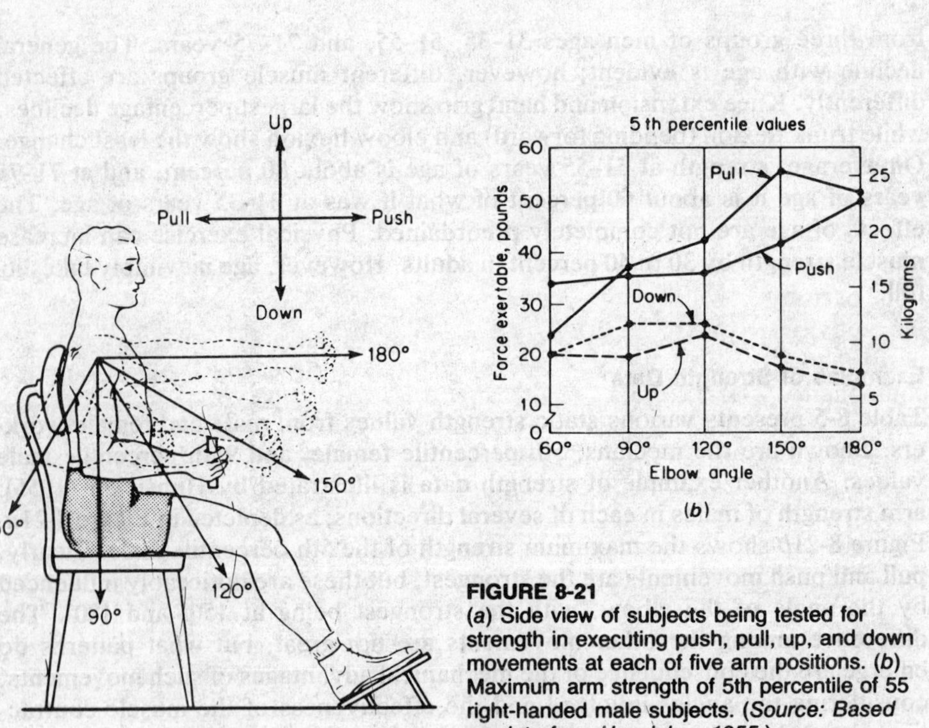

FIGURE 8-21
(*a*) Side view of subjects being tested for strength in executing push, pull, up, and down movements at each of five arm positions. (*b*) Maximum arm strength of 5th percentile of 55 right-handed male subjects. (*Source: Based on data from Hunsicker, 1955.*)

Endurance

If one considers the endurance of people to maintain a given static muscular force, we can all attest from our own experience that such ability is related to the magnitude of the force. This is shown dramatically in Figure 8-22, which depicts the general pattern of endurance time as a function of static force requirements of the task. It is obvious that people can maintain their maximum static effort very briefly, whereas they can maintain a force of around 25 percent or less of their own maximum for a somewhat extended period (10 min or more). The implication of this relationship is fairly obvious—if it is necessary to require individuals to maintain static force over a time, the force required should be well below each individual's own static force capacity.

In the case of repetitive dynamic work, the combination of force and frequency of repetition determines the length of time that the activity can be endured. When subjects perform rhythmic, maximal isometric contractions in time to a metronome, the force generated gradually decreases because of fatigue, but levels off at a value that can be maintained for a long time. Figure 8-23 presents the force (as a percentage of maximum isometric strength) that can be maintained over time in such activities as a function of the number of contractions per minute. (The data are averages from various muscle groups.) At 10 contractions per minute, about 80 percent of the maximum isometric strength could be applied without impairment. At 30 contractions per minute, however, only 60 percent of the maximum could be sustained.

Discussion of Strength and Endurance

The evidence seems to indicate that strength and endurance are substantially correlated. In a study by Jackson, Osburn, and Laughery (1984) dealing with this issue, the subjects were administered certain strength tests and, in addition, performed work sample tests that simulated certain mining and oil field job activities. They found that the strength tests predicted quite well the performance on the work sample tests which involved an "absolute endurance component." Such results suggest that, in the selection of people to perform

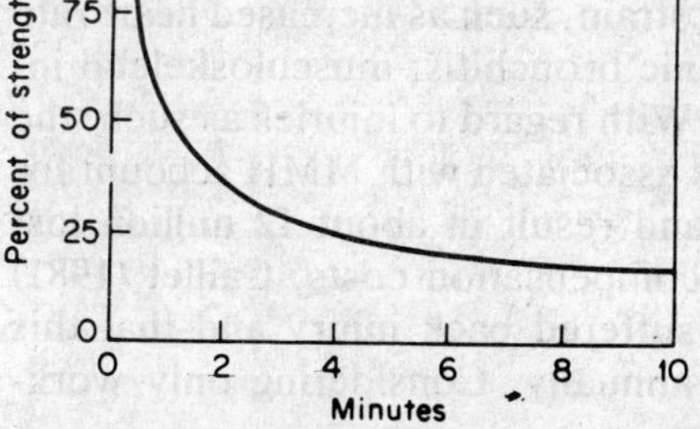

FIGURE 8-22
Endurance time as a function of static force requirements. (*Source: Kroemer, 1970, Fig. 4, as adapted from various sources.*)

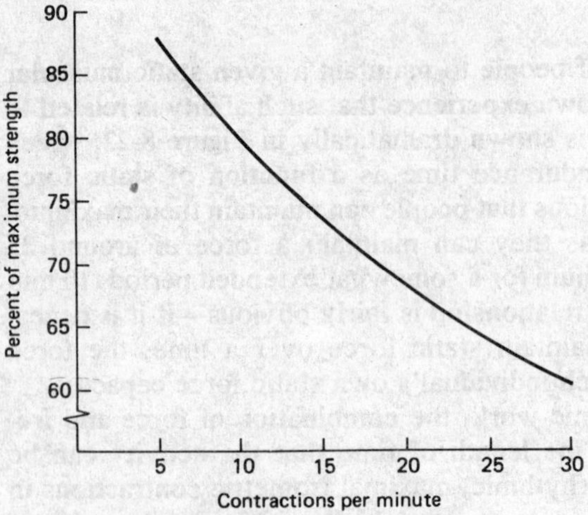

FIGURE 8-23
Percent of maximum isometric strength that can be maintained during rhythmic contractions. Each point is an average for finger, hand, arm, and leg muscles combined. (*Source: Molbech, 1963; as presented by Astrand and Rodahl, 1986, Fig. 3-26. Reproduced with permission of McGraw-Hill.*)

demanding physical work requiring endurance, it would be possible to use strength tests rather than endurance tests (that take more testing time).

Note that a given level of strength and endurance is not fixed forever. Exercise can increase these within limits, these increases sometimes being in the range of 30 to 50 percent above beginning levels.

MANUAL MATERIALS HANDLING

Many jobs and activities in life require manual materials handling (sometimes abbreviated MMH). This includes a wide variety of activities such as loading and unloading boxes or cartons, removing materials from a conveyer belt, stacking items in a warehouse, etc. The physical movements and associated demands involved in such activities are so varied that we can touch on only certain basic aspects. Remember that such factors as individual differences, physical condition, sex, etc., markedly influence the abilities of individuals to perform such activities.

Health Effects of Manual Materials Handling

Various short- and long-term health effects can be attributed to MMH. Some of these are (National Institute for Occupational Safety and Health, 1981) lacerations, bruises, and fractures; cardiovascular strain, such as increased heart rate and blood pressure; muscular fatigue; chronic bronchitis; musculoskeletal injury, especially to the spine; and back pain. With regard to injuries as such, the National Safety Council reports that injuries associated with MMH account for about 25 percent of all industrial injuries and result in about 12 million lost workdays per year and over $1 billion in compensation costs. Caillet (1981) estimates that 70 million Americans have suffered back injury and that this number will increase by 7 million people annually. Considering only work-related back strains and sprains, almost 50 percent of such injuries occur while lifting objects. An additional 9 percent occur while pushing or pulling objects, and about 6 percent occur while holding, wielding, throwing, or carrying objects (Klein, Roger, Jensen, and Sanderson, 1984).

Ayoub and Mital (1989) conclude that despite improved medical care, increased automation in industry, and more extensive use of preemployment exams, only a marginal decline in work injuries has occurred since 1972, but that the costs of such injuries has increased at an alarming rate.

Approaches to Assessing MMH Capabilities

There are three distinct approaches taken for assessing MMH capabilities and for setting recommended workload limits. We briefly discuss each. For more detailed discussion, see Ayoub and Mital (1989) or NIOSH (1981).

Biomechanical Approach Biomechanical approaches view the body as a system of links and connecting joints corresponding to segments of the body such as upper arm (link), elbow (joint), and forearm (link). Principles of physics are used to determine the mechanical stresses on the body and the muscle forces needed to counteract the stresses.

Muscle attachments are close to the joints where they act, but the forces that have to be countered are usually applied some distance from the joint. An example is holding a weight in your hand with your elbow at a 90° angle. The weight of your forearm and the weight in your hand must be countered by the bicep muscle operating around the elbow. The bottom line is that large muscle forces are required to counter relatively small external loads. Holding a 9-lb (4-kg) weight in our hand as described above requires the bicep muscle to exert a force of 74 lb (33.75 kg) to counter the forces.

Researchers have developed various computerized models for determining the forces and torques acting on the body during manual materials handling tasks (for reviews see Kroemer, Snook, Meadows, and Deutsch, 1988; Chaffin, 1987; Chaffin and Andersson, 1991). Figure 8-24 shows an example of a low-back biomechanical model for analyzing compressive (F_{comp}), muscle (F_{musc}), and shear (F_{shear}) forces during static lifting. Fortunately, the explanation of the model is beyond the scope of this book.

In many MMH tasks such as lifting, carrying, pushing, and pulling, significant forces are produced in the lower back, L5/S1 disc (i.e., the disc located between the fifth lumbar and first sacral vertebrae), the site of most MMH back injuries. During MMH tasks, considerable compressive forces can be generated on the vertebral bodies in the L5/S1 region. The strength of the vertebral body itself to withstand compressive forces appears to be the critical variable in determining the compressive limits of the spine. The discs appear to be better able to tolerate compressive forces than do the vertebrae.

The biomechanical approach has been limited to analyzing infrequent MMH tasks. The goal is to limit task demands to be within the strength capacity and

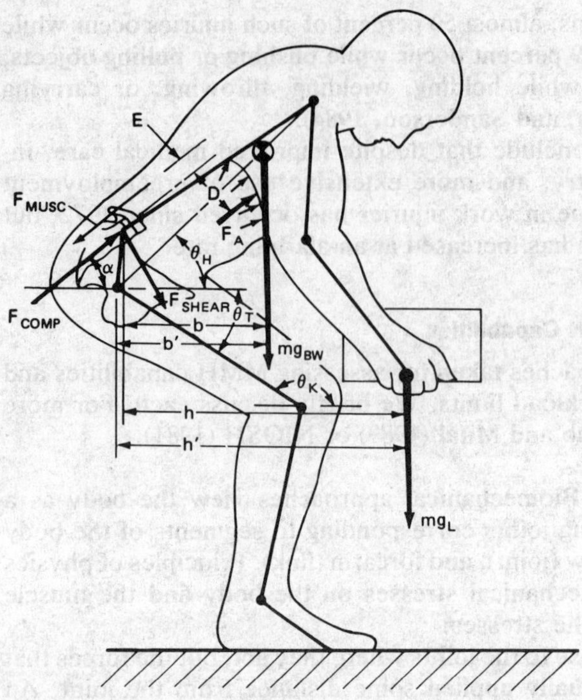

FIGURE 8-24
An example of a simple low-back model of lifting for static coplanar lifting analysis. (*Source: Chaffin and Andersson, 1991, Figure 6.22, as adapted from Chaffin, 1975.*)

compressive force tolerance of the body. It is usually the infrequent lift or push that creates excessive forces. Tasks that are performed repeatedly are somewhat more likely to be done at a level that is within appropriate biomechanical limits.

Physiological Approach The physiological approach is best suited to MMH tasks that are done frequently and over some duration of time. For example, in many instances, people are required to perform an MMH task throughout an entire 8-h shift. The physiological approach is concerned with energy consumption and the stresses acting on the cardiovascular system.

Various researchers have developed equations for predicting the energy costs of various types of MMH tasks (for a review, see Genaidy and Asfour, 1987). The equations, or models, take into account such variables as body weight, weight of load, gender, vertical start and end positions of a lift, dimensions of the load, and frequency of handling. All of the models have limitations and care must be exercised when applying them.

We discussed the physiological costs of work earlier in this chapter and included some examples related to MMH tasks.

Psychophysical Approach The underlying premise of the psychophysical approach is that people integrate and combine both biomechanical and physiological stresses in their subjective evaluation of perceived stress. The use of psychophysics in assessing a lifting task requires that subjects adjust the weight of a load (or in some instances, the frequency of handling a load) to the maximum amount they can sustain without strain or discomfort and without becoming unusually tired, weakened, overheated, or out of breath. The maximum selected is called the *maximum acceptable weight of load (MAWL)*. Ayoub and Mital (1989) consider the psychophysical approach to be one of the most appropriate for setting criteria for MMH tasks.

One disadvantage of the approach is that special controls are necessary to get valid data. The people tested must be representative of the population to which the data will be applied. Repeated testing over days or weeks is recommended, training periods may be required, and even using objects with false bottoms is desirable to reduce visual cues as to the load being handled. MAWLs determined on the basis of a 20 to 45-min trial period overestimate the actual weight people would handle during an 8-h shift. The actual acceptable weight decreases by about 3.4 percent per hour for males and 2 percent per hour for females, from the MAWL determined during the initial trial period (Mital, 1983).

As in the case of the prior approaches, models have been developed to predict MAWL. Genaidy, Asfour, Mital, and Waly (1990) review such models which are equations taking into account such variables as the height of a lift, frequency of lifting, object dimensions, and working time.

Discussion Each of these three approaches results in different recommended load limits that we discuss later in this chapter. In addition, the effect of particular variables may be different depending on the approach used to assess the stress on the body. For example, the stability and distribution of the load being lifted can affect the MAWL, but not necessarily the level of energy expenditure. One lifting posture may be beneficial in terms of the level of biomechanical stresses produced, but may significantly increase the level of energy expenditure. It is therefore important to keep in mind the criteria used and its relevance to the task at hand when applying data and recommendations in the area of manual materials handling.

Lifting Tasks

Lifting tasks make up a large proportion of MMH tasks and are involved in far more back injuries than other types of MMH tasks. For these reasons, we will review a few of the variables that influence the level of stress placed on the body during lifting. Although we deal with these variables one at a time, keep in mind that there are powerful interactions among them.

Horizontal Position of the Load The horizontal position of a load in relationship to the L5/S1 disc is one of the most significant factors affecting the compressive forces experienced at the L5/S1 disc. This is shown in Figure 8-25 for various loads. The compressive force at the L5/S1 disc is about 400 lb

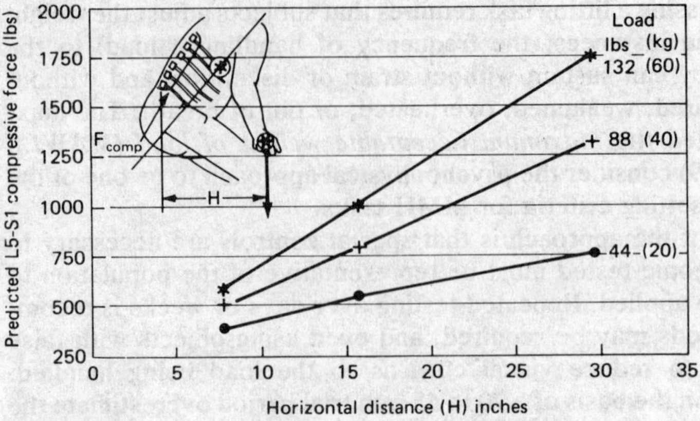

FIGURE 8-25
Effect of weight of load and horizontal distance between the load center of gravity and the L5/S1 disc on the predicted compressive force on the L5/S1 disc. (*Source: Adapted from NIOSH, 1981, Figs. 3.4 and 3.5.*)

(182 kg) when a 44 lb (20 kg) weight is held about 8 in (20 cm) in front of the L5/S1 disc. Move that weight out to 30 in (75 cm) in front of the L5/S1 disc and the compressive force increases substantially to about 750 lb (341 kg).

Height and Range of Lift Common practice provides for categorizing height of lift as follows: (1) floor to knuckle, (2) knuckle to shoulder, and (3) shoulder to reach (from shoulder to maximum reach above). The evidence indicates very strongly that reaches above the shoulder are most demanding in terms of physiological criteria and are least acceptable to workers (Mital, 1984).

Further, Davies (1972) reports that the energy cost of lifting objects from the floor to about 20 in (51 cm) is about half again as much as lifting the same weight from about 20 to 40 in (51 to 102 cm). This is because of the additional effort of raising and lowering the body. On the basis of Davies' analysis, the most efficient lift range is between 40 and 60 in (102 and 152 cm). This suggests that, where feasible, workplaces should be designed to provide for the primary lifting to be within this range.

Method of Lifting from the Floor It has been rather common custom (even perhaps an article of faith) to recommend that lifting from the floor be carried out from a squatting position, with knees and hips bent and the back reasonably straight (sometimes called the *straight-back bent-knee method, squat lift,* or *leg lift*). The squat method generally is in contrast to the *stoop method* (or *back lift*), in which the legs tend to be straight, with the back bending forward and doing most of the lifting. Another lift style is called *free-style,* which resembles a semisquat posture where the load may be rested on the thigh during lifting.

Despite common custom, Ayoub and Mital (1989) conclude that the free-style posture is the least stressful. Garg and Saxena (1979), for example, compared the energy costs of these methods of lifting as a function of maximum acceptable workload. Figure 8-26 shows that energy costs were lowest for the free-style lift, especially at the higher levels of workload.

In general, comparing squat and stoop lifting, squat lifting with a straight back results in lower biomechanical stresses on the lower back (Anderson and Chaffin, 1986). There is one caveat, however: the load being lifted must fit between the knees. If the load is so large that it must be lifted in front of the knees, the horizontal distance from the load to the L5/S1 disc becomes excessive and greatly increases the compressive forces on the low back. In a stoop lift, persons can stand over the load with their toes touching the load, thereby minimizing the horizontal distance and the compressive force. Loads to be lifted should be packaged in small containers to permit squat lifting with the load between the knees.

With respect to energy expenditure, a stoop lift is less stressful than a squat lift, as shown in Figure 8-26. This is because with a squat lift the entire body must be raised each time the load is lifted. In addition, many people do not have the strength in their legs to lift both the load and their body at one time and hence they cannot effectively use the squat lift.

Frequency of Lifting Everyday experience tells us that we can tolerate occasional exertion, as in lifting, much better than frequent exertions. This has been confirmed by numerous investigations. We present two illustrations. Mital (1984) had women and men with considerable industrial work experience perform various types of lifting tasks. Using the psychophysical approach he determined the MAWL that the subjects believed they could lift at various frequencies (lifts per min) if the rate were continued over a regular 8-h workday. Figure 8-27 presents that data as a percentage of the MAWL selected for a frequency of 1 lift per min. At 12 lifts per minute, for example, the MAWL was only 70 to 80 percent of what it was for 1 lift per min.

Genaidy and Asfour (1989) had men squat-lift various loads from the floor to table height at various lifting frequencies. They continued at the task until they

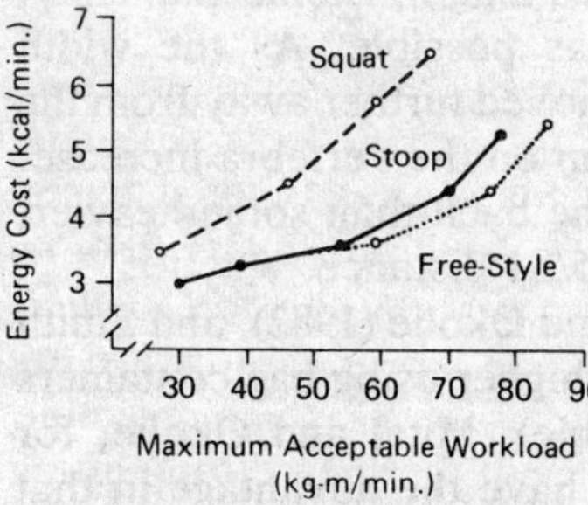

FIGURE 8-26
Energy costs of lifting a load from the floor as related to maximum acceptable workload based on the judgments of the subjects. (*From Garg and Saxena, 1979, Fig. 2, p. 899. Reprinted with permission of American Industrial Hygiene Association Journal.*)

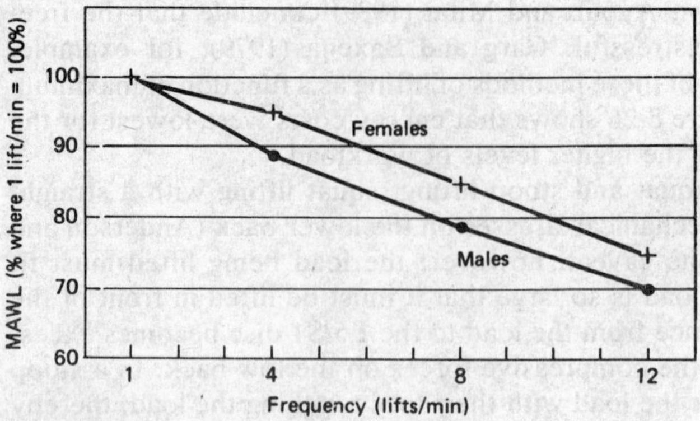

FIGURE 8-27
Relationship between frequency of lifting and maximum acceptable weight of load (MAWL) over an 8-h work day. The values for 1 lift/min were used as the base and equal 100 percent. (*Source: Mital, 1984, Figs. 3 and 4.*)

could no longer lift the load at the prescribed frequency. This was taken as endurance time. Figure 8-28 presents the results. There is a significant decline in endurance time as the frequency of lifting is increased. It should also be pointed out that as frequency increases, fewer total lifts are made during the endurance time. For example, consider a 9-lb (20-kg) load, at 4 lifts per min. Endurance time is 292 min, and 1171 lifts are made. At 10 lifts per min, endurance time is only 27 min, and therefore, only 272 lifts are made. Another example of the tortoise and hare—speed isn't everything.

Object Characteristics Various characteristics of the object being lifted affect the biomechanical, physiological, and psychophysical stresses experienced. Such characteristics include the following:

• *Object size:* The length (side to side) and width (fore and aft) of the object being lifted influence the MAWL, energy expenditure, and spinal stresses. In terms of energy expenditure, Ayoub and Mital (1989) recommend any increase in container volume should be accomplished by first increasing its height, then its width up to a limit of 20 in (50 cm), and finally its length. Biomechanically, the width of the container should be as small as possible. As the width increases, the center of gravity (CG) of the load is moved further away from the L5/S1 disc and compressive and sheer forces action on the vertebra increase. Light, large objects can create greater forces on the back than some heavier, but smaller objects due to the difference in CG–L5/S1 distance.

• *Object shape:* Garg and Saxena (1980), Mital and Okolie (1982), and Smith and Jiang (1984) have all found that the MAWL is higher using bag containers (collapsible) than when using boxes (noncollapsible). Mital and Okolie, for example, reported an 18 percent difference. Bags have the advantage in that

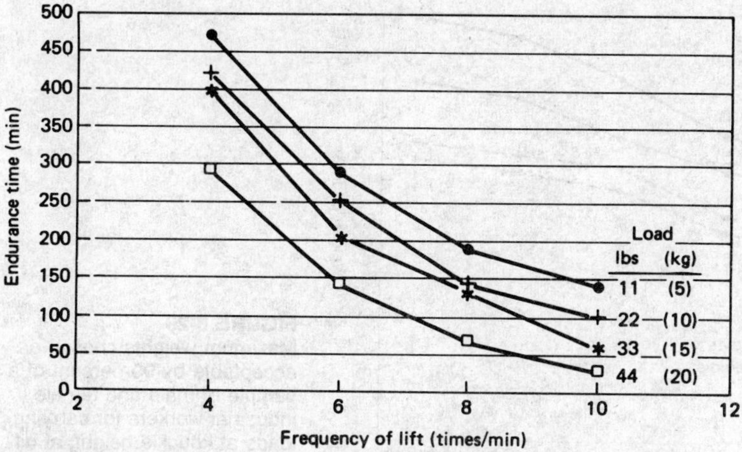

FIGURE 8-28
Endurance time for squat-lifting various loads at various lift frequencies. Lift was from floor to table height. (*Source: Genaidy and Asfour, 1989, Fig. 1.*)

they can be carried closer to the body, thereby reducing the CG–L5/S1 distance.

- *Load distribution and stability:* Ayoub and Mital (1989) point out that when the CG of a load falls outside the centerline of the body, the load causes a lateral bending moment on the lumbar spine resulting in rotation of each vertebra on its adjacent vertebra. Such torsional stresses from asymmetric loading reduce the MAWL, although they do not affect the level of energy expenditure. A shifting CG, as occurs when handling liquids when the container is not full, can reduce MAWLs by as much as 31 percent (Mital and Okolie, 1982).
- *Handles:* The general and consistent finding from the literature is that handling loads with handles is safer and less stressful than handling such loads without handles. The MAWL for boxes without handles is about 4 to 12 percent less than when handles are provided (Garg and Saxena, 1980).

Carrying Tasks

As with lifting tasks, there are marked differences in the weights people find acceptable, depending on the frequency with which the activity is carried out. Aside from individual differences, there are systematic differences between males and females. An example, Figure 8-29 presents data for males and females on the maximum weights reported as acceptable by 90 percent of the subjects for two distances of carry. The effects of frequency are particularly noticeable, with carrying distance having an additional effect on the level of acceptable loads. Figure 8-29 presents data for carrying at knuckle height.

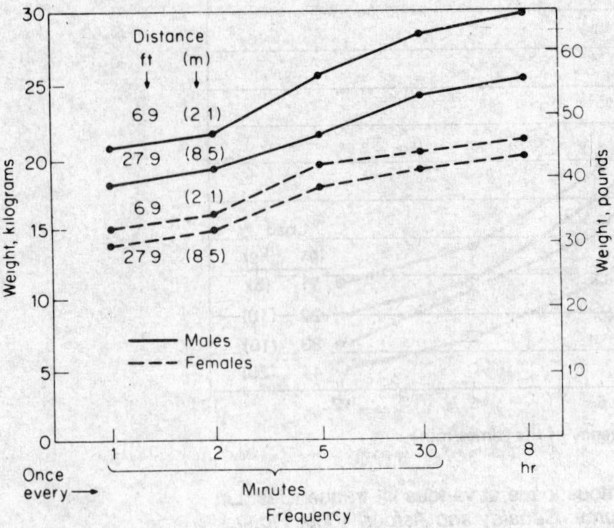

FIGURE 8-29
Maximum weights chosen as acceptable by 90 percent of a sample of male and female industrial workers for carrying loads at knuckle height, at different frequencies, for two distances. (*Source: Snook, 1978, Table 10, p. 980.*)

When the load is carried at elbow height, the acceptable weight is somewhat lower.

Dutta and Taboun (1989), using a physiological approach, studied the effects of various factors on lifting and carrying loads. Figure 8-30 presents oxygen consumption data as a function of the load handled, the distance carried (at 2.5 mi/h), and the frequency of handling. An increase in any of these variables results in an increased level of energy expenditure (i.e., increased oxygen consumption). The largest effect is associated with increasing the frequency of handling.

Pushing Tasks

Additional data on manual handling tasks from Snook deal with acceptable levels of weights in pushing tasks. The data shown in Figure 8-31 also reveal systematic differences in the maximum acceptable loads depending on the frequency with which the tasks are performed, with appreciable differences in distances pushed, but only slight sex differences (in fact, for the shortest distance there was no difference by sex). This example represents data for pushing at average shoulder height. Pushing at elbow or knuckle height appears not to influence the levels reported as acceptable.

RECOMMENDED LIMITS FOR MMH TASKS

Setting limits for MMH tasks is a tricky affair. The limits depend on the criteria one selects. Each of the three approaches to assessing MMH stress have provided recommended limits for such tasks. Unfortunately, the limits recommended are not consistent, and even within one approach, different authors

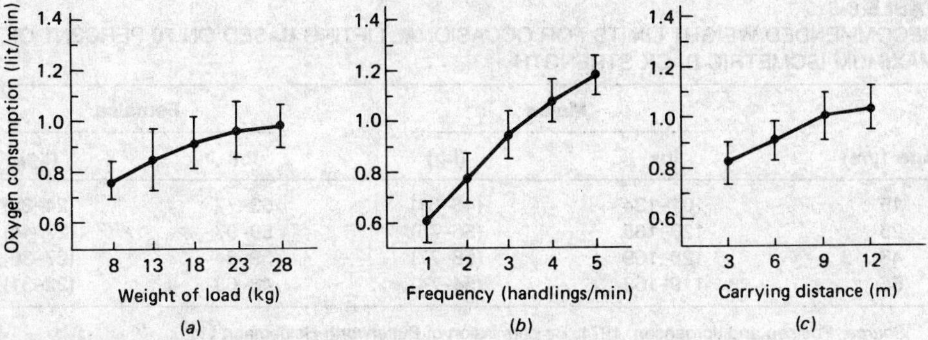

FIGURE 8-30
Effect of weight of load (*a*), frequency of handling (*b*), and carrying distance (*c*) on oxygen consumption for a lifting and carrying task. Data are averaged across values of the other variables. Walking speed was 2.5 mi/h (4 km/h). (*Source: Dutta and Taboun, 1989, Figs. 7, 9, and 11.*)

make somewhat different recommendations. In general, recommended limits based on biomechanical criteria are best applied to infrequent MMH tasks. Limits based on physiological criteria are best applied to repetitive and longer-duration tasks. The psychophysical approach can be used in either situation depending on how the data were gathered. We briefly review the three types of recommended limits. Most of the recommendations, however, apply to lifting tasks because most MMH back injuries occur during that activity.

Biomechanical Recommended Limits

Biomechanical recommended limits focus principally on two factors: strength and compressive forces on the spine. Poulsen and Jorgensen (1971), for example, propose weight limits for occasional lifts (1 to 2 lifts per h) based on 70 percent of the maximum isometric back strength and 50 percent for repeated lifting. Table 8-6 presents the recommendations for occasional lifting.

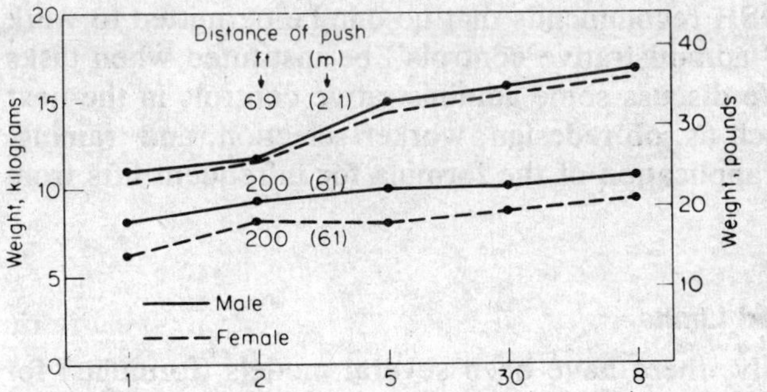

FIGURE 8-31
Maximum weights chosen as acceptable by 90 percent of a sample of female and male industrial workers for a pushing task at shoulder height, at different frequencies, for two distances. (*Source: Snook, 1978, Tables 5, 6, 7, 8, pp. 975–978.*)

TABLE 8-6
RECOMMENDED WEIGHT LIMITS FOR OCCASIONAL LIFTING BASED ON 70 PERCENT OF MAXIMUM ISOMETRIC BACK STRENGTH

	Males		Females	
Age (yrs)	lbs	(kg)	lbs	(kg)
15	101–134	(46–61)	53–77	(24–35)
25	123–165	(56–75)	59–97	(27–44)
45	128–169	(58–77)	59–84	(27–38)
55	119–163	(54–74)	48–68	(22–31)

Source: Poulsen and Jorgensen, 1971. By permission of Butterworth-Heinemann Ltd.

There is considerable disagreement as to the cause of low-back pain and the genesis of ruptured or slipped discs. Despite this, there is consensus that the compressive force acting on the L5/S1 disc is a major factor and a primary measure of the stress to the low back during lifting. The strength of the vertebrae to resist compressive forces is in large part a function of the age of the person, cross-sectional area of the vertebrae, and the bone-mineral content of the body. From various sources of data, NIOSH (1981) concludes that tasks generating more than 1430 lb (650 kg) of compressive force on the low back are hazardous to all but the healthiest of workers. In terms of specifying a recommended level, however, NIOSH proposes a much lower level of 770 lb (350 kg) as an upper-limit design standard.

Based on these criteria of compressive force on the low back (L5/S1 disc), NIOSH (1981) developed a guideline (in the form of a formula) for determining the acceptability of a lifting task. The formula takes into account the following variables: horizontal position of the load forward of the ankles, vertical position of the load at the start of the lift, vertical distance traveled during the lift, and the average frequency of lifting. The complete formula is presented in Appendix C. Two limits are defined in the NIOSH guidelines: the *action limit (AL)* and the *maximum permissible limit (MPL)*, which equals 3 times the AL.

The criteria used to set the action limit and maximum permissible limit are outlined in Table 8-7. NIOSH recommends that no one be permitted to work above the MPL and that "administrative controls" be instituted when tasks exceed the action limit. We discuss some administrative controls in the next section of this chapter, such as job redesign, worker selection, and training. Figure 8-32 illustrates the application of the formula for infrequent lifts from floor to knuckle height.

Physiological Recommended Limits

As we discussed previously, there have been several models (formulas) for predicting energy expenditure for repetitive lifting tasks. Unfortunately, the

TABLE 8-7
CRITERIA FOR SETTING NIOSH LIFTING RECOMMENDATIONS*

Criteria	Action limit	Maximum permissible limit
Increase in injuries	Moderate	Significant
Compressive force on L5-S1 disc	770 lb (350 kg)	1430 lb (650 kg)
Energy expenditure	3.5 kcal/min	5.0 kcal/min
Percent with strength capability:		
Males	99%	25%
Females	75%	<1%

* Maximum permissible limit = 3 × action limit.

various formulas predict different levels of energy expenditure for the same set of conditions. In applying any of the formulas, some limit on the rate of energy expenditure must be set. Some investigators suggest 33 percent of maximum aerobic power (MAP) as the maximum average 8-h level of energy expenditure. The MAP for healthy young adult males is about 3 L of oxygen/min, or 15 kcal/min. One-third of this is 5 kcal/min.

FIGURE 8-32
Levels of risk for lifting tasks as related to horizontal location of load and weight lifted for infrequent lifts from floor to knuckle height. (*This figure is derived from a guideline formula developed by NIOSH, 1981.*)

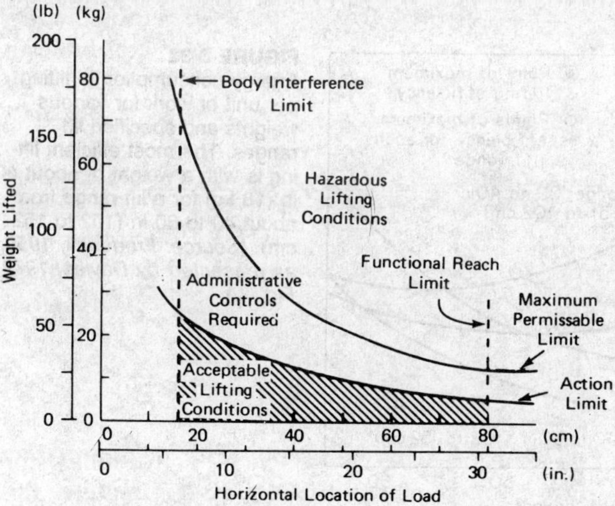

To illustrate physiological models, we present an early model developed by Frederick (1959). The formula for computing energy expenditure is:

$$E = F \times a \times W \times C/1000$$

Where E = energy expenditure (kcal/h)
F = frequency of lifting (lifts/h)
a = vertical lifting range (ft)
W = weight to be lifted (lb)
C = energy consumption (g·cal/ft·lb) from Figure 8-33

The formula only applies for lifts within a range specified in Figure 8-33. Figure 8-33 also indicates the points of maximum efficiency for each lifting range. In general, the point of maximum efficiency occurs at higher weights, the lower the lifting range. Frederick, by the way, recommends that energy expenditure should not exceed 3.33 kcal/min (200 kcal/h) for an average male working a full shift. This is somewhat lower than recommendations made by others.

Psychophysical Recommended Limits

The maximum acceptable weight of load (MAWL) for a particular lifting task is considered the recommended load limit for that situation. Snook and Ciriello (1991) present extensive tables of MAWL values for male and female industrial workers for lifting and lowering various sized boxes. Table 8-8 presents 50th percentile MAWL values for infrequent lifting and lowering of a 13 in (34 cm) (fore-aft dimension) box a distance of 20 in (51 cm).

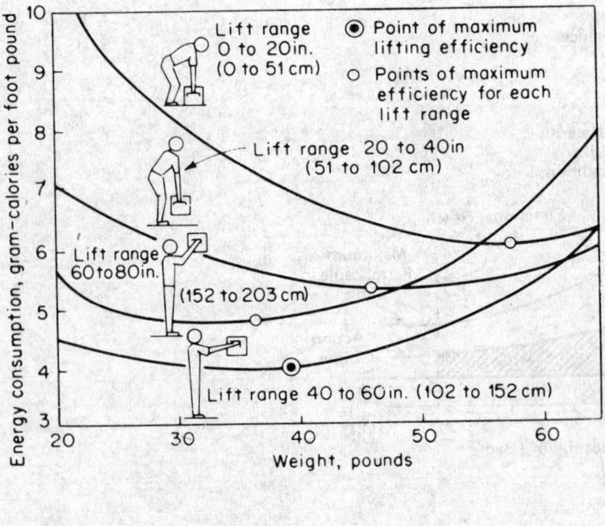

FIGURE 8-33
Energy consumption in lifting per unit of work for various weights and specified lift ranges. The most efficient lifting is with a weight of about 40 lb (18 kg) for a lift range from about 40 to 60 in (102 to 152 cm). (*Source: Frederick, 1959, as presented by Davies, 1972, Fig. 3.*)

TABLE 8-8
MAXIMUM ACCEPTABLE WEIGHTS FOR MALES AND FEMALES PERFORMING INFREQUENT LIFTING AND LOWERING TASKS

	1 per 30 min		1 per 8 h	
Task	**lb**	**(kg)**	**lb**	**(kg)**
		Lifting		
Floor to knuckle				
Males	86	(39)	101	(46)
Females	44	(20)	59	(27)
Knuckle to shoulder				
Males	73	(33)	79	(36)
Females	34	(17)	42	(19)
		Lowering		
Knuckle to floor				
Males	92	(42)	117	(53)
Females	44	(20)	59	(27)
Shoulder to knuckle				
Males	77	(35)	95	(43)
Females	40	(18)	48	(22)

Source: Adapted from Snook and Ciriello, 1991.

REDUCING THE RISK OF MMH OVEREXERTION

Reducing the risk of overexertion requires attacking the problem on multiple fronts. We briefly discuss job design, work selection, and work training as tools for reducing such risks.

Job Design

Of course, the best solution to an MMH-related problem is to eliminate the need for manually handling the materials. This can often be accomplished by providing mechanical lifting aids such as lift tables, hoists, cranes, etc. The work area can sometimes be modified so that all materials are provided at the work level. Where MMH cannot be eliminated, Ayoub and Mital (1989) suggest, among other things, the following to reduce the demands of the job:

1 Decrease the weight of the objects handled (e.g., order supplies in smaller-quantity containers).

2 Use two or more people to move heavy or large objects.

3 Change the activity; for example, pull or, better yet, push rather than carry.

4 Minimize horizontal distances between start and end of the lift (i.e., minimize carrying distance).

5 Stack materials no higher than shoulder height.

6 Keep heavy objects at knuckle height.
7 Reduce frequency of lifting.
8 Incorporate rest periods.
9 Incorporate job rotation to less strenuous jobs.
10 Design containers with handles that can be held close to the body.

Worker Selection

When it is not feasible or it is impractical to redesign a job to reduce MMH-related stress, a strategy of preemployment screening should be used. The idea is that once an employee's capacity has been estimated, an attempt can be made to assign him or her to a job that does not exceed that capacity.

Various types of tests have been used, including: back x-rays (the value of x-rays for predicting future back injuries is much debated), strength testing (by itself, this is not a sufficient measure of worker capacity, but it can be valuable for selecting workers for jobs requiring high levels of force production), and medical exams (of questionable predictive value, but recommended if a person is required to exceed biomechanical or physiological recommended limits).

Whatever preemployment screening procedures are used, they should have the following characteristics (NIOSH, 1981): safe to administer; give reliable, quantitative values; be related to specific job requirements; be practical to administer; and should predict risk of future injury or illness.

Training

There is considerable controversy concerning the effectiveness of how-to-lift or safe-lift training programs for reducing MMH-related injuries. Despite widespread understanding of safe-lift principles* among industrial workers, the number of MMH-related injuries has not decreased significantly. Actually, there are several lifting techniques, each having its merits for some situations (Jones, 1985). As Ayoub and Mital (1989) conclude: "The most useful rule regarding safe lifting is that there is no single, correct way to lift." The best lifting techniques must be tailored to the specifics of the task. The problem is that many tasks involve changing parameters and expecting workers to assess every lift and select the safest technique before performing it may not be practical. This is not to say that such training is of no value, but rather to stress that it is not a cure-all for MMH-related injuries. Such training coupled with MMH hazard recognition training, refresher courses, and reinforcement by supervisors probably can reduce MMH-related risks.

Strength, flexibility, and fitness training appears to have a positive effect on injury rates (Congleton, 1983; Snook and White, 1984). Despite these findings,

* Maintain a straight back, use leg muscles, keep the load as close to the body as possible, use smooth body motion, avoid jerking, and turn the feet rather than twisting the torso.

however, Snook and White consider the effectiveness of such training as a technique for back injury prevention to be debatable. They conclude this because most of the studies cited as evidence suffer from various epidemiologic problems and do not incorporate adequate experimental controls.

Ayoub and Mital (1989, p. 273) provide a prudent conclusion worth quoting:

> Enough evidence is available in support of training program effectiveness to warrant its further employment, *provided those programs are conducted in conjunction with ergonomic job design and employee selection procedures*. (emphasis added)

DISCUSSION

Physical work and manual materials handling will continue to be part and parcel of the working world. Undoubtedly, all heavy physical labor and stressful manual materials handling tasks cannot be eliminated. However, much can be done to reduce the level of risk to which many workers are exposed. Identifying potentially hazardous jobs, applying sound human factors principles to redesigning the work, and training and selecting the workers will go a long way toward reducing injuries and illnesses associated with heavy physical labor.

REFERENCES

American Industrial Hygiene Association (AIHA) (1971). *Ergonomic guide to assessment of metabolic and cardiac costs of physical work*. Akron, OH: AIHA.

Anderson, C., and Chaffin, D. (1986). A biomechanical evaluation of five lifting techniques. *Applied Ergonomics*, 17(1), 2–8.

Astrand, I., Astrand, P., Hallback, I., and Kilbom, A. (1973). Reduction in maximal oxygen uptake with age. *Journal of Applied Physiology*, 35, 649.

Astrand, I., Fugelli, P., Karlsson, C., Rodahl, K., and Vokac, Z. (1973). Energy output and work stress in coastal fishing. *Scandinavian Journal of Clinical Laboratory Investigations*, 31, 105.

Astrand, P., and Rodahl, K. (1977). *Textbook of work physiology* (2d ed.). New York: McGraw-Hill.

Astrand, P., and Rodahl, K. (1986). *Textbook of work physiology* (3d ed.). New York: McGraw-Hill.

Ayoub, M., and Mital, A. (1989). *Manual materials handling*. London: Taylor & Francis.

Blink, B. (1962). The physical working capacity in relation to working time and age. *Ergonomics*, 5, 25–28.

Borg, G. (1985). *An introduction to Borg's RPE scale*. Ithaca, NY: Movement Publications.

Brouha, L. (1967). *Physiology in industry*. New York: Pergamon.

Bureau of Labor Statistics (1982). *Back injuries associated with lifting* (Bulletin 2144). Washington, DC: U.S. Department of Labor.

Caillet, R. (1981). *Low back pain syndrome*. Philadelphia: FA Davis.

Chaffin, D. (1975). Ergonomics guide for the assessment of human static strength. *American Industrial Hygiene Association Journal,* 35, 505–510.

Chaffin, D. (1987). Biomechanical strength models in industry. In American Conference of Governmental Industrial Hygienists (ed.), *Ergonomic interventions to prevent musculoskeletal injuries in industry*. Chelsea, MI: Lewis Publishers.

Chaffin, D., and Andersson, G. (1991). *Occupational biomechanics* (2d ed.). New York: Wiley.

Chaffin, D., Herrin, G., and Keyserling, W. (1978). Preemployment strength testing. *Journal of Occupational Medicine,* 20(6), 403–408.

Congleton, J. (1983). *Design and evaluation of a neutral posture chair*. Ph.D. dissertation, Texas Tech University, Lubbock, TX.

Datta, S., and Ramanathan, N. (1971). Ergonomics comparison of seven modes of carrying loads on the horizontal plane. *Ergonomics,* 14(2), 269–278.

Davies, B. (1972). Moving loads manually. *Applied Ergonomics,* 3(4), 190–194.

Dutta, S., and Taboun, S. (1989). Developing norms for manual carrying tasks using mechanical efficiency as the optimization criterion. *Ergonomics,* 32(8), 919–943.

Edholm, O. (1967). *The biology of work*. New York: McGraw-Hill.

Fleishman, E., Gebhardt, D., and Hogan, J. (1984). The measurement of effort. *Ergonomics,* 27(9), 947–954.

Frederick, W. (1959). Human energy in manual lifting. *Modern Materials Handling,* 14(3), 74–76.

Freivalds, A. (1986). The ergonomics of shovelling and shovel design—An experimental study. *Ergonomics,* 29(1), 19–30.

Garg, A., and Saxena, U. (1979). Effects of lifting frequency and technique on physical fatigue with special reference to psychophysical methodology and metabolic rate. *American Industrial Hygiene Association Journal,* 40, 894–903.

Garg, A., and Saxena, U. (1980). Container characteristics and maximum acceptable weight of lift. *Human Factors,* 22, 487–495.

Genaidy, A., and Asfour, S. (1987). Review and evaluation of physiological cost prediction models for manual materials handling. *Human Factors,* 29(4), 465–476.

Genaidy, A., and Asfour, S. (1989). Effects of frequency and load of lift on endurance time. *Ergonomics* 32(1), 51–57.

Genaidy, A., Asfour, S., Mital, A., and Waly, S. (1990). Psychophysical models for manual lifting tasks. *Applied Ergonomics,* 21(4), 295–303.

Gordon, E. (1957). The use of energy costs in regulating physical activity in chronic disease. *A.M.A. Archives of Industrial Health,* 16, 437–441.

Grandjean, E. (1988). *Fitting the task to the man* (4th ed.). London: Taylor & Francis.

Hunsicker, P. (1955). *Arm strength at selected degrees of elbow flexion* (Tech Rept. 54-548). Wright Air Development Center: U.S. Air Force.

Jackson, A., Osburn, H., and Laughery, K. (1984). Validity of isometric strength tests for predicting performance in physically demanding tasks. *Proceedings of the Human Factors Society 28th Annual Meeting*. Santa Monica, CA: Human Factors Society, pp. 452–454.

Jensen, R. (1985). Events that trigger disabling back pain among nurses. *Proceedings of the Human Factors Society 29th Annual Meeting*. Santa Monica, CA: Human Factors Society, pp. 799–801.

Jensen, R. (1986). Work-related back injuries among nursing personnel in New York. *Proceedings of the Human Factors Society 30th Annual Meeting*. Santa Monica, CA: Human Factors Society, pp. 244–248.

Jensen, C., Schultz, G., and Bangerter, B. (1983). *Applied kinesiology and biomechanics* (3d ed.). New York: McGraw-Hill.

Jones, D. (1985, February). Back injury prevention—Are programs adequate? *Professional Safety*, 18–24.

Jones, N., Morgan-Campbell, E., Edwards, R., and Robertson, D. (1975). *Clinical exercise testing*. Philadelphia: W. B. Saunders.

Khalil, T. (1973). An electromyographic methodology for the evaluation of industrial design. *Human Factors*, 15(3), 257–264.

Kilbom, A. (1990). Measurement and assessment of dynamic work. In J. Wilson, and E. N. Corlett (eds.), *Evaluation of human work*. London: Taylor & Francis, pp. 520–541.

Klein, B., Roger, M., Jensen, R., and Sanderson, L. (1984). Assessment of workers' compensation claims for back sprain/strains. *Journal of Occupational Medicine*, 26, 443–448.

Kodak, Eastman Company (1986). *Ergonomic design for people at work*, vol. 2. New York: Van Nostrand Reinhold.

Kroemer, K. (1970). Human strength: Terminology, measurement, and interpretation of data. *Human Factors*, 12(3), 297–313.

Kroemer, K., Kroemer, H., and Kroemer-Elbert, K. (1986). *Engineering Physiology*. Amsterdam: Elsevier.

Kroemer, K., Snook, S., Meadows, S., and Deutsch, S. (eds.) (1988). *Ergonomic models of anthropometry, human biomechanics, and operator-equipment interfaces: Proceedings of a workshop*. Washington, DC: National Academy Press.

Lehmann, G. (1958). Physiological measurements as a basis of work organization in industry. *Ergonomics*, 1, 328–344.

Lind, A., and Petrofsky, J. (1979). Amplitude of the surface EMG in fatiguing isometric contractions. *Muscle and Nerve*, 2, 257–264.

Maas, S., Kok, M., Westra, H., and Kemper, H. (1989). The validity of the use of heart rate in estimating oxygen consumption in static and in combined static/dynamic exercise. *Ergonomics*, 32(2), 141–148.

Michael, E., Hutton, K., and Horvath, S. (1961). Cardiorespiratory response during prolonged exercise. *Journal of Applied Physiology*, 16, 997–999.

Mital, A. (1983). The psychophysical approach in manual lifting: A verification study. *Human Factors*, 25, 485–491.

Mital, A. (1984). Comprehensive maximum acceptable weight of lift database for regular 8-hour work shifts. *Ergonomics*, 27(11), 1127–1138.

Mital, A., and Okolie, S. (1982). Influence of container shape, partitions, frequency, distance, and height level on the maximum acceptable amount of liquid carried by males. *American Industrial Hygiene Association Journal*, 43, 813–819.

Molbech, S. (1963). Average percentage force at repeated maximal isometric muscle contractions at different frequencies. *Communications from the Testing and Observations Institute of the Danish National Association for Infantile Paralysis*, no. 16. Cited in Astrand, P., and Rodahl, K. (1986).

Murrell, K. (1965). *Human performance in industry*. New York: Reinhold.

National Institute for Occupational Safety and Health (NIOSH) (1981, March). *Work practices guide for manual lifting* [DHHS (NIOSH) Publication No. 81-122]. Washington, DC: Superintendent of Documents.

Passmore, R., and Durnin, J. (1955). Human energy expenditure. *Physiological Reviews*, 35, 801–875.

Petrofsky, J., Glaser, R., and Phillips, C. (1982). Evaluation of the amplitude and frequency components of the surface EMG as an index of muscle fatigue. *Ergonomics*, 25, 213–223.

Poulsen, E., and Jorgensen, K. (1971). Back muscle strength, lifting and stooped working postures. *Applied Ergonomics*, 2, 133–137.

Pytel, J., and Kamon, E. (1981). Dynamic strength as a predictor for maximal and acceptable lifting. *Ergonomics*, 24(9), 663–672.

Roebuck, J., Kroemer, K., and Thomson, W. (1975). *Engineering anthropometry methods*. New York: Wiley-Interscience.

Rohmert, W. (1973*a*). Problems in determining rest allowances, part 1: Use of modern methods to evaluate stress and strain in static work. *Applied Ergonomics*, 4(2), 91–95.

Rohmert, W. (1973*b*). Problems in determining rest allowances, part 2: Determining rest allowances in different human tasks. *Applied Ergonomics*, 4(2), 158–162.

Simonson, E. (1971). Recovery and fatigue. Significance of recovery processes for work performance. In E. Simonson (ed.), *Physiology of work capacity and fatigue*. Springfield, IL: Thomas.

Smith, J., and Jiang, B. (1984). A manual materials handling study of bag lifting. *American Industrial Hygiene Association Journal*, 45, 505–508.

Snook, S. (1978). The design of manual handling tasks. *Ergonomics*, 21(12), 963–985.

Snook, S., and Ciriello, V. (1991). The design of manual handling tasks: Revised tables of maximum acceptable weights and forces. *Ergonomics*, 34(9), 1197–1214.

Snook, S., and Irvine, C. (1969). Psychophysical studies of physiological fatigue criteria. *Human Factors*, 11, 291–299.

Snook, S., and White, A. (1984). Education and training. In M. Pope, J. Frymoyer, and G. Andersson (eds.), *Occupational low back pain*. New York: Praeger.

Strobbe, T. (1982). *The development of a practical strength testing program in industry*. Ph.D. dissertation, Pennsylvania State University, University Park, PA.

Suggs, C., and Splinter, W. (1961). Some physiological responses of man to workload and environment. *Journal of Applied Physiology*, 16, 413–420.

Taylor, H., Buskirk, E., and Henschel, A. (1955). Maximal oxygen intake as an objective measure of cardiorespiratory performance. *Journal of Applied Physiology*, 8, 73.

Viitasalo, J., Era, P., Leskinen, A., and Heikkinen, E. (1985). Muscular strength profiles and anthropometry in random samples of men aged 31–35, 51–55, and 71–75 years. *Ergonomics*, 28(11), 1563–1574.

Vos, H. (1973). Physical workload in different body postures while working near to, or below ground level. *Ergonomics*, 16(6), 817–828.

CHAPTER 9

MOTOR SKILLS

People receive information, process it, and make responses. This is the basic input-mediation-output model of human behavior. As we saw in earlier chapters, the process is actually more complex than this. In this chapter, we concentrate on the response part of the model. We begin by presenting information about our joints and the range of movement we have around them. Motor responses and motor skills, however, involve much more than just flexibility in our joints. We therefore discuss in this chapter the speed and accuracy and control of such movements. The literature in the area of motor skills is vast, coming from psychology, engineering, biology, neuroscience, kinesiology, and physical education. We cannot hope to cover this area in one chapter, so we have selected a few topics and representative research to provide some understanding of the basic concepts and a flavor for the field.

BIOMECHANICS OF HUMAN MOTION

By virtue of our skeletal structure, muscles, and nervous system, we are capable of an incredibly wide repertoire of physical activities. We are capable of simple activities like threading a needle or throwing a ball or complex gymnastic moves such as a back somersault with a full twist. *Biomechanics* deals with the various aspects of physical movements of the body. It uses laws of physics and engineering concepts to describe motion undergone by the various body segments and the forces acting on those body parts during normal daily activities (Frankel and Nordin, 1980). At a more specific level, *kinesiology* is the study of human motion as it relates to the construction of the skeletal-muscular system. Kinesiology describes motions of the body segments and identifies the muscle actions responsible for those motions. Chaffin and Andersson (1984) consider kinesiology to be a subdiscipline of biomechanics.

Types of Body Movements

Body movements occur around movable joints. A joint is a point in the skeletal system where two or more bones meet or articulate. There are two general types of joints that are principally used in physical activities, namely synovial joints and cartilaginous joints.

The *synovial joints* include hinge joints (e.g., fingers and knees), pivot joints (e.g., elbows, which are also hinge joints), and ball-and-socket joints (e.g., shoulders and hips). The primary *cartilaginous joints* are the vertebrae of the spine which collectively make possible considerable rotation and forward bending of the trunk of the body. Each type of joint allows certain types of movements. Some of the basic types that are performed by the body members are described below (Jensen, Schultz, and Bangerter, 1983):

- *Flexion:* A movement of a segment of the body causing a decrease in the angle at the joint, such as bending the arm at the elbow. Bending sideways, such as with the trunk or neck is called *lateral flexion.*
- *Extension:* A movement in the opposite direction of flexion which causes an increase in the angle at the joint, such as straightening the elbow. *Hyperextension* is extension of a body segment to a position beyond its normal extended position, such as arching the back.
- *Abduction:* A movement of a body segment in a lateral plane away from the midline of the body, such as raising the arm sideways.
- *Adduction:* A movement of a body segment toward the midline as when moving the arm from the outward horizontal position downward to the vertical position.
- *Rotation:* A movement of a segment around its own longitudinal axis. A body segment may be rotated inward toward the midline of the body; such a movement is called *medial rotation*. Rotation of the hand and forearm are given special names: *pronation* is rotation that results in a palm-down position; *supination* is rotation that results in a palm-up position.
- *Circumduction:* A circular or conelike movement of a body segment, such as swinging the arm in a circular movement about the shoulder.

Range of Movement

Range of motion is the amount of movement through a particular plane that can occur in a joint; it is expressed in degrees of movement. Figure 9-1 presents some of the basic body movements discussed above for specific joints. In each case, the mean and the 5th to 95th percentile range of movement is given. The data are based on a sample of 100 male college students.

The range of movement in any joint is dependent on the bone structure of the joint, amount of bulk (muscle or other tissue) near the joint, and the elasticity of the muscles, tendons, and ligaments around the joint. For example, as the

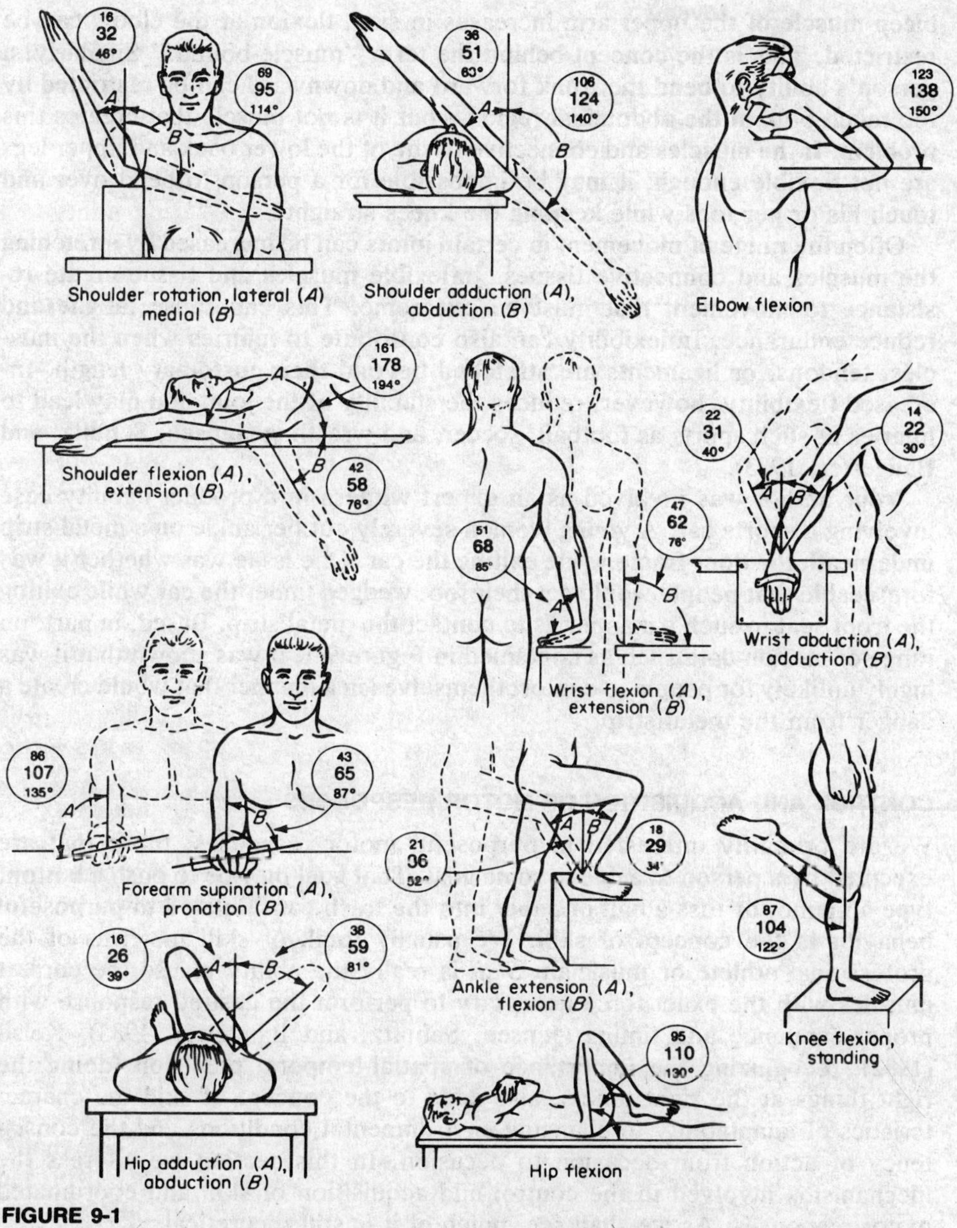

FIGURE 9-1
Range (in degrees) of rotation and movement of certain upper and lower extremities, based on a sample of 100 male college students. The three values (in degrees) given for each are the 5th percentile, the mean, and the 95th percentile. (*Source: Houy, 1983, Table 1. Copyright by the Human Factors Society, Inc. and reproduced by permission.*)

bicep muscle of the upper arm increases in size, flexion at the elbow can be restricted. This is the concept behind the term "muscle-bound." Similarly, a person's ability to bend the trunk forward and downward can be restricted by too much bulk in the abdominal region—but it is not muscle that creates this problem. If the muscles and connective tissue of the lower back and upper legs are not flexible enough, it may be impossible for a person to bend over and touch his or her toes while keeping the knees straight.

Often the range of movement in certain joints can be increased by stretching the muscles and connective tissues. Inflexible muscles and tissue create resistance to movement that must be overcome. This can cause fatigue and reduce endurance. Inflexibility can also contribute to injuries when the muscles, tendons, or ligaments are stretched beyond their customary length. Increased flexibility, however, reduces the stability of the joint and may lead to injuries in such sports as football, soccer, and wrestling (Jensen, Schultz, and Bangerter, 1983).

Your author was involved as an expert witness in a product liability case involving a sports car. A young woman severely cut her ankle on a metal strip underneath the door frame while exiting the car. One issue was whether it was foreseeable that people would get their foot wedged under the car while exiting the front seat in such a manner as to contact the metal strip. Based, in part, on range-of-motion data such as contained in Figure 9-1, it was shown that it was highly unlikely for people to contort themselves in a manner that would create a danger from the metal strip.

CONTROL AND ACQUISITION OF MOTOR RESPONSES

We are primarily interested in purposeful motor responses, those that are executed by a person to achieve some goal. That goal may be to push a button, type a memo, or toss a ball of paper into the trash can. Related to purposeful behavior is the concept of skill. We usually think of skill in terms of the professional athlete or musician. *Skill* is really the ability to use the correct muscles with the exact force necessary to perform the desired response with proper sequence and timing (Jensen, Schultz, and Bangerter, 1983). Kelso (1982), recognizing the importance of spatial-temporal precision (doing the right things at the right times), also adds to the concept of skill the characteristics of adaptability to changing environmental conditions and the consistency of action from occasion to occasion. In this section we address the mechanisms involved in the control and acquisition of skill and coordinated motor responses. As we shall see, much of it is still theoretical.

Types of Responses

Before discussing the control and acquisition of motor responses, we describe certain classes of motor movements. Each of these types of movements has been the focus of specialized research, some of which we will discuss in this and the next chapter. The following are some of the major classes of motor movements:

- *Discrete movements* involve a single reaching movement to a stationary target, such as reaching for a control or pointing to a word on a computer screen. Discrete movements may be made with or without visual control.
- *Repetitive movements* involve a repetition of a single movement to a stationary target or targets. Examples include hammering a nail or tapping a cursor key on a computer keyboard.
- *Sequential movements* involve discrete movements to a number of stationary targets regularly or irregularly spaced. Examples include typewriting or reaching for parts in various stock bins.
- *Continuous movements* involve movements that require muscular control adjustments of some degree during the movement, as in operating the steering wheel of a car or guiding a piece of wood through a band saw.
- *Static positioning* consists of maintaining a specific position of a body member for a period of time. Strictly speaking, this is not a movement, but rather the absence of movement. Examples include holding a part in one hand while soldering, or holding a needle to thread it.

These various types of movements may be combined in sequence so that they blend into one another. For example, placing the foot on an automobile accelerator pedal is a discrete movement, but this may be followed by a continuous movement of adjusting the amount of pressure on the pedal to the conditions of traffic or static positioning if the accelerator must be maintained in a specific position for a period of time as might occur on the open road.

Sensory Feedback

Motor responses can be influenced by internal and external information. The classic model of motor skills involves a closed-loop, servocontrol system in which the system compares the actual motor response, as represented by internal and external information, to some desired goal or set point. The difference between the goal and the actual response is computed as an error. This error information is then conveyed to a higher level of control, where a new set of output commands are generated to reduce the error. This model has been successful for describing continuous movements such as tracking (see Chapter 10). Winstein and Schmidt (1989) point out, however, that there are no real biological correlates of such a system and that feedback information, although certainly capable of modifying centrally organized actions, is not essential for purposeful movements to occur with precision.

In this section of the chapter we briefly discuss the role of information in modifying motor responses.

Role of Sensory Information Sensory information can be classified into two broad classes: that which is available during or after the motor response, called *feedback;* and that which is available prior to the action that regulates and triggers coordinated responses, called *feedforward.* Feedforward information is usually visual in nature. Let us concentrate on feedback information at this time.

Winstein and Schmidt (1989) propose that there is a continuum of varying degrees of conscious activity that can occur between the reception of incoming sensory information and a change in motor behavior. At one end are automatic reflexes where no conscious mediation is involved between reception and response. These are very fast responses, but they cannot be modified; they are hard-wired into the system. At the other end of the continuum are voluntary behaviors where all phases of the response are mediated by conscious control. These types of behaviors are relatively slow, but the output can be modified to fit the specific demands of the task. Between these extremes are all sorts of reflexlike responses that, although fast, are still modifiable to some extent to fit the specific demands of the task. Many investigators believe that a major by-product of learning and skill development is that the behavior becomes more automatic and under less conscious control, thereby reducing the demands on attention required by the behavior.

Sources of Motor Feedback There are various types of receptors in the human body that can sense and transmit information that contributes to our sense of movement and the position of our limbs, or *kinesthesis.* (The term kinesthesis has begun to replace the term *proprioception* as the preferred term in this area.) These receptors are located in muscles, tendons, joints, and cutaneous (skin) tissue. The primary receptors in the muscles are the *muscle spindles,* highly specialized receptors that sense muscle length and rate of change of length. The primary stimulus for the muscle spindle occurs when the muscle is stretched. The response is an almost immediate increase in the muscle's output acting against the stretching action. In addition, the muscle spindle can be innervated by higher-level processes to bring about slower, more conscious control of the muscle's output. It is now generally accepted that muscle spindles play a greater role in controlling motor responses than was once believed (Kelso, 1982; Winstein and Schmidt, 1989).

At one time it was believed that information about the position of the joints originated from receptors located around the joints. It was thought that the receptors would fire nerve impulses in proportion to the angular velocity of joint rotation and would adopt different steady firing rates depending on the particular joint angle—a nice straightforward system. Today, however, joint receptors are no longer considered capable of encoding static joint angle, and it is the tension in the joint that influences the firing rate whether the joint is moved or not. Precisely what information joint receptors code is still open to debate. This illustrates that our knowledge about how we perceive the position of our limbs and control their movement is still evolving and many fundamental questions have still not been completely answered.

Receptors in the tendons (called *Golgi tendon organs*) appear to be sensitive to muscle tension generated through muscle contractions. These receptors inhibit muscle contraction and, together with muscle spindles, may play a role in regulating muscle action. Receptors in the skin may also play a role in sensing joint movement, rather than position. They may augment feedback from other receptors, especially from the hands, feet, and face (Winstein and Schmidt, 1989).

Vision, of course, is a primary source of feedback information regarding our motor responses. There is evidence that visual information may even control motor responses on a nonconscious level (Bridgeman, Kirch, and Sperling, 1981) as well as on a conscious control level. Vision is also important for the timing of motor responses.

Discussion Controlling motor responses is an extremely complex process involving a variety of different receptors providing information that is processed at various levels of consciousness. There is much we do not fully understand, but advances in neuroscience should open new windows of understanding in the future. The role of sensory feedback in the control of motor responses has taken somewhat of a back seat to the newer concept of motor programs, which we discuss next.

Higher-Level Control of Motor Responses

Schmidt (1982*a*) makes a distinction between two types of movements: fast and slow. Fast movements take 200 ms or less, such as kicking a ball or striking an emergency stop button on a machine. We seem to plan such movements in advance and then "let them fly." Little or no conscious control seems to be possible for a short period of time when the movement is executed—even when we sense that something has changed and the movement is now inappropriate. An example is a batter who swings at an obviously bad pitch. On the other hand, slow responses, those taking longer than about 200 ms to execute, provide ample opportunities for conscious control guided by various forms of sensory feedback. Threading a needle is an example. Two major classes of theories, open- and closed-loop, have been put forward to explain fast, highly practiced motor responses.

Closed-Loop Theories Closed-loop theories, as we discussed earlier, emphasize the role of sensory feedback and consider all behavior, fast or slow, to be controlled by comparing the effect of the action to some representation of what the action should be. Any discrepancy is considered error and the response is altered to reduce the error. The most well developed closed-loop theory is that of Adams (1971, 1976). He postulates a comparison between a perceptual trace, or memory, of the movement to the sensory feedback created by the ongoing response. The error serves as a stimulus for subsequent corrective movements. Closed-looped theories may be adequate to explain slow movements, but Schmidt (1982*a*) and others raise serious doubts about whether a closed-loop theory can explain fast movements. People need about 200 ms to perceive and use information to alter their responses. This is about the minimum reaction time for processing visual information. Consider the finger movements of a concert pianist. It would be difficult or impossible to visually control the series of successive movements required. By the time one processed the information from the first movement, several others would have occurred and the person would lag behind the series. A pause would have to be introduced to allow processing of the feedback information, yet no pause is required.

Open-Loop Theories An alternative to the closed-loop theories are open-loop theories. Open-loop theories stress the development of higher-level programs, called *motor programs,* which contain all the information necessary for movement patterning. It is believed that these motor programs are structured before the movement sequence begins. Evidence in support of this comes from a study by Klapp, Anderson, and Berrian (1973), wherein they found that reaction time (time from signal onset to the beginning of the response) was longer the more complex the response was going to be. That is, it appears to take longer to structure a motor program for a complex movement than for a simple movement. This result is inconsistent with closed-loop theories, which propose that the later component of a movement is not readied until feedback from earlier responses has been processed.

There is ample evidence for the existence of innate motor programs. For example, when a baby as young as 1 month is supported upright and placed on a motorized treadmill, the baby exhibits an adultlike alternating stepping response (Thelen, 1987). Grillner (1985) even speculates that innate movement patterns form the basis of learned motor programs and that learning a new movement involves learning to combine and sequence in a new fashion parts of the neuronal apparatus used to control innate movements.

Errors Associated with Motor Programs Motor behavior is goal-directed. When hitting a golf ball off the tee, the goal is to hit the ball to the middle of the fairway at some specific distance. This goal must be translated into a set of muscle actions to execute the proper swing. Schmidt (1982*a*) identifies two types of error involved in the use of motor programs. The first he calls *errors of selection.* Here the person calls up the wrong motor program although that program is executed perfectly. Continuing with our golf example, the player may have selected a motor program more appropriate for swinging an iron than a wood. Correcting an error of selection requires the person to perceive, during the course of the movement, that such an error was made and it requires that a new motor program be executed; one cannot simply fix the old one. This process takes time, a minimum of about 120 to 200 ms.

The second type of error is called an *error of execution*. The correct motor program is called up, but something goes wrong with the execution. Perhaps the muscles are a little more fatigued than was anticipated or the situation is not as expected. Evidence suggests that corrections can be made to an ongoing motor program at a reflex level. For example, at a restaurant your author knows of, they use mugs made of lightweight plastic that look just like heavy glass. Often when first lifting the mug a person, apparently anticipating heavy glass, jerks it off the table. In this case, reflex-based corrections can be made, but they require about 30 to 80 ms to initiate—sometimes just fast enough to prevent a spill, sometimes not.

From the foregoing discussion, we can see that sensory information plays several roles in controlling motor responses. First, it serves a feedforward role in providing information about the environment and conditions within the body needed to set up the appropriate motor program. Second, it serves a feedback function to correct errors of selection. Finally, it serves as feedback to initiate reflexlike corrections for errors of execution.

Some Problems for Motor Programs Schmidt (1982*a*) recognizes two problems with the simplistic notion that different motor programs are required for different responses. First, if we had to have a different motor program for every response and variation of response, there would be no room in our brain to store them. Consider throwing a ball. Imagine if we needed separate programs for throwing overhand or sidearm; high in the air or along the ground; or for throwing hardballs, tennis balls, or Ping-Pong balls. Schmidt identified a second problem: How do we make a movement we never made before? Where does the program come from if the movement pattern has never been done?

One theoretical approach for handling these problems is to move away from the idea that the motor program specifies every detail in the motor response and instead adopt a multilevel or hierarchical view of programming. In this view, abstract descriptions of an action reside at the higher levels and are transformed into specific patterns of movement at lower levels (Glencross, 1980; Keele, 1981; Summers, 1989; Shaffer, 1982; Schmidt, 1982*a*). These higher-level programs are often referred to as *generalized motor programs*.

Generalized Motor Programs A generalized motor program is one that is hypothesized to contain a set of instructions that can be generalized across situations. This is done by supplying the program with specific parameters that tailor the response to the situation. There is still much debate about what information is fixed or coded in generalized programs and what information or parameters can be provided to the programs to tailor them. There is some evidence that the sequencing and relative timing of movements are invariant features in a program. When a person attempts to speed up or slow down the execution of a motor program, all the movements speed up or slow down proportionately.

Schmidt (1982*b*, 1984) postulates that parameters that can be specified in a generalized motor program include overall speed of response, muscles that execute the response, force of the response, and size of the response. An interesting test that supports the notion of generalized motor programs was conducted by Raibert (1977). Subjects produced writing samples (a complex motor skill) with their right hand, right arm, left hand, mouth gripping the pen, and pen taped to the bottom of their right foot. As shown in Figure 9-2, the samples are remarkably similar despite the fact that different muscles were used to produce each sample. This suggests that a generalized motor program was used to control the response.

Schema If we accept the foregoing theory of generalized motor programs as a fundamental mechanism for controlling skilled motor behavior, we must consider how the person determines the values of the parameters needed to execute the program. Schmidt (1975*a*, 1975*b*, 1982*b*) postulates that people formulate abstract rules, called *schema,* based on past experience to determine what parameter values might work in the present situation. The theory proposes that when we make a response, we briefly store the following four types of information:

1 *Parameters:* The person stores the parameter values (such as force, speed, etc.) that were used for the program just completed.

2 *Movement outcome:* The person stores what happened in the environment as a result of the response. For example, where did the golf ball land?

3 *Sensory consequences:* The person also stores what the response "felt" like, looked like, sounded like, etc. as it was being executed. A good golf swing feels smooth; one hears the club cut through the air and sees the club head striking the ball. All this and more is assumed to be stored after the response.

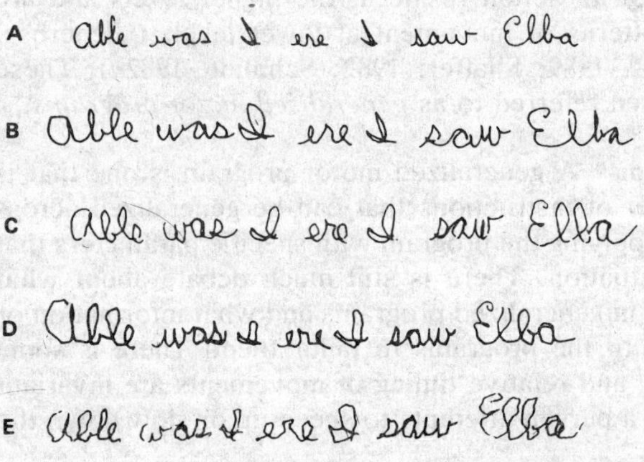

FIGURE 9-2
Samples of writing produced by the same person using (*A*) right hand, (*B*) right arm, (*C*) left hand, (*D*) mouth gripping the pen, and (*E*) pen taped to bottom of right foot. The samples are remarkably similar despite different muscles being used. This supports the existence of generalized motor programs. (*Source: Raibert, 1977.*)

4 *Initial conditions:* The person also remembers the initial conditions that existed before the response was initiated. Back on the golf course, such things as the direction and force of the wind, position of the feet relative to the ball, and the evenness of the ground would be initial conditions that might be stored.

Using these four types of information the person, theoretically, develops or modifies rules or schema that relate the various pieces of information together. Once the schema are updated with the new information, the information is forgotten; only the updated rule is kept. (This theoretical position reduces the problem of where all the information from all our responses is stored; only a few rules are stored, not all the individual pieces of information.)

Schmidt believes that there are two primary types of schema developed. The first he calls *recall schema*. These rules relate initial conditions to the specific values of parameters used. For example, how much additional force is required to swing the golf club with a given headwind. The more varied the experiences of the performer, the better will be the schema developed. A second type of schema, called *recognition schema,* relate the sensory consequences with the outcome of the movement in the environment. In a sense, you learn what it feels like to properly execute the response and what it feels like when you are making an error, such as topping the ball or taking a 6 in divot out of the grass. Using these schema, the performer generates the expected sensory consequences of the movement pattern before starting the movement. Comparing the expected sensations with the actual sensations provides a measure of error that can be used to correct the movement during its execution if time permits.

One practical application of this theory is in the field of motor skill learning. The theory would predict that to learn a motor skill one should practice the movement pattern under widely varying conditions that require the learner to produce many variations of the basic response pattern. This should facilitate the development of recall schema and should aid the performer to accommodate new, unpracticed, initial conditions. This prediction was supported by research carried out by McCracken and Stelmach (1977) for a simple movement task. Subjects moved a slider on a track so that the movement time was maintained at 200 ms. One group practiced by always moving the same distance (low variability), while the other group practiced by moving four different distances in random order (high variability). After training, the subjects performed the task at a distance neither group had practiced. Figure 9-3 presents the results. Accuracy was defined as the absolute error in response time from the desired 200 ms. During practice the low-variability group produced more accurate movements than did the high-variability group. This is not too surprising. However, when they transferred to the new distance, the low-variability group produced larger errors than did the high-variability group. One week later, the difference between the groups was diminished, but still apparent. The advantage of variability in the conditions of practice appears to be even stronger when training children than when training adults (Moxley, 1979).

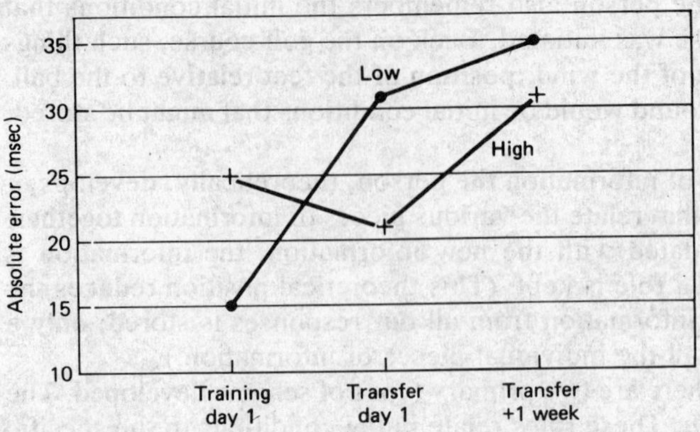

FIGURE 9-3
Absolute error in reproducing a fixed movement time as a function of the type of practice. Subjects were trained at either one distance (*Low*) or four distances (*High*). Both groups then tried to maintain the fixed movement time for a movement at a distance neither group practiced (*Transfer Day 1*). One week later both groups again performed at the new distance (*Transfer + 1 week*). Results demonstrated the value of varied motor practice on performance. (*Source: McCracken and Stelmach, 1977. Reprinted with permission of the Helen Dwight Reid Educational Foundation. Published by Heldref Publications, Washington, D.C. Copyright 1977.*)

Discussion

The control of motor responses is complex and not fully understood. This is an area in which theory is constantly being developed and refined. There is a distinct ''computer'' flavor to these theories. We see concepts such as generalized programs, rule- (schema-) based decision making, and hierarchical levels of control. Using a computer model to understand motor behavior has provided interesting and valuable theories, but there are important differences between computers and the human nervous system. We must be on guard not to be ''swept away'' by the computer analogy.

The study of motor skills has all sorts of practical applications, including how to teach and refine skilled performance, diagnose and treat motor-related handicaps, and understand common behavior and errors. The cartoon in Figure 9-4, for example, cries out for an explanation based on the concepts of motor programs and levels of processing.

SPEED OF MOVEMENTS

In many situations people have to make some type of physical response on the basis of stimuli they receive from the environment (from visual displays, auditory signals, events, etc.). In some circumstances it is critical that the response be made as quickly as possible, for example, applying the brakes of an automobile when an emergency occurs or throwing a baseball to second base

FIGURE 9-4
An example of skilled performance gone awry. (*Reprinted with permission of UFS, Inc.*)

when the runner on first attempts to steal. It is in such situations that a premium is placed on rapid response time. An understanding of what components comprise response time and what variables affect each component can aid in the design of tasks and displays for situations where response time is critical.

Total response time can be partitioned into two major components: reaction time and movement time. *Reaction time* is the time from onset of a signal calling for a response until the beginning of the response. The time from the beginning of the response to its completion is *movement time*. In some simple situations it is difficult to separate the two. For example, the time taken to press a button upon which you are resting your finger in response to a tone is considered reaction time, although it involves a minute amount of movement time (the actual time to depress the button). We discuss reaction and movement time and present some factors that influence each.

Reaction Time

Reaction time is used both as a tool for uncovering complex mental events and as a practical measure of performance. When reaction time is used as a tool for exploring mental processes, small changes, on the order of 10-20 ms, can have significant theoretical implications. As a practical measure of performance, however, it is a rare situation where 10-20 ms would have any real practical significance. Two types of reaction time can be distinguished based on the nature of the task: simple reaction time and choice reaction time.

Simple Reaction Time Simple reaction time is the time to initiate a response when only one particular stimulus can occur and the same response is always required. Usually, the person is aware that the stimulus will occur within a short time and within a specific spatial area. There are relatively few situations, outside of a controlled laboratory experiment, where simple reaction time is relevant. In most real-life situations there can be several different stimuli that may require different responses and the occurrence of the stimulus is not so closely anticipated.

Under ideal conditions, simple reaction time is typically between 150 to 200 ms (0.15 to 0.20 s), with 200 ms being a fairly representative value. The following are some variables that can influence simple reaction time, but for the most part each changes reaction time by less than about 100 ms.

1 *Stimulus modality:* Simple auditory or tactual reaction time is approximately 40 ms faster than simple visual reaction time.

2 *Stimulus detectability:* Teichner and Krebs (1972) studied the effects of flash intensity, duration, and size on simple visual reaction time. When the flash was very dim (less than 3 cd/m^2; see Chapter 16 for discussion of luminance), reaction time to very small and very brief flashes could be as much as 500 ms. However, when brightness was above 3 cd/m^2, above threshold for cone vision, simple reaction time improved to 200 ms, even for very small and very brief flashes.

3 *Spatial frequency:* Simple reaction time to detect the presence of a vertical grating increases about 80 ms from low-spatial-frequency gratings (1 cycle per degree) to high-frequency gratings (16 cpd) (Lupp, Hauske, and Wolf, 1976). (See Chapter 4 for a discussion of spatial frequency.)

4 *Preparedness or expectancy of a signal:* When a stimulus occurs infrequently, or when the person is not expecting it, simple reaction time increases. This was illustrated in a study conducted by Warrick, Kibler, and Topmiller (1965), in which typists at their regular jobs were asked to press a button located to the side of their typewriter whenever a buzzer sounded. The buzzer sounded only once or twice a week over a 6-month period. The reaction time to the "unexpected" signals averaged about 100 ms above that obtained when the subjects had received a warning 2 to 5 s before the buzzer.

5 *Age:* Simple reaction time changes little with age from about 15 to about 60. There is substantial slowing at younger ages and only very moderate slowing after age 60 (Keele, 1986).

6 *Stimulus location:* Stimuli presented in the peripheral field of view (45° from the fovea) are responded to about 15 to 30 ms slower than are centrally presented stimuli (Keele, 1986).

Choice Reaction Time Choice reaction time is an issue when one of several possible stimuli are presented, each of which requires a different response. In general, the more possible stimuli and responses there are, the slower the reaction time. We discussed this in Chapter 3 in reference to the Hick-Hyman law. Reaction time increases linearly with the logarithm (to the base 2) of the number of choices. This is shown in Figure 9-5, which plots data from a representative experiment carried out some time ago (Woodworth, 1938). The data are graphed with number of alternatives (or choices) on a linear scale and again using a log scale. The data more closely approximate a straight line when a log scale is used for number of alternatives. One consequence of this logarithmic relationship is that doubling the number of alternatives (1 to 2, 2 to 4,

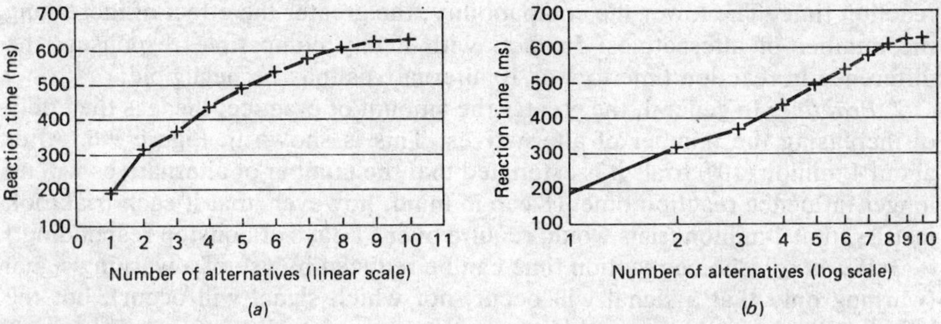

FIGURE 9-5
Logarithmic relationship between number of alternatives (choices) and choice reaction time. Number of alternatives is shown on a linear scale (*a*) and on a logarithmic scale (*b*). Panel (*b*) shows a linear relationship between reaction time and the log of the number of alternatives. (*Source: Based on data from Woodworth, 1938.*)

4 to 8, 3 to 6, 5 to 10) increases choice reaction time by a constant. A representative value found in many studies is that choice reaction time increases about 150 ms for each doubling of the number of alternatives, this being added to simple reaction time of about 200 ms.

Actually, choice reaction time is a function of the probability of a stimulus occurring. It is faster for more probable events. When the number of alternatives is increased, the probability of any one alternative decreases, resulting in an increase in choice reaction time. If different stimuli occur with different probabilities, reaction time will be shorter for the more probable ones and longer for the less probable ones. One interpretation of this is offered by Keele (1986). A person expects a high-probability stimulus and prepares a response for it. If it occurs, the appropriate response is already retrieved from memory and reaction time is fast. If the expected stimulus fails to occur, the person must retrieve the appropriate response from memory. The time taken to retrieve another response is in part a function of the number of responses possible, but there are other factors that influence choice reaction time as well. A few of these are:

1 *Compatibility between stimuli and responses:* Compatibility refers to the degree to which relationships are consistent (or natural) with human expectations. We discuss compatibility at greater length in Chapter 10. The more natural the relationship between stimuli and responses, the faster the choice reaction time. Consider a simple example of five lights arranged in a row (A-B-C-D-E) and five buttons (1-2-3-4-5) arranged one under each light. Choice reaction time will be fastest when the correct responses to the lights are compatible (i.e., for A press 1; for B press 2, and so on). Choice reaction time will be much slower if the correct button-press response to each light was selected randomly. In addition to generally increasing choice reaction time, compatibility also changes the effect that the number of alternatives has on choice reaction time. The lower the compatibility, the greater the effect of increasing the number of alternatives. In fact, with highly compatible responses, the difference in reaction time to 2 or 10 alternatives may be negligible.

2 *Practice:* In general, the greater the amount of practice, the less the effect of increasing the number of alternatives. This is shown in Figure 9-6. After about 1 million (10^6) trials it is estimated that the number of alternatives will no longer influence reaction time. (Keep in mind, however, that if each trial took just 1 s, that 1 million trials would require over 11 days of nonstop responding.)

3 *Warning:* Choice reaction time can be reduced by use of a warning signal (warning only that a signal will occur, not which signal will occur), but the effect depends on the interval between the warning and the signal. For example, Posner, Klein, Summers, and Buggie (1973), using a two-choice task, found an optimum interval of 200 ms at which time choice reaction time improved by about 50 to 80 ms compared to no warning. This is shown in Figure 9-7 for both compatible and incompatible stimulus-response arrangements. What is not shown in Figure 9-7 is that as reaction time decreases, errors increase.

4 *Type of movement:* As we discussed earlier with reference to motor programs, the more complex the movement, the greater will be the reaction time. However, the effect is rather small and only of theoretical importance.

5 *More than one stimulus:* When two successive signals require a separate response, it has been shown that when the second signal occurs before the response is made to the first signal, the response to the second signal is delayed. This is shown rather vividly in Figure 9-8. The first signal was a visual presentation of a digit; either 1, 2, or 5 choices were possible. The response was to press corresponding buttons with the left hand. At various intervals after the

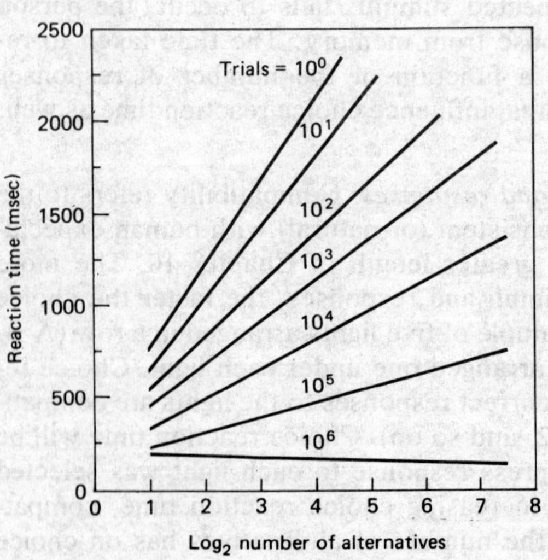

FIGURE 9-6
Choice reaction time as a function of number of alternatives and amount of practice. Based on data from several studies. (*Source: Teichner and Krebs, 1974. Copyright 1974 by the American Psychological Association. Reprinted by permission.*)

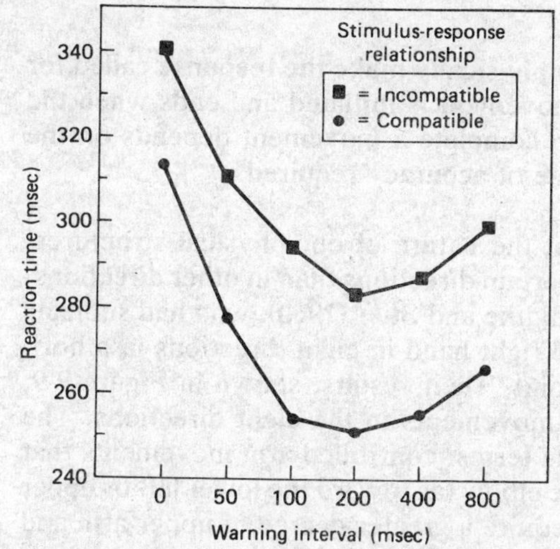

FIGURE 9-7
Choice reaction time as a function of warning signal interval. A warning tone occurred at various intervals before an X mark on the right or left of a screen appeared. In the compatible condition the subject pressed the right key if the X appeared on the right side, and the left key if the X appeared on the left. In the incompatible condition, the X on the left required pressing the right button and vice versa. A warning interval of about 200 ms optimally improves reaction time but also results in increased errors (not shown). (*Source: Posner, Klein, Summers, and Buggie, 1973. Reprinted by permission of Psychonomic Society, Inc.*)

onset of the digit, one of two tones was presented and the subject responded by pressing one of two keys with the other hand. The delay in responding to the second signal is always greater the more choices there were in the first signal. Reaction time to the second signal is more or less normal if the second signal occurs after the first response, as was always the case in Figure 9-8 when the stimuli were separated by about 1000 ms. The reaction time to the first signal is largely unaffected by the appearance of the second one.

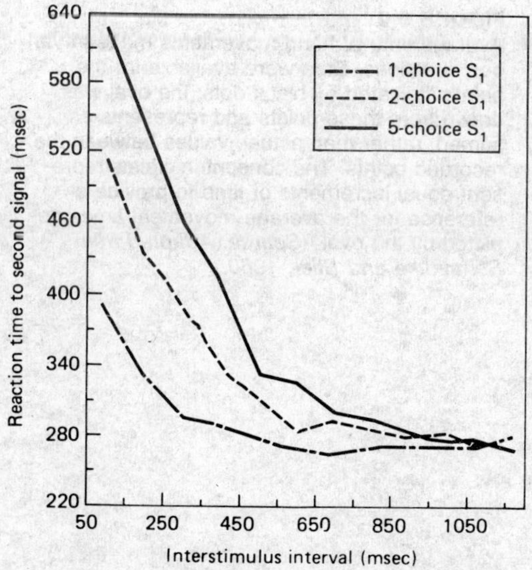

FIGURE 9-8
Reaction time to a second signal soon after the occurrence of a first signal. The first signal required a keypress to either 1, 2, or 5 possible digits. The second signal, a tone, required one of two possible responses with the other hand. When the second signal occurred very shortly after the first signal and before the first response, reaction time to the second signal increased. The more choices there are in the first signal, the greater the delay in responding to the second signal. (*Source: Karlin and Kestenbaum, 1968.*)

Movement Time

Movement time is the time required to physically make the response called for by the stimulus. It begins when the movement is initiated and ends when the movement ends. The time required to complete a movement depends on the nature of the movement and the degree of accuracy required.

Direction of Movement Because of the nature of our physical structures, motions can be made more rapidly in certain directions than in other directions. We use as an example data from Schmidtke and Stier (1960), who had subjects make positioning movements with the right hand in eight directions in a horizontal plane from a center starting point. Their results, shown in Figure 9-9, show the average times to make the movements in the eight directions. The pattern suggests that, in biomechanical terms, controlled arm movements that are primarily based on a pivoting of the elbow (as toward the lower left or upper right) take less time than those that require a greater degree of upper-arm and shoulder action (as toward the lower right or upper left). Also evidence from other sources indicates that such movements are also more accurate. This has implications for design of workplaces, for example, for the placement of part bins for assembly tasks.

Distance and Accuracy Required It seems intuitive that movement time would be affected by the distance moved and the precision demanded by the size of the target to which one is moving. The longer the distance and/or the smaller the target, the longer the movement will take. Fitts (1954; Fitts and Peterson, 1964) discovered that movement time was a logarithmic function of

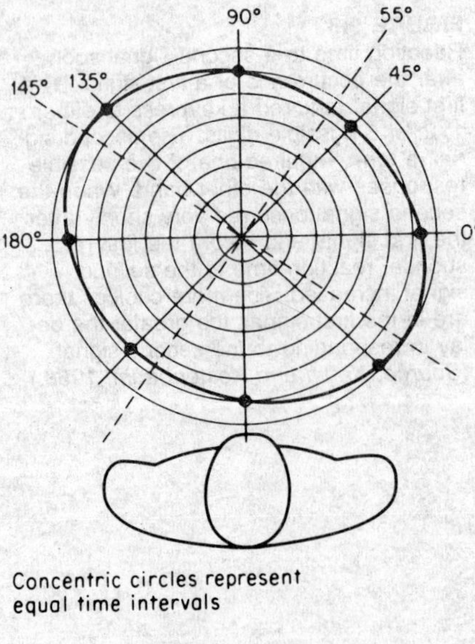

FIGURE 9-9
Average time of hand movements made in various directions. Data were available for the points indicated by black dots; the oval was drawn from these points and represents assumed, rather than actual, values between the recorded points. The concentric circles represent equal increments of time to provide a reference for the average movement times depicted by the oval. (*Source: Adapted from Schmidtke and Stier, 1960.*)

distance when target size was held constant, and that movement time was also a logarithmic function of target size when distance was held constant. He formulated this into what is now called Fitts' law. Mathematically, Fitts' law is stated as follows:

$$MT = a + b \log_2\left(\frac{2D}{W}\right)$$

where MT = movement time
a and b = empirically derived constants
D = distance of movement from start to target center
W = width of the target

(The distance-width ratio is multiplied by 2 simply to avoid negative logarithms when the target width is greater than the distance to the center of the target.) Log to the base 2 is used; movement time is related to the information (in bits) contained in the movement (see Chapter 3).

The empirical constants a and b in the equation depend on the type of movement involved. To illustrate, Figure 9-10 presents data from movements of the arm, wrist (hand), and finger. The slope of the line decreases from arm to finger. The smaller the slope, the smaller the effect of the distance-width ratio. Bailey and Presgrave (1958) had subjects make simple arm movements at a rate they felt they could maintain over a workday. They found a slope of about 180 ms, which is considered to be the upper bound estimate for simple arm movements (Keele, 1986).

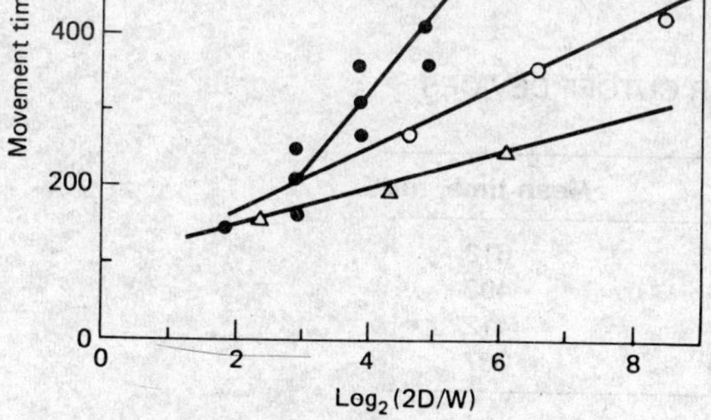

FIGURE 9-10
Example of Fitts' law for arm, wrist, and finger movements. The slopes of the functions are: for arm, 105 ms; for wrist, 43 ms; and for finger, 26 ms. (*Source: Based on data from Langolf, Chaffin, and Foulkes, 1976. Reprinted with permission of the Helen Dwight Reid Educational Foundation. Published by Heldref Publications, Washington, D.C. Presented in Keele, 1986, Fig. 30.18. Copyright © John Wiley & Sons, Inc. Reprinted with permission.*)

Fitts' law has been shown to hold for movements of the head (Radwin, Vanderheiden, and Lin, 1990) and feet (Drury, 1975); for movements made underwater (Kerr, 1978); and for movements made with remote manipulators (McGovern, 1974).

Crossman and Goodeve (1983) and Keele (1968) propose models that account for Fitts' law. They suggest that a movement to a target consists of a series of submovements, each taking a fixed amount of time. A submovement is an impulse response to visual feedback that attempts to reduce any error that is detected. The required precision of the movement (target width) affects the number of corrections to be made, but not the time for these corrections. The time required is postulated to be about 250 ms. If these models are correct, we would predict that Fitts' law would not hold for movements that cannot be controlled by visual feedback, that is, movements that take less than about 200 ms. In fact, Fitts' law does breakdown for movements that take less than 200 ms (Glencross and Barrett, 1989). Gian and Hoffmann (1988) call movements taking less than about 200 ms *ballistic movements*. They find that movement time for such movements is linearly related to the square root of the distance moved. It should be pointed out that other theorists have developed models to account for Fitts' law and its breakdown with fast movements that make different assumptions about the role of visual feedback (Meyer, Smith, and Wright, 1982; Beggs and Howarth, 1972*a,b*).

Discussion

We see that a variety of factors can influence response time. In one applied study, for example, Pattie (1973) investigated the time to activate four types of possible emergency cutoff switches as they might be used on agricultural tractors: clutch, toggle switch, horn rim, and rim blow. These devices reflect differences in the location, the type of response to be made, and even the body member (foot versus hand). The results, given in Table 9-1, show very distinct differences, ranging from 337 to 613 ms.

Even under the best circumstances there is still a lag in responding to situations in our environment. However, even if time is, as is said, of the essence, we should not throw up our hands in complete despair of the time lag

TABLE 9-1
MEAN TIMES TO ACTIVATE FOUR POWER CUTOFF DEVICES ON A TRACTOR

Device	Mean time, ms
Clutch	613
Toggle switch (underneath steering wheel)	498
Horn rim (on steering wheel)	412
Rim blow (on under edge of steering wheel)	337

Source: Pattie, (1973).

in human responses. There are, indeed, ways of aiding and abetting people in responding rapidly to stimuli, for example, by using sensory modalities with shortest reaction time, presenting stimuli in a clear and unambiguous manner, minimizing the number of alternatives from which to choose, giving advance warning of stimuli if possible, using body members that are close to the cortex to reduce neural transmission time, using control mechanisms that minimize response time, and training the individuals.

ACCURACY OF MOVEMENTS

Often accuracy of a response is of greater importance than the speed of the response, within limits. An example is when drivers unintentionally put their foot on the accelerator rather than the brake. If this is done when the car is started and put into gear, the driver experiences full, unexpected acceleration with an apparently complete failure of the braking system. An accident often results. Schmidt (1989) provides evidence that such unintended acceleration accidents are probably due to the inconsistency in foot trajectory generated by the processes that generate muscular forces and timing of such movements.

Below we discuss both the accuracy of movements not controlled by visual feedback and the effect of tremor on continuous control. We do not discuss the accuracy of movements controlled by visual feedback because such movements are exceptionally accurate given sufficient time to perform them. Fitts' law describes the relationship between accuracy and movement time for such responses.

Movements Not Controlled by Visual Feedback

Movements in which the person cannot see the movement either because attention must be directed elsewhere or the environment is dark are examples of blind-positioning movements. Such movements are obviously not controlled by visual feedback. In addition, as we discussed previously, movements that are made very quickly (under about 200 ms) are also not controlled by visual feedback, but rather are preprogrammed before the movement is initiated. We discuss a few factors that affect the accuracy of blind-positioning movements and very fast movements not under visual control.

Target Location A typical type of blind-positioning movement is one in which a person moves a hand (or foot) as in reaching for a control device when the eyes are otherwise occupied. A classic study by Fitts (1947) probably provides the best available data relating to the accuracy of such movements in free space. Fitts arranged targets around a subject at 0°, 30°, 60°, and 90° angles left and right of straight ahead in three tiers: shoulder level, 45° above shoulder level (i.e., above head level), and 45° below shoulder level (i.e., below waist level). Blindfolded subjects were given a sharp pointed marker which they used

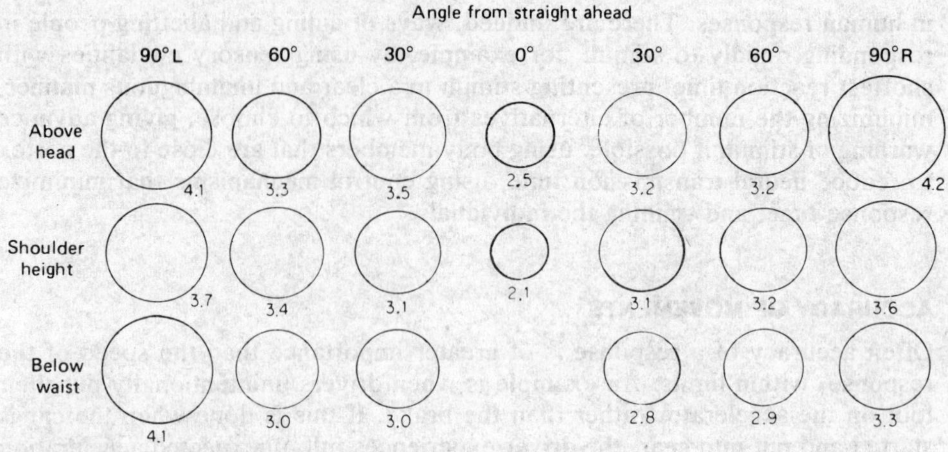

FIGURE 9-11
Relative accuracy for blind positioning movements to various locations. The position of the circles represents the location of the targets: from 90° left to 90° right of straight ahead; at three tiers, shoulder height, 45° above, and 45° below. The size of each circle represents the relative inaccuracy of aiming; larger circles are more inaccurate. The average error (in inches) from the center of the target is given below each circle. (*Source: Based on data from Fitts, 1947, as presented by USAF, AFHRL, Human Engineering Division.*)

to touch each target when they tried to reach for it. Figure 9-11 shows the results. The size of each circle is proportional to the average error score for that location target; the smaller the size, the better the accuracy. We can see that blind-positioning movements are most accurate in the dead-ahead positions and least accurate at the side positions; also the center and lower tiers are slightly more accurate than the upper tier. Control devices or other devices that are to be reached for blindly should, therefore, be located in, or near, the straight-ahead position and at or below shoulder level. In addition, more space should be allocated between controls placed at the sides of an operator than between those placed in front of the person to reduce the chance of accidentally contacting the wrong control while making a blind reach.

Distance of Movement Brown, Knauft, and Rosenbaum (1947) had subjects slide a knob along a fixed straight path. They presented the subjects with a light at the desired end position for the sliding movement. The light was extinguished and the subjects moved the knob, in the dark, to where they thought the desired end position was located. The results are shown in Figure 9-12 for movements made toward and away from the body. In general, short distances (about 1 in or less) were overestimated, resulting in the subjects overshooting the target. Long distances were underestimated, resulting in undershooting the target. How long is long seems to depend on whether the movement is away or toward the body. Overshooting short distances and undershooting long distances is sometimes called the *range effect*.

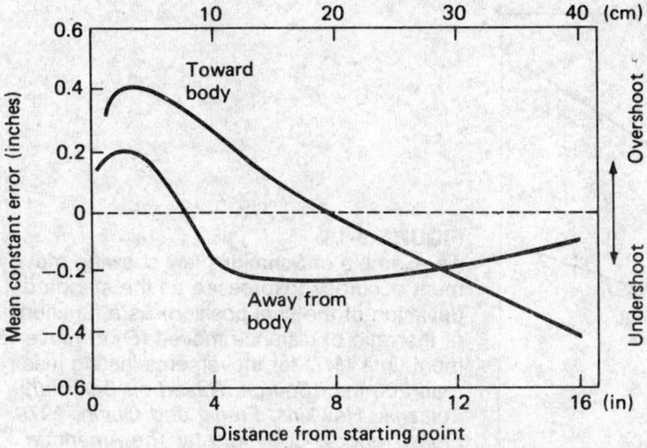

FIGURE 9-12
Movement accuracy (expressed as mean constant error) for restricted blind-positioning movements for different distances made toward or away from the body. (*Source: Brown, Knauft, and Rosenbaum, 1947.*)

Speed of Movement Recall that Fitts' law does not hold for movements executed so fast that they cannot benefit from visual feedback. Schmidt, Zelaznik, Hawkins, Frank, and Quinn (1979) found that for such movements, accuracy was a linear function of the distance moved and the movement time. The mathematical formulation of this relationship has been called Schmidt's law:

$$W = a + b\left(\frac{D}{MT}\right)$$

where W = standard deviation of the end point dispersion, a measure of accuracy, also known as effective target width
a and b = empirically derived constants
D = distance
MT = movement time

Schmidt's law provides a good description of accuracy for movements lasting 140 to 200 ms over distances up to about 12 in (30 cm). The constants a and b differ depending on whether errors are measured parallel or perpendicular to the line of travel, as shown in Figure 9-13. In general, movements lasting longer than 200 ms are considerably more accurate for a given D/MT value than are faster movements, probably because visual feedback is used to correct errors in the case of slower movements.

Schmidt's law breaks down under conditions that involve forces that exceed 60 to 70 percent of maximum capability. Such movements are actually more accurate, especially in the direction perpendicular to the line of travel. For example, swinging a bat in baseball becomes less accurate the faster the swing or the more massive the bat, but only to a point. Beyond that point, however, increasing the speed or the bat's mass should produce an increase in accuracy (Keele, 1986).

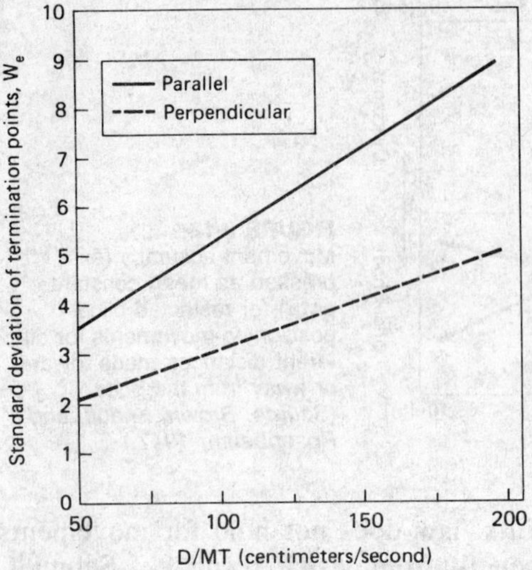

FIGURE 9-13
An example of Schmidt's law showing movement accuracy (expressed as the standard deviation of the final position) as a function of the ratio of distance moved (*D*) to movement time (*MT*) for movements lasting less than 200 ms. (*Source: Based on Schmidt, Zelaznik, Hawkins, Frank, and Quinn, 1979, Figs. 7 and 9. Copyright by The American Psychological Association, adapted by permission. As presented in Keele, 1986, Fig. 30.22. Copyright © John Wiley & Sons, Inc. Reprinted with permission.*)

Continuous Control and Tremor

Some tasks require high accuracy and continuous freehand control, such as in sign painting, drawing, etc. Such movements can be carried out with minimum tremor (and greater accuracy) when the movement is on the horizontal plane and in a lateral (left-right) direction with the forearm pivoted at the elbow. The advantage of this direction of movement is illustrated by the results of a study by Mead and Sampson (1972), in which they had subjects move a stylus (15 in long with a 4-in, 90° bend at the tip) along a narrow groove in any one of the four positions shown in Figure 9-14. As the subject moved the stylus in the directions indicated in that figure, any time the stylus touched the side of the groove, an "error" was electronically recorded. These errors, which can be viewed as measures of tremor, are also shown in Figure 9-14 and indicate that tremor was greatest during an in-out arm movement in the vertical plane (in which the tremor was up-down) and that it was least during a right-left arm movement in the horizontal plane (in which the tremor was in-out).

Continuous control movements such as those involved in tracking tasks are discussed further in Chapter 10.

Static Muscular Control

Static muscular control is not a movement as such, but rather is the absence of a displacement of the body or body member. In such control, however, the muscles are very definitely involved, in that certain sets of muscles typically operate in opposition to each other to maintain equilibrium of the body or body member. Thus, if a body member, such as the hand, is being held in a fixed position, the various muscles controlling hand movement are in a balance that

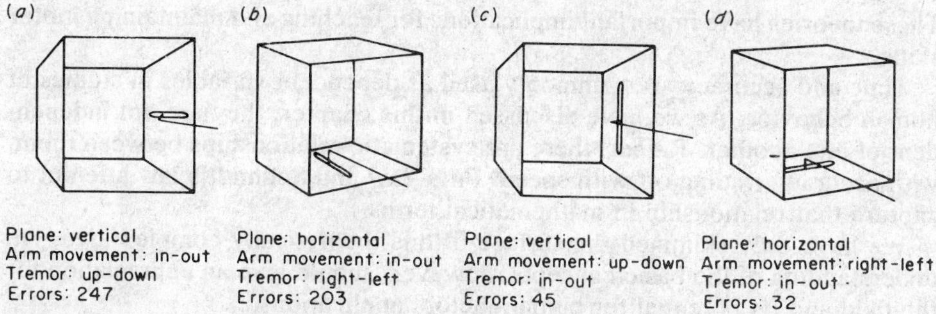

FIGURE 9-14
Directions and planes of arm movements with stylus as used in study of hand steadiness, with direction of hand tremor and number of "errors" (number of times the stylus touched the side of the groove) for each condition. (*Source: Adapted from Mead and Sampson, 1972, Fig. 1 and Table 1.*)

permits no net movement one way or the other. The tensions set up in the muscles to bring about this balance, however, require continued effort, as most of us who have attempted to maintain an immobile state for any time can testify. In fact, it has been stated that maintaining a static position produces more wear and tear on people than some kind of adjustive posture.

Deviations from static postures are of two types: those called *tremor* (small vibrations of the body member) and those characterized by a gross drifting of the body or body member from its original position.

Tremor in Maintaining Static Position Tremor is of particular importance in work activities in which a body member must be maintained in a precise and immovable position (as in holding an electrode in place during welding). An interesting aspect of tremor is that the more a person tries to control it, the worse it usually is. The following four conditions help to reduce tremor:

1 Use of visual reference.

2 Support of body in general (as when seated) and of body member involved in static reaction (as hand or arm).

3 Hand position. There is less hand tremor if the hand is within 8 in above or below the heart level.

4 Friction. Contrary to most situations, mechanical friction in the devices used can reduce tremor by adding enough resistance to movement to counteract, in part, the energy of the vibrations of the body member.

DISCUSSION

The topic of motor skills has benefited from multidisciplinary research. Theories abound for explaining and predicting such behavior. Newer theories recognize the importance of higher-level information processing and control.

These theories have important implications for teaching and maintaining motor skills.

Time and accuracy are commonly used as dependent variables in studies of human behavior. As we have discussed in this chapter, they are not independent of one another. Rather, there are systematic relationships between them, with accuracy trading off with speed. Fitts' law and Schmidt's law attempt to capture that relationship in mathematical terms.

We have only skimmed the surface of this theoretically complex area. An understanding of the basic concepts, however, should give an appreciation for the field and its potential for human factors applications.

REFERENCES

Adams, J. (1971). A closed-loop theory of motor learning. *Journal of Motor Behavior*, 3, 111–150.

Adams, J. (1976). Issues for a closed-loop theory of motor learning. In G. Stelmach (ed.), *Motor control: Issues and trends*. New York: Academic.

Bailey, G., and Presgrave, R. (1958). *Basic motion time study*. New York: McGraw-Hill.

Beggs, W., and Howarth, C. (1972*a*). The accuracy of aiming at a target. Some further evidence for a theory of intermittent control. *Acta Psychologica*, 36, 171–177.

Beggs, W., and Howarth, C. (1972*b*). The movement of the hand towards a target. *Quarterly Journal of Experimental Psychology*, 24, 448–453.

Bridgeman, B., Kirch, M., and Sperling, A. (1981). Segregation of cognitive and motor aspects of visual information using induced motion. *Perception and Psychophysics*, 29, 336–342.

Brown, J., Knauft, E., and Rosenbaum, G. (1947). The accuracy of positioning reactions as a function of their direction and extent. *American Journal of Psychology*, 61, 167–182.

Chaffin, D., and Andersson, G. (1984). *Occupational biomechanics*. New York: Wiley.

Crossman, E., and Goodeve, P. (1983). Feedback control of hand movements and Fitts' law. *Quarterly Journal of Experimental Psychology*, 35A, 251–278.

Drury, C. (1975). Application of Fitts' law to foot-pedal design. *Human Factors*, 17, 368–373.

Fitts, P. (1947). A study of location discrimination ability. In P. Fitts, (ed.), *Psychological research on equipment design*, Research Report 19. Army Air Force, Aviation Psychology Program. Columbus, OH: Ohio State University.

Fitts, P. (1954). The information capacity of the human motor system in controlling the amplitude of movement. *Journal of Experimental Psychology*, 47, 381–391.

Fitts, P., and Peterson, J. (1964). Information capacity of discrete motor responses. *Journal of Experimental Psychology*, 67, 103–112.

Frankel, V., and Nordin, M. (1980). *Basic biomechanics of the skeletal system*. Philadelphia: Lea & Febiger.

Gian, K-C., and Hoffmann, E. (1988). Geometrical conditions for ballistic and visually controlled movements. *Ergonomics*, 31(5), 829–839.

Glencross, D. (1980). Levels and strategies of response organization. In G. Stelmach, and J. Reguin (eds.), *Tutorials in motor control*. Amsterdam: North-Holland.

Glencross, D., and Barrett, N. (1989). Discrete movements. In H. Holding (ed.), *Human skills* (2d ed.). Chichester, England: Wiley, pp. 107–146.

Grillner, S. (1985). Neurobiological bases of rhythmic motor acts in vertebrates. *Science,* 228, 143–149.

Houy, D. (1983). Range of joint movement in college males. *Proceedings of the 27th Human Factors Society Annual Meeting.* Santa Monica, CA: Human Factors Society, pp. 575–579.

Jensen, C., Schultz, G., and Bangerter, B. (1983). *Applied kinesiology and biomechanics.* New York: McGraw-Hill.

Karlin, L., and Kestenbaum, R. (1968). Effects of number of alternatives on the psychological refractory period. *Quarterly Journal of Experimental Psychology,* 20, 167–178.

Keele, S. (1968). Movement control in skilled motor performance. *Psychological Bulletin,* 70, 387–403.

Keele, S. (1981). Behavioral analysis of movement. In V. Brooks (ed.), *Handbook of physiology: Section 1: The nervous system,* vol. 2: *Motor control,* part 2. Baltimore: American Physiological Society.

Keele, S. (1986). Motor control. In K. Boff, L. Kaufman, and J. Thomas (eds.), *Handbook of perception and human performance,* vol. 2: *Cognitive processes and performance.* New York: Wiley.

Kelso, J. (1982). *Human motor behavior: An introduction.* Hillsdale, NJ: Lawrence Erlbaum Associates.

Kerr, R. (1978). Diving, adaptation, and Fitts' law. *Journal of Motor Behavior,* 10, 255–260.

Klapp, S., Anderson, W., and Berrian, R. (1973). Implicit speech in reading, reconsidered. *Journal of Experimental Psychology,* 100, 368–374.

Langolf, G., Chaffin, D., and Foulkes, J. (1976). An investigation of Fitts' law using a wide range of movement amplitudes. *Journal of Motor Behavior,* 8, 113–128.

Lupp, U., Hauske, G., and Wolf, W. (1976). Perceptual latencies to sinusoidal gratings. *Vision Research,* 16, 969–972.

McCracken, H., and Stelmach, G. (1977). A test of the schema theory of discrete motor learning. *Journal of Motor Behavior,* 9, 193–201.

McGovern, D. (1974). Factors affecting control allocation for augmented remote manipulation. Unpublished doctoral dissertation, Stanford University, Stanford, CA.

Mead, P., and Sampson, P. (1972). Hand steadiness during unrestricted linear arm movements. *Human Factors,* 14, 45–50.

Meyer, D., Smith, J., and Wright, C. (1982). Models for the speed and accuracy of aimed movements. *Psychological Review,* 89, 449–482.

Moxley, S. (1979). Schema: The variability of practice hypothesis. *Journal of Motor Behavior,* 11, 65–70.

Pattie, C. (1973, May). Simulated tractor overturnings: A study of human responses in an emergency situation. Unpublished doctoral dissertation, Purdue University, Lafayette, IN.

Posner, M., Klein, R., Summers, J., and Buggie, S. (1973). On the selection of signals. *Memory and Cognition,* 1, 2–12.

Radwin, R., Vanderheiden, G., and Lin, M. (1990). A method for evaluating head-controlled computer input devices using Fitts' law. *Human Factors,* 32, 423–438.

Raibert, M. (1977). Motor control and learning by the state space model. Unpublished doctoral dissertation, Massachusetts Institute of Technology, Cambridge, MA.

Schmidt, R. (1975*a*). A schema theory of discrete motor skill learning. *Psychological Review,* 82, 225–260.

Schmidt, R. (1975*b*). *Motor skills.* New York: Harper & Row.

Schmidt, R. (1982*a*). More on motor programs and the schema concept. In J. Kelso (ed.), *Human motor behavior: An introduction.* Hillsdale, NJ: Lawrence Erlbaum Associates, pp. 189–235.

Schmidt, R. (1982*b*). *Motor control and learning: A behavioral emphasis.* Champaign, IL: Human Kinetics.

Schmidt, R. (1984). The search for invariance in skilled movement behavior. *Research Quarterly for Exercise and Sport,* 56, 188–200.

Schmidt, R. (1989). Unintended acceleration: A review of human factors contributions. *Human Factors,* 31, 345–364.

Schmidt, R., Zelaznik, H., Hawkins, B., Frank, J., and Quinn, J. Jr. (1979). Motor output variability: A theory for the accuracy of rapid motor acts. *Psychological Review,* 86, 415–451.

Schmidtke, H., and Stier, F. (1960). Der Aufbau komplexer Bewegungsablaufe aus Elementarbewegungen. *Forschungsberichte des Landes Nordrhein-Westfalen,* 822, 13–32.

Shaffer, L. (1982). Rhythm and timing in skill. *Psychological Review,* 89, 109–122.

Summers, J. (1989). Motor programs. In H. Holding (ed.), *Human skills* (2d ed.). Chichester, England: Wiley, pp. 49–69.

Teichner, W., and Krebs, M. (1972). Laws of simple visual reaction time. *Psychological Review,* 79, 344–358.

Teichner, W., and Krebs, M. (1974). Laws of visual choice reaction time. *Psychological Review,* 81, 75–98.

Thelen, E. (1987). We think, therefore, we move. *Cahiers de Psychologie Cognitive,* 7, 195–198.

Warrick, M., Kibler, A., and Topmiller, D. (1965). Response time to unexpected stimuli. *Human Factors,* 7(1), 81–86.

Winstein, C., and Schmidt, R. (1989). Sensorimotor feedback. In H. Holding (ed.), *Human skills* (2d ed.). Chichester, England: Wiley, pp. 17–47.

Woodworth, R. (1938). *Experimental psychology* (rev. ed.). New York: Henry Holt.

CHAPTER **10**

HUMAN CONTROL OF SYSTEMS

The types of systems and mechanisms people control in their jobs and everyday lives vary tremendously from simple light switches to complex power plants and aircraft. Whatever the nature of the system, the basic human functions involved in its control remain the same. The human receives information, processes it, selects an action, and executes the action. The action taken then serves as the control input to the system. In the case of most systems, there typically is some form of feedback to the person regarding the effects of the action taken.

In this chapter we discuss three topics that involve all the human functions of control. In a sense this chapter can be viewed as a transition from the display-oriented chapters (4, 5, 6, and 7) to the chapters related more to controls (11 and 12). Compatibility, our first topic, deals with the relationships between controls and displays and affects the ease and adequacy with which people can select and carry out appropriate actions from several alternatives. Tracking, our second topic, is a special form of human control that is involved when the control required is continuous and must conform to some external input signal. It is a task that often involves complex information processing and decision-making activities to direct proper control of a system, and this task is greatly influenced by the displays and dynamics of the system being controlled. Our third topic is supervisor control. As more and more processes become automated, humans assume less direct control of the physical processes. Rather, they supervise the automated process, planning and monitoring its function and perhaps intervening when things go awry. This sort of control function is evident in, for example, chemical processing plants, nuclear power plants, jet aircraft, and factories using computer-assisted manufacturing (CAM) techniques.

COMPATIBILITY

We introduced the concept of *compatibility* in Chapter 3 and distinguished four types: conceptual, spatial, movement, and modality compatibility. (It would be a good idea to review that material now.) In this chapter we deal with spatial and movement compatibility as they relate to the relationships between controls and displays.

Compatibility refers to the degree to which relationships are consistent with human expectations. As such, expectations have profound impact on human performance. Where compatibility relationships are designed into the system, (1) learning is faster, (2) reaction time is faster, (3) fewer errors are made, and (4) user satisfaction is higher. Although people can and do learn to use systems that are out of sync with their expectations, they do so at a price. Such systems place a greater information processing demand on an operator. Especially under conditions of stress, a person may make the most "natural" (or compatible) response; if this is not the correct response, an error or accident could result.

Two things should be kept in mind about the use of compatibility relationships: (1) Some are stronger than others; that is, some expectations are shared by greater proportions of a population than are others. (2) In some circumstances it may be necessary to violate one compatibility relationship to take advantage of another one in the design of some systems; sometimes it is impossible to avoid this. This last fact is illustrated by the results of a study by Bergum and Bergum (1981) in which 93 percent of subjects expected upward movement of a pointer on a vertical scale to represent an increase, while 71 percent of the same group expected the numbers on the scale to increase from top to bottom!

Spatial Compatibility

There are many variations of spatial compatibility; most deal either with the *physical similarities* between displays and their corresponding controls or with the *arrangement* of displays and their controls.

Physical Similarity of Displays and Controls Sometimes there exists the opportunity to design related displays and controls so there is reasonable correspondence of their physical features, and perhaps also of their modes of operation. Such a case was well illustrated years ago in a classic study by Fitts and Seeger (1953). In this study three different displays and three different controls were used in all possible combinations. The displays consisted of lights in various arrangements. As a light went on, the subject was to move a stylus along a corresponding channel to turn the light off. The three displays and controls are illustrated in Figure 10-1. Different groups of subjects used each of the nine combinations of stimulus-response panels; performance was measured in terms of reaction time, errors, and information lost. The results,

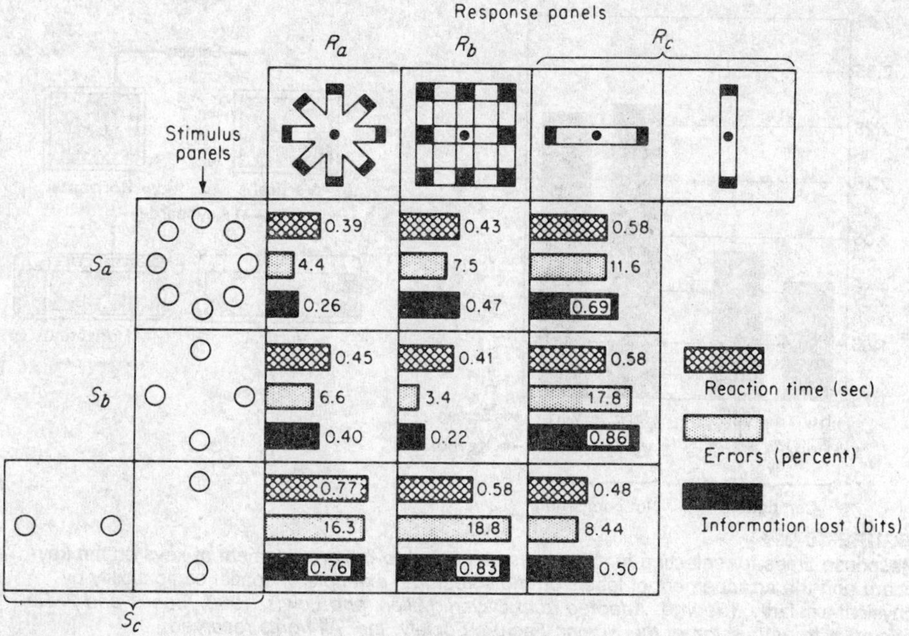

FIGURE 10-1.
Illustrations of signal (stimulus) panels and response panels used by Fitts and Seeger. The values in any one of the nine squares are the average performance measures for the combination of stimulus panel and response panel in question. The compatible combinations are S_a-R_a, S_b-R_b, and S_c-R_c, for which results are shown in the diagonal cells. (*Source: Fitts and Seeger, 1953.*)

also presented in Figure 10-1, showed that performance was best when the stimulus panel physically resembled the response panel (i.e., combinations S_a-R_a, S_b-R_b, and S_c-R_c).

Consider a more modern example of spatial compatibility by physical similarity. Function keys are arranged on computer keyboards in two basic configurations, either as a row (or rows) across the top of the keyboard or in columns on one side as shown in Figure 10-2. The meanings of function keys depend on the software program being used and in some cases the same key will have different meanings in the same program depending on the task being performed. To aid users, some programs present labels for function keys on the screen. The labels can then be changed as the meaning of the keys change. Bayerl, Millen, and Lewis (1988) compared configurations for presenting function key labels on the screen when using keyboards with different function key arrangements as shown in Figure 10-2. The mean time taken to locate and press function keys (response time) for each combination of screen and keyboard layout is also shown in Figure 10-2. When the labels are arranged in a manner physically similar to the arrangement of the keys on the keyboard, response time is much faster.

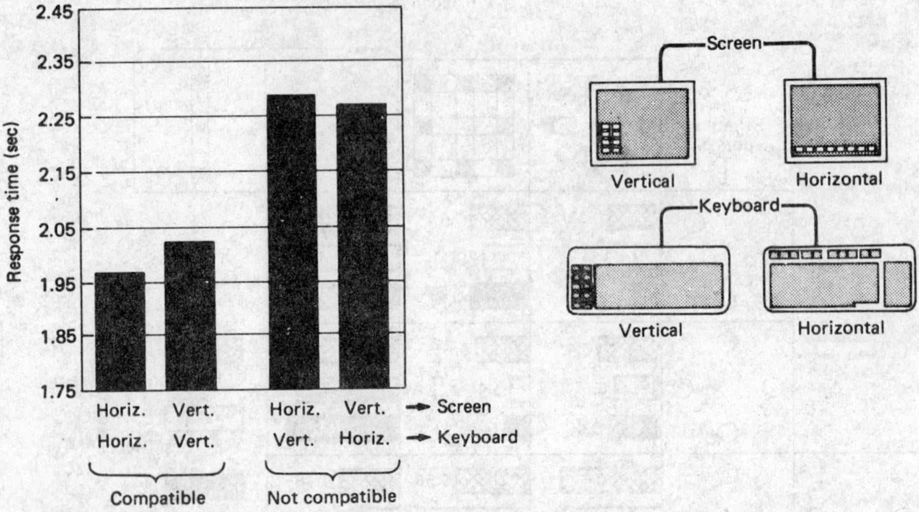

FIGURE 10-2.
Response times for selecting function keys as related to the arrangement of keys on the keyboard and the arrangement of labels on the screen: an example of spatial compatibility by physical similarity. (*Source: Adapted from Bayerl, Millen, and Lewis, 1988, Figs. 1 and 2. Reprinted with permission of the Human Factors Society, Inc. All rights reserved.*)

Physical Arrangement of Displays and Controls Both experiments and rational considerations lead one to conclude that, for optimum use, corresponding displays and controls should be arranged in corresponding patterns. This aspect of compatibility has been put to the test by a down-to-earth gadget used morning, noon, and night (and sometimes in between)—the arrangement of the burner controls on a four-burner stove. Several studies have investigated this problem. Chapanis and Lindenbaum (1959) and Ray and Ray (1979) presented various arrangements of controls and burners (such as those shown in Figure 10-3) to subjects and asked them to turn on specific burners. The number of errors was recorded. The results from both studies are shown in Table 10-1. As can be seen, the results of the two studies are in complete agreement with respect to the relative rankings of those arrangements tested by both. Taking both studies together, we see that arrangement I, with the offset burners, was clearly the best. When burners are aligned, arrangement II appears to be superior.

In a further study dealing with burner controls, Shinar and Action (1978) asked subjects to indicate which of the unmarked controls they thought controlled each of the burners (the burners were aligned rather than being offset). Referring to Figure 10-3, the most frequently chosen arrangement was III, being chosen by 31 percent of the subjects. Arrangement II, which resulted in fewer errors than arrangement III in the other studies, was chosen by only 25

TABLE 10-1
PERCENTAGE OF ERRORS IN EXPERIMENTAL USE OF BURNER CONTROLS ON STOVES SHOWN IN FIGURE 10-3

Design	Chapanis and Lindenbaum (1959)	Ray and Ray (1979)
I	0	Not tested
II	6	9
III	10	16
IV	11	19
V	Not tested	12

Source: Chapanis and Lindenbaum (1959) and Ray and Ray (1979).

percent of the subjects. This suggests that the arrangements people choose may not always result in the optimum levels of performance. It is probably best, therefore, to use actual performance rather than rely on subjective preference or choice to decide on arrangements between controls and displays.

Osborne and Ellingstad (1987) extended prior research on arranging stove burners and controls by considering the effect of adding to the stove top what they called *sensor lines,* that is lines drawn from controls to the displays to

FIGURE 10-3.
Control-burner arrangements of a simulated stove used in experiments by Chapanis and Lindenbaum, and by Ray and Ray. (*Source: Adapted from Chapanis and Lindenbaum, 1959.*)

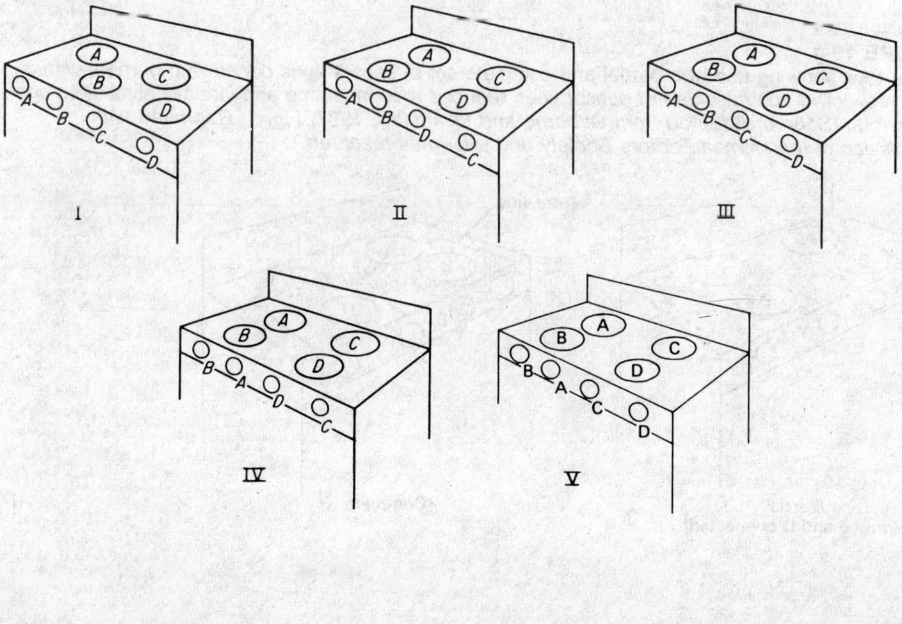

which they are linked. They added either a partial or a complete set of sensor lines as shown in Figure 10-4. The arrangement is the same as arrangement II in Figure 10-3. They also included the offset arrangement (I) of Figure 10-3 for comparison. The results showed that adding sensor lines (either a partial or complete set) essentially eliminated control errors. Although response time was reduced (by 113 ms) by adding sensor lines, responses were still faster to the offset arrangement (917 ms) than they were to the arrangements with partial or complete sensor lines (997 and 980 ms, respectively). So if you or someone you know has a stove where the burners are not offset, add sensor lines to the stove top to reduce response time and virtually eliminate errors. (Yet another example of human factors working for you.)

Movement Compatibility

In several different types of circumstances movement compatibility becomes important. The following are some examples:

- Movement of a control device to *follow* the movement of a display (moving a level to the right to follow a right movement of a blip on a radarscope)
- Movement of a control device to *control* the movement of a display (tuning a radio to a particular wavelength)
- Movement of a control device that produces a specific system response (turning a steering wheel to the right to turn right)
- Movement of a display indication without any related response (the clockwise turn of the hands of a clock)

People's expectations regarding movement relationships are often referred to as *population stereotypes,* with some stereotypes being stronger than others.

FIGURE 10-4.
Stove tops showing use of a partial and complete set of sensor lines connecting burners with their respective controls. Use of sensor lines reduced response time and virtually eliminated control errors. (*Source: Adapted from Osborne and Ellingstad, 1987, Fig. 1. Reprinted with permission of the Human Factors Society, Inc. All rights reserved.*)

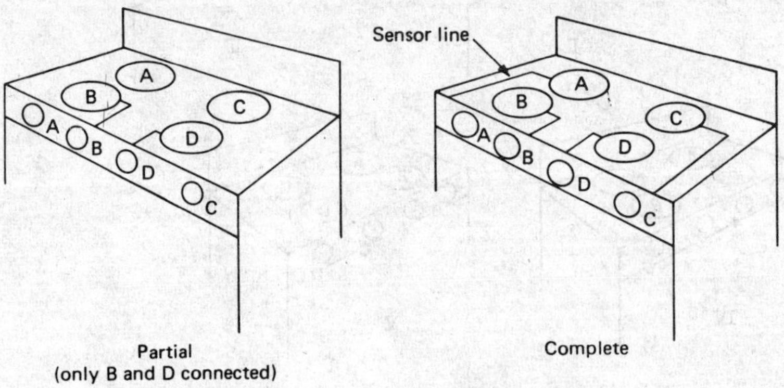

Movement compatibility relationships depend in part on features of the controls and displays as well as their physical orientation to the user (such as whether they are in the same or different planes relative to the user). On the basis of research and experience, certain principles of movement compatibility have merged, some of which are discussed now.

Rotary Controls and Rotary Displays in Same Plane In the case of fixed rotary scales with moving pointers, the principle that has become firmly established is that a clockwise turn of the control should be associated with a clockwise turn of the pointer, and that such rotation generally should indicate an increase of the value in question. Conversely, a counterclockwise movement of the control and display pointer should be associated with a decrease of the value.

For moving scales with fixed pointers Bradley (1954) postulated that the following principles generally would be desirable:

1 The scale should rotate in the same direction as its control knob (i.e., there should be a direct drive between control and display).
2 The scale numbers should increase from left to right.
3 The control should turn clockwise to increase settings.

Unfortunately, it is not possible to incorporate all three of these principles into a single fixed-pointer moving-scale display and rotary control assembly. As shown in Figure 10-5, only two of the principles can be followed at the same time in these types of situations. Since one of the principles must therefore be sacrificed, which should it be? To answer that question, Bradley tested various control-display assemblies, including the four shown in Figure 10-5. The following criteria were used to evaluate the designs: starting errors (an initial movement in the wrong direction), setting errors (incorrect settings), and rank-order preferences of the subjects. Some of the results are shown at the bottom of Figure 10-5.

From the results it is clear that direct-drive linkage between the control and display (assemblies *A* and *B*) is the most important principle to preserve. The next most important principle appears to be that the scale numbers increase from left to right. Of lesser importance is that a clockwise control movement results in an increased setting.

Rotary Controls and Linear Displays in the Same Plane In the case of rotary controls and fixed-scale linear displays, the control can be placed above, below, to the left, or to the right of the display. Various compatibility principles have been defined for such circumstances. The first is *Warrick's principle* (Warrick, 1947), which can be described as the expectation that the pointer on the display will move in the same direction as that side of the control which is nearest to it. This principle applies only when the control is located to the side of the display.

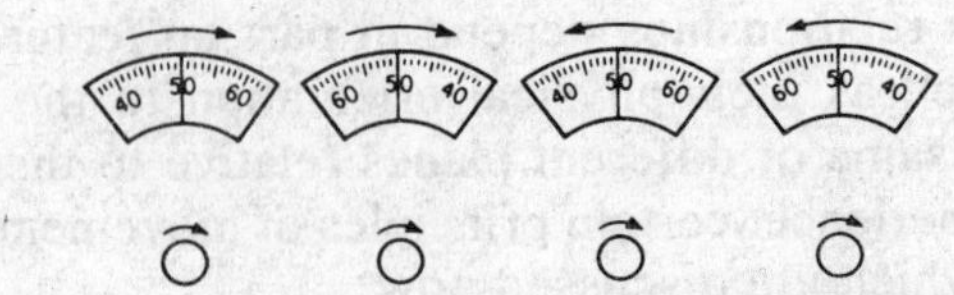

Assembly	A	B	C	D
Drive	Direct	Direct	Reversed	Reversed
Scale numbers increase	Left to right	Right to left	Left to right	Right to left
With clockwise knob movement setting will:	Decrease	Increase	Increase	Decrease

	A	B	C	D
Starting errors	13	11	87	106
Setting errors	0	9	1	8
Preference (number of times ranked "first")	31	22	17.5	1.5

FIGURE 10-5.
Some of the moving-display and control-assembly types used in a study by Bradley. The various features relate to three desirable characteristics given below the diagrams; crosshatching indicates an undesirable feature. With the usual display orientation all three desirable features are not possible. Some data on three criteria are given at the bottom of the figure, indicating the general preferability of *A*. (*Source: Adapted from Bradley, 1954.*)

Brebner and Sandow (1976) formulated two other principles that apply to the use of rotary controls with linear displays (specifically to vertical displays). The first is called the *scale-side principle*, that is, the expectation that the pointer will move in the same direction as the side of the control knob which is on the same side as the scale markings on the display. This principle operates when the control is at the top, bottom, or side of a vertical display. The second principle is called *clockwise-for-increase principle* which states that people will turn a rotary control clockwise to increase the value on the display no matter where the control is located relative to the display.

Warrick's principle, the clockwise-for-increase principle, and the scale-side principle were compared and contrasted in experiments by Brebner and Sandow (1976) and by Petropoulos and Brebner (1981) with various arrangements of displays and controls. Subjects were shown slides and were asked to indicate in which direction they would turn the knob to "move the indicator to 15." Figure 10-6 presents some of the conditions tested, the predictions based on the three principles, and the results obtained from the two studies. There was a major discrepancy between the two studies with regard to arrangement *C* which may have been due to differences in methodology. In general, however, it appears that when principles clash, as in arrangements *A* and *C*, the strength of the stereotype is weaker than when the principles are congruent, as in arrangements *B* and *D*.

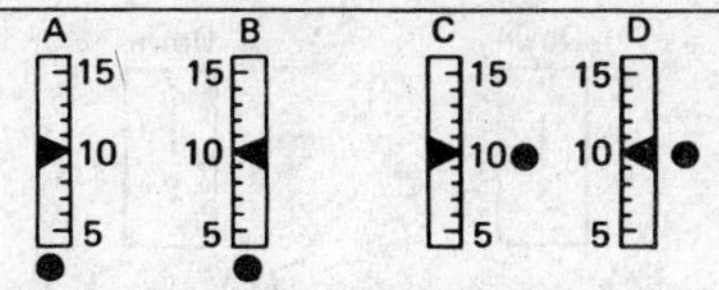

	A	B	C	D
Predictions:				
Warrick's principle	NA	NA	C	C
Scale-side principle	CC	C	CC	C
Clockwise-for-increase	C	C	C	C
Results: Percent choosing:				
Clockwise	43 (45)	80 (72)	73 (57)	86 (85)
Counterclockwise	57 (55)	20 (28)	27 (43)	14 (15)

NA = not applicable C = clockwise CC= counterclockwise

FIGURE 10-6.
Four configurations of rotary controls and vertical linear scales. Shown are the predicted stereotypes based on three principles. The percentages choosing each direction of rotation to move the pointer to 15 are shown for two studies: Brebner and Sandow (1976) and, in parentheses, Petropoulos and Brebner (1981).

Rotary knobs can be placed above the displays to which they relate, but this is generally considered to be a poor idea because the operator's hand will block the display while using the control. Thus, if the control is placed below the display (probably the best location since the hand will not block the display), arrangement *B* in Figure 10-6 should be used. If the control is placed to the side, arrangement *D* is recommended with the understanding that a left-hander can block the display when operating the control.

Movement of Displays and Controls in Different Planes Sometimes control devices may be in a different plane from the displays with which they are associated. Several studies have been conducted to investigate stereotypes associated with rotary and stick-type controls and linear displays positioned in different planes. Holding (1957), for example, found that with rotary controls, people's responses tended to be based on two principles: (1) a general clockwise-for-increase principle and (2) a helical, or screwlike, tendency in which clockwise rotation is associated with movement away from—and counterclockwise movement is associated with movement toward—the individual.

Spragg, Finck, and Smith (1959) investigated the four combinations of control and display movements shown in Figure 10-7 in connection with a tracking task. For the horizontally mounted stick, on the vertical plane, the superiority of the up-up relationship (control movement up associated with display movement up) over the up-down relationship is evident in the results shown in Figure 10-7. For a vertically mounted stick, on the horizontal lateral-cutting plane, there was less difference between the forward-up and forward-down relationships.

Based on these studies and others, Grandjean (1988) recommended the movement compatibility relationships shown in Figure 10-8 for rotary and stick-type controls and linear displays located in various planes.

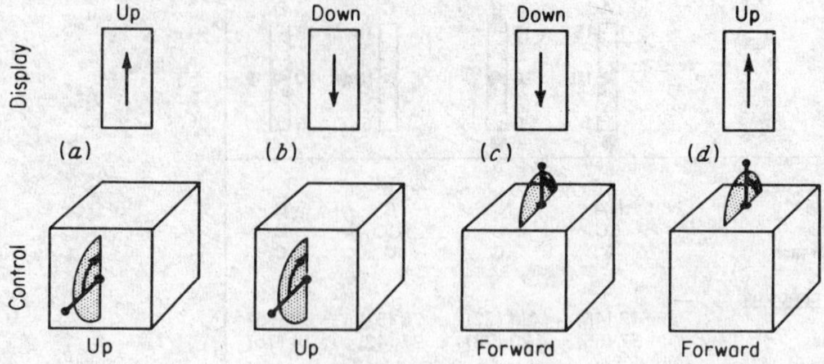

Average Tracking Score	239	149	221	227

FIGURE 10-7.
Tracking performance with horizontally mounted and vertically mounted stick controls and varying control-display relationships. (*Source: Adapted from Spragg, Finck, and Smith, 1959, data based on trials 9 to 16.*)

Simpson and Chan (1988) express some reservations about applying some of the results and recommendations in Figures 10-7 and 10-8 to three-dimensional tasks in which a control lever controls the up-down movement of a physical machine component. Examples include controlling the up-down movement of booms, scoops, or drill presses. Using a three-dimensional mock-up of a machine Simpson and Chan had subjects move a lever to control various machine components. The control lever could operate in three orientations: up-down, left-right, and fore-aft. They found that to move a machine component up, there was a strong stereotype for moving the control forward; to move the component down, the stereotype was to move the control lever aft. A weaker stereotype was found for moving the control down to move the component up,

FIGURE 10-8.
Recommended movement relationships for rotary and stick-type controls and linear displays located in various planes. (*Source: Grandjean, 1988, Fig. 112.*)

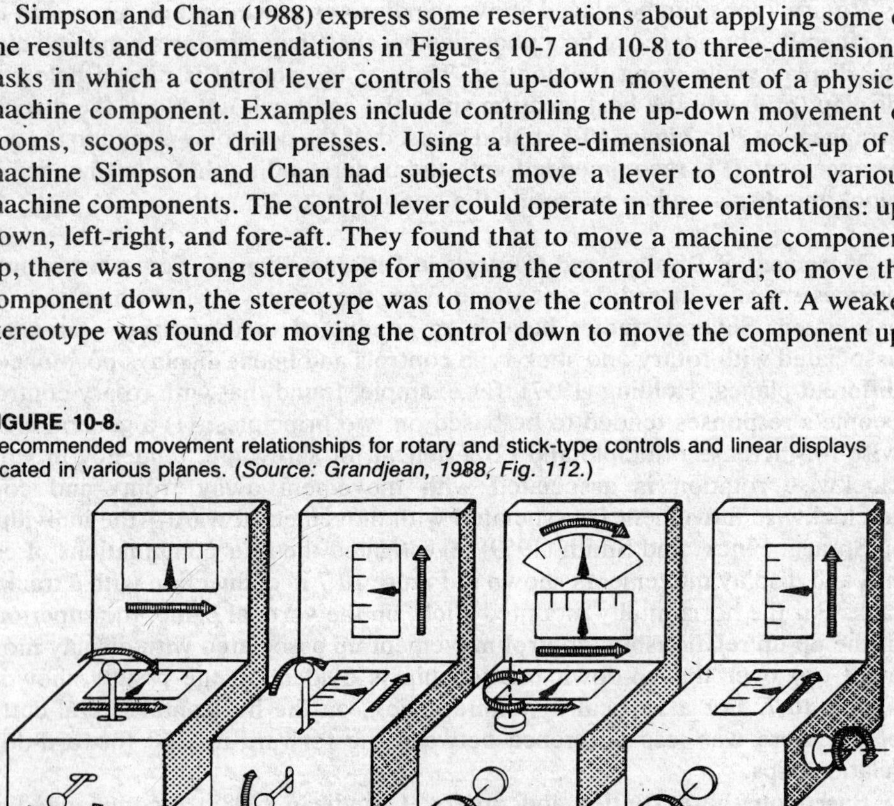

and up to move the component down. In fact, less than 50 percent of the subjects adopted the up-up/down-down relationship recommended in Figures 10-7 and 10-8. Apparently, these subjects (mostly engineers and machine operators) adopted a seesaw model wherein pushing down the control lever raised the machine component. Given the relative strengths of these stereotypes, it would perhaps be safer to use a fore-aft control movement to raise and lower machine components rather than an up or down control movement.

Movement Relationships of Rotary Vehicular Controls In the operation of most vehicles there is no "display" to reflect the "output" of the system; rather, there is a "response" of the vehicle. In such instances, if the wheel control is in a horizontal plane, an operator tends to assume an orientation toward the forward point of the control, as shown in Figure 10-9*a* (Chapanis and Kinkade, 1972). If the wheel is in a vertical plane, the operator tends to orient to the top of the control, as shown in Figure 10-9*b*.

A rather horrendous control problem one of the authors has seen is with shuttle cars for underground coal mines that have a control wheel that controls left and right turns. This wheel is on the right-hand side of the car relative to the driver when going in one direction, but when going in the other direction the driver moves to face the direction of travel, so that the wheel is then on the driver's left. The control relationships of this are so complicated that new drivers have major problems in learning how to control the cars.

FIGURE 10-9.
The most compatible relationships between the direction of movement of horizontally and vertically mounted rotary controls and the response of vehicles. (*Source: Adapted from Chapanis and Kinkade, 1972, Figs. 8-6 and 8-7.*)

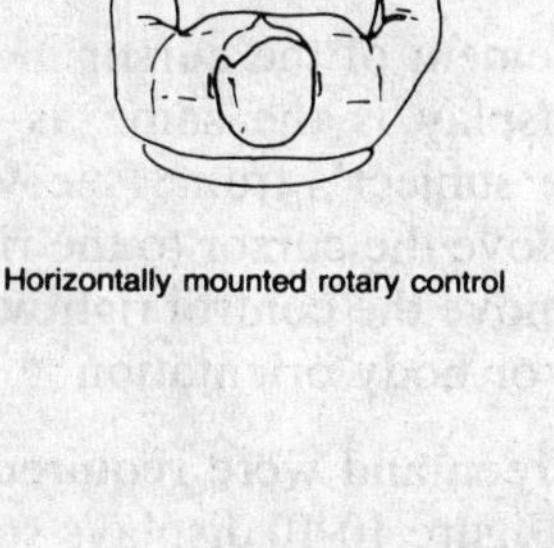

(*a*) Horizontally mounted rotary control

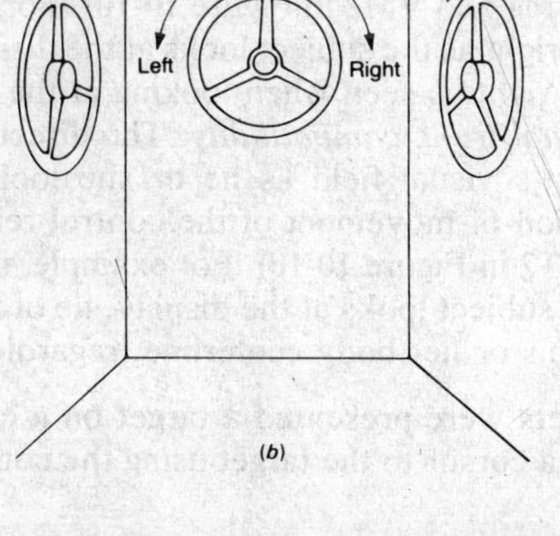

(*b*)

Movement Relationships of Power Switches In the United States the most universally accepted stereotype for power switches is up for on and down for off. (In Great Britain, on the other hand, the stereotype is just the opposite.) But what happens when the power switch is mounted to operate left-right or toward-away; which way is on? Lewis (1986) conducted a little study to find out. The following results were found (the percentage of subjects choosing the option is contained in parentheses):

Up = on (97%)
Right = on (71%)
Away = on (52%)

Lewis suggests that we stick with vertical power switches and discourage other orientations.

Orientation of the Operator and Movement Relationships In our previous discussions, we tacitly assumed that the operator faced the display and the control was located essentially in front of his or her body. There are situations, however, in which the operator looking at the display is not looking in the direction of the control, but might be looking 90° or 180° from the control. An example is adjusting a car's right-side mirror with a remote control located either on the dashboard in front of the driver (in which case, the mirror being controlled is 90° to the right of the control), or on the driver's door (in which case, the mirror is 180° to the right of the control). Worringham and Beringer (1989) investigated display-control relationships in such situations. They distinguished three types of directional compatibility:

Control-display compatibility: A control movement in a given direction produces a parallel movement of the cursor on the display, independent of operator position or orientation (see CD1 and CD2 in Figure 10-10). For example, in CD1, to make the cursor move in a given direction the operator moves the control in the same direction, and it does not matter which direction the operator is facing.

Visual-motor compatibility: The direction of motion of the cursor in the subject's visual field as he or she looks at the display is the same as the direction of the motor response if he or she were looking at the controlling limb (see VM1 and VM2 in Figure 10-10). For example, in VM1, to move the cursor to the right as the subject looks at the display, he or she would move the control to the right as seen when looking at the control.

Visual-trunk compatibility: The direction of movement of the cursor in the subject's visual field as he or she looks at the display is the same as the direction of movement of the control relative to the subject's trunk (see VT1 and VT2 in Figure 10-10). For example, in VT1, to move the cursor to the right as the subject looks at the display, he or she would move the control rightward from his or her body centerline, regardless of head or body orientation.

Subjects were presented a target on a computer screen and were required to move a cursor to the target using the control lever. Figure 10-10 displays some

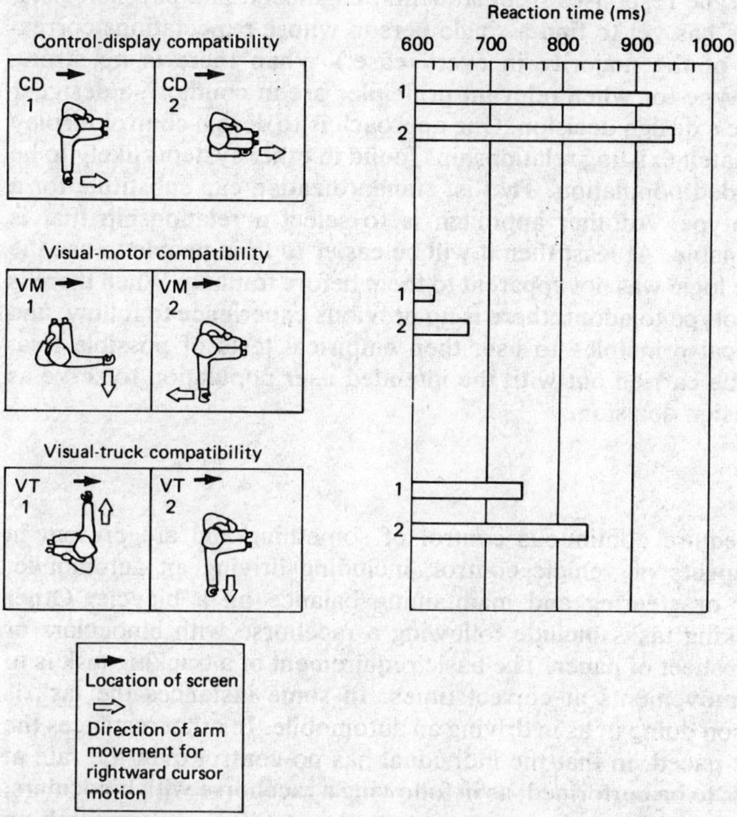

FIGURE 10-10.
Relationships between direction of arm movement and cursor movement for various conditions investigated by Worringham and Beringer. See text for explanation of control-display, visual-motor, and visual-trunk compatibilities. Shown also are the mean reaction times (time to first movement) found in each situation. (*Source: Adapted from Worringham and Beringer, 1989, Fig. 2.*)

of the orientations investigated and the mean reaction times (time to first movement) found. Similar results were found for movement time. Quite clearly, visual-motor compatibility resulted in the fastest reaction times. Although not shown in Figure 10-10, even faster reaction times were found in situations in which the operator could look at both the display and control at the same time and the cursor moved in the same direction as the control moved. These situations all involved visual-motor compatibility as well as control-display compatibility.

Discussion

Although fairly clear-cut population stereotypes do exist for certain control-display relationships, these are by no means universal. (After years of soliciting population stereotype responses from students, engineers, and psychologists, one of the authors has yet to find a single person whose expectations correspond with those of the majority in every case.) When there is no strong population stereotype (or when relevant principles are in conflict), a designer still needs to make a design decision. One approach is to design control-display relationships to match existing relationships found in other systems likely to be used by the intended population. That is, standardization can substitute for a population stereotype. Another approach is to select a relationship that is logical and explainable. At least then it will be easier to train people to use the system even if the logic was not apparent to them before training. When there is no clear-cut stereotype to adopt, there is no previous experience to follow, and there are no logical principles to use, then empirical tests of possible relationships should be carried out with the intended user population to serve as the basis for a design decision.

TRACKING

Tracking tasks require continuous control of something and are present in practically all aspects of vehicle control, including driving an automobile, piloting a plane, or steering and maintaining balance on a bicycle. Other examples of tracking tasks include following a racehorse with binoculars or tracing a line on a sheet of paper. The basic requirement of a tracking task is to execute correct movements at correct times. In some instances the task is paced by the person doing it, as in driving an automobile. In other instances the task is externally paced, in that the individual has no control over the rate at which the task has to be performed, as in following a racehorse with binoculars.

The analysis of tracking tasks can be extremely complex, and research on tracking has been primarily engineering-oriented, focusing on mathematical representations of the responses of well-trained operators. We do not deal extensively with such topics, but the interested reader can consult Kelley (1968), Poulton (1974), and Wickens (1986) for more extensive treatments of the field. In this chapter we discuss briefly some background regarding certain concepts involved in tracking and some inklings of the many variables associated with tracking performance.

Inputs and Outputs in Tracking

In a tracking task, an input, in effect, specifies the desired output of the system; for example, the curves in a road (the input) specify the desired path to be followed by an automobile (the output).

Inputs Inputs in a tracking task can be constant (e.g., steering a ship to a specified heading or flying a plane to an assigned altitude) or variable (e.g., following a winding road or chasing a maneuvering butterfly with a net). Such input typically is received directly from the environment and sensed by mechanical sensors or by people. If it is sensed mechanically, it may be presented to operators in the form of signals on some display. The input signal is sometimes referred to as a *target* (and in certain situations it actually is a target), and its movement is called a *course*. Whether the input signal represents a real target with a course or some other changing variable such as desired changes in temperature in a production process, it usually can be described mathematically and shown graphically as a function of time. While in most instances the mathematical or graphic representations do not depict the real geometry of the input in spatial terms, such representations still have utility for characterizing the input.

Inputs can take on complex patterns over time as anyone who has ever chased an elusive butterfly can tell you. Despite this, several elementary patterns can be distinguished which, when combined, form the most complex inputs. The elementary inputs are the step input, ramp input, and sine input. Figure 10-11 shows these three forms of input. A *sine input* is characterized by a sinusoidal (i.e., sine) wave. A *step input* specifies a discrete change in value, such as an instruction from a control tower for an aircraft to change altitude or the need to increase the temperature control of an oven at some time. A *ramp input* specifies a constant rate of change of some variable. Since velocity is the rate of change in position, maintaining a constant velocity (i.e., speed) of a car would be an example of tracking a ramp input.

Outputs The output is usually brought about by a physical response with a control mechanism (if by an individual) or by the transmission of some form of energy (if by a mechanical element). In some systems the output is reflected by some indication on a display, sometimes called a *follower,* or a *cursor;* in other systems it can be observed by the outward behavior of the system, such as the movement of an automobile. In either case it is frequently called the *controlled element*.

Pursuit and Compensatory Displays in Tracking

The input (the target) and the output (the controlled element) can be presented on either a *pursuit* display or a *compensatory* display, as illustrated in Figure 10-12. With a pursuit display both indications move, each showing its own location in space in relationship to the other. If the relative movement of the two elements is represented by a single value (a single dimension), they can be shown on a scale such as those in the upper part of Figure 10-12, whereas if this movement needs to be depicted in two dimensions (as in chasing a moving aircraft), the display might look like that in the lower left part of Figure 10-12. The task of the operator is to align the moving cursor with the moving target. In a compensatory display, using either one or two dimensions (shown to the right in Figure 10-12), only one of the two indicators moves and the other is fixed. The task for the operator is to get the moving indicator (cursor) to align with the fixed indicator (target).

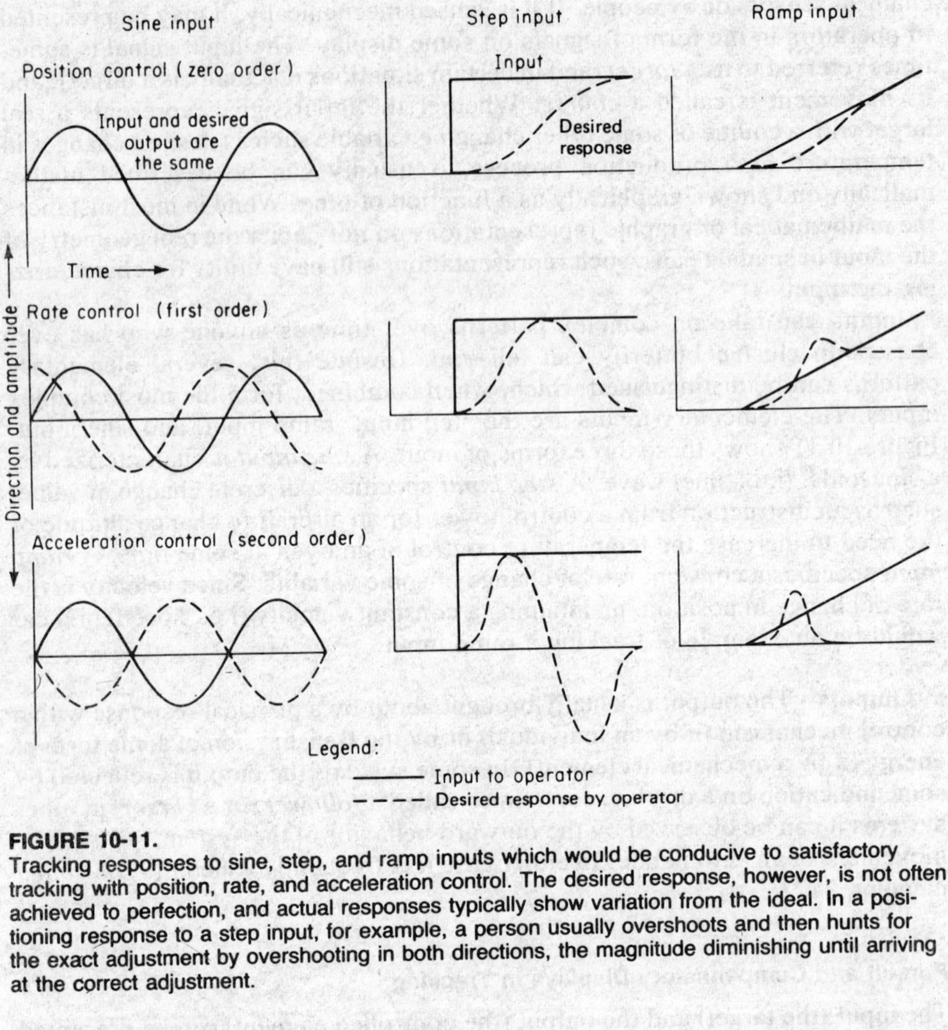

FIGURE 10-11.
Tracking responses to sine, step, and ramp inputs which would be conducive to satisfactory tracking with position, rate, and acceleration control. The desired response, however, is not often achieved to perfection, and actual responses typically show variation from the ideal. In a positioning response to a step input, for example, a person usually overshoots and then hunts for the exact adjustment by overshooting in both directions, the magnitude diminishing until arriving at the correct adjustment.

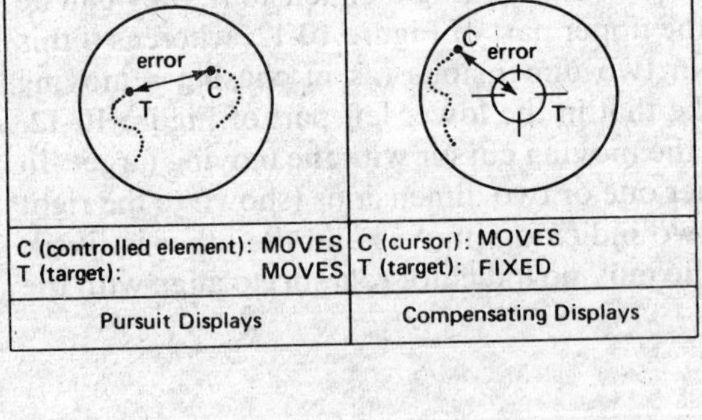

FIGURE 10-12.
Illustrations of compensatory and pursuit-tracking displays. A compensatory tracking display shows only the *difference* (error) between the target *T* and the controlled element *C*. A pursuit display shows the location (or other value represented) of both the target and the controlled element.

When the two indicators are superimposed, by using either a compensatory or pursuit display, the controlled element is said to be *on target;* any difference represents error, and the function of the operator is to manipulate the controls to eliminate or minimize that error. With a pursuit display, the operator can determine whether the error is due to target movement or the movement of the controlled element, and further, she or he can see the target's course independent of any movements of the controlled element. With a compensatory display, however, only the absolute error, or difference between the target and controlled element, is shown. If the error begins to increase, the display does not indicate whether the target has moved, the controlled element has moved, or both have moved. A practical advantage of compensatory displays, however, is that they conserve space on an instrument panel because they do not need to represent the whole range of possible values or locations of the two elements.

Control Order of Systems

Control order refers to what we might call the hierarchy of control relationships between the movement of a control and the output it controls. The nature of control orders and their implications for tracking tasks are very intricate, and we do not go too deeply into this topic. To give some impression of control orders, however, let us begin with the concept of *position*. This refers to the output of the system, with the objective of the tracking task being to control the system so that the output corresponds as closely as possible to the input. (The input, of course, specifies what the output should be.) Although we can readily envision the "position" of a vehicle in space, let us think of it in a broader context as a measure of the output. It might be shown as a moving pointer on a scale, a blip on a radar screen, an index of the revolutions per minute of a machine, or something else.

Position (Zero-Order) Control In a *position-control* tracking task, the movement of the control device controls the position *directly,* such as moving a spotlight to keep it on an actor on a stage, following the movements of an athlete with a camera, or (in a laboratory task) following a moving curved line with a pen or other device. If the system involves a display, there is a direct relationship between the control movement and the display movement it produces.

Rate (First-Order) Control With a *rate-control* system, the direct effect of the operator's movement is to control the rate at which the output is being changed. The accelerator of an automobile is a rate or first-order control device since it controls the speed (rate of change of position) of the automobile. In particular, the distance by which the pedal is depressed controls the speed, although there is a time lag between depressing the pedal and the speed controlled by the pedal. In turn, the speed of the automobile controls its position along the road. The operation of certain machine guns is controlled by hand cranks that control the rate at which the gun changes direction, and would involve rate control.

Acceleration (Second-Order) Control *Acceleration* is the rate of change in the rate of movement of something. The operation of the steering wheel of an automobile is an example of acceleration control since the angle at which the wheel is turned controls the angle of the front wheels. In turn, the direction in which the automobile wheels point determines the rate at which the automobile turns. Thus, a given rotation of the steering wheel gives the automobile a corresponding acceleration toward its turning direction. Poulton (1974, p. 324) makes the point that the control of certain chemical plants or nuclear reactors may approximate an acceleration or second-order system.

Higher-Order Control Certain control systems can be considered as *higher-order* systems, such as third- or fourth-order control. A third-order control, for example, would involve the direct control of the rate of change of the acceleration (jerk), that then controls the rate, that finally controls the position of whatever is being controlled. The control of a large ship could approximate a third-order or even a fourth-order control because of the linkages between the person steering and the actual movement and position of the ship and the mass (that is, the weight) of the ship. In continuous control processes that have a series of control linkages (such as a ship), the sequence of chain-reaction effects can be described in terms of mathematical functions, such as a change in the *position* of one variable changing the *velocity* (rate) of the next, the *acceleration* of the next, etc.

Control Responses with Various Control Orders

The operators of many systems are expected to make control responses to bring about the desired operation of the system as implied by the input (such as a road to be followed, the flight path of an aircraft, or the appropriate temperature control over time of an industrial process). In the absence of any scheme for helping the operator, such control can be complicated with higher-level control orders. Examples of the appropriate responses for ramp, step, and sine inputs with position, rate, and acceleration control systems are shown in Figure 10-11. In each case the dotted line represents the response over time (along the horizontal) that would be required for satisfactory tracking of the input in question. In general, the higher the order of control, the greater is the number of controlled movements that need to be made by an operator in response to any single change in the input, as illustrated in Figure 10-11. With respect to tracking performance, people do about as well with a zero-order control system as they do with a first-order control system, but performance really deteriorates when a second-order control system is used. Tracking error can increase from 40 to 100 percent (Wickens, 1986).

As an example, consider a step input (the target jumps forward and stops) and the responses to it from using a joystick (the middle column of diagrams in Figure 10-11). In the case of a position control (zero order), the response is simply to move the stick forward the proper distance and hold it. With a rate control (first order), the stick is moved forward to impart a particular velocity to the controlled element. As the controlled element approaches the position of the stationary target, the stick must be returned to the null position (zero velocity) to stop the controlled element before the element overshoots the position of the target. In the case of an acceleration control (second order), things really get interesting, as shown at the bottom of Figure 10-11. With a step input, the control is moved forward, which imparts an acceleration (increasing speed) to the controlled element. The stick must then be brought back to the null position (zero acceleration) which leaves the controlled element moving at a constant velocity. To avoid overshooting the target, the stick must be moved in the opposite direction to decelerate the controlled element. The stick cannot be held there, however, or else the controlled element will stop and start accelerating in the opposite direction. The stick, therefore, must be returned to the null position so that no acceleration or deceleration is being imparted to the controlled element. And all this is needed just to move the controlled element forward and stop it! It is no wonder that performance deteriorates with second-order control systems.

Human Limitations in Tracking Tasks

Humans, unfortunately, are not very good at tracking tasks, especially when they involve higher-order control parameters. Wickens (1984) identifies specific information processing limitations that affect tracking performance, i.e., processing time, bandwidth, and anticipation.

Processing Time People do not process information instantaneously, hence there is a time delay between a change in a target and the initiation of the responses required to track the target. The magnitude of the time delay is dependent on the order of the system being controlled (McRuer and Jex, 1967). The time delay with zero- and first-order systems is about 150 to 300 ms. For a second-order system, the delay is on the order of 400 to 500 ms. These delays are harmful to tracking performance because the operator, in essence, is always chasing the target and is always somewhat behind it unless some means are provided to help the operator. We discuss a few of these techniques later in this chapter.

Bandwidth *Bandwidth* refers to the upper limit of frequency with which corrective decisions can be made, and hence the term defines the maximum frequency of a random input that can be successfully tracked. This bandwidth is normally between 0.5 and 1.0 Hz (Elkind and Sprague, 1961). Since two corrections are required per cycle, a 1-Hz limit corresponds to a limit of roughly two corrections per second. This limit appears to be a central processing limit rather than a motor response limit because people have no difficulty in tracking predictable courses as high as 2 to 3 Hz (Pew, 1974).

Anticipation Often operators must track targets by using systems that have time lags or that respond sluggishly to control inputs (ships and planes are examples). This requires that the operator anticipate future errors based on present conditions and then make control responses that are expected to reduce that anticipated future error. Unfortunately, humans are not very good at anticipating future outputs, especially for slow, sluggish systems. Part of this difficulty is due to the limitations inherent in working memory. Making the calculations necessary to predict the future state of a higher-order system can stress all but the most experienced operators.

Factors That Influence Tracking Performance

The effectiveness of human control of tracking operations is influenced by a wide variety of factors such as the nature of the displays and of the controls and the features of the tracking system. A few such factors are discussed briefly below.

Preview of Track Ahead In some tracking situations an individual has some preview of the track ahead, as in driving an automobile on a winding road that can be seen ahead. (In a blinding snowstorm or heavy fog, the individual does not have such a preview.) In general terms some preview of the input assists an operator in a tracking task. Although the nature of the task presumably can influence the possible benefit of a preview, Kvälseth (1979) indicates that such preview is most beneficial if the preview shows the portion of the track that immediately precedes the present position, rather than showing a lagged preview with a gap between what is previewed and the present position. The duration of the preview seems to be of less consequence than the opportunity to have at least some preview. Kvälseth (1978*a*) indicates that performance improves steadily as the preview span approaches approximately 0.5 s, beyond which the usefulness of the preview does not increase much. The primary advantage of preview is that it enables the operator to compensate for time lags. The 0.5-s preview corresponds well with the response time (lag) of operators using higher-order systems. When there are other time lags in the system, larger previews are expected to be beneficial.

Type of Display: Pursuit versus Compensatory In reviewing the research relative to the possible merits of visual analog pursuit versus compensatory displays, Poulton (1974, p. 166) concluded that when there is a choice between the two, a conventional pursuit (true motion) display is preferable to a compensatory (relative motion) display. His argument was based on a summary of numerous studies, as shown in Table 10-2. In the case of the seven studies in

TABLE 10-2
SUMMARY OF COMPARISON OF NUMEROUS STUDIES OF PURSUIT AND COMPENSATORY TRACKING

Control order	Number of studies	Difference in results reliably in favor of: Pursuit	Compensatory	Results inconclusive
Zero-order	45	29	0	16
Higher-order	34	14	7*	13

*These studies used inappropriate experimental methods, and the results should probably be disregarded.
Source: Poulton, 1974.

which compensatory tracking was reported to be best and three of the studies involving zero-order control included in the inconclusive column, Poulton argued that the results could be attributed to inappropriate experimental designs and therefore the results should probably be disregarded. As an illustration of the advantage of pursuit displays, Figure 10-13 shows the results of one study comparing the two display types (Briggs and Rockway, 1966).

One reason pursuit displays are generally better than compensatory displays is that the operator can see the separate effects of target and controlled-element movements on the error generated. This makes it easier to predict the target's course and to learn the consequences of various control actions on the movement of the controlled element. When the target is fixed (as, for example, when a ship is docking) and the displayed error is a consequence of only the movement of the controlled element, the advantage of a pursuit display is diminished (Wickens, 1984).

Another advantage of pursuit displays is that they involve greater movement compatibility. If a target suddenly moves to the left, the pursuit display shows a movement to the left and the correct response is a leftward movement to chase the target—a compatible arrangement. With a compensatory display, however, the left-moving target is displayed as a right-moving error, while the correct

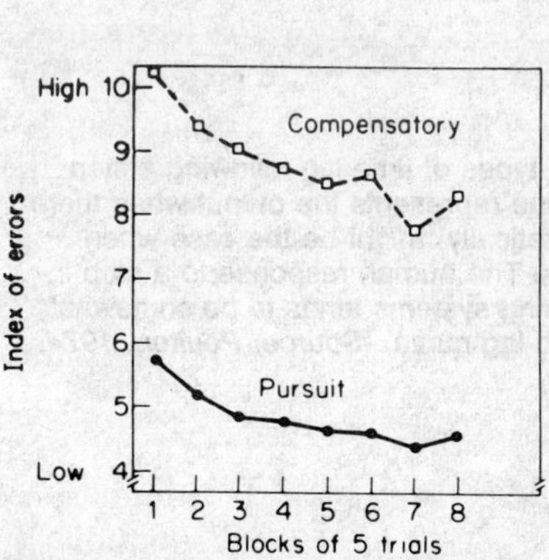

FIGURE 10-13.
Comparison of errors of subjects when using a compensatory versus a pursuit display in a tracking task. (*Source: Adapted from Briggs and Rockway, 1966, Fig. 2. Copyright © 1966 by the American Psychological Association. Reprinted by permission.*)

response remains a leftward control movement—obviously a less compatible arrangement than with a pursuit display.

Kvälseth (1978*b*) compared pursuit and compensatory displays by using digital displays (like counters) rather than the traditional analog displays. With digital displays, it is more difficult for users to visualize the target's course, since only a series of changing numbers is shown. As might be expected under such conditions, no difference in performance was found between the two types of displays.

Although pursuit displays generally result in better tracking performance than compensatory displays (at least with visual analog presentations), practical considerations may sometimes argue for the use of compensatory displays, especially because they may occupy less space on control panels.

Time Lags in Tracking In a tracking task when an error is detected by the operator, three actions need to take place: (1) the operator must choose and execute a corrective response, (2) the system being controlled must then respond to the control input, and (3) the result must be displayed to the operator. Each of these activities takes time; i.e., each introduces a *time lag* into the task. In general, time lags tend to degrade operator performance. This is due in part to the greater demands placed on working memory and the increased need to anticipate future events when lags are present.

There are several types of time lags. *Response lag* is the time taken by the operator to make a response to an input. *Control-system lag* is the time between a control action of an operator and the response of the system under control. For sluggish, higher-order systems, such as a ship or large aircraft, it may be several seconds before the system responds to an operator's control actions. *Display-system lag* is the delay between the response of the system being controlled or a change in the target and the display of that response or change.

Control-system lag has been a major focus of research on time lags. There are three basic types of control lag, these being illustrated in Figure 10-14 for a step input. Of these, *transmission time lag* simply delays the effect of a

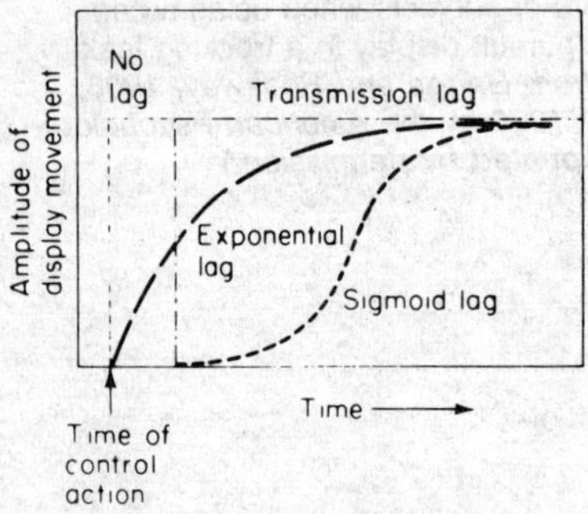

FIGURE 10-14.
Illustration of three types of time lag following a step input. The dotted line represents the output when there is no lag, as (theoretically) might be the case when turning on a switch. The human response to a step input with most tracking systems tends to be somewhat similar to a sigmoid lag curve. (*Source: Poulton, 1974, Fig. 11.9, p. 206.*)

person's response; the output follows the control response by a constant time interval. An *exponential lag* refers to a situation in which the output is represented by an exponential function following a step input. And a *sigmoid time lag* is represented by the S-shaped curve of Figure 10-14.

The effects of various types of lags are intricately related to the various features of the tracking system and are not discussed here in detail. As an example of the effects of lags on performance, Kao and Smith (1978) present data that indicate that in a compensatory tracking task involving bimanual (two-handed) versus unimanual (one-handed) control, increasing time lags, up to 0.8 s, resulted in performance degradation for both forms of control; but with a longer display lag (1.5 s) bimanual performance was more adversely affected than unimanual performance. The results are shown in Figure 10-15.

Aside from some of the intricacies of lags with various types of tracking systems, Poulton (1974, p. 373) points out that all three types of control lag increase the error in tracking. Although such lags are generally undesirable, there are sometimes ways to minimize these effects. For example, Poulton (p. 378) indicates that a design engineer may be able to reduce the effective order of a system from acceleration control to rate control by introducing an approximately exponential lag. As another illustration of methods of minimizing the effects of lag, Rockway (1954) has shown that, with long (as opposed to short) delays, the control-response ratio (C/R ratio) can have some effect on tracking performance. The *C/R ratio* is the ratio of the amount of movement of the control device as related to the amount of corresponding movement of the display indication (see Chapter 11). In his study, Rockway found that high C/R ratios (1:3 and 1:6) resulted in performance degradation with long lags, whereas low C/R ratios (1:15 and 1:30) resulted in the maintenance or even improvement of performance with such lags. Thus, in some circumstances there are schemes that can be used to minimize the possible effects of time lags.

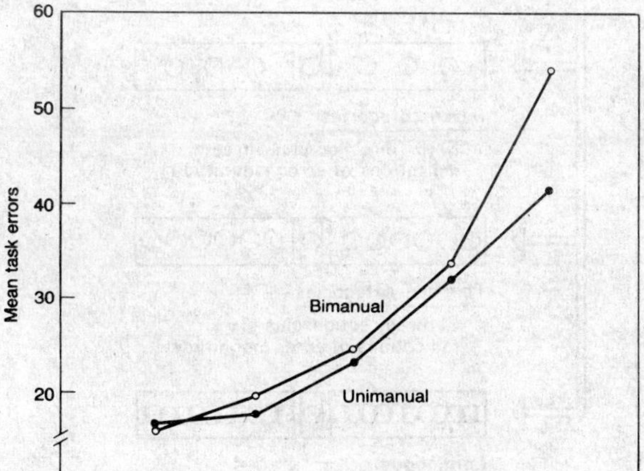

FIGURE 10-15.
Illustration of the effects of various display lags in a compensatory tracking task performed with bimanual and unimanual control. The difference for the 1.5 s display lag is statistically significant. (*Source: Kao and Smith, 1978, Fig. 2, p. 666.*)

Specificity of Displayed Error in Tracking In certain compensatory tracking systems the error (the difference between input and output) can be presented in varying degrees of *specificity*. Some such variations are shown in Figure 10-16 as they were used in a tracking experiment by Hunt (1961), including 3 categories of specificity (left, on target, and right); 7 categories; 13 categories; and continuous. The accuracy of tracking performance under these conditions (for two levels of task difficulty) indicates quite clearly that performance improved with the number of categories of information (greater specificity), this improvement taking a negatively accelerated form for tasks of both levels of difficulty. Although the results of other studies are not entirely consistent with these, the evidence suggests that in a tracking task, performance is facilitated by the presentation of more specific, rather than less specific, display information.

Paced versus Self-Paced Tracking Most tracking tasks are self-paced, in that the person has control over the rate of the output, as in driving an automobile. In such an activity the driver usually can select the speed. With some tracking tasks, however, the pace is not under the individual's control. Poulton (1974, p.8), for example, refers to the fact that an airline pilot, when

FIGURE 10-16.
Compensatory displays used in study of the effects of specificity of feedback error information and tracking performance. The feedback error was presented by the use of lights (3, 7, 13, and continuous). (*Source: Adapted from Hunt, 1961.*)

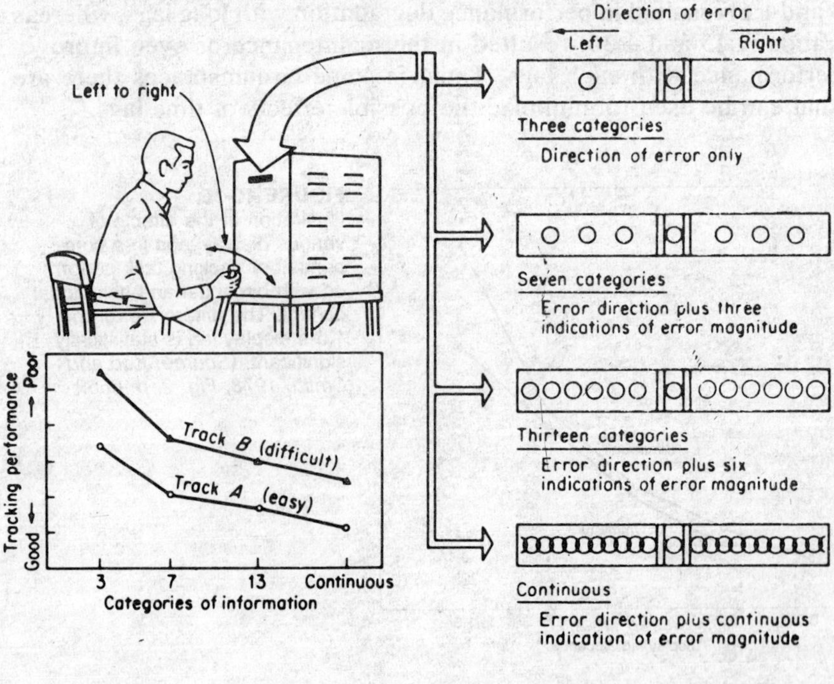

coming in to land, has to hold airspeed within close limits as well as keep the plane within a closely defined glide path. Tracking is easiest when the task is self-paced and increases in difficulty as the degree of external pacing is increased.

Procedures for Facilitating Tracking Performance

As we can see, humans have built-in time lags and limited bandwidths and are poor at anticipating future system states; in addition, control order, preview, and time lags influence tracking performance. These limitations and influences are especially important with higher-order systems (second-, third-, or fourth-order systems). Often the characteristics of such systems exceed the capabilities of the operators. Something, therefore, needs to be done to compensate for this disparity. A few such procedures are discussed below.

Aiding One such procedure is the use of *aiding*. Aiding was initially developed for use in gunnery tracking systems and is most applicable to tracking situations of this general type, in which the operator is following a moving target with some device. Its effect is to modify the output of the control in order to help the tracker. In *rate aiding* a single adjustment of the control affects both the rate and the position components of the tracking system. Suppose we are trying to keep a high-powered telescope directed exactly on a high-flying aircraft by using a rate-aided system. When we fall behind the target, our control movement to catch up again automatically speeds up the *rate* of motion of our telescope (and thus, of course, its position). Similarly, if our telescope gets ahead of the target, a corrective motion automatically slows down its rate (and influences its position accordingly). Such rate aiding simplifies the problem of quickly matching the rate of motion of the following device to that of the target and thus improves tracking performance. In *acceleration aiding* the control movement controls three variables of the controlled element, namely, acceleration, rate, and position.

The operational effect of aiding is to shift from the operator the mental operations that approximate those of differentiation, integration, and algebraic addition which are required in some tracking tasks, so the operator's choice is primarily that of amplification. (In effect, the operator simply has to figure out the ratio of the movement of the control to that of the controlled element.) The operational effects of aiding depend on a number of factors, such as the nature of the input signal, the control order, and whether the system is a pursuit or compensatory type. Thus, aiding should be used selectively, in those control situations in which it is uniquely appropriate.

Predictor Displays Still another procedure for simplifying the control of high-order systems (especially large vehicles such as submarines) is the use of predictor displays, as proposed primarily by Kelley (1962, 1968). In effect, predictor displays use a fast-time model of the system to predict the future excursion of the system (or controlled variable) and display this excursion to the operator on a scope or other device. The model repetitively computes predictions of the real system's future, based on one or more assumptions about what the operator will do with the control (e.g., return it to a neutral position, hold it where it is, or move it to one or another extreme). The predictions so generated are displayed to enable the operator to reduce the difference between the predicted and desired output of the system. A distinguishing feature of a predictor display is that it shows the present state of the system as well as the predicted future state.

An example of a predictor display is seen in Figure 10-17, which shows the present and predicted depth errors of a submarine over a 10-s period. A computer-generated stylized representation of an aircraft predictor display is shown in Figure 10-18. This display presents the present and predicted positions of the aircraft 8 s in the future during a landing approach.

Predictor displays offer particular advantages for complex control systems in which the operator needs to anticipate several seconds in advance, such as with submarines, aircraft, and spacecraft. The advantages in such situations have been demonstrated by the results of experiments such as the one by Dey (1972) simulating the control of a VTOL (vertical takeoff and landing) aircraft. The values of one index of the deviations from a desired "course" with and without the use of a predictor display were 2.48 and 7.92, respectively. The evidence from this and other experiments shows a rather consistent enhancement of control performance with a predictor display.

Quickening Quickening is closely related to predictor displays in that the display presents information about where the system will likely be in the future given the action (or inaction) taken by the operator. Thus, the operator learns quickly (hence the name *quickening*) the future consequences of a control action. The distinction between a quickened display and a predictor display, however, is that a quickened display does not present information about the present position or state of the system, whereas a predictor display does. A quickened display looks like a regular display except it shows what the situation will be in the future rather than what the situation is at present. An

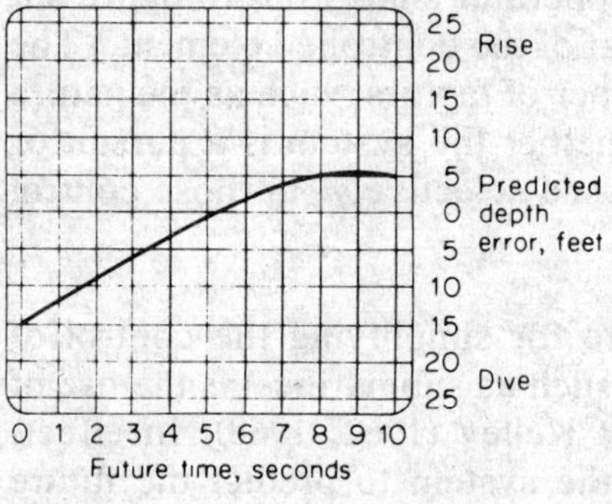

FIGURE 10-17.
Example of a predictor display for a submarine. The display shows the predicted depth error, in feet, extrapolated to 10 s, assuming that the control device would be immediately returned to a neutral position. (*Source: Kelley, 1962.*)

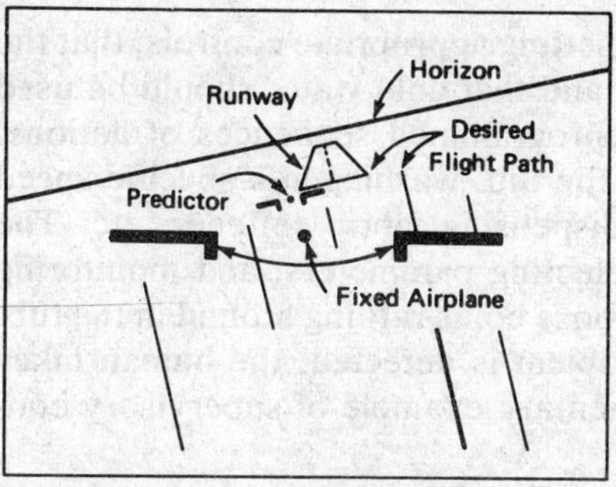

FIGURE 10-18.
Stylized representation of an aircraft predictor display showing the present and predicted position of an aircraft 8 s in the future during a landing approach. (*Source: Jensen, 1981, Fig. 1. Copyright by the Human Factors Society, Inc. and reproduced by permission.*)

operator using a quickened display can be lulled into a false sense of security. The fact that the quickened display indicates that the controlled element is on target does not necessarily mean that *at that moment* the controlled element is indeed on target. It is only an indication that in the future the element is likely to be on target—within the accuracy constraints of the equations used to predict the future state. Quickening can be of some benefit under specific conditions, but usually predictor displays will result in superior tracking performance.

Discussion The relative utility of the various methods of facilitating performance of tracking tasks becomes inextricably intertwined with the type of input (step, ramp, sine wave, etc.), with the control order (zero, first, second, etc.), and with the type of display (pursuit or compensatory). Aiding can be useful in certain circumstances but has limited general applicability. In connection with quickening, Poulton (1974, pp. 180–185) raised serious questions about the research strategies used in some studies and also referred to certain disadvantages of quickened displays. In general Poulton concluded that true motion predictor displays are likely to be far easier and safer to use for control systems of high order than quickened displays.

Note that quickened and predictor displays are not an either-or proposition. They can be combined in various proportions in a single display (Roscoe, Corl, and Jensen, 1981), and performance can be improved under some circumstances by combining them (Jensen, 1981).

SUPERVISORY CONTROL

Our discussion of tracking focused on the human as a direct controller of the system, sensing the difference between the target and the controlled element and making motor responses to reduce that difference. With the continued improvement in computer technology and advances in expert systems (see Chapter 3), the computer has become an intelligent intermediary between the human and the operation of the system. Take a very simple example that captures the idea of an intermediary between human and process: a washing machine. The human tells the machine, by setting appropriate controls, that the wash load is composed of delicate fabrics and that cold water should be used for wash and rinse. An intermediary (preprogrammed sequences of actions) then carries out the specific task of filling the tub, washing at a specific speed and for a given length of time, rinsing, dispensing fabric softener, etc. The human's role involves setting the goal, selecting parameters, and monitoring the process for possible problems such as loose coins rattling around in the tub, or suds overflowing the machine. If a problem is detected, the human takes corrective action. This represents an elementary example of supervisory control.

The term *supervisory control* in the context of human-computer systems derives from the close analogy of it to characteristics of a supervisor's interaction with subordinate human staff members (Sheridan, 1987). A supervisor plans and sets goals and then gives general directives to his or her staff members, who, in turn, translate the directive into detailed actions. The staff also collects and transforms detailed information into summary form for the supervisor. The degree of intelligence of the staff members determines the level of involvement of the supervisor in the process. In human-computer systems, an "intelligent" computer serves the function of the subordinate staff member, receiving information from the human and directing lower-level task actions, receiving information from lower-level processes and summarizing the information for the human.

Table 10-3 summarizes 10 levels of human-computer interaction relevant to supervisory control. As can be seen, the levels imply greater and greater autonomy for the computer and a smaller and smaller role for the human. This list makes two things clear. First, automation is not an all-or-none affair; there are degrees of automation. Second, there are interesting and challenging design

TABLE 10-3
A TAXONOMY OF HUMAN-COMPUTER INTERACTION RELEVANT TO SUPERVISORY CONTROL

1	Human does the whole job up to the point of turning it over to the computer to implement.
2	Computer helps by determining the options.
3	Computer helps determine options and suggests one, which human need not follow.
4	Computer selects action and human may or may not do it.
5	Computer selects action and implements it if human approves.
6	Computer selects action, informs human in plenty of time to stop it.
7	Computer does whole job and tells human what it did.
8	Computer does whole job and tells human what it did only if the human explicitly asks.
9	Computer does whole job and tells human what it did and what it, the computer, decides the human should be told.
10	Computer does whole job if it decides it should be done, and if so, tells human if it decides the human should be told.

Source: Sheridan, 1980, Table 40.10. Reprinted with permission from *Technology Review*, copyright 1980.

decisions to be made regarding the roles of humans and computers in system design. The question of how to allocate functions between humans and computers is a topic we discuss in Chapter 22.

Supervisory Roles

Sheridan (1987) lists five major categories of behaviors encompassed in the role of supervisor: planning, teaching, monitoring, intervening, and learning. The supervisor plans, in part, by deciding on goals, strategies, and contingency actions. Teaching involves deciding how to instruct the computer system to carry out the plans. Monitoring includes deciding what information to receive, receiving it, and determining the meaning of it. Intervening occurs when the human decides that automatic control will not produce satisfactory performance. Learning relates to the process by which the human modifies his or her knowledge of how the system operates and incorporates that new knowledge into future operations.

There is very little data available on the entire process of supervisory control; however, specific components have received attention, notably monitoring and troubleshooting. Moray (1986) provides a good overview of these topics.

A somewhat different approach to characterizing supervisory control involves the behavior taxonomy of Rasmussen (1983). Rasmussen distinguishes three types of behavior: skill-based, rule-based, and knowledge-based. *Skill-based behavior* is the highly practiced, nearly effortless behavior, characteristic of skilled manual control (for example, riding a bicycle or playing the piano). At the other end is *knowledge-based behavior,* which involves high-level situation assessment and evaluation. Such behavior is typified during unfamiliar situations for which control rules based on previous experience are unavailable or inappropriate. Knowledge-based behavior involves setting goals, selecting strategies, and thinking through or testing the consequences of a plan. In the middle between skill-based and knowledge-based behavior is *rule-based behavior*. Here behavior is controlled by stored rules or "know-how" (for example, instructions given by someone, a cookbook recipe, rules of thumb developed through prior experience). These three types of behaviors should be considered not as distinct categories but rather on a continuum, with each type blending into the next type. In supervisory control, the human assumes more of a knowledge-based role while the intelligent computer provides the skill-based behavior and may provide rules for rule-based behavior.

Rasmussen also discusses the role information plays with each type of behavior. At the skill-based level, information serves as *signals*. These signals indicate whether performance is on or off target or indicates when to initiate a behavior. The signals have no meaning beyond the moment. Rule-based behavior uses information as *signs*. Signs must be interpreted to assess their meaning. The interpretation allows the person to modify or select rules for controlling behavior. The signs may be interpreted differently in different situations. For example, during landing of an aircraft, a meter reading of 150 may indicate normal conditions, but during takeoff, the same reading may invoke a different rule that initiates a sequence of skill-based behaviors (switches may be thrown, throttles advanced, etc.). At the knowledge-based level, information serves as *symbols*. Symbols are necessary for reasoning and represent information about the conceptual model of the system that the person has in his or her head (called a *mental model*). The meter reading of 150 during takeoff may be a symbol of a leaking hydraulic line or a valve stuck in the open position. In the context of supervisory control, information received by the operator from the intelligent computer is more in the nature of signs and symbols than signals. How to properly present such information is, and will continue to be, a challenge for human factors specialists.

Future Implications of Supervisory Control

Supervisory control holds the promise of freeing us from the tedious manual control of systems and elevating us to more dignified higher-level cognitive behavior for which we, as humans, are uniquely suited. One immediate problem that supervisory control can create, however, is that the operator may be less capable of intervening in the control of the system during an emergency (such as when the intelligent computer breaks down). The operator may not know what the intelligent computer was doing or what strategy it was employing when the emergency occurred. Further, the operator may lack the necessary skills required to take control because such skills are not practiced when the system is working properly. The possibility of human error during an emergency, therefore, may well increase under supervisory control.

Sheridan (1980, 1987) raises some long-term social implications of supervisory control:

1 Unemployment.

2 Desocialization: The trend in supervisory control is toward fewer people per work team, and eventually one person will be adequate in most situations. People will talk more with computers than with people.

3 Remoteness from the product: Supervisory control removes people from hands-on interaction with the work product.

4 Deskilling: Skilled workers may resent the transition to supervisory controller because of fear that when and if called on to take over and do the job manually, they may not be able to. They may feel loss of professional identity built up over an entire working life.

5 Intimidation: The stakes are higher if the person does something wrong.

6 Discomfort in the assumption of power: Some people may have anxiety about taking responsibility for a complex process or may become arrogant in their role.

7 Technological illiteracy: People may lack the technological understanding of how the computer does what it does. They may come to resent this and the elite class that does understand it.

8 Mystification: People could become mystified and superstitious about the power of the computer, even seeing it as a kind of "big brother" authority.

9 Sense of not being productive.

10 Eventual abandonment of responsibility: Individuals may eventually feel that they are no longer responsible for what happens; the computers are.

These implications are frightening. It is, therefore, incumbent upon system designers and human factors specialists to insure that steps are taken in the design of systems and the creation of supervisory control jobs to reduce or eliminate these possibilities. Humans serving in supervisory control roles must be given sufficient social contacts with other people and a feeling of worth by maintaining their old skills and identifying new skills. They must be made to feel comfortable with their new responsibilities, come to understand what the computer does and not be mystified by it, and realize that they are in charge of setting goals and criteria by which the system operates (Sheridan, 1987).

DISCUSSION

The basic functions of those who are to control a system or mechanism are rooted in the nature of the process or operation in question. However, specific design features of the system or mechanism predetermine the specific operational demands imposed on the controller, in particular the inputs that must be received, the types of decisions to be made, and the control actions to be taken. Considerations of human factors during the design stage can contribute to the design of systems and mechanisms that can be controlled adequately by human beings and that can contribute to the well-being of those involved in subsequent control operations.

REFERENCES

Bayerl, J., Millen, D., and Lewis, S. (1988). Consistent layout of function keys and screen labels speeds user responses. *Proceedings of the Human Factors Society 32d Annual Meeting*. Santa Monica, CA: Human Factors Society, pp. 344–346.

Bergum, B. D., and Bergum, J. E. (1981). Population stereotypes: An attempt to measure and define. *Proceedings of the Human Factors Society 25th Annual Meeting*. Santa Monica, CA: Human Factors Society, pp. 662–665.

Bradley, J. V. (1954). *Desirable control-display relationships for moving-scale instrument* (Tech. Rept. 54-423). Dayton, OH: U.S. Air Force, Wright Air Development Center.

Brebner, J., and Sandow, B. (1976). The effect of scale side on population stereotype. *Ergonomics*, 19(5), 571–580.

Briggs, G. E., and Rockway, M. R. (1966). Learning and performance as a function of the percentage of pursuit tracking component in a tracking display. *Journal of Experimental Psychology*, 71, 165–169.

Chapanis, A., and Kinkade, R. G. (1972). Design of controls. In H. P. Van Cott and R. G. Kinkade (eds.), *Human engineering guide to equipment design*. Washington, DC: Government Printing Office.

Chapanis, A., and Lindenbaum, L. (1959). A reaction time study of four control-display linkages. *Human Factors*, 1(4), 1–7.

Dey, D. (1972). The influence of a prediction display on a quasi-linear describing function and remnant measured with an adaptive analog-pilot in a closed loop. *Proceedings of Seventh Annual Conference on Manual Control* (NASA SP-281). Washington, DC: National Aeronautics and Space Administration.

Elkind, J. I., and Sprague, L. T. (1961). Transmission of information in simple manual control systems. *IEEE Transactions on Human Factors in Electronics*, HFE-2, 58–60.

Fitts, P. M., and Seeger, C. M. (1953). S-R compatibility: Spatial characteristics of stimulus and response codes. *Journal of Experimental Psychology*, 46, 199–210.

Grandjean, E. (1988). *Fitting the task to the man*. London: Taylor & Francis.

Holding, D. H. (1957). Direction of motion relationships between controls and displays in different planes. *Journal of Applied Psychology*, 41, 93–97.

Hunt, D. P. (1961). The effect of the precision of informational feedback on human tracking performance. *Human Factors*, 3, 77–85.

Jensen, R. S. (1981). Prediction and quickening in perspective flight displays for curved landing approaches. *Human Factors*, 23, 355–363.

Kao, H. S. R., and Smith, K. U. (1978). Unimanual and bimanual control in a compensatory tracking task. *Ergonomics*, 21(9), 661–669.

Kelley, C. R. (1962, March). Predictor instruments look to the future. *Control Engineering*, 86.

Kelley, C. R. (1968). *Manual and automatic control*. New York: Wiley.

Kvälseth, T. O. (1978*a*). Effect of preview on digital pursuit control performance. *Human Factors*, 20(3), 371–377.

Kvälseth, T. O. (1978*b*). Human performance comparisons between digital pursuit and compensatory control. *Ergonomics*, 21(6), 419–425.

Kvälseth, T. O. (1979). Digital man-machine control systems: The effects of preview lag. *Ergonomics*, 22(1), 3–9.

Lewis, J. (1986). Power switches: Some user expectations and preferences. *Proceedings of the Human Factors Society 30th Annual Meeting*. Santa Monica, CA: Human Factors Society, pp. 895–899.

McRuer, D. T., and Jex, H. R. (1967). A review of quasi-linear pilot models. *IEEE Transactions on Human Factors in Electronics*, 8, 231.

Moray, N. (1986). Monitoring behavior and supervisory control. In K. Boff, L. Kaufman, and J. Thomas (eds.), *Handbook of perception and human performance*, vol. 2: *Cognitive processes and performance*. New York: Wiley, pp. 40-1 to 40-51.

Osborne, D., and Ellingstad, V. (1987). Using sensor lines to show control-display linkages on a four burner stove. *Proceedings of the Human Factors Society 31st Annual Meeting*. Santa Monica, CA: Human Factors Society, pp. 581–584.

Petropoulos, H., and Brebner, J. (1981). Stereotypes for direction-of-movement of rotary controls associated with linear displays: The effects of scale presence and position, of pointer direction, and distances between the control and the display. *Ergonomics*, 24, 143–151.

Pew, R. W. (1974). Human perceptual-motor performance. In B. Kantowitz (ed.), *Human information processing*. Hillsdale, NJ: Erlbaum Associates.

Poulton, E. C. (1974). *Tracking skill and manual control*. New York: Academic.

Rasmussen, J. (1983). Skills, rules, and knowledge; signals, signs, and symbols, and other distinctions in human performance models. *IEEE Transactions on Systems, Man and Cybernetics,* SMC-13, 257–266.

Ray, R. D., and Ray, W. D. (1979). An analysis of domestic cooker control design. *Ergonomics,* 22, 1243–1248.

Rockway, M. R. (1954). *The effect of variations in control-display ratio and exponential time delay on tracking performance* (Tech. Rept. 54-618). U.S. Air Force, Wright Air Development Center.

Roscoe, S. N., Corl, L., and Jensen, R. S. (1981). Flight display dynamics revisited. *Human Factors,* 23, 341–353.

Sheridan, T. (1980, October). Computer control and human alienation. *Technology Review,* 61–73.

Sheridan, T. (1987). Supervisory control. In G. Salvendy (ed.), *Handbook of human factors*. New York: Wiley, pp. 1243–1268.

Shinar, D., and Acton, M. B. (1978). Control-display relationships on the four-burner range: Population stereotypes versus standards. *Human Factors,* 20(1), 13–17.

Simpson, G., and Chan, W. (1988). The derivation of population stereotypes for mining machines and some reservations on the general applicability of published stereotypes. *Ergonomics,* 31(3), 327–336.

Spragg, S. D. S., Finck, A., and Smith, S. (1959). Performance on a two-dimensional following tracking task with miniature stick control, as a function of control-display movement relationship. *Journal of Psychology,* 48, 247–254.

Warrick, M. J. (1947). Direction of movement in the use of control knobs to position visual indicators. In P. M. Fitts (ed.), *Psychological research on equipment design* (Research Rept. 19). Columbus, OH: Army Air Force, Aviation Psychology Program.

Wickens, C. D. (1984). *Engineering psychology and human performance*. Columbus, OH: Merrill.

Wickens, C. (1986). The effects of control dynamics on performance. In K. Boff, L. Kaufman, and J. Thomas (eds.), *Handbook of perception and human performance,* vol. 2: *Cognitive processes and performance*. New York: Wiley.

Worringham, C., and Beringer, D. (1989). Operator orientation and compatibility in visual-motor task performance. *Ergonomics,* 32(4), 387–400.

CHAPTER **11**

CONTROLS AND DATA ENTRY DEVICES

Human beings have demonstrated amazing ingenuity in designing machines for accomplishing things with less wear and tear upon themselves. Some of these machines perform more efficiently the functions that were once performed solely by hand, whereas others accomplish things that previously could only be dreamed of. Since most machines require human control, they must include control devices such as wheels, pushbuttons, or levers. It is through these control devices that people make their presence known to machines—that is, in addition to kicking the side of a candy machine. Computers are becoming part and parcel of our everyday lives, both on and off the job. Fortunately, computers cannot yet read our minds. In order to "communicate" with computers we have to use various devices to input commands and data. There is an amazing array of control devices and techniques available for inputting commands to computers, such as keyboards, mice, voice, and gestures. These devices and techniques, as well as the more traditional controls, must be selected and designed to be suitable for the desired control actions in terms of sensory, cognitive, psychomotor, and anthropometric characteristics of the intended users.

In this chapter we examine some of the more common types of control devices and some of the factors which influence their use by humans. We discuss hand controls, foot controls, data entry devices, and some other, more esoteric, control innovations. It is not our intent to recite a comprehensive set of recommendations regarding specific design features of such devices, although a partial set of such recommendations is given in Appendix B.

FUNCTIONS OF CONTROLS

The primary function of a control is to transmit control information to some device, mechanism, or system. The type of information so transmitted can be divided into two broad classes: discrete and continuous information. *Discrete* information is information that can represent only one of a limited number of conditions such as on-off; high-medium-low; boiler 1, boiler 2, boiler 3; or alphanumerics such as A, B, and C or 1, 2, and 3. *Continuous* information, on the other hand, can assume any value on a continuum, such as speed (as 0 to 60 km/h), pressure [as 1 to 100 lb/in^2 (0.07 to 7.03 kg/cm^2)], position of a valve (as fully closed to fully open), or amount of electric current (as 0 to 10 A). Visual display terminals (VDTs) often require the user to transmit to the system a special class of continuous information which we call *cursor positioning*. The information relates to the physical location of the cursor on the screen and is important for tasks such as selecting a word for editing, drawing a picture on the screen, or selecting a system function from a list of functions shown on the screen.

The information transmitted by a control may be presented in a display, or it may be manifested in the nature of the system response. The distinction between controls and displays is becoming more and more blurred with the advent of advanced technology. For example, push buttons are available that are, themselves, miniature dot-matrix displays (3.8 × 2.5 cm, or 1.5 × 1.0 in) which can be programmed to display messages or symbols (Micro Switch, 1984). VDTs can display push buttons and slider controls which can be activated by using touch-screen technology, an example of computer-generated controls.

Generic Types of Controls

There are numerous types of control devices available today. Certain types of controls are best suited for certain applications. One simple way to classify controls is based on the type of information they can most effectively transmit (discrete versus continuous) and the force normally required to manipulate them (large versus small). The amount of force required to manipulate a control is a function of the device being controlled, the mechanism of control, and the design of the control itself. Electric and hydraulic systems typically require small forces to actuate controls, whereas direct mechanical linkage systems may require large forces. Table 11-1 lists some of the more common types of controls and the types of applications for which they tend to be most suited. Certain of these types of controls are illustrated in Figure 11-1. Appendix B contains additional information about many of these control devices, including operational characteristics and specific design features.

FACTORS IN CONTROL DESIGN

Although certain types of controls might be considered more appropriate for one application than for another, the overall utility of the control can be greatly influenced by such factors as ease of identification, size, control-response ratio, resistance, lag, backlash, deadspace, and location. We discuss some of these factors here. Discussion of location of controls is deferred to Chapter 14.

TABLE 11-1
COMMON TYPES OF CONTROLS CLASSIFIED BY TYPE OF INFORMATION TRANSMITTED AND FORCE REQUIRED TO MANIPULATE

	Type of information transmitted		
		Continuous	
Force required to manipulate control	Discrete	Traditional	Cursor positioning
Small	Push buttons (including keyboards) Toggle switches Rotary selector switches Detent thumb wheels	Rotary knobs Multirotational knobs Thumb wheels Levers (or joysticks) Small cranks	Joysticks Trackballs Mice Digitizing tablets Touch tablets Light pens Touch screens
Large	Detent levers Large hand push buttons Foot push buttons	Handwheels Foot pedals Large levers Large cranks	

Identification of Controls

Although the correct identification of controls is not critical in some circumstances (as in operating a pinball machine), there are some situations, however, in which their correct and rapid identification is of major consequence—even of life and death. For example, McFarland (1946) cites cases and statistics relating to aircraft accidents that had been attributed to errors in identifying control devices. For example, confusion between landing gear and flap controls was reported to be the cause of over 400 Air Force accidents in a 22-month period during World War II. More recently (1989) a commercial jet crashed in New York's East River. Apparently, the copilot hit the disengage button during takeoff rather than a button that would have automatically accelerated the plane to a preset speed for takeoff. Two people died in the ensuing crash. It is with such circumstances in mind that control identification becomes important.

Gamst (1975) reports that, in some railroad locomotives, engineers intending to turn off the signal lights grasp the wrong control and actually turn off the fuel pumps instead, thus killing all the diesel engines in the train. Moussa-Hamouda and Mourant (1981) interviewed 405 automobile drivers with cars equipped with fingertip-reach controls attached to the steering column. Of the 405 drivers, 44 percent reported instances of inadvertent operation, for example, intending to operate the turn signal and instead turning on the windshield wipers.

The identification of controls is essentially a coding problem; the primary coding methods include shape, texture, size, location, operational method, color, and labels.

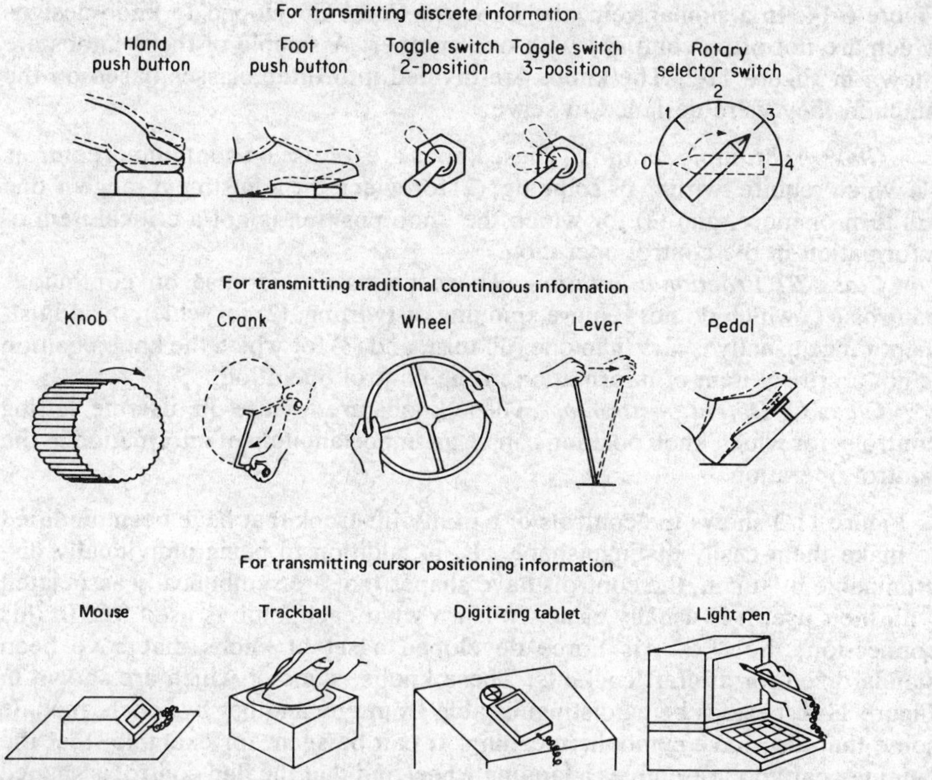

FIGURE 11-1
Examples of some types of control devices classified by the type of information they best transmit and the force required to activate them.

As discussed in Chapter 3, the choice of a coding method and the specific codes depends on the detectability, discriminability, compatibility, meaningfulness, and standardization of the codes selected. The utility of a coding method or specific set of codes typically is evaluated by such criteria as the number of discriminable differences that people can make (such as the number of shapes they can identify), bits of information transmitted, accuracy of use, and speed of use.

Shape Coding of Controls The discrimination of shape-coded controls essentially involves tactual sensitivity. One procedure used to select controls that are not confused with one another is to present all possible pairs of controls to blindfolded subjects who, by touch alone, indicate whether the controls are the same or different. It is then possible to determine which shapes were confused with other shapes. Using a procedure similar to this, Jenkins (1947*b*) identified two sets of eight knob shapes, such that the knobs within each group were rarely confused with one another. These two sets of knobs were presented in

Figure 6-14. In a similar vein, the U.S. Air Force developed 15 knob designs which are not often confused with one another. A sample of these knobs are shown in Figure 11-2. The knobs are divided into three classes based on the function they were designed to serve:

- *Class A: Multiple rotation.* These knobs are for use on continuous controls (1) which require twirling or spinning, (2) for which the adjustment range is one full turn or more, and (3) for which the knob position is not a critical item of information in the control operation.
- *Class B: Fractional rotation.* These knobs are for use on continuous controls (1) which do not require spinning or twirling, (2) for which the adjustment range usually is *less* than one full turn, and (3) for which the knob position is not a critical item of information in the control operation.
- *Class C: Detent positioning.* These knobs are for use on discrete setting controls, for which knob position can be an important item of information in the control operation.

Figure 11-3 shows the controls of a utility lift-truck that have been modified to make them easily distinguishable. If, in addition to being individually discriminable by touch, the controls have shapes that are symbolically associated with their use, it is usually easier to learn what each knob is used for. In this connection, the U.S. Air Force developed a set of knobs that have been standardized for aircraft cockpits. These knobs, some of which are shown in Figure 11-4, besides being distinguishable from one another by touch, include some that also have symbolic meaning. It can be seen, for example, that the landing-gear knob resembles a landing wheel and that the flap control is shaped like a wing.

Texture Coding of Controls In addition to shape, control devices can vary in surface texture. This characteristic was studied (along with certain other variables) in a series of experiments with flat cylindrical knobs such as those

FIGURE 11-2
Examples of knob designs for three classes of use that are seldom confused by touch. The diameter or length of these controls should be between 0.5 and 4.0 in (1.3 to 10 cm), except for class C, where 0.75 in (1.9 cm) is the minimum suggested. The height should be between 0.5 and 1 in (1.3 to 2.5 cm). (*Source: Adapted from Hunt, 1953.*)

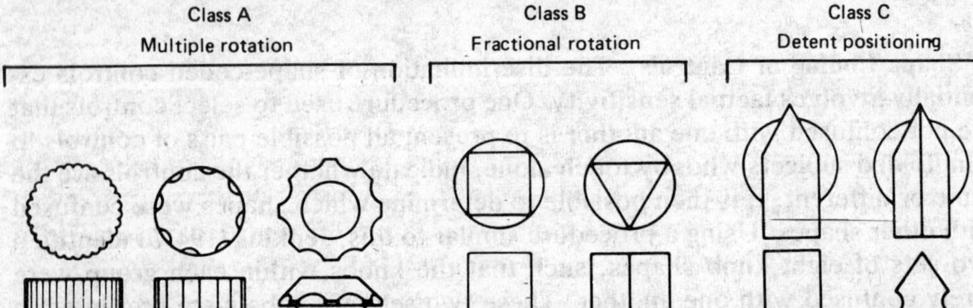

FIGURE 11-3
Controls of a utility lift truck that have been modified to make them easily distinguishable.

shown in Figure 11-5 (Bradley, 1967). In one phase of the study, knobs of this type [2-in (5.1-cm) diameter] were used, and subjects were presented with individual knobs through a curtained aperture and were asked to identify the particular design they felt. The smooth knob was not confused with any other, and vice versa; the three fluted designs were confused with one another but not with other types; and the knurled designs were confused with one another but not with other designs. Also with gloved hands and with smaller-sized knobs (in a later phase of the study) there was some cross-confusion among classes, but this was generally minimal. The investigator proposes that three surface characteristics can thus be used with reasonably accurate discrimination, namely,

FIGURE 11-4
Examples of some standardized shape-coded knobs for United States Air Force aircraft. A number of these have symbolic associations with their functions, such as a wheel representing the landing-gear control. (*Source: Air Force System Command, 1980.*)

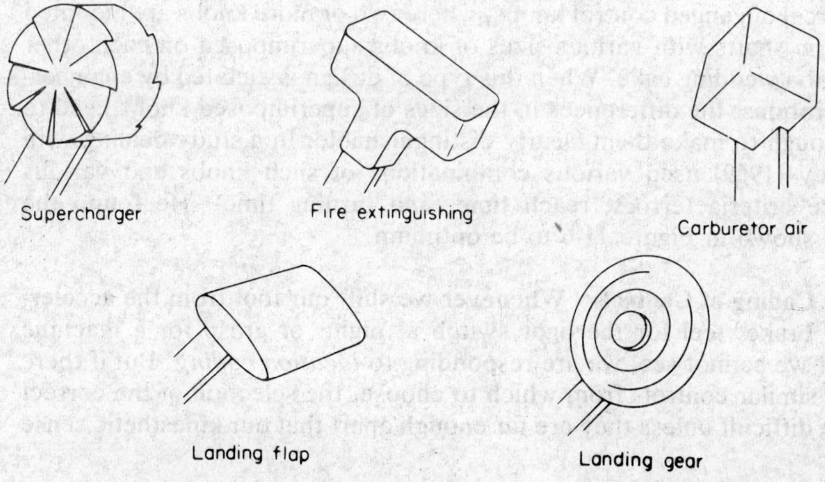

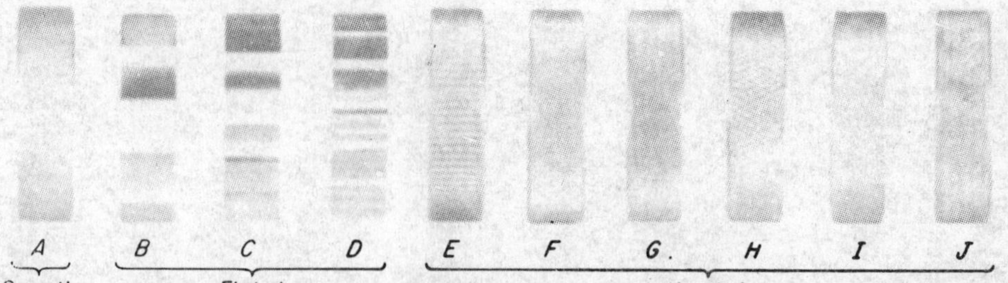

FIGURE 11-5
Illustration of some of the knob designs used in study of tactual discrimination of surface textures. Smooth: *A*; fluted: *B* (6 troughs), *C* (9), *D* (18); and knurled: *E* (full rectangular), *F* (half rectangular), *G* (quarter rectangular), *H* (full diamond, *I* (half diamond), and *J* (quarter diamond). (*Source: Bradley, 1967.*)

smooth, fluted, and knurled. Dirt and grime could also hinder discrimination, especially between smooth and knurled surfaces.

Size Coding of Controls Size coding of controls is not as useful for coding purposes as shape, but there may be some instances where it is appropriate. When such coding is used, the different sizes should, of course, be discriminable one from the others. Part of the study by Bradley reported above dealt with the discriminability of cylindrical knobs of varying diameters and thicknesses. It was found that knobs that differed by ½ in (12.7 mm) in diameter and by ⅜ in (9.5 mm) in thickness could be identified by touch very accurately, but smaller differences sometimes resulted in confusion. Incidentally, Bradley proposes that a combination of three surfaces textures (smooth, fluted, and knurled), three diameters [¾, 1¼, and 1¾ in (19.1, 31.8, and 44.4 mm)], and two thicknesses [⅜ and ¾ in (9.5 and 19.1 mm)] could be used in all combinations to provide 18 tactually identifiable knobs.

Aside from the use of coding for individual control devices, size coding is part and parcel of ganged control knobs, where two or more knobs are mounted on concentric shafts with various sizes of knobs superimposed on each other like layers of a wedding cake. When this type of design is dictated by engineering considerations, the differences in the sizes of superimposed knobs need to be great enough to make them clearly distinguishable. In a study dealing with this, Bradley (1969) used various combinations of such knobs and various performance criteria (errors, reach time, and turning time). He found the dimensions shown in Figure 11-6 to be optimum.

Location Coding of Controls Whenever we shift our foot from the accelerator to the brake, feel for the light switch at night, or grasp for a machine control that we cannot see, we are responding to *location coding*. But if there are several similar controls from which to choose, the selection of the correct one may be difficult unless they are far enough apart that our kinesthetic sense

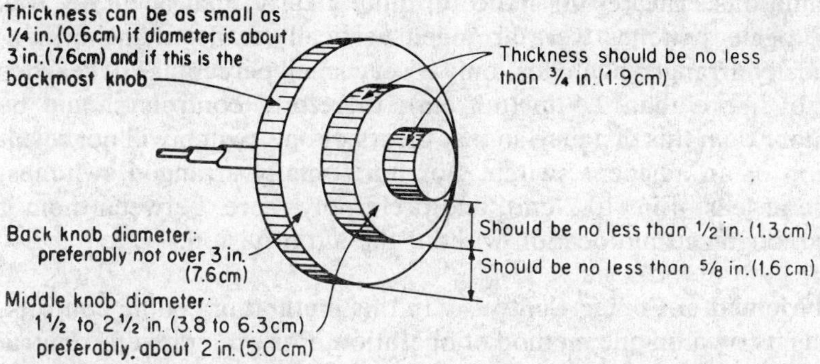

FIGURE 11-6
Dimensions of concentrically mounted knobs that are desirable in order to allow human beings to differentiate knobs by touch. (*Source: Adapted from Bradley, 1969.*)

makes it possible for us to discriminate. Some indications about this come from a study by Fitts and Crannell as reported by Hunt (1953). In this study blindfolded subjects were asked to reach for designated toggle switches on vertical and horizontal panels, the switches being separated by 1 in (2.5 cm). The major results are summarized in Figure 11-7, which shows the percentage of reaches that were in error by specified amounts when the panels were in horizontal and vertical positions. The results have been simplified by combin-

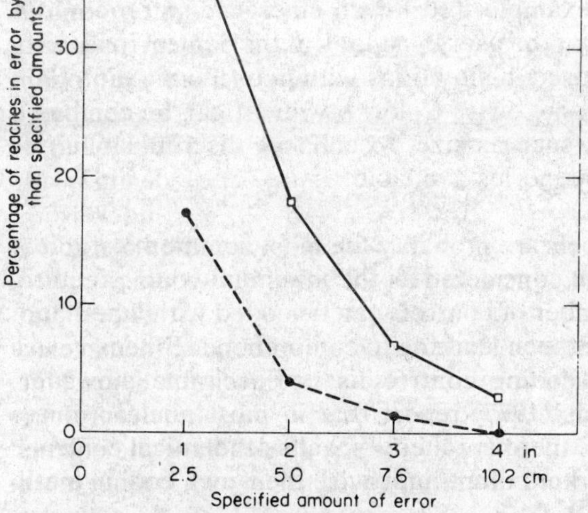

FIGURE 11-7
Accuracy of blind reaching to toggle switches (nine in a row on switch box) with switch box positioned horizontally and vertically. Each data point represents the mean of several conditions. (*Source: Adapted from Fitts and Crannell, as presented by Hunt, 1953.*)

ing several conditions. The results indicate, quite clearly, that accuracy was greatest when toggle switches were arranged vertically rather than horizontally. For vertically arranged switches, only a very small percentage of reaches were in error by more than 2.5 in (6.3 cm); therefore, controls should be separated by more than this distance so that errors of one switch will not result in the activation of an adjacent switch. For horizontally arranged switches, there should be at least 4 in (10.2 cm), and preferably more, between them if they are to be recognized by location without the aid of vision.

Operational Method of Coding Controls In this method of coding controls, each control has its own unique method of operation. For example, one control might be a push-pull variety and another of a rotary variety. Each can be activated *only* by the movement that is unique to it. Moussa-Hamouda and Mourant (1981) compared various operational coding schemes for operating automobile windshield wiper/washer systems from fingertip-reach controls mounted on a stalk attached to the steering column. Figure 11-8 shows the various combinations tested, the average time required to operate the various configurations, and the percentage of responses that resulted in activation of the wrong function. The authors recommended the rotate-away or rotate-toward configurations over the others. It is apparent that operational coding of controls would be inappropriate if there were any premium on control operation time or operating errors were of considerable importance. When operational coding is used, it is desirable that compatible relationships be utilized and that control operations be standardized. By and large, this method of coding should be avoided except in those individual circumstances in which it seems to be uniquely appropriate.

Color Coding of Controls Color coding was discussed in Chapter 4 in reference to visual displays. Color can be useful for identifying controls as well. A moderate number of coding categories can be used, and colors can often be picked that are meaningful, for example, red for an emergency stop control. One disadvantage, of course, is that the user must look at the control to identify it. Thus, color coding cannot be used in situations with poor illumination or in which the control is likely to become dirty. Color, however, can be combined with some other coding method, such as size, to enhance discriminability or increase the number of coding categories available.

Label Coding of Controls Labels are probably the most common method of identifying controls and should be considered as the minimum coding requirement for any control. A large number of controls can be coded with labels and, if properly chosen, do not require much learning to comprehend. Extensive use of labels as the only means of coding controls is not desirable, however. Seminara, Gonzalez, and Parsons (1977) report that in most nuclear power plant control rooms, for example, there are literally walls of identical controls, distinguished only by labels. Workers often improvise their own coding meth-

Stalks		Wiper (2 speed) and Washer Functions	Mean Performance Time (s)	Percentage of Inadvertent Operation Errors
Rotate Away		Wiper-rotates toward instrument panel Washer-button on end of stalk	0.90	4.6
Button		Wiper-button on end of stalk Wiper speed-rocker switch Washer-hand switch moves toward steering column	1.10	9.5
Hand Switch		Wiper-hand switch moves toward steering column Wiper speed-rocker switch Washer-button on end of stalk	1.00	4.9
Rotate Toward		Wiper-rotates toward driver Washer-button on end of stalk	0.88	3.5
Side Switch		Wiper-side switch pushes toward column Washer-button on end of stalk	1.19	4.3

FIGURE 11-8
Comparison of various configurations of automobile windshield wiper/washer system controls. Shown are the operational coding methods used for the various functions, the average response time, and the percentage of responses that were incorrect. (*Source: Adapted from Moussa-Hamouda and Mourant, 1981.*)

ods including attaching uniquely shaped draft beer dispenser handles to the controls. Labels take time to read and thus should not be the sole coding method where speed of operation is important. Labels should be placed above the control (so that the hand will not cover them when the operator is reaching for the control) and should be visible to the operator before reaching for the control.

Discussion of Coding Methods In the use of codes for identification of controls, two or more code systems can be used in combination. Actually, combinations can be used in two ways. First, *unique combinations* of two or more codes can be used to identify separate control devices, such as the various combinations of texture, diameter, and thickness mentioned before (Bradley, 1967). Second, there can be completely *redundant codes*, such as identifying each control by a distinct shape *and* by a distinct color. Such a scheme probably would be particularly useful if accurate identification were especially critical. In discussing codes, we would be remiss if we failed to make a plug for standardization in the case of corresponding controls used in various models of the same type of equipment, such as automobiles and tractors. When individuals are likely to transfer from one model to another of the same general equipment type, the same system of coding should be used if at all possible. Otherwise, it is probable that marked "habit interference" will result and that people will revert to their previously learned modes of response.

Some general considerations that affect the choice of a control code should be kept in mind, such as whether the operators will have the luxury of being able to look at the control to identify it, whether they will have to continuously monitor a display, whether they will work in the dark, or whether they have a visual handicap. If vision is restricted, then only shape, size, texture, location, or operational method should be considered.

A second consideration is maintenance. When designing a system, including the controls, one must take into consideration the maintenance of that system. If shape, size, texture, or color coding is used, one must keep a stock of different spare controls and run the risk that the wrong control will be replaced if one is damaged. Thus, if there is a significant chance that controls will be damaged and have to be replaced, perhaps location or operational method coding should be considered.

We have been discussing coding methods mostly in terms of accuracy and speed of identification. Some coding methods, such as shape, size, and texture, may, however, influence the operation of the controls themselves. Carter (1978) compared the speed and accuracy of using knobs of different sizes and shapes. The study was aimed at designing underwater diving equipment, so the tests took place in cold water with subjects wearing gloves, and the knobs were located on different parts of the body. The relative speed and accuracy of operating the knobs was somewhat dependent on the location of the knobs, but there were clear differences in performance with knobs that differed in shape, size, or both. It is important to select control codes that are discriminable, compatible, and meaningful, but it is also important to consider the possible effect of the specific codes on the speed and accuracy with which the control is operated. Incidentally, controls that require the user to grip them tightly should not use shape codes with sharp edges or protruding points.

Control-Response Ratio

In continuous control tasks or when a quantitative setting is to be made with a control device, a specified movement of the control will result in a system response. The system response may be represented on a display or it may not, as in turning the steering wheel of a car. We call the ratio of the movement of the control device to the movement of the system response the *control-response ratio* (C/R ratio). In the past, the C/R ratio has been called the *control-display ratio* (C/D ratio). It was felt that control *display* was really not appropriate when no real display was present and that control *response* is a more general term. Movement of the control, display, or system may be measured in linear distance (in the case of levers, vertical dials, etc.) or in angles or number of revolutions (in the case of knobs, wheels, circular displays, etc.). A very *sensitive* control is one which brings about a marked change in the controlled element (display) with a slight control movement; its C/R ratio would be low (a small control movement is associated with a large display movement). The reciprocal of the C/R ratio is referred to as *gain* and is really the same thing as sensitivity. A control with a low C/R ratio has a high gain. Examples of low and high C/R ratios are shown in Figure 11-9.

C/R Ratios and Control Operation The performance of human beings in the use of continuous control devices which have associated display movements can be affected by the C/R ratio. This effect is not simple, but rather is a function of the nature of the human motor activities involved in using such controls. In a sense, there are two types of human motions in such tasks. There is a *gross*-adjustment movement (travel time or a sluing movement) in which the operator brings the controlled element (say, the display indicator) to the approximate desired position. This gross movement is followed by a *fine*-adjustment movement, in which the operator makes adjustments to bring the controlled element precisely to the desired location. (Actually, these two movements may not be individually identifiable, but there is typically some change in motor behavior as the desired position is approached.)

Optimum C/R Ratios The determination of an optimum C/R ratio for any continuous control or quantitative-setting control task needs to take into account these two components of human motions. Where the control in question

FIGURE 11-9
Generalized illustrations of low and of high control-response ratios (C/R ratios) for lever and rotary controls. The C/R ratio is a function of the linkage between the control and display.

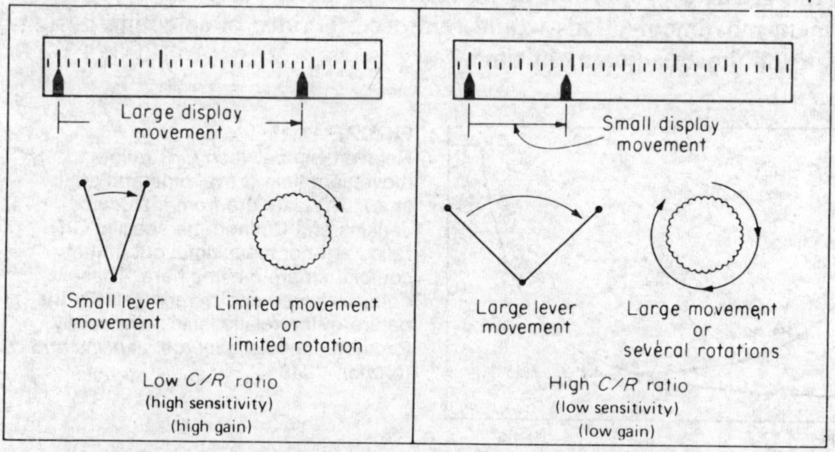

is a knob, the general nature of the relationships is reflected by the results of studies by Jenkins and Connor (1949) and is illustrated in Figure 11-10. That figure illustrates the essential features of the relationships, namely, that travel time drops off sharply with decreasing C/R ratios and then tends to level off, and that adjustment time has the reverse pattern. Thus, if you were trying to find a radio station using a knob with a high C/R ratio (where it takes a couple of turns of the control to move the pointer on the dial a small distance), it would take you some time to get the pointer across the dial to the general location of the station (travel time). Once it was there, however, you could quickly "home in" on the station without overshooting it. With a low C/R ratio (where a small turn of the control sends the pointer across the dial), you could get the pointer near the station in no time, but it would take you a considerable time to "home in" because you would tend to overshoot the station. In determining the optimum C/R ratio, one must often trade off travel and adjustment time. The optimum ratio is somewhere around the point of intersection, as shown in Figure 11-10. Within this general range, the combination of travel and adjustment time usually would be minimized. Note that when a joystick or a lever is used instead of a control knob, the curve for travel time is almost a horizontal straight line. This is because it is almost as fast to move a stick through 180° as through a smaller angle. The optimum C/R ratio is then based on minimizing adjustment time only.

Not everyone, however, agrees on the importance of C/R ratio in the design of joystick controls. Buck (1980), for one, found that with joysticks neither travel nor adjustment time was affected by the C/R ratio. Rather, the physical widths of the target, measured on the display and in the control (but not their ratio), were the primary factors affecting performance. Arnaut and Greenstein (1987), using touch tablets and trackballs to move cursors on a VDT screen, found evidence supporting Buck's conclusion that the size of the target as well as the amplitudes of the control and cursor movements on the display, and not just the ratio of the two, are important factors in designing controls. Buck also found that the greatest proportion of performance time was spent in making fine-adjustment movements. This would reinforce the idea of selecting design parameters to minimize adjustment time.

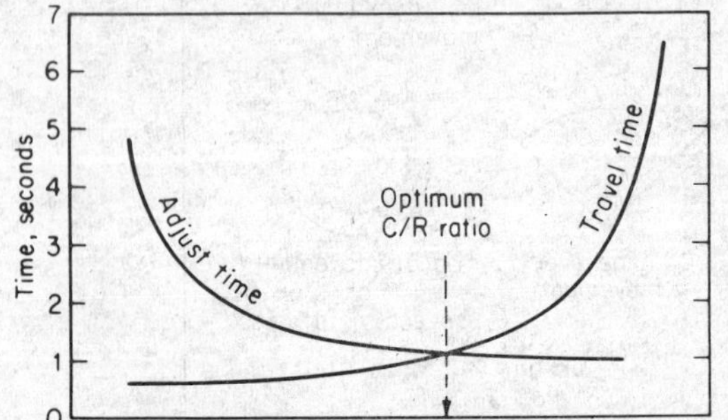

FIGURE 11-10
Relationship between C/R ratio and movement time (travel time and adjust time). The data are from a study by Jenkins and Connor; the specific C/R ratios are not meaningful out of that context, so are omitted here. These data, however, depict very typically the nature of the relationships, especially for knob controls. (*Source: Jenkins and Connor, 1949.*)

The numeric value for the optimum C/R ratio is a function of the type of control (knob, lever, crank, etc.), size of the display, tolerance permitted in setting the control, and other system parameters such as control order and lag. Unfortunately, there are no formulas for determining optimum C/R ratios. Rather this ratio should be determined experimentally for the control and display parameters being contemplated. To give some ballpark figures, Jenkins and Connor (1949) reported optimum C/R ratios for knobs in the range of 0.2 to 0.8; Chapanis and Kinkade (1972) reported optimum C/R ratios for levers in the range of 2.5 to 4.0.

Resistance in Controls

The major part of the feedback or "feel" of a control is due to the various types of resistance associated with the control. Control manipulation takes principally two forms: the amount of displacement of the control and the amount of force applied to the control. The operator of the control can sense both movement and force through the proprioceptive and kinesthetic senses. Force and movement, therefore, are the primary sources of control feedback.

To use most controls requires both force to overcome resistance and some displacement of the control (such as depressing a brake pedal or turning a steering wheel). Some controls, however, require only force or only displacement. A pure displacement control would have virtually no resistance to movement, and the only type of feedback would be the amount of movement of the body member. These are called *free-position,* or *isotonic,* controls. A *pure force,* or *isometric*, control, on the other hand, involves no displacement. The output is related to the amount of force or pressure that is applied to the control (these are sometimes called *stiff sticks*). Pure force controls are becoming more common with advances being made in electronics and servo control mechanisms.

It is difficult to specify which type of feedback, force or displacement, is most useful. However, some research and experience, plus a bit of conventional wisdom, suggest that in many circumstances a combination of the two is useful, such as in the operation of a steering wheel. In special circumstances, however, one type or the other may have an advantage. For example, from the scant amount of literature available, Poulton (1974) concludes that pure force controls are superior to a combination force-displacement (spring-centered) control when they are used with higher-order tracking control systems (see Chapter 10) to track a relatively fast-moving, gyrating target.

Types of Resistance All control devices have some resistance, although in the case of free-positioning (isotonic) controls this is virtually negligible. Resistance can, of course, serve as a source of feedback, but the feedback from some types of resistance can be useful to the operator while the feedback from other types of resistance can have a negative effect. In some circumstances the designer can design or select those controls with resistance characteristics that will minimize possible negative effects and that possibly can enhance performance. This can be done in a number of ways, such as with servomechanisms.

with hydraulic or other power, and with mechanical linkages. The primary types of resistance are elastic (or spring-loaded), static and coulomb friction, viscous damping, and inertia. We discuss each in turn.

- *Elastic resistance:* Such resistance (as in spring-loaded controls) varies with the *displacement* of a control device (the greater the displacement, the greater the resistance). The relationship may be linear or nonlinear. A major advantage of elastic resistance is that it can serve as useful feedback, combining both force and displacement in a redundant manner. Poulton indicates that for tracking tasks, spring-loaded controls are best for compensatory tracking in a position control system with a slow ramp target to track (see Chapter 10). With elastic resistance, the control automatically returns to the null position when the operator releases it; hence it is ideal for "dead-man" switches. By designing the control so that there is a distinct gradient in resistance at critical positions (such as near the terminal), additional cues can be provided. In addition, such resistance permits sudden changes in control direction but reduces the likelihood of undesirable activation caused by accidental contact with the control.
- *Static and coulomb friction: Static friction,* the resistance to initial movement, is maximum at the initiation of a movement but drops off sharply. *Coulomb (sliding) friction* continues as a resistance to movement, but this friction force is not related to either velocity or displacement. This can be illustrated by pressing a pencil eraser firmly against a flat horizontal surface with one hand while trying to slide the eraser slowly by pushing it horizontally with the other hand. You have to exert some force before you can get the pencil (or control) to move; i.e., you have to overcome the static friction. Once the pencil is moving, however, a smaller amount of force, to overcome the coulomb friction, will keep it moving no matter how fast it is moving or how far you move it. If you try to change directions, you will notice a delay and some irregularity in the movements. With a couple of possible exceptions, static and coulomb friction tend to cause degradation in human performance. This is essentially a function of the fact that there is no systematic relation between such resistance and any aspect of the control movement (such as displacement, speed, or acceleration); thus, it cannot produce any meaningful feedback to the user of the control movement. On the other hand, such friction has the advantages of reducing the possibility of accidental activation and of helping to hold the control in place.
- *Viscous damping:* Viscous damping feels like moving a spoon through thick syrup. The amount of force required to move the spoon (or control) to overcome viscous damping is related to the velocity of the control movement. The faster you move the control, the more viscous damping you have to overcome. Viscous damping generally has the effect of resisting quick movement and of helping to execute smooth control, especially in maintaining a prescribed *rate* of movement. The feedback, being related to *velocity* and not displacement, probably is not readily interpretable by operators. Such resistance, however, does minimize accidental activation.

• *Inertia:* This is the resistance to movement (or change in direction of movement) caused by the mass (weight) of the mechanism involved. It varies in relation to *acceleration.* A force exerted on an inertia control will have little effect at first because it produces only an acceleration of the control (the greater the force, the greater the acceleration). The acceleration has to operate for a little while before the control will begin to move very much. Once the control begins to move, however, it is hard to stop. Inertia opposes quick control movements and thus probably would be a disadvantage in tracking. As Poulton points out, the human arm has enough inertia of its own. Inertia does, however, aid in smooth control and can aid in turning a crank at a fixed speed (Poulton). It also reduces the possibility of accidental activation of the control.

Combining Resistances Almost all controls that move involve some forms of resistance, often more than one type. The effects of more than one type of resistance on control operation can be complex. Howland and Noble (1955), for example, show how time on target with a position control is affected by various combinations of control resistance. They used a knob control manipulated with the fingertips, and relatively small amounts of resistance were used. Table 11-2 summarizes the results. Elastic resistance alone resulted in the best performance, even better than no resistance at all. The addition of inertia by itself, or in combination with the other types of resistance, always resulted in a decrement in performance. The worst combination was elastic resistance plus inertia.

Deadspace

Deadspace in a control mechanism is the amount of control movement *around the null position* that results in no movement of the device being controlled. It is almost inevitable that some deadspace will exist in a control device. Dead-

TABLE 11-2
EFFECTS OF VARIOUS COMBINATIONS OF CONTROL RESISTANCE ON TRACKING PERFORMANCE

	Type of control resistance present			
Condition	**Elastic***	**Viscous damping†**	**Inertia‡**	**Time on target, %**
1	Yes	No	No	60
2	Yes	Yes	No	52
3	No	Yes	No	44
4	No	No	No	41
5	No	No	Yes	35
6	No	Yes	Yes	35
7	Yes	Yes	Yes	34
8	Yes	No	Yes	27

*Elastic resistance: 0.02 in · lb/°
†Viscous damping: 0.02 (in · lb)/(° · s)
‡Inertia: 0.002 (in · lb)/(° · s^2)
Source: Howland and Noble, 1955.

space of any consequence usually affects control performance, with the amount of effect being related to the sensitivity of the control system. Rockway (1957) observed that tracking performance deteriorated with increases in deadspace (in degrees of control movement that produced no movement of the controlled device). But he found the deterioration was less with the less sensitive systems than with more sensitive systems. This, of course, suggests that deadspace can, in part, be compensated for by building in less sensitive C/R relationships.

Rogers (1970) found that increasing the deadspace in a control resulted in an almost linear increase in the time needed to acquire a target. In the experiment a ball had to be spun at a minimum velocity to move the response marker. Deadspace corresponded to the minimum velocity required.

Poulton (1974) suggests that deadspace may be more detrimental in compensatory tracking tasks than in pursuit, but with higher-order control systems a small amount of deadspace may be helpful when the control is spring-centered. This is because the control may not always spring back to exactly the same position when it is released.

Backlash

The best way to think of backlash is to imagine operating a joystick or lever with a loose hollow cylinder fitted over it. When the cylinder is moved to the right, for example, the stick touches it on the left. If you are moving the cylinder to the right and then reverse directions, the stick does not start to return to the left until it comes up against the right side of the cylinder. Until the stick contacts the cylinder, the operator's control movements have no effect on the system. In essence, then, backlash is deadspace *at any control position.*

Typically, operators cannot cope very well with backlash. This effect was illustrated by the results of an investigation by Rockway and Franks (1959) using a control task under varying conditions of backlash and of control gain (the reciprocal of C/R ratio). The results show that performance deteriorated with increasing backlash for all control gains but was most accentuated for high gains. The implications of such results are that if a high control gain is strongly indicated (as in, say, high-speed aircraft), the backlash needs to be minimized in order to reduce system errors; or conversely, if it is not practical to minimize backlash, the control gain should be as low as possible—also to minimize errors from the operation of the system.

DESIGN OF SPECIFIC HAND-OPERATED CONTROLS

The various factors discussed above relate to all types of controls, although some factors are more relevant to tracking controls than to other types. We now present a few examples of research investigating various design parameters of specific types of controls. These studies illustrate how specific design features affect the adequacy with which people can use the control for a particular purpose. Within the scope of this chapter, we can only present a smattering of the range of research carried out on control devices. We discuss briefly some aspects of cranks and handwheels, rotary knobs, and joysticks. We discuss data entry devices, cursor positioning devices, and foot-operated controls in later sections of this chapter.

Cranks and Handwheels

Cranks and handwheels frequently are used as a means of applying force to perform various types of functions, such as moving a carriage on a cutting tool or lifting objects. We present the results of some studies which illustrate the effect of size and friction on various aspects of operator performance. It should become clear that the optimum configuration depends on the aspect of operator performance one wishes to optimize.

The work output of subjects using three sizes of cranks and five levels of resistance was investigated by Katchmer (1957) with a view to identifying fairly optimum combinations. The cranks were adjusted to waist height for each subject, and the subjects were instructed to turn the crank at a rapid rate until they felt they could no longer continue the chore (or until 10 min had passed). A summary of certain data from the study is presented in the graphs of Figure 11-11. In total foot-pounds of work (Figure 11-11*a*), the highest work levels occurred for the moderate torque loads of 30 and 50 in·lb (0.54 and 0.9 kg·mm) using the 5-in (13-cm) and 7-in (18-cm) radius cranks. Torque had no effect on the work output when the small 4-in (10-cm) crank was used. Looking at the average time that subjects worked before quitting (Figure 11-11*b*), we see that, again, the larger the radius of the crank, the longer the subjects would continue to turn it, except under the higher torque conditions. But now, in terms of time,

FIGURE 11-11
Average performance measures for subjects using 4-, 5-, and 7-in radius cranks under 10, 30, 50, 70, and 90 in·lb torque resistance. (*Source: Adapted from Katchmer, 1957.*)

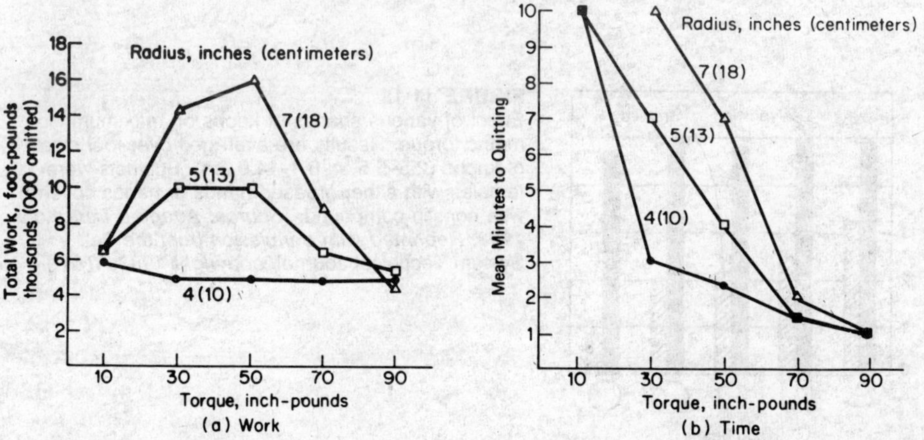

the optimum torque is 10 in·lb (0.18 kg·mm), not 30 to 50 in·lb (0.54 and 0.9 kg·mm), as was the case with work output.

These graphs dramatically illustrate the point that a particular design may be optimum for one criterion but not for another (in this case, the different criteria being tolerance time and total work output). It is important, therefore, that the aspect or aspects of human performance one wants to maximize be specified before design parameters are chosen.

Knobs for Producing Torque

Often, control knobs are used to apply fairly high levels of torque to equipment, for example, in turning on or off a water faucet, tightening a clamp mechanism, or turning a door knob. Kohl (1983) measured the maximum isometric force that females could exert on smooth aluminum knobs of various shapes and sizes [diameters of 2.5 to 5.5 in (6.4 to 14.0 cm)]. Subjects performed the task under two conditions: with greased hands and with hands covered with a nonslip compound. Figure 11-12 presents results after performance is averaged across the various knob sizes. It is no surprise that greasy hands reduced the amount of force (torque) that could be applied to the knobs. There was little effect of knob shape when the nonslip compound was used; however, with greased hands, torque *decreased* as the number of sides on the knob increased. As might be expected, the larger the knob diameter, the more torque that could be developed, but this effect diminished as the number of sides on the knob increased. Kohl cautioned that the triangular knob caused discomfort and therefore recommended the square shape.

Bullinger and Muntzinger (1984) performed a similar experiment, varying the shape of knobs having a diameter of 80 mm (3.15 in). The subjects in this experiment were males, and they performed the task with clean knobs and with knobs soiled with dirt and oil. Figure 11-13 shows the knob shapes used and the average maximum torque produced with each. Larger differences in torque production were found between shapes in this study than in Kohl's study. The

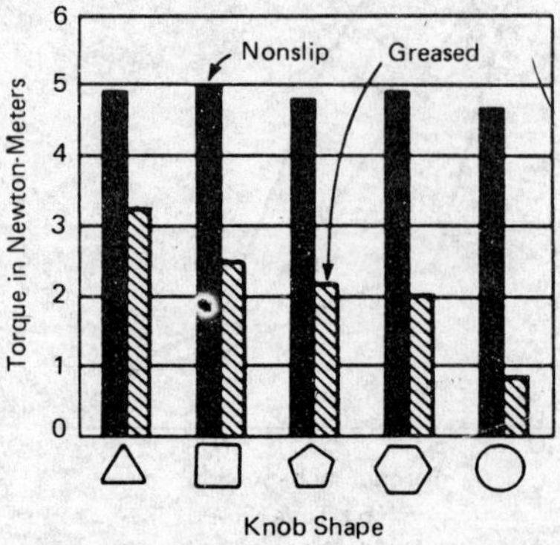

FIGURE 11-12
Effect of various shapes of knobs on maximum isometric torque. Results are averaged over four sizes of knobs (2.5–5.5 in; 6.4–14.0 cm). Subjects were females with either greased hands or hands covered with nonslip compound. (*Source: Adapted from Kohl, 1983. Reprinted with permission from the* Bell System Technical Journal, *copyright 1983 AT&T.*)

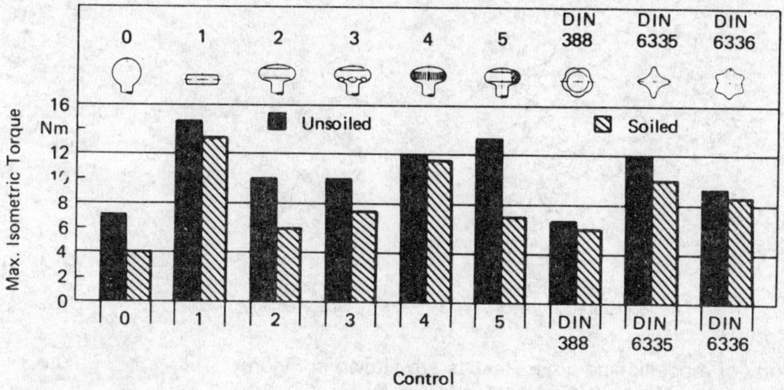

FIGURE 11-13
Maximum isometric torque as a function of knob shape. Knobs were either clean or soiled with dirt and oil. Subjects were males. Control 5 is coated with resin. (*Source: Bullinger and Muntzinger, 1984.*)

control producing the highest average torque (i.e., control 1, a rectangular bar) was not recommended because of the discomfort caused by the edges digging into the palm of the hand. Control 4, in soiled conditions, and control 5, in clean conditions, were recommended.

Both of these studies illustrate the need to consider multiple criteria in selecting an optimal design. In both studies the control producing the highest level of torque was not recommended because of the discomfort caused by its shape. A good design must consider performance, user comfort, and user satisfaction.

Stick-Type Controls

At first, one might think that the length of a joystick or lever would dramatically affect the speed or accuracy with which it could be used. This does not seem to be true, however. Jenkins and Karr (1954), for example, found that varying the length of a joystick from 12 to 30 in (30.5 to 76.2 cm) was relatively unimportant as long as the C/R ratio was around 2.5 or 3.0. Hartman (1956), using a tracking task, found only moderate effects on performance by varying joystick length from 6 to 27 in (15 to 69 cm). There was about a 10 percent advantage in using 18-in (46-cm) sticks relative to other lengths.

Mehr (1973) reports on a study that compared time to reposition a cursor given various types of joysticks. The joysticks used are shown in Figure 11-14. The first joystick was operated in two modes: as a *displacement joystick,* which remained in position when released, and as a *spring-return joystick*. The finger- and thumb-operated joysticks were force-operated isometric controls. The spring-return joystick and both isometric joysticks operated as first-order rate controls, while the displacement joystick operated as a zero-order position control. Figure 11-15 shows the mean repositioning time as a function of the

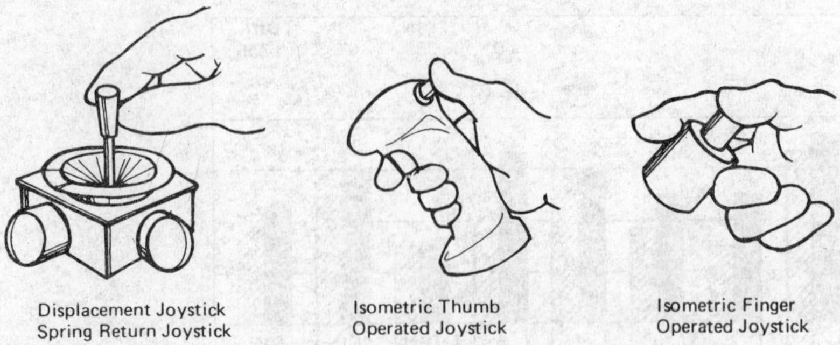

FIGURE 11-14
Joysticks used in study of repositioning time. Results are shown in Figure 11-15. (*Source: Adapted from Mehr, 1973.*)

length, measured in steps, of the repositioning movement (the display area was 1000 × 750 steps). The results showed that repositioning time was fastest with the finger-operated isometric control. The difference between the displacement and spring-return joysticks is difficult to explain because the presence or absence of the spring was confounded with whether the control was a zero- or first-order control. Notice that with the rate-controlled joysticks the repositioning time did not increase much, if at all, for movements longer than 500 steps.

Multifunction Hand Controls

Advances in technology have made some systems more complex and have endowed them with more functions. An excellent example is a high-performance fighter aircraft. A pilot must be able to control various aircraft, communication, and weapon functions in a demanding and hostile environment. Multifunction hand controls are becoming common in such aircraft. Figure 11-16, for example, shows the left-hand control of an F-18 aircraft. The control itself is actually two slide controls, one controlling the throttle of each engine, that can be operated together or separately or can be locked in any position. In addition, four control functions are operated by the thumb, two more are

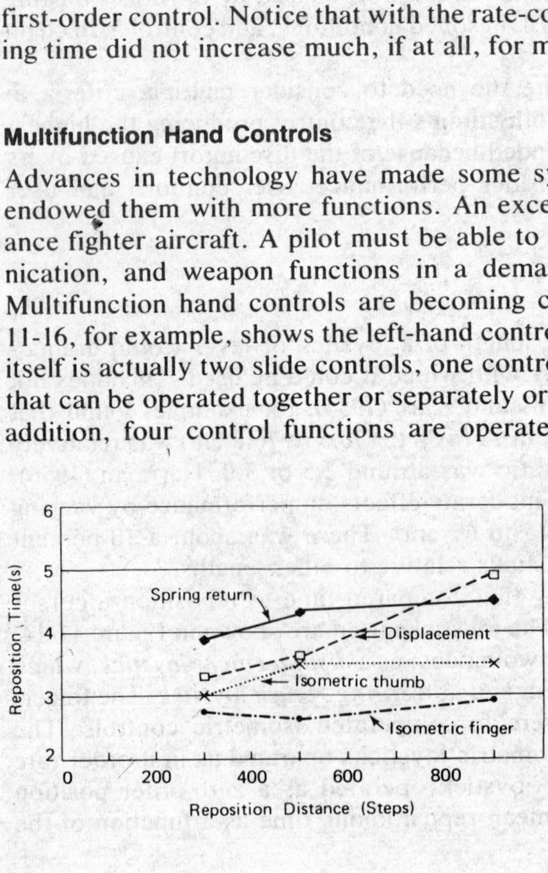

FIGURE 11-15
Repositioning time for the various types of joysticks shown in Figure 11-14. Spring return and both isometric joysticks were first-order rate controls. Displacement joystick remained in position when released and was a zero-order position control. All devices used optimized designs. (*Source: Adapted from Mehr, 1973.*)

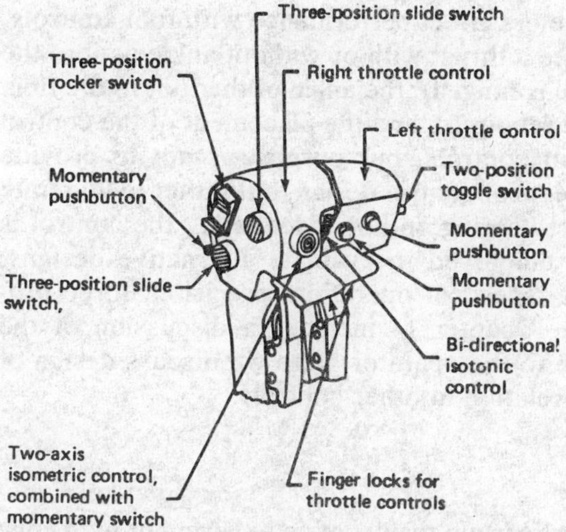

FIGURE 11-16
The left-hand control of the F-18 aircraft. (*Adapted from: Wierwille, 1984, Fig. 2.*)

operated by the index finger, and five more are operated by the remaining three fingers. (By the way, the right-hand control, not shown here, contains seven control functions in addition to controlling the pitch and roll of the aircraft.)

Wierwille (1984) points out that the design of such multifunction hand controls is based on several principles: (1) The operator should not have to observe the control to operate it; (2) the hand should remain in contact with the primary controls throughout critical operations of the system; and (3) auxiliary controls should be able to be activated without loss of physical contact with the primary controls. Unfortunately, little research has been conducted on multifunction controls, so we do not even know, for example, the maximum number of functions that can be efficiently controlled by an operator. As Wierwille points out, comprehensive testing, therefore, is mandatory whenever a multifunction control is used in a system.

FOOT CONTROLS

Far and away, hand controls are more widely employed than foot controls. One reason for this, as Kroemer (1971) points out, is that the general tenor of human factors handbooks implies that the feet are slower and less accurate than the hands. Kroemer hastens to add that this belief is neither supported nor discredited by experimental results. Besides this, foot controls often restrict the posture of the user. Having to maintain one foot (or both) on a control makes it more difficult to shift around in the seat or change position of the leg. Anyone who has taken a long drive on the open highway can verify this. Operating foot controls from a standing position requires people to balance all their weight on the other foot. Despite all this, foot controls will continue to have a place in the control tasks of seated operators.

Several important design parameters affect performance with foot controls, such as whether the controls require a thrust with or without ankle action, the location of the fulcrum (if the pedal is hinged), the angle of the foot to the tibia bone of the leg, the load (the force required), and the placement of the control relative to the user. As with hand controls, our purpose is not to provide detailed design guidance on all aspects of control design, but rather to illustrate how specific features have a direct bearing on how adequately the control is used. Various criteria have been employed to evaluate alternative designs, such as reaction time, travel time, speed of operation, precision, force produced, and subjective preference. Chapter 14 includes a discussion of the placement of foot controls relative to the operator; here we discuss design of foot controls and their placement relative to other controls.

Pedal Design Considerations

For illustrative purposes we present certain results of an experiment by Ayoub and Trombley (1967) in which one of the factors varied was the location of the fulcrum of the pedal. In one set of experimental conditions, the pedal, to be activated, had to be moved through an arc of 12°, no matter where the fulcrum was located (constant angle). The closer the fulcrum is to the heel, the farther the ball of the foot must travel to achieve a 12° arc. In a second set of conditions, the pedal, to be activated, had to be moved a constant distance [0.75 in (1.9 cm)]. In this case, the closer the fulcrum is to the heel, the smaller the arc will be to achieve the activation distance.

The mean travel times for various fulcrum locations are shown in Figure 11-17. It can be seen that for a constant angle of movement (of 12°), the optimum location of the fulcrum is forward of the ankle (about one-third of the distance between the ankle and the ball of the foot), whereas for a constant distance of movement the optimum location of the fulcrum is at the heel. However, as Kroemer points out, the somewhat inconsistent results of various experiments with pedals (in part based on differences in experimental conditions) preclude any general statements as to what pedal allows the fastest activation or highest frequency in repetitive operation under a variety of specified conditions.

Foot Controls for Discrete Control Action Foot control mechanisms usually are used to control one or a couple of functions. There are some indications, however, that the feet can be used for more varied control functions. In one investigation, for example, Kroemer developed an arrangement of 12 foot positions (targets) for subjects as illustrated in Figure 11-18. Using procedures that need not be described here, he was able to measure the speed and accuracy with which subjects could hit the various targets with their feet. By and large, he found that after a short learning period the subject could perform the task with considerable accuracy and with very short travel time (averaging about 0.1 s). Although forward movements were slightly faster than backward or lateral movements, these differences were not of any practical consequence. His

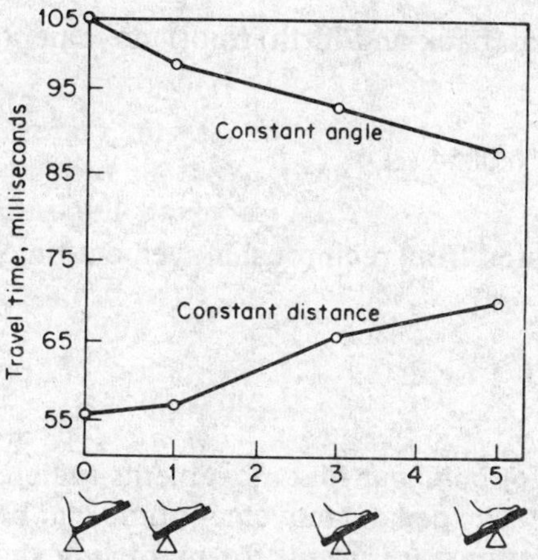

FIGURE 11-17
Mean travel time in pedal movement as related to location of the fulcrum for conditions of constant angle of movement (12°) and constant distance of movement. (0.75 in or 1.9 cm) (*Source: Adapted from Ayoub and Trombley, 1967, Fig. 13. Copyright 1967 by American Institute of Industrial Engineers, Inc., 25 Technology Park, Atlanta, Norcross, Ga. 30071, and reproduced with permission.*)

results strongly suggest that it might be possible to assign control tasks to the feet which previously have been considered in the domain of the hands. Anyone who can play, or has watched someone play, the pedals of an organ can attest to the precision and timing of foot movements that can be attained with practice.

The time required to move a foot from one pedal to another is a function of the distance traveled and the size of the pedals. Fitt's law (Fitts, 1954), developed to predict hand movement time (see Chapter 9), states that movement time is a direct function of task difficulty and that task difficulty is directly related to the movement distance and inversely related to the size of the target. Thus the smaller the target and the longer the distance to be moved, the more difficult the task and the longer the movement time required. Drury (1975) used a modification of Fitts' law (Welford, 1968) to predict movement times for operating foot pedals. Drury's index of (task) difficulty (ID) was defined as

$$ID = \log_2 \left[\frac{A}{W + S} + 0.5 \right]$$

where A = movement amplitude (distance) to centerline of target
W = target width
S = shoe sole width

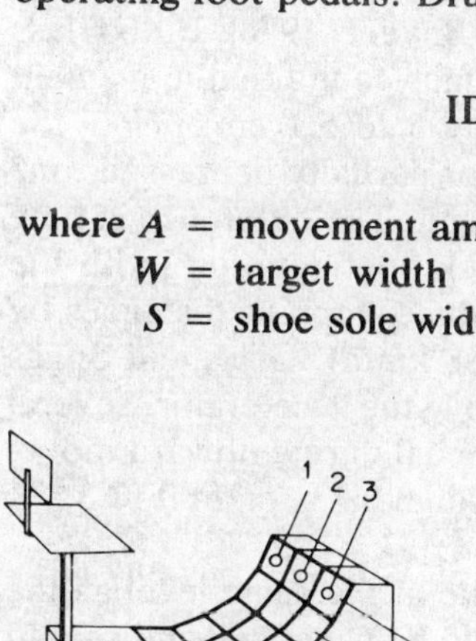

FIGURE 11-18
Arrangement of foot targets used in experiment by Kroemer in studying the speed and accuracy of discrete foot motions. (*Source: Kroemer, 1971, Fig. 7.*)

The equation developed to predict reciprocal (back-and-forth) (foot) movement time (RMT) was

$$\text{RMT} = 0.1874 + 0.0854\ (\text{ID})$$

Single-movement time (SMT) can be estimated from reciprocal movement time (RMT) by

$$\text{SMT} = \text{RMT}/1.64$$

These equations appear to be accurate for coplanar foot movements (where pedals are in the same plane). For noncoplanar pedals, movement time can be double, as discussed below. Drury's equations are useful for predicting the effects of varying the size of a pedal and/or the distance between pedals on movement time. As it turns out, with typical spacing between pedals, the size of the pedal (over a reasonable range) has only a minimal effect on movement time. The farther apart the pedals, however, the greater the effect on movement time of varying the size of the pedals. Remember, reaction time must be added to movement time to obtain overall performance time estimates.

Automobile Brake and Accelerator Pedals The most commonly used foot controls are the brake and accelerator controls of automobiles. Aside from their individual characteristics, an important factor in their use is their relative positions. In most automobiles the accelerator is lower than the brake pedal, thus requiring the lifting of the foot from the accelerator, its lateral movement, and then the depression of the brake. Several investigators, however, have found that movement time from accelerator to brake is shorter when the accelerator and brake are at the same level (Davies and Watts, 1970; Glass and Suggs, 1977; Morrison, Swope, and Halcomb, 1986; Snyder, 1976). Both Glass and Suggs, and Morrison, Swope, and Halcomb, in fact, found that movement time was optimum when the brake was 1 to 2 in (2.5 to 5.1 cm) *below* the accelerator. In this configuration with the accelerator partially depressed, one could merely slide the foot to the brake without lifting it. The difference in time between the optimum placement and a typical truck configuration [with the brake 4 in (10.2 cm) above the accelerator] could reduce stopping distances by approximately 6 ft (1.8 m) with travel at 55 mi/h (88 km/h). Glass and Suggs also report that when the brake was even with or above the accelerator, several subjects caught their foot on the edge of the brake pedal during upward movement from the accelerator. This resulted in an additional 0.3 s, or 24 ft (7.3 m), when travel was at 55 mi/h (88 km/h).

At first blush it may appear that placing the brake at the same level as the accelerator (coplanar) or below the accelerator would be beneficial and should be done. But there is more to pedal placement than foot-movement time. Casey and Rogers (1987) point out that inadvertent activation of the accelerator is more likely with coplanar pedals. In fact, the National Highway Traffic Safety Administration (1983) believed that the "runaway car" phenomenon reported by nearly 500 owners of Audi 5000 cars (i.e., drivers putting the transmission into gear with their foot apparently on the brake would experience a sudden and unexpected acceleration) was due to "insufficient height differential between gas and brake pedals" and the drivers actually having their foot on the accelerator rather than the brake. (See Schmidt, 1989, for a motor skills and human factors analysis of such accidents in automobiles.) Further, Casey and Rogers point out that, with the accelerator pedal placed at a comfortable distance from the driver, coplanar or brake-below-accelerator configurations would not permit short-legged people to extend their leg the additional distance necessary to fully depress the brake pedal. Adequate pedal travel distance is also necessary to effectively modulate braking forces. The lesson from all this is that design recommendations must be evaluated from a total systems perspective, rather than just in terms of one measure of performance.

DATA ENTRY DEVICES

Our insatiable appetite for information has brought about a proliferation of machines and devices for storing, transmitting, and analyzing that information we hold so dear. These machines, such as the typewriter, telephone, computer, and calculator, require alphabetic or numerical information, or both, as input. Several types of devices have been used for the input of alphanumeric information, such as knobs, levers, thumb wheels, and push-button keyboards. With very few exceptions, keyboard entry devices are generally superior to other data entry devices such as thumb wheels, knobs, and levers. Miner and Revesman (1962), for example, found that 10-key keyboards, such as those used on hand calculators, resulted in a 33 percent reduction in entry time and a 74 percent reduction in errors compared to using either knobs or levers to enter numeric data. Therefore, we focus our attention on keyboards because of their clearcut advantages for data entry.

The speed and accuracy with which alphanumeric information can be entered by use of a keyboard, or any other device for that matter, is dependent on the quality of the data presented to the operator. As would be expected on the basis of common sense, data entry speed and accuracy are generally greater when the data presented to the operator are clear and legible. As discussed by Seibel (1972), speed and accuracy will also be greater if (1) the operator is familiar with the format of information to be entered; (2) upper- and lowercase characters are used for written text; and (3) long messages or strings of digits are divided into chunks, that is, 123 456 rather than 123456.

Seibel also points out that there is no consistent difference in data entry rate between the usual alphanumeric and straight numeric data entry. In addition, messages composed of random sequences of just 10 of the possible alphabetic characters are not entered more rapidly or more accurately than random sequences of all 26 characters.

Chord versus Sequential Keyboards

Most data entry keyboards are *sequential* in the sense that individual characters are entered in a specific sequence, there being a specific key or other device for every character. On the other hand, in *chord* keyboards a single input unit (e.g., a number, letter, or word) requires the simultaneous activation of two or more keys. Chord keyboards, although far less common than sequential keyboards, are used with stenotype machines, pianos, and some mail-sorting machines.

Noyes (1983) presented a history of the developments and applications of chord keyboards. Most of the development and evaluation of chord keyboards took place during the 1960s and 1970s. The principal advantage of a chord keyboard is its suitability for one-handed alphanumeric data entry. Studies that have compared chord keyboards to sequential keyboards have generally found that the chord keyboard is equal to or better than the sequential keyboard (Bowen and Guinness, 1965; Conrad and Longman, 1965). Although chord keyboards seem to offer advantages over sequential keyboards, chord keyboards can be more difficult to learn; in particular, it is more difficult to learn the coding system (i.e., the combination of keys that represent individual items).

Noyes concludes that there probably is no real need for a general-purpose chord keyboard since sequential keyboards adequately fulfill the majority of our everyday requirements. There may, however, be special applications where chord keyboards could be useful, especially where one-handed alphanumeric data entry is required.

Keyboard Arrangement

Since most keyboards are sequential, we discuss briefly some aspects of alphabetic and numeric keyboard arrangements.

Alphabetic Keyboards The standard typewriter keyboard (called QWERTY because of the sequence of letters in the topline) appears to have been intentionally designed to slow the operator's keystroke rate. The most commonly occurring letters (at least in English) are generally delegated to the weakest, slowest fingers, and frequently occurring letter combinations are on the same hand. You might ask, "How could they have been so stupid?" The answer is, they weren't. The QWERTY keyboard was, in fact, intentionally designed to slow the rate of keying. Back then, mechanical typewriters could not function as fast as people could stroke the keys, and keys often jammed. QWERTY slowed the operators and thus reduced the frequency of jammed keys. By the time technology caught up with human keying capabilities, it was too late to change. Various alternative arrangements of the conventional typewriter keyboard have been proposed. The most notable was by Dvorak (1943) over a quarter of a century ago. Figure 11-19 shows the QWERTY keyboard and the

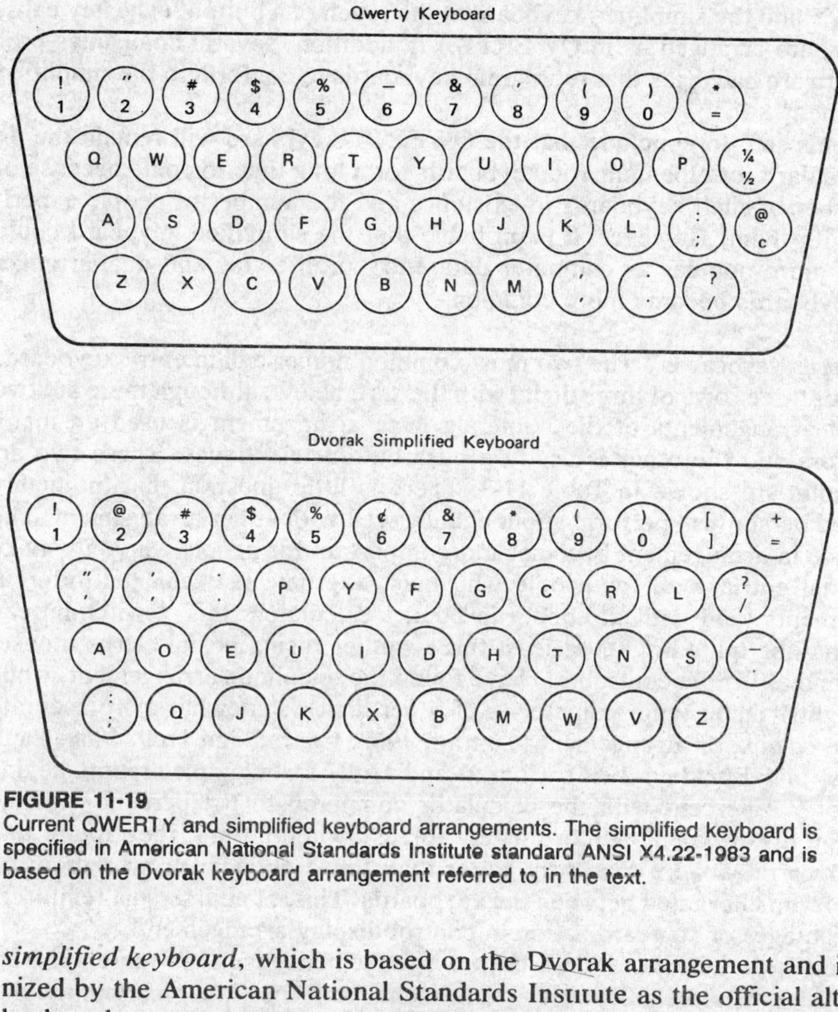

FIGURE 11-19
Current QWERTY and simplified keyboard arrangements. The simplified keyboard is specified in American National Standards Institute standard ANSI X4.22-1983 and is based on the Dvorak keyboard arrangement referred to in the text.

simplified keyboard, which is based on the Dvorak arrangement and is recognized by the American National Standards Institute as the official alternative keyboard arrangement to QWERTY (ANSI X4.22-1983). (In the original Dvorak keyboard, the numeric keys across the top row were in this order: 7-5-3-1-9-0-2-4-6-8.)

Overall, the research clearly shows that the Dvorak arrangement is superior to the QWERTY keyboard. Some claim 10 to 20 percent improvements in speed with less hand and finger fatigue. Norman (1983), however, contends that a 5 to 10 percent improvement is more realistic and questions how much of the improvement is actually realized on the job. Norman concludes that the superiority of the Dvorak arrangement is not great enough to justify switching from QWERTY and retraining millions of people to type with the new arrangement. The real question is not whether QWERTY should be scrapped, but rather whether both arrangements should be equally available on keyboards. In fact, there is a popular personal computer that is designed to switch between

QWERTY and the simplified keyboard at the touch of a button. (The key caps, however, are arranged as in QWERTY.) In addition, several companies produce software packages that reprogram keyboards to conform to the simplified arrangement.

It seems safe to conclude that the QWERTY keyboard will remain the de facto standard for alphanumeric keyboards for a long time to come because of the number of such keyboards in existence and the number of people experienced with them. However, it is probable that the simplified keyboard could become more popular as computer data entry proliferates and dual-arrangement keyboards become more common.

Numeric Keyboards The two most common numeric data entry keyboards consist of three rows of three digits with the zero below, although there are two different arrangements of the numerals. One arrangement is used on many calculators and the other is used on push-button telephones. These two arrangements are shown in Table 11-3. There is little question that the highly practiced person will perform about equally well with either arrangement. The difference in arrangement becomes more important for people who only make occasional entries and for people who must alternate between the different arrangements (and with the boom in pocket calculators this "switch-hitter" group may be quite large). At least three studies (using postal clerks, housewives, and air traffic controllers) have found the telephone arrangement, while only slightly faster (approximately 0.05 s per digit), somewhat more accurate than the calculator arrangement (Conrad, 1967; Conrad and Hull, 1968; Paul, Sarlanis, and Buckley, 1965). Conrad and Hull, for example, report keying errors of 8.2 percent with the calculator compared to 6.4 percent with the telephone arrangement. Unfortunately for us switch-hitters, they found that these error rates were markedly better than those obtained from a group of subjects who alternated between the keyboards. This, of course, just reinforces the advantages of standardization in control display arrangements.

Although the telephone push-button arrangement results in more accurate performance than the calculator arrangement, how does it compare to the dial telephone? Pollard and Cooper (1978) found that even after almost 2 years of extensive experience with the push-button telephone, operators still made more errors than a control group of dial telephone operators (2.96 versus 1.60 percent). Data entry time, however, is significantly longer with a dial telephone.

TABLE 11-3
COMMON NUMERIC KEYBOARDS

Calculator	Telephone
7 8 9	1 2 3
4 5 6	4 5 6
1 2 3	7 8 9
0	0

Keyboard Feel

We cannot hope to discuss all the varied keyboard factors that can impact data entry performance. One variable, however, often mentioned in product review articles is the *feel* of the keyboard. Although it is not a scientific term, keyboard feel is a function of, among other things, key travel and resistance characteristics, auditory activation feedback, and hysteresis. *Hysteresis* refers to the tendency of a key switch to remain in the closed position even after a partial release of downward pressure; it is defined as the difference in travel distance between the closing and opening points of a switch. Too little hysteresis can produce key bounce and cause inadvertent insertions of extra characters; too much hysteresis can interfere with high-speed typing.

Brunner and Richardson (1984) compared three keyboards that differed in terms of a number of keying parameters, as shown in Table 11-4. The snap-spring keyboard was characterized by a rapid buildup of resistance on the downstroke with a sharp drop-off in midtravel at the point of switch closure. The elastomer keyboard exhibited a rapid buildup of resistance in the initial part of the downstroke which disappeared at the point of switch closure and then increased again during keystroke overtravel. Two versions of this keyboard were used. One had no auditory feedback when the switch closed, and the other had an electronic click timed to switch closure. The linear spring keyboard exhibited light initial resistance which doubled at the point of switch closure.

Among both occasional and expert typists, the linear-spring keyboard was least preferred and considered most fatiguing to use, presumably because it presented the least actuation resistance and the most ambiguous auditory feedback (it clicked when the switch closed and again when it opened). Figure 11-20 shows the mean interkeystroke interval (a measure of typing speed) and the mean number of insertion errors made with the keyboards. As can be seen, the elastomer keyboards (with and without auditory feedback) resulted in the fastest keying times and low numbers of insertion errors. The linear spring keyboard with low resistance and negligible hysteresis resulted in the greatest number of insertion errors.

TABLE 11-4
KEYBOARD CHARACTERISTICS USED BY BRUNNER AND RICHARDSON

Characteristics	Snap spring	Elastomer	Linear spring
Key travel	3.8 mm	3.2 mm	3.8 mm
Switch closure	2.45 mm	1.78 mm	1.92 mm
Switch opening	1.45 mm	1.78 mm	1.895 mm
Hysteresis	1.0 mm	0.0 mm	0.025 mm
Auditory feedback	Snap at open and closure	*a.* None *b.* Click at closure	Up stop and down stop

Source: Brunner and Richardson, 1984.

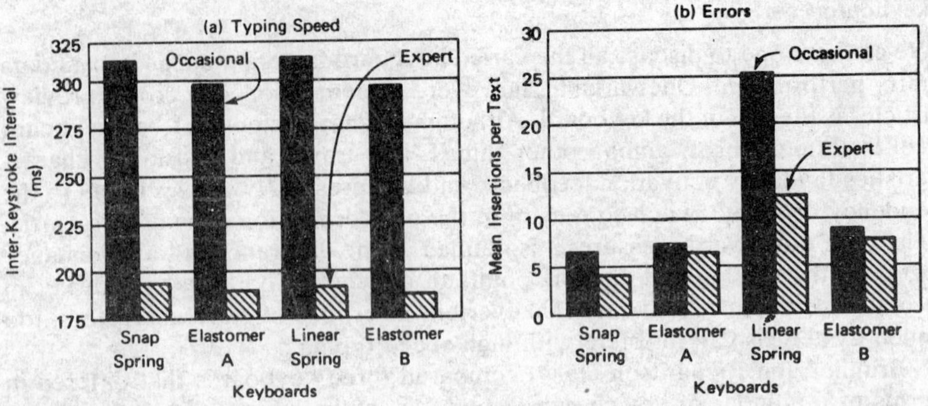

FIGURE 11-20
Effects of various types of keyboards on mean interkeystroke interval and mean number of insertion errors for occasional and expert typists. Keyboards are described in Table 11-4. (*Source: Adapted from Brunner and Richardson, 1984, Tables 1 and 3.*)

Membrane Keypads

Membrane keypads are becoming increasingly popular for use in many consumer products from microwave ovens to pocket calculators. Membrane switches usually consist of mechanical contacts separated by a very thin nonconductive layer. The nonconductive layer has spaces in it corresponding to the keys; when the top layer is depressed, the contacts positioned above and below the space close. Membrane keypads are inexpensive, afford considerable design flexibility, and are sealed to protect against contamination.

Such keypads, however, present some interesting human factors problems. First, key travel distance is virtually nonexistent. Whereas conventional push-button keyboards may have to be depressed 100 to 200 mil (1 mil = 1/1000 in) for activation, the typical membrane keypad can be activated by depressing it 10 to 20 mil. Thus, the familiar keystroke feedback is absent. Second, to help reduce accidental activation, membrane keypads often require more force to activate them than do conventional push-buttons. For example, push buttons may require 50 to 90 g to activate, while membrane keypads may require 200 to 500 g. Third, the actual contact areas (where the spaces in the nonconductive inner layer are positioned) are often difficult to locate. Designers print graphics on the keypads or, as we will see, use other techniques to aid users in positioning their fingers on the contact areas.

Membrane switch technology has been applied most often to situations that require infrequent or intermittent data entry, such as entering a few numbers on a pocket calculator to balance a checkbook. Loeb (1983), however, investigated the use of a membrane keypad (with auditory feedback) for typing tasks, using subjects differing in typing ability. Figure 11-21 presents the average number of words typed per minute [(number of words typed − number of words in error)/total time in minutes] for the various groups of subjects. There was very little difference between the conventional and membrane keyboards

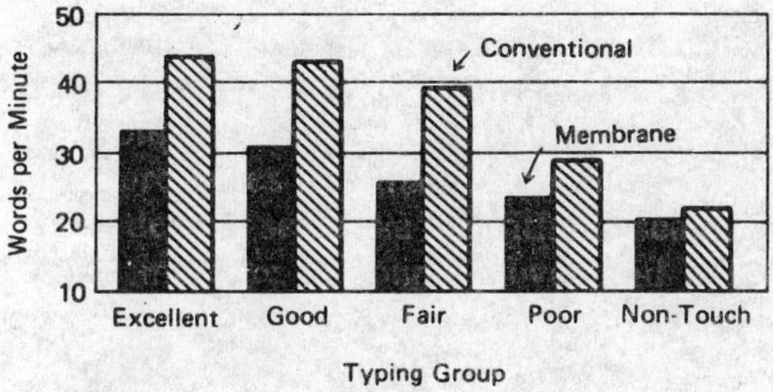

FIGURE 11-21
Comparison of typing performance using a conventional keyboard and a membrane keyboard. Words per minute (wpm) = (number of words typed − number of words in error)/total time in minutes. Categories of typists were defined as follows: Excellent = >60 wpm; Good = 50–60 wpm; Fair = 40–49 wpm; Poor = 26–39 wpm; Nontouch = 16–25 wpm. (*Source: Adapted from Loeb, 1983, Table II. Reprinted with permission from the Bell System Technical Journal, copyright 1983 AT&T.*)

for the non-touch typists. The difference, however, was more pronounced for the touch typists (20 to 35 percent). Loeb did note that the advantage of the conventional keyboard over the membrane keyboard was reduced substantially, but not completely, as a function of limited practice on the membrane keyboard. It is possible, therefore, that with enough practice the membrane keyboard may be as good as a conventional keyboard for typing, at least as far as productivity is concerned.

The task of key pressing involves locating the key and activating it. Both actions require feedback to the user. Roe, Muto, and Blake (1984) conducted a study investigating various techniques for providing membrane keypad users with such feedback. Three feedback techniques were used in all possible combinations: (1) *auditory tone* for activation feedback, (2) *embossing* (each key outlined with a raised rim) for finger-position feedback, and (3) *snap domes* (a raised metal dome over each contact area which is depressed to activate the key) to supply good finger-position feedback and possibly some activation feedback. The results indicated that some form of activation feedback is especially important for reducing errors (virtually all the errors were errors of omission). Keypads incorporating the tone supplied the best activation feedback and resulted in the fewest number of errors. With the tone present, adding finger-position feedback with embossing, domes, or both resulted in a further, but equal, decrease in errors. With no tone present, the use of domes, with or without embossing, resulted in fewer errors than using only embossing. Without the tone it appeared, therefore, that embossing alone failed to provide activation feedback; the domes, however, did provide such feedback, although not as effectively as did the tone. The conclusion, therefore, is that both activation and finger-position feedback are important in membrane keypad use and should be designed into the device to aid the user.

Split and Tilted Keyboards

Almost all alphabetic keyboards on typewriters and computers arrange the keys in straight parallel rows. This configuration requires that the hands bend outward at the wrist when typing. This is shown in Figure 11-22. There is

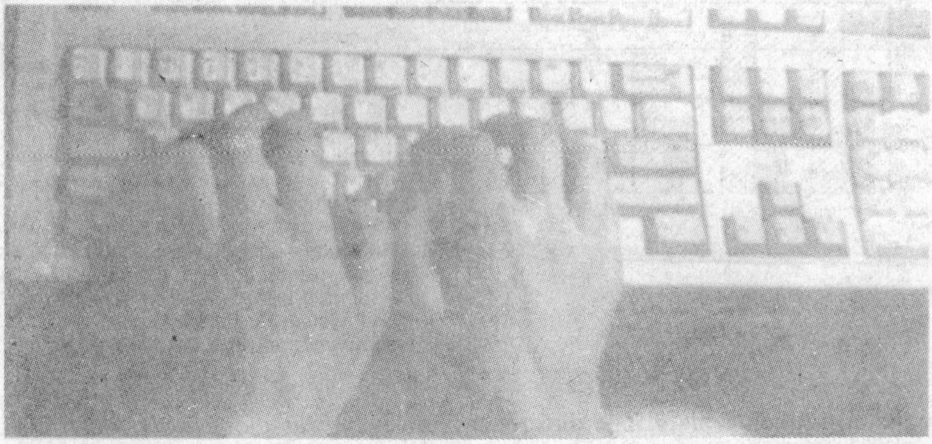

FIGURE 11-22
Typical keyboard arrangement of straight parallel rows of keys. This arrangement forces an unnatural outward bending of the hands at the wrists.

evidence that typing with the wrists in such a position can lead to discomfort and tenosynovitis (an inflammation of the tendon sheaths) in the wrist. As early as 1926, Klockenberg proposed splitting the keyboard into two parts so that the hands could be kept in a more natural, straight-wrist, position. Several such designs have been proposed and tested (for example, Kroemer, 1972; Nakaseko, Grandjean, Hunting, and Gierer, 1985; Ilg, 1987). The basic features of one such design are shown in Figure 11-23. The two halves of the keyboard are angled out and sloped laterally down. Some designs place the keys in cupped areas to conform to the different lengths of the fingers. The results from the various studies are encouraging. People seem to get accustomed to the designs quickly and report less discomfort using them. The degree of bending at the wrist is markedly reduced. The angle between the halves of the keyboard of the various designs range from about 25 to 30°. Although split keyboards are not common today, they may gain in popularity and may be just what the doctor ordered for those experiencing wrist problems.

Handwritten and Gestural Data Entry

New developments in computer hardware and software have made possible the recognition of handwritten input by a computer. The few implementations of this technology to date have used a touch screen or tablet upon which the user writes. The computer recognizes the scribbling and displays the corresponding characters on the screen. A couple of computer companies have begun to sell notebook-size machines, but so far only in Japan (Sanger, 1990). The reason, of course, is because Japanese is a mixture of three alphabets, including 7000 Chinese-based ideographs (picturelike symbols), the value of a computer that can recognize handwritten input would be of more value there than in other countries that have more limited alphabets. Mahach (1989), for example, evalu-

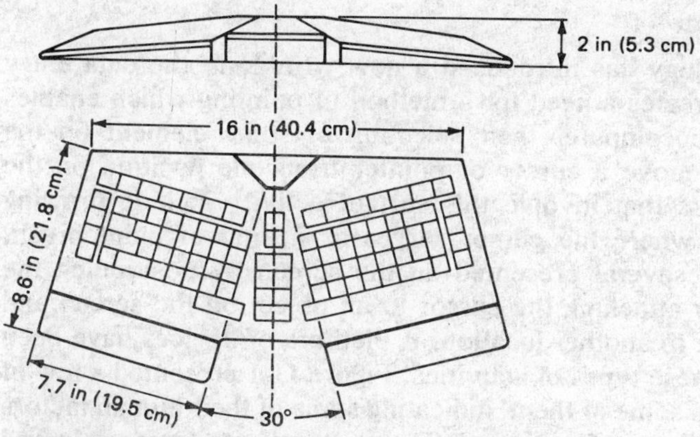

FIGURE 11-23
Representative design of a split and tilted keyboard. The configuration allows a more natural hand position with less outward bending of the hands at the wrists than in typical parallel row keyboards. (*Source: Adapted from Ilg, 1987, Fig. 8.*)

ated a handwriting recognition device for English text entry and found it to be slower than using a keyboard and far more error-prone.

A takeoff on handwritten input is the use of handwritten gestures to communicate with a computer. Gestural inputs have been explored most often within the framework of a text-editing task. The gestures are written on a touch tablet or screen. The gestures are similar to proofreaders' or copy editors' symbols. Figure 11-24 demonstrates a few possible gestural inputs. Wolf (1988) compared a prototype gestural interface to a keyboard for editing a spreadsheet. Subjects completed the task with the gestural interface in 69 to 76 percent of the time it took them to do it with the keyboard and preferred it to the keyboard.

Handwritten and gestural data entry still have a way to go before they become fully operational, but they do show promise for certain specific applications.

FIGURE 11-24
Illustrations of gestural inputs to edit a spreadsheet. The commands do the following: change *Californio* to *California*; insert 1 column between June and July; erase the Ohio row; and sum the June column.

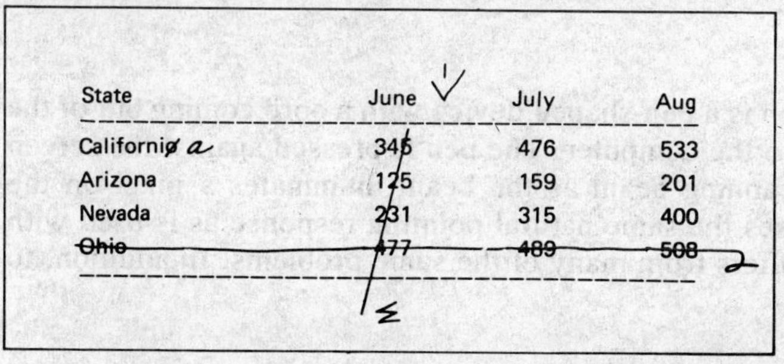

Cursor Positioning Devices

Computer-based technology has introduced a new wrinkle in the data entry field. This technology creates a need for a method of pointing which enables users to indicate to the computer their selection of some element on the computer display or to move a cursor or pointer from one position on the screen to another. Tasks that involve this sort of activity include drawing pictures on the screen where the cursor serves as a kind of paint brush, selecting a choice from several presented on the screen by positioning the cursor over the item, or attaching the cursor to an object on the screen and moving it and the object to another location. A plethora of devices have been developed to facilitate these types of activities. Figure 11-1 presented a few of these. We briefly discuss some of them, indicating some of their human factors advantages and disadvantages. For more information refer to Greenstein and Arnaut (1988).

Touch Screen Touch-screen devices use either a touch-sensitive overlay on the computer screen or beams (e.g., infrared or acoustic) projected across the screen that are interrupted when the screen is touched. The user simply points a finger at something on the screen and touches (or almost touches) the screen. This technique is easy to learn and requires the most natural response for the user—if you want something, point at it. It is, however, not well suited for drawing on the screen. There are some other drawbacks: the finger obscures what is being pointed at; holding up the arm to the screen for long periods is fatiguing; and the finger leaves a smudge on the screen which reduces character legibility. In addition, pointing is not very accurate. Beringer (1983) and Beringer and Peterson (1985) found, for example, that when people use a touch screen to select an item, they tend to hit below the target by ⅛ to ¼ in (3.2 to 6.4 mm) and that accuracy was best for targets at the bottom of the screen and worst for those at the top. This appears to be due to the positioning of the arm and to parallax.

The size of keys (touch targets) displayed on the screen affects a user's performance. Beaton and Weiman (1984) varied the horizontal and vertical size and separation of touch keys in a target selection task. Only vertical size affected errors, with fewer errors being made with the larger keys [0.4 in (10.16 mm)] than with the smaller keys [0.2 in (5.08 mm)]. Figure 11-25 presents one configuration preferred by subjects that also resulted in the smallest number of errors.

Light Pen A light pen is a pen-shaped device with a cord coming out of the end, and it is attached to the computer. The pen is pressed against the screen and senses the CRT scanning beam as the beam illuminates a pixel on the screen. The light pen uses the same natural pointing response as is used with the touch screen and suffers from many of the same problems. In addition, to

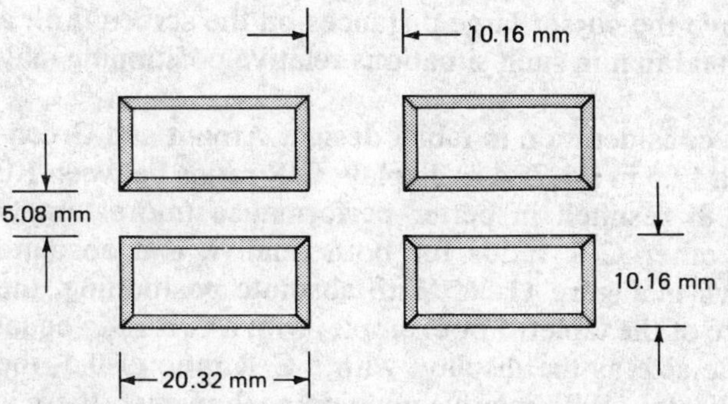

FIGURE 11-25
Dimensions of touch-screen keys preferred by subjects and yielding the smallest number of errors. (*Source: Beaton and Weiman, 1985*)

point, the user must find the pen and pick it up. The cord on the pen can also become tangled on the keyboard or monitor. Usually, pointing resolution is much better with a light pen than with a touch screen.

Graphics Tablet A graphics tablet is a device with a flat surface that can be placed on a desktop or held in the lap of the user. Movement of a finger or a stylus on the tablet's surface controls the cursor on the computer screen. Some graphics tablets, also called *digitizing tablets*, use a special stylus or puck to electromagnetically or acoustically sense position on the tablet's surface. Some of these digitizing tablets are as large as a drafting table. Graphics tablets can also be touch-sensitive (these are sometimes called *touch pads*) and can be used with a finger or ordinary stylus. Compared to a touch screen, the graphics tablet separates the input from the system response. The user has to watch the screen while touching the tablet, but the position of the tablet reduces arm fatigue. Drawing pictures with a stylus and tablet is similar to drawing with a pencil on paper. With digitizing tablets the user can rest the hand on the tablet without causing spurious inputs. This is not the case with all touch pads, however.

Graphics tablets can employ absolute or relative cursor positioning. In *absolute positioning*, when a user places a finger on the tablet, the cursor on the screen will move from its present position to a position that corresponds to the location of the finger on the tablet. In essence, the tablet is a representation of the screen. Touching the four corners of the tablet would jump the cursor to the four corners of the screen. In *relative positioning,* the cursor remains in its current position no matter where the finger is placed on the tablet. It is only movement of the finger that causes the cursor to move correspondingly. If one were to just touch each of the four corners of a relative positioning tablet without moving the finger on the tablet, the cursor would not move on the screen at all. In general, absolute positioning is faster (Arnaut and Greenstein, 1986) and more accurate (Ellingstad et al., 1985) than relative positioning. As Greenstein and Arnaut (1988) point out, however, when the tablet is small in comparison to the display, absolute positioning requires that small movements on the tablet be made to move the cursor large distances on the screen (i.e., a small C/R ratio is required) and that in such situations relative positioning may be preferable.

C/R ratio is an important consideration in tablet design. Arnaut and Greenstein (1986) found that with a 12.5 in (31.75 cm) display, C/R ratios between 1.0 and 1.25 (gain of 1.0 to 0.8) resulted in better performance (more targets selected per second) than other C/R ratios for both relative and absolute positioning tablets, as shown in Figure 11-26. With absolute positioning, the C/R ratio determines the size of the tablet. For example, with a C/R ratio equal to 1.0, the tablet is the same size as the display; with a C/R ratio of 0.5, the tablet is half the size of the display. With relative positioning, however, there is no relationship between C/R ratio and the size of the tablet. If a movement across the entire tablet surface does not move the cursor across the entire display, then the user simply has to repeat the movement.

It is possible to program a tablet to include a C/R ratio that is proportional to the velocity of the finger's movement on the tablet. A rapid finger movement of a given distance would result in a greater movement of the cursor than would an equal but slower movement of the finger. This velocity-dependent C/R ratio is also called a *lead-lag compensation system.* Because gross movements tend to be rapid and fine positioning tends to be done more slowly, adding a lead-lag compensation system provides low C/R ratios (high gain) for gross movements and high C/R ratios (low gain) for fine tuning. Becher and Greenstein (1986) found faster target acquisition times with the velocity-dependent C/R ratio than with a constant C/R ratio, but there was a slight increase in error rate.

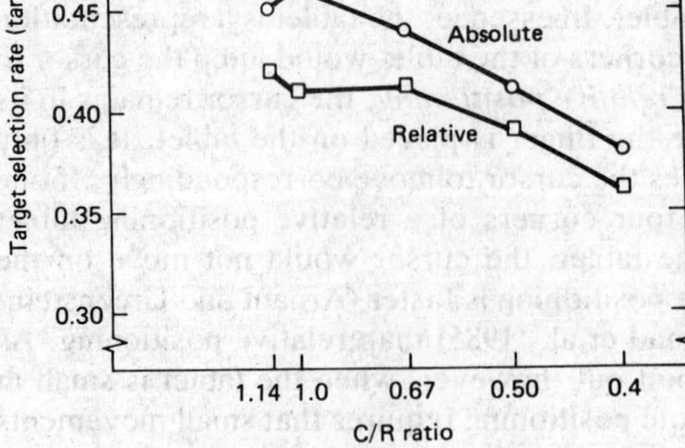

FIGURE 11-26
Target selection performance with absolute and relative positioning touch pads as a function of C/R ratio or gain. (*Source: Arnaut and Greenstein, 1986, Fig. 2. Reprinted with permission of the Human Factors Society, Inc. All rights reserved.*)

Mouse A mouse is a hand-held device with one or more buttons that is rolled on a desktop and simultaneously controls the movement of a cursor on the screen. As with graphics tablets, the C/R ratio of the mouse can be changed or made velocity-dependent. A mouse, however, is also a relative positioning device. In general, a mouse is easy and fast to use. One disadvantage of the mouse is that a clear space near the computer screen is required to operate it; mice do not work well in a rat's nest of papers and open books. Freehand drawing is not as natural as drawing with a pencil-shaped stylus because the mouse is moved with arm and wrist movements rather than finger movements.

Other Cursor Positioning Devices In addition to the more esoteric devices mentioned above, keyboards, joysticks, and trackballs are also used for cursor positioning. Keyboards (cursor keys for up, down, left, and right) are notoriously slow and are not at all good for drawing. Their advantage is that the hands do not have to leave the keyboard during data entry. Joysticks and trackballs can be good pointing devices but are not well suited for drawing.

Comparison of Various Cursor Positioning Devices Albert (1982) compared the devices listed in Table 11-5 on speed and accuracy of positioning a cursor. The rankings of the devices on these criteria are also shown in Table 11-5. The trackball, the most accurate device, was 100 percent more accurate than the touch screen, the least accurate device. The fastest device, the touch screen, was more than 8 times faster than the slowest device, the keyboard. These are impressive differences and illustrate the need to select the best device for the task at hand. Notice that there is a clear speed-accuracy tradeoff. The more accurate devices tended to be the slower devices; thus, the selection of the best device must take into account the relative importance on the task of speed and accuracy.

TABLE 11-5
COMPARISON OF VARIOUS DEVICES ON SPEED AND ACCURACY OF CURSOR POSITIONING

	Ranking	
Device	**Speed**	**Accuracy**
Touch screen	1 (fastest)	6.5 (worst)
Light pen	2	6.5
Digitizing tablet	3	2
Trackball	4	1 (best)
Force joystick	5	3
Position joystick	6	4
Keyboard	7 (slowest)	5

Source: Adapted from Albert, 1982, figs. 1 and 2. Copyright by the Human Factors Society, Inc. and reproduced by permission.

Card, English, and Burr (1978) compared performance using a mouse, isometric joystick, step keys (up, down, left, and right), and text keys (character, word, line, and paragraph) in a text-editing task. The results are shown in Table 11-6. The mouse was found to be the fastest and also to have the lowest error rate. This is in contrast to the other studies mentioned which found speed-accuracy trade-offs among the devices tested.

Epps (1987) compared the following devices on seven graphics-editing tasks: absolute-positioning touch pad (C/R ratio = 1.0), mouse (C/R ratio = 0.77), trackball (velocity-dependent C/R ratio), relative-positioning touch pad (velocity-dependent C/R ratio), rate-controlled displacement joystick, and rate-controlled force joystick. Subjects consistently performed best in terms of task completion time with the trackball and the mouse. Performance with the absolute touch pad was close to that with the trackball and mouse for moderate cursor positioning accuracy tasks, but degraded as positioning accuracy requirements increased.

Research on cursor positioning devices is still in its infancy. Specific design parameters of the devices and their relationship to the display have hardly been addressed. Such features as C/R ratio, physical size, methods for confirming a selection, and feedback on the selection made have not been systematically investigated. As technology advances, higher-resolution screens and devices will become possible with greater accuracy demands placed on the user. Researchers will have to scramble to keep up with the advancing technology.

SPECIAL CONTROL DEVICES

The hands and feet are traditionally the parts of the body delegated to the task of operating controls. Microcomputer technology, however, has opened new vistas for control devices and applications. Our purpose here is to introduce some devices that are being developed or are in current use. The emphasis is on those devices which involve challenging human factors problems.

Teleoperators

The technological explosion of recent decades has made it possible to explore outer space and the depths of the ocean, thus exposing humans to new and

TABLE 11-6
COMPARISON OF TEXT-EDITING DEVICES

Device	Time, s	Error, %
Mouse	1.66	5
Joystick	1.83	11
Step keys	2.51	13
Text keys	2.26	9

Source: Card, English, and Burr, 1978.

hazardous environments. These and other developments require some types of extension of human control functions across distance or through some physical barrier, as in the control of lunar vehicles or of underwater activities. The control systems for performing such functions have been called *teleoperators* (Corliss and Johnson, 1968). Teleoperators are general-purpose human-machine systems that augment the physical skills of the operator. A person is always in the teleoperator control link on a real-time basis. Robots are not considered teleoperators because they operate autonomously and sometimes (as in the science fiction movie *2001: A Space Odyssey*) counter to human interests.

An example of a manipulator teleoperator is shown in Figure 11-27. Similar types of devices are commonly used today for handling dangerous materials (such as radioactive, explosive, or biologically toxic materials) or working in hostile environments such as involved in undersea work. One of the more spectacular examples is the 50-ft (15.2-m) remote manipulator arm of the U.S. space shuttle. The arm has been successfully used to retrieve wayward satellites and fix malfunctioning equipment on the shuttle itself.

Human Factors Considerations The design of various types of teleoperators must, of course, take into account the biomechanics of human movement, anthropometrics, and the nature of the kinesthetic and proprioceptive information people use to control their movements. It is important for efficient teleoperator control that the device be designed to be compatible with the operator. Rice, Yorchak, and Hartley (1986) discuss human factors consider-

FIGURE 11-27
Experimental remote-handling equipment used in various studies.
(*Source: Courtesy of USAF, Air Force Human Resources Lab.*)

ations involved in use of teleoperators in space. In this regard, particular consideration needs to be given to the nature of the feedback given to the operator. In many teleoperator applications, the operators are denied direct physical contact with the objects being manipulated by the teleoperator device; thus the teleoperators lose their usual kinesthetic feedback. Feedback regarding the amount of force being exerted by the teleoperator can be given to the operator through the use of servo systems or direct mechanical linkage. Information on slip and shear force, however, is more difficult to convey in teleoperator systems (Feldman, 1982).

More and more teleoperator applications deny the operator direct visual access to the effector end of the teleoperator. In such cases, television cameras are commonly used to extend the operator's vision. Generally, two cameras are better than one. Whether they should be set up to give two different (orthogonal) views of the task area or one stereo image is still being debated (Rice, Yorchak, and Hartley, 1986). Use of TV, however, does introduce major problems. TV distorts distance and depth cues; and when the cameras can be moved, zoomed, etc., operators can become visually disoriented. The use of remote sensing (visual and kinesthetic) can also introduce time delays between the operator's control action and the feedback received about the device's response. This is especially true if the operator is on earth controlling a manipulator in space where time delays can be from 2 to 8 s.

An especially tricky aspect of teleoperators is the design of the controls used by the operator to direct the actions of the effector. A manipulator has multiple joints, each of which can be extended and rotated in a number of different planes. Designing a control configuration that will allow smooth, coordinated movements is a real challenge. Corliss and Johnson (1968) feel that teleoperator controls should be "built along anthropomorphic lines." The problem with this is that teleoperators can perform movements impossible for humans to perform, such as rotating the wrist 360°. Often computers are enlisted to aid the operator, such as by using preprogrammed movement sequences that can be invoked by the operator as needed.

Speech-Activated Control

The use of speech to converse with computers is a stock item in science fiction books and movies. The advantages are obvious: The operator is not tied to a console, the operator's hands are free to do other chores (such as pour a cup of coffee), and, at least in science fiction, only minimal training of operators is required. Although such casual conversations with machines are not yet upon us, recent advances in speech recognition have spawned new and novel solutions to system problems.

Applications for Speech Recognition Speech recognition systems are usually substituted in place of keyboard entry or pencil-and-paper recording of data. The best applications appear to be where data entry must occur simultaneously with other activities. Entering destination data while sorting packages is an example; recording part numbers while assembling components or dialing a car phone while driving are others. In addition, speech recognition is well suited to situations where the operator moves around to collect data. Inspectors on automobile assembly lines, for example, use speech to record defects as they walk around the car. One advantage of such a system is that the data are immediately fed into the computer, rather than having to be transcribed from handwritten notes. Speech recognition systems are also well suited for use by the handicapped for such activities as controlling motorized wheelchairs, or turning on and off lights, coffee pots, etc. (Noyes, Haigh, and Starr, 1989). McCauley (1984) discusses other applications including aircraft cockpits and interactive training systems. Noyes and Frankish (1989) discuss applications of speech recognition in the office. Poock (1986) summarizes speech recognition development efforts in various countries.

Types of Speech Recognition Systems Speech recognition systems can be classified along two dimensions. The first is whether the system is *speaker-dependent* or *speaker-independent*. Speaker-dependent systems, by far the most common, are trained to recognize a particular person's pronunciation of a limited vocabulary of words, phrases, or sentences. Some speaker-dependent systems, however, are robust enough that they can provide 85 to 90 percent recognition accuracy of utterances from people other than the one on which it was trained. Speaker-independent systems can recognize words spoken by anyone. The second dimension relates to the constraints placed on how the speaker must utter the words. Speech recognition systems are classed as *isolated-word, connected-word,* or *continuous-speech* systems. In isolated-word systems, clearly the most common, the speaker must set off each utterance from the next with a pause. An utterance can be a word, phrase, or short sentence. This pause requirement is relaxed somewhat in connected-word systems; the speaker can string together several utterances without pausing between them, but each word in the string must be enunciated accurately. In continuous-speech systems, there are no pause constraints on the speaker; entire sentences can be uttered as in natural speech. Almost all commercial speech recognition systems today are of the speaker-dependent, isolated-word variety. Some speaker-independent systems are being used in telephone applications, but their vocabulary is very limited, often to the digits 0 to 9.

Connected-word and continuous-speech systems in use today also tend to be limited to small vocabularies and short, simple sentences (Reddy and Zue, 1983). Continuous-speech systems require about 10 times as much computational power as isolated-word systems (Pfauth and Fisher, 1983) and face some monumental recognition problems. As an illustration, consider the following sentence: "You gave the cat your dinner." Because of the phenomenon called *coarticulation,* the "t" of "cat" combines with the "y" of "your" and makes the sentence sound like "you gave the catcher dinner."

Current speaker-dependent, isolated-word systems recognize words in one of two fundamental ways. The most common involves comparison of the frequency spectrum of the voice signal (see Chapter 7) with the spectral characteristics of each vocabulary item stored in the computer until an acceptable match is found. The second approach focuses on isolating certain key features of an utterance, e.g., stops and fricatives, and comparing these to the stored vocabulary items. For a more detailed, but readable, explanation, see Levinson and Liberman (1981). A problem in practice is that background noise can render a valid command unrecognizable because of the distortion to the sound pattern transmitted to the system.

Human Factors Issues There are several human factors–related issues inherent in the speaker-dependent, isolated-word type of system (McCauley, 1984; Hapeshi and Jones, 1988). First, the system can handle only a limited predefined vocabulary, with most currently available systems having vocabularies in the range of 20 to 200 items. The larger the vocabulary, generally, the lower the recognition accuracy and the slower the processing, all other things being equal. Several things can be done to increase vocabulary size and recognition accuracy. By placing constraints on what items can be said in what order, very large vocabularies can be generated. For example, the first word or phrase in a voice command could be limited to one of, say, 50 preestablished items. When it is recognized, a second set of 50 items, composed of those items which are allowed to follow the first item, is searched. This process can be repeated, thus allowing for vocabularies of several hundred item strings (Coler et al., 1977). In addition, the specific vocabulary items can be selected to be maximally discriminable from one another (Green et al., 1983). A large package delivery service in the United States uses speech recognition to sort parcels. To minimize recognition errors, they use the names of professional football teams—Colts, Bears, Jets, Giants—to indicate the destinations of the parcels (Modern Materials Handling, 1983).

A second problem with isolated-word systems is that each word must be set off with pauses. If the speaker fails to insert a pause after each word or inserts a pause before the end of the word, the system very likely will fail to recognize it. The need to control speech pauses makes it mandatory that the operators of isolated-word systems be trained to discipline their voices and speech. Further, a sigh, cough, or mumble may be erroneously accepted as a valid response by the system. Speakers must be trained to control such extraneous verbal behavior.

In practice, not only must the operator be trained, but so must the system itself. As pointed out in Chapter 7, people have different speech patterns, inflections, frequency modulations, etc.; not everyone sounds the same. A speaker-dependent speech recognition system must develop a complete vocabulary for each operator. This is usually done by having the operator repeat each item in the system's vocabulary a few times to develop a good representation of that item, for that speaker, in the computer's memory. Most systems today require around four repetitions (McCauley, 1984). This sort of training is very tedious and restricts the size of the vocabulary that can, in practical terms, be used. Imagine having to repeat distinctly each of 300 words 4 times!

People's voices change owing to fatigue and stress. The speech recognition system, trained to recognize a rested, calm (perhaps even bored) voice, may not recognize the speaker in a stress situation or after the voice becomes fatigued. Imagine controlling a piece of machinery with a speech recognition system. Your clothes become entangled, and you are being pulled into the machine. You yell "Stop!" Unless you yelled "stop" in the same way as when you first trained the system, it is very likely that the machine will not respond. And yelling louder will not help.

Performance of Speech Recognition Systems In most reports, speech recognition systems turn out to be as good as or better than keyboard data entry, although error rates may be higher depending on the specific system, vocabulary, and task situation. Error rates of 1 percent are claimed by some systems, but this is usually under ideal conditions. Error rates can be, and usually are, higher in actual working situations. Speech recognition is not, however, the panacea for all data entry problems. Morrison et al. (1984), for example, found that doing text editing with a speech recognition system (in which the commands were spoken and the text was typed) was preferred less than was straight keyboard entry. Switching modalities during a command was inherently disruptive.

We can conclude that speech recognition has progressed to where it is a valuable method of data entry and is especially well suited to particular applications, but it will probably never totally replace the keyboard as the primary device for data entry.

Eye-Activated Control

Advances in equipment used to track eye and head movements have made it possible to convert eye movements to control input. To date, most eye-activated control applications have been in the military field. Other applications could be in control of teleoperators, selection of instruments on a panel, and visual search tasks. Calhoun, Arbak, and Boff (1984) describe one way such a system would operate. The user, wearing a special helmet to monitor head and eye position, would direct his or her line of sight to, say, a switch. The switch would then change (e.g., its color, brightness, etc.) to give feedback that the signal had been received by the system. To complete the switch activation, the user would make some type of input, perhaps a verbal response to a speech recognition system or a separate button press to acknowledge or accept the selection.

Accuracy of eye positioning has been reported to be approximately ± 10 minutes of visual arc at the center of the visual field (Chapanis and Kinkade, 1972; Coluccio and Mason, 1970). This is very close to the 10 to 20 minutes of arc reported by Calhoun, Arbak, and Boff for successive eye fixations in an actual eye control system. Calhoun, Arbak, and Boff did find that accuracy was poorer for fixations in the lower part of the visual field. Eye control has some disadvantages, including the following: there is potential for overburdening of the visual system, especially in a visual dual-task situation; eye control is difficult in a vibrating or acceleratory environment because either one will produce eye movements which the operator cannot prevent; distractions which might cause operators to shift their gaze will degrade performance; the time required to sense and calculate the line of sight causes a corresponding delay in the feedback to the operator that an object has been selected; and the operator must hold the gaze for some short time to reduce the likelihood of inadvertent activation. Calhoun, Arbak, and Boff report median total activation times (from onset of some signal, through gaze fixation, to confirming selection with an accept switch) of 1.5 to 1.7 s. Calhoun, Janson, and Arbak (1986) had pilots select switches on a control panel while performing a compensatory tracking task and found no differences in time to select a switch or tracking performance between manual switch selection and eye-controlled selection with a separate manual accept. Delays inherent in the speech recognition system, however, resulted in slower selection times and greater tracking error when voice was used to accept the eye-controlled selection.

DISCUSSION

We have discussed various aspects of controls and their influence on operator performance. One might think that the selection or design of controls has "all been done"; however, numerous examples of inappropriately designed or selected controls can still be found today, for example, in nuclear power plants, heavy-equipment cabs, and numerous consumer products. A wide variety of controls and control mechanisms are available off the shelf; but as we saw in this chapter, the best design often depends on the particular demands of the task and the criteria used to judge performance. For those who view controls as somehow less glamorous than other areas of human factors, we hope the discussions of cursor positioning devices, teleoperators, and speech recognition systems illustrate that exciting things are still happening in this area.

REFERENCES

Air Force System Command (1980, June). *Design handbook 1-3, human factors engineering* (3d ed.). U.S. Air Force.

Albert, A. (1982). The effect of graphic input devices on performance in a cursor positioning task. *Proceedings of the Human Factors Society 26th Annual Meeting*. Santa Monica, CA: Human Factors Society, pp. 54–58.

Arnaut, L., and Greenstein, J. (1986). Optimizing the touch tablet: The effects of control-display gain and method of cursor control. *Human Factors*, 28(6), 717–726.

Arnaut, L., and Greenstein, J. (1987). An evaluation of display/control gain. *Proceedings of the Human Factors Society 31st Annual Meeting*. Santa Monica, CA: Human Factors Society, pp. 437–441.

Ayoub, M. M., and Trombley, D. J. (1967). Experimental determination of an optimal foot pedal design. *Journal of Industrial Engineering*, 17, 550–559.

Beaton, R., and Weiman, N. (1984). *Effects of touch key size and separation on menu-selection accuracy* (Tech. Rep. No. TR 500-01). Beaverton, OR: Tektronix, Human Factors Research Laboratory.

Becker, J., and Greenstein, J. (1986). A lead-lag compensation approach to display/control gain for touch tablets. *Proceedings of the Human Factors Society 30th Annual Meeting*. Santa Monica, CA: Human Factors Society, pp. 332–336.

Beringer, D. (1983). The pilot-computer direct-access interface: Touch panels revisited. In R. Jensen (ed.), *Proceedings of the 2d Symposium on Aviation Psychology*. Columbus: Ohio State University, Aviation Psychology Laboratory.

Beringer, D., and Peterson, J. (1985). Underlying behavioral parameters of the operation of touch-input devices: Biases, models, and feedback. *Human Factors*, 27(4), 445–458.

Bowen, H. M., and Guinness, G. V. (1965). Preliminary experiments on keyboard design for semiautomatic mail sorting. *Journal of Applied Psychology*, 49(3), 194–198.

Bradley, J. V. (1967). Tactual coding of cylindrical knobs. *Human Factors*, 9(5), 483–496.

Bradley, J. V. (1969). Desirable dimensions for concentric controls. *Human Factors*, 11(3), 213–226.

Brunner, H., and Richardson, R. (1984). Effects of keyboard design and typing skill on user keyboard preferences and throughput performance. In D. Attwood and C. McCann (eds.), *Proceedings of the 1984 International Conference on Occupational Ergonomics*. Rexdale, Ontario, Canada: Human Factors Association of Canada, pp. 267–271.

Buck, L. (1980). Motor performance in relation to control-display gain and target width. *Ergonomics*, 23, 579–589.

Bullinger, H., and Muntzinger, W. (1984). Development of ergonomically designed controls. In D. Attwood and C. McCann (eds.), *Proceedings of the 1984 International Conference on Occupational Ergonomics*. Rexdale, Ontario, Canada: Human Factors Association of Canada, pp. 447–451.

Calhoun, G., Arbak, C., and Boff, K. (1984). Eye-controlled switching for crew station design. *Proceedings of the Human Factors Society 28th Annual Meeting*. Santa Monica, CA: Human Factors Society, pp. 258–262.

Calhoun, G., Janson, W., and Arbak, C. (1986). Use of eye control to select switches. *Proceedings of the Human Factors Society 30th Annual Meeting*. Santa Monica, CA: Human Factors Society, pp. 154–158.

Card, S., English, W., and Burr, B. (1978). Evaluation of mouse, rate-controlled isometric joystick, step keys, and text keys for text selection on a CRT. *Ergonomics*, 21, 601–613.

Carter, R. (1978). Knobology underwater. *Human Factors*, 20, 641–647.

Casey, S., and Rogers, S. (1987). The case against coplanar pedals in automobiles. *Human Factors*, 29(1), 83–86.

Chapanis, A., and Kinkade, R. (1972). Design of controls. In H. P. Van Cott and R. Kinkade (eds.), *Human engineering guide to equipment design* (rev. ed.). Washington, DC: Government Printing Office.

Coler, C., Plummer, R., Huff, E., and Hitchock, M. (1977). Automatic speech recognition research at NASA Ames Research Center. In M. Curran, R. Breaux, and E. Huff (eds.), *Voice technology for interactive real-time command/control systems applications, proceedings of a symposium*. Moffett Field, CA: NASA Ames Research Center.

Coluccio, T., and Mason, K. (1970). Measurement of target position in a realtime display by use of the Oculometer eye direction monitor. Paper presented at the Human Factors Society annual meeting.

Conrad, R. (1967). Performance with different push-button arrangements, *Het PTT-Bedrijfdeel,* 15, 110–113.

Conrad, R., and Hull, A. J. (1968). The preferred layout for numerical data-entry keysets. *Ergonomics,* 11(2), 165–173.

Conrad, R., and Longman, D. (1965). A standard typewriter versus chord keyboard—An experimental comparison. *Ergonomics,* 8, 77–88.

Corliss, W., and Johnson, E. (1968). *Teleoperator controls* (NASA SP-5070). Washington, DC: NASA Office of Technology Utilization.

Davies, B. T., and Watts, J. M., Jr. (1970). Further investigations of movement time between brake and accelerator pedals in automobiles. *Human Factors,* 12(6), 559–561.

Drury, C. (1975). Application of Fitts' Law to foot-pedal design. *Human Factors,* 17, 368–373.

Dvorak, A. (1943). There is a better typewriter keyboard. *National Business Education Quarterly,* 12, 51–58.

Ellingstad, V., Parng, A., Gehlen, G., Swierenga, S., and Auflick, J. (1985, March). *An evaluation of the touch table as a command and control input device.* Vermillion, SD: University of South Dakota, Department of Psychology.

Epps, B. (1987). A comparison of cursor control devices on a graphics editing task. *Proceedings of the Human Factors Society 31st Annual Meeting.* Santa Monica, CA: Human Factors Society, pp. 442–446.

Feldman, M. (1982, June). Evolving toward projected man. *Nuclear News,* 25(8), 127–132.

Fitts, P. (1954). The information capacity of the human motor system in controlling the amplitude of movement. *Journal of Experimental Psychology,* 47, 381–391.

Gamst, F. (1975). Human factors analysis of the diesel-electric locomotive cab. *Human Factors,* 17, 149–156.

Glass, S., and Suggs, C. (1977). Optimization of vehicle-brake pedal foot travel time. *Applied Ergonomics,* 8, 215–218.

Green, T., Payne, S., Morrison, D., and Shaw, A. (1983). Friendly interfacing to simple speech recognizers. *Behaviour and Information Technology,* 2, 23–38.

Greenstein, J., and Arnaut, L. (1988). Input Devices. In M. Helander (ed.), *Handbook of human-computer interaction.* New York: Elsevier, pp. 495–519.

Hapeshi, K., and Jones, D. (1988). The ergonomics of automatic speech recognition interfaces. *International Reviews of Ergonomics,* 2, 251–290.

Hartman, B. O. (1956). *The effect of joystick length on pursuit tracking* (Rept. 279). Fort Knox, KY: U.S. Army Medical Research Laboratory.

Howland, D., and Noble, M. (1955). The effect of physical constraints on a control on tracking performance. *Journal of Experimental Psychology,* 46, 353–360.

Hunt, D. P. (1953). *The coding of aircraft controls* (Tech. Rept. 53-221). U.S. Air Force, Wright Air Development Center.

Ilg, R. (1987). Ergonomic keyboard design. *Behaviour and Information Technology,* 6(3), 303–309.

Jenkins, W. (1947*b*). The tactual discrimination of shapes for coding aircraft-type controls. In P. Fitts (ed.), *Psychological research on equipment design* (Research Rept. 19). Columbus, OH: Ohio State University Army Air Force, Aviation Psychology Program.

Jenkins, W., and Connor, M. B. (1949). Some design factors in making settings on a linear scale. *Journal of Applied Psychology*, 33, 395–409.

Jenkins, W., and Karr, A. C. (1954). The use of a joy-stick in making settings on a simulated scope face. *Journal of Applied Psychology*, 38, 457–461.

Katchmer, L. T. (1957, March). *Physical force problems; 1. Hand crank performance for various crank radii and torque load combinations* (Tech. Memo 3-57). Aberdeen, MD: Aberdeen Proving Ground, Human Engineering Laboratory.

Klockenberg, E. (1926). *Rationalisierung der Schreibmaschine und ihrer Bedienung*. Berlin: Springer.

Kohl, G. (1983). Effects of shape and size on knobs on maximal hand-turning forces applied by females. *The Bell System Technical Journal*, 62, 1705–1712.

Kroemer, K. H. E. (1971). Foot operation of controls. *Ergonomics*, 14(3), 333–361.

Kroemer, K. (1972). Human engineering the keyboard. *Human Factors*, 14, 51–63.

Levinson, S., and Liberman, M. (1981, April). Speech recognition by computer. *Scientific American*, 244, 64–76.

Loeb, K. (1983, July–August). Membrane keyboards and human performance. *The Bell System Technical Journal*, 6, 1733–1749.

Mahach, K. (1989). A comparison of computer input devices: Linus pen, mouse, cursor keys, and keyboard. *Proceedings of the Human Factors Society 33d Annual Meeting*. Santa Monica, CA: Human Factors Society, pp. 330–334.

McCauley, M. (1984). Human factors in voice technology. In F. Muckler (ed.), *Human factors review—1984*. Santa Monica, CA: Human Factors Society.

McFarland, R. A. (1946). *Human factors in air transport design*. New York: McGraw-Hill.

Mehr, M. (1973). Two-axis manual positioning and tracking controls. *Applied Ergonomics*, 4, 154–157.

Micro Switch (1984). *Programmable display pushbutton PD series*. Freeport, IL.

Miner, F. J., and Revesman, S. L. (1962). Evaluation of input devices for a data setting task. *Journal of Applied Psychology*, 46, 332–336.

Modern Materials Handling. (1983, April 6). Voice recognition back again, and better, pp. 52–55.

Morrison, D., Green, T., Shaw, A., and Payne, S. (1984). Speech-controlled text-editing: Effects of input modality and of command structure. Paper submitted to *International Journal of Man-Machine Studies*.

Morrison, R., Swope, J., and Halcomb, C. (1986). Movement time and brake pedal placement. *Human Factors*, 28(2), 241–246.

Moussa-Hamouda, E., and Mourant, R. (1981). Vehicle fingertip reach controls—Human factors recommendations. *Ergonomics*, 12, 66–70.

Nakaseko, M., Grandjean, E., Hunting, W., and Gierer, R. (1985). Studies on ergonomically designed alphanumeric keyboards. *Human Factors*, 27(2), 175–187.

National Highway Traffic Safety Administration (1983, September). Recall campaign 1978–1983 model Audi 5000 vehicles equipped with automatic transmission, installation of brake pedal plate, Audi campaign code FR (Recall 83-V-095). Unpublished recall campaign. Cited in Casey and Rogers (1987).

Norman, D. (1983, August 23). *The DVORAK revival: Is it really worth the cost?* La Jolla; Institute for Cognitive Science (C-015), University of California, San Diego.

Noyes, J. (1983). Chord keyboards. *Applied Ergonomics*, 14, 55–59.

Noyes, J., and Frankish, C. (1989). A review of speech recognition applications in the office. *Behaviour and Information Technology*, 8(6), 475–486.

Noyes, J., Haigh, R., and Starr, A. (1989). Automatic speech recognition for disabled people. *Applied Ergonomics,* 20(4), 293–298.

Paul, L., Sarlanis, K., and Buckley, E. (1965). A human factors comparison of two data entry keyboards. Paper presented at *Sixth Annual Symposium of the Professional Group on Human Factors in Electronics,* IEEE.

Pfauth, M., and Fisher, W. (1983, September). Voice recognition enters the control room. *Control Engineering,* 30(9), 147–150.

Pollard, D., and Cooper, M. (1978). An extended comparison of telephone keying and dialing performance. *Ergonomics,* 21, 1027–1034.

Poock, G. (1986). Speech recognition research, applications and international efforts. *Proceedings of the Human Factors Society 30th Annual Meeting.* Santa Monica, CA: Human Factors Society, pp. 1278–1283.

Poulton, E. (1974). *Tracking skill and manual control.* New York: Academic.

Reddy, R., and Zue, V. (1983, November). Recognizing continuous speech remains an elusive goal. *IEEE Spectrum,* 20(11), 84–87.

Rice, J., Yorchak, J., and Hartley, C. (1986). Planning for unanticipated satellite servicing teleoperations. *Proceedings of the Human Factors Society 30th Annual Meeting.* Santa Monica, CA: Human Factors Society, pp. 870–874.

Rockway, M. (1957, September). *Effects of variation in control deadspace and gain on tracking performance* (Tech. Rept. 57–326). Dayton, OH: U.S. Air Force, Wright Air Development Center.

Rockway, M., and Franks, P. (1959, January). *Effects of variations in control backlash and gain on tracking performance* (Tech. Rept. 58–553). Dayton, OH: U.S. Air Force, Wright Air Development Center.

Roe, C., Muto, W., and Blake, T. (1984). Feedback and key discrimination on membrane keypads. *Proceedings of the Human Factors Society 28th Annual Meeting.* Santa Monica, CA: Human Factors Society, pp. 277–281.

Rogers, J. (1970). Discrete tracking performance with limited velocity resolution. *Human Factors,* 12, 331–339.

Sanger, D. (1990, March 26). Japan's scratch-pad computers. *The New York Times.*

Schmidt, R. (1989). Unintended acceleration: A review of human factors contributions. *Human Factors,* 31(3), 345–364.

Seibel, R. (1972). Data entry devices and procedures. In H. P. Van Cott and R. G. Kinkade (eds.), *Human engineering guide to equipment design.* Washington, DC: Government Printing Office.

Seminara, J., Gonzalez, W., and Parsons, S. (1977). *Human factors review of nuclear power plant control room design* (EPRI NP-309). Palo Alto, CA: Electric Power Research Institute.

Snyder, H. (1976). Braking movement time and accelerator-brake separation. *Human Factors,* 18, 201–204.

Welford, A. (1968). *Fundamentals of skill.* London: Methuen.

Wierwille, W. (1984). The design and location of controls: A brief review and an introduction to new problems. In H. Schmidtke (ed.), *Ergonomic data for equipment design.* New York: Plenum.

Wolf, C. (1988). A comparative study of gestural and keyboard interfaces. *Proceedings of the Human Factors Society 32d Annual Meeting.* Santa Monica, CA: Human Factors Society, pp. 273–277.

CHAPTER 12

HAND TOOLS AND DEVICES

Some people still believe that the use of tools for pounding, digging, scraping and cutting is what distinguishes humans from apes. Actually, there is considerable evidence (Washburn, 1960) that such hand tools and other devices were used by prehuman primates almost a million years ago. Today hand tools are crafted for uncounted specific applications as well as for general-purpose activities. Until recently, however, human factors had largely ignored the design of hand tools and other hand-held devices, concentrating on more sophisticated equipment. Perhaps the attitude was that a million years of evolutionary experience would produce tools and devices uniquely adapted for human use. Actually, a million years is no guarantee of proper design. Indeed, many hand tools and devices are not designed for efficient, safe operation by humans—especially for repetitive tasks.

Improperly designed tools and devices have several undesirable consequences, including accidents and injuries. Aghazadeh and Mital (1987) surveyed various state agencies in the United States regarding hand tool industrial injuries. Hand tool–related injuries comprise about 9 percent of all work-related compensable injuries. Aghazadeh and Mital estimate that there are over 260,000 hand tool–related injuries in the United States each year and that the associated medical costs alone come to some $400 million. Power tools account for only about 21 percent of all hand tool related–injuries. The most common tools implicated in injuries were knives, wrenches, and hammers. Such injury data, for the most part, represent single-incident traumatic events such as smashing a finger or cutting the palm of the hand. Other more insidious consequences of improper tool design are cumulative traumas (some of which we discuss later), such as carpal tunnel syndrome, tenosynovitis, "trigger finger," ischemia, vibration-induced white finger, and even tennis elbow. Incidence rates of cumulative-trauma disorders can be quite high in industries and jobs requiring repetitive use of hand tools. For example, according to the Bureau of Labor Statistics, in 1986, 480 of every 10,000 meatpacking workers suffered from cumulative-trauma disorders. More specifically, Armstrong et al. (1982) found the overall incidence of cumulative trauma in a poultry processing plant to be almost 13 cases per 200,000 work hours (200,000 work hours is equivalent to roughly 100 people working for 1 yr, or 200 people working for 0.5 yr, and so on). In one department, thigh skinning, the incidence rate was almost 130 cases per 200,000 work hours! The Department of Labor reports that in 1988 cumulative (or repetitive) trauma accounted for 48 percent of all industrial *illnesses* in the United States. This is an often-cited statistic, but it needs to be put in perspective. Illnesses are only 3.7 percent of all industrial *incidences* (illnesses + injuries). Therefore, cumulative traumas account for only about 1.8 percent of all industrial illnesses *and* injuries. In many cases, cumulative-trauma disorders do not show up on accident injury reports but often lead to reduced work output, poorer-quality work, increased absenteeism, and single-incident traumatic injuries.

The proper design of hand tools and devices requires the application of technical, anatomic, kinesiological, anthropometric, physiological, and hygienic considerations. Tools cannot be designed in isolation. Often a redesign of the work space, for example, changing the height of the work surface or repositioning the workpiece relative to the worker, can compensate for a less than optimally designed tool.

An important aspect of hand tool use that is often overlooked is training. It is usually assumed that workers know how to use hammers, pliers, saws, wrenches, etc. Actually, nothing could be further from the truth. As a rule, workers do not always know the proper way to use tools. A good training program in tool usage can often reduce the incidence of injuries and cumulative-trauma disorders.

In this chapter we concentrate on the human factors considerations in hand tool and device design. Our emphasis is on factors that contribute to cumulative trauma and both the quality and quantity of productivity. We start with a discussion of the anatomy and functioning of the human hand. This allows us to formulate some general human factors principles of hand tool and device design. To close the chapter, a few examples of applying human factors principles to specific tools and devices are presented.

HUMAN HAND

The human hand is a complex structure composed of bones, arteries, nerves, ligaments, and tendons, as shown in Figure 12-1. The fingers are flexed by muscles in the forearm. The muscles are connected to the fingers by tendons which pass through a channel in the wrist. This channel is formed by the bones of the back of the hand on one side and the transverse carpal ligament (flexor retinaculum) on the other. The resulting channel is called the *carpal tunnel*.

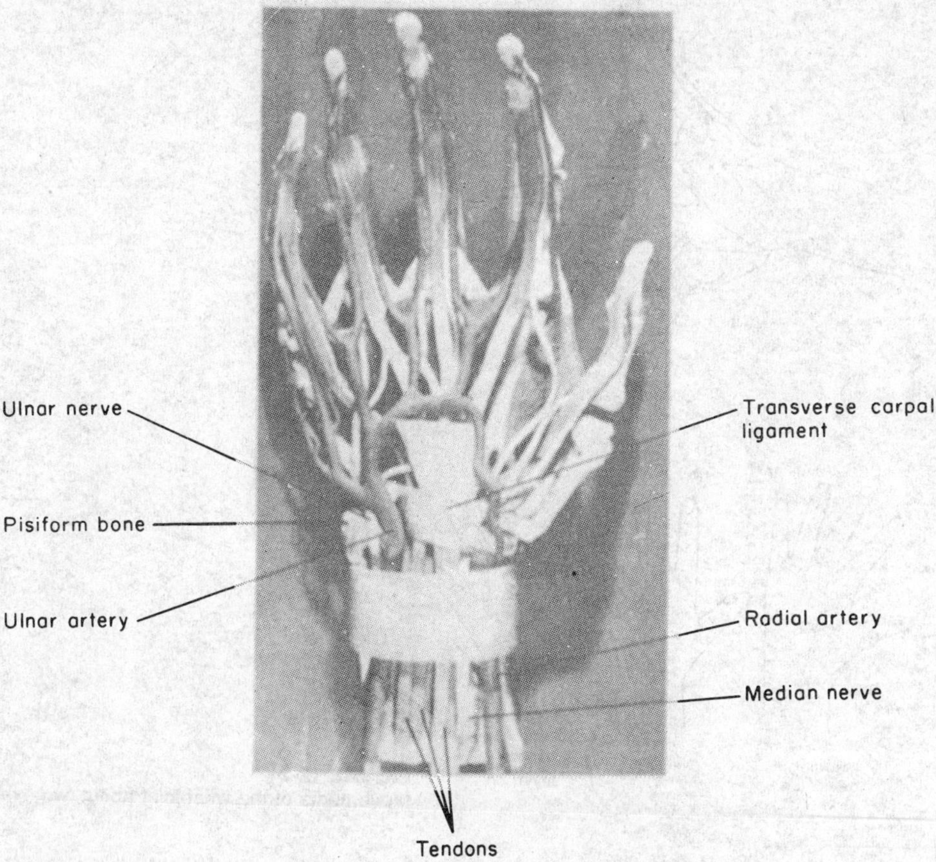

FIGURE 12-1
The anatomy of the hand as seen from the palm side. Shown are the blood vessels, tendons, and nerves underneath the ligaments of the wrist. (*Source: Tichauer, 1978, Fig. 50.*)

Through this tunnel passes a whole host of vulnerable anatomic structures including the radial artery and median nerve. Running over the outside of the transverse carpal ligament are the ulnar artery and ulnar nerve. This artery and this nerve pass beside a small bone in the wrist called the *pisiform bone*. All this may seem a bit clinical, but a clear understanding of the relationships of these structures is important to appreciate the consequences of improper hand tool design.

The bones of the wrist connect to the two long bones of the forearm—the *ulna* and the *radius*. The radius connects to the thumb side of the wrist, and the ulna connects to the little-finger side of the wrist. The configuration of the wrist joint permits movements in only two planes, each one at an approximately 90° angle to the other. The first plane allows palmar flexion or, when it is performed in the opposite direction, dorsiflexion, as shown in Figure 12-2*a*. The second movement plane, shown in Figure 12-2*b*, consists of either ulnar deviation or radial deviation of the hand.

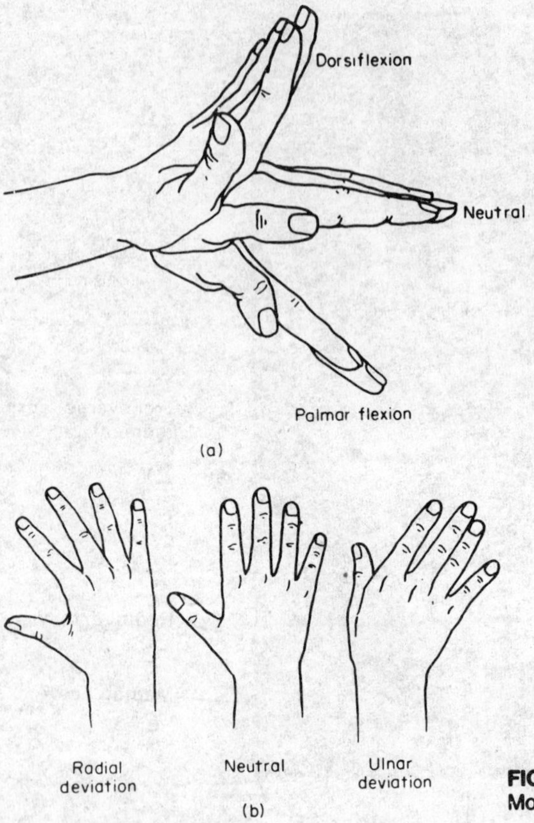

FIGURE 12-2
Movements of the wrist joint about two axes.

The ulna and radius of the forearm connect to the *humerus* of the upper arm, as shown in Figure 12-3. The *bicep muscle* connects to the radius. When the arm is extended, the bicep muscles will pull the radius strongly against the humerus. This can cause friction and heat in the joint. The bicep is both a flexor of the forearm and an outward rotator of the wrist. This can be seen by bending the arm 90° at the elbow and rotating the wrist outward. Notice how the bicep muscle contracts and bulges. Thus any movement that requires a strong pull and simultaneous inward rotation of the hand should be avoided. As pointed out by Tichauer and Gage (1977), good practical design can be observed in the operation of a corkscrew. A wine server pulls hard at the cork while rotating the right forearm outward.

PRINCIPLES OF HAND TOOL AND DEVICE DESIGN

It is not possible to list or discuss all the many principles of hand tool design here. The major principles are discussed as they relate to the biomechanics of the human hand. Also, it is not our intention to present detailed design specifications for specific hand tools. The interested reader can refer to Fraser (1980), Eastman Kodak Co. (1983), and Freivalds (1987) for such data.

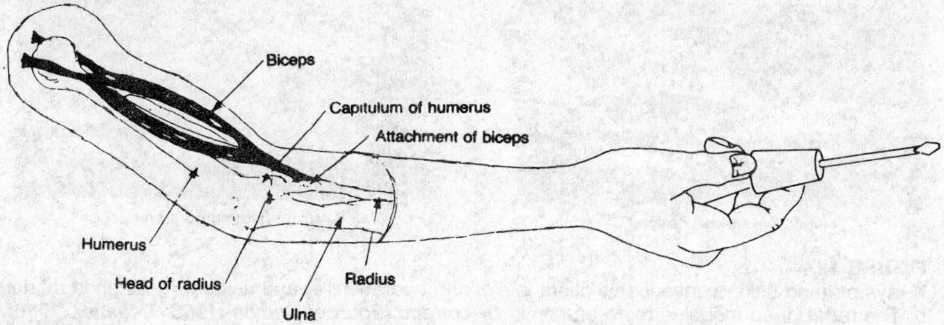

FIGURE 12-3
The elbow joint showing the connection of the bicep to the radius. (*Source: Tichauer, 1978, Fig. 31.*)

Maintain a Straight Wrist

The flexor tendons of the fingers pass through the carpal tunnel of the wrist. When the wrist is aligned with the forearm, all is well. However, if the wrist is bent, especially in palmar flexion or ulnar deviation (or both), problems occur. The tendons bend and bunch up in the carpal tunnel. Continued use will cause *tenosynovitis,* an inflammation of the tendons and their sheaths. A common type of motion which can lead to tenosynovitis is clothes wringing (Tichauer, 1978), in which, for example, the wringing is done by a clockwise movement of the right fist and counterclockwise action of the left. This same type of motion is involved in inserting screws in holes, manipulating rotating controls such as found on steering handles of motorcycles, and looping wire while using pliers.

In addition to tenosynovitis, a condition called *carpal tunnel syndrome* can develop. Carpal tunnel syndrome is a disorder caused by injury of the median nerve where it passes through the carpal tunnel of the wrist. The symptoms include numbness, loss of feeling and grip, and finally muscle atrophy and loss of hand functions. This condition occurs 3 to 10 times more often in women than in men (Dionne, 1984), reflecting either a physiological-anatomical difference or only the fact that women are more likely to engage in work that involves repetitive bent-wrist motions. Incidence rates of carpal tunnel syndrome among workers depend, of course, on the particular job being performed. Silverstein, Fine, and Armstrong (1987) found, among a sample of manufacturing industrial workers, that 5.6 percent of those involved in jobs requiring both high force application and short cycle time (high repetitive motion) tested positive for carpal tunnel syndrome. Armstrong (1983) cites studies showing rates of carpal tunnel syndrome of 3.4 and 25.6 cases per 200,000 work hours.

A key rule in hand tool use is to avoid ulnar deviation. The x-ray of a hand holding conventional pliers in Figure 12-4*a* shows a classic case of ulnar deviation. Redesigning the pliers, as shown in Figure 12-4*b,* so that the handles bend rather than the wrist, allows a more natural alignment of wrist and forearm. Tichauer (1976) reports the results of a comparison of two groups of

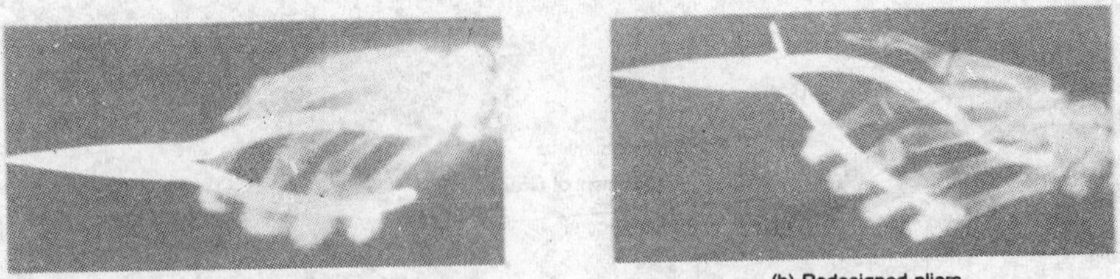

(a) Conventional design (b) Redesigned pliers

FIGURE 12-4
X-rays of hand using conventional pliers in a wiring operation, *a*, and using a redesigned model, *b*. The redesigned model is more anatomically correct. (*Source: Damon, 1965; Tichauer, 1966; photographs courtesy of Western Electric Company, Kansas City.*)

40 electronics assembly trainees, one using conventional straight pliers and the other using the redesigned bent pliers, during a 12-week training program. Figure 12-5 shows the incidence of carpal tunnel syndrome, tenosynovitis, and tennis elbow in the two groups over the 12 weeks of training. There was a sharp increase in symptoms in the tenth and twelfth weeks for the group using the straight pliers while no such increase occurred in the group using the redesigned bent pliers.

The idea of bending the tool, not the wrist, has been applied to other things by John Bennett (Emanuel, Mills, and Bennett, 1980), who patented the idea of bent handles (19° ± 5°) for all tools and sports equipment. Examples of this

FIGURE 12-5
Comparison of two groups of trainees using different pliers. Shows percent of workers with tenosynovitis, epicondylitis (tennis elbow), and carpal tunnel irritation. (*Source: Adapted from Tichauer, 1976. Reprinted with permission from* Industrial Engineering *magazine, 1976, copyright American Institute of Industrial Engineers, Inc., 25 Technology Park/Atlanta, Norcross, GA 30092.*)

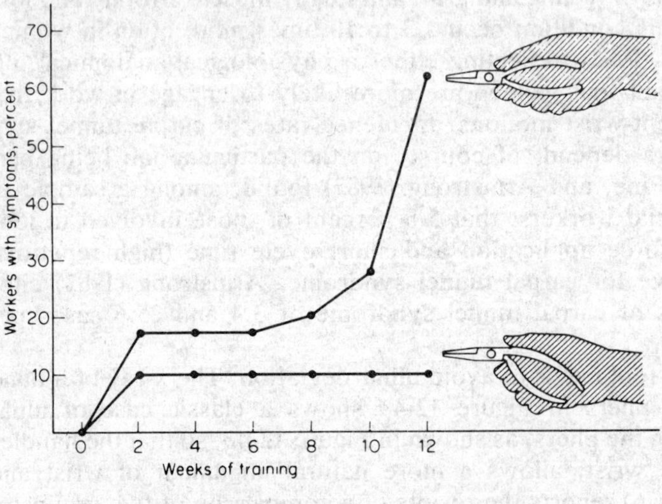

design are shown in Figure 12-6 for a push broom and for a hammer. Schoenmarklin and Marras (1989*a*, 1989*b*) found that among novices, the use of hammers with handles bent 20 or 40 degrees resulted in less total ulnar deviation, especially at impact (but more radial deviation at the starting position of the hammer stroke) than did straight-handled hammers. And further, the bent handles did not affect hammering performance, forearm muscle fatigue, or discomfort ratings when compared to straight-handled hammers. Knowlton and Gilbert (1983) found similar results for ulnar deviation using a professional carpenter as a subject. In addition, they found that hand grip strength measured before and after driving twenty 3.5-in (8.9-cm) nails into a block of wood showed much greater decrements when the straight-handled hammer was used. Emanuel, Mills, and Bennett (1980), although not reporting the results of any systematic evaluation of bent-handle tools, did report the experiences of several individuals and organizations that have used tools with such handles, all with positive results. The U.S. Forest Service, for example, has tested the concept on a set of 19 types of tools such as knives, axes, hoes, shovels, and shears. Preliminary results indicate some decrease in fatigue and in general a preference for the bent handle over the straight handle. Krohn and Konz (1982) also found that people preferred a hammer handle with a slight bend (10°) over a traditional straight-handled hammer. The experiences of individuals with existing injuries or wrist problems suggest that the bent-handle design might be particularly useful for certain handicapped persons. As an aside, a softball bat with a bent handle has been manufactured and is legal for the game.

In addition to ulnar deviation, other types of wrist bending can also cause problems. Radial deviation, particularly if combined with pronation and dorsiflexion, increases pressure between the head of the radius and the capitulum of the humerus in the elbow (see Figure 12-3). Many wire brushes have to be held in this way when they are used overhead. This can lead to *tennis elbow* (*epicondylitis*), an inflammation of the tissue in the elbow region.

FIGURE 12-6
Examples of the Bennett handle that helps the user keep the wrist straight while using the tool. (*Source: Emanuel, Mills, and Bennett, 1980, Fig. 1.*)

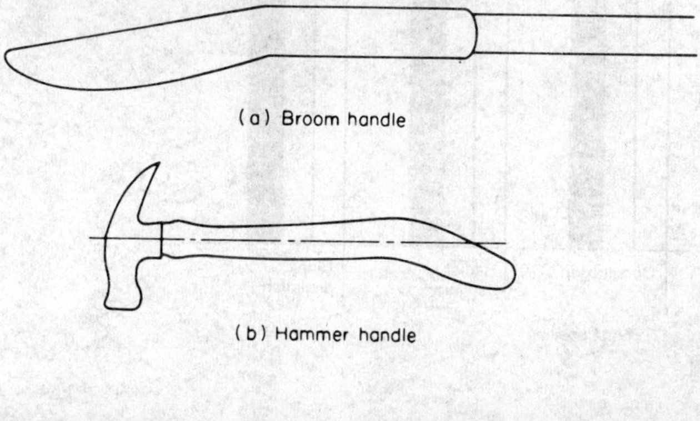

We have seen that ulnar deviation can cause tenosynovitis and that radial deviation can lead to tennis elbow. If that were not enough, Terrell and Purswell (1976) report that grip strength is reduced if the wrist is bent in any direction. Figure 12-7 shows the grip strength as a percentage of the maximum grip strength (achieved with the wrist in a neutral position with supinated forearm). Reductions in grip strength may increase the likelihood that the user will lose control of the tool or drop it, leading to an injury or poor-quality work. Attempts to maintain a strong grip will increase fatigue.

Figure 12-8 shows a man using a hacksaw in a manner that violates the straight-wrist principle. The man's right hand is in ulnar deviation while the left hand is in dorsiflexion with a touch of radial deviation. After a few hours of sawing like this, he will be in no shape to play the piano!

Avoid Tissue Compression Stress

Often in the operation of a hand tool or device, considerable force is applied with the hand as in squeezing pliers or scraping paint with a paint scraper. Such actions concentrate considerable compressive force on the palm of the hand.

FIGURE 12-7
Grip strength as a function of wrist and forearm position. Grip strength is the average maximal grip sustained for 3 s, expressed as a percentage of neutral supinated grip strength. (*Source: Based on data from Terrell and Purswell, 1976, Table 1.*)

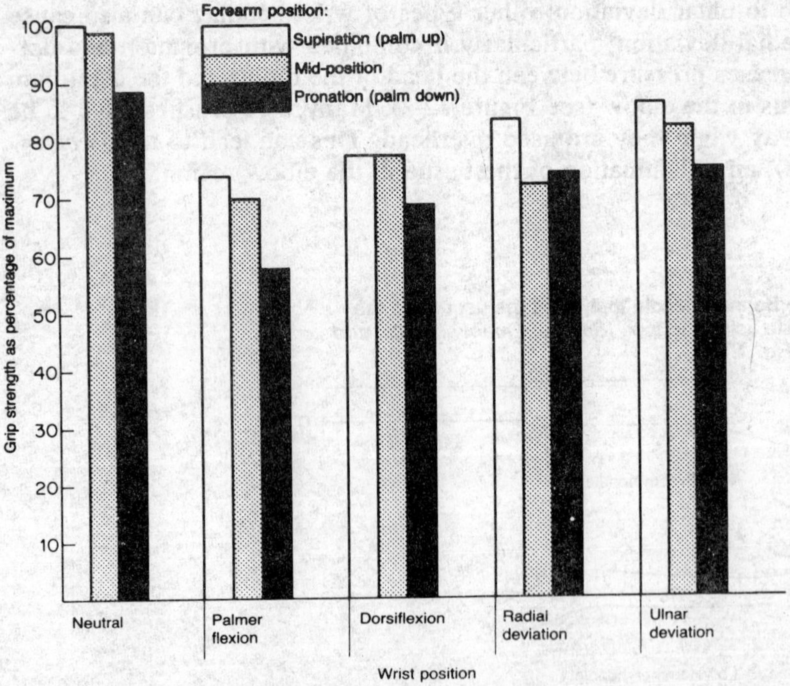

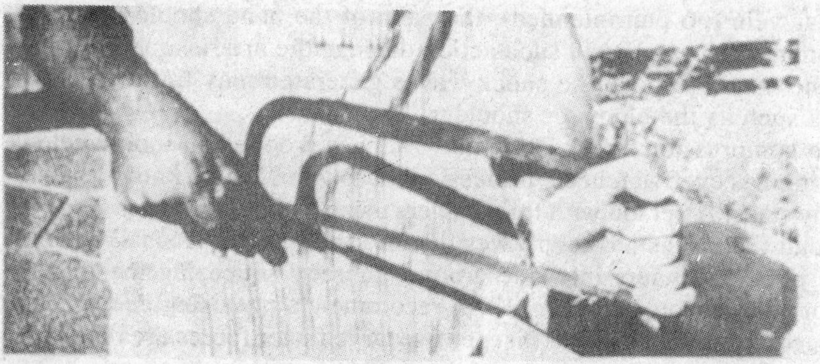

FIGURE 12-8
Man using a hacksaw and violating the straight-wrist principle. The right hand is in ulnar deviation while the left hand is in dorsiflexion. (*Source: Greenberg and Chaffin, 1977, Fig. 35.*)

Particularly pressure-sensitive areas are those overlying critical blood vessels and nerves, specifically the ulnar and radial arteries. Figure 12-9*a* shows a hand holding a conventional paint scraper. The handle digs into the palm and obstructs blood flow through the ulnar artery. This obstruction of blood flow, or *ischemia,* leads to numbness and tingling of the fingers. Tichauer (1978) reports that affected workers will often take temporary breaks from work to relieve the symptoms. Thrombosis of the ulnar artery has also been reported (Tichauer, 1978).

If possible, handles should be designed to have large contact surfaces to distribute the force over a larger area and to direct it to less-sensitive areas, such as the tough tissue between thumb and index finger. Figure 12-9*b,* for example, shows an improved paint scraper handle that rests on the tissue between thumb and finger, thus preventing pressure on the critical areas of the palm.

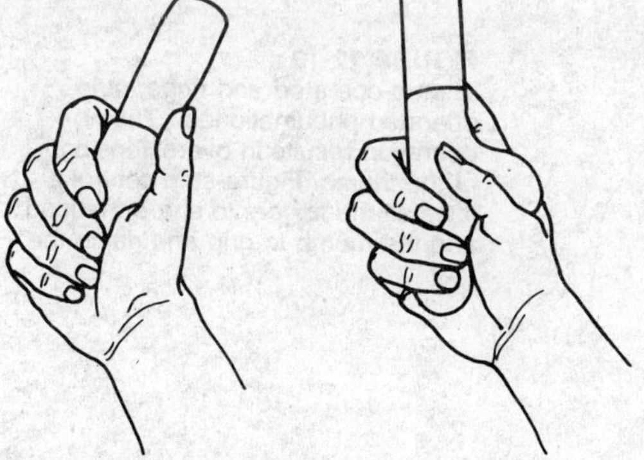

FIGURE 12-9
A conventional paint scraper that presses on the ulnar artery, and a modified handle which rests on the tough tissues between thumb and index finger, thus preventing pressure on the critical areas of the hand. (*Source: Tichauer, 1967.*)

In a similar vein (no pun intended), the palm of the hand should never be used as a hammer. Not only will such action damage the arteries, nerves, and tendons of the hand, but also the shock waves generated may travel to other body regions such as the elbow or shoulder.

Related to compression stress is the use of finger grooves on tool handles. As anyone who has ever watched a professional basketball game knows, hands come in many sizes. A person with thick fingers using a tool with finger grooves often finds that the ridges of the grooves dig into the fingers. A small-handed person may put two fingers into one groove, thereby squeezing the fingers together. For this reason Tichauer (1978) recommends not using deep finger grooves or recesses in tool handles if repetitive large finger forces are required.

Avoid Repetitive Finger Action

Occasionally, if the index finger is used excessively for operating triggers, a condition known as *trigger finger,* a form of tenosynovitis, develops. The afflicted person typically can flex but cannot extend the finger actively. The finger must be passively straightened, and when it is, an audible click may be heard. The condition seems to occur most frequently if the handle of the tool or device is so large that the distal phalanx (segment) of the finger has to be flexed while the middle phalanx must be kept straight (Tichauer, 1978).

As a rule, frequent use of the index finger should be avoided, and thumb-operated controls should be used. The thumb is the only finger that is flexed, abducted, and opposed by strong, short muscles located entirely within the palm of the hand. One must be careful, however, not to hyperextend the thumb such as shown in Figure 12-10*a*. This causes pain and inflammation. Preferable to thumb controls is the incorporation of a *finger-strip* control, as shown in Figure 12-10*b,* which allows several fingers to share the load and frees the thumb to grip and guide the tool.

The grip strength of the hand is related to the size of the object being gripped, as shown in Figure 12-11*a*. Maximum grip strength, for both males

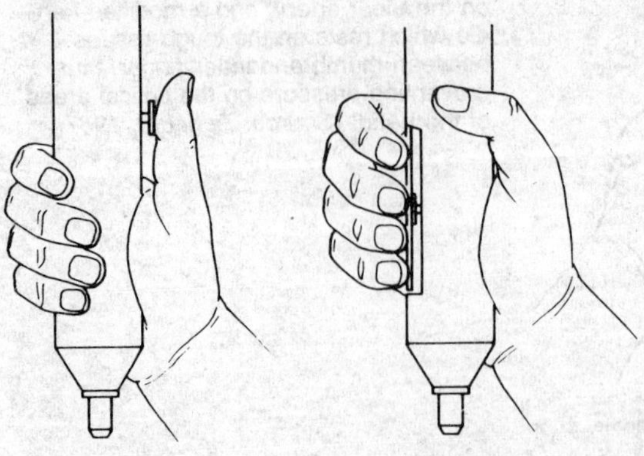

FIGURE 12-10
Thumb-operated and finger-strip-operated pneumatic tool. Thumb operation results in overextension of the thumb. Figure-strip control allows all the fingers to share the load and the thumb to grip and guide the tool.

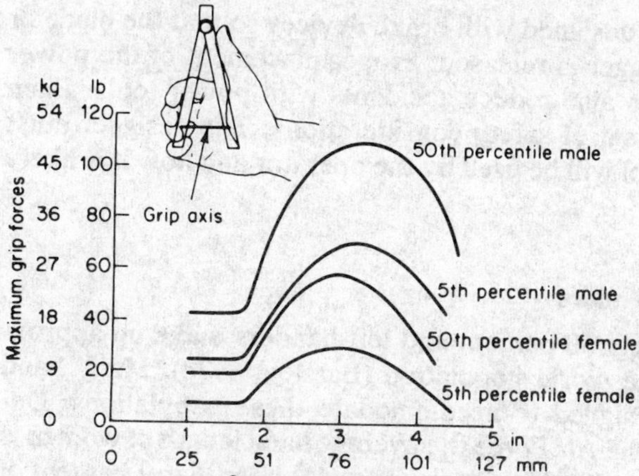

(a) Grip strength

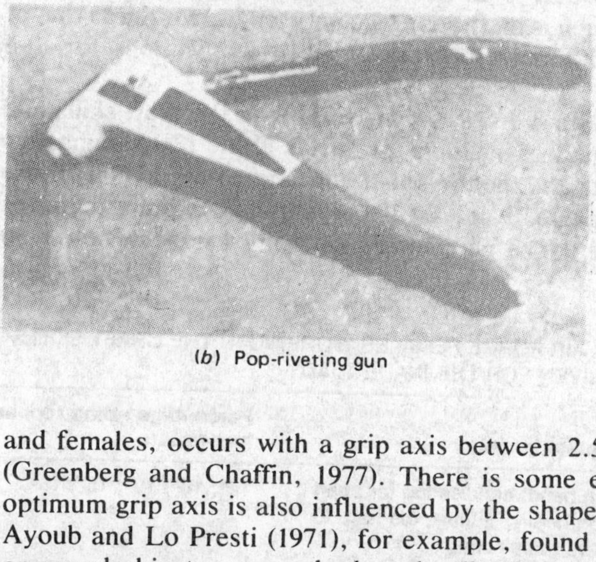
(b) Pop-riveting gun

FIGURE 12-11
(*a*) Maximum grip strength for various handle openings. Subjects were 50 male and 50 female electronics manufacturing employees. (*b*) A pop-riveting gun which to be operated must be grasped and squeezed in the fully open position. The outside edges of the handles are 6 in (150 mm) apart. (*Source: Greenberg and Chaffin, 1977, Figs. 27 & 28.*)

and females, occurs with a grip axis between 2.5 and 3.5 in (64 and 89 mm) (Greenberg and Chaffin, 1977). There is some evidence, however, that the optimum grip axis is also influenced by the shape of the object being gripped. Ayoub and Lo Presti (1971), for example, found that maximum grip strength on round objects occurred when the diameter of the object was about 1.6 in (41 mm). Therefore, if a tool handle is round, the optimum grip axis may be smaller than shown in Figure 12-11*a*. Figure 12-11*b* shows a pop-riveting gun which must be gripped in the fully open position—a distance of 6 in (152 mm) between the outside edges—and squeezed. This is nearly impossible for most women to perform.

Design for Safe Operation

Designing tools and devices for safe operation would include eliminating pinching hazards and sharp corners and edges. This can be done by putting guards over pinch points or stops to prevent handles from fully closing and pinching the palm of the hand. Sharp corners and edges can be rounded. Power tools such as saws and drills can be designed with brake devices so that the blade or bit stops quickly when the trigger is released. Proper placement of the power switch for quick operation can also reduce accidents with power tools. Each type of tool presents its own set of safety considerations. The designer must consider, in detail, how the tool will be used by the operator and how it is likely to be *misused* by the user.

Remember Women and Left-Handers

Women make up approximately 50 percent and left-handers make up approximately 8 to 10 percent of the world population (Barsley, 1970). Many hand tools and devices are not designed to accommodate these populations. Ducharme (1977) reports that in the Air Force the average hand length of women is about 2 cm (0.8 in) shorter than that of the average male. Less than 1 percent of Air Force men have a hand that is as short as the average woman's hand. Further, grip strength of women is on the average only about two-thirds that of men (Konz, 1979). These differences obviously have implications for tool design.

Ducharme surveyed 1400 Air Force women working in the craft skills and asked them to rate the adequacy of the tools they used. Table 12-1 summarizes some of the major complaints and the tools that engendered them. Only those tools that were judged inadequate by at least 10 percent of the women in one or more craft fields are included. The percentages of women in each field who

TABLE 12-1
TOOLS RATED INADEQUATE BY AIR FORCE FEMALES WORKING IN THE CRAFT SKILLS AND SOME OF THE REASONS GIVEN FOR THE INADEQUACY

Tool	Rated inadequate because	Percentage rating tool as inadequate
Wire strippers	Hard to hold in hand; handles too far apart to squeeze; too heavy; clumsy; too hard to squeeze; fingers get pinched	12[a]; 18[b]; 19[c]; 15[d]; 18[e]
Crimping tool	Handles too far apart; too hard to squeeze; not able to manipulate	14[a]; 14[c]; 25[d]
Soldering iron	Too heavy; handle too large; clumsy, too bulky	17[a]; 15[c]
Soldering gun	Too heavy; cannot reach trigger; hard to hold in hand	15[a]
Twist wire pliers	Too large to grip; handles too far apart	29[b]
Metal shears	Too large; need two hands to cut	22[f]
Rivet cutter	Too hard to squeeze; awkward	17[f]
Caulking gun	Hard trigger; awkward	11[g]

[a]Communication-electronic systems field.
[b]Missile electronics maintenance field.
[c]Avionics systems field.
[d]Aircraft accessory maintenance field.
[e]Mechanical/electrical field.
[f]Metalworking field.
[g]Structural/pavements field.
Source: Ducharme, 1977.

considered each tool inadequate are also shown in Table 12-1. From scanning the complaints it is obvious that almost all reported inadequacies can be traced to the fact that women have smaller hands and less grip strength than most men. With women assuming a greater role in traditionally male-dominated occupations, the design of hand tools and devices must reflect the anthropometric and ergonomic differences between men and women.

Another neglected group is left-handers. Actually, there are degrees of handedness from strong right-handed through ambidextrous to strong left-handed. Tools should be designed to be used in the operator's preferred hand. Often, designs preclude the use of tools by left-handers. For example, the drill shown in Figure 12-12 requires left-handers to operate the drill with the right hand. If a threaded fastener were provided in the right side of the drill housing, the handle could be moved for left-handed use.

Laveson and Meyer (1976) point out that some serrated knives have the beveled (cutting) edge on the left side. The left-hander presses down and to the left; the beveled edge then fails to cut, since the pressure is on the wrong side. A serrated bread knife, when used by a left-hander, will often tear rather than cut the bread. The difficulty could be corrected by double-edge beveling.

VIBRATION

Typically in the human factors literature, the topic of *vibration* refers to whole-body vibration such as heavy-equipment operators' experience as they drive over bumpy roads. We discuss this type of vibration in Chapter 19. Our focus

FIGURE 12-12
Drill with handle designed for right-handed operation. (*Source: Greenberg and Chaffin, 1977, Fig. 25.*)

here, however, is on hand vibration induced by such power tools as chain saws, pneumatic drills, grinding tools, and chipping hammers.

Hand-Arm Vibration Syndrome (HAVS)

In the early 1900s reports began to appear that linked hand-held vibratory tools with vascular spasm in the hands (Taylor, 1974). Since then numerous other reports and investigations have been published on what has become known as *vibration-induced white finger (VWF)*, *vibration syndrome*, or simply *hand-arm vibration syndrome (HAVS)*.* For an excellent, and readable, summary of the literature, see NIOSH (1989).

HAVS is a complex phenomenon whose exact pathology is not fully understood, but it probably involves damage to the nerves and smooth muscles of the blood vessels in the hand. Although the National Institute for Occupational Safety and Health (NIOSH) has linked regular, prolonged use of vibratory hand tools in certain occupations to vibration syndrome, little is known about what specific vibration parameters are the most necessary to control (NIOSH, 1983).

The primary symptom of HAVS is a reduction in blood flow to the fingers and hand. This is caused by the smooth muscles of the blood vessels in the hand and fingers constricting and thereby reducing the flow of blood. Workers afflicted with HAVS will have *vascular attacks* wherein the hand or fingers blanch (turn white). The feeling is the same as when the hand or foot "falls asleep." The person complains of "pins and needles" (i.e., tingling), numbness, or pain. These attacks seem to be brought on by cold. They are especially prevalent among workers who use vibratory tools in cold environments. The attacks can last minutes or hours.

Workers afflicted with HAVS appear to have reduced blood flow in the hands under normal conditions in addition to experiencing vascular attacks. Koradecka (1977), for example, examined pneumatic drill operators, manual grinder operators, and control groups working under similar microclimatic conditions but not exposed to vibration. He found that the exposed groups exhibited a 38 to 67 percent mean reduction in hand skin blood flow compared with control subjects.

One consequence of this reduced blood flow is a reduction in skin temperature. Figure 12-13 shows the mean finger skin temperature of vibration-exposed and control subjects. The vibration-exposed group starts out with a lower skin temperature, but when the hand is heated [10 min in 113°F (45°C) water], both groups attain the same temperature. Cooling the hands, however, restores the initial difference which remains throughout recovery. This lower skin temperature is often associated with a decrease in sensitivity and fine finger dexterity and a loss of grip.

* Several other names have also been used in the literature: *Raynaud's phenomenon of occupational origin*, *dead hand*, *dead finger*, *white finger*, *occupational vasomotor traumatic neurosis*, and *traumatic vasospastic disease*.

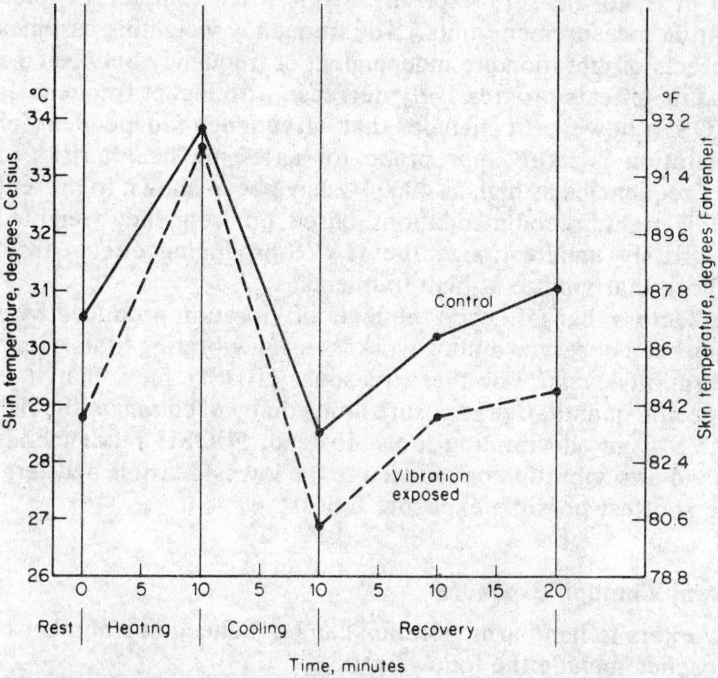

FIGURE 12-13
Skin temperature of the fourth finger of the left hand during thermal testing of a group of workers occupationally exposed to hand vibration and a control group not exposed. (*Source: Koradecka, 1977, Fig. 13.*)

Although most cases of HAVS are not debilitating, advanced cases have been known to lead to gangrene of the fingertips (Walton, 1974). The prevalence of HAVS, of course, varies with the type and amount of exposure and other individual differences. NIOSH (1989) reviews over 50 studies of workers using vibrating tools. The prevalence of HAVS among exposed workers ranges from 6 to 100 percent with the average being 50 percent. Minimum daily exposures for several hours each day for months or years are usually required before the first signs and symptoms appear.

Vibration induced from hand-held tools has also been implicated in numerous other diseases, including neuritis, decalcification and cysts of the radial and ulnar bones, and pain and stiffness in the joints (see Wasserman et al., 1982, for a review).

A Question of Standards

There have been numerous standards, guidelines, and recommendations regarding safe exposure limits to hand-arm vibration (e.g., ISO, 1986; ACGIH, 1988; ANSI, 1986; NIOSH, 1989). With the exception of the NIOSH recommendations, other guidelines and standards use a frequency-weighted acceleration, expressed in m/s², to measure exposure levels. (See Chapter 19 for a discussion of vibration measurement units.) The frequency weighting assumes that the harmful effects of vibration are independent of frequency between 6.3 and 16 Hz, but that the effects progressively decrease with higher frequencies up to 1500 Hz. NIOSH, however, concludes that a frequency-*in*dependent (or unweighted) acceleration is more appropriate for assessing health risks to exposed workers. Frequencies as high as 4000 Hz have been shown to present a risk. If NIOSH is right, recommendations based on frequency-weighted acceleration may grossly underestimate the HAVS-producing effects from tools, especially those that vibrate at high frequencies.

There are many factors that influence the level of vibration produced by a tool, and HAVS has been observed among workers using vibrating tools over a wide range of vibration levels. For these reasons, NIOSH feels that it is premature to set specific quantitative exposure limits that will eliminate the risk of developing HAVS from all vibrating tools. Instead, NIOSH recommends that exposure to hand-arm vibration be reduced to the lowest feasible acceleration levels and the shortest possible exposure times.

Controlling Hand-Arm Vibration Exposure

The exposure of workers to hand-arm vibration can be reduced in a number of ways. Some approaches include the following:

1 Select tools with the lowest level of vibration. Figure 12-14 shows vibration levels from individual tools within particular classes of tools used in car body repair work. Substantial reductions in levels of vibration can be achieved by simply purchasing tools with low vibration levels. Electric motors and rotary gasoline engines create less vibration than reciprocating gasoline engines.

2 Properly maintain tools and keep cutting tools sharpened.

3 Use vibration-reduction (damping) gloves.

4 Minimize grip force needed to hold and control the tool.

5 Alternate work tasks that do not require use of vibrating tools with those that do.

6 Limit daily use of vibrating tools.

7 Provide long rest breaks while using vibrating tools.

8 Limit the number of days per week that vibrating tools are used by a worker.

Pyykko and associates (Pyykko, 1974; Pyykko et al., 1978; Pyykko et al., 1982; Pyykko et al., 1986) in Finland followed 66 forestry workers using gasoline-powered chain saws from 1972 until 1983. In 1972, the prevalence of HAVS symptoms in this group was 34 percent and reached a peak in 1975 of 38 percent. Antivibration chain saws were introduced in the mid-1970s, and by 1983, only 5 percent of the 66 workers showed HAVS symptoms. Antivibration chain saws measured in 1977 and 1983 showed levels of about 16 to 17 m/s². Non-antivibration chain saws in 1970 measured from 49 to 105 m/s².

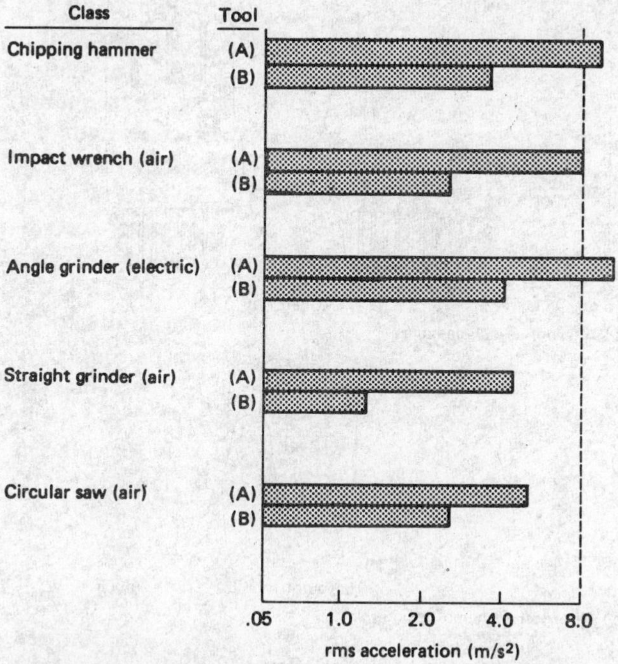

FIGURE 12-14
Weighted hand-arm vibration accelerations in various tools used in car body repair work. This illustrates the wide range of vibration levels both between and within classes of power tools. (*Source: Adapted from Hansson, Eklund, Kihlberg, and Ostergren, 1987, Fig. 1.*) By permission of Butterworth-Heinemann Ltd.

GLOVES

A chapter on hand tools and devices would not be complete without at least a brief discussion of gloves. Gloves are often used in conjunction with hand tools for protection against abrasions, cuts, punctures, and temperature extremes. Gloves come in an amazing number of varieties. In general, however, they can be distinguished in terms of the material used for construction (e.g., cotton, leather, vinyl, neoprene, asbestos, and even metal), the cut (i.e., clute cut and gunn cut, as shown in Figure 12-15), the design of the thumb (i.e., wing or straight, as shown in Figure 12-15), and the type of wristband (e.g., knit, band top, gauntlet, or extended length).

Effect on Manual Performance

All the research on the effects of wearing gloves on manual performance have concentrated on comparing gloves of different materials. No studies exist which evaluate glove design independent of the material. We do not know,

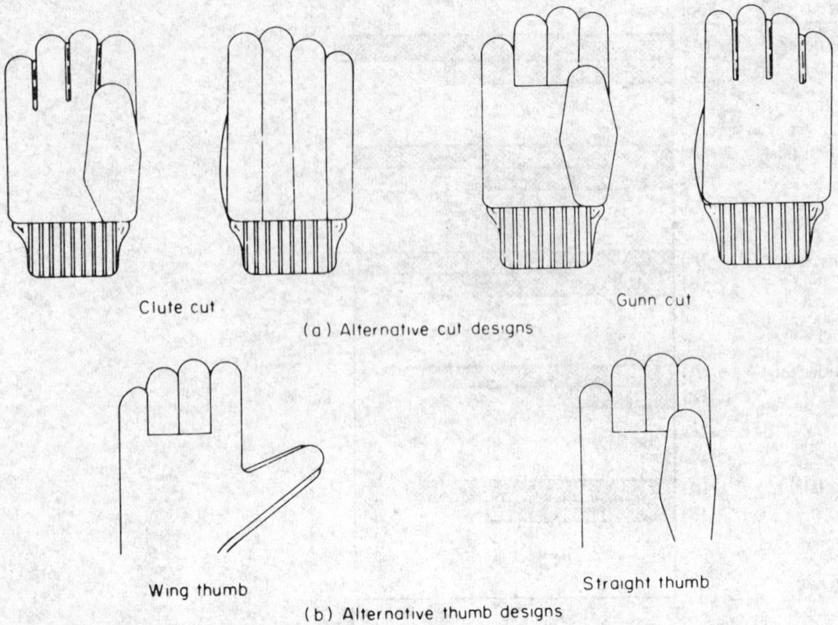

FIGURE 12-15
Common styles of work gloves distinguished by cut and thumb design.

therefore, whether there is any difference in performance while gunn or clute cut gloves are worn.

Research on the effects of various glove materials on manual performance indicates that the nature of the task influences the relative superiority of the various materials. On some tasks there will be no differences in performance when different gloves are worn, while on other tasks large differences will be found. In general, however, performance on tasks requiring fine motor control and tactile feedback will be adversely affected by wearing gloves, as compared to bare-handed performance. An exception was noted in Chapter 6 wherein surface irregularities can be detected more accurately with thin cloth gloves than with bare fingers.

A study by Weidman (1970) is a good example of research in this area. Weidman compared four gloves [leather with canvas back, heavy-duty terry cloth, neoprene, and polyvinyl chloride (PVC)] and a bare-hand condition on performance of a maintenance-type task. The entire task was divided into five subtasks:

1 Using a key on a key ring to open a lock to get into a box
2 Replacing 16 television tubes by hand
3 Using a socket wrench and open-end wrench to remove 20 bolts
4 Using a screwdriver to remove four screws and replace two tubes
5 Using a trigger-type solder gun to remove two leads

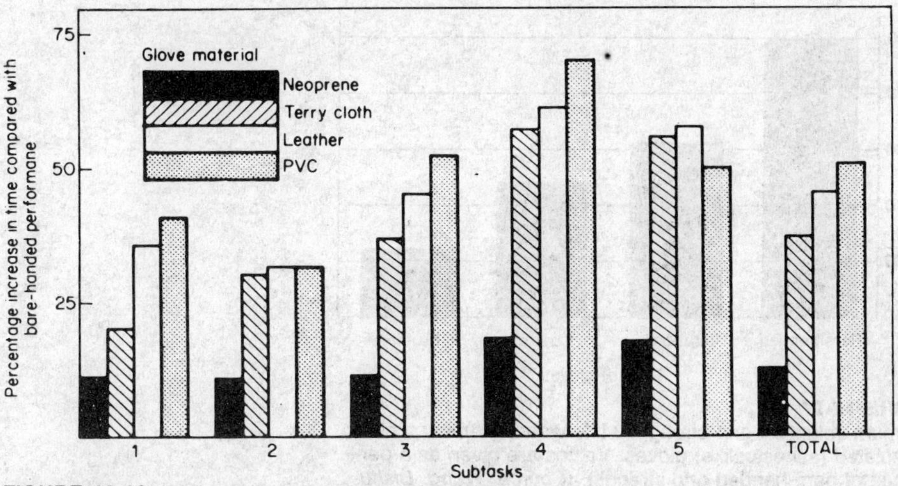

FIGURE 12-16
Performance (time to complete) on a maintenance-type task while wearing gloves constructed of five different materials. Performance is shown as percentage increase compared to bare-handed performance. (See text for a description of the subtasks.) (*Source: Weidman, 1970.*)

The results are shown in Figure 12-16 as percentage increase in task completion times compared to the bare-hand condition. For subtasks 1 and 5 there was no statistically significant decrement, or differential decrement, caused by any type of glove. This was caused by the overall short time required to complete these subtasks (less than 50 s). For subtasks, 2, 3, and 4 and total time, the results showed that performance did not differ between neoprene and bare-hand conditions. Leather, terry cloth, and PVC, however, were consistently inferior to both neoprene and bare hands but did not differ from each other.

Grip Strength and Gloves

A rather consistent finding in the literature is that grip strength is reduced when gloves are worn. Figure 12-17, for example, compares maximum voluntary grip strength measured bare-handed and wearing either rubber, cotton, or insulated (asbestoslike) gloves. Interestingly, Sudhakai, Schoenmarklin, Lavender, and Marras (1988) found no differences in forearm muscle activity (based on EMG recordings) between bare-handed and glove-handed grips, despite lower measured grip strength while wearing gloves. For some reason, a certain amount of the muscle force generated is lost between the hand and the glove. Some partial explanations of this include: the thickness of a glove changes the size of the grip axis (see Figure 12-11); the pliability of a glove reduces the force transmitted to the grip strength dynamometer; and the thickness of material between the fingers reduces the force that can be exerted.

A loss of grip strength while wearing gloves may lead to negative consequences, including dropping a tool, poorer control of a tool, lower quality work, and increased muscle fatigue.

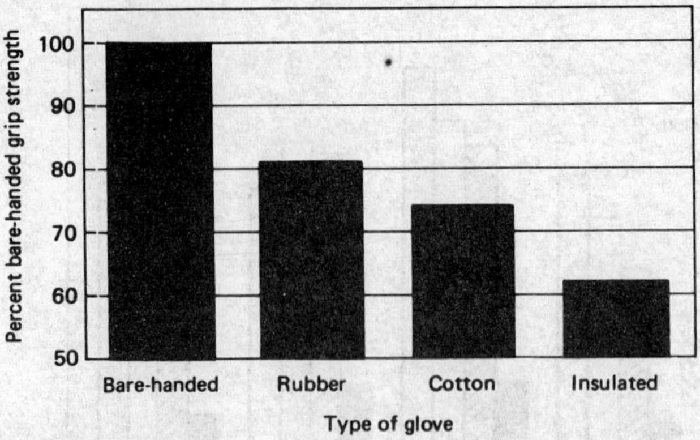

FIGURE 12-17
Maximum voluntary grip strength while wearing rubber, cotton, or insulated (asbestoslike) gloves. Values are given as a percentage of bare-handed grip strength. (*Source: Wang, Bishu, and Rodgers, 1987, Fig. 1.*)

Other Considerations

Irritants entrapped unintentionally in the glove may work into the skin, producing in some cases benign tumors. Chemicals might be soaked up in the glove material and cause dermatitis or other tissue damage.

These potential hazards speak to a need for careful evaluation of the work environment and the selection of suitable gloves. Knitted wristbands can reduce the intrusion of potential irritants. Neoprene- or PVC-coated gloves can reduce absorption of chemicals. The selection of proper glove material must also take into account the type of hazards (cuts, punctures, abrasions and thermal) likely to be encountered on the job. Some materials are better suited to handle certain hazards than are others (Coletta et al., 1976). Gloves must be given as careful thought as the design or selection of the hand tools.

EVALUATION OF ALTERNATIVE DESIGNS

In recent years there has been an increase in the application of human factors to hand tool and device design. In this section we present a few examples. The emphasis is not on the innovation per se but on the methodology used to develop and evaluate the new designs. In Chapter 2 we discussed laboratory and real-life research as well as types of criteria that can be used to measure possible effects of new designs. In several of the examples to follow, combinations of laboratory and real-life research methods were employed to provide a comprehensive evaluation of alternative designs. We illustrate the kinds of information sought and the insights provided by the various methodologies employed.

Toothbrush Design

Guilfoyle (1977) provides an informative description of the human factors development and evaluation of what ultimately became the Reach toothbrush shown in Figure 12-18. Although maybe not the most sophisticated of tools, the toothbrush is probably one of the most widely used hand-held tools in the world. Since the first bone shaft and hog-hair bristle toothbrush in 1780, the only major development had been the introduction of nylon filament bristles in the 1930s.

DuPont approached Applied Ergonomics, Inc., to design an improved toothbrush. The first step was to review relevant literature and conduct a series of consultations with dentists. From this preliminary investigation they discovered that no human factors research had been applied to toothbrush design and that the primary objective of a toothbrush was removal of plaque from the teeth. A secondary objective was gingival (gum) massage. Since there was little available literature on the public's dental care habits, a questionnaire was distributed to 300 adults. The information obtained helped the designers focus on basic dental care problems.

After analyzing the results of the questionnaire, the design team began collecting basic physical dimensions on existing toothbrushes and on the anthropometric characteristics of consumers (measurements of hands, teeth, and mouths). To obtain more information on how users handle a toothbrush, a series of time-motion film studies were made of people brushing their teeth. The design team studied how much time was devoted to brushing different mouth areas and the stroke directions used. The team also examined the way people manipulated the brush.

FIGURE 12-18
Reach toothbrush designed by Applied Ergonomics, Inc., and manufactured by Johnson and Johnson. Innovative features include a small bristle head and an angled (12°) and shaped handle with contoured thumb rest area.

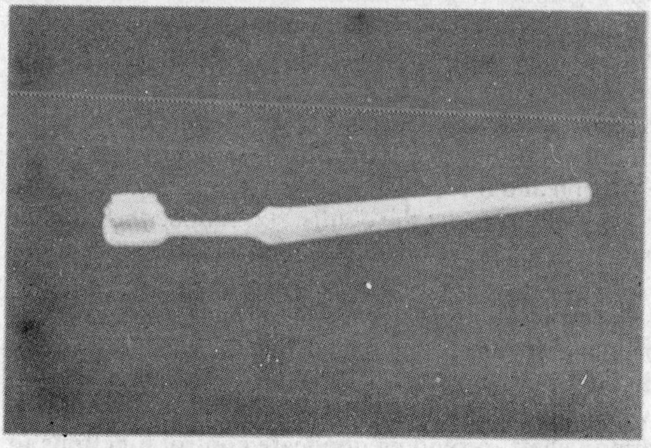

The time-motion studies made it clear that a contoured grip was undesirable because it hindered hand movements. A round handle was briefly considered but was dropped because it would not resist rotational forces. Another reason the round handle was dropped illustrates the need to consider the entire system into which a tool or device will fit. In this case, the diameter of the handle would have to be small enough to fit into the standard bathroom toothbrush holder. To do so would have made the handle too weak.

In addition to the time-motion studies, other laboratory studies related brushing time and bristle diameter to plaque removal. Based on the information obtained from all the varied sources, two prototype toothbrushes were designed and a sufficient number were manufactured for testing. Both prototypes had individual characteristics that required testing.

These two prototypes were then tested against two other commercially available toothbrushes for removing plaque. Both prototypes proved superior. Interviews with the test subjects indicated a preference for the bristle head of one prototype and the handle design of the other. The features were blended into the final product. Plaque removal tests with the final design demonstrated superior removal compared to other conventionally designed brushes. Some of the features incorporated into the Reach toothbrush include a small bilevel bristle head to concentrate brushing on a small area, an angled (12°) and shaped handle for easier manipulation, and a contoured thumb area which makes brushing easier.

This design and evaluation process illustrates the use of past literature, expert (dentists') opinions, questionnaire data, objective laboratory studies (time-motion and plaque-removal studies), and user opinion data to arrive at an improved tool design.

Multiple-Function Dental Syringe

Those of us who perhaps have not properly used our toothbrushes may have been on the receiving end of this tool. When a dentist fills a cavity, the dental assistant often works alongside, evacuating (with a suction vacuum) tooth debris being drilled by the dentist as well as providing air, water, and an air-water spray. Currently two instruments, a three-way air-water-spray syringe, and a separate suction device are used, as shown in Figure 12-19*a*. Evans et al. (1973) reasoned that reducing the number of instruments handled by the dental assistant by combining the suction and three-way syringe into a single four-way device would (1) free a second hand to assist the dentist, (2) reduce the number of instruments in the mouth, (3) increase visibility, and (4) reduce patient discomfort.

The four-way device is shown in Figure 12-19*b*. The air and water streams are each operated by depressing small spring-loaded levers with the thumb. Spray is achieved by activating both levers simultaneously. Suction is controlled by rotating thumb wheels located to the sides of the levers. The thumb wheels are joined by a common shaft, providing duplicate control functions to the operator, thus accommodating left-handed as well as right-handed oper-

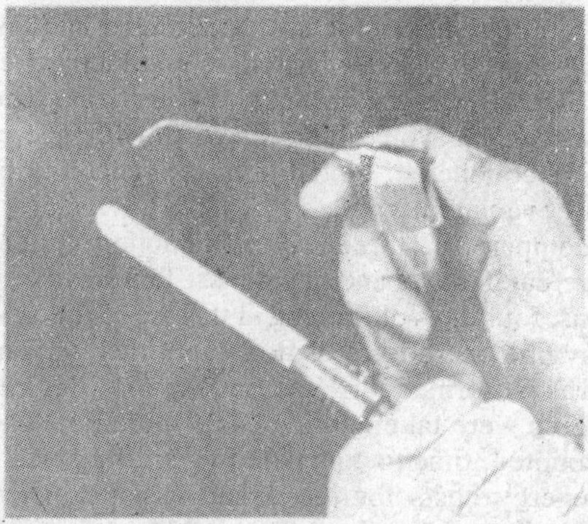

(*a*)

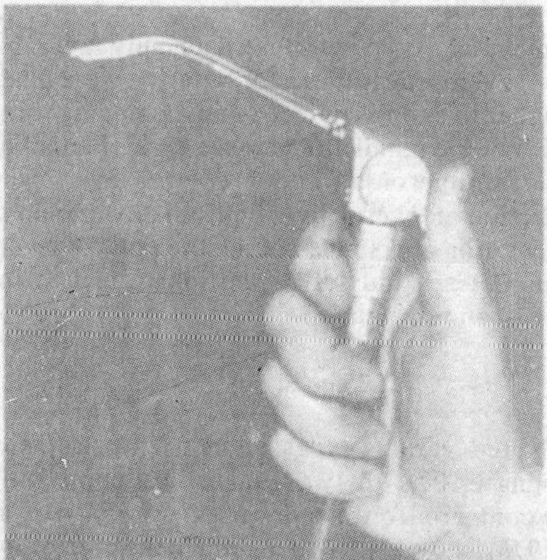

(*b*)

FIGURE 12-19
(*a*) The conventional three-way syringe (air, water, and spray) with separate suction used in oral dentistry. (*b*) The four-way (air, water, spray, and suction) device designed and evaluated by Evans et al. (*Source: Evans, Lucaccini, Hazell, and Lucas, 1973, Figs. 1 and 3.*)

ators—a plus. The tip has a 45° angle bend and rotates to provide access to difficult-to-reach areas of the mouth.

The evaluation of the new instrument consisted of mechanical performance tests, mock clinical tests (simulated operations), and actual field tests (Evans et al., 1973). We briefly review each to illustrate the criteria used and the insights gained. The mechanical performance tests were conducted to determine whether the device satisfied basic requirements for air, water, and suction pressure. The mock (or simulated) clinical trials consisted of conducting restorative dental operations on a dental manikin instead of a patient. The purpose was to be sure that the instrument would function satisfactorily and safely before the live field tests were conducted. In addition, the reliability of the device under sustained use was evaluated.

Field tests were conducted to compare the three-way plus suction system with the new four-way device by actual dentists and dental assistants working on patients. At each of five locations, four days of training in the use of the new device were given. On the fifth day one operation was done with the conventional system, and one operation was performed by using the four-way device. Videotape and motion picture records were taken during each operation. The number of hand movements was counted, time to complete the cutting portion of the operation was recorded, energy costs for the dental assistant were calculated from samples of expired air, and finally operator preferences and suggestions were obtained. Here we see again the use of multiple criteria for evaluation and again the inclusion of subjective user-preference data. The results are shown in Table 12-2.

The interviews with dentists and dental assistants showed a highly favorable attitude toward the new device, but several suggestions for redesign were made including adding a plastic cover to the tip end to protect against chipping tooth enamel; reducing the size and length of the handle to increase ease of handling (especially for female users); replacing the on-off action control levers with variable controls; changing the direction of lever activation from down to up to conform to thumb motion; and separating and shape-coding the control levers to minimize inadvertent activation. These kinds of suggestions attest to the value of user inputs in system design and evaluation.

Writing Instruments

Kao, in a series of studies, evaluated various types of writing instruments (1976, 1979) and alternative designs (1977). In two studies, ballpoint pens, fountain pens, and pencils were compared on a number of criteria. An attitudinal survey of users (Kao, 1976) showed that ballpoint pens were most preferred, followed (in order) by pencils, and finally fountain pens. The survey

TABLE 12-2
COMPARISON OF TWO MULTIPLE FUNCTION DENTAL SYRINGES

Criteria	Four-way	Three-way plus suction
Hand movements	42	65
Cutting time, s	84	91
Energy expenditure, kcal/(min·kg)	0.023	0.023

Source: Evans et al., 1973.

addressed several subjective perceptions of the users including legibility, writing ease, and writing speed. In 1979, Kao followed up his subjective evaluation with laboratory tests. The dependent variables were objective measures of writing speed and writing pressure. He included felt-tip pens in the evaluation.

The results are important both for practical reasons (e.g., whether to invest $10 in a fountain pen or to pick up a $1.98 ballpoint pen) and as an illustration of trade-offs that we often discover when evaluating alternative designs using multiple criteria. The results are shown in Figure 12-20. The time measure represents the time required to write 10 lowercase *a*'s. The results indicate that for faster writing time the ballpoint pen is the best instrument and the fountain pen the worst. On the other hand, for more comfortable and less fatiguing handwriting performance, the felt-tip pen is clearly the best and the ballpoint is now the worst. So here we have a trade-off: speed versus fatigue. Given the relatively small differences in time between the instruments and large differences in pressure, Kao (1979) concludes that probably the felt-tip pen is the best all-round type of instrument for handwriting.

Some people might think that finding trade-offs among the designs being evaluated would make the evaluator's life more difficult—and it does. The consolation is knowing that a more complete understanding of the pros and cons of each design was achieved by using multiple criteria. If Kao had assessed only writing time, he might have erroneously concluded that the ballpoint pen was the most efficient instrument.

Kodak Disc Camera

The Eastman Kodak Company introduced the disc camera system in 1982, and the Human Factors Department of Kodak was a part of the product development program from the start. More than 50 experiments were conducted that

FIGURE 12-20
Comparison of writing speed and writing pressure with four writing instruments. Time represents the time required to write 10 lowercase *a*'s. (*Source: Kao, 1979, Figs. 1 and 2.*)

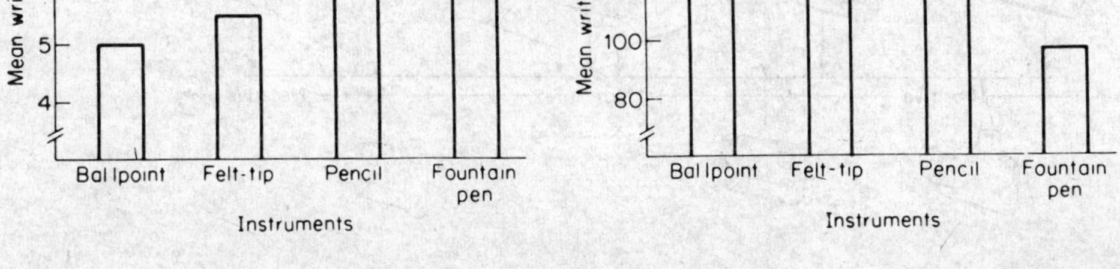

utilized over 2000 subjects and assessed all aspects of the design and operation of the camera (Faulkner, 1982). The initial efforts centered on basic questions of customer needs, understanding of the circumstances under which people take pictures, and the typical problems people encounter. In one phase of the program, 30,000 prints taken by amateur photographers were examined to determine, among other things, the picture-taking frequency as a function of camera-to-subject distance and ambient light level (Faulkner and Rice, 1982). In another phase various camera configurations were tested for holdability (Caplan, 1982). The cameras differed in terms of the location of the shutter release, whether a cover and/or handle was provided, and other factors. Subjects simulated taking pictures with the various camera configurations, and the frequency with which they covered the lens or flash with their fingers was recorded. The worst configuration would have been expected to result in 14 times the number of degraded pictures as the best configuration owing to fingers blocking the lens and flash.

In a series of three experiments, subjects' preferences for various camera configurations were assessed (Faulkner, Rice, and Heron, 1983). The overall ratings from one of the studies are shown, along with the camera configurations tested, in Figure 12-21. Using various statistical procedures, the investigators identified three relatively distinct clusters among their subjects. The three subject clusters produced somewhat different ratings of the various configurations. One cluster of subjects appeared to be more concerned with superficial features such as pop-up flashes and lens covers. Another cluster seemed more concerned with ease of holding the camera in terms of comfort and steadiness. The third cluster rated the cameras more in terms of the location of the shutter release. No one configuration completely satisfied all three groups.

The development of the Kodak disc camera is a good example of a systematic application of human factors to the design of a hand-held consumer product. Unfortunately, the small size of the negatives provided poor quality pictures and Kodak discontinued production of the disc camera.

FIGURE 12-21
Preference ratings for various configurations of disc cameras. (*Source: Adapted from Faulkner, Rice, and Heron, 1983. Copyright by the Human Factors Society, Inc. and reproduced by permission.*)

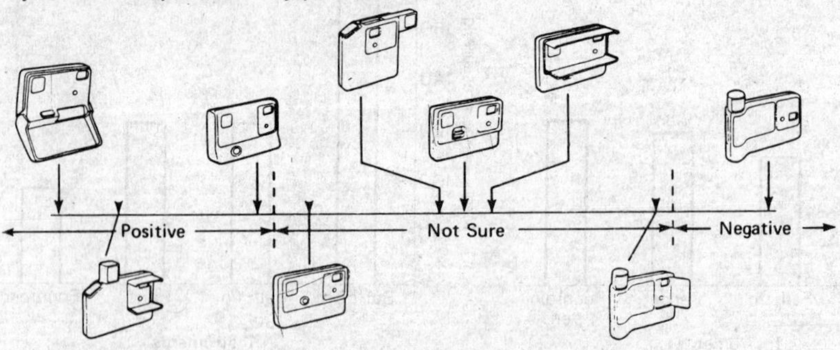

DISCUSSION

In bygone centuries people used various tools and hand devices to accomplish certain objectives, such as making things or performing certain tasks, including clubbing their attackers. Various types of hand tools and other hand-held devices still serve—and will continue to serve—many purposes. In the future we can look forward to new hand tools designed for specialized applications. To be of real value, these new tools must be designed with full cognizance of the operational system (and its limitations and constraints) into which the device will fit. In addition, the designer must consider the human factors, ergonomic, and biomechanical aspects of the user-tool interface.

The evaluation of new tool designs will require creativity to devise methodologies for reliable and valid measurement of relevant criteria. Combinations of various research strategies, including laboratory, simulation, and real-life field testing, will be required. Although hand tools have been around for a million years, there is still considerable work to be done to adapt them optimally for human use.

REFERENCES

Aghazadeh, F., and Mital, A. (1987). Injuries due to handtools. *Applied Ergonomics*, 18(4), 273–278.

American Conference of Government Industrial Hygienists (ACGIH) (1988). *TLVs: Threshold limit values and biological exposure indices for 1988–1989*. Cincinnati, OH: ACGIH, pp. 83–88.

American National Standards Institute (ANSI) (1986). *Guide for the measurement and evaluation of human exposure to vibration transmitted to the hand* (ANSI S3.34). New York: ANSI.

Armstrong, T. (1983). *Ergonomics guides: An ergonomics guide to carpal tunnel syndrome*. Akron, OH: American Industrial Hygiene Association.

Armstrong, T., Foulke, J., Joseph, B., and Goldstein, S. (1982). Investigation of cumulative trauma disorders in a poultry processing plant. *American Industrial Hygiene Association Journal*, 43(2), 103–116.

Ayoub, M., and Lo Presti, P. (1971). The determination of an optimum size cylindrical handle by use of electromyography. *Ergonomics*, 4(4), 503–518.

Barsley, M. (1970). *Left-handed man in a right-handed world*. London: Pitman.

Caplan, S. (1982). Designing new cameras for improved holdability. *Proceedings of the Human Factors Society 26th Annual Meeting*. Santa Monica, CA: Human Factors Society, pp. 195–198.

Coletta, G., Arons, I., Ashley, L., and Drennan, A. (1976). *The development of criteria for firefighter's gloves*, vol. 1: *Glove requirements* (No. 77–134–A). Cincinnati: Department of Health, Education, and Welfare, NIOSH.

Damon, F. (1965). The use of biomechanics in manufacturing operations. *The Western Electric Engineer*, 9(4).

Dionne, E. (1984, March). Carpal tunnel syndrome: Part I—The problem. *National Safety News*, pp. 42–45.

Ducharme, R. (1977). Women workers rate "male" tools inadequate. *Human Factors Society Bulletin*, 20(4), 1–2.

Eastman Kodak Co. (1983). *Ergonomic Design for People at Work,* vol. 1. Belmont, CA: Lifetime Learning Publications.

Emanuel, J., Mills, S., and Bennett, J. (1980). In search of a better handle. *Proceedings of the Symposium: Human Factors and Industrial Design in Consumer Products.* Medford, MA: Tufts University.

Evans, T., Lucaccini, L., Hazell, J., and Lucas, R. (1973). Evaluation of dental hand instruments. *Human Factors,* 15, 401–406.

Faulkner, T. (1982). Human factors in disc photography. *Third National Symposium on Human Factors and Industrial Design in Consumer Products.* Santa Monica, CA: Consumer Products Technical Group of the Human Factors Society, pp. 279–281.

Faulkner, T., and Rice, T. (1982). Human factors, photographic space, and disc photography. *Proceedings of the Human Factors Society 26th Annual Meeting.* Santa Monica, CA: Human Factors Society, pp. 190–194.

Faulkner, T., Rice, T., and Heron, W. (1983). The influence of camera configuration on preference. *Human Factors,* 25, 127–141.

Fraser, T. (1980). *Ergonomic principles in the design of hand tools,* Occupational Safety and Health Series no. 44. Geneva, Switzerland: International Labour Office.

Freivalds, A. (1987). The ergonomics of tools. In D. Oborne, (ed.), *International reviews of ergonomics,* vol. 1. London: Taylor & Francis.

Greenberg, L., and Chaffin, D. (1977). *Workers and their tools.* Midland, MI: Pendell Publishing.

Guilfoyle, J. (1977). Look what design has done for the toothbrush. *Industrial Design,* 24, 34–38.

Hansson, J., Eklund, L., Kihlberg, S., and Ostergren, C. (1987). Vibration in car repair work. *Applied Ergonomics,* 18(1), 57–63.

International Organization for Standards (ISO) (1986). *Mechanical vibration—Guidelines for the measurement and the assessment of human exposure to hand-transmitted vibration* (ISO 5349–1986). Geneva: ISO.

Kao, H. (1976). An analysis of user preference toward handwriting instruments. *Perceptual and Motor Skills,* 43, 522.

Kao, H. (1977). Ergonomics in penpoint design. *Acta Psychologica Taiwanica,* 18, 49–52.

Kao, H. (1979). Differential effects of writing instruments on handwriting performance. *Acta Psychologica Taiwanica,* 21, 9–13.

Knowlton, R., and Gilbert, J. (1983). Ulnar deviation and short-term strength reductions as affected by a curve-handled ripping hammer and a conventional claw hammer. *Ergonomics,* 26, 173–179.

Konz, S. (1979). *Work design.* Columbus, OH: Grid Inc.

Koradecka, D. (1977). Peripheral blood circulation under the influence of occupational exposure to hand-transmitted vibration. In D. Wasserman and W. Taylor (eds.), *Proceedings of the International Occupational Hand-Arm Vibration Conference* (No. 77-170). Cincinnati, OH: Department of Health, Education, and Welfare, NIOSH.

Krohn, R., and Konz, S. (1982). Bent hammer handles. *Proceedings of the Human Factors Society 26th Annual Meeting.* Santa Monica, CA: Human Factors Society, pp. 413–417.

Laveson, J., and Meyer, R. (1976). Left out "lefties" in design. *Proceedings of the Human Factors Society 20th Annual Meeting.* Santa Monica, CA: Human Factors Society, pp. 122–125.

National Institute for Occupational Safety and Health (NIOSH) (1983). *Current intelligence bulletin 38: Vibration syndrome* (DHHS-NIOSH Publ. 83–110). Cincinnati, OH: NIOSH.

National Institute for Occupational Safety and Health (NIOSH) (1989). *NIOSH criteria for a recommended standard: Occupational exposure to hand-arm vibration* (DHHS-NIOSH Publ. 89–106). Cincinnati, OH: NIOSH.

Pyykko, I. (1974). The prevalence and symptoms of traumatic vasospastic disease among lumberjacks in Finland: A field study. *Work & Environmental Health,* 11, 118–131.

Pyykko, I., Sairanen, E., Korhonen, O., Farkkila, M., and Hyvarinen, J. (1978). A decrease in the prevalence and severity of vibration-induced white fingers among lumberjacks in Finland. *Scandinavian Journal of Work and Environmental Health,* 12(4), 237–241.

Pyykko, I., Korhonen, O., Farkkila, M., Starck, J., and Aatola, S. (1982). A longitudinal study of the vibration syndrome of Finnish forestry workers. In A. Brammer and W. Taylor (eds.), *Vibration effects on the hand and arm in industry.* New York: Wiley, pp. 157–167.

Pyykko, I., Korhonen, O., Farkkila, M., Starck, J., Aatola, S., and Jantti, V. (1986). Vibration syndrome among Finnish forest workers, a follow-up from 1972 to 1983. *Scandinavian Journal of Work and Environmental Health,* 12(4), 307–312.

Schoenmarklin, R., and Marras, W. (1989*a*). Effect of handle angle and work orientation on hammering: I. Wrist motion and hammering performance. *Human Factors,* 31(4), 397–411.

Schoenmarklin, R., and Marras, W. (1989*b*). Effect of handle angle and work orientation on hammering: II. Muscle fatigue and subjective ratings of body discomfort. *Human Factors,* 31(4), 413–420.

Silverstein, B., Fine, L., and Armstrong, T. (1987). Occupational factors and carpal tunnel syndrome. *American Journal of Industrial Medicine,* 11, 343–358.

Sudhakai, L., Schoenmarklin, R., Lavender, S., and Marras, W. (1988). The effects of gloves on grip strength and muscle activity. *Proceedings of the Human Factors Society 32d Annual Meeting.* Santa Monica, CA: Human Factors Society, pp. 647–650.

Taylor, W. (1974). The vibration syndrome: Introduction. In W. Taylor (ed.), *The vibration syndrome.* New York: Academic.

Terrell, R., and Purswell, J. (1976). The influence of forearm and wrist orientation on static grip strength as a design criterion for hand tools. *Proceedings of the Human Factors Society 20th Annual Meeting.* Santa Monica, CA: Human Factors Society, pp. 28–32.

Tichauer, E. (1966). Some aspects of stress on forearm and hand in industry. *Journal of Occupational Medicine,* 8(2), 63–71.

Tichauer, E. (1967). Ergonomics: The state of the art. *American Industrial Hygiene Association Journal,* 28, 105–116.

Tichauer, E. (1976, February). Biomechanics sustains occupational safety and health. *Industrial Engineering.* 8(2), 46–56.

Tichauer, E. (1978). *The biomechanical basis of ergonomics.* New York: Wiley.

Tichauer, E., and Gage, H. (1977). Ergonomic principles basic to hand tool design. *American Industrial Hygiene Association Journal,* 38, 622–634.

Walton, K. (1974). The pathology of Raynaud's phenomenon of occupational origin. In W. Taylor (ed.), *The vibration syndrome.* New York: Academic.

Wang, M.-J., Bishu, R., and Rodgers, S. (1987). Grip strength changes when wearing three types of gloves. *Proceedings of INTERFACE 87*. Santa Monica, CA: Human Factors Society, pp. 349–354.

Washburn, S. (1960). Tools and human evolution. *Scientific American,* 203, 3–15.

Wasserman, D., Taylor, W., Behrens, V., Samueloff, S., and Reynolds, D. (1982). *Vibration white finger disease in U.S. workers using pneumatic chipping and grinding hand tools, I: Epidemiology* (DHHS-NIOSH Publ. 82–118). Cincinnati, OH: NIOSH.

Weidman, B. (1970). *Effect of safety gloves on simulated work tasks* (AD 738981). Springfield, VA: National Technical Information Service.

PART FOUR

WORKPLACE DESIGN

13 Applied Anthropometry, Work-Space Design, and Seating

14 Arrangement of Components within a Physical Space

15 Interpersonal Aspects of Workplace Design

CHAPTER **13**

APPLIED ANTHROPOMETRY, WORK-SPACE DESIGN, AND SEATING

In our everyday experience we use all sorts of physical equipment and facilities, many of which we find are not suitable to us because of their design features. Examples include bathroom sinks that are too low, chairs that are uncomfortable to sit in, shelves that are too high to reach, pants that fit in the waist but are too tight across the buttocks, and equipment that needs repair but provides inadequate space for inserting a hand while holding the necessary tool. These examples illustrate failures to design equipment and facilities to fit the physical dimensions of the people who will use them.

One consequence of automation and the information revolution is that jobs are changing. Today we spend more of our time sitting, sitting at computer terminals, sitting at control panels, sitting in libraries and classrooms, and sitting in front of the television. Poorly designed seats and workstations can cause pain and even injury to the back, muscle aches and pains in the shoulders and neck, and circulatory problems in the legs. In this chapter we discuss how things can be designed to fit the physical dimensions of people with special emphasis on designing seats and seated workstations.

ANTHROPOMETRY

Anthropometry deals with the measurement of the dimensions and certain other physical characteristics of the body such as volumes, centers of gravity, inertial properties, and masses of body segments. We confine our discussion to measurement of dimensions because such data are fundamental to a wider range of design problems. There are two primary types of body measurement: static and dynamic (functional). What is sometimes called *engineering anthropometry* is concerned with the application of both types of data to the design of the things people use. We briefly discuss static and dynamic anthropometry before discussing how such data are used in the design of work spaces and equipment.

Static Dimensions

Static dimensions are measurements taken when the body is in a fixed (static) position. They consist of skeletal dimensions (between the centers of joints, such as between the elbow and the wrist) or of contour dimensions (skin-surface dimensions such as head circumference. Many different body features can be, and have been, measured. The NASA *Anthropometric Source Book* (vol. 2, 1978), for example, illustrates 973 such measurements and presents data on certain of these measurements from 91 worldwide surveys. Many of these measurements, of course, have very specific applications, such as in designing helmets, earphones, or gloves. However, measurements of certain body features have rather general utility, and summary data on some of these features are presented for illustrative purposes. These data are representative of the U.S. civilian population 20 to 60 years of age (Kroemer, 1989). The specific body features measured are shown in Figure 13-1. For each measure and body weight, data for the 5th, 50th, and 95th percentiles are given in Table 13-1. (See Chapter 2 for a definition of percentiles.)

Body measurements vary as a function of age (see Figure 13-2), sex, and for different ethnic populations. In connection with age, for example, stature and related dimensions (lengths and heights) generally increase until the late teens or early twenties, remain relatively constant throughout early adulthood, and decline from early-to-middle adulthood into old age (Stoudt, 1981). (An exception is the length of the ears, which continue to grow throughout life.)

Differences in anthropometry between males and females are dramatically illustrated in Figure 13-3*a*, which shows the range of overlap of the 5th and 95th percentile males and females for selected heights and lengths. Figure 13-3*b* shows a similar display comparing 5th to 95th percentile U.S. Air Force blacks and whites and Japanese civilians on selected heights. Notice in Figure 13-3*b* the degree of overlap of blacks and whites for stature compared to the overlap for crotch height. This discrepancy is because blacks tend to have longer legs and shorter torsos than whites.

Differences in anthropometric dimensions between people working in different occupations is common. For example, Sanders (1977) found truck drivers to be taller and heavier than the general civilian population. Ayoub et al. (1982) found that underground coal miners had larger circumferences (torso, arms, legs) than did military personnel but did not differ significantly in terms of linear dimensions (heights and lengths). Occupational differences can result from a host of factors including imposed height and/or weight restrictions, amount of physical activity involved in the work, and self-selection on the part of applicants for practical or sociological reasons.

At the turn of the twentieth century in Western cultures it would have been difficult to find men over 6 ft (1.8 m) tall, yet today they seem to be in

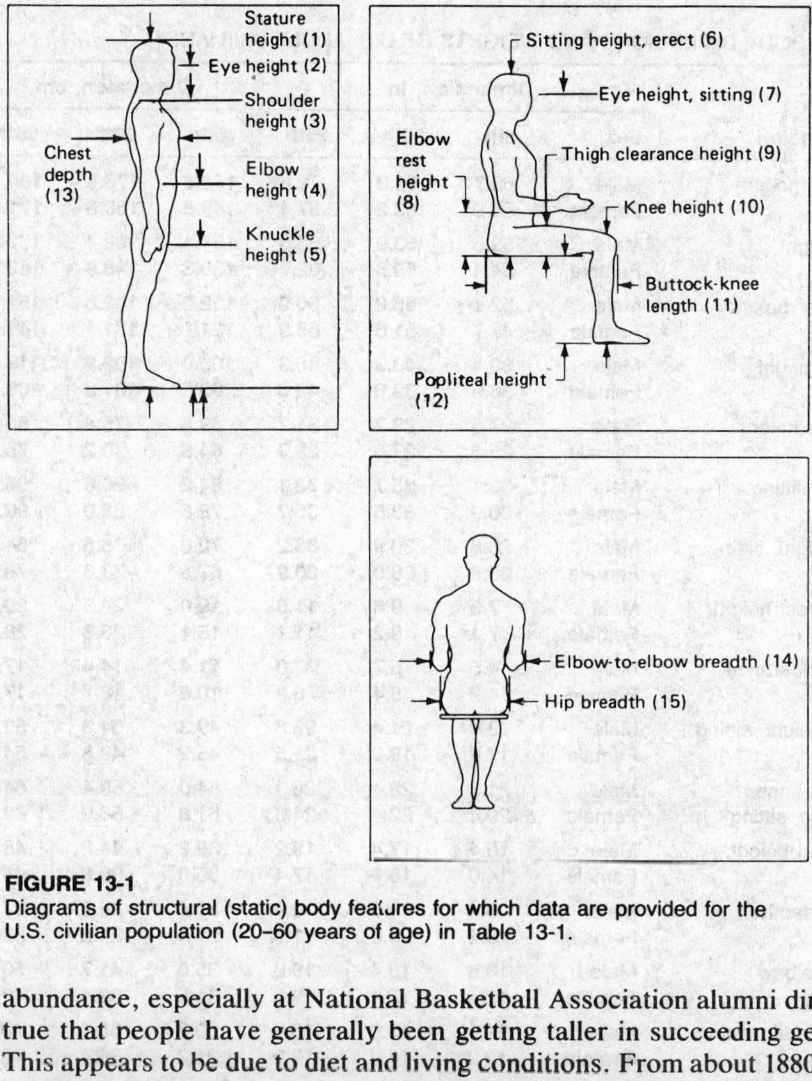

FIGURE 13-1
Diagrams of structural (static) body features for which data are provided for the U.S. civilian population (20–60 years of age) in Table 13-1.

abundance, especially at National Basketball Association alumni dinners. It is true that people have generally been getting taller in succeeding generations. This appears to be due to diet and living conditions. From about 1880 to 1960 in virtually all European countries, the United States, Canada, and Australia the average adult stature increased about 1 cm (0.5 in) per decade (Tanner, 1978). The evidence now, however, is that this trend appears to have leveled off since about 1960 (Cameron, 1979; Roche, 1979) with little increase expected within these countries in the foreseeable future. There is still a chance, then, that you will be able to look eye-to-eye with your kids when they grow up.

Dynamic (Functional) Dimensions

These dimensions are taken under conditions in which the body is engaged in some physical activity. In most physical activities (whether one is operating a steering wheel, assembling a mousetrap, or reaching across the table for the

TABLE 13-1
SELECTED BODY DIMENSIONS AND WEIGHTS OF U.S. ADULT CIVILIANS*

		Dimension, in			Dimension, cm		
Body dimension	**Sex**	**5th**	**50th**	**95th**	**5th**	**50th**	**95th**
1. Stature (height)	Male	63.7	68.3	72.6	161.8	173.6	184.4
	Female	58.9	63.2	67.4	149.5	160.5	171.3
2. Eye height	Male	59.5	63.9	68.0	151.1	162.4	172.7
	Female	54.4	58.6	62.7	138.3	148.9	159.3
3. Shoulder height	Male	52.1	56.2	60.0	132.3	142.8	152.4
	Female	47.7	51.6	55.9	121.1	131.1	141.9
4. Elbow height	Male	39.4	43.3	46.9	100.0	109.9	119.0
	Female	36.9	39.8	42.8	93.6	101.2	108.8
5. Knuckle height	Male	27.5	29.7	31.7	69.8	75.4	80.4
	Female	25.3	27.6	29.9	64.3	70.2	75.9
6. Height, sitting	Male	33.1	35.7	38.1	84.2	90.6	96.7
	Female	30.9	33.5	35.7	78.6	85.0	90.7
7. Eye height, sitting	Male	28.6	30.9	33.2	72.6	78.6	84.4
	Female	26.6	28.9	30.9	67.5	73.3	78.5
8. Elbow rest height, sitting	Male	7.5	9.6	11.6	19.0	24.3	29.4
	Female	7.1	9.2	11.1	18.1	23.3	28.1
9. Thigh clearance height	Male	4.5	5.7	7.0	11.4	14.4	17.7
	Female	4.2	5.4	6.9	10.6	13.7	17.5
10. Knee height, sitting	Male	19.4	21.4	23.3	49.3	54.3	59.3
	Female	17.8	19.6	21.5	45.2	49.8	54.5
11. Buttock-knee distance, sitting	Male	21.3	23.4	25.3	54.0	59.4	64.2
	Female	20.4	22.4	24.6	51.8	56.9	62.5
12. Popliteal height, sitting	Male	15.4	17.4	19.2	39.2	44.2	48.8
	Female	14.0	15.7	17.4	35.5	39.8	44.3
13. Chest depth	Male	8.4	9.5	10.9	21.4	24.2	27.6
	Female	8.4	9.5	11.7	21.4	24.2	29.7
14. Elbow-elbow breadth	Male	13.8	16.4	19.9	35.0	41.7	50.6
	Female	12.4	15.1	19.3	31.5	38.4	49.1
15. Hip breadth, sitting	Male	12.1	13.9	16.0	30.8	35.4	40.6
	Female	12.3	14.3	17.2	31.2	36.4	43.7
X. Weight (lbs and kg)	Male	123.6	162.8	213.6	56.2	74.0	97.1
	Female	101.6	134.4	197.8	46.2	61.1	89.9

* Body dimensions are depicted in Figure 13-1.
Source: Kroemer, 1989. (Courtesy of Dr. J. T. McConville, Anthropology Research Project, Yellow Springs, OH 45387, and Dr. K. W. Kennedy, USAF-AMRL-HEG, OH 45433.)

FIGURE 13-2
Some anthropometric changes that occur with age. (*By permission of Johnny Hart and North American Syndicate, Inc.*)

salt) the individual body members function in concert. The practical limit of arm reach, for example, is not the sole consequence of arm length; the limit is also affected by shoulder movement, partial trunk rotation, possible bending of the back, and the function to be performed by the hand. Figure 13-4 gives some impression of the possible interactions by various body members of a forklift truck operator. This figure is an illustration of what is sometimes called *somatography* in that it shows the three views of the operator: the frontal, side, and top.

Discussion

Far more static anthropometric data exists than does dynamic anthropometric data even though functional measures are more representative of actual human activity. Although there is no systematic procedure for translating static anthro-

FIGURE 13-3
Comparisons of 5th- and 95th-percentile heights and lengths: (*a*) comparison of males and females; (*b*) comparison of U.S. Air Force blacks and whites and Japanese civilians. (*Source: NASA, 1978*).

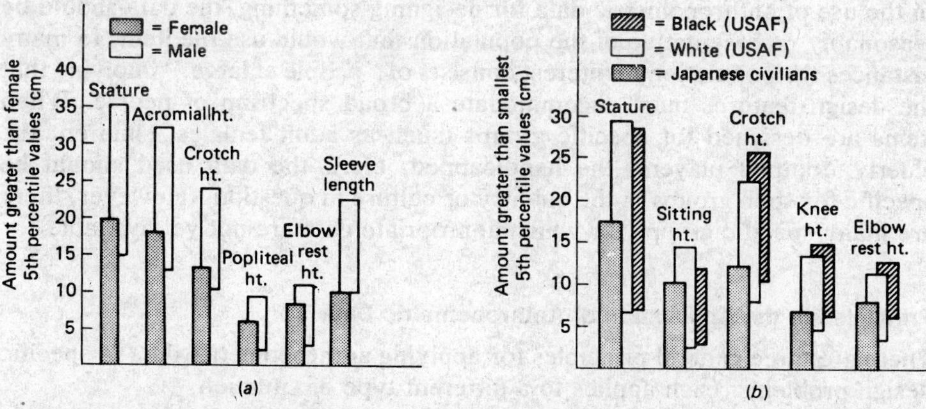

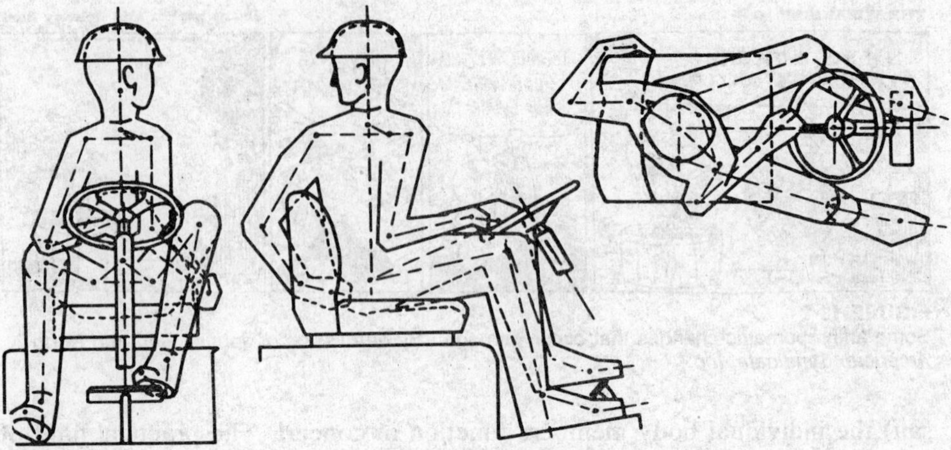

FIGURE 13-4
Three views of an operator of a forklift truck that illustrate the interactions of the body members. Such views represent what is sometimes called *somatography*. (*From North, 1980.*)

pometric data into dynamic measurements, Kroemer (1983) offers the following rules of thumb that may be helpful:

• Heights (stature, eye, shoulder, hip): reduce by 3 percent.

• Elbow height: no change, or increase by up to 5 percent if elevated at work.

• Knee or popliteal height, sitting: no change, except with high-heel shoes.

• Forward and lateral reaches: decrease by 30 percent for convenience, increase by 20 percent for extensive shoulder and trunk motions.

Obviously, these conversions are only very rough estimates, which can be affected by body postures, working conditions, etc.

USE OF ANTHROPOMETRIC DATA

In the use of anthropometric data for designing something, the data should be reasonably representative of the population that would use the item. In many instances the population of interest consists of "people at large," implying that the design features must accommodate a broad spectrum of people. When items are designed for specific groups (such as adult females, children, the elderly, football players, the handicapped, etc.), the data used should be specific for such groups in the country or culture in question. (However, there are many specific groups for which appropriate data are not yet available.)

Principles in the Application of Anthropometric Data

There are three general principles for applying anthropometric data to specific design problems; each applies to a different type of situation.

Design for Extreme Individuals In designing certain features of our built physical world, one should try to accommodate all (or virtually all) the population in question. In some circumstances a specific design dimension or feature is a limiting factor that might restrict the use of the facility for some people; that limiting factor can dictate either a *maximum* or *minimum* value of the population variable or characteristic in question.

Designing for the *maximum* population value is the appropriate strategy if a given maximum (high) value of some design feature should accommodate all (or virtually all) people. Examples include heights of doorways, sizes of escape hatches on military aircraft, and strength of supporting devices (such as a trapeze, rope ladder, or workbench). In turn, designing for the *minimum* population value is the appropriate strategy if a given minimum (low) value of some design feature has to accommodate all (or virtually all) people. Examples include the distance of a control button from the operator and the force required to operate the control.

Usually there are reasons for accommodating most, but not 100 percent, of the population. For example, it is not reasonable to have all doorways 9 ft (2.7 m) high to accommodate circus giants. Thus, it frequently is the practice to use the 95th male and 5th female percentiles of the distributions of relevant population characteristics as the maximum and minimum design parameters.

Designing for Adjustable Range Certain features of equipment or facilities can be designed so they can be adjusted to the individuals who use them. Some examples are automobile seats, office chairs, desk heights, and footrests. In the design of such equipment, it frequently is the practice to provide for adjustments to cover the range from the 5th percentile female to the 95th percentile male of the relevant population characteristic (sitting height, arm reach, etc.). The use of such a range is especially relevant if there could be technical problems in trying to accommodate the very extreme cases (i.e., 100 percent of the population); frequently the technical problems involved in accommodating the extreme cases are disproportionate to the advantages gained in doing so. Note that using a range from the 5th percentile female to the 95th percentile male will result in accommodating 95 percent of a 50/50 male/female population, not 90 percent, because of the overlap between male and female body dimensions. Generally, designing for an adjustable range is the preferred method of design, but of course, it is not always possible.

Designing for the Average First of all, there is no "average" individual. A person may be average on one or two body dimensions, but because there are no perfect correlations it is virtually impossible to find anyone who is average on more than a few dimensions. Often designers design for the average as a cop-out so that they do not have to deal with the complexity of anthropometric data. This is not to say that one should never design for the average. On the contrary, a thorough analysis of the situation may prove that an average value is acceptable. Such a situation would probably involve noncritical work where it is not appropriate to design for an extreme and where adjustability is impractical. For example, a checkout counter at a supermarket built for the average customer would probably inconvenience the majority of customers less than one built for either jockey Willy Shoemaker or basketball player Wilt Chamberlain. Designing for the average should only be done after careful consideration of the situation and never as an easy way out.

Discussion of Anthropometric Design Principles The discussion of the above principles refers to the application of anthropometric data for single dimensions (such as height or arm reach). The design problem becomes more sticky when one needs to take into account combinations of several dimensions. For example, the setting of limits such as the 95th and 5th percentiles on each of several dimensions can eliminate a fairly high percentage of the population. For instance, Bittner (1974) found in one situation the 95th and 5th percentile limits on each of 13 dimensions would exclude 52 percent of the population instead of the 10 percent implied by the 95th and 5th percentile limits of the individual dimensions. This occurs because body dimensions are not perfectly correlated with each other. For example, people with short arms do not necessarily have short legs. Not all the people excluded because they are outside the 5th to 95th percentile range on one variable will be the same people who are excluded on another measure. It is important, therefore, to consider the relationships (correlations) between body dimensions in the design of things based on combinations of dimensions.

Adding the 5th- or 95th-percentile values of body segments (e.g., fingertip to elbow and elbow to shoulder lengths) will not produce the corresponding percentile value for the combined dimension (e.g., fingertip to shoulder length). Once again this is because of the imperfect correlation between body dimensions. Robinette and McConville (1981) demonstrated this by building a 5th-percentile female out of 5th-percentile body segments (ankle height, ankle to crotch, crotch to buttock, etc.). The resulting female was 6 in (15.6 cm) shorter than the actual 5th-percentile value for stature. To derive composite measures requires taking into account the imperfect correlations between the segments being added (or subtracted) by using regression analysis.

In the application of anthropometric data, it is sometimes the practice to use physical models, such as the articulated model shown in Figure 13-5. Such models usually represent a specific percentile of the population. In addition, there are computer programs available that assess the adequacy of tentative work-space designs in terms of anthropometric considerations. We discuss some of these in Chapter 22.

It must be granted that tables of anthropometric data do not make for enthralling bedtime reading (although they might help those who have insomnia). But when we consider the application of such data to the design of the physical facilities and objects we use, we can see that such intrinsically uninteresting data play a live, active role in the real, dynamic world in which we live and work.

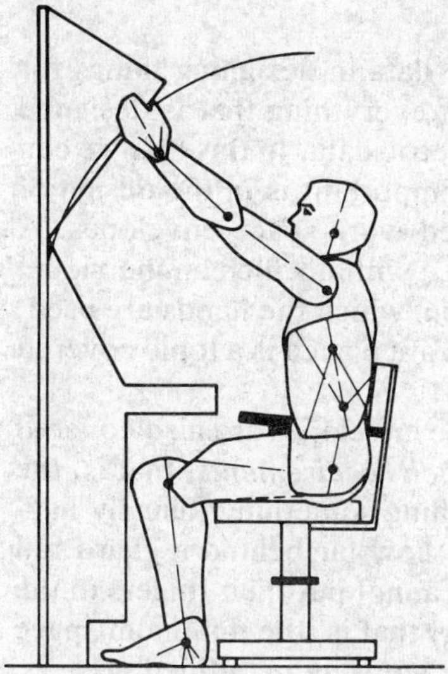

FIGURE 13-5
An articulated anthropometric scale model, such as used in the design of work spaces. (*Source: Meyer, 1979, Fig. 5.*)

In the application of anthropometric data to specific design problems, there can be no nicely honed set of procedures to follow, because of the variations in the circumstances in question and in the types of individuals for whom the facilities would be designed. Using anthropometric data in design involves art as well as science. As a general approach, however, the following suggestions are offered:

1 Determine the body dimensions important in the design (e.g., sitting height as a basic factor in seat-to-roof dimensions in automobiles).

2 Define the population to use the equipment or facilities. This establishes the dimensional range that needs to be considered (e.g., children, women, U.S. civilians, different age groups, world populations, different races).

3 Determine what principle should be applied (e.g., design for extreme individuals, for an adjustable range, or for the average).

4 When relevant, select the percentage of the population to be accommodated (for example, 90 percent, 95 percent) or whatever is relevant to the problem.

5 Locate anthropometric tables appropriate for the population, and extract relevant values.

6 If special clothing is to be worn, add appropriate allowances (some of which are available in the anthropometric literature).

7 Build a full-scale mock-up of the equipment or facility being designed and, using the mock-up, have people representative of large and small users walk through representative tasks. All the anthropometric data in the world cannot substitute for a full-scale mock-up.

WORK SPACES

There are many possible uses for anthropometric data in designing things for people. It would not be presumptuous to say that everything that is designed for people represents a potential use of anthropometric data. In this text we can illustrate only a few such uses. One of the most important is in the design of work spaces, including what are sometimes called work-space envelopes. A *work-space envelope* is the three-dimensional space within which an individual works. Typically this boils down to the space within which the hands are used. Where, within this envelope, controls and displays are placed is a topic covered in Chapter 14 on physical space and arrangement.

Two topics related to the concept of work-space envelopes are also covered in this section. The first might be called *out-of-reach requirements,* that is, the distances required to prevent a person from reaching something (usually hazardous) over a barrier. For example, at the zoo, how far behind a guard rail should the hawk cage be placed to insure people cannot put their fingers in the cage? The second topic is *clearance requirements,* that is, the minimum space needed to move through a tight space or perform work in a confined area.

Work-Space Envelopes for Seated Personnel

The limits of the work-space envelope for seated personnel are determined by functional arm reach, which in turn is influenced especially by the direction of arm reach and the nature of the manual activity (i.e., the task or function) to be performed. Functional arm reach is also influenced by such factors as the presence of any restraints and by the apparel worn. Some examples of the effects of these variables are given for illustration.

Effects of Direction of Reach and Presence of Restraints on Work-Space Envelope The effect of these variables on the work-space envelope is illustrated by the results of a study by Roth, Ayoub, and Halcomb (1977). They measured the functional arm reach of subjects at various lateral angles from a dead-ahead seated position (from −45° left to +120° right) and at various levels (ranging from −60° to +90°) from a seat reference point (SRP). Measurements were taken of the "grip-center" reach point at 114 such locations, under both restrained and unrestrained conditions. In the restrained conditions the shoulders of the subjects were held back against the seat back, whereas in the unrestrained condition the subjects could move their shoulders. The subjects were selected to be reasonably representative of adults in terms of height and weight.

Some of the results are shown in Figure 13-6, in particular the envelope for the 5th percentiles of males and females for each condition. Each line represents the 5th percentile value at specified levels above a seat reference point. (Data are given for only certain levels.) On the basis of the data for either sex, one could envision a three-dimensional space, the outer surface of which would

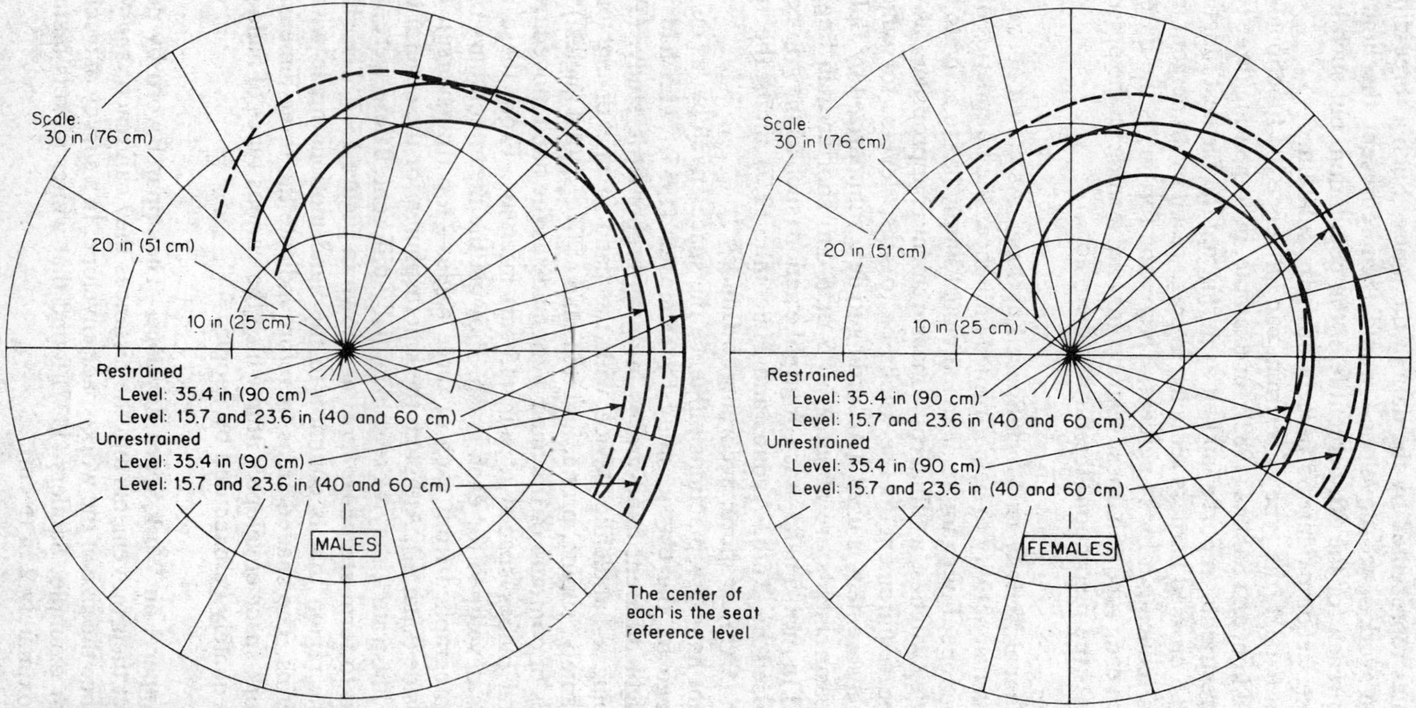

FIGURE 13-6
Functional arm reach for the 5th percentiles of males and females at specified levels above a seat reference point. The three-dimensional space represented by these data can be considered as forming a work-space envelope that would accommodate 95 percent of the population. (*Source: Roth, Ayoub, and Halcomb, 1977, for description of procedures used in obtaining data. Appreciation is expressed to Dr. M. M. Ayoub and his associates for special tabulations used in the preparation of this figure.*)

include the values represented by the data in the figure. Such a three-dimensional space—a work-space envelope—would then represent the limits of convenient arm reach for the 5th percentile of the population, but such limits would, of course, accommodate 95 percent of the population.

As one might expect, the type of restraint can influence the functional arm reach. Garg, Bakken, and Saxena (1982), for example, in comparing three types of vehicle seat restraints, report a lap belt as permitting the farthest functional arm reach, with a crossed harness reducing average reach by 14 percent and a parallel harness reducing it by 24 percent. The reason for this is the different degrees to which each type of harness restricts forward leaning by the person in the seat. The less the restriction, the further the reach.

Effects of Manual Activity on Work-Space Envelope As stated above, the nature of the manual activity to be performed influences the boundaries of the work-space envelope. For example, if an individual simply has to activate push buttons or toggle switches, a "fingertip" measurement is appropriate, as contrasted with the requirement to use knobs or to grasp levers, for which a "thumb tip" measurement is used. [According to data summarized by Bullock (1974) such measurements are about 2 in (5 or 6 cm) shorter than fingertip measurements.] In turn, a hand-grasp or griplike action (such as the grip-center measurement used by Roth, Ayoub, and Halcomb, 1977) limits the reach further, by 2 in (5 cm) or more, according to Bullock.

Even different hand-grasp actions influence the space envelope, as demonstrated years ago by the classic study by Dempster (1955), who had male subjects use eight different hand grasps in an anthropometric study. (These involved grasping a handlelike device with the hand in one of eight fixed orientations, namely, supine, prone, inverted, and at five spatial angles.) Photographic traces of contours of the hand were made as the hand moved over a series of frontal planes spaced at 6-in (15-cm) intervals. From these a *kinetosphere* was developed for each grasp, showing the mean contours of the tracings as photographed from each of three angles—top (transverse), front (coronal), and side (sagittal). Although the kinetospheres for the several types of grip are not illustrated, they were substantially different. These were, however, combined to form *strophospheres* as shown in Figure 13-7. The shaded areas define the region that is common to the hand motions made with the various hand grips. The shaded areas therefore indicate the three-dimensional space in the work-space envelope within which the various types of hand grips could most adequately be executed by people.

Effects of Apparel on Work-Space Envelope The apparel worn by people can also restrict their movements and the distances they can reach, and therefore can influence the size of the work-space envelope. In a survey of truck and bus drivers, for example, Sanders (1977) found that winter jackets restricted reach by approximately 2 in (5 cm).

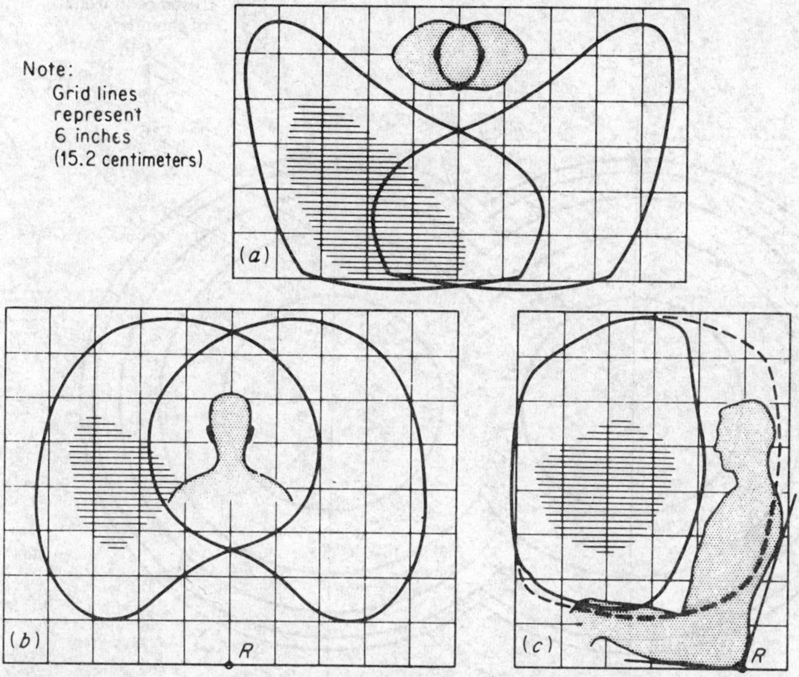

FIGURE 13-7
Strophosphere resulting from superimposition of kinetospheres of range of hand movements with a number of hand-grasp positions in three-dimensional space. The shaded areas depict the region common to all hand motions (prone, supine, inverted, and several different angles of grasps), probably the optimum region, collectively, of the different types of hand manipulations. (*Source: Adapted from Dempster, 1955.*)

Work-Space Envelopes for Standing Personnel

Stand with your back and heels against a wall and reach forward as far as you can. Generally, you are limited in how far forward you can reach in a standing position because you begin to topple over as your center of gravity (cg) extends beyond the limits of the base of support provided by your feet. Step away from the wall and try it again. Pushing the pelvis backward acts as a counterbalance and keeps your cg over your feet as you reach out farther. Standing reach, then, is essentially a matter of body equilibrium, and the reach envelope will be modified by any factor that affects this. Reach is diminished if a weight is carried in the reaching hand or if an obstacle is placed behind a person that limits counterbalancing activities. Reach is increased by increasing the base of the feet.

Pheasant (1986) describes the zone of convenient reach as the space in which an object may be reached without undue exertion, i.e., within arm's reach. Figure 13-8 presents the standing zone of convenient reach for a 5th-percentile female as a function of the distance in front of the shoulder. This zone requires

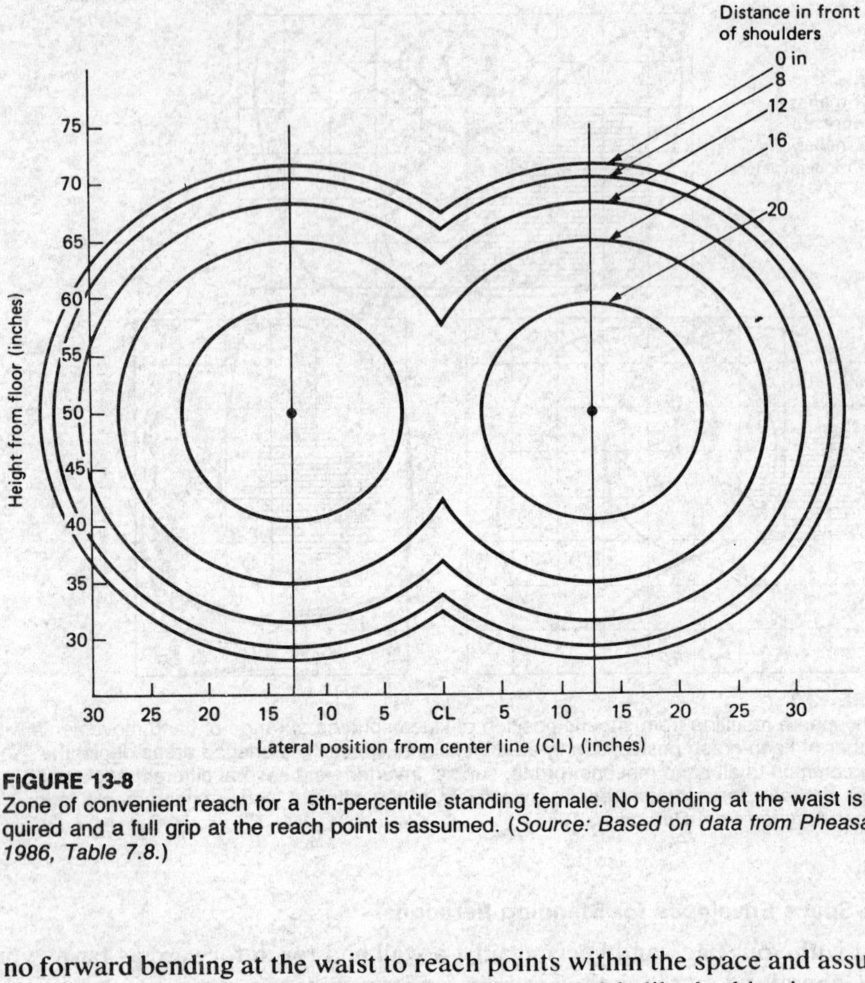

FIGURE 13-8
Zone of convenient reach for a 5th-percentile standing female. No bending at the waist is required and a full grip at the reach point is assumed. (*Source: Based on data from Pheasant, 1986, Table 7.8.*)

no forward bending at the waist to reach points within the space and assumes a full grip at the reach point. (Looking at Figure 13-8 is like looking into two ice-cream cones joined together at the top.)

Discussion of Work-Space Envelopes Work-space envelopes consist of the three-dimensional spaces that are reasonably optimum for seated or standing persons who perform some type of manual activity. Thus (for example) control devices and other objects to be used usually should be located within such space. The reasonable limits of such space are determined by functional arm reach, which is influenced by such variables as direction of arm reach, the nature of the manual activity, the use of restraints, apparel worn, the angle of the backrest, and personal variables such as age, sex, ethnic group, and handicaps. Whenever feasible, such spaces should be designed with consideration for the personal characteristics of the population to use the facility. It is fairly standard practice to design such space for the 5th percentile of the using population, thus making it suitable for 95 percent of the population.

In connection with work-space envelopes for special populations, Rozier (1977), for example, reports that with amputees who have a below-elbow prosthesis there is an average decrease in usable work space of 45 percent, and with an above-elbow prosthesis the average decrease is 83 percent. Thus, with such special populations the design of the work space requires particular (and sometimes individual) attention.

Out-of-Reach Requirements

In many situations objects that are not to be touched must be placed near where people work or play. Examples include *objets d'art* in a museum or robots in a factory (look but don't touch). In such situations, usually a barrier of some height is placed between the person and the object. The question is how far from the barrier should the object (or target, from the person's perspective) be placed so that people cannot touch it. The answer depends on the height of the barrier and the height of the object. Thompson (1989) using males of the 99th percentile in stature measured maximum reach for various combinations of these two variables. Figures 13-9*a* and 13-9*b* show two representative reach conditions used in the study. Figure 13-9*c* presents the out-of-reach requirements (mean plus 1 standard deviation for 99th-percentile males). This data should be used when designing for the extreme, and then some.

Clearance Requirements

People sometimes have to work in, move through, or even just fit into some restricted or awkward spaces. This is especially true for some types of maintenance work. For illustrative purposes examples of the clearances required for certain types of work situations are given in Figure 13-10. Note that in most cases, heavy clothing adds 4 to 6 in (10 to 15 cm) to the requirements and in the case of escape hatches 10 in (25 cm).

The clearance requirements of people can be stretched to include some unusual circumstances, such as the sleeping space needed by long-haul truck drivers, many of whom drive in pairs so that one can sleep while the other drives. The Federal Motor Carrier Safety Regulations presently specify the minimum space for sleeper berths in trucks. (Such berths are sort of shelves behind the driver.) An investigation of the dimensions of the sleeping postures of 239 drivers was carried out by Sanders (1980). He reports the following dimensions of the preferred and prostrate (face-down) postures of the 95th-percentile drivers. The legal specifications are also given.

Sleeper berths	Length	Width
Preferred position	1.98 m (78 in)	0.84 m (33 in)
Prostrate position	2.04 m (80 in)	0.86 m (34 in)
Legal specifications	1.90 m (75 in)	0.61 m (24 in)

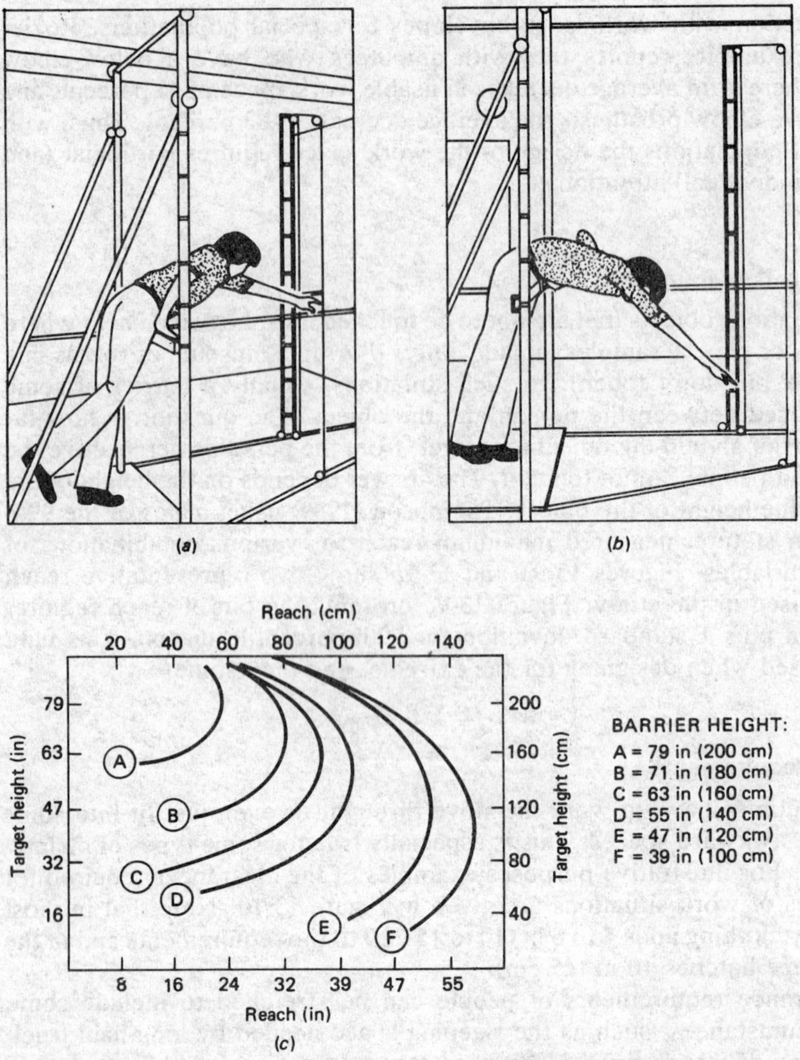

FIGURE 13-9
Out-of-reach requirements as a function of barrier and target height. The graph (*c*) presents the mean reach plus 1 standard deviation for 99th-percentile stature males. The data are based on a sample size of 30 and the subjects wore shoes while making the reaches. (*Source: Adapted from Thompson, 1989, Figs. 1, 3–8.*)

We can see that the postures of the subjects would not fit within the federal specifications, especially for the width of the berth. Such data can then serve as the basis for setting legal standards as well as for the more general objectives of designing for human use.

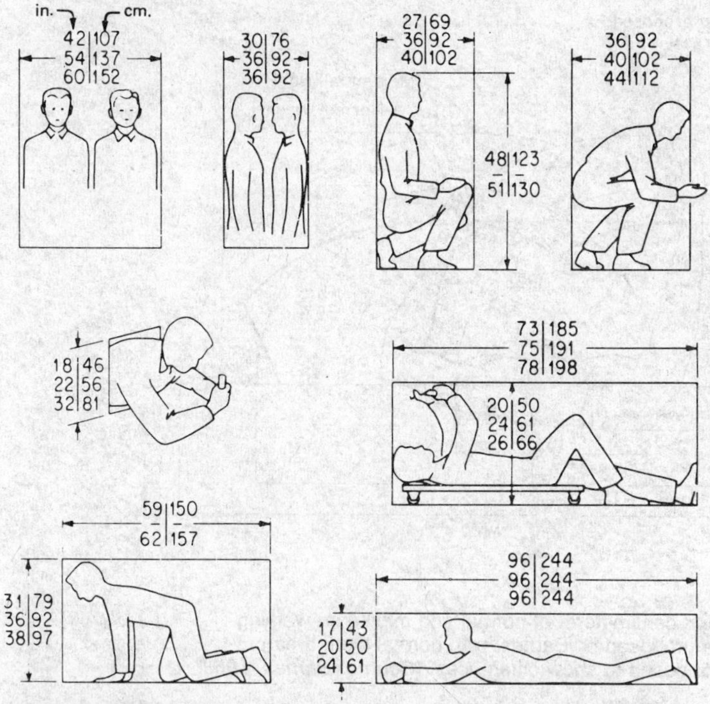

FIGURE 13-10
Clearances for certain work spaces that individuals may be required to work in or pass through. *Note:* The three dimensions given (inches at left, centimeters at right) are (from top to bottom in each case) minimum, best (with normal clothing), and with heavy clothing (such as arctic). (*Source: Adapted from Rigby, Cooper, and Spickard, 1961.*)

DESIGN OF WORK SURFACES

Within the three-dimensional envelope of a workplace, specific design decisions need to be made about various features of the workplace, including the location and design of whatever work surfaces are involved, such as benches, desks, tables, and control panels. This is not the place to provide guidelines for designing the myriad work surfaces that exist in this world, but some samples may be helpful to illustrate how anthropometric and biomechanical data can be used.

Horizontal Work Surface Area

The horizontal work surface area to be used by seated and "sit-stand" workers generally should provide for manual activities to be within convenient arm's reach. Certain *normal* and *maximum* areas were proposed by Barnes (1963) and earlier by Farley (1955) and have been used rather widely. These areas as defined by Barnes are shown in Figure 13-11 and are described as follows:

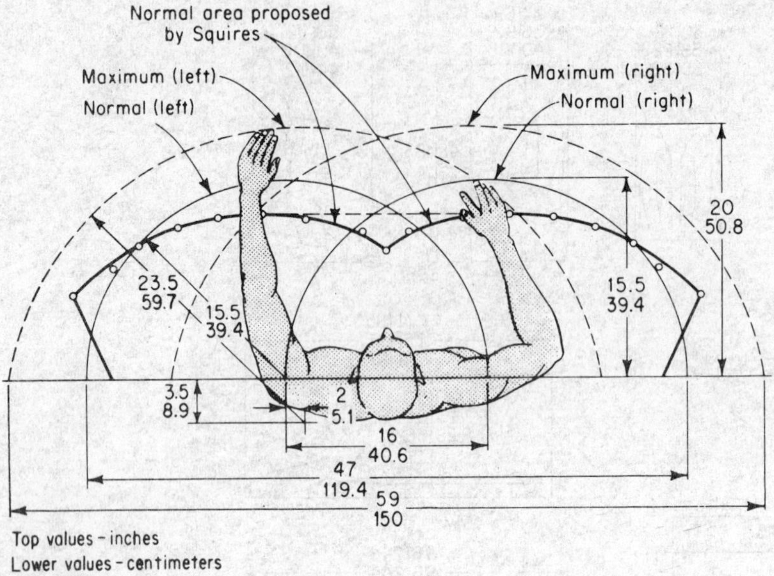

FIGURE 13-11
Dimensions (in inches and centimeters) of normal and maximum working areas in horizontal plane proposed by Barnes, with normal work area proposed by Squires superimposed to show differences. (*Source: Barnes, 1963; Squires, 1956.*)

1 *Normal area.* This is the area that can be conveniently reached with a sweep of the forearm while the upper arm hangs in a natural position at the side.

2 *Maximum area.* This is the area that could be reached by extending the arm from the shoulder.

Related investigations by Squires (1956), however, have served as the basis for proposing a somewhat different work-surface area that takes into account the dynamic interaction of the movement of the forearm as the elbow also is moving. It is for this reason that Das and Grady (1983*b*) recommend it over that proposed by Farley or Barnes. The area proposed by Squires is superimposed over the area proposed by Barnes in Figure 13-11. It is a bit more shallow than that proposed by Barnes but covers more area than that proposed by Farley. An additional argument in favor of Squires' model is that it requires less forward extension of the forearm, thus minimizing the stress on the elbow joint (Tichauer, 1978).

Although most office activities such as reading and writing are carried out on horizontal surfaces such as desks and tables, Eastman and Kamon (1976) propose that (where feasible) a slanted surface be used. In their study they found that subjects using slanted surfaces (12° and 24°) had better posture, showed less trunk movement, and reported less fatigue and less discomfort than when using horizontal surfaces. Bridger (1988) found similar results, as depicted in Figure 13-12. When using a slanted surface (15°) subjects sat with less bending of the neck, a more upright trunk, and less trunk flexion than when using a horizontal work surface. The evidence seems clear that using slanted work surfaces for visual tasks, such as reading, offers considerable advantages in terms of posture over traditional horizontal work surfaces. Now we have to work on how to keep the materials being read from sliding off the surface.

Work-Surface Height: Seated

Some of those who have experienced backaches, neck aches, and shoulder pains at work can testify that the height of a work surface can indeed bring on such aches and pains. If a work surface is too low, the back may be bent over too far; and if it is too high, the shoulders must be raised above their relaxed posture, thus triggering shoulder and neck discomfort.

When discussing work-surface height some confusion may be introduced if a distinction is not made between work-surface height and working height. *Work-surface height* is simply the height of the upper surface of a table, bench, desk, counter, etc. measured from the floor. Even this simple notion becomes a little complex when slanted work surfaces are referred to; usually the height of the front edge and the angle of the surface are specified. *Working height*, however, depends on what one is working on. When writing on paper, the working height is the same as the work-surface height. When using a keyboard (typewriter or

FIGURE 13-12
Comparison of typical postures when horizontal or slanted (15°) work surfaces are used for reading. The slanted surface results in less bending of the neck, more upright trunk, and less trunk flexion than does the horizontal surface. (*Source: Adapted from Bridger, 1988, Fig. 4. Reprinted with permission of the Human Factors Society, Inc. All rights reserved.*)

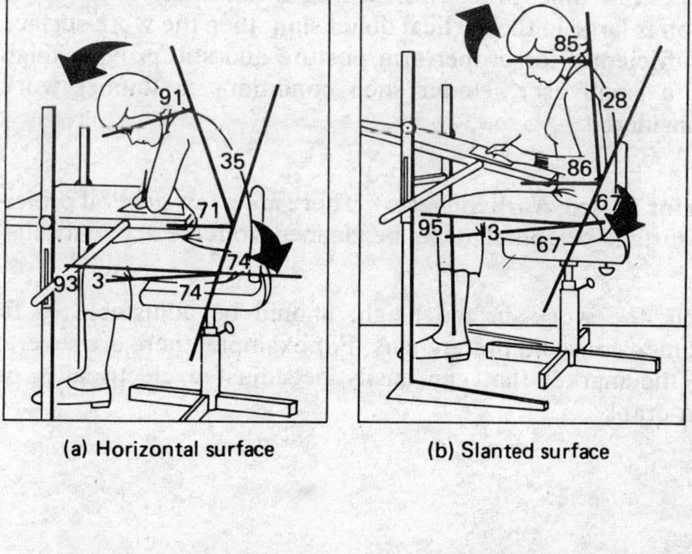

(a) Horizontal surface

(b) Slanted surface

computer) the working height is taken as the height of the home row of keys (the "asdfghjkl" row on a standard keyboard). When washing vegetables in a sink, the working height is actually below the work-surface height.

Seated Work-Surface Height and Arm Posture In recent years some investigators have recommended reducing work-surface heights, generally to permit relaxed postures of the upper arms with respect to working height. Working with relaxed upper arms and elbows at about 90° provides comfort and helps maintain straight wrists, which can be beneficial when performing repetitive tasks such as typing or electronic assembly. On the basis of a European survey, Bex (1971) reports that the most common heights of desks have, in fact, been reduced from about 30 in (76 cm) in 1958 to about 28.5 in (72 cm) in 1970. Based on his own and other anthropometric data, he argues for a further reduction of fixed desk heights to about 27 in (68.6 cm).

Seated Work-Surface Height and Thigh Clearance Basing seated work-surface height on arm posture alone can lead to problems. Work-surface height is also influenced by seat height (discussed later), the thickness of the work surface, and the thickness of the thighs. The clearance between the seat and the underside of the work surface should accommodate the thighs of the largest user. ANSI (Human Factors Society, 1988) recommends 26.2 in (66.5 cm) as the minimum height for the underside of a nonadjustable seated work surface. Such work surfaces can cause problems for smaller users. They usually have to raise their seats so that their elbow height is equal to the working height. In so doing, often their feet cannot touch the floor. When this occurs, a foot rest is needed to support their feet. With adjustable-height work surfaces, small users can adjust the height so that the working height is at elbow height with their feet on the floor. ANSI (Human Factors Society, 1988) recommends a range of height adjustments for the underside of the work surface of 20.2 to 26.2 in (51.3 to 66.5 cm). This works fine unless the work surface is unusually thick or the object being worked on is large in the vertical dimension; then the work surface cannot be lowered sufficiently for proper arm posture and still provide thigh clearance—even for a small user. Under such conditions, a thinner work surface should be considered.

General Principles for Seated Work Surfaces There are a few general principles related to work-surface heights that can be gleaned from the above discussion:

1 If at all possible the work-surface height should be adjustable to fit individual physical dimensions and preferences. For example, there are several computer tables on the market that can easily be adjusted electrically, or manually by use of a crank.

2 The work surface should be at a level that places the working height at elbow height.

3 The work surface should provide adequate clearance for a person's thighs under the work surface.

Seated Work Surface Height and Nature of the Task Table 13-2 presents some recommendations for work-surface heights from various sources based on representative anthropometric data. The recommendations for fine and precision work (Ayoub, 1973) are somewhat at odds with the second general principle of placing working heights at elbow height. Specially, the recommendations would place the working surface upon which fine and precision work is done at 6 and 2 in (15 and 5 cm) above elbow height respectively. In such instances there should be provision for arms to rest on the work surface. The higher surfaces are recommended so that the work is within close visual range which would be important for such work.

Work-Surface Height: Standing

The critical features for determining work-surface heights for standing workers are in part the same as for seated workers, i.e., elbow height and the type of work being performed. Figure 13-13 shows recommended heights for precision work, light work, and heavy work as related to elbow height (Grandjean, 1988). For light and heavy work the recommended work-surface heights are below elbow height, whereas that for precision work is slightly above (generally to provide elbow support for precise manual control). We recommend, however, that precision tasks be performed sitting down.

TABLE 13-2
RECOMMENDATIONS FOR SEATED WORK-SURFACE HEIGHTS FOR VARIOUS TYPES OF TASKS

	Male		Female	
Type of task (seated)	**in**	**cm**	**in**	**cm**
Fine work (e.g., fine assembly)[1]	39.0–41.5	99–105	35.0–37.5	89–95
Precision work (e.g., mechanical assembly)[1]	35.0–37.0	89–94	32.5–34.5	82–87
Light assembly[1]	29.0–31.0	74–78	27.5–29.5	70–75
Coarse or medium work[1]	27.0–28.5	69–72	26.0–27.5	66–70
Reading and writing[2]	29.0–31.0	74–78	27.5–29.0	70–74
Range for typing desks[2]	23.5–27.5	60–70	23.5–27.5	60–70
Computer keyboard use[3]	23.0–28.0	58–71	23.0–28.0	58–71

Sources: [1] Ayoub, 1973; [2] Grandjean, 1988; [3] Human Factors Society, 1988.

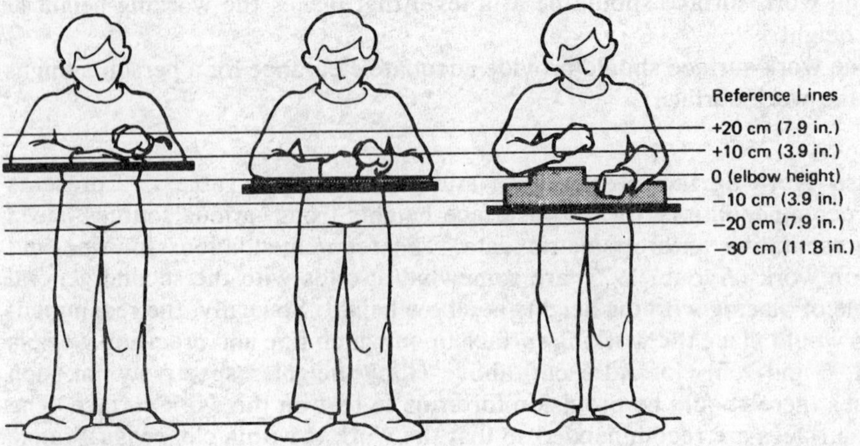

FIGURE 13-13
Relationship between elbow height (from floor) and recommended work-surface height for three types of work performed while standing. The zero horizontal reference line is the elbow height of the individual, and the other lines represent levels above and below. Average elbow height reported by Grandjean for Europeans is 105 cm (41.3 in) for males and 98 cm (38.6 in) for females. (*Adapted from Grandjean, 1988, Fig. 32.*)

Even though the work-surface height should be influenced in part by the nature of the task, individual preferences for work-surface height also are relevant. This was reflected by the results of a survey by Ward and Kirk (1970) of the preferences of British homemakers with respect to work-surface heights in the performance of three different types of tasks. The percentages preferring work surfaces at certain levels (relative to the elbow) in the performance of these tasks are given in Table 13-3. Notice that in Figure 13-13 the recommended work-surface height for heavy work is *below* elbow height yet in Table 13-3, 50 percent of the subjects preferred a work surface *above* elbow height for exerting pressure. Balancing anthropometric considerations with user preferences involves compromise but speaks quite clearly for height-adjustable work surfaces.

In general, fixed-height work surfaces for use while standing should be designed for the largest users. Platforms can then be supplied to smaller users to lift them to the proper height. Even better, of course, is the use of adjustable-height work surfaces that can be raised or lowered electrically, hydraulically, or manually.

Over the years there have been recommendations for standing work-surface heights, but the recommendations depend on the distribution of elbow heights assumed in the population. Table 13-4 presents recommended heights for fixed and adjustable work surfaces. The recommendations are based on the elbow height data from Table 13-1 and the recommendations from Figure 13-13. The fixed-height work-surface recommendations assumes that a platform will be available to lift smaller users to the proper level.

TABLE 13-3
PERCENTAGE OF SUBJECTS EXPRESSING PREFERRED WORK-SURFACE HEIGHTS IN PERFORMING THREE KITCHEN TASKS

Type of task	Level relative to elbow		
	Lower	Even	Above
Working above surface (peeling vegetables, slicing bread, etc.)	54	14	32
Working on surface (spreading butter, chopping ingredients, etc.)	16	11	73
Exerting pressure (ironing, rolling pastry, etc.)	41	9	50

Source: Ward and Kirk, 1970.

Work Surfaces for Standing or Sitting

It is sometimes desirable to provide the opportunity to perform a job in either a standing or a sitting posture or to permit workers to alternate their postures. In such instances the work-surface height should permit a relaxed position of the upper arm, and the chair and footrest should permit such a posture, as illustrated in Figure 13-14.

SCIENCE OF SEATING

Whether at work, at home, at horse races, on buses, or elsewhere, the members of the human race spend a major fraction of their lives sitting down. As we know from experience, the chairs and seats we use cover the gamut of comfort and discomfort. Improper design of chairs and seats also can affect the work performance of people and can contribute to backaches and back problems. In this section we discuss a few important principles of seat design, followed by

TABLE 13-4
RECOMMENDED STANDING WORK-SURFACE HEIGHTS FOR THREE TYPES OF TASKS*

Type of task (standing)	Sex	Fixed height		Adjustable height	
		in	cm	in	cm
Precision work (with elbows supported)	Males	49.5	126	42.0–49.5	107–126
	Females	45.5	116	37.0–45.5	94–116
Light assembly work	Males	42.0	107	34.5–42.0	88–107
	Females	38.0	96	32.0–38.0	81–96
Heavy work	Males	39.0	99	31.5–39.0	80–99
	Females	35.0	89	29.0–35.0	74–89

* Assumes platform will be available to lift smaller users to proper level.

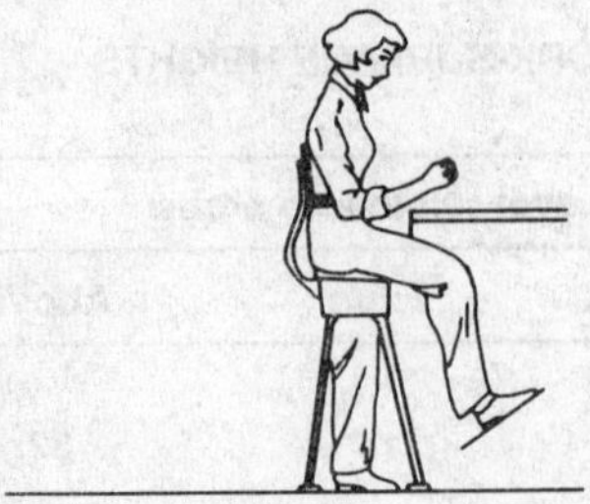

FIGURE 13-14
A work station that permits the worker to stand or sit and still maintain an appropriate relationship with the work surface. (*Adapted from Das and Grady, 1983a, Fig. 4.*)

more specific recommendations, and finally close by illustrating a few seat designs for specific purposes.

General Principles of Seat Design

A lot has been written of late on seat design with designers coming up with all sorts of novel design ideas. Our purpose here is not to explore the frontiers of seat design, but rather to relate some of the basic information and principles that form the solid ground upon which future designs should build. Unfortunately, some of what appears to be solid ground may, in time, prove to be sandy loam. Your author is often asked what is the best chair on the market. The answer is that it is the chair that you find most comfortable after sitting in it for a few hours. The problem is that different people find different chairs best for them. There is no one best chair for everyone. The best approach is to provide two or three good designs from which people can choose. The following are a few principles of seat design.

Promote Lumbar Lordosis When standing erect, the lumbar portion of the spine (the small of the back, just above the buttocks) is naturally curved inward (concave), that is, it is *lordotic*. Natural lumbar lordosis aligns the vertebrae of the spine in a near vertical axis through the thigh and pelvis, as shown in Figure 13-15. However, when one is sitting with the thighs at 90°, the lumbar region of the back flattens out and may even assume an outward bend (convex), that is, it becomes *kyphotic*, as shown in Figure 13-15. This occurs because the hip joint rotates only about 60°, forcing the pelvis to rotate backward about 30° to achieve the 90° thigh angle. Lumbar kyphosis results in increased pressure on the discs located between the vertebrae of the spine.

Andersson et al. (1979) found that the use of a 2-in (4-cm) thick lumbar support had a marked impact on maintaining lumbar lordosis with a seat backrest angle of 90°. When the backrest angle was reclined to 110° (and the lumbar support was used), the lumbar spine resembled closely the lumbar curve of a person standing. Interestingly, the exact location of the support within the lumbar region did not significantly influence any of the angles measured in the lumbar region. The lumbar support, therefore, does not have to perfectly match the inward arch of the lower back.

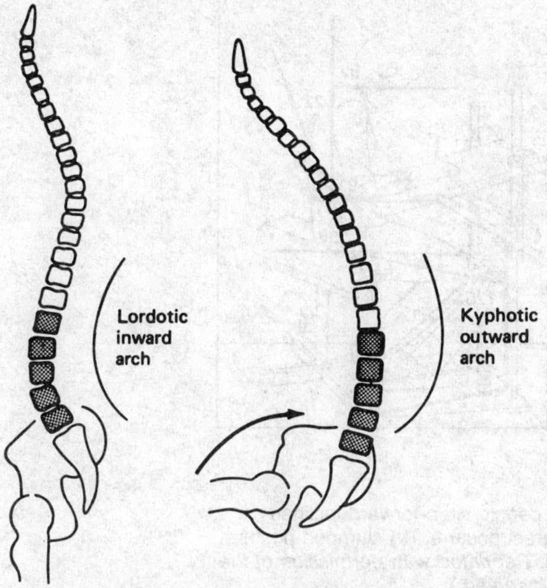

FIGURE 13-15
Posture of the spine when (*a*) standing and (*b*) sitting. Lumbar portion of spine is lordotic when standing and kyphotic when sitting. The shaded vertebrae are the lumbar portion of the spine. (*Source: Grandjean, 1988, Fig. 47.*)

Another approach to preventing the flattening or kyphosis of the lumbar spine is to provide a forward-tilting seat (Mandal, 1985). A forward-tilting seat opens the angle between the hip and upper torso, thus producing a more relaxed posture. For example, Keegan (1953) found by the use of x-rays that the most relaxed, or most normal, sleeping posture when people sleep on their side, is one in which there is about a 135° angle with the spine. This is very similar to the posture assumed by people in the weightless environment of space (see Chapter 19).

Bendix (1986) did find a lordotic effect of forward-tilting seats, but the effect was small and depended on other factors, such as seat height and table top slant. Bridger (1988) found some evidence supporting the value of a forward-tilting seat but also noted that some people assumed a slumped posture even with the tilted chair. Figure 13-16 shows typical erect and slumped postures observed by Bridger when subjects used a forward-tilting seat. Figure 13-16 can be compared to Figure 13-12 where a horizontal seat is used. The value of a forward-tilting chair seems to be enhanced if a slanted work surface is also used.

Minimize Disc Pressure The discs between the vertebrae can be damaged by excessive pressure. Unsupported sitting, i.e., not using a backrest, increases disc pressure considerably over that experienced while standing. Nachemson and Elfstrom (1970), for example, found that unsupported sitting in an upright, erect posture (forced lordosis) resulted in a 40 percent increase in pressure compared to standing. Unsupported sitting in a forward slumped posture increased pressure 90 percent compared to standing.

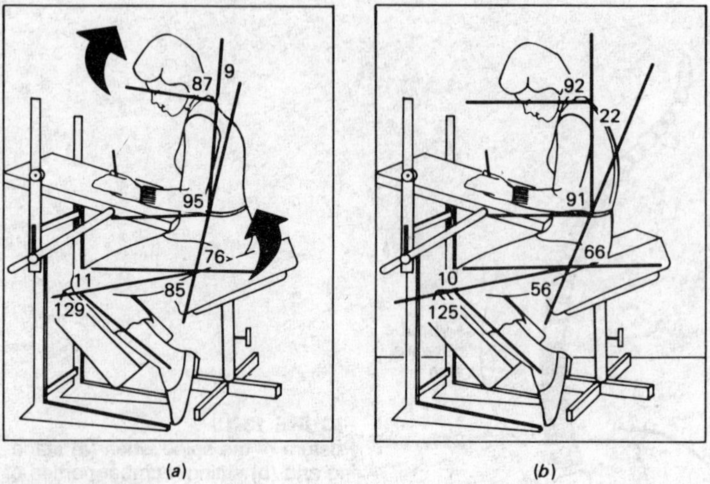

FIGURE 13-16
Two common postures observed when people sit in forward-inclined seats with a slanted work surface: (*a*) erect posture; (*b*) slumped posture. (*Source: Bridger, 1988, Figs. 4d and 5. Reprinted with permission of the Human Factors Society, Inc. All rights reserved.*)

Andersson (1987) reports on several factors that reduce disc pressure. Use of a reclined backrest has a marked effect with considerable reductions in pressure by reclining backrests from vertical (90°) to just 100 to 110°. Use of a lumbar support also reduces disc pressure as does the use of arm rests, but with a reclined backrest the effect of arm rests is negligible.

Minimize Static Loading of the Back Muscles Andersson (1987) reports that muscular activity as measured by electromyography (EMG) is similar when standing or sitting. In fact, EMG activity decreases when sitting in a forward slumped posture, even though, as discussed above, this posture produces maximum pressure on the discs. There are ways, however, to relax the muscles without sacrificing the discs. Andersson and Ortengren (1974) found a reduction in muscular activity in the back when the backrest was reclined up to 110°, beyond which little additional relaxation was found. The effects of a lumbar support on EMG activity have been mixed (Andersson, 1987).

Reduce Postural Fixity Grieco (1986) discusses the problem of *postural fixity,* that is, sitting in one position for long periods without significant postural movement. This is especially common when using a computer where the hands remain on the keyboard and the eyes are fixed on the screen. The human body is simply not made to sit in one position for long periods of time. The discs between the vertebrae depend on changes in pressure to receive nutrients and remove waste products. Discs have no blood supply; fluids are exchanged by osmotic pressure. Sitting in one posture—no matter how good it is—will result in reduced nutritional exchanges and in the long term may promote degenerative processes in the discs.

Postural fixity also promotes static loading of the back and shoulder muscles, which can result in aches and cramping. Postural fixity also causes restriction in blood flow to the legs, which can cause swelling (edema) and discomfort.

Chair design can reduce postural fixity some by allowing the user to rock in the chair and assume a variety of postures. The best defense against postural fixity, however, is to periodically stand up and work the kinks out by flexing and bending the back and legs. Dainoff and Dainoff (1986) even suggest that attached to every office chair should be a tag that says *Caution: Prolonged sitting may be hazardous to your health*.

Provide for Easy Adjustability Adjustable furniture is fundamental to good human factors design. Studies have shown that providing adjustable seats increases productivity (Springer, 1982) and reduces complaints of shoulder and back pain (Shute and Starr, 1984). The problem is that workers are usually not aware of the adjustability features available on their chairs and rarely use the ones they know about. In a survey of 2000 air traffic controllers, Kleeman and Prunier (1980) found only about 10 percent adjusted their seats during the day and more than half were not even aware of some of the adjustments that were available. The personal experience of one of your authors confirms this among newspaper employees. One feels like a hero when showing someone how to adjust their backrest angle or backrest height to achieve a more comfortable posture.

One factor that contributes to people's reluctance to adjust their furniture is that the adjustments are not always easy. Lueder (1986*a*) provides the following guidelines for increasing the ease of making adjustments:

1 Controls can be easily reached and adjusted from the standard seated work position.

2 Labels and instructions on the furniture are easy to understand.

3 Controls are easy to find and interpret.

4 Tools are not necessary.

5 Controls provide immediate feedback (for example, seats that adjust in height by rotating the seat pan delay feedback because the user must get up and down repeatedly to determine the correct position).

6 The direction of operation of controls is logical and consistent.

7 Few motions are required to use the controls.

8 Adjustments require the use of only one hand.

Even following the above guidelines is not enough. Employees must be educated on how to adjust their chairs, why it is important, the principles behind the adjustment, and the proper posture to attain. Dropping a booklet on an employee's desk is no answer (McLeod, Mandel, and Malven, 1980). Individual and small group training sessions with hands-on instruction are necessary.

Specific Design Recommendations

Our purpose is not to provide a detailed design guide for all types of chairs. We present a few basic recommendations, drawing from the general principles of seat design discussed above and the ANSI standard for chairs used at computer terminals (Human Factors Society, 1988)

Seat Height and Slope One design principle that has been generally accepted is that the seat height should be low enough to avoid excessive pressure on the underside of the thigh. Such pressure can reduce blood circulation to the lower legs (although it is recognized that postural fixity is a greater contributor to reduced circulation than pressure from too high a seat). In general, to reduce excessive pressure the seat should be lower than the distance from the floor to the underside of the thigh when the person is seated (i.e., popliteal height). It is fairly common practice to design seats for the 5th percentile of popliteal heights, which for males and females is 15.4 in and 14 in (39 cm and 36 cm), respectively. Given that heels add an inch or more, many fixed-height general-purpose chairs have seat heights around 17 or 18 in (43 or 46 cm). This design recommendation, however, is being questioned (Lueder, 1986*b*; Mandal, 1982). Low seats result in more lumbar flexion (an undesirable condition) than do high seats. This is because in a high seat the thighs tend to slope downward, thus increasing the angle between the thighs and upper torso. In fact, people tend to set adjustable seats about 1.5 to 2 in (4 to 5 cm) higher than their popliteal height taking into account shoes and footrests if used (Sauter and Arndt, 1984; DeGroot and Vellinga, 1984). The extra height is not enough to place undue pressure on the underside of the thighs. Fixed-height general-purpose seats, therefore, should be 18 to 19 in (46 to 48 cm) high, with a rounded, waterfall front seat edge and a softer layer of cushioning on the outside edge of the seat.

Where seats are adjustable in height, they should accommodate the 5th-percentile female to the 95th-percentile male. ANSI recommends a minimum range of 16 to 20.5 in (40.6 to 52 cm) based on a compressed seat [i.e., loaded with 100 lbs (45.4 kg)]. Often workstation seats cannot be adjusted high enough for long-legged people. In trying to accommodate short-legged individuals, apparently the upper range of adjustability has been sacrificed.

The tilt of the seat pan has been discussed in regard to several of the general principles presented above. There is no doubt that biomechanical advantages can accrue if forward-tilting seats are used. The advantages, however, do not accrue to everyone. Such seats appear to be most beneficial when used with a slanted work surface. Tilted seats must also be set higher than a standard seat. It takes a little time to get use to sitting in a forward-tilting seat, especially when the angle exceeds about 6° (Lueder, 1991). Adjustable seats should permit some degree of forward tilt as well as a slight backward tilt. ANSI recommends seat-pan angles of 0 to 10° backward tilt. Mandal (1982) recommends 10 to 15° of forward tilt in seats that are used with a table for reading, writing, or data entry. Striking a compromise between these recommendations, Lueder (1986*a*), suggests 5 to 15° forward tilt to 5° backward tilt.

Seat Depth and Width These features obviously depend in part on the type of seat being considered. In the design of multipurpose seats for the public, however, a couple of guidelines should be observed: (1) The depth should be set to be suitable for small persons (to provide clearance for the calf of the leg and to minimize thigh pressure), and (2) the width should be set to be suitable for large persons. On the basis of comfort ratings for chairs of various designs, Grandjean et al. (1973) recommend that for multipurpose chairs depths not exceed 16.8 in (43 cm) and that the width of the seat surface be not less than 15.7 in (40 cm), although such a seat width [or perhaps one a bit wider, say 17 in (43 cm)] would do the trick for individual seats. If people are to be lined up in a row or seats are to be adjacent to each other, elbow-to-elbow breadth values need to be taken into account, with even 95th percentile values around 19 and 20 in (48 and 51 cm) producing a moderate sardine effect. (And for bundled-up football spectators with thermos, the sardine effect is amplified further.) In any event, these are approximate minimum values for chairs with arms on them (for well-fed friends you should have even wider lounge chairs). ANSI recommends the following for chairs used at computer workstations:

Seat depth: 15 to 17 in (38 to 43 cm)
Seat width: 18.2 in (45 cm)

Contouring and Cushioning The human body has evolved in such a way that, when the body is seated, the primary weight of the body can best be supported by the ischial tuberosities (sometimes called the *sitting bones*) of the buttocks. Although the primary weight should be supported by these bones, Rebiffé (1969) expresses the opinion that seats should provide for a distribution of the weight over the entire buttocks, with the pressure decreasing from these bones to the periphery of the buttocks. Such a weight distribution can be aided by the use of contoured seats. Although the contouring of seats can help to produce such a distribution of weight over the buttocks, it can have the disadvantage of restricting movement and promoting postural fixity. The specific amount of contouring involves a trade-off between supporting postures, distributing pressures, and providing freedom of movement.

The density and thickness of the seat-pan cushion affects pressure distribution, as does contouring. The soft, "cushy" chair that feels so comfortable when you first sit in it probably won't after a period of time. One sinks into the seat, posture is restricted, blood circulation is reduced, skin temperature increases, and pain may result. It is generally recommended that seat cushion thickness range from 1.5 to 2 in (4 to 5 cm) and that the covering "give" and "breathe" (Lueder, 1986*a*).

Seat Back Parameters The following are a few recommendations for seat backs on chairs used with computer workstations:

Seat back angle: ANSI recommends a minimum range of 90° to 105° with the seat pan, but considering the general principles discussed above, up to 120° would be preferable.

Seat back width: ANSI recommends at least 12 in (30.5 cm) in the lumbar region.

Seat back height: Lueder (1986) recommends a minimum of 19.5 in (50 cm). As a backrest reclines, the lumbar support moves upward relative to the lumbar spine (Andersson, 1987). Ideally then, the seat back should lower as the backrest reclines so that the lumbar support remains in the same position relative to the spine. (In fact, there are a few chairs on the market that actually do this.)

Lumbar support: ANSI recommends a support that is 6 to 9 in (15.2 to 22.9 cm) in height and 12 in (30.5 cm) wide positioned 6 to 10 in (15.2 to 25.4 cm) above the seat reference point. The lumbar support should protrude forward about 2 in (5 cm) from the back of the seat.

Seat Designs for Specific Purposes

Designing seats for specific purposes combines elements of science and art. We have stressed the science (anthropometry, biomechanics, and comfort), but the art (appearance and styling) cannot be ignored as important criteria for judging the overall value of a design. The science-art distinction was made obvious to one of your authors. He was developing generic human factors design guidelines for Walt Disney Imagineering (WDI) for use in the design of rides and attractions at Disneyland. A part of the guidelines dealt with ride operator workstations, including seat design. After reviewing the seat design specifications, the design engineers at WDI stated that the information would be very useful to them but asked, "How does one make it look like a mushroom?" You see, ride operator workstations at Disneyland are often thematically coordinated to the attraction. For example, the operator of the Alice in Wonderland ride might sit on a seat that looks like a mushroom. Human factors guidelines can only go so far; however, the comfort, biomechanics, and utility of a seat (science) should not be compromised just for appearance's sake.

We present a few examples of seats for specific purposes to illustrate the influence of purpose on seat design. (Mushroom seats will not be discussed.)

Computer Workstation Chairs The tremendous surge in the use of computers has triggered considerable research and concern regarding the design of chairs for use at computer workstations. We have reviewed basic design recommendations for such seats in the previous section. Our purpose here is to present a few new ideas that are showing up in the marketplace.

As mentioned above, people do not usually bother to adjust their seats, and with the many possible adjustments that could be incorporated into a chair, the average user could become overwhelmed by the choices (see Figure 13-17). Some manufacturers have developed so-called *dynamic seat designs,* in which fewer adjustments by the user are required. Lueder (1991) lists three approaches to providing dynamic design. The first involves designs in which the seat motions are synchronized automatically. For example, the seat-pan angle

FIGURE 13-17
With the many possible adjustments that could be incorporated into a chair, the average user could become overwhelmed by the choices. (*Source: Duffy, copyright © 1988 Universal Press Syndicate. Reprinted with permission. All rights reserved.*)

might automatically reposition when the backrest angle is adjusted. The second approach involves a "floating" backrest that follows and supports the user as he or she moves in the chair. As the user leans forward or back, the backrest moves with the person. The third approach uses new flexible seat materials that bend with the user. The backrest and pan might be a single molded unit, but each will flex, twist, and bend somewhat independently in response to user movements. The efficiency of these chairs has not been documented (Dainoff and Dainoff, 1986), but they seem to be doing well in the marketplace (but then again, so did Hula Hoops).

One interesting design (Cyborg by Rudd International Inc.) attempts to reduce postural fixity by making the user move in the seat. Through the use of a hydraulic cylinder in the seat pan, the seat pan slowly sinks backward and down as a person continues to sit in the seat. The seat moves through an arc of only about 4° in about 1.5 min. When weight is lifted from the seat pan, even slightly, as for example by reaching across the desk, the cylinder recycles, the seat raises back up, and the process begins again. The mechanism and range of seat positions are shown in Figure 13-18. This movement, according to the

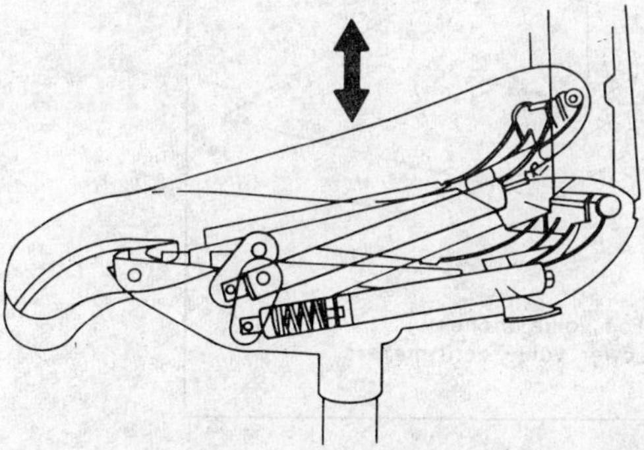

FIGURE 13-18
Cut-away view of the seat pan of a Cyborg chair showing the mechanism and range of dynamic change in the seat-pan angle from shifting one's body weight while sitting. The manufacturer claims this reduces postural fixity by forcing the user to change posture in response to the changing seat-pan angle. (*Source: Rudd International, Inc., Washington, DC.*)

manufacturer, "forces" the user to change position to compensate for the continuously changing seat-pan angle, and hence reduces postural fixity. Again, the benefits of this approach has not been scientifically validated. (But it sure beats sitting on tacks to reduce postural fixity.)

Multipurpose Chairs In a study by Grandjean et al. (1973), some 25 men and 25 women were asked to rate the comfort of 11 parts of the body when testing 12 different designs of multipurpose chairs. In addition, each subject compared every chair with every other one and rated the overall comfort by the paired-comparison method. The contours of the two most preferred chairs are shown in Figure 13-19, along with design recommendations made on the basis of the analysis of the results of all the data. The recommendations include foam rubber of 2 to 4 cm (about 0.75 to 1.5 in) on the entire seat.

Chairs for Resting and Reading The desirable features of chairs for relaxation and reading are, of course, different from those of chairs used for more

FIGURE 13-19
Contours of the two multipurpose chairs (of 12) judged to be most comfortable by 50 subjects and the design features recommended for multipurpose chairs based on the study. (*Source: Grandjean et al., 1973, Figs. 2, 6, 13.*)

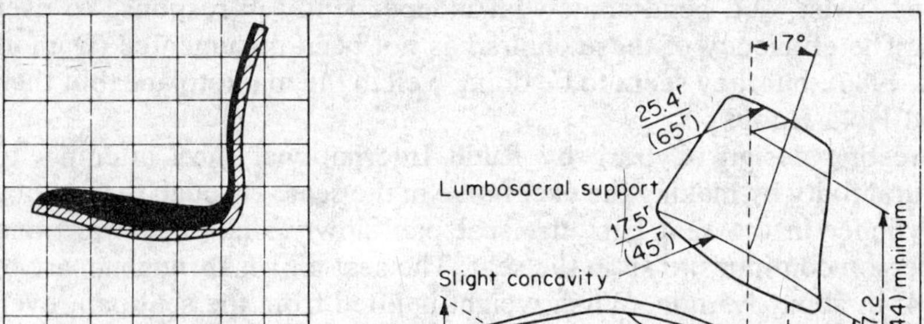

TABLE 13-5
RANGES OF DIMENSIONS OF SEATS FOR READING AND RESTING PREFERRED BY SUBJECTS

Dimension	Reading	Resting
Seat inclination, deg	23–24	25–26
Backrest inclination, deg	101–104	105–108
Seat height, cm	39–40	37–38
Seat height, in	15.3–15.7	14.6–15.0

Source: Grandjean, Boni, and Krestzschmer, 1969.

active use. A study was carried out by Grandjean, Boni, and Krestzschmer (1969) in which they used a "seating machine" for eliciting judgments of subjects about the comfort of various seat designs. The seating machine consisted of features that could be adjusted to virtually any profile. Without summarizing all the results, they found that the angles and dimensions shown in Table 13-5 were preferred by more subjects than others for the two purposes of reading and resting. Profiles for two such chairs are shown in Figure 13-20. Note in these the distinct angles of the backs and seats to provide full back support with particular support for the lower (lumbar) sections of the spine.

VIDEO DISPLAY TERMINAL (VDT) WORKSTATIONS

We close this chapter with a discussion of VDT workstations. This will give us an opportunity to present some guidelines that were not covered previously. More importantly, however, we discuss some difficulties in applying current VDT workstation guidelines and standards.

Designing a VDT workstation requires more than just putting a computer on a standard office desk and providing an adjustable chair. Manufacturers have recognized the special needs of computer users and have designed tables that

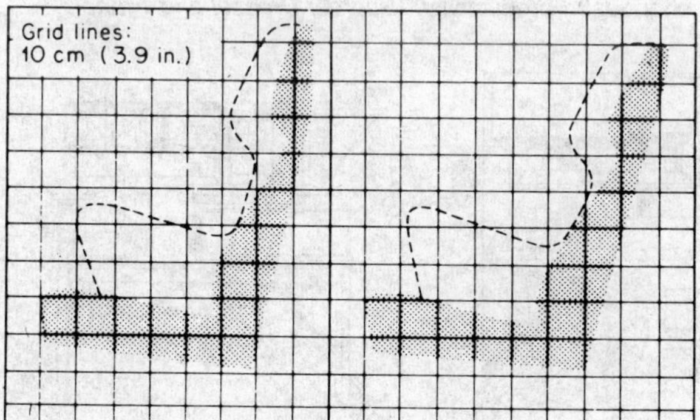

FIGURE 13-20
Profiles of seats proposed for reading and resting. The dotted lines correspond to the armrests and a possible outer contour. The shaded area shows the surface of the seat, including upholstering to be 6 cm (2.5 in) thick. (*Source: Grandjean, Boni, and Krestzschmer, 1969, Fig. 2, p. 310.*)

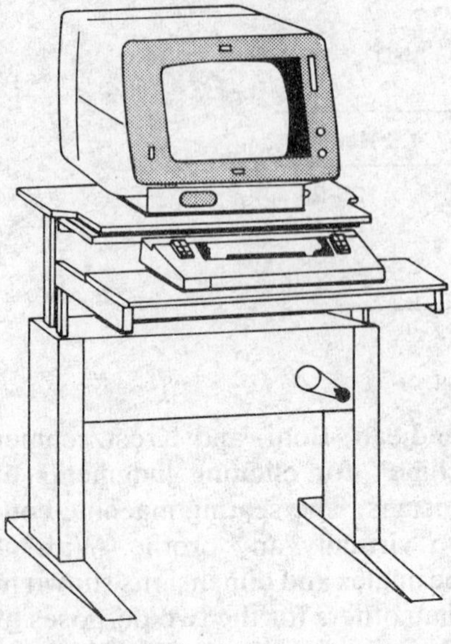

FIGURE 13-21
An example of an adjustable VDT table in which the heights of the keyboard and terminal can be adjusted independently. The terminal surface can be tilted, and the keyboard surface can also be pulled forward and tilted. (*Source: Steelcase, Inc., Grand Rapids, MI.*)

are height-adjustable; some even have separately adjustable keyboard and terminal surfaces. An example of such a table is shown in Figure 13-21. In addition to table and chair, a VDT workstation might include a footrest, wrist rest, task light, and document holder.

Workstation design must also consider storage, desk surface area, and privacy needs. The importance of these and other workstation design factors was demonstrated in a study conducted by Dressel and Francis (1987). They compared the three workstations shown in Figure 13-22 in terms of productivity and worker satisfaction. Three groups of workers in a government installa-

FIGURE 13-22
Three workstations evaluated by Dressel and Francis. The control (C) workstation was government-issue steel furniture. The E1 workstation was entirely new furniture and included an "ergonomic" chair. The E2 workstation just added additional steel furniture to the C workstation. The results compare before-and-after changes in productivity. (*Source: Dressel and Francis, 1987, Fig. 1.*)

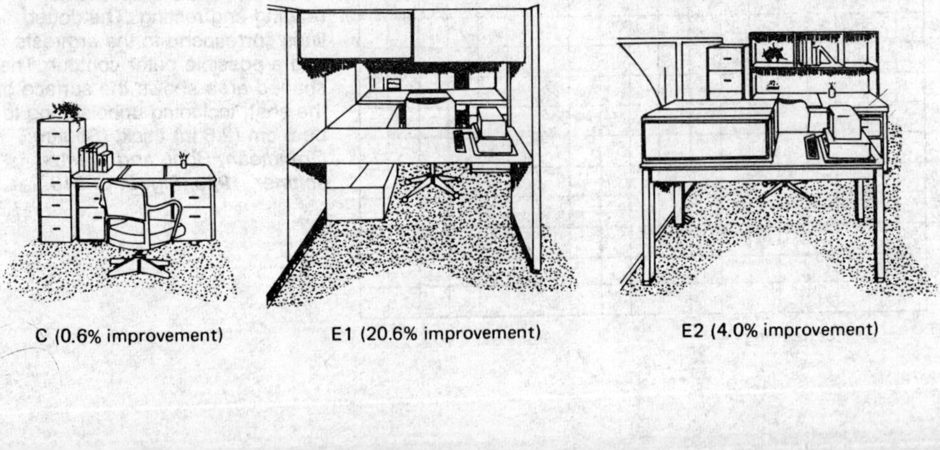

C (0.6% improvement) E1 (20.6% improvement) E2 (4.0% improvement)

tion were studied for 10 months prior to and 11 months after the installation of the two experimental workstations (E1 and E2). The control group (C) remained in the original workstation throughout the entire 21 months. The E1 workstation included all new furniture and an "ergonomic" chair. The E2 workstation simply added more government-issue steel furniture to the control workstation. The E1 furniture resulted in a significant improvement in productivity and satisfaction. Dressel and Francis calculated that the savings in labor from the increased productivity would pay for the E1 furniture in about one year.

The Use of Guidelines and Standards

There are many guidelines and standards for designing VDT workstations. In the United States there is the ANSI standard (Human Factors Society, 1988), but there are also standards in Germany, Sweden, and other countries (Helander and Rupp, 1984; Abernethy, 1988). In addition, there are guidelines developed by companies in the computer industry such as AT&T (1983) or IBM (1984), and finally, just about every book on the subject offers a set of guidelines (e.g., Grandjean, 1987; Lueder, 1986*a*).

Dainoff and Dainoff (1986) discuss some of the shortcomings of standards and guidelines. First, there are major differences in coverage, structure, and degree of detail between the various sources. Standards are based on consensus agreement among scientists, manufacturers, employers, and labor representatives. As such, they can become watered down in the process of satisfying such diverse interests.

Second, there is the problem of obsolescence. New technologies (e.g., flat-panel displays) may not meet standards based on older technologies (e.g., cathode-ray tubes), or the standard may inhibit development of new technologies because they collide with the established standard.

Third, and most disturbing, is that the specific contents of many standards and guidelines simply are not in agreement and the extent of the disagreement can be quite large. Helander and Rupp (1984), for example, reported that recommendations on the acceptable range of height for keyboard support surfaces contained in various sources often did not even overlap.

Finally, many recommendations and standards are based on research investigating a single factor such as keyboard angle and do not adequately take into account the interaction of other factors. Acceptable keyboard angles probably are a function of the keyboard support height and seat height. Applying each recommendation in isolation without considering interactions may result in unnecessarily large ranges of adjustability for specific features.

Use Preferences

We can add to the shortcoming of standards discussed above. The recommendations often do not correspond to how users actually set up the components of their workstation. There have been several laboratory studies of user preferences (e.g., Cushman, 1984; Rubin and Marshall, 1982). The data, however, are somewhat suspect in that tasks of very brief duration were performed by the subjects. Far better are studies conducted in actual office environments over long periods of VDT use (e.g., Grandjean, Hünting, and Piderman, 1983; Cornell and Kokot, 1988). The results of such studies are clear. When VDT users are given fully adjustable furniture, they do not follow the recommendations. Generally, they set everything higher than what is set forth in the standards. Often the range of adjustment recommended in the standard would not be adequate to accommodate the range of user preferences.

Grandjean, Hünting, and Piderman (1983) present the typical VDT user posture as shown in Figure 13-23. VDT users tend to lean back in their chair and extend their arms up and forward. The keyboard often is set higher and at a steeper angle than is recommended in most standards. Leaning back in the seat lowers the eye height and users often set the screen higher than recommended. Leaning back also increases the distance from the eyes to the screen and users sit further from the screen than would be recommended given the size of the characters on the screen. A question still being debated is whether design recommendations should be based on user preferences or on anthropometric and biomechanical considerations. Giving the nod to user preference assumes that users know what they are doing. There is some evidence that they do (Grandjean, 1987), but the numbers of users complaining of skeletal-muscular aches and pains may be evidence that they do not.

Another interesting finding is that there is little correlation between anthropometric body measurements and the preferred workstation dimensions (Grandjean, Hünting, and Piderman, 1983; Cornell and Kokot, 1988). Appar-

FIGURE 13-23
Typical VDT user posture: (*a*) shows relevant body angles (means) assumed by VDT users; (*b*) artist rendition of typical posture—leaning back in the chair. (*Source: Grandjean, 1987, Figs. 83 and 85.*)

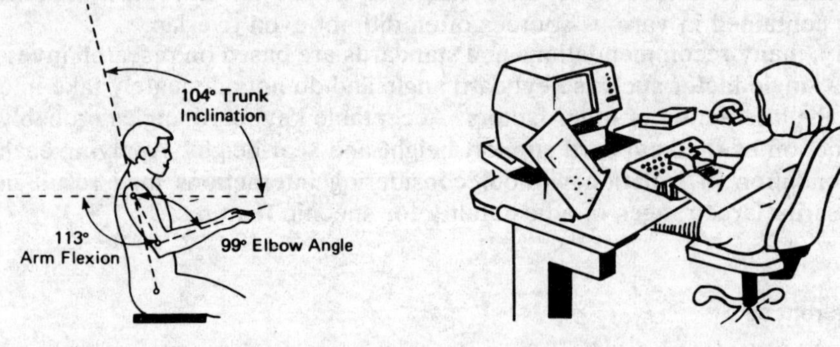

TABLE 13-6
RECOMMENDATIONS FOR VDT WORKSTATION ADJUSTMENT RANGES FOR KEYBOARD AND TERMINAL

Dimension	Human Factors Society 1988	Lueder 1986*a*
Keyboard height (floor to home row)	23–28 in (58.5–71 cm)	24–32 in (61–81.5 cm)
Keyboard angle	0–25°	0–25°
Screen position		
Angle below horizon	0–60°	
Support surface height		24.4–35 in (62–89 cm)
Screen angle to vertical		
Ideal		+/− 20°
Minimum		+/− 7°

ently, users have not learned about the relationship between the dimensions of workstations and such things as popliteal height, elbow height, and sitting eye height.

Recommendations

We see no need to generate yet another set of recommendations for VDT workstation design. We therefore present in Table 13-6 a summary of the ANSI standard (Human Factors Society, 1988) and another recent guideline (Lueder, 1986a). Table 13-6 concentrates on the height and angle of the keyboard and terminal and assumes fully adjustable equipment.

One thing that comes across loud and clear is the need for adjustable furniture for people who use VDTs on a more or less continuous basis. We should make the distinction between workstations that are easily adjustable by the user and those that require tools and perhaps the removal of a few screws and bolts to affect adjustments. The latter would probably be acceptable where only one person used a particular workstation. Were several people share a workstation or workers on different shifts occupy the same workstation, then serious consideration should be given to providing the more easily adjustable type of furniture.

DISCUSSION

The design of workplaces includes the work envelope, work surfaces (such as desks, tables, etc.), and seats (if used) as well as the design and location of equipment used. (The next chapter deals with some aspects of the specific arrangement of equipment components.) The design of workplaces clearly is

FIGURE 13-24
A couple of human factors problems in a nuclear power plant. In the photograph on the left, the control console is arranged so that certain controls could be accidentally activated by the knee or the hand used to brace the operator when reaching for the farthest controls. In the photograph on the right, certain of the displays are at a height that makes reading or lamp replacement a problem. Special stepladders are required. (*Copyright 1979, Electric Power Research Institute, EPRI Report NP-1118, "Human Factors Methods for Nuclear Control Room Design." Reprinted with permission.*)

rooted in the use of anthropometric data. Examples of work situations that were not well designed are easy to find. A couple of extreme examples are shown in Figure 13-24. (You may have run across some poorly designed workplaces, too.)

REFERENCES

Abernethy, C. (1988). HCI standards: Origins, organizations and comment. In D. Oborne (ed.), *International Review of Ergonomics*, 2, 31–54.

Andersson, G. (1987). Biomechanical aspects of sitting: An application to VDT terminals. *Behaviour and Information Technology*, 6(3), 257–269.

Andersson, G., Murphy, R., Ortengren, R., and Nachemson, A. (1979). The influence of backrest inclination and lumbar support on the lumbar lordosis in sitting. *Spine*, 4, 52–58.

Andersson, G., and Ortengren, R. (1974). Myoelectric back muscle activity during sitting. *Scandinavian Journal of Rehabilitation Medicine*, Supplement, 3, 73–90.

AT&T Bell Laboratories (1983). *Video display terminals*. Short Hills, NJ.

Ayoub, M. M. (1973). Work place design and posture. *Human Factors*, 15(3), 265–268.

Ayoub, M., Bethea, N., Bobo, M., Burford, C., Caddel, D., Intaranont, K., Morrissey, S., and Salan, J. (1982). *Mining in low coal*, vol. 2: *Anthropometry*. (OFR 162(2)-83). Pittsburgh, PA: Bureau of Mines.

Barnes, R. M. (1963). *Motion and time study* (5th ed.). New York: Wiley.

Bendix, T. (1986). Seated trunk posture at various seat inclinations, seat heights, and table heights. *Human Factors*, 26, 695–703.

Bex, F. H. A. (1971). Desk heights. *Applied Ergonomics*, 2(3), 138–140.

Bittner, A. C., Jr. (1974). Reduction in user population as the result of imposed anthropometric limits: Monte Carlo estimation, (TP-74-6). Point Mugu, CA: Naval Missile Center.

Bridger, R. (1988). Postural adaptations to a sloping chair and work surface. *Human Factors*, 30(2), 237–247.

Bullock, M. I. (1974). The determination of functional arm reach boundaries for operation of manual controls. *Ergonomics*, 17(3), 375–388.

Cameron, N. (1979). The growth of London school children 1904–1966: An analysis of secular trend and intracounty variation. *Annals of Human Biology*, 6, 505–525.

Cornell, P., and Kokot, D. (1988). Naturalistic observation of adjustable VDT stand usage. *Proceedings of the Human Factors Society 32d Annual Meeting*. Santa Monica, CA: Human Factors Society, pp. 496–500.

Cushman, W. (1984). Data entry performance and operator preferences for various keyboard heights. In E. Grandjean (ed.), *Ergonomics and health in modern offices*. London: Taylor & Francis.

Dainoff, M., and Dainoff, M. (1986). *People and productivity: A manager's guide to ergonomics in the electronic office*. Toronto, Ont., Canada: Holt, Rinehart & Winston.

Das, B., and Grady, R. M. (1983*a*). Industrial workplace layout design: An application of engineering anthropometry. *Ergonomics*, 26(5), 433–447.

Das, B., and Grady, R. M. (1983*b*). The normal working area in the horizontal plane: A comprehensive analysis between Farley's and Squires' concepts. *Ergonomics*, 26(5), 449–459.

DeGroot, J., and Vellinga, R. (1984). Practical usage of adjustable features in terminal furniture. *Proceedings of the 1984 International Conference on Occupational Ergonomics*. Rexdale, Ont., Canada: The Human Factors Society of Canada, pp. 308–312.

Dempster, W. T. (1955, July). *Space requirements of the seated operator* (Tech. Rept. 55–159). Wright-Patterson Air Force Base, OH:U.S. Air Force, WADC.

Dressel, D., and Francis, J. (1987). Office productivity: Contributions of the workstation. *Behaviour and Information Technology*, 6(3), 279–284.

Eastman, M. C., and Kamon, E. (1976). Posture and subjective evaluation at flat and slanted desks. *Human Factors*, 18(1), 15–26.

Electrical Power Research Institute (1979). *Human Factors Methods for Nuclear Control Room Design* (NP-1118). Palo Alto, CA: Electrical Power Research Institute.

Farley, R. R. (1955). Some aspects of methods and motion study as used in development work. *General Motors Engineering Journal*, 2, 20–25.

Garg, A., Bakken, G. M., and Saxena, U. (1982). Effect of seat belts and harnesses on functional arm reach. *Human Factors*, 24(3), 367–372.

Grandjean, E. (1987). *Ergonomics in computerized offices*. London: Taylor & Francis.

Grandjean, E. (1988). *Fitting the task to the man* (4th ed.). London: Taylor & Francis.

Grandjean, E., Boni, A., and Krestzschmer, H. (1969). The development of a rest chair profile for healthy and notalgic people. *Ergonomics,* 12(2), 307–315.

Grandjean, E., Hünting, W., Wotzka, G., and Shärer, R. (1973). An ergonomic investigation of multipurpose-chairs. *Human Factors,* 15(3), 247–255.

Grandjean, E., Hünting, W., and Piderman, M. (1983) VDT workstation design: Preferred settings and their effects. *Human Factors,* 25(2), 161–175.

Grieco, A. (1986). Sitting posture: An old problem and a new one. *Ergonomics,* 29(3), 345–362.

Helander, M., and Rupp, B. (1984). An overview of standards and guidelines for visual display terminals. *Applied Ergonomics,* 15(3), 185–195.

Human Factors Society (1988). *American national standard for human factors engineering of visual display terminal workstations* (ANSI/HFS 100-1988). Santa Monica, CA: Human Factors Society.

International Business Machines (1984). *Human factors of workstations with visual displays.* San Jose, CA: IBM.

Keegan, J. J. (1953). Alterations of the lumbar curve. *Journal of Bone and Joint Surgery,* 35, 589–603.

Kleeman, W., and Prunier, T. (1980). Evaluation of chairs used by air traffic controllers of the U.S. Federal Aviation Administration. In R. Easterby, K. Kroemer, and D. Chaffin (eds.), *NATO symposium on anthropometry and biomechanics: Theory and application.* London: Plenum Press.

Kroemer, K. (1983). Engineering anthropometry: Work space and equipment to fit the user. In D. Oborne and M. Gruneberg, *The physical environment at work.* London: Wiley.

Kroemer, K. (1989). Engineering anthropometry. *Ergonomics,* 32(7), 767–784.

Lueder, R. (1986*a*). Work station design. In R. Lueder (ed.), *The ergonomics payoff: Designing the electronic office.* Toronto, Ont., Canada: Holt, Rinehart & Winston.

Lueder, R. (1986*b*). Seat height revisited. *Human Factors Society Bulletin,* 29(11), 4–5.

Lueder, R. (1991). Seating a new worker. In J. Sweere (ed.), *Chiropractic family practice.* Rockville, MD: Aspen.

Mandal, A. C. (1982). The correct height of school furniture. *Human Factors,* 24(3), 257–269.

Mandal, A. (1985). *The seated man.* Denmark: Dafnia Publications.

McLeod, W., Mandel, D., and Malven, F. (1980). The effects of seating on human tasks and perceptions. In H. Poydar (ed.), *Proceedings of the Symposium of Human Factors and Industrial Design in Consumer Products.* Medford, MA: Department of Engineering Design, Tufts University, pp. 117–126.

Meyer, R. P. (1979). Articulated anthropometric modes. *C P News* (Newsletter of the Consumer Products Technical Group of the Human Factors Society), 4(2). Santa Monica, CA: Human Factors Society.

Nachemson, A., and Elfstrom, G. (1970). Intravital dynamic pressure measurements in lumbar discs. *Scandinavian Journal of Rehabilitation Medicine,* Supp., 1.

National Aeronautics and Space Administration (NASA). (1978). *Anthropometric source book,* vol. 1: *Anthropometry for designers;* vol. 2: *A handbook of anthropometric data;* vol. 3: *Annotated bibliography* (NASA Ref. Pub. 1024). Houston, TX: NASA.

North, K. (1980). Ergonomics methodology—An obstacle or promoter for the implementation of ergonomics in industrial practice? *Ergonomics,* 23(8), 781–795.

Pheasant, S. (1986). *Bodyspace: Anthropometry, ergonomics and design.* London: Taylor & Francis.

Rebiffé, P. R. (1969). Le siège du conducteur: Son adaptation aux exigences fonctionnelles et anthropométriques. *Ergonomics,* 12(2), 246–261.

Rigby, L. V., Cooper, J. I., and Spickard, W. A. (1961, October). *Guide to integrated system design for maintainability* (Tech. Rept. 61–424). U.S. Air Force, ASD.

Robinette, K., and McConville, J. (1981). *An alternative to percentile models.* SAE Tech. Paper Series No. 810217. Warrendale, PA: Society of Automotive Engineers.

Roche, A. (1979). Secular trends in stature, weight and maturation. *Monographs of the Society for Research in Child Development,* Serial no. 179; 44(3-4), 3–27.

Roth, J. T., Ayoub, M. M., and Halcomb, C. G. (1977). Seating, console and workplace design: Seated operator reach profiles. *Proceedings of the Human Factors Society 21st Annual Meeting.* Santa Monica, CA: Human Factors Society, pp. 83–87.

Rozier, C. K. (1977). Three-dimensional work space for the amputee. *Human Factors,* 19(6), 525–533.

Rubin, T., and Marshall, C. (1982). Adjustable VDT workstations: Can naive users achieve a human factors solution? *International Conference on Man-Machine Systems,* IEEE Conference Publication No. 121. Piscatanay, NJ: IEEE.

Sanders, M. S. (1977). *Anthropometric survey of truck and bus drivers: Anthropometry, control reach and control force.* Westlake Village, CA: Canyon Research Group.

Sanders, M. S. (1980). Sleep envelopes and sleeper berth requirements. *Human Factors,* 22(3), 313–317.

Sauter, S., and Arndt, R. (1984). Ergonomics in the automated office; gaps in knowledge and practice. In G. Salvendy (ed.), *Human-computer interaction.* Amsterdam: Elsevier.

Shute, S. J., and Starr, S. J. (1984). Effects of adjustable furniture on VDT users. *Human Factors,* 26(2), 157–170.

Springer, T. (1982). *Visual display terminals: A comparative evaluation of alternatives.* Bloomington, IL: State Farm Mutual Automobile Insurance Co.

Squires, P. C. (1956). *The shape of the normal work area* (Rept. 275). New London, CT: Navy Department, Bureau of Medicine and Surgery, Medical Research Laboratory.

Stoudt, H. W. (1981). The anthropometry of the elderly. *Human Factors,* 23(1), 29–37.

Tanner, J. (1978). *Foetus into man.* London: Open Books.

Thompson, D. (1989). Reach distance and safety standards. *Ergonomics,* 32(9), 1061–1076.

Tichauer, E. R. (1978). *The biomechanical basis of ergonomics.* New York: Wiley.

Ward, J. S., and Kirk, N. S. (1970). The relation between some anthropometric dimensions and preferred working surface heights in the kitchen. *Ergonomics,* 13(6), 783–797.

CHAPTER **14**

ARRANGEMENT OF COMPONENTS WITHIN A PHYSICAL SPACE

An important facet of design is arranging components within some physical space. We arrange controls and displays on a control panel; a telephone, pen holder, and books on a desk; desks and equipment in an office; offices, classrooms, and restrooms in a building; and buildings in an industrial park. All of these examples, each at a different level of design, involve arranging components within some physical space. We use the term *component* to refer to any physical entity that must be placed within a defined space. The component is the lowest level entity being arranged in the space. When arranging furniture in an office a desk is a component, but the stapler, pen holder, and books on the desk are not.

In this chapter we discuss some basic principles and methods for arranging components and then concentrate on information relevant to arranging controls and displays within a work space or on a control panel.

PRINCIPLES OF ARRANGING COMPONENTS

Ideally, we would like to place each component in an optimum location for serving its purpose. This optimum would be predicated on human capabilities and characteristics, including sensory capabilities, and anthropometric and biomechanical characteristics. The optimum location would facilitate performance of the activities carried out in the space. Unfortunately, it is usually not possible to place each component in its optimum location. One can only fit so many displays in the optimum viewing area; only so many offices can be close to the restrooms. Placing a control in the optimum location for fast response may separate it from the display to which it is related. To bring order to such potential chaos requires setting priorities and making trade-offs. These priorities and trade-offs, however, are not chiseled in stone like the Ten Commandments, but rather, must be determined by the design team. Factors such as those discussed later usually play a role in making the determinations.

Before we touch on a few methods that are used to figure out what should go where, let us set down a few general guidelines (in addition to the idea of the optimum location mentioned above) that may be helpful. Depending on the circumstance, these guidelines can be concerned with one or both of two separate but interrelated phases: that concerned with the *general location* of components (either individual components usually used by themselves or groups of related components) and that concerned with the *specific arrangement* of components within their "general" location (that is, the arrangement of components within a group of related components in that general location).

Importance Principle

This principle states that important components be placed in convenient locations. Importance refers to the degree to which the component is vital to the achievement of the objectives of the system. The determination of importance usually is a matter of judgment made by people who are experts in the operation of the system.

Frequency-of-Use Principle

This principle states that frequently used components be placed in convenient locations. For example, the activation control of a punch press should be conveniently located because it is used very frequently, or a copying machine should be near a typist.

Functional Principle

The *functional principle* of arrangement provides for the grouping of components according to their function, such as the grouping of displays, controls, or machines that are functionally related in the operation of the system. Thus, temperature indicators and temperature controls might well be grouped, and electric power distribution instruments and controls usually should be in the same general location.

Sequence-of-Use Principle

In the use of certain items, sequences or patterns of relationship frequently occur in the operation of equipment or in performing some service or task. In applying this principle, the items would be so arranged as to take advantage of such patterns.

Figure 14-1*a*, for example, shows an original control panel from an automobile plant. When manually operated, the sequence of actions is shown by the numbers (1 to 16) and arrows. The panel is not entirely arranged according to the sequence-of-use principle. Figure 14-1*b* shows the panel redesigned making greater use of the principle. The redesign incorporates a more orderly and logical sequence of actions. As an aside, notice that step 3 requires raising the carriage (a part of the machine) and step 16 requires lowering the carriage. In the original design (Figure 14-1*a*), the two controls were far apart. In the redesign, the controls are placed together; this is an example of the application of the functional principle discussed above.

Discussion

In putting together the various components of a system, no single guideline can, or should, be applied consistently across all situations. But, in a very general way, and in addition to the optimum premise, the notions of importance and frequency probably are particularly applicable to the more basic phase of locating components in a general area in the work space; in turn, the sequence-of-use and functional principles tend to apply more to the arrangement of components within a general area.

FIGURE 14-1
Control panel in an automobile plant: (*a*) original design; (*b*) redesign using sequence-of-use principle. (*Source: Kochhar and Barash, 1987, Figs. 7 and 8. By permission of Butterworth-Heinemann Ltd.*)

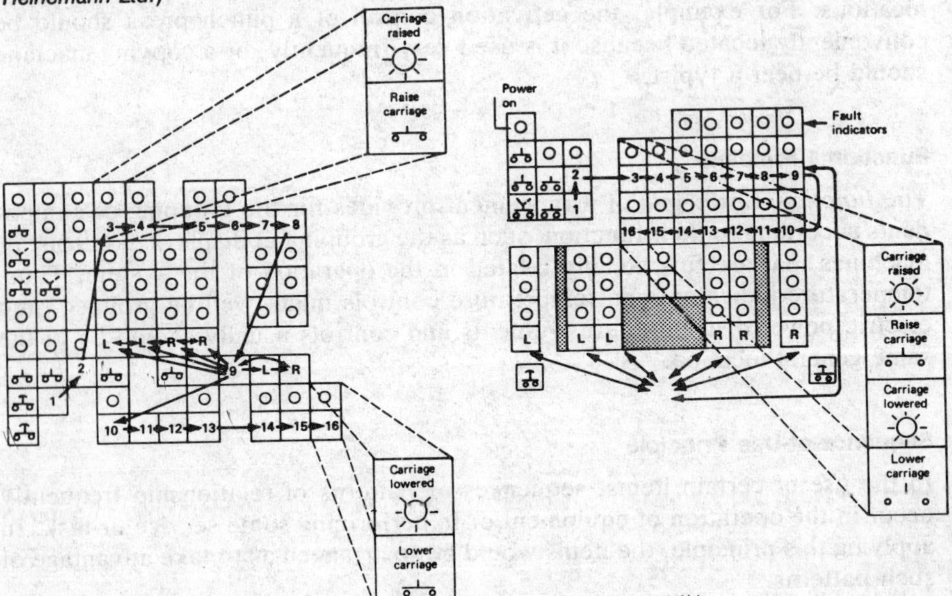

The application of these various principles of arrangement of components generally has had to be predicated on rational, judgmental considerations, since there has been little empirical evidence available regarding the evaluation of these principles. However, some data from at least one study cast a bit of light on this matter. The study in question, by Fowler et al. (1968), consisted of the evaluation of various control panel layouts in which the controls and displays had been arranged following each of the four principles described above. The panels included 126 standard military controls and displays. The arrangement of these following the four principles need not be described; but it should be added that, for each principle, three control panels were developed, varying in terms of three "levels" of application of the principle (based on a scoring scheme), these being high, medium, and low. The 200 male college student subjects used the various arrangements in a simulated task. Their performance was measured in terms of time and errors, the results for the time criterion being shown in Figure 14-2. This figure (and corresponding data regarding errors) showed a clear superiority for the sequence-of-use principle. Even the arrangements that were based on the low and medium application levels of this principle were better than, or equal to, the high levels for the other principles.

Although the sequence-of-use principle came out on top in this study, this principle obviously could be applied only in circumstances (as in this study) in which the operational requirements actually do involve the use of the components in question in rather consistent sequences. Where this is the case, this principle certainly should be followed. Where it is not the case, however, consideration should be given to the functional principle for arranging components.

METHODOLOGIES FOR ARRANGING COMPONENTS

Arranging components in a work space, be they tables and control panels in a room or controls and displays on a console, requires the availability of relevant data and the use of certain methods in applying the data.

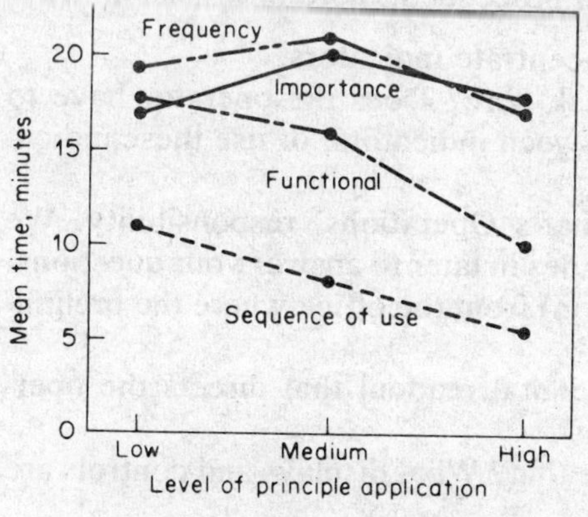

FIGURE 14-2
Time required to carry out a standard simulated task in the use of controls and displays arranged on the basis of four principles. In the case of each principle, three control panels were used, these varying in terms of the rated "level" or degree to which the principle has been applied in the panel design. (*Source: Adapted from Fowler, Williams, Fowler, and Young, 1968.*)

Types of Data for Use in Arranging Components

The types of data that are relevant for use in arranging components generally fall into the following broad classes:

- *Basic data about human beings*. Anthropometric and biomechanical data are especially relevant, but other types of data may also be useful, such as data on sensory, cognitive, and psychomotor skills. Such data generally come from research undertakings and are published in various source books.
- *Task analysis data*. These are data about the work activities of people who are (or would be) involved in the specific system or work situation in question. (For our purposes we refer to the work activities as *tasks* regardless of the level of specificity of the activity.) See Chapter 22 for more discussion of task analysis.
- *Environmental data*. This category covers any relevant environmental features of the situation, such as illumination, noise, vibration, motion, heat, traffic and congestion, etc. See Part 5 for a discussion of these topics.

The task data must be developed by the use of some systematic task analysis procedure. Certain methods of developing, organizing, and presenting task data are central to the processes of arranging components. Some of these methods are discussed and illustrated below.

How Not to Do It

Although it may appear intuitively obvious that a thorough understanding of the tasks to be performed is absolutely essential to arrange components in a work space, it is not always followed in practice. A consultant (one of your authors) had the occasion to assist in the layout of displays on a 32 × 8 ft (9.7 × 2.4 m) nuclear power plant console that consisted of four interrelated subconsoles. The consultant met with two engineers to discuss the console. A scale drawing of a preliminary design had been made, but no mock-up was available, nor was there any intention of having one made. The consultant asked the two engineers to "walk him through" the console so he could become familiar with it. The conversation proceeded, more or less as follows:

Engineers: These are the six *XYZ* concentrate indicators.

Consultant: What is the operator's task here? Does the operator have to read a specific value, compare values between indicators, or use these indicators while manipulating controls?

Engineers: We really don't know—that's Operations' responsibility. We will see that someone from Operations comes in later to answer your questions.

[The Operations office was 25 mi (40 km) from the office where the preliminary design was developed.]

Engineers: Let's continue. This is a digital readout that directs the operator's sequence of actions.

Consultant: How does the operator use that? What displays and controls are referred to by the readout? How much time does the operator have to respond to the display?

Engineers: We really don't know—that is Operations' responsibility.

Consultant (now becoming a little irritated): Exactly what did the two of you have to do with this console anyway?

Engineers: We designed it! Each of us started at a different end, and we laid out the displays going across. We then compared the layouts and selected the parts from each we liked the best.

Consultant: I think this would be a good time for a coffee break.

Gathering Basic Task Data

When a modification of an existing system is being developed, data relating to an existing model may be appropriate. Such data can be obtained by various methods, such as the use of film; observation; the use of eye-movement recordings (with eye cameras or other related devices); and interviews with experienced personnel (including questions to elicit their opinions, such as about frequency or importance of various activities or about the desirable arrangement of components).

However, in the case of new systems or facilities (without current counterparts), information about the activities to be performed (such as frequency, sequence, and other activity parameters) needs to be inferred from whatever tentative drawings, plans, procedures, or concepts are available.

Types of Task Data

Although a thorough task analysis yields all sorts of valuable information for designing a work space, we concentrate on two broad classes related to arranging components within the space. We assume that an analysis has been made to determine what components will be needed to do the tasks. That is, we assume that all the controls, displays, tables, chairs, telephones, bookshelves, and what-have-you have been enumerated. The objective now is to arrange these items in the space allocated.

The broad classes of task-related information useful in arranging components are (1) information on the use of the components individually and (2) information on the relationships between components as they are used.

Information Dealing with Components Individually

For each component, information is collected to determine the frequency with which each component is (or would be) used and how important or critical its use is considered to be. Ratings by experts are usually relied on to assess importance and sometimes are used to assess frequency as well. If data are available for an existing system, however, usually it is preferable to use such data for determining task frequency.

Often, importance and frequency are not perfectly correlated. That is, although frequently used components are often important, some important components are not used frequently—but when they are needed, they are very important. Because of this, it is often easier to deal with a composite frequency-importance index rather than with separate indices of frequency and importance. Usually the frequency data are converted to a scale comparable to the scale used to rate importance. For example, if importance is rated on a 5-point scale from "unimportant" to "extremely important," the frequency data can be converted to a 5-point scale from, perhaps, "seldom used" to "very frequently used." Several options are available for combining the two ratings: (1) add the values; (2) multiply the values; (3) differentially weight frequency and importance, and then add them; or (4) differentially weight frequency and importance, and then multiply them. The different methods may result in a somewhat different rank ordering of components. If there are marked differences in the rankings, a choice must be made on the basis of judgment, since there are no absolute guidelines to follow.

An example of a different composite index is the *control accessibility index* developed by Banks and Boone (1981) to arrange control devices in a work space. Control devices, of course, need to be physically accessible to operators (as contrasted with displays, for example, which can be out of reach). Of the several variables that can affect the relative accessibility of controls, Banks and Boone developed their quantitative index of accessibility based on (1) frequency of use, (2) relative position of controls with respect to the operator, and (3) the operator's reach envelope. The index I_a was derived in part from ratings made by subjects who used mock-ups of three control panels for experimental purposes.

The details of the procedures and the resulting index need not be reported, but the results reflect the feasibility of deriving a quantitative index of accessibility for use in locating controls within a person's work-space envelope or in its general vicinity.

Information Dealing with Relationships between Components

Relationships between components, be they people or things, are called *links*.

Types of Links Links fall generally into three classes; communication links, control links, and movement links. Communication and control links can be considered as functional. Movement links generally reflect sequential movements from one component to another. Some versions of the three types follow:

1 Communication links

a Visual (person to person or equipment to person)

b Auditory, voice (person to person, person to equipment, or equipment to person)
c Auditory, nonvoice (equipment to person)
d Touch (person to person or person to equipment)

2 Control links
a Control (person to equipment)

3 Movement links (movements from one location to another)
a Eye movements
b Manual movements, foot movements, or both
c Body movements

The kinds of information usually collected about links include how often the components are linked (e.g., how often display *a* is viewed immediately before or after display *b* or how often person 1 talks to person 2), in what sequence the links occur, assuming a fixed sequence exists (e.g., whether the operator views display *a* then *e* then *f* or whether the sequence is *f* then *a* then *e*), and the importance of the links. These sorts of data can be gathered over time, across specific tasks, or over several repetitions of a single task.

Summarizing Link Data Link data are often summarized in a *link table*. An example of a link table is shown in Figure 14-3 for components in a hypothetical

FIGURE 14-3
Illustration of a chart of the link values of the relationships between pairs of components in a hypothetical minicomputer laboratory. (*Source: Cullinane, 1977, Fig. 1.*)

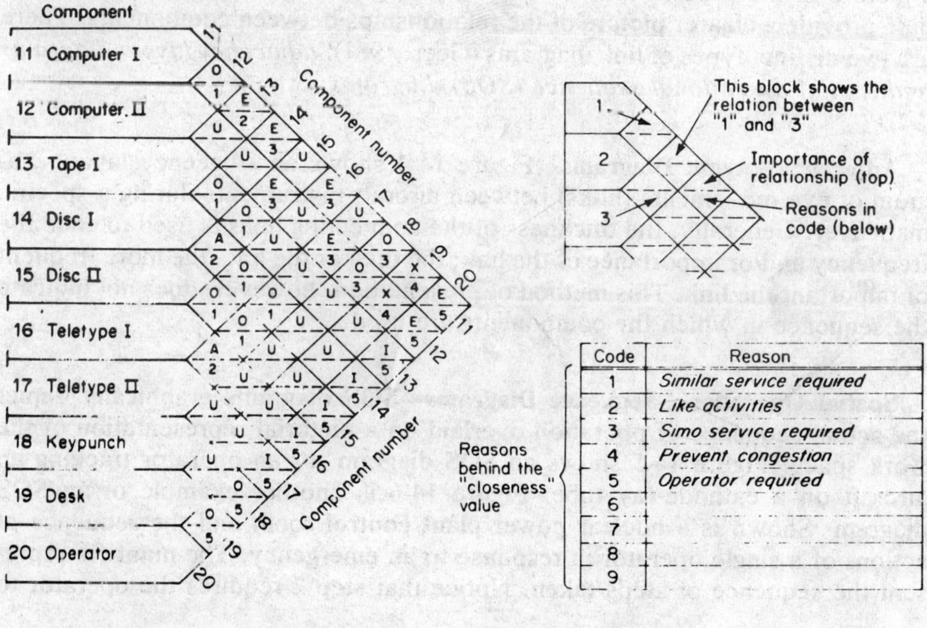

minicomputer laboratory. In this example the relationship of each component with any other one (i.e., each link) was rated by using the following scale:

A: It is *absolutely essential* for the activities to be located close together.
E: It is *essential* for two activities to be located close together.
I: It is *important* that the two activities be located close together.
O: *Ordinary* closeness is acceptable for the two activities being considered.
U: A link does not exist, and it is *unimportant* whether two activities are placed together.
X: It is *undesirable* for two activities to be placed together.

These are shown in the upper triangular part of the square formed by the downsloping line of one component and the up-sloping line of another (these being *A, E, I, O, U*, or *X*). The number in the lower triangular part of the square is a code explaining the reasons for the relationship. (In the example in Figure 14-3 the codes used for this purpose are in the inset in the lower right-hand corner.) To illustrate the use of the table, notice that for the operator (20), closeness to the desk (19) and keypunch (18) is acceptable (code O); but it is essential (code E) that the operator be located near the two computers (11 and 12).

Graphic Representation of Link Data

The link table shown in Figure 14-3 is one method of presenting link data; however, it is somewhat mind-boggling and is not very effective for developing a picture of the problem at hand. Other graphical methods of presenting link data provide a clearer picture of the relationships between components. There are two distinct types of link diagrams (Geer, 1981): *adjacency layout diagrams and spatial operational-sequence (SOS) diagrams*.

Adjacency Layout Diagrams Figure 14-4 shows an adjacency layout diagram of eye movements (links) between aircraft instruments during a specific maneuver. Generally, the thickness of the connecting lines is used to code the frequency and/or importance of the link; the thicker the line, the more frequent or important the link. This method of presentation, however, does not indicate the sequence in which the components are used.

Spatial Operational-Sequence Diagrams SOS diagrams graphically depict the actual sequence of operation overlaid on a pictorial representation of the work space. Figure 14-5 shows an SOS diagram for an operator tracking an aircraft on a cathode-ray tube. Figure 14-6 is another example of an SOS diagram. Shown is a nuclear power plant control room and the sequence of actions of a single operator in response to an emergency. The numbers represent the sequence of steps taken. Notice that step 2 requires the operator to

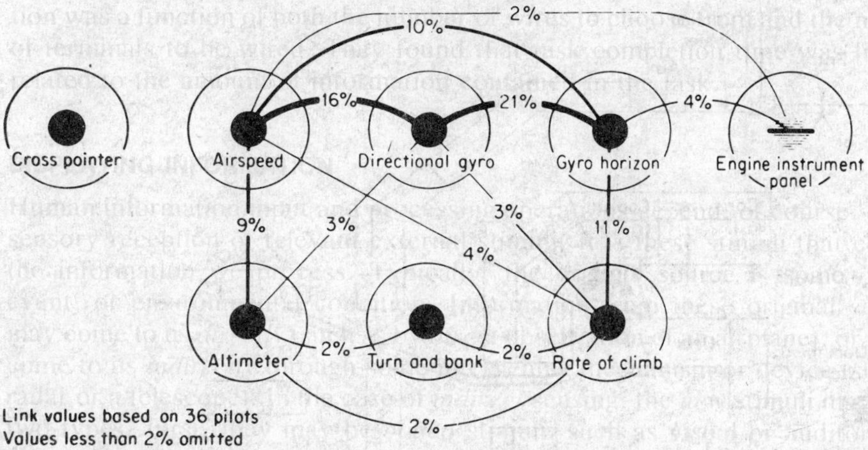

FIGURE 14-4
An adjacency layout diagram of eye-movement link values between aircraft instruments during a climbing maneuver. (*Source: Jones, Milton, and Fitts, 1949.*)

actually leave the control room to check a display. Obviously, SOS diagrams are most useful when there is a set sequence of operations performed with the components.

Arranging Components by Using Link Data

Probably the most common method of arranging components by using link data is through trial and error. The designer physically arranges scale drawings of the components, trying to maximize criteria that often conflict with one an-

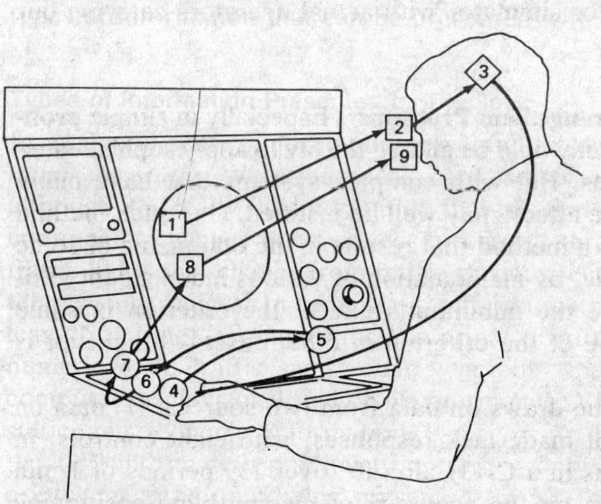

FIGURE 14-5
An example of a spatial operational-sequence diagram showing the sequence of sensory, decision, and control activities involved in tracking an aircraft. (*Adapted from Geer, 1981.*)

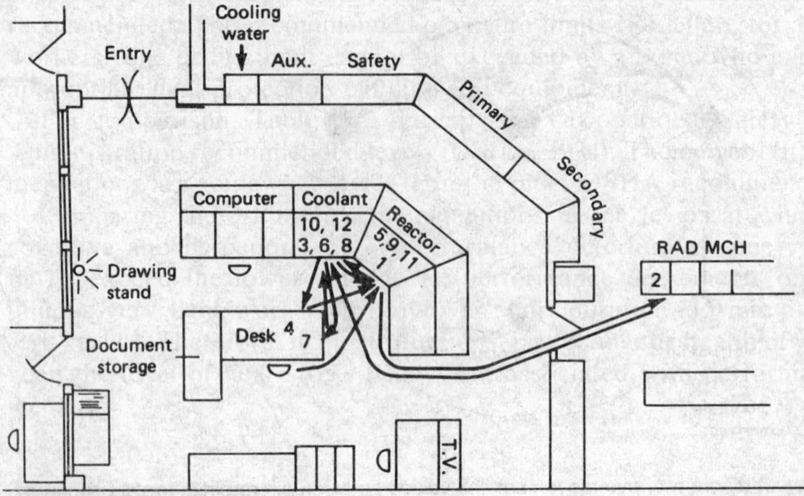

FIGURE 14-6
A spatial operational-sequence diagram of one nuclear power plant control room operator's actions in response to an emergency. The numbers represent the sequence of steps taken. (*Source: Seminara, Gonzales, and Parsons, 1977, Fig. 4-13. Copyright 1977, Electrical Power Research Institute. Reprinted with permission.*)

other. Trying to keep the most frequently used components in the most advantageous locations while arranging components to minimize the length of frequent or important links—and all the while keeping in mind sequences of operations and functional grouping—is probably as much art as it is science. There are, however, quantitative analysis techniques to aid the designer. No matter what technique is used, it is important that the resulting arrangement be validated by using a mock-up or simulator with actual operators carrying out actual or simulated tasks.

Quantitative Solutions to Arrangement Problems Especially in simple problems of arranging components, it would be gilding the lily to apply sophisticated quantitative analytical methods. But with complex systems that have many components, some quantitative attack may well be justified. One such method is *linear programming*. This is a method that results in the optimizing of some criterion, or dependent variable, by manipulation of various independent variables. The optimum would be the minimum value of the criterion in some cases, and the maximum value of the criterion in other cases—whichever is desired.

An example of this technique draws on data from two sources: (1) data on frequencies with which a pilot made task responses, with eight controls, in flying simulated cargo missions in a C-131 aircraft, over 139 periods of 1 min (Deininger, 1958); and (2) data on the accuracy of manual blind-positioning responses in various areas, based on the study by Fitts (1947). For this particular purpose accuracy data for 8 of the 20 target areas were used. The accuracy scores are average errors in inches in reaching to targets in the eight areas. Linear programming, when used in the analysis of the data by Huebner and Ryack (1961), involved the derivation of a *utility cost* rating for each of eight controls in each of the eight areas shown in Figure 14-7. Each one was computed by multiplying the *frequency* of responses involving each control (from Deininger) by the *accuracy* of responses in each area (from Fitts). For example, the values for the *C* control (cross-pointer set) for the eight areas range from 42.8 to 69.8, each such value being the product of the *frequency* value for the control (in this case 20) and the mean *accuracy* scores for the areas shown in Figure 14-7. For any possible arrangement of controls one can derive a *total cost*, this being the sum of the utility costs of the several controls in their locations specified by that arrangement. By linear programming it was possible to identify the particular arrangement for which the total cost was minimum. The resulting optimum arrangement is shown in Figure 14-7, with the individual controls simply being identified by a letter code. Note that each control is not in its *own* optimum location, but that *collectively* the derived arrangement is optimum in terms of the criterion of total cost. (Control *C*, mentioned above, was thus placed in area 6, even though its utility cost in that location was less.)

Another example of a quantitative approach for designing a facility is described by Bonney and Williams (1977), this dealing with the arrangement of control devices and other items at the "pulpit" of a bar and rod mill. Figure 14-8*a* shows the existing layout, the items being located on a wall panel, a left desk, a right desk, and the floor. Figure 14-8*b* shows the arrangement based on a computer program that was intended to optimize the collective locations of the items. Without going into details, the computerized arrangement was estimated to reduce limb movements from 35 to 31 m and provided for 98 percent

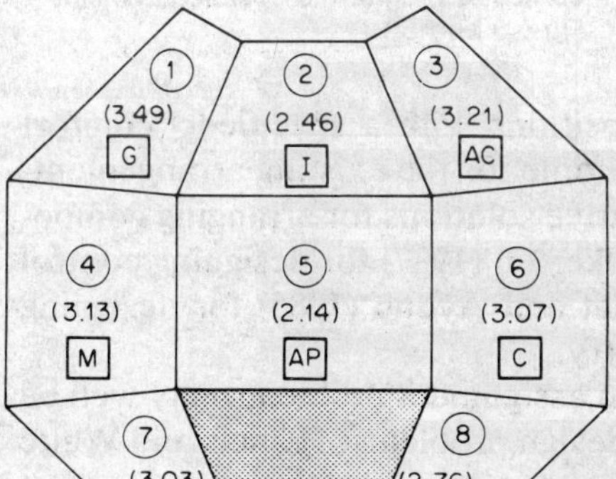

FIGURE 14-7
Perspective drawing of eight areas used in application of linear programming to the arrangement of eight aircraft controls. Mean accuracy scores of blind-positioning movements are given in parentheses for the eight target areas. The letters in boxes are symbols for the eight controls, their locations shown here representing the optimum linear programming solution. (*Sources: Perspective drawing, Huebner and Ryack, 1961; mean accuracy scores, based on data from Fitts, 1947.*)

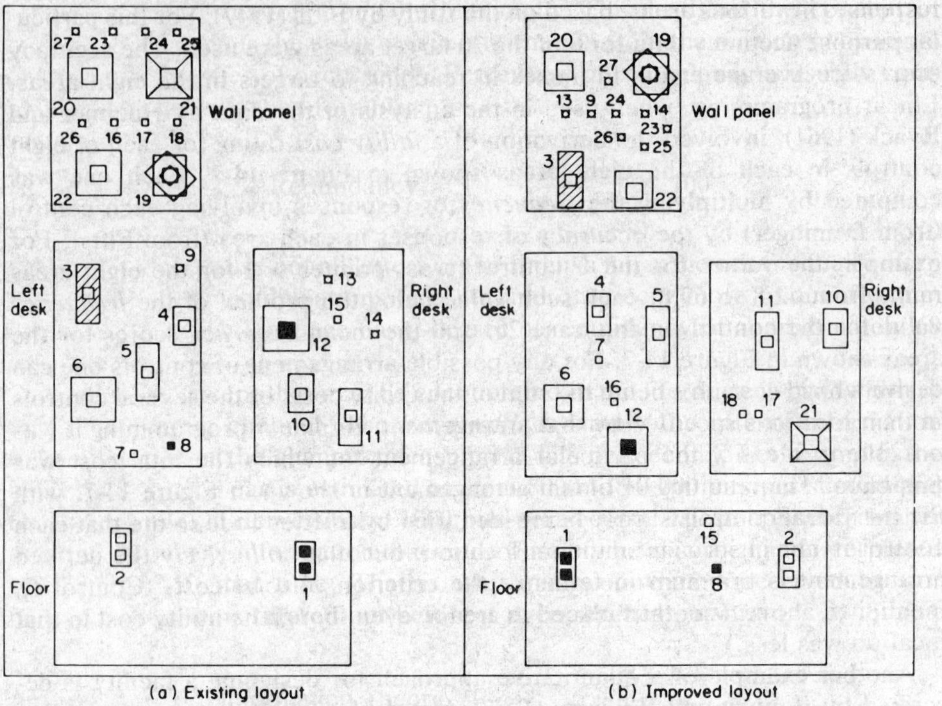

Key

1. pedal 1	6. lever 4	11. lever 6	16. button 4	21. lever 9	26. lamp 7
2. pedal 2	7. button 1	12. lever 7	17. button 5	22. lever 10	27. lamp 8
3. lever 1	8. button 2	13. lamp 2	18. button 6	23. lamp 4	
4. lever 2	9. lamp 1	14. lamp 3	19. knob 1	24. lamp 5	
5. lever 3	10. lever 5	15. button 3	20. lever 8	25. lamp 6	

FIGURE 14-8
(*a*) Existing arrangement of controls and other items of the "pulpit" of a bar and rod mill in Great Britain. (*b*) An improved arrangement developed with a computer program to optimize the location of the items. (*Source: Bonney and Williams, 1977, Figs. 3 and 5.*)

of the operational time in postures and positions with a specified "comfort rating," as compared with 50 percent of the time for the existing arrangement.

Other, more recent, examples of quantitative solutions for arranging components include work done by Palmiter and Elkerton (1987) for designing control panels in automobile factories, and by Pulat and Ayoub (1985) for designing control panels in the process control industry.

Although quantitative solutions for design assignment problems may well be justified in connection with some complex design problems, Francis and White (1974, Chap. 6) point out that in many circumstances such procedures are not warranted, and simpler approaches can produce substantially similar results. In connection with the design problem represented in Figure 14-7, for example, an approach such as the following might be used:

1 Put the control with highest frequency in the area with the lowest error (area 5).

2 Put the control with the second highest frequency in the area with the second lowest error (area 2).

3 Continue, to the eighth control.

GENERAL LOCATION OF CONTROLS AND DISPLAYS WITHIN WORK SPACE

As indicated above, it is reasonable to assume that any given component in a system or facility would have some reasonably optimum location, predicated on whatever sensory, anthropometric, biomechanical, or other considerations are relevant. Although the optimum locations of some specific components probably would depend on situational factors, some generalizations can be made about certain classes of components. Certain of these are discussed here.

Visual Displays

The normal line of sight is usually considered to be about 15° below the horizon. Visual sensitivity accompanied by moderate eye and head movements permits fairly convenient visual scanning of an area around the normal line of sight. The area for most convenient visual regard (and therefore generally preferred for visual displays) has generally been considered to be defined by a circle roughly 10° to 15° in radius around the normal line of sight.

However, there are indications that the area of most effective visual regard is not a circle around the line of sight but rather is more oval. Such an indication is found in the research of Haines and Gilliland (1973), who measured subjects' times of response to a small light that flashed in various locations of the visual field. Figure 14-9 shows the mean response times to the lights in the various regions of the visual field. Each boundary on that figure indicates the region within which mean response time can be expected to be about the same (the *isoresponse* time region). The generally concentric lines tend to form ovals—but slightly lopsided ovals, being flatter above the line of sight than below.

The subjects in this study detected the lights even toward the outer fringes of this area, but the primary implications of this (and some other) research are that critical visual displays should be placed within a reasonably moderate oval around the normal line of sight.

Hand Controls

The optimum location of hand control devices is, of course, a function of the type of control, the mode of operation, and the appropriate criterion of performance (accuracy, speed, force, etc.). Certain preceding chapters have dealt with some tangents to this matter, such as the discussion of the work-space envelope in Chapter 13.

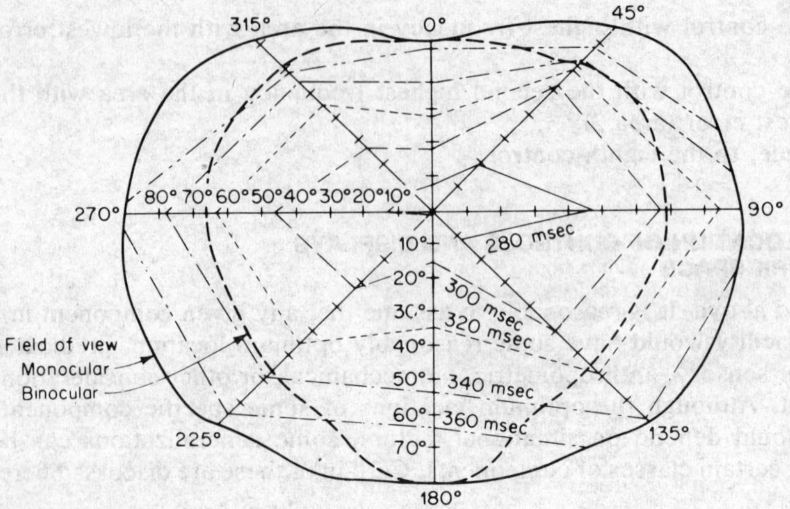

FIGURE 14-9
Isoresponse times within the visual field. Each line depicts an area within which the mean response time to lights is about the same. (*Source: Adapted from Haines and Gilliland, 1973, Fig. 2. Copyright 1973 by the American Psychological Association and reproduced by permission.*)

Controls That Require Force Many controls are easily activated, and so a major consideration in their location is essentially the ease of reach. Controls that require at least a moderate force to apply (such as certain control levers and hand brakes in some tractors), however, bring in another factor—force that can be exerted in a given direction with, say, the hand in a given position. Investigations by Dupuis (1957) and by Dupuis, Preuschen, and Schulte (1955), dealt with this question, specifically the pulling force that can be exerted, when a person is seated, when the hand is at various distances from the body (actually, from a seat reference point). Figure 14-10, which illustrates the results, shows the serious reduction in effective force as the arm is flexed when it is pulled toward the body. The maximum force that can be exerted by pulling is about 57 to 66 cm forward from the seat reference point, and this span, of course, defines the optimum location of a lever control (such as a hand brake) if the pulling force is to be reasonably high.

The best location for cranks and levers (especially those to be operated continuously to and fro) is in front of the sitting or standing operator, so that the handle travels at about waist height in the sagittal plane (i.e., the vertical plane from front to back) passing through the shoulder.

Controls on Panels Many controls are positioned on panels or in areas forward of the person who is to use them. Because of the anthropometric and biomechanical characteristics of people, controls in certain locations can be operated more effectively than those in other locations. Figure 14-11 shows one

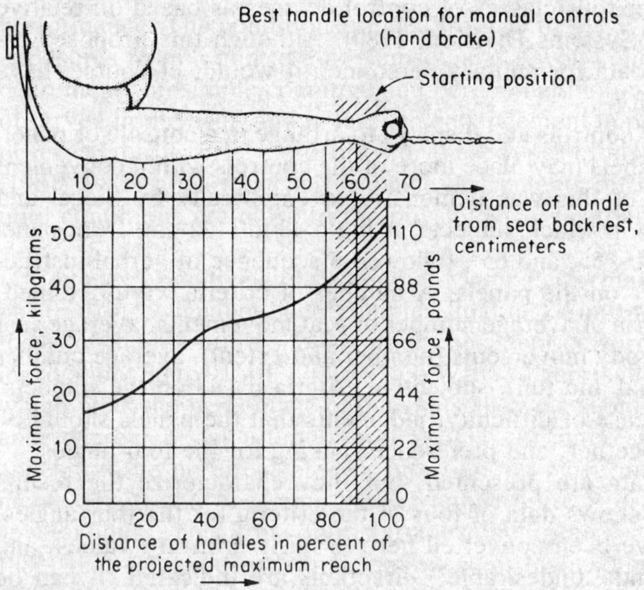

FIGURE 14-10
Relationship between maximum pulling force (such as on a hand brake) and location of control handle. (*Source: Dupuis, Preuschen, and Schulte, 1955.*)

FIGURE 14-11
Preferred vertical surface areas and limits for different classes of manual controls. (*Source: Adapted from Aeronautical Systems Division, 1980.*)

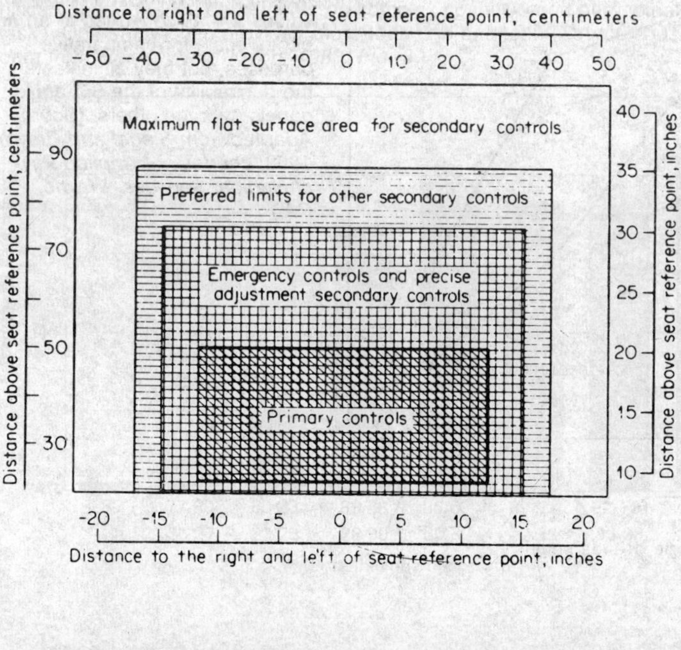

set of preferred areas for four classes of control devices as based on relative priorities (Aeronautical Systems Division, 1980). Although this proposed arrangement is based on data for military personnel, it would, of course, have more general applicability.

When there are many controls and displays to arrange in a console or panel, the use of angled side panels may place more of the controls within convenient access. The advantage of this was demonstrated empirically by Siegel and Brown (1958) in a study in which subjects, using a 48-in (122-cm) front panel with side panels at 35, 45, 55, and 65°, followed a sequence of verbal instructions to use the controls on the panels. A number of criteria were obtained, including objective criteria of average number of seat movements, average seat displacement, average body movements (number and extent), average number of arm extensions (partial and full), subjective criteria based on the subjects' responses of degree of ease or difficulty, judgments that the panels should be wider apart or closer together, and preference ranking for the four angles.

Only some of the data are presented, but they characterize the results generally. Figure 14-12 shows data of four of the criteria for the four angles. The criterion scales have been converted here to fairly arbitrary values, and only the "desirable" and "undesirable" directions are indicated. It can be seen, however, that all four criteria were best for the 65° side panels. The consistency across all criteria (these and the others) was quite evident.

Although Figures 14-11 and 14-12 provide general guidelines for locating controls, including consideration of relative priorities, sometimes the opera-

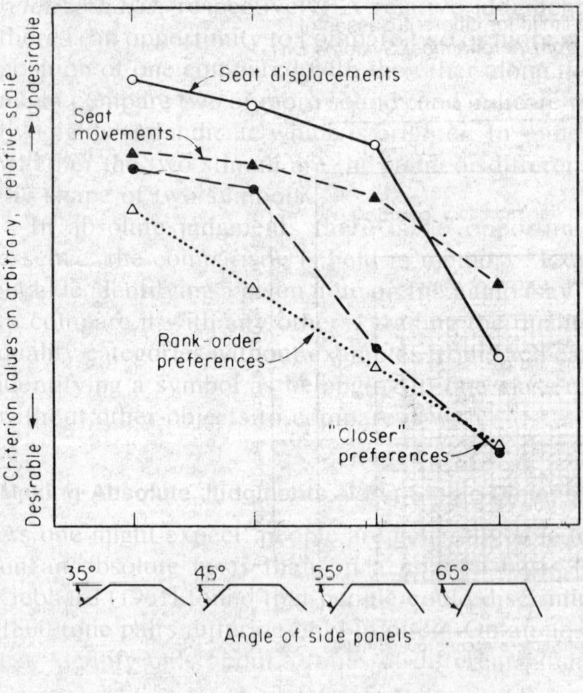

FIGURE 14-12
Representation of criteria from study relating to angles of side panels of console. The criterion values are all converted to an arbitrary scale for comparative purposes, but they all indicate the desirability of the 65° angle panels over the others. (*Source: Adapted from Siegel and Brown, 1958, courtesy of Applied Psychological Services, Wayne, PA.*)

tional requirements of certain specific types of controls impose impossible constraints on their location. This was illustrated, for example, by the results of a study by Sharp and Hornseth (1965) in which seated subjects operated each of three types of controls (knobs, toggle switches, and pushbuttons) at each of 12 locations in each of three consoles (far, middle, and close). The controls of the close console were positioned for convenience of reach of individuals of small build (about the 5th percentile of males), and only data for this console are given here.

Figure 14-13*a* shows the time in seconds to activate the three types of controls when they are located at various angles from the center position. This indicates that activation time was minimum for all three controls at about 25° from center. Figure 14-13*b* in turn shows the areas of the shortest performance times for each type of control (e.g., the area within which times were within 5 percent of minimum times for the control in question). The smaller area for the toggle switch suggests that the selection of a location for such devices may be more critical than for the other devices if time is of the essence. However, as suggested by the investigators, a well-designed system preferably should not impose response-time requirements on operators in which differences of 0.1, 0.2, or 0.3 s are crucial.

Two-Hand Controls Some operations require the simultaneous use of controls by both hands. For example, in the operation of some metal-forming presses, the operators have to press two push buttons for safety reasons to keep the hands away from the press when it is activated. In some instances these push buttons ("palm" buttons operated by the palm of the hand) are at eye level. This location was suspected of being responsible for a high rate of

FIGURE 14-13
Data on time to activate push button and toggle switch and time to reach knob, when the controls are in various positions. Data are for *left* hand and for close console. Part (*a*) gives mean times for controls positioned 25 in (63.5 cm) above seat reference level. Part (*b*) gives contours of the area for which times were *shortest* for the particular control (e.g., times within 5 percent of the minimum time for the control). (*Source: Adapted from Sharp and Hornseth, 1965.*)

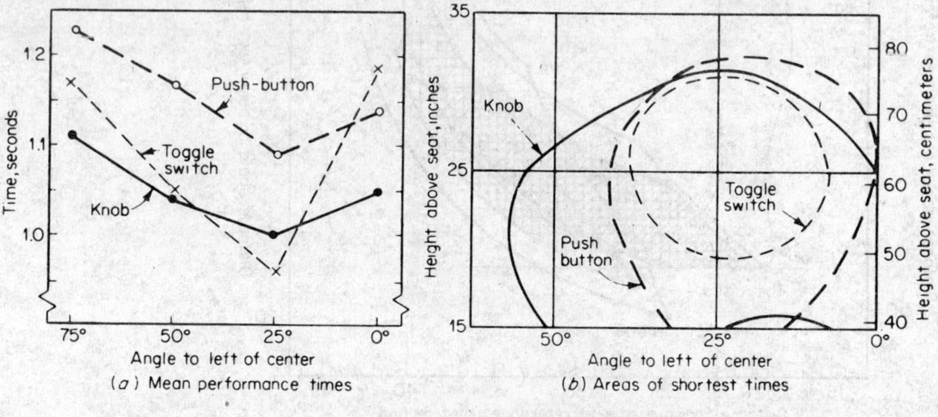

muscular strain and sprain injuries in an automobile plant (Nemeth and Blanche, 1982). In an investigation triggered by this suspicion, the experimenters had subjects perform a metal-processing task which required the use of dual palm buttons at eye level and at waist level. Electromyograph recordings of three muscle groups showed more than four times as much muscle activity with the eye-level buttons as with the waist-level buttons. Such a difference undoubtedly would later be translated to a lower level of muscular strain with waist-level buttons.

Foot Controls

Since only the most loose-jointed among us can put their feet behind their heads, foot controls generally need to be located in fairly conventional areas, such as those depicted in Figure 14-14. These areas, differentiated as optimal and maximal, for toe-operated and heel-operated controls, have been delineated on the basis of dynamic anthropometric data. The maximum areas indicated require a fair amount of thigh or leg movement or both, and preferably they should be avoided as locations for frequent or continual pedal use. Inci-

FIGURE 14-14
Optimal and maximal vertical and forward pedal space for seated operators. (*Source: Adapted from Aeronautical Systems Division, 1980.*)

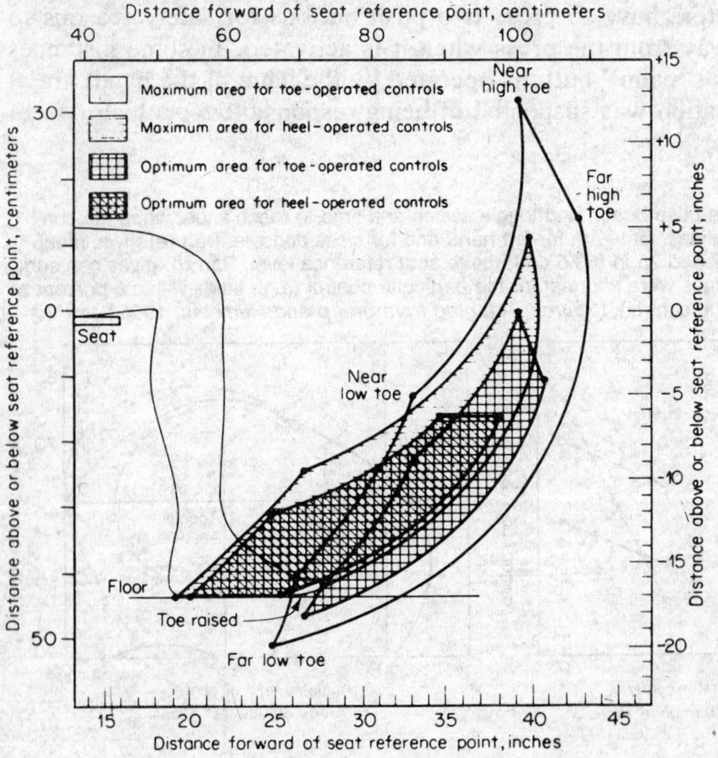

dentally, Figure 14-14 is predicated on the use of a horizontal seat pan; with an angular seat pan (and an angled backrest) the pedal locations need to be manipulated accordingly (although such adjustments have been published, they are not given here).

The areas given in Figure 14-14 generally apply to foot controls that do not require substantial force. For applying considerable force, a pedal preferably should be fairly well forward. This point is illustrated by Figure 14-15, which shows the mean maximum brake-pedal forces for 100 Air Force pilots when the leg was in a "normal" position and in an extended position (with the leg essentially forward from the seat). The figure shows that the maximum brake-pedal forces were clearly greater for the extended position for three seat levels used in the experiment. (The seat level was adjusted so the "floor-to-eye" distances were standardized.) As an aside, Figure 14-15 also shows that the forces for both leg positions were greatest when the foot angle on the pedal was between about 15 and 35° from the vertical.

SPECIFIC ARRANGEMENT OF CONTROLS AND DISPLAYS WITHIN WORK SPACE

As indicated in the above discussion of the general arrangement of components, certain general areas are most suitable (or even required) for locating various types of components, such as visual displays, hand controls, and foot controls. Within the constraints imposed by these considerations, the next

FIGURE 14-15
Mean maximum brake-pedal forces exerted by 100 Air Force pilots in two leg positions and for various brake-pedal angles and three floor-to-eye distances. (The seat level was adjusted for three conditions so the eye level was at the three specified distances above the floor.) (*Source: Adapted from Hertzberg and Burke, 1971, Fig. 6.*)

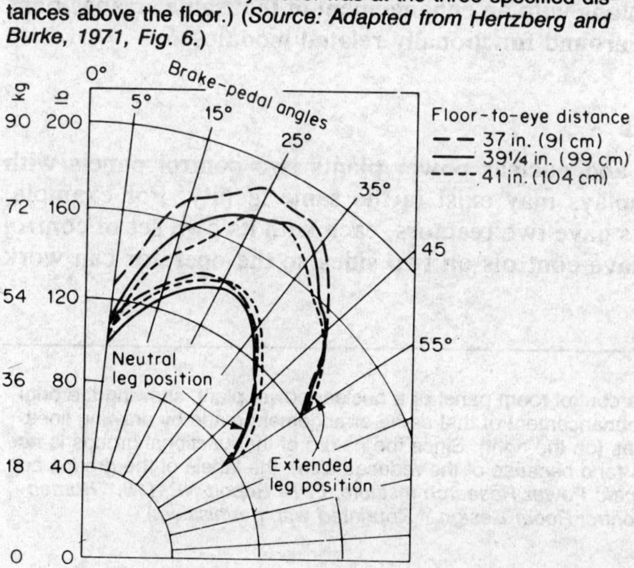

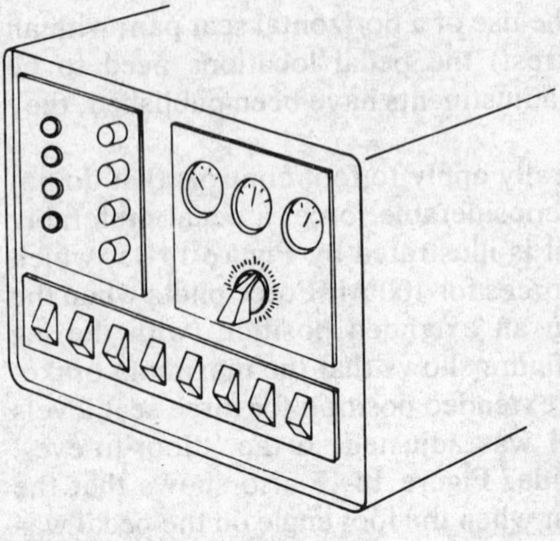

FIGURE 14-16
Example of controls and displays grouped by function, in which the different groups are clearly indicated.

process is that of arranging components within the areas appropriate for them. In this process the arrangement of groups of components can be based on the principles of sequence or function. Where there are common sequences, or at least frequent relationships, in the use of components, the layout usually should be such as to facilitate the sequential process—as in hand movements, eye movements, etc. Where there are no fixed or common sequences, the components should be grouped on the basis of function. In such instances the various groups should be clearly indicated by borders, color, or other means. An example of such groups is shown in Figure 14-16. When relevant, compatibility principles should be followed in arranging components.

Even in existing systems it may be possible to add borders around groups of functionally related components to clarify their relationships. Such a procedure is illustrated in Figure 14-17, which shows the original design of a control panel of a nuclear power plant, along with an enhancement of that same arrangement made by drawing borders around functionally related modules.

Mirror-Image Arrangements

In some process control and nuclear power plants two control panels with identical controls and displays may exist in the same facility. For example, some nuclear power plants have two reactors, each with its own set of control panels. Some machines have controls on two sides so the operator can work

FIGURE 14-17
Simplified illustration of part of a control room panel of a nuclear power plant, showing the original design (on the left) and an enhancement of that same arrangement made by drawing lines around related groups of modules (on the right). Since the nature of the functional groups is not relevant for illustrative purposes (and because of the reduced size), the labels of the groups are not given. (*Copyright 1979, Electric Power Research Institute, EPRI Report NP-118, "Human Factors Methods for Nuclear Control Room Design." Reprinted with permission.*)

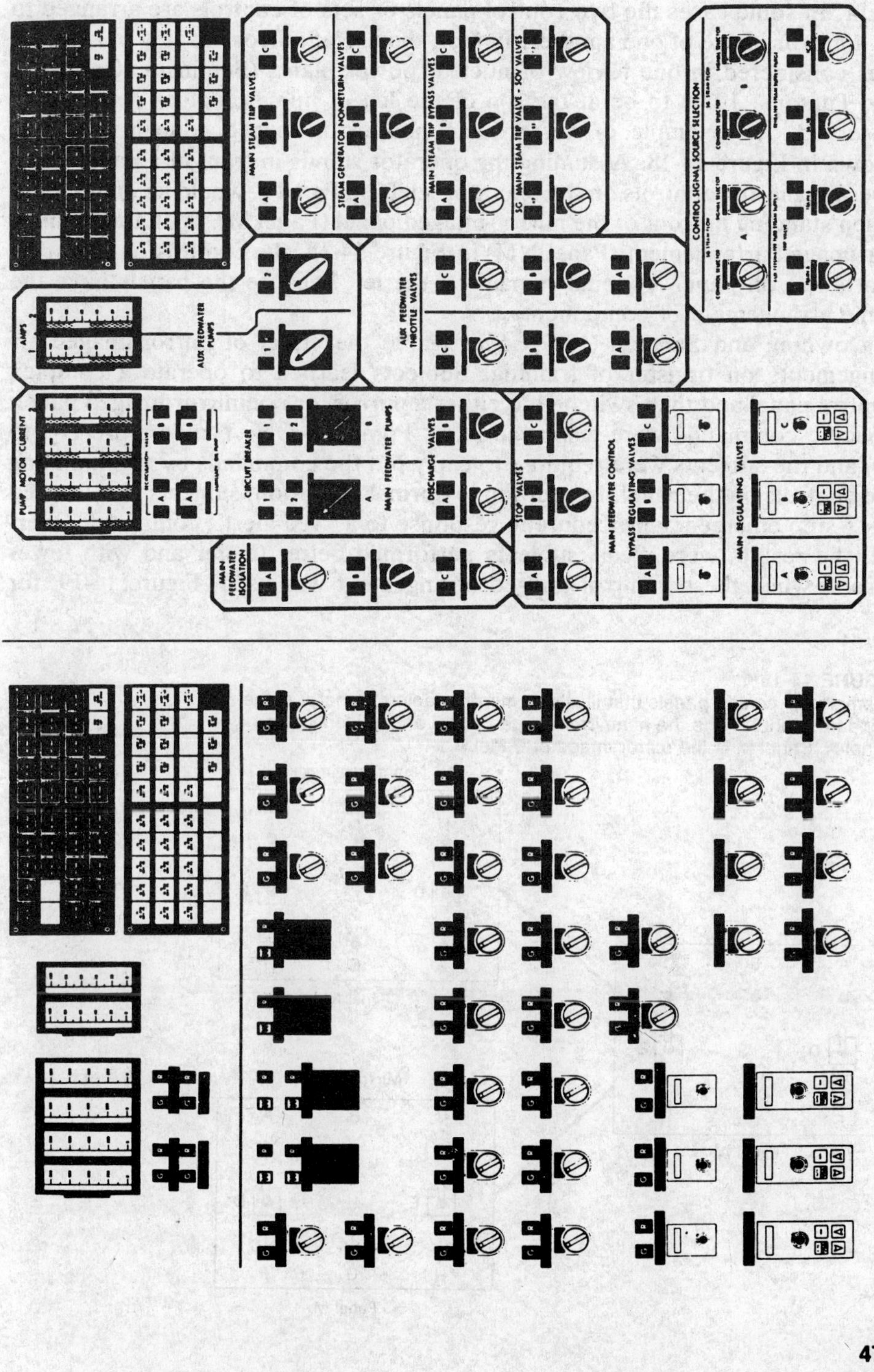
MAIN STEAM TRIP VALVES
STEAM GENERATOR NONRETURN VALVES
MAIN STEAM TRIP BYPASS VALVES
SG MAIN STEAM TRIP VALVE - SOV VALVES
CONTROL SIGNAL SOURCE SELECTION
AMPS
AUX FEEDWATER PUMPS
AUX FEEDWATER THROTTLE VALVES
PUMP MOTOR CURRENT
CIRCUIT BREAKER
MAIN FEEDWATER PUMPS
DISCHARGE VALVES
STOP VALVES
MAIN FEEDWATER CONTROL
BYPASS REGULATING VALVES
MAIN REGULATING VALVES
MAIN FEEDWATER ISOLATION

from either side. This is common, for example, on underground mining equipment. In some cases the two control panels or sets of controls are arranged to be a mirror image of one another. In fact, the use of mirror imaging of controls was considered in one review of nuclear power plants (Seminara, Gonzales, and Parsons, 1977) to be at the top of the list of human factors deficiencies observed. An example of mirror- and nonmirror-imaged control panels is shown in Figure 14-18. Assuming the operator stands in front of a panel when operating it, the controls on the operator's left on Panel A would be on the right when standing in front of the mirror-imaged panel (Panel M). With the nonmirror-imaged arrangement (Panel NM) in Figure 14-18, the controls on the operator's left on Panel A would remain on the left because the panels have the same arrangement of components.

Downing and Sanders (1987) investigated the effect of mirror-imaged arrangements on transfer of training. Subjects learned to operate a complex control panel and then switched to either a mirror- or nonmirror-imaged panel. Specific commands were displayed (e.g., Pump 1—ON; Temperature D-3 to 40) and the subjects were required to carry out the commands by manipulating the controls on the panel. In addition to normal operation, subjects also learned a six-step emergency procedure in response to a "red alert" command. Overall, the results were clear, subjects performed better (faster and with fewer errors) when the nonmirror-imaged arrangement was used. Figure 14-19, for

FIGURE 14-18
Examples of control panels that illustrate mirror imaging. Panel A is the standard. Panel NM is the nonmirror-imaged panel and is identical to Panel A. Panel M is the mirror image of Panel A.

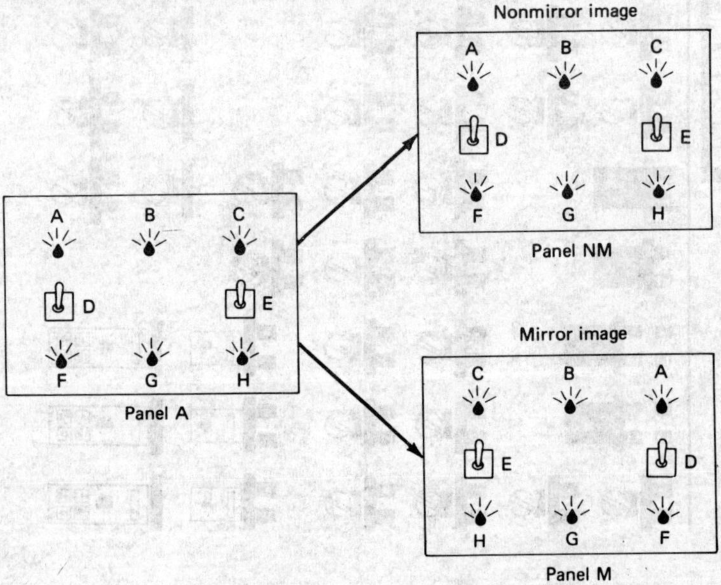

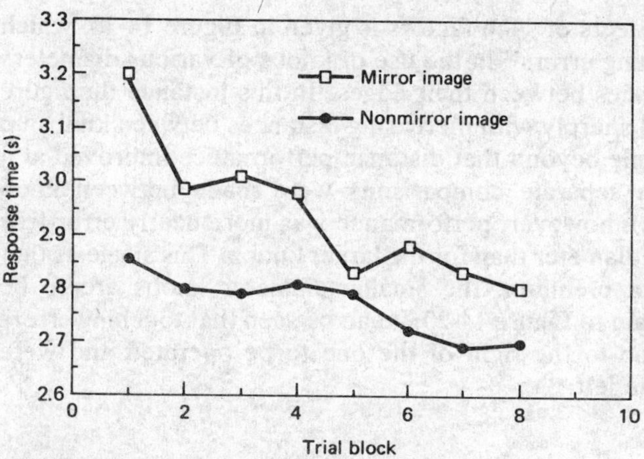

FIGURE 14-19
Response time to commands when subjects, after working on one control panel, switch to either a mirror-imaged or a non-mirror-imaged panel. (*Source: Downing and Sanders, 1977, Fig. 4. Reprinted with permission of the Human Factors Society, Inc. All rights reserved.*)

example, shows the response time for individual commands as a function of trial blocks (78 trials were grouped into 8 trial blocks of 9 or 10 trials each). Mean response times with the nonmirror-imaged arrangement were faster than with the mirror-imaged panel, especially during the early trial blocks. The difference between mirror- and nonmirror-imaged configurations grew smaller as subjects continued to practice and became more accustomed to the mirror-imaged panel. The emergency response and completion times, however, remained faster for the nonmirror-imaged condition than for the mirror imaged condition even after 74 responses to specific commands. Subjects switching to the mirror-imaged arrangement required, on average, almost 6 s of additional time to complete the six-step emergency procedure than did those switching to the nonmirror-imaged arrangement. The results of this study suggest that mirror imaging should be avoided if possible in designing pairs of control panels that will likely be used by the same operators.

SPACING OF CONTROL DEVICES

Although we have talked about minimizing the distances between components, such as the sequential links between controls, there are obvious lower-bound constraints that need to be respected, such as the physical space required in the operation of individual controls to avoid touching other controls. Whatever lower-bound constraints there might be would be predicated on the combination of anthropometric factors (such as of the fingers and hands) and on the precision of normal psychomotor movements made in the use of control devices.

An illustration of the effects of such factors is given in Figure 14-20, which shows inadvertent "touching errors" in the use of knobs of various diameters as a function of the distances between their edges. In this instance the figure shows that errors dropped sharply with increasing distances between knobs up to about 1 in (2.5 cm), while beyond that distance performance improved at a much slower rate. When separate comparisons were made between knob centers (rather than edges), however, performance was more nearly error-free for knobs of ½-in (1.2 cm) diameter than for the larger knobs. This suggests that when panel space is at a premium, the smaller-diameter knobs are to be preferred. By referring again to Figure 14-20, it can be seen that touching errors were greatest for the knob to the right of the one to be operated and were minimal for the one on the left.

FIGURE 14-20
Frequency of inadvertent touching errors for knobs of various diameters as a function of the distance between knob edges. Areas of circles in inset indicate relative frequency with which the four surrounding knobs were inadvertently touched when knob *x* was being operated. (*Source: Bradley, 1969, Fig. 2.*)

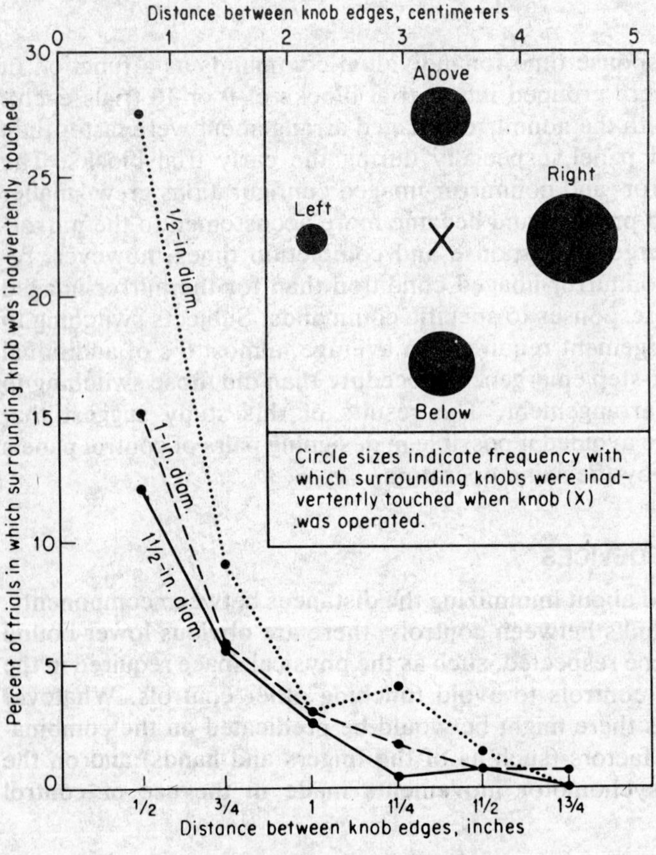

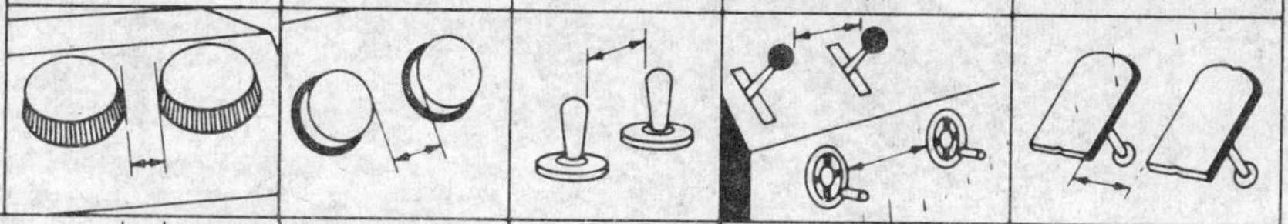

Number of body members and type of use		Knobs	Push buttons	Toggle switches	Cranks, levers	Pedals
1, randomly	in	2(1)	2(1/2)	2(3/4)	4(2)	6(4)
	cm	5(2.5)	5(1.3)	5(1.8)	10(5)	15(10)
1, sequentially	in		1(1/4)	1(1/2)		4(2)
	cm		2.5(.6)	2.5(1.3)		10(5)
2, simultaneously	in	5(3)			5(3)	
	cm	12.7(7.6)			12.7(7.6)	
2, randomly, sequentially	in		1/2 (1/2)	3/4 (5/8)		
	cm		1.3(1.3)	1.8(1.6)		

FIGURE 14-21
Recommended separation (in inches and centimeters) between adjacent controls. Preferred separations are given for certain types of use with corresponding minimum separation in parentheses. (*Source: Adapted from Chapanis, 1972.*)

On the basis of various studies of control devices Chapanis (1972, Chap. 8) set forth certain recommended distances (preferred and minimal) between pairs of similar devices, as given in Figure 14-21.

GENERAL GUIDELINES IN DESIGNING INDIVIDUAL WORKPLACES

In designing workplaces some compromises are almost inevitable because of competing priorities. In this regard, however, appropriate link values can aid in the trade-off process. Some general guidelines for designing workplaces that involve displays and controls are as follows (Van Cott and Kinkade, 1972, Chap. 9):

- *First priority:* Primary visual tasks
- *Second priority:* Primary controls that interact with primary visual tasks
- *Third priority:* Control-display relationships (put controls near associated displays, compatible movement relationships, etc.)
- *Fourth priority:* Arrangement of elements to be used in sequence
- *Fifth priority:* Convenient location of elements that are used frequently
- *Sixth priority:* Consistency with other layouts within the system or in other systems

DISCUSSION

The built features of our environments include many "things" that we use: machines, computers, desks, other office equipment, storage spaces, kitchen cabinets, control panels, hospital laboratories, etc. The physical location and

FIGURE 14-22
The control room of the Three Mile Island nuclear power plant. (*Photograph courtesy of Dr. Thomas B. Malone, Essex Corporation.*)

arrangement of such items (and of their specific features) clearly can affect the effectiveness and safety with which they are used as well as the comfort and physical well-being of the users. In many circumstances (especially where very few items or components are to be used) there is no particular problem in developing an arrangement that is reasonably satisfactory. In some situations, however (especially where there are many components), the development of a satisfactory arrangement can be very tricky. An example of a very complex situation is given in Figure 14-22, this being the control room of the Three Mile Island nuclear power plant. In connection with this installation, the chairperson of the presidential committee investigating the accident at Three Mile Island is quoted as saying, "It looks to me like this is very bad human engineering." Subsequent investigations confirmed this opinion. No single picture can illustrate all the specific human factors deficiencies of the control system, but Figure 14-22 may at least give some impression of the problems of controlling such a facility and of the importance of locating and arranging the many components of systems so as to simplify the work demands of the people involved and to enhance the likelihood of satisfactory work performance.

REFERENCES

Aeronautical Systems Division (1980, June 25). *Human factors engineering* (3d ed., rev. 1). (AFSC DH 1-3). Wright-Patterson Air Force Base, OH.

Banks, W. W., and Boone, M. P. (1981). A method for quantifying control accessibility. *Human Factors*, 23(3), 299–303.

Bonney, M. C., and Williams, R. W. (1977). CAPABLE: A computer program to layout controls and panels. *Ergonomics*, 20(3), 297–316.

Bradley, J. V. (1969). Optimum knob crowding. *Human Factors*, 11(3), 227–238.

Chapanis, A. (1972). Design of controls. In H. P. Van Cott and R. G. Kinkade (eds.), *Human engineering guide to equipment design* (rev. ed.). Washington, DC: Government Printing Office.

Cullinane, T. P. (1977). Minimizing cost and effort in performing a link analysis. *Human Factors,* 19(2), 151–156.

Deininger, R. L. (1958). *Process sampling, workplace arrangements, and operator activity levels.* Unpublished report, Engineering Psychology Branch, U.S. Air Force, WADD. Wright-Patterson Air Force Base, OH.

Downing, J., and Sanders, M. (1987). The effects of panel arrangement and locus of attention on performance. *Human Factors,* 29(5), 551–562.

Dupuis, H. (1957). Farm tractor operation and human stresses. Paper presented at the meeting of the American Society of Agricultural Engineers, Chicago: December 15–18, 1957.

Dupuis, H., Preuschen, R., and Schulte, B. (1955). *Zweckmäbige gestaltung des schlepperführerstandes.* Dortmund, Germany: Max Planck Institutes für Arbeitphysiologie.

Electric Power Research Institute (1979). *Human factors methods for nuclear control room design,* vol. 1: *Human factors enhancement of existing nuclear control rooms* (EPRI NP-1118). Palo Alto, CA: Lockheed Missiles and Space Co., Inc.

Fitts, P. M. (1947). A study of location discrimination ability. In P. M. Fitts (ed.), *Psychological research on equipment design* (Res. Rept. 19). Army Air Force, Aviation Psychology Program. Columbus, OH: Ohio State University.

Fowler, R. L., Williams, W. E., Fowler, M. G., and Young, D. D. (1968, December). An investigation of the relationship between operator performance and operator panel layout for continuous tasks (Tech. Rept. 68–170). U.S. Air Force AMRL (AD-692 126). Wright-Patterson Air Force Base, OH.

Francis, R. L., and White, J. A. (1974). *Facility layout and location: An analytical approach.* Englewood Cliffs, NJ: Prentice-Hall.

Geer, C. W. (1981). *Human engineering procedure guide* (AMRL-TR-81-35). Wright-Patterson Air Force Base, OH.

Haines, R. F., and Gilliland, K. (1973). Response time in the full visual field. *Journal of Applied Psychology,* 58(3), 289–295.

Hertzberg, H. T. E., and Burke, F. E. (1971). Foot forces exerted at various aircraft brake pedal angles. *Human Factors,* 13(5), 445–456.

Huebner, W. J., Jr., and Ryack, B. L. (1961, March). *Linear programming and work place arrangement: Solution of assignment problems by the product technique* (Tech. Rept. 61–143). U.S. Air Force, Air Research and Development Command, WADD. Wright-Patterson Air Force Base, OH.

Jones, R. E., Milton, J. L., and Fitts, P. M. (1949). *Eye fixations of aircraft pilots: IV. Frequency, duration and sequence of fixations during routine instrument flight* (U.S. Air Force Tech. Rept. 5975).

Kochhar, D., and Barash, D. (1987). Status displays in automated assembly. *Applied Ergonomics,* 18(2), 115–124.

Nemeth, S. E., and Blanche, K. M. (1982). Ergonomic evaluation of two-hand control location. *Human Factors,* 24(5), 567–571.

Palmiter, S., and Elkerton, J. (1987). Evaluation metrics and a tool for control panel design. *Proceedings of the Human Factors Society 31st Annual Meeting.* Santa Monica, CA: Human Factors Society, pp. 1123–1127.

Pulat, B., and Ayoub, M. (1985). A computer-aided panel layout procedure for process control jobs—LAYGEN. *IEEE Transactions,* 17, 84–93.

Seminara, J., Gonzales, W., and Parsons, S. (1977). *Human factors review of nuclear power plant control room design* (EPRI Report NP-309). Palo Alto, CA: Electric Power Research Institute.

Sharp, E., and Hornseth, J. P. (1965, October). *The effects of control location upon performance time for knob, toggle switch, and push button* (Tech. Rept. 65–41). AMRL. Wright-Patterson Air Force Base, OH.

Siegel, A. I., and Brown, F. R. (1958). An experimental study of control console design. *Ergonomics,* 1, 251–257.

Van Cott, H. P., and Kinkade, R. G. (1972). Design of individual workplaces. In H. P. Van Cott and R. G. Kinkade (eds.), *Human engineering guide to equipment design* (rev. ed.). Washington, DC: Government Printing Office.

CHAPTER 15

INTERPERSONAL ASPECTS OF WORKPLACE DESIGN

The previous two chapters dealt with the details of workplace layout and the arrangement of components within a physical space. The criteria relied heavily on anthropometric and biomechanical considerations. In this chapter, however, we take a somewhat different approach to workplace design and discuss issues that deal more with the interpersonal aspects of design. The criteria rely more on social and affective (attitudinal and emotional) consequences. We focus on two environments with which most people are familiar. The first, office environments, are clearly workplaces. The second, places where we live, are not commonly considered workplaces but nonetheless offer interesting examples of interpersonal design considerations. Our purpose is to illustrate some of the issues pertinent to built environments, and some relevant research findings, rather than presenting an exhaustive review of the literature.

EVALUATING THE BUILT ENVIRONMENT

The varied features of the built environment can have profound effects on people. In addition to the effects of the physical features of such environments, people are influenced by such nonphysical features as social, cultural, technological, economic, and political factors characteristic of the environment. The built environment and its associated nonphysical factors affect a wide gamut of human experience, including performance in work and other aspects of life, social behavior, attitudes, satisfaction and dissatisfaction, mental health, quality of life, and physical health and well-being.

Both field and laboratory studies are used to assess the affective response of people to various aspects of the built environment. A common laboratory methodology is to show to subjects drawings, pictures, or scale models of built environments, such as a living room or an office, and ask them to rate it along various dimensions, such as friendliness, warmth, complexity, etc. Features of the environment, such as placement of windows, ceiling height, or color of walls, are varied in order to assess whether ratings change when the feature changes.

We discuss some of these studies later; however, such studies have some potentially serious limitations. Looking at a drawing or scale model of an environment can be a vastly different experience from being in the actual full-scale version of it. Further, opinions and attitudes about an environment formed from a one- or two-minute exposure may be quite different from the opinions and attitudes about that environment developed after living or working in it for several weeks or months. In laboratory studies either subjects are not instructed about what activities will take place in the space or they are asked to imagine the environment will be used for some purpose. The purpose or functions carried out in an environment can affect people's opinions and attitudes about the environment. We may, for example, prefer rooms with lower ceilings when we eat than when we are watching television.

Field studies and surveys, of course, have the advantage of involving prolonged exposure to the environment under actual "real-life" conditions. But it is often difficult to isolate the specific aspect(s) of the environment contributing to the attitudes and behaviors being measured. As in other areas of human factors, a combination of laboratory and field studies probably is the best method of unraveling the varied effects of the built environment on people's attitudes and behaviors.

Discussion

The design of buildings and other features of the built environment is a different process because good design should be directed toward satisfying several goals (i.e., criteria) simultaneously. However, it usually is not feasible to fulfill all the desired objectives with a single design. In this regard, Bennett (1977) makes the point that various design objectives are not equal and tend to form a hierarchy. In planning a private home, for example, a person might consider such factors as location, convenience to transportation, type of neighborhood, size and number of rooms, style of architecture, and type of building materials, along with cost. The relative values the individual places on these and other features usually would place the values in some rough hierarchy; so if not all the desirable features can be fulfilled, those low in the hierarchy usually would be dropped or modified. The decision made under such circumstances represent the trade-offs that plague human factors designers in the design of many items.

THE OFFICE AS A BUILT ENVIRONMENT

To many people the office environment is like a second home where they spend almost half their waking hours. In the United States there were more people working in offices in 1980 than in factories. There were 44.8 million white-collar workers compared to only 29.8 million blue-collar workers (Kleeman, 1981). With all those people spending all that time in offices, it is no wonder that office planning and design has become a big business.

In previous and upcoming chapters we discuss aspects of office design such as illumination, speech communications, noise, and workstation design. These are areas traditionally addressed by human factors specialists and other professionals. In recent years, our attention has widened to consider other, more global aspects of the office environment including overall size, arrangement of furnishings, and even the value of windows. We have come to recognize that such aspects of the office environment can affect the way people interact and their attitudes toward their jobs.

To provide some context for our discussion of office design, we briefly discuss the nature of office work. Our discussion is not exhaustive, but does provide some information that can influence the design of the office.

Office Activities

As with any design effort, it is essential that we understand the functions and tasks to be performed by the people using whatever it is we are designing. This is equally true for designing offices. There have been several attempts to develop taxonomies of office tasks. Table 15-1 is representative and illustrates the range of tasks typically performed in offices.

One major activity performed in offices is information development and gathering; one criterion for good office design is to facilitate that function. To gain some insight, Hodges and Angalet (1968) interviewed 1500 engineers, scientists, and technical personnel employed by 84 companies, institutes, and

TABLE 15-1
A TAXONOMY OF OFFICE TASKS

Cognitive	*Physical*
Information development and gathering	Filing and retrieval
Information storage and retrieval	Writing
Reading and proofreading	Mail handling
Data analysis and calculating	Traveling
Planning and scheduling	Copying and reproducing
Decision making	Collating and sorting
Social	Pickup and delivery
Telephoning	Typing and keying
Dictating	Keeping calendars
Conferring	Using equipment
Meeting	
Procedural	
Completing forms	
Checking documents	

Source: Galitz, 1980, Fig. 4-1.

universities. Table 15-2 presents the findings concerning where information was first sought when it was needed. About 30 percent of the respondents' informational needs were satisfied without search, and 50 percent were sought within the local work environment. This suggests that informal and personal information sources be documented and incorporated into the office design and that formal information sources (personal files, department files, etc.) be close to people so they can use them more productively (Galitz, 1980).

The percentage of time spent on office tasks varies with the position in the organization. Table 15-3, for example, presents some data from a study of 1200 white-collar workers from 11 companies at 40 different locations (Knopf, 1982). As one moves down the organizational ladder from executive to secretary, the percentage of time spent in face-to-face communications decreases, while time spent in document-based communications increases. The same study reported that executives spend 47 percent of their time in meetings, both formal and informal. An office with executives should be designed, therefore, to provide space where meetings can be held without distraction and with adequate privacy.

What Kind of Office?

Arguments have been made for the pros and cons of various types of offices—large and small, landscaped and conventional. A few studies have dealt with this matter, although the results have been somewhat ambiguous, if not conflicting. Part of the problem lies in the proliferation of terms used to describe various office designs. At the risk of stepping on a few toes, let us suggest the

TABLE 15-2
WHERE OFFICE WORKERS GO FIRST FOR INFORMATION

Situation	Source	Frequency, %
No search required	Recall	19.0
	Received with task	10.5
	Asked a colleague	14.5
	Searched own collection	13.0
	Internal company consultant	9.5
Local work environment	Departmental files	5.5
	Assigned to subordinate	4.5
	Respondent's own action	2.5
	Asked supervisor	1.0
	Library or librarian	10.0
	Manufacturer or supplier	6.0
External to work environment	Customer	2.0
	External consultant	1.0
	DOD information systems	1.0

Source: Hodges and Angalet, 1968. Copyright by the Human Factors Society, Inc. and reproduced by permission.

TABLE 15-3
PERCENTAGES OF MAJOR TYPES OF COMMUNICATIONS IN OFFICES AS A FUNCTION OF POSITION IN THE ORGANIZATION

Type of communication	Executive	Manager	Knowledge worker	Secretary
Face-to-face	53	47	23	Negligible
Document	27	29	42	55
Telephone	16	9	17	20
Other	4	15	18	25

Source: Knopf, 1982.

following distinctions. A *large* office refers to size rather than to how the space is partitioned. A large office may be an *open-plan* or a *cellular* office (those divided into many smaller offices for individuals or small groups). An open-plan office is not absolutely open, but it can be subdivided by movable furniture and screens. One type of open-plan office is the "bull pen" office with desks arranged in neat rows, often as far as the eye can see. Another type of open-plan office is the landscaped, or *Bürolandschaft,* office which originated in Germany (Brookes, 1972). Such an office consists of one large, open—but "landscaped"—area which is planned and designed about the organizational processes to take place within it. The people who work together are physically located together, the geometry of the layout reflecting the pattern of the work groups. The areas of the various work groups are separated by plants; low, movable screens; cabinets; shelves; etc., as shown in Figure 15-1.

In a nationwide survey of 1047 office workers (Steelcase, 1978) almost one-third reported they were working in an open-plan office other than a bull pen office. Of the respondents only 15 percent reported working in bull pen offices.

Flexibility is one of the major advantages cited for open-plan offices. Space can be reconfigured easily, and workstations can be rearranged at little cost to meet changing needs or work patterns. In addition, more of the available space is utilized in an open-plan office because less space is devoted to hallways and walls. We briefly discuss a few studies that have compared open-plan offices (usually the term as used in such studies does not include bull pen offices) with traditional cellular offices to illustrate other possible benefits and some potential limitations of open-plan offices.

Social Behavior in Offices There are some indications that small-office environments are more conducive to the development of social affinity for others than large offices. This was reflected, for example, by the results of a sociometric study by Wells (1965) in which office personnel in large and small offices were each asked to indicate their choices of individuals beside whom they would like to work. The data summarized in Table 15-4 show the percentages of the choices made by individuals (in open and small areas) of other persons in their own section or department. Although there was greater internal

FIGURE 15-1
Example of a landscaped office in which individual offices and work groups are separated by plants, screens, cabinets, and shelves. This office is in the Administration Services Building at Purdue University.

cohesion among personnel working in the smaller areas than in the open areas, also there were more isolates (individuals not chosen by anyone).

A factor often related to the type of office plan is the degree to which a company permits personalization of the workplace. *Personalization* refers to the deliberate adornment, decoration, modification, or rearrangement of an environment by its occupants to reflect their individual identities (Sommer, 1974). One survey of office workers (BOSTI, 1981) found that 61 percent had added personal items such as pictures, clocks, plants, rugs, etc. to their work spaces. If one includes personal, work-related items (e.g., awards, pencil cups, etc.), three out of four office workers personalize their work space. Personalization is usually tolerated more in closed-plan offices than in open-plan offices. Because the entire space is visible in open-plan offices, planners often want the office to look the way they designed it to look and consider personalization of an office space as "blight," spoiling the aesthetic environment (Sundstrom, 1986). The research evidence on the value of personalization for job satisfaction is circumstantial. People regard personalization as important

TABLE 15-4
SOCIOMETRIC CHOICES OF OFFICE WORKERS IN OPEN AND SMALL OFFICE AREAS

	Choices of members of own section, %	Reciprocal choices within own section, %
Open areas	64	38
Small areas	81	66

Source: Wells, 1965.

and feel that being able to personalize a workplace indicates that the company recognizes their individuality.

Another factor of office plan design that impacts interpersonal behavior is status markers. *Status markers* are those aspects of design that differentiate people of different rank in an organization. Louis Harris and Associates (1978) in a survey of office workers asked, "Which characteristics make the working areas of senior executives seem different from other areas of the office?" The following six items were mentioned by 20 percent or more of the respondents:

- Style of desks, tables, and chairs
- Amount of privacy
- Overall amount of personal space
- Paintings, posters, and wall decorations
- Materials used for desks, tables, and chairs
- Type of floor covering

It is usually easier to provide status markers in a closed-plan office than in an open-plan office. In fact, one of the purported advantages of the open plan is the reduction of status markers and a more egalitarian work environment. There is no evidence, however, that an absence of status markers by itself is beneficial, let alone whether it is even possible.

Disturbances and Distractions in Offices There appear to be indications that open-plan offices contribute to an increased number of distractions and disturbances, some work-related, others of a social nature. Mercer (1979), for example, observed samples of workers in three office environments; traditional cellular offices, open-plan offices, and what was called an *action office* (apparently similar to landscaped offices). Some of the results related to the number of disturbances and distractions occurring per hour in the three environments are presented in Table 15-5. Disturbances tended to be higher in the action and open-plan offices than in the traditional ones. Work-related distractions were considerably higher in the open plan than in the other two types of offices. What is also somewhat amazing is that in all three offices there was an average of one disturbance or distraction every 2 min! No wonder so many office

TABLE 15-5
NUMBER OF DISTURBANCES AND DISTRACTIONS OCCURRING PER HOUR IN THREE TYPES OF OFFICE PLANS

	Type of office		
	Action	**Open plan**	**Traditional**
Disturbances			
Work-related	11.4	8.1	7.1
Social	5.5	5.3	4.5
Total	16.9	13.4	11.6
Distractions			
Work-related	8.9	23.4	10.3
Social	2.4	2.3	4.8
Total	11.3	25.7	15.1
Total	28.2	39.1	26.7

Source: Adapted from Mercer, 1979, Table 2.

workers are needed to get the work done. Complaints about disturbances and distractions are fairly common in open-plan offices, as Nemecek and Grandjean (1973) found when they surveyed 519 workers in 15 open-plan offices in Switzerland. They found that 69 percent of the respondents complained about disturbances in concentration and 11 percent mentioned that confidential conversations were impossible. To reinforce the importance of these sorts of findings, Steelcase (1978) found that 41 percent of their nationwide sample of office workers indicated that the characteristic most important in helping to get the work done well was the ability to concentrate without noise or other distractions.

Health in Office Environments Hedge (1984) reported that complaints of frequent headaches were almost twice as prevalent among open-plan office employees as among traditional-plan office employees. Problems of eye irritation, coughs, colds, and sore throats, however, appeared more related to whether the office was air-conditioned than to the type of office plan. All these symptoms were more prevalent in air-conditioned offices.

Designing Open-Plan Offices A key factor in the success of an open-plan office is whether the layout is based on actual communication patterns and work group needs. Surveys should be conducted to assess who talks to whom, who works with whom, what type of interactions take place, privacy needs, and what facilities (equipment, storage, etc.) are needed by each person to do the work. Given such information, an intelligent office layout can be formulated.

Wichman (1984) offers the following specific design recommendations to enhance the utility of an open-plan office:

• *Use sound-absorbing materials on all major surfaces wherever possible.* Noise is often more of a problem than expected.

• *Leave some elements of design for the workstation user.* Workstations are usually overly designed and inflexible. People need to have control over their environments; so leave some opportunities for changing or rearranging things.

• *Provide both vertical and horizontal surfaces for the display of personal belongings.* People like to personalize their workstations.

• *Install telephones that ring "silently."* In small workstations phones that flash a light for the first two "rings" before emitting an auditory signal dramatically reduce disturbances.

• *Provide all private work areas with a system to signal the willingness of the occupant to be disturbed.* There is a need to express changing needs for privacy and community.

• *Provide several easily accessible islands of privacy.* This would include small rooms with full walls and doors that can be used for conferences and for private or long-distance telephone calls.

• *Have clearly marked flow paths for visitors.* For example, hang signs from the ceiling showing where secretaries and department boundaries are located. The visual stimulus configuration of an open-plan office is often too complex for visitors to quickly develop a cognitive map.

• *Design workstations so it is easy for drop-in visitors to sit down while speaking.* This will tend to reduce disturbances to other workers.

• *Plan for ventilation air flow.* Most traditional offices have ventilation ducting. This is usually not the case with open-plan cubicles, so they become dead-air cul-de-sacs that are extremely resistant to post hoc resolution.

• *Overplan for storage space.* Open-plan systems with their emphasis on tidiness seem to chronically underserve the storage needs of people.

Discussion The attitudes and opinions of people about landscaped offices are something of a mixed bag. People who work in such offices have been reported by certain investigators (Brookes, 1972; Brookes and Kaplan, 1972; McCarrey et al., 1974) to complain of noise, visual bustle, lack of privacy and confidentiality of communications, and lack of "territory definition." However, such offices are generally viewed as developing greater solidarity and offering greater personal contact. In addition, most people react favorably to the aesthetic aspects of such offices. The implications regarding productive efficiency are ambiguous. Brookes (1972), for example, states, "It looks better but it works worse," whereas McCarrey et al. (1974) report opinions tending to imply increased productivity.

At present it is probably not feasible to say that the reported advantages of landscaped offices, do, or do not, outweigh the disadvantages. However, probably landscaped offices would be preferred over large, open, bull pen types.

Windows or No Windows?

Because of factors such as cost and energy conservation, there has been renewed interest in windowless buildings. Windows are no longer necessary to provide light and ventilation, but there is some concern that they may be of some value in fulfilling what Manning (1965) refers to as a "psychological need" for some "contact with the outside world." The use of deep buildings, where only the outer ring of offices has windows, and underground workplaces (Wise and Wise, 1984) have further focused attention on the psychological value of windows.

In general, office workers seem to prefer working in buildings with windows. Without windows, workers often complain about lack of daylight, poor ventilation, not knowing the weather outside, and not having a view (Collins, 1975). Ruys (1970), for example, in a survey of workers in small windowless offices found that 87 percent expressed dissatisfaction with the lack of windows.

Although many office workers, when asked, express dissatisfaction with a lack of windows, there is no evidence of reduced work performance or of workers being "driven crazy" because of a lack of windows. Also studies of windowless schools and factories where large groups of people interact in the windowless space have not always revealed negative attitudes. In the case of schools, for example, windows are often seen as a source of distraction, and removing them is often seen as benefiting the educational process (Chambers, 1963; Tikkanen, 1970). As Boyce (1981) points out, there is probably no single, general conclusion about attitudes toward windowless buildings. Factors such as the size of the windowless space, the level of activity within the space, and workers' expectations all mediate attitudes toward such environments. One thing seems clear, however: Where windows are available, it is usually the executives who have them. And this probably says more about the value of windows than all the surveys that have been done to date.

Office Furnishings and Arrangements

Some fairly definite indications of the subjective reactions of people to office furnishings and arrangements can be reported. For one thing, people tend to report feelings of spaciousness being greatest in offices that are moderately furnished (in terms of number of chairs, desks, etc.), as contrasted with those that are more empty or overfurnished (Imamoglu, 1973).

In addition, people react more favorably to offices that have living things (as plants) and aesthetic objects (as posters) and are tidy than to those that lack these features (Campbell, 1979). These features make visitors feel more welcome and comfortable, as reflected by the responses of subjects presented with slides of offices with various combinations of features (see Table 15-6).

Another feature of offices that influences the reaction of visitors and interpersonal relationships is the arrangement of chairs for visitors. When a visitor is seated opposite the desk of the person being visited, the desk tends to serve

TABLE 15-6
RATINGS OF RESPONDENTS REGARDING SLIDES OF OFFICES WITH VARIOUS FEATURES

	Mean ratings (9 = favorable; 1 = unfavorable)					
Quality	**Plants**	**No plants**	**Posters**	**No posters**	**Tidy**	**Messy**
Comfort of visitor	6.0	4.6	5.7	4.9	6.1	4.5
Invitingness of office	5.9	4.3	5.6	4.7	6.1	4.1
How welcome visitor feels	6.0	4.7	5.8	5.0	6.2	4.5

Source: Campbell, 1979.

as a barrier. In general, visitors have a greater sense of friendliness when both chairs are on the same side of the desk, as illustrated in Figure 15-2*a* as contrasted with 15-2*b* (Campbell, 1979; Zweigenhaft, 1976).

Office Automation

We would be remiss if we did not at least mention office automation and its possible impact on office design. The computer and video display terminal (VDT) are probably the most visible aspects of office automation, but there are other aspects as well. Helander (1985) presents a taxonomy of office automation tools based on the type of interaction and office task being automated as shown in Figure 15-3. Helander points out that, at present, there are no automated aids for the face-to-face social skills of supervising, counseling, persuading, and negotiating. Automated aids are making their biggest impact in the areas of perceptual motor skills, rule-based decision making, and analysis and problem solving.

The implications for office design of various automated tools such as voice or electronic mail, electronic bulletin boards, local area networks, and computerized data bases are difficult to foresee. Already the boundaries of the office are expanding; sources of information and methods of gathering information are changing; and interaction patterns are broadening. As just one example consider *computer-supported cooperative work (CSCW)* in which a group of people meet at the same time and same place to cooperatively work on a task (Cornell, Luchetii, Mack, and Olson, 1989; Mack, 1989). In a CSCW environment, each participant has a keyboard and display, as well as access to a large (often rear projection) public display. Participants can work in small teams and share information with other teams on the large screen, or they can input information directly to the large screen as the entire group interacts. These systems raise all sorts of interesting interpersonal issues. For instance, who

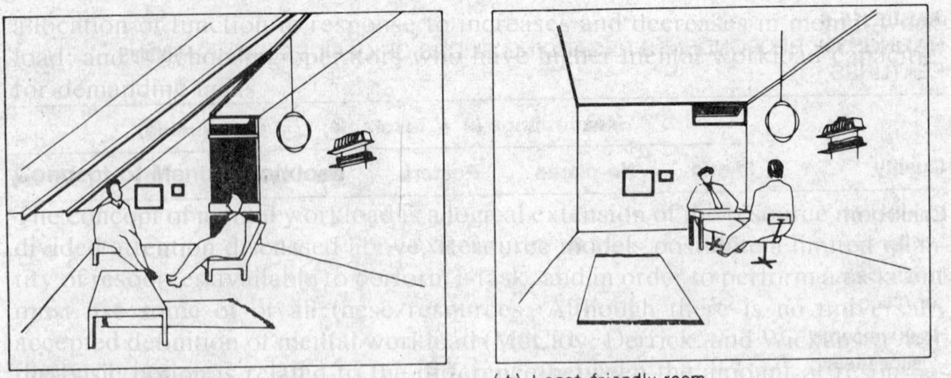

(*a*) Most friendly room (*b*) Least friendly room

FIGURE 15-2
Examples of the drawings of rooms used in a study relating to the perceived dimensions of rooms. These examples are of rooms that were appraised to be most friendly and least friendly. (*Source: Adapted from Wools and Canter, 1970, Figs. 4 and 5.*)

controls the cursor on the big screen? Can one modify work someone else placed in public view? Will an omnipresent big screen reduce interpersonal communication? How do computer "timid" people fare in such an environment?

Discussion

Large or small offices? Landscaped offices? Windows or no windows? Although research data about these and other aspects of offices are still quite skimpy, the research that is available (such as that discussed above) gives an impression of ambiguity, inconsistency, and lack of support for at least certain expectations or hypotheses. In reflecting about this disturbing state of affairs, we need to keep in mind the fact that in part the measures of the effects of some design features consist of subjective reactions of people, such as preferences, attitudes, and aesthetic impressions. Overt manifestations in terms of behavioral criteria such as work performance are difficult to document, but at the same time the favorable disposition of workers to certain types of working situations indicates that they probably have some long-range hidden values. Further, although people might prefer a particular environment situation, probably they have a fair quota of resiliency or adaptability that makes it possible to adjust to a variety of circumstances.

DWELLING UNITS

There are many aspects of dwelling units that have human factors implications. There has been a considerable body of literature to address such issues as space requirements in the kitchen and bathroom and indoor climate [see, for example, Grandjean (1973)]. As in our discussion of the office environment, we

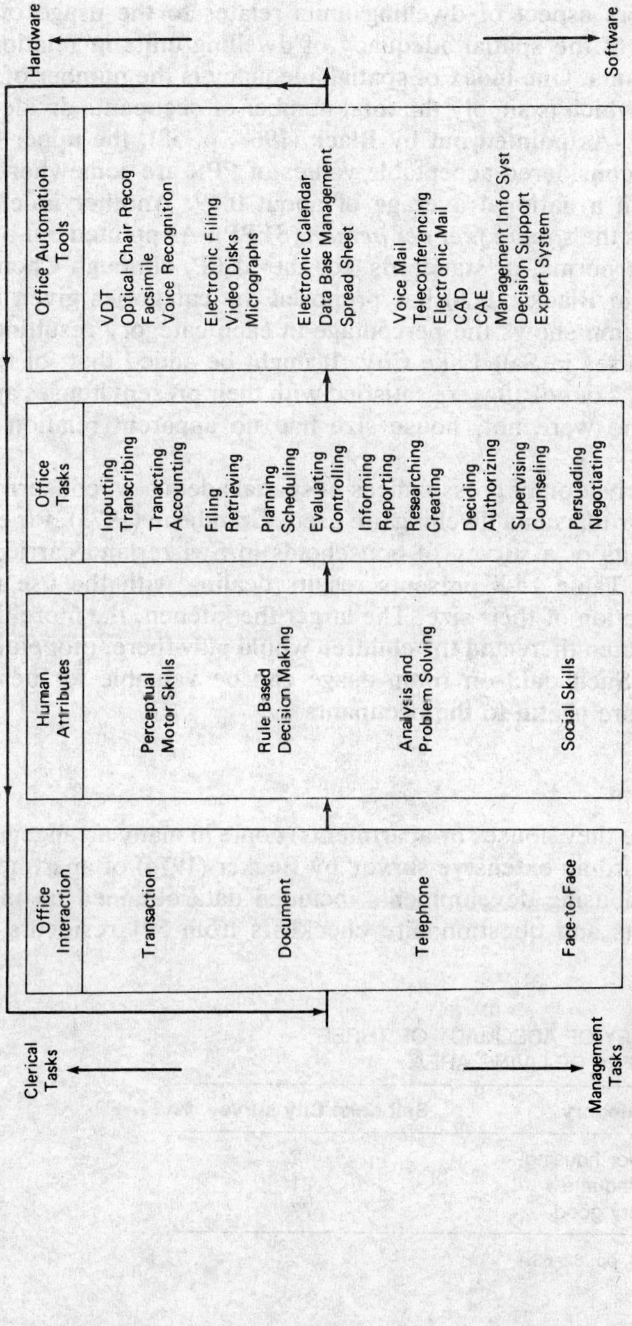

FIGURE 15-3
Taxonomy of office automation tools based on the type of office interaction, human attribute, and office tasks involved. (*Source: Helander, 1985, Fig. 1. Copyright by the Human Factors Society, Inc. and reproduced by permission.*)

focus on architectural and interior design features. Even within this limited domain, we touch only briefly on a few such aspects.

Room Usage in Dwellings

One human factors aspect of dwelling units relates to the usage of rooms, including indices of the spatial adequacy of dwelling units in relation to the number of occupants. One index of spatial adequacy is the number of *persons per room* (PPR), which is simply the total number of occupants divided by the number of rooms. As pointed out by Black (1968, p. 58), the upper limits of what are usually considered acceptable values of PPR are somewhere around 1.00 or 1.20, with a national average of about 0.69. Another index that is sometimes used is the *square feet per person* (SFPP). As pointed out by Black, there are no U.S. norms, or standards, for the SFPP, although Chombart de Lauwe (as cited in Black, 1968) has proposed the categories given in Table 15-7; the last column shows the percentage in each category resulting from a survey of 121 houses in Salt Lake City. It might be added that, of the home owners surveyed, 7 out of 10 were satisfied with their present houses and in the case of those who were not, house size had no apparent relation to their dissatisfaction.

Besides the derivation of gross indices of spatial adequacy, concern has also been paid to how rooms in a dwelling are used. Grandjean (1973), for example, presents the results of a survey of households in Switzerland carried out by Bächtold (1964). Table 15-8 presents results dealing with the use made of kitchens as a function of their size. The larger the kitchen, the more likely the meals would be eaten there and the children would play there, (hopefully not at the same time). Such data on room usage can be valuable for designing a dwelling to be more useful to the occupants.

Room Arrangement

Dwelling units (be they houses or apartments) come in many arrangements and sizes. In this regard an extensive survey by Becker (1974) of apartment occupants of public housing developments included data obtained by interviews with 257 residents and questionnaire checklists from 591 residents. Certain

TABLE 15-7
RESULTS OF SURVEY OF ADEQUACY OF THREE CATEGORIES OF SFPP OF LIVING AREA

SFPP	Category	Salt Lake City survey, %
<130	Poor housing	2
131–215	Adequate	21
>215	Very good	77

Source: Black, 1968, pp. 62–63.

TABLE 15-8
USE MADE OF KITCHENS IN RELATION TO THEIR SIZE, GIVEN AS PERCENTAGE OF THOSE PARTICIPATING IN SURVEY

	Kitchen area		
Activity	**Up to 6 m²**	**7 m²**	**8 m² or more**
Breakfast	49	66	85
Midday meal	42	48	71
Evening meal	40	53	74
Washing, ironing	42	74	78
Children playing and working	2	8	25
Spending leisure time	11	15	26

Source: Bächtold, 1964 as presented in Grandjean, 1973.

questions dealt with room size and arrangement of the living-dining-eating spaces. The responses to these questions are summarized:

• Many variations of room size and arrangement were considered to be equally satisfactory by the residents, but (as would be expected) satisfaction tended to be greater with larger rooms.

• More residents preferred a separate dining area (39 percent) or separate dining area and large eat-in kitchen (28 percent) to combined living-dining area (19 percent) or living room with large eat-in kitchen (18 percent). A couple of these arrangements are illustrated in Figure 15-4 along with proposed designs that would permit several options for arrangement of eating and living space. Although most residents preferred a separate dining area, 83 percent said they would not bc willing to give up any of their living room space for a separate dining area or an all-purpose room. In other words, residents presumably found living room space so essential (whatever the size) that they could not conceive of reducing it.

In addition to dining and living areas, Grandjean (1973) lists the following guidelines for the placement of bedrooms within a dwelling:

• Bedrooms should be separated from the living part of the dwelling (i.e., living room, dining room, kitchen).

• Bedrooms should be near a bathroom.

• Bathrooms should be accessible from children's rooms without having to pass through the living room.

• To make it easier to watch small children, at least one child's bedroom should be close to that of the parents.

• Because the parents' bedroom is used so little during the day, it may face north.

• Since children's rooms are used so much in the daytime, they should face south.

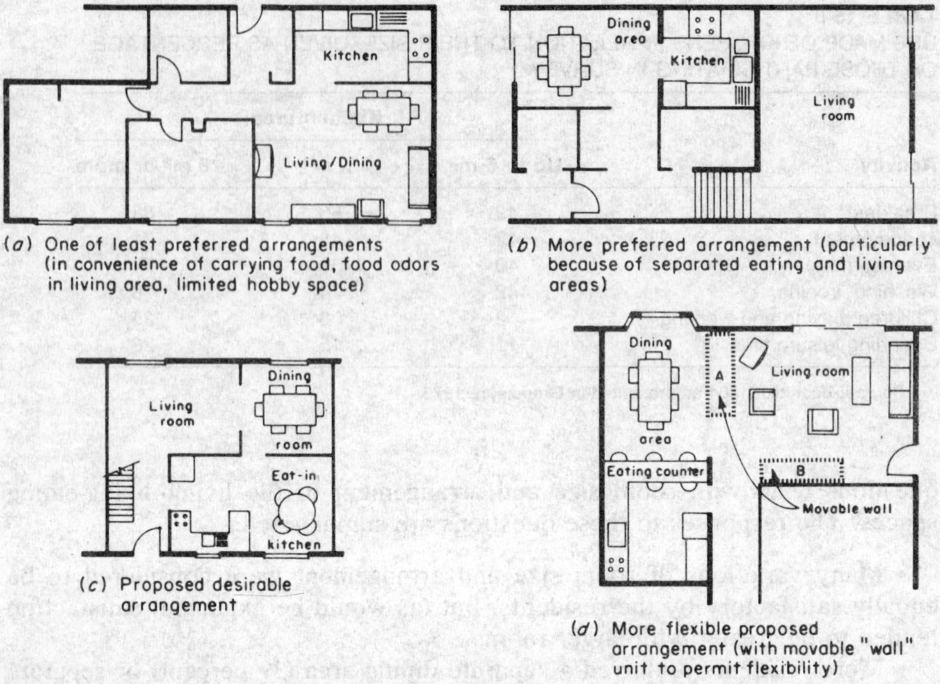

FIGURE 15-4
Order of preference of residents of a couple of multifamily housing units regarding dining and eating arrangements, and a couple of proposed designs, one of which (*d*) would permit several options for eating and arrangement of living space (this one having a movable wall unit which could be placed in various locations, such as at *A* or *B*). (*Source: Adapted from Becker, 1974, Figs. 5a, 5b, 5c, and 5d.*)

Interior Design Features

Numerous aspects of interior design can influence people's perceptions of a dwelling or living space. For illustrative purposes we briefly discuss a few.

Ceiling Height and Slope. The usual height for rooms in dwelling units is 8 ft (2.4 m). In a study by Baird, Cassidy, and Kurr (1978), subjects were placed in a room that had an adjustable ceiling. After the ceiling was set at a given height, the subject was asked to give a preference rating on a scale ranging from -10 to $+10$, these ratings being given in two hypothetical frames of reference—one in which the subject would be involved in no activity and another in which the subject would be dining. The mean ratings, shown in Figure 15-5, indicate that the preferred heights were about 2 ft (0.6 m) above the conventional height.

In another phase of this study the authors found that subjects tended to prefer sloping ceilings (in particular those with 3/12 and 5/12 pitches) and wall corners that were greater than 90° rather than less than 90°.

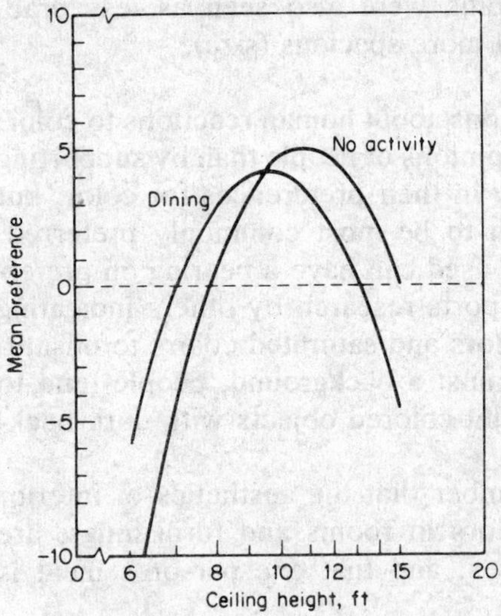

FIGURE 15-5
Mean preferences for ceiling height of rooms for "no activity" and for dining. Note that the ceiling height is shown on a logarithmic scale. (*Source: Adapted from Baird, Cassidy, and Kurr, 1978, Fig. 3, p. 725. Copyright 1978 by the American Psychological Association. Reprinted by permission.*)

Windows People seem to prefer spaces with windows to those without, but how do windows affect our subjective perceptions of a room? Kaye and Murray (1982) asked 176 students to rate watercolor perspective drawings of a living room on 30 adjectives, such as *cluttered, interesting, confined, gloomy,* and *colorful.* Using a statistical technique called *factor analysis,* the authors identified four underlying factors that seemed to account for the ratings on the individual adjectives. The four factors are listed in Table 15-9 along with some of the adjectives most associated with each. The presence or absence of a window in the picture affected the social-aesthetic, mood, and size factors. The presence of a window made the rooms appear more friendly, inviting, and

TABLE 15-9
FOUR FACTORS UNDERLYING RATINGS FOR VARIOUS ROOM ARRANGEMENTS USING AN ADJECTIVE RATING FORM

Factor	Representative adjectives
Social-aesthetic	Bright, happy, inviting, friendly, exciting
Physical organization	Cluttered, organized, accidental
Mood	Unattractive, drab, dead, closed, gloomy
Size	Large, spacious

Source: Kaye and Murray, 1982.

exciting (social-aesthetic). Windowed rooms were also seen as less drab, gloomy, and closed (mood) and larger and more spacious (size).

Color There have been many speculations about human reactions to color; however, the topic is dominated more by opinions of people than by supporting research evidence. People differ markedly in their preferences for color, but generally blue, green, and red hues seem to be most commonly preferred. However, the context in which colors are used can have a bearing on preferences. Aside from hue, Bennett (1977) reports research by others indicating that people tend to prefer light to dark colors and saturated colors to unsaturated colors. When objects are placed against a background, people tend to prefer high-contrast combinations, i.e., light-colored objects with dark background colors or vice versa.

In all this, however, we need to remember that the aesthetics of interior design, with various combinations of colors in rooms and furnishings, are essentially based on subjective preferences, and that one person's meat is another person's poison.

Kunishima and Yanase (1985) presented to subjects color slides of empty living rooms with different wall colors. A total of 60 slides varying in hue, brightness, and saturation were presented. Subjects rated each room on 16 bipolar-adjective scales (e.g., spacious-crowded, calm-restless). Through factor analysis, three underlying factors were identified that seemed to account for the ratings on the individual bipolar scales: activity (gay, fresh, happy); evaluation (elegant, refined, comfortable, good); and warmth (warm, stuffy). The authors found that hue had its greatest effect on the warmth factor with reds rated high, greens neutral, and blues low. Increased brightness resulted in higher ratings on the activity factor. Saturation affected the evaluation factor and to a lesser degree the activity factor. As saturation increased, the rating on the evaluation factor decreased sharply while the rating on the activity factor increased somewhat.

The relationship between hue and perceived warmth has been reported by many investigators. Greene and Bell (1980), however, carried the idea a step further. They assessed whether different colors actually affected people's perception of thermal comfort. After all, if people felt warmer in the presence of "warm" colors (reds and oranges), why not lower the thermostat and still achieve the same level of thermal comfort? Unfortunately, being in a room of a given temperature, whether it was painted red (warm color) or blue (cool color), resulted in the same degree of thermal comfort.

In assessing the influence of color in our lives, one has the impression of being caught between Scylla (accepting the many common beliefs and pronouncements about the effects of color) and Charybdis (rejecting such beliefs and notions, assuming that color is of only nominal consequence in human life—except possibly for the fate of paint stores). An in-between frame of reference actually seems to be warranted, but hard data about the various aspects of color are still limited.

SPECIAL-PURPOSE DWELLINGS

We briefly discuss two types of special-purpose dwellings: college dormitories and housing for the elderly and handicapped. College dormitories are an example of how the design of buildings can influence the social behavior of people. For the handicapped and many elderly people, circumstances argue for special attention to housing features. These circumstances include physical limitations and health conditions, the fact that such persons spend more time in their housing units, and the additional cost constraints realized for many such people.

College Dormitories

Let us consider the effects of dormitory design on social behavior. As one phase of a study reported by Baum and Valins (1977), the investigators compared residents of suite-type and corridor-type dormitories in terms of the residential locations of their friends. (The individual room in both types of dormitories were nearly the same size, and the average square footage per person, including lounges and bathrooms, was about the same.) The comparison showed quite clearly that more of the friends of residents of suite-type dormitories lived within the same dormitory (an average of 61 percent) as contrasted with friends of residents of corridor-type dormitories (an average of 27 percent). These results strongly suggest that suite-type dormitories are more conducive to the formation of friendships.

Another phase of this study dealt with residents of long versus short corridors. In general, those living on short corridors had a sense of greater privacy and of less crowding, and they experienced less aggressiveness on the part of other residents than did those living on long corridors. These and other results from this and other studies rather consistently reflect differences in the social behavior and attitudes of residents of different type of dormitories.

Housing for the Handicapped and Elderly

It has been estimated that by the year 2010 almost one-quarter of the U.S. population will be 55 years old or older and 40 percent of the elderly population will be 75 years of age or older (Newman, Zais, and Struyk, 1984). Roughly 45 percent of the current elderly population reported some sort of health limitation which, to some degree, restricted their activities (U.S. Public Health Service, 1981). The types of disabilities that are receptive to design of facilities and products for the elderly and handicapped include difficulty in interpreting information; loss of sight; loss of hearing; poor balance; lack of coordination; limited stamina; difficulty in moving the head; loss of upper-extremity skills; difficulty with handling and fingering; difficulty bending, kneeling, etc.; reliance on walking aids; and inability to use lower extremities.

Numerous publications discuss designing for the elderly and handicapped (see, for example, Altman, Lawton, and Wohlwill, 1984; Hale, 1979; Howell,

1980; Koncelik, 1982; Parsons, 1981; and Steinfeld, 1979*a*, 1979*b*). And it is not feasible here to discuss the many facets of the problem. Rather, we simply highlight a few of the major aspects that have important human factors implications.

Some of the specific features of housing facilities cited as being especially relevant for the aged are the following (Browning and Saran, 1978; Lang, 1978; Parsons, 1978; Rohles, 1978): safety and security; bathing facilities; handles and railings; warning devices; wide doorways and halls (for possible wheelchairs); adequate provision for mobility in general; furniture; facilities for recreation and entertainment (including TV, hobby facilities, etc.). Adequate provision for such features would apply equally to private homes and to various types of facilities for the elderly and retired, such as retirement housing developments, retirement homes, public housing, and convalescent homes.

BEYOND THE DWELLING UNIT

Multifamily housing developments involve human factors considerations over and above those associated with the individual dwelling units comprising them. These factors are associated with such features as the height, size, and arrangement of apartment buildings and with certain associated outdoor features. Becker (1974), for example, in a survey of housing developments in urban and suburban areas throughout New York State, found that the exterior appearance of a development was "very important" to most residents (67 percent). Variations in the shape, pattern, and form of buildings that increased their individuality were much appreciated; straight, rectilinear, and symmetric forms were strongly disliked. In addition, the barrier around a development can influence the affective response to the development, as illustrated in Figure 15-6. The wire fence at the left gives an institutional impression; yet probably it would not serve as an impenetrable barrier for someone intent on breaching it. The hedgelike fence at the right, however, creates a psychological boundary and a feeling of enclosure and so, while not communicating rejection to nonresidents, would tend to discourage strangers from entering. The burgeoning problems of urban centers are undoubtedly one of the major challenges of current life. The many facets of these problems leave few inhabitants unscathed. A partial inventory would include problems associated with health, recreation, mobility, segregation, education, congestion, physical housing, crime, and loss of individuality. The current manifestations of these problems lend some validity to the forebodings of Ralph Waldo Emerson and Henry Thoreau, who viewed with deep misgivings the encroachment of civilization on human life, especially in the form of large population centers. The tremendous population growth, however, makes it inevitable that many people must live in close proximity to others (and thus requires the existence of urban centers). Given this inevitability, however, we propose to operate on the hypothesis that, by proper design, urban centers can be created which might make it possible to achieve

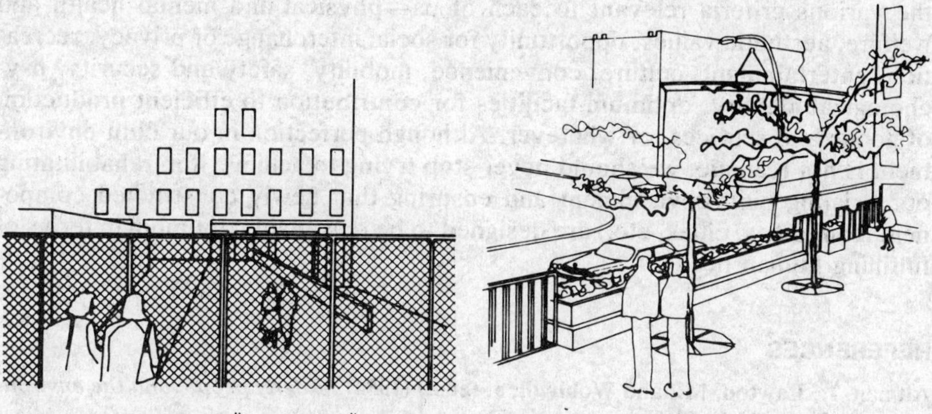

Chain-like fence gives "institutional" impression.

Hedge-like fence creates psychological boundary, with "soft" territorial definition.

FIGURE 15-6
Illustration of two types of barriers for a multifamily housing development. The one at the left gives an institutional feeling, whereas the one at the right has a more pleasing appearance yet still serves as a psychological boundary and gives a feeling of enclosure. (*Source: Adapted from Becker, 1974, Figs. 7 and 8.*)

the fulfillment of a wide spectrum of reasonable human values—perhaps even those that Emerson and Thoreau and maybe those that we might esteem.

As Proshansky, Ittelson, and Rivlin (1976) point out, cities are not just social, political, economic, and cultural systems, but geographical and physical systems as well. The interaction of these various features of cities with the inhabitants clearly influences the quality of life, but these influences are, of course, not the same for all inhabitants. The crowding and density in slum areas are, of course, in marked contrast to those features in spacious luxury apartments or in posh suburban areas.

It is not feasible within the limits of these pages to discuss the problems of urban communities and to speculate about their possible solution. The sociologists, social psychologists, and clinical psychologists clearly could contribute through research to the understanding of the impact of such communities on the lives of the inhabitants. In addition, however, it is believed that the human factors disciplines have much to contribute to the solution of some of the human problems of urban communities.

DISCUSSION

It is probable that utopia in the built environment that forms our living spaces is a will-o'-the-wisp, an unattainable goal. In other words, it is doubtful whether human beings' living space in the buildings and other structures they use and in the communities they build could ever provide for the broad-scale fulfillment of the various criteria relevant to each of us—physical and mental health and welfare, aesthetic values, opportunity for social interchange or privacy, recreation, entertainment, culture, convenience, mobility, safety and security, psychological identity, optimum facilities for contribution to efficient production of goods and services, or whatever. Although perfection in our built environment is not realistic, we should never stop trying to achieve it in rehabilitating our existing built environment and ensuring that newly constructed components (buildings, cities, etc.) are designed to be reasonably optimum in terms of fulfilling human needs.

REFERENCES

Altman, I., Lawton, M., and Wohlwill, J. (eds.) (1984). *Elderly people and the environment*. New York: Plenum.

Bächtold, R. (1964). Der moderne Wohnungs—und Siedlungsbau als soziologisches Problem. Dissertation, University of Fribourg, Switzerland.

Baird, J., Cassidy, B., and Kurr, J. (1978). Room preference as a function of architectural features and user activities. *Journal of Applied Psychology*, 63(6), 719–727.

Baum, A., and Valins, S. (1977). *Architecture and social behavior*. Hillside, NJ: Lawrence Erlbaum Associates.

Becker, F. D. (1974). *Design for living: The resident's view of multi-family living*. Ithaca, NY: Center for Urban Development Research, Cornell University.

Bennett, C. (1977). *Spaces for people: Human factors in design*. Englewood Cliffs, NJ: Prentice-Hall.

Black, J. C. (1968). Uses made of spaces in owner-occupied houses. Ph.D. thesis, University of Utah, Salt Lake City.

BOSTI (1981). *The impact of office environment on productivity and quality of working life: Comprehensive findings*. Buffalo, NY: Buffalo Organization for Social and Technological Innovation.

Boyce, P. (1981). *Human factors in lighting*. New York: Macmillan.

Brookes, M. J. (1972). Office landscape: Does it work? *Applied Ergonomics*. 3(4), 224–236.

Brookes, M. J., and Kaplan, A. (1972). The office environment: Space planning and affective behavior. *Human Factors*, 14(5), 373–391.

Browning, H. W., and Saran, C. (1978). A proposed plan of home safety for the older adult. *Proceedings of the Human Factors Society 22d Annual Meeting*. Santa Monica, CA: Human Factors Society pp. 592–596.

Campbell, D. E. (1979). Interior office design and visitor response. *Journal of Applied Psychology*, 64(6), 648–653.

Chambers, J. (1963). A study of attitudes and feelings towards windowless classrooms. Ed.D. dissertation, University of Tennessee.

Collins, B. (1975). *Windows and people: A literature survey*. NBS building science series. Washington: National Bureau of Standards.

Cornell, P., Luchetii, R., Mack, L., and Olson, G. (1989). CSCW anecdotes and directions. *Proceedings of the Human Factors Society 33d Annual Meeting*. Santa Monica, CA: Human Factors Society, pp. 867–871.

Galitz, W. (1980). *Human factors in office automation*. Atlanta: Life Office Management Association.

Grandjean, E. (1973). *Ergonomics of the home*. London: Taylor & Francis.

Greene, T., and Bell, P. (1980). Additional considerations concerning the effects of "warm" and "cool" colours on energy conservation. *Ergonomics*. 23, 949–954.

Hale, G. (1979). *The source book for the disabled*. London: Imprint Books.

Hedge, A. (1984). Ill health among office workers: An examination of the relationship between office design and employee well-being. In E. Grandjean (ed.), *Ergonomics and health in modern offices*. London: Taylor & Francis.

Helander, M. (1985). Emerging office automation systems. *Human Factors*, 27(1), 3–20.

Hodges, J., and Angalet, B. (1968). The prime technical information source—The local work environment. *Human Factors*, 10(4), 425–430.

Howell, S. (1980). *Designing for aging: Patterns of use*. Cambridge, MA: MIT Press.

Imamoglu, V. (1973). The effect of furniture density on the subjective evaluation of spaciousness and estimation of size of rooms. In R. Küller (ed.), *Architectural Psychology: Proceedings of the Lund Conference*. Stroudsburg, PA: Dowden, Hutchinson & Ross, pp. 341–352.

Kaye, S., and Murray, M. (1982). Evaluation of an architectural space as a function of variations in furniture arrangement, furniture density, and windows. *Human Factors*, 24(5), 609–618.

Kleeman, W., Jr. (1981). *The challenge of interior design*. Boston: CBI Publishing.

Knopf, C. (1982). *Proceedings of Probe Research Seminar on Voice Technology*. New Brunswick, NJ: Probe Research, Inc.

Koncelik, J. (1982). *Aging and the product environment*. Stroudsburg, PA: Dowden, Hutchinson & Ross.

Kunishima, M., and Yanase, T. (1985). Visual effects of wall color in living rooms. *Ergonomics*, 28(6), 869–882.

Lang, C. (1978). Seniors plan senior housing. *Proceedings of the Human Factors Society 22d Annual Meeting*. Santa Monica, CA: Human Factors Society. pp. 545–549.

Louis Harris and Associates, Inc. (1980). *The Steelcase national study of office environments: Do they work?* Grand Rapids, MI: Steelcase.

McCarrey, M. W., Peterson, L., Edwards, S. and Von Kulmiz, P. (1974). Landscape office attitudes. *Journal of Applied Psychology*, 59(3), 401–403.

Mack, L. (1989). Technology for computer-supported meetings. *Proceedings of the Human Factors Society 33d Annual Meeting*. Santa Monica, CA: Human Factors Society, pp. 857–861.

Manning, P. (ed.) (1965). *Office design: A study of environment* [SfB (92): UDC 725.23]. Liverpool, England: Pilkington Research Unit, Department of Building Science, University of Liverpool.

Mercer, A. (1979). Office environments and clerical behavior. *Environment and Planning B*. 6, 29–39.

Nemecek, J., and Grandjean, E. (1973). Results of an ergonomic investigation of large-space offices. *Human Factors*, 15(2), 111–124.

Newman, S., Zais, J., and Struyk, R. (1984). Housing older America. In I. Altman, M. Lawton, and J. Wohlwill (eds.), *Elderly people and the environment*. New York: Plenum.

Parsons, H. (1978). Bedrooms for the aged. *Proceedings of the Human Factors Society 22d Annual Meeting*. Santa Monica, CA: Human Factors Society. pp. 550–557.

Parsons, H. (1981). Residential design for the aging (for example, the bedroom). *Human Factors*, 23(1), 39–58.

Proshansky, H. M., Ittelson, W. H., and Rivlin, L. G. (1976). *Environmental psychology*. New York: Holt.
Rohles, F. H., Jr. (1978). Habitability of the elderly in public housing. *Proceedings of the Human Factors Society 22d Annual Meeting*. Santa Monica, CA: Human Factors Society. pp. 693–697.
Ruys, T. (1970). Windowless offices. Master's thesis, University of Washington, Seattle.
Sommer, R. (1974). *Tight spaces: Hard architecture and how to humanize it*. Englewood Cliffs, NJ: Prentice-Hall.
Steelcase (1978). *The Steelcase national study of office environments: Do they work?* Grand Rapids, MI: Steelcase, Inc.
Steinfeld, E. (1979*a*). *Access to the built evironment: A review of literature* (HUD-PDR-405). Washington, DC: Department of Housing and Urban Development.
Steinfeld, E. (1979*b*). *Accessible buildings for people with walking and reaching limitations* (HUD-PDR-397). Washington, DC: Department of Housing and Urban Development.
Sundstrom, E. (1986). *Work places: The psychology of the physical environment in offices and factories*. Cambridge, MA: Cambridge University Press.
Tikkanen, K. (1970). *Significance of windows in classrooms*. Master's thesis, University of California at Berkeley.
U.S. Public Health Service (1981). Current estimates from the national health interview survey, United States, 1980. *Vital and Health Statistics,* ser. 10, no. 139. Washington, DC: National Center for Health Statistics.
Wells, B. W. P. (1965). The psycho-social influence of building environment: Sociometric findings in large and small office spaces [SfB (92): UDC 301.15]. *Building Science,* 1, 153–175.
Wichman, H. (1984). Shifting from traditional to open offices: Problems suggested design principles. In H. Hendrick and O. Brown, Jr. (eds.), *Human factors in organizational design and management*. London: Elsevier.
Wise, J., and Wise, B. (1984). Humanizing the underground workplace: Environmental problems and design solutions. In H. Hendrick and O. Brown, Jr. (eds.), *Human factors in organizational design and management*. London: Elsevier.
Wools, R., and Canter, D. (1970). The effect of the meaning of buildings on behavior. *Applied Ergonomics,* 1(3), 144–150.
Zweigenhaft, R. L. (1976). Personal space in the faculty office: Desk placement and the student-faculty interaction. *Journal of Applied Psychology,* 61(4), 524–532.

PART FIVE

ENVIRONMENTAL CONDITIONS

16 Illumination

17 Climate

18 Noise

19 Motion

CHAPTER 16

ILLUMINATION

In many aspects of life we depend on the sun as our source of illumination, as in driving in the daylight, playing golf, and picking tomatoes. When human activities are carried on indoors, or at night, however, it is usually necessary to provide some form of artificial illumination. There does not, however, appear to be much of a difference between artificial illumination and natural daylight for performance of simple visual tasks, such as threading needles or proofreading (Santamaria and Bennett, 1981). The design of artificial illumination systems, as we shall see, does have an impact on the performance and comfort of those using the environment as well as on the effective responses of the people to the environment. Illuminating engineering is both an art and a science. The scientific aspects include the measurement of various lighting parameters and the design of energy-efficient lighting systems. The artistic side comes into play in combining light sources to create, for example, a particular mood in a restaurant, to highlight a display in a store, or to complement a particular color scheme. It is not our intention here to make anyone an illuminating engineer. Rather, our aim is to familiarize you with the basic concepts in the field and to illustrate the importance of proper illumination from a human factors point of view. We leave the artistic aspects to the artists and concentrate on the scientific aspects. In recent years, much has been written about proper ambient lighting for areas where people work with video display terminals (VDTs). In recognition of the special problems involved in designing a lighting environment for VDT use, we have devoted the last section of this chapter to those issues.

THE NATURE OF LIGHT

Light, according to the Illuminating Engineering Society (IES), is "radiant energy that is capable of exciting the retina (of the eye) and producing a visual sensation" (IES Nomenclature Committee, 1979). The entire electromagnetic spectrum consists of waves of radiant energy that vary from about 10^{-15} m to about 10^4 m in length. This tremendous range includes cosmic rays; gamma rays; x-rays; ultraviolet rays; the visible spectrum; infrared rays; radar; FM, TV, and radio broadcast waves; and power transmission—as illustrated in Figure 16-1.

The visible spectrum ranges from about 380 to 780 nm. The *nanometer* (formerly referred to as a *millimicron*) is a unit of wavelength equal to 10^{-9} (one-billionth) m. Light can be thought of as the aspect of radiant energy that is visible; it is then basically psychophysical in nature rather than purely physical or purely psychological.

Color

Variations in wavelength within the visible spectrum give rise to the perception of color, the violets being around 400 nm, blending into the blues (around 450 nm), the greens (around 500 nm), the yellow-oranges (around 600 nm), and the reds (around 700 nm and above). The eye cannot really see colors unless the ambient luminance (i.e., amount of ambient light reaching the eye) is above about 3 candela per square meter (cd/m^2). The concept of luminance is discussed later in this chapter.

Just as the ear is not equally sensitive to all frequencies of sound, the eye is not equally sensitive to all wavelengths of light. Unlike the ear, however, the eye is composed of two basic receptors, rods and cones (see Chapter 4), each of which has its own sensitivity function. At high levels of illumination, the rods and cones both function (*photopic vision*) and the eye is most sensitive to light wavelengths around 550 nm (green). As illumination levels decrease, however, the cones cease to function, the rods take over the entire job of seeing (*scotopic*

FIGURE 16-1
The radiant energy (electromagnetic) spectrum, showing the visible spectrum. *(Source: Adapted from* Light and color, *1968, p. 5.)*

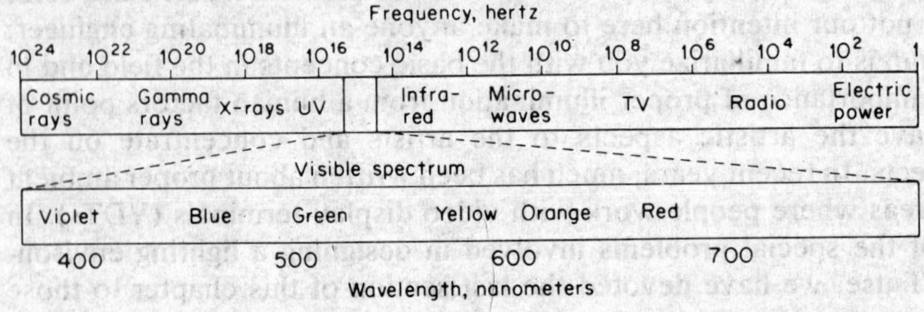

vision), and the eye becomes most sensitive to wavelengths around 500 nm (blue-green). This shift in sensitivity from photopic to scotopic vision is called the *Purkinje effect*. One practical application of this effect is that targets can be made blue-green to increase the probability of detection at night.

Light comes to us from two sources: *incandescent bodies* (''hot'' sources, such as the sun, luminaires, or a flame) and *luminescent bodies* (''cold'' sources, i.e., the objects we see in our environment which reflect light to us). A hot light source that includes all wavelengths in about equal proportions is called *white light*. Most light sources, as luminaires, have spectra that include most wavelengths but that tend to have more energy in certain areas of the spectrum than in others. These differences make the lights appear yellowish, reddish, bluish, etc. Illuminating engineers speak of the color temperature of a light source to describe its color appearance. The *color temperature* of a light source is the temperature at which the walls of a furnace must be maintained so that light from a small hole in it will have the same (or, in the case of fluorescent lamps, similar) color appearance as that of the light source. The noon sun, i.e., daylight, has a color temperature of 5500 K. Lower temperatures are more reddish, and higher temperatures are more bluish. Color temperature specifies only the color appearance of a light source and not the actual spectral composition of the light. Different spectral compositions can have the same color appearance but produce different colors from a surface illuminated by them. Therefore, color temperature is an incomplete and unreliable indicator of how objects will appear (in terms of color) when they are illuminated by the source. Color temperature does have value, however, for comparing the color appearance of similar types of light sources such as incandescent or fluorescent lights.

As light from a hot source falls upon an object, some specific combination of wavelengths is absorbed by the object. The light that is so reflected is the effect of the interaction of the spectral characteristics of the light source with the spectral absorption characteristics of the object. If a colored object is viewed under white light, it is seen in its *natural* color. If it is viewed under a light that has a concentration of energy in a limited segment of the spectrum, the light may alter the apparent color of the object. We have probably all experienced, at one time or another, the difficulty of finding a familiar colored car in a parking lot illuminated by yellow sodium lights. The spectral composition of the yellow lights changes the perceived colors of the cars.

The light that is reflected from an object—and that produces our sensations of the object's color—can be described in terms of three characteristics: *dominant wavelength, luminance,* and *saturation*. The dominant wavelength gives rise to our sensation of *hue*, or color. Luminance, sometimes referred to as *value*, is associated with the relative amount of light reflected and gives rise to our sensation of *lightness*, or *brightness*. Although there is a general relationship between the luminance of light and the subjective response of lightness, not all colors that reflect equal total amounts of light energy are necessarily perceived as equal in lightness. This is due to the fact that the eye is differentially sensitive to various wavelengths, as we discussed earlier. Saturation is the predominance of a narrow range of wavelengths, or the degree of difference of a color from a gray with the same luminance. Saturation is also referred to as *purity*, or *chroma*.

Color Systems It is very difficult to describe a particular color given the variations in hue, saturation, and lightness that are possible. Although difficult, it is often necessary to communicate such information without actually reproducing the color. For example, a design specification may have to describe the colors for, say, visual codes or labels. To assist in this communication process, various color systems have been developed to serve as standards for describing colors.

Dating at least to the seventeenth century, there have been dozens of attempts to order colors in some form of three-dimensional space (Hesselgren, 1984). Fortunately, the systems in use today have several features in common. Most systems are based on a color cone such as the one shown in Figure 16-2. Neutral-gray samples are located on the central vertical axis, going from black at the bottom to white at the top. The hues are arranged around the axis in the same sequence as the hues in the visual spectrum (rainbow), with purples being added where the reds and blues meet. The distance out from the central vertical axis represents the degree of saturation, with the intense hues on the periphery and the more "washed out" hues nearer the axis. The major color systems differ in the spacing of the gray samples between the black and white endpoints

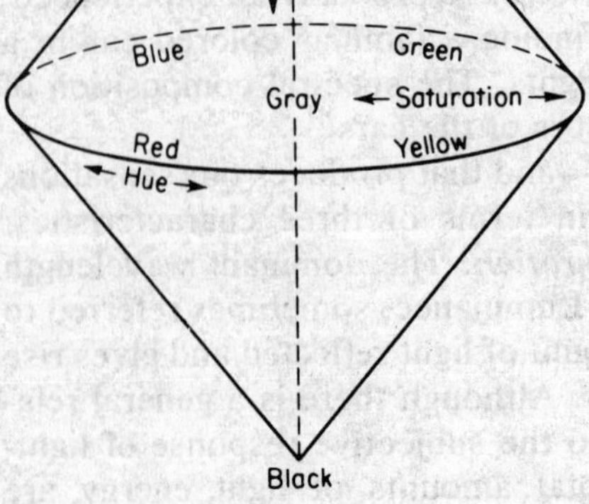

FIGURE 16-2
The color cone. Hue is shown on the circumference, lightness (from light to dark) on the vertical, and saturation on the radius from circumference to the center.

of the vertical axis, in the spacing of the hues around the hue circle, and in the location of the samples in each hue plane.

Several color systems consist of color plates or chips arranged in order based on the color cone. These systems include the Ostwald system (Container Corporation of America, 1942), the DIN color system (Richter, 1955), the natural color system (Hard and Sivik, 1981), and the Munsell system (Munsell Color Co., 1973). The Munsell system is probably the most widely known. It is intended to present the colors so that they are subjectively evaluated as differing from each other by equal amounts of the three basic color attributes. The system contains over 1000 color samples, each given a unique notation based on its hue (dominant wavelength), value (luminance), and chroma (saturation). For example, a strong red would be designated 7.5R/4/12. For a more detailed discussion of various color order systems, see Nimeroff (1968).

Another widely used system for describing color is the Commission Internationale de l'Eclairage (CIE) colorimetric system (CIE, 1971). The system is based on the fact that any color can be matched by a combination of three spectral wavelengths of light in the red, green, and blue regions. The CIE system determines the relative proportions of three imaginary light sources needed to produce a color. These imaginary light sources are mathematically derived and are aptly named X, Y, and Z. The X source corresponds, roughly, to the red region of the spectrum, the Y source to the green region, and the Z source to the blue region. By using techniques beyond the scope of this book, the relative proportion of each of these sources that produces a particular color is determined. [See Hardesty and Projector (1973) for computational procedures and examples.] These proportions, designated x, y, and z, are called the *color coordinates* of the color being matched. Since the three coordinates must add to 1.0, only two coordinates (x and y) are used to specify a color (z can be computed by subtraction). Figure 16-3 shows the CIE 1931 chromaticity diagram produced by combinations of the x and y color coordinates. All pure colors (that is, those which consist of a single wavelength) lie on the outer edge of the figure (the edge is called the *spectrum locus*). Figure 16-3 shows some of the wavelengths from 400 to 770 nm on the spectrum locus. Also indicated in the figure is the equal-energy point E, which represents a perfectly white, colorless surface. The saturation of a color increases as one moves toward the spectrum locus and away from the equal energy point. The color names in Figure 16-3 describe the general color regions of the diagram; the colors actually blend from one to the other.

Note that several CIE chromaticity diagrams exist. The CIE 1931 diagram, the most frequently referenced, is for use with colors that occupy visual angles up to 4°. The CIE 1964 diagram is used for colors subtending larger visual angles. Both these diagrams cover only hue and saturation, which together make up the quality of *chromaticity*. Brightness is not addressed by either diagram.

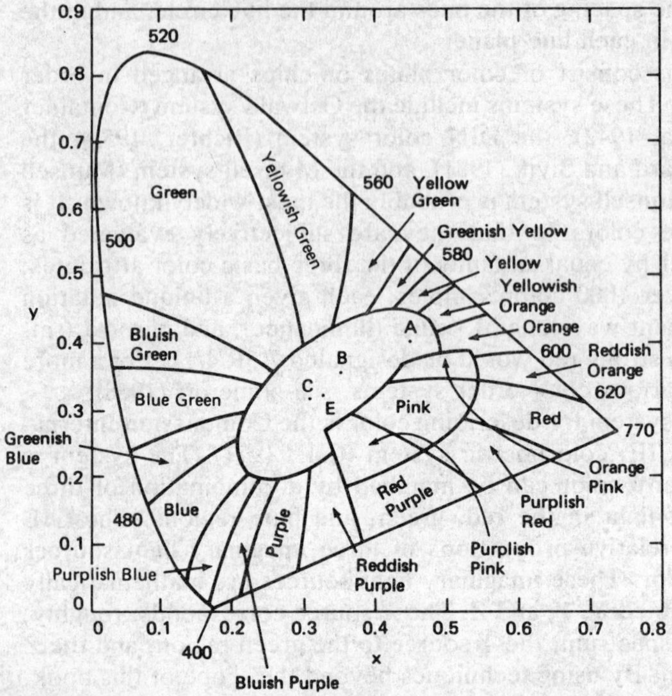

FIGURE 16-3
The CIE 1931 chromaticity diagram with descriptions of the various color regions. Here A, B, and C represent the positions of standard sources roughly equivalent to gas-filled incandescent lamps, noon sunlight, and average daylight, respectively; E represents the equal-energy point corresponding to perfect white, colorless light. See text for discussion of the x and y coordinates. (*Source: Adapted from Hardesty and Projector, 1973.*)

Measurement of Light

There are many concepts and terms that relate to the measurement of light (*photometry*). We define a few of them here and show their interrelationships. One source of confusion to the photometric novice is the multitude of different measurement units used in the field. Part of this problem stems from the existence of two systems of measurement, the U.S. Customary System (USCS) and the International System of units (or SI units). We bow to the rest of the world and the scientific community and encourage the use of SI units. The USCS units are mentioned, however, and conversions are given to aid those of us who were weaned on footcandles and foot-lamberts.

The fundamental photometric quantity is *luminous flux,* which is the rate at which light energy is emitted from a source. The unit of luminous flux is the *lumen* (lm). Luminous flux is a somewhat esoteric concept which is similar to other flow rates such as gallons per minute. Time is implied in the unit of luminous flux. Thus one would say that light is emitted from a 100-W incandescent lamp at a rate of 1740 lm. The *luminous intensity* of a light source is measured in lumens emitted by the source per unit solid angle.* The unit of luminous intensity is the *candela* (cd). A 1-cd source emits 12.57 lm.

Consider a source of some luminous intensity emitting luminous flux in all directions. Imagine the source as being placed inside (at the center of) a sphere. The amount of light striking any point on the inside surface of the sphere is called *illumination*, or *illuminance*. It is measured in terms of luminous flux per unit area, as, for example, lumens per square foot (lm/ft^2) or lumens per square meter (lm/m^2). To obscure matters a bit, special names have been given to units of illuminance. One lumen per square foot is called a *footcandle* (fc), a USCS unit, whereas 1 lumen per square meter is called a *lux* (lx), an SI unit. One footcandle equals 10.76 lx; however, an accepted practice for some purposes is to consider 1 fc to equal 10 lx and to forget the fraction (Kaufman and Christensen, 1984).

Figure 16-4 illustrates the relationship among candelas, footcandles, and lux. Our 1-cd source emits 12.57 lm, and the total surface area of a sphere equals 12.57 times the radius squared (or $4\pi r^2$). At a radius of 1 m, therefore, our 1-cd

*A solid angle is measured in steradians (sr). There are 12.57 sr in a sphere.

FIGURE 16-4
Illustration of the distribution of light from a light source following the inverse-square law. (*Source: General Electric Company, 1965, p. 5.*)

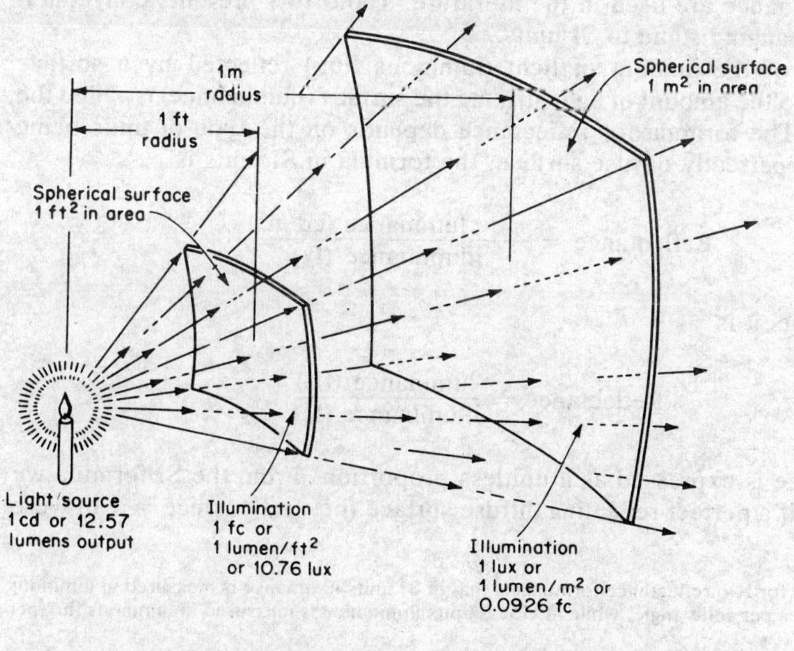

source is distributing 12.57 lm evenly over the 12.57 m^2 of surface area. Thus, the amount of light on any point is 1 lm/m^2, or 1 lx. The same logic holds at a radius of 1 ft for the footcandle.

The amount of illumination striking a surface from a point source follows the *inverse-square law:*

$$\text{Illuminance (lx)} = \frac{\text{candlepower (cd)}}{D^2}$$

where D is the distance from the source in meters. At 2 m a 1-cd source would produce ¼ lx, and at 3 m it would produce 1/9 lx.

We now have light striking a surface (*illuminance*), but we must consider what happens to light that strikes a surface. Some of it is absorbed, and some of it is reflected. The light that is reflected from the surface of objects is what allows us to "see" objects—their configuration and color. The amount of light per unit area leaving a surface is called *luminance*. The light leaving the surface may be reflected by the surface or emitted by the surface, as would occur with a fluorescent light panel. The amount of light can be measured in terms of luminous flux (lumens) or luminous intensity (candelas). When the amount of light is measured in lumens and area is in square feet, the USCS unit of luminance is the *foot-lambert* (fL). When the amount of light is measured in candelas and the area is in square meters, the SI unit of luminance is *candela per square meter* (cd/m^2). A piece of white paper lying on a table illuminated by 300 lx will have a luminance of about 70–80 cd/m^2. Actually numerous measures of luminance are used in the literature. Table 16-1 presents conversion factors for changing some to SI units.

The ratio of the amount of light (luminous flux) reflected by a surface (luminance) to the amount of light striking the surface (illuminance) is called the *reflectance*. The formula for reflectance depends on the type of units being used.* For a perfectly diffuse surface, the formula in SI units is

$$\text{Reflectance} = \frac{\pi \times \text{luminance (cd/m}^2)}{\text{illuminance (lx)}}$$

In USCS units it is

$$\text{Reflectance} = \frac{\text{luminance (fL)}}{\text{illuminance (fc)}}$$

Reflectance is expressed as a unitless proportion. From the SI formula we can see that if a perfect reflecting diffuse surface (i.e., reflectance = 1.0) were

*The reason for two reflectance formulas is that in SI units luminance is measured in luminous flux per unit area per solid angle, while in USCS units luminance is measured in luminous flux per unit area.

TABLE 16-1
CONVERSION FACTORS FOR VARIOUS UNITS OF LUMINANCE

Unit	To convert to cd/m², multiply by:
Nit	1.0
Stilb	1×10^4
Apostilb	0.3183
Millilambert	3.183
Foot-lambert	3.426
Candles per square inch	1.55×10^3
Candles per square foot	10.76

illuminated by 1 lx, the surface luminance would be $1/\pi$ cd/m². When the surface being considered is not perfectly diffuse, reflectance is replaced by the luminance factor. The *luminance factor* is the ratio of the luminance of a surface viewed from a particular position and lit in a specified way to the luminance of a diffusely reflecting white surface viewed from the same direction and lit in the same way.

LAMPS AND LUMINAIRES

The term *lamp* is a generic term for an artificial source of light. A *luminaire*, however, is a complete lighting unit consisting of a lamp or lamps together with the parts designed to distribute the light, to position and protect the lamps, and to connect the lamps to the power supply (IES Nomenclature Committee, 1979). A trip to a lighting store will quickly convince you that lamps come in hundreds of configurations and sizes and that luminaire designs are limited only by the taste of the designer, ranging from simple globes to elaborate crystal chandeliers.

In the next two sections we introduce some of the generic types of lamps and luminaires and discuss some of the implications of choosing one or another.

Lamps

There are two main classes of lamps: *incandescent filament lamps*, in which light is produced by electric heating of a filament or by combustion of gases within a thin mesh mantle, and *gas-discharge lamps*, in which light is produced by the passage of an electric current through a gas.

Gas-discharge lamps are of three types: high-intensity discharge (HID) lamps including mercury, metal halide, and high-pressure sodium lamps; low-pressure sodium lamps; and fluorescent lamps. In the case of fluorescent lamps, the radiation created by the electric current passing through the gas is invisible to the eye (it is ultraviolet radiation). This radiation, however, is used to excite a phosphor coating on the inner surface of the bulb, which in turn produces visible radiation. Different phosphors can create different colors of illumination.

Lamp Color The various types of lamp differ markedly in terms of their characteristic spectral distribution, or what we plain folk call *color*. Figure 16-5 shows representative distributions for two types of lamps. These differences can affect task performance and the subjective impressions of the people in the illumination environment. The CIE (1974) has developed the general *color rendering index* (CRI) for quantifying the accuracy with which any specific light source renders colors. As pointed out by Boyce (1981), however, the CRI gives only gross indications of color-rendering accuracy. Nonetheless, the CRI is a good indicator of expected performance in color judgment tasks that involve a wide range of colors. As the CRI increases, color judgment errors tend to decrease, as can be seen in Figure 16-6, which presents results of a color judgment task under various types of light sources with different CRIs. Boyce (1988) recommends avoiding lamps with CRIs below about 60 when accurate color judgments are important. Collins and Worthey (1985), for example, tested meat and poultry inspectors under five different light sources (incandescent, cool-white fluorescent, deluxe cool-white fluorescent, high-pressure sodium discharge, and low-pressure sodium discharge). They found that more inspection errors were made when the CRI of the light source was less than 60.

Table 16-2 summarizes color characteristics of several common types of lamps. In addition to indicating each lamp's approximate color temperature and CRI, Table 16-2 also indicates the colors that tend to be strengthened and those that tend to be grayed by each source. Where color judgment is important, the selection of the proper lamp can be critical.

Energy Considerations In today's world, energy conservation has become a prime goal in system design, and this is especially true in design of lighting systems. The efficiency of a light source, called *lamp efficacy*, is measured in terms of the amount of light produced [lumens (lm)] per unit of power consumed [watts (W)]. Figure 16-7 shows some typical lamp efficacy values for various types of light sources.

Switching from incandescent to fluorescent lights can result in real dollar savings. For example, compared to the incandescent lamp, the fluorescent tube represents a 41 percent energy savings, lasts 20 times longer, and gives 30 percent more light to boot! And who said there was no such thing as a free lunch?

Incidentally, don't be fooled into thinking all incandescent bulbs of the same wattage give off the same amount of light—they don't! Bulb manufacturers are now required to put the tested lumen output on their packages. For example, a long-life 100-W bulb may give off as little as 1470 lm compared to perhaps 1740 lm for a standard 100-W bulb. Proper choice of light source often can result in substantial energy and cost savings—a worthy goal for any human factors design effort.

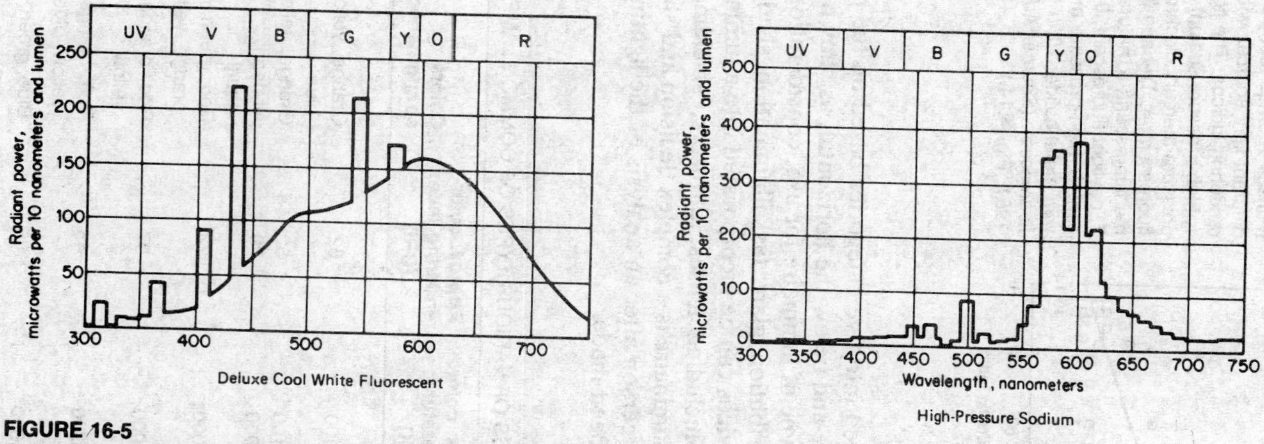

FIGURE 16-5
Representative spectral distributions (chromaticity diagrams) for high-pressure sodium, and deluxe cool-white fluorescent lamps. (*Source: IES, 1976, Figs. 4-13, 4-23.*)

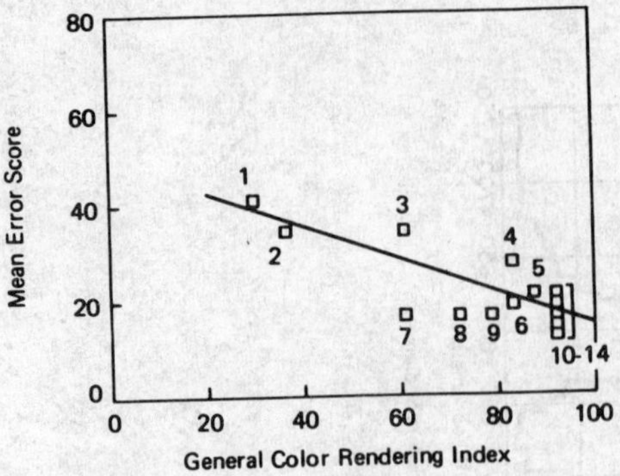

FIGURE 16-6
Mean error scores on the Farnsworth-Munsell 100-hue test performed under different light sources with different color rendering indices. The light sources are: 1, high-pressure sodium discharge; 2, high-pressure mercury discharge; 3, Homelite fluorescent; 4, Tri-band fluorescent; 5, Kolor-rite fluorescent; 6, natural fluorescent; 7, daylight fluorescent; 8, Plus-White fluorescent; 9, high-pressure mercury discharge with metal halide; 10 to 14, artificial daylight fluorescent. (*Source: Adapted from Boyce, 1981*, Fig. 4.14.)

Luminaires

Luminaires are classified into five categories based on the proportion of light (lumens) emitted above and below the horizontal, as shown in Figure 16-8. In selecting a particular type of luminaire for use, consideration must be given to the pattern of light distribution, glare, task illumination, shadowing, and energy efficiency. Various devices can be incorporated into a luminaire to control the distribution of light, including lenses, diffusers, shielding, and reflectors. Choice of an efficient luminaire is a complex decision and should be made by a qualified, experienced person after an analysis of the lighting needs and physical environment has been made.

TABLE 16-2
COLOR CHARACTERISTICS OF VARIOUS TYPES OF COMMON LAMPS

Type of lamp	Approx. color temperature (K)	Approx. color-rendering index (CRI)	Colors strengthened	Colors grayed
Fluorescent				
Cool white	4400	62	Orange, yellow blue	Red
Warm white	3100	52	Orange, yellow	Red, blue, green
Deluxe cool-white	4000	89	All nearly equal	None very much
Deluxe warm-white	3000	73	Red, green orange, yellow	Blue
Incandescent	3000	97	Red, orange yellow	Blue
Color-improved mercury	3900	45	Red, orange yellow, blue	Green
Clear metal halide	4600	65	Blue, green	Red
High pressure	2200	21	Yellow, orange green	Red, blue

Source: Wotton, 1986.

FIGURE 16-7
Lamp efficacy ranges for common light sources. (*Source: IES, 1976, Fig. 3-4.*)

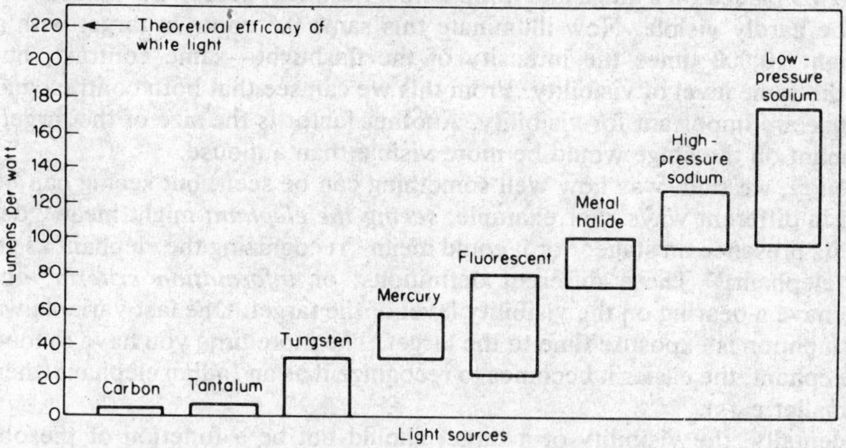

THE CONCEPT OF VISIBILITY

Visibility refers to how well something can be seen by the human eye. Visibility, therefore, involves human judgment. There is no device that can measure visibility directly; a human must always be involved in its determination. One key factor influencing visibility of a target is how well it stands out from its background, that is, its *contrast*, which was defined in Chapter 4. Although contrast is related to visibility, it is not the same as visibility. DiLaura

FIGURE 16-8
Types of luminaires based on the proportion of lumens emitted above and below the horizontal. (*Source: IES, 1976, Fig. 5-6.*)

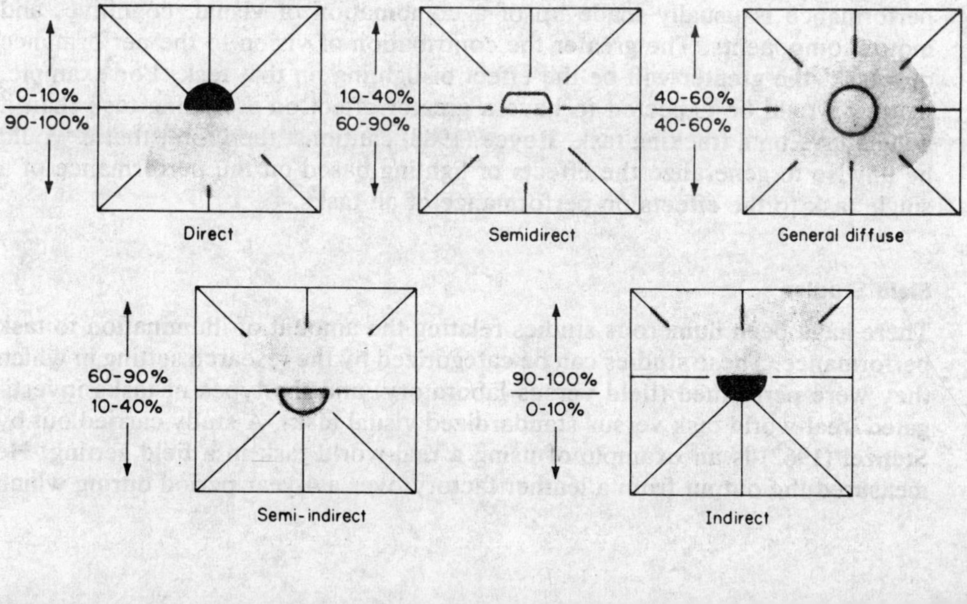

(1978) provides a simple illustration of this. Imagine a target with a contrast equal to 0.5 placed on a stage and lit up with a flashlight held in the balcony. It would be hardly visible. Now illuminate this same 0.5 contrast target with a stage light 10,000 times the intensity of the flashlight—same contrast, but hardly the same level of visibility. From this we can see that both contrast and luminance are important for visibility. Another factor is the size of the target; an elephant on the stage would be more visible than a mouse.

Visibility, we said, was how well something can be seen, but seeing can be defined in different ways. For example, *seeing the elephant* might mean "detecting its presence on stage," or it could mean "recognizing the elephant as an Indian elephant." These different definitions, or *information criteria*, obviously have a bearing on the visibility level of the target. One last variable we should mention is exposure time to the target. The more time you have to look at the elephant, the easier it becomes to recognize it as an Indian elephant (they have smaller ears).

Incidentally, the visibility of a target should not be a function of the observer; visibility should be a characteristic of the task itself.

EFFECTS OF LIGHTING ON PERFORMANCE

We have already related bits and pieces of information about the performance effects of lighting, including accuracy of color judgment under various types of lamps. As pointed out by Boyce (1981), lighting itself cannot produce work output. What lighting can do is to make details easier to see and colors easier to discriminate without producing discomfort or distraction. Workers can then use this increased ease of seeing to increase output, if they have the motivation and ability to do so.

Lighting influences the performance of different tasks in different ways. Lighting parameters can only effect the visual aspects of a task, but task performance is usually made up of a combination of visual, cognitive, and motor components. The greater the contribution of vision to the performance of a task, the greater will be the effect of lighting on that task. For example, lighting would be expected to have a greater effect on a reading task than it would have on a tracking task. Boyce (1988) cautions, therefore, that it would be unwise to generalize the effects of lighting based on the performance of a single task to the effects on performance of all tasks.

Field Studies

There have been numerous studies relating the amount of illumination to task performance. These studies can be categorized by the research setting in which they were performed (field versus laboratory) and the types of tasks investigated (real-world task versus standardized visual task). A study carried out by Stenzel (1962) is an example of using a real-world task in a field setting. He measured the output from a leather factory over a 4-year period during which the illumination was changed from 350 to 1000 lx. Output increased when the illumination level was increased. Knave (1984) presents an example of a study using a standardized visual task (visual acuity eye chart) in a field setting (a foundry). He reports that increasing illumination improved acuity as measured by a low-contrast eye chart.

Field studies often lack sufficient experimental controls to rule out alternative explanations for the changes in performance under the lighting conditions tested. Often, more than just the lighting changes. Such things as supervision, the task itself, and the physical design of the workplace are often altered along with the change in the illumination. Further, the change in illumination often involves multiple lighting changes. Not only might the illumination be increased, but also different types of lamps and luminaires may be used with different spectral characteristics, along with changes in the number and position of the lamps. Under such conditions it is not possible to isolate the specific causal factor that changed performance. All we can say is that performance was not the same under the conditions tested.

Laboratory Studies

In an attempt to overcome the problems inherent in field studies, investigators have moved into the more controlled laboratory setting. Here greater control can be exerted over outside influences, and the specific lighting characteristics of interest can be manipulated independently of other lighting parameters. Here, too, investigators have used real-world and standardized visual tasks. An example of using real-world tasks in the laboratory is the work done by Merritt et al. (1983) in which they investigated the effect of amount of illumination on detection of important features in underground mining, including holes and rubble on the floor and a crack appearing in a rock wall.

Examples of laboratory studies using a standardized visual task are the classic studies of Blackwell (Blackwell, 1959, 1961, 1964, 1967; Blackwell and Blackwell, 1968, 1971). The task was to detect the presence of a uniformly luminous disk which subtended a visual angle of 4 minutes (approximately 1.1 mm at a distance of 1 m) presented on a uniformly luminous screen. What Blackwell found was that when the background luminance was decreased (analogous to decreasing the amount of illumination on the task), the contrast of the just-barely-visible disk had to be increased to make it just barely visible again.

Laboratory studies are not without problems, however. First, subjects know they are in an experiment and are usually motivated to perform well. This high motivation, however, could mask the effects of subtle lighting changes. Standardized visual tasks performed in the laboratory are usually simple tasks, often of a reading variety, which do not require depth perception, texture perception, perception of moving objects, or peripheral perception. This limits the generalizability of the findings. Because laboratory studies purposely exclude extraneous variables which might influence performance, there is always the possibility that the effect found in the pristine laboratory will "wash out" when it is brought into the real world, where other factors can influence performance.

Models of Visual Performance

Various models have been put forth that mathematically relate lighting parameters to visual performance. Models of threshold visual performance, where one can just barely see or recognize the object of interest, are concerned with the probability of the target being detected or recognized. The work of Blackwell generated such a model, called the *visibility level model*.

On the other hand, models of supratheshold visual performance, where just seeing or recognizing the target is not a problem, are concerned with questions of time; for example, how quickly can something be read? Rea (1986), for example, used a numerical verification task in which subjects were required to indicate discrepancies between two printed lists of five-figure numbers. One list (the response list) was always printed in high-contrast ink; the contrast of the other list (the reference list) was varied. Rea, in his model, relates contrast and adaptation luminance to the time taken to read the reference list.

Boyce (1988) points out that none of the visual performance models is capable of predicting task performance except where the nonvisual components of performance approach zero, and none can, at present, deal with visual performance of visually complex tasks. He concludes that we still have a long way to go with respect to modeling the effects of lighting on visual performance. Despite the limitations of existing models, it is still possible to make some general conclusions regarding the effect of lighting on visual performance.

General Conclusions

Although field and laboratory studies have their unique problems, when similar findings emerge from studies carried out in both settings, we can have more confidence in the conclusions. Fortunately, such is the case with the effects of amount of illumination on performance. Although we cannot review all the studies in this area, Figure 16-9 presents representative results relating illumination (illuminance) level to time required to complete several industrial tasks (Bennett, Chitlangia, and Pangrekar, 1977). The general conclusion from such research is that increasing the level of illumination results in smaller and smaller improvements in performance until performance levels off. The point where this leveling off occurs is different for different tasks. Generally, the more difficult the task, i.e., the smaller the detail or the lower the contrast, the higher the illumination level at which this occurs.

Boyce (1981) notes that larger improvements in visual performance can be achieved by changing features of the task (i.e., increasing its size or contrast) than by increasing the illumination level, at least over any range of practical interest. Further, it is rarely possible to make performance on a visually

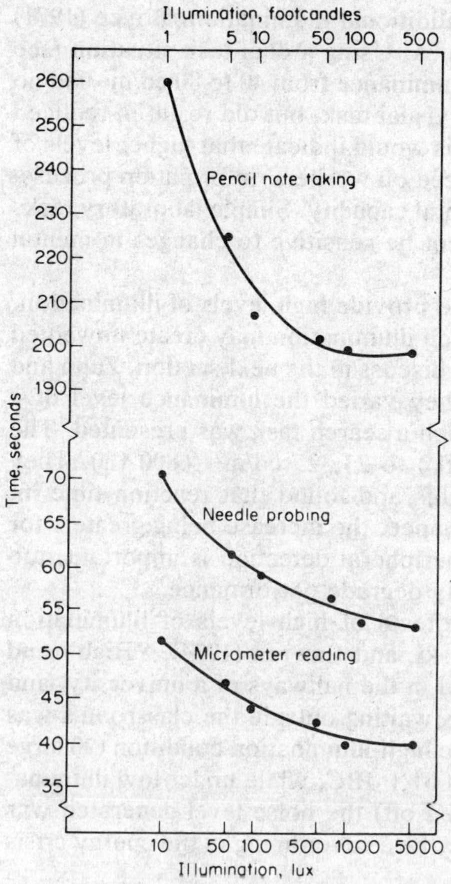

FIGURE 16-9
Relationship between amount of illumination and task completion time for three representative industrial tasks. (*Source: Bennett, Chitlangia, and Pangrekar, 1977, Figs. 1, 3, and 4.*)

difficult task (small size, low contrast) reach the same level as that on a visually easy task simply by increasing the illumination level.

This fact has lead to some controversy over just how important the amount of illumination is for task performance. Results from several sources (including Bennett, Chitlangia, and Pangrekar, 1977; Hughes and McNelis, 1978) indicate that the age of the observer has a major impact on the results, with age taking its toll; the amount of illumination may be more important for older persons. Ross (1978), after reviewing numerous studies, concludes that above 34 to 68 cd/m² (10 to 20 fL) of background luminance, other variables, notably age and print quality, are more important determiners of performance than amount of illumination. Ross further concludes that increasing illumination above 500 lx (50 fc) results in little additional improvement in task performance. This can be seen in Figure 16-9.

Although performance levels off with increased illumination, it may be that people must expend additional effort at lower illumination levels to maintain their performance. In essence, although performance levels off, the difficulty of the task may continue to decrease with additional illumination. Boyce (1971) presents some evidence to support this idea. Using a dual-task situation (see Chapter 3), he found that increasing task luminance from 40 to 80 cd/m^2 had no effect on the performance of the primary visual task, but did result in reduced errors on the secondary auditory task. This would indicate that higher levels of illumination may reduce the demands placed on workers' information processing systems and hence increase spare mental capacity. Simple laboratory tasks used to assess illumination effects may not be sensitive to changes in mental processing demands.

It may not always be wise, however, to provide high levels of illumination. Besides being wasteful of energy, too much illumination may create unwanted effects. Glare is one such effect which we discuss in the next section. Zahn and Haines (1971) offer another example. They varied the luminance level of a centrally located diffuse screen upon which a search task was presented. The luminance varied from 29.1 cd/m^2 (8.5 fL) to 23,325 cd/m^2 (6800 fL). They measured reaction time to peripheral lights and found that reaction time increased with increased central task luminance, the increase being greatest for the more peripheral locations. Thus, if peripheral detection is important, too much central task luminance may actually degrade performance.

Another example of the unpleasant effects of high levels of illumination comes from a study by Sanders, Gustanski, and Lawton (1974). "High" and "low" levels of illumination were created in the hallways of a university, and the noise level generated by the students waiting outside the classrooms was measured. The mean noise level under the high-illumination condition (20 large overhead fluorescent light panels lit) was 61.1 dB*C*, while under low illumination (two-thirds of the panels were turned off) the noise level generated was only 50.3 dB*C*. Here then, it appears, may be a silver lining in the energy crisis cloud—peace and quiet.

HOW MUCH IS ENOUGH?

The problem of determining the level of illumination that should be provided for various visual tasks has occupied the attention of illuminating engineers, psychologists, and others for many years. Over the years there has been a succession of recommendations, each claiming to provide adequate illumination for visual comfort and task performance. The recommended levels, however, continually increase. Current recommended levels are about 5 times greater than the levels recommended 30 years ago for the same tasks. The recommended levels increase as the efficiency of lighting improves.

One method for determining required illumination levels is based on the work of Blackwell which we discussed previously. This method is endorsed by the CIE (1972) and was used by the Illuminating Engineering Society (IES) until 1981. Because of uncertainties in the Blackwell approach, however, the IES in 1981 adopted a much simpler approach for determining minimum levels of illumination. The approach is based on consensus among lighting experts as to the level of illumination they considered adequate for the situation under study. The IES recognizes that the approach is an oversimplified solution to a difficult and complex problem; but if it is used with intelligence and judgment, this approach should aid in the design of effective lighting systems (Kaufman and Christensen, 1984). It should be pointed out, however, that the IES recommendations prescribe higher levels of illumination than do European guidelines. The following are a few examples:

Type of work	German DIN	IES
Precise assembly work	1000 lx	3000 lx
Very precise machine tool work	1000	7500
General office work	500	750

The first step in the IES procedure is to identify the type of activity to be performed in the area for which illumination recommendations are sought. The *IES Lighting Handbook* (Kaufman and Christensen, 1984) contains extensive tables listing various types of tasks, each referencing one of the illumination categories shown in Table 16-3. Although we cannot reproduce the lists of tasks here, the column "Type of activity" in Table 16-3 contains enough information to give a sense of what is involved in each category. Note that categories A through C do not involve visual tasks.

Table 16-3 lists, for each category, a range (low-middle-high) of illuminances. To decide which of the three values to choose, one must compute a correction factor by using Table 16-4. For each characteristic in Table 16-4 (workers' ages, speed or accuracy, and reflectance of task background) a weight is assigned (−1, 0, +1). These weights are then added algebraically to obtain the total weighting factor (TWF), (for example, +1 − 1 = 0). However, since categories A through C do not involve visual tasks, the *speed or accuracy* correction is not used for these categories and the average reflectance of the room, walls, and floor are used rather than task background reflectance. Table 16-5 is then used to select the low, middle, or high values in the category based on the computed TWF.

The IES (1982) also recommends minimum illumination levels for safety (as opposed to efficient visual performance of tasks). These values are shown in Table 16-6. Related to safety is the question of how much lighting is necessary to permit people to find their way out of a building during an emergency. Jaschinski (1982) and Boyce (1985) simulated such an escape task under various levels of emergency lighting. They both found that escape time increased as the light level decreased. Down to about 1 lx there was only slight increases in escape time. Further decreases in illumination level, however, resulted in rapidly increasing escape times. It is recommended that emergency lighting be maintained above 1 lx, preferably above 2 lx, and where many elderly people are apt to be present, a minimum of 4-lx emergency lighting is recommended.

TABLE 16-3
RECOMMENDED ILLUMINATION LEVELS FOR USE IN INTERIOR LIGHTING DESIGN

Category	Range of illuminances, lx (fc)	Type of activity
A	20–30–50* (2–3–5)*	Public areas with dark surroundings
B	50–75–100* (5–7.5–10)*	Simple orientation for short temporary visits
C	100–150–200* (10–15–20)	Working spaces where visual tasks are performed only occasionally
D	200–300–500† (20–30–50)†	Performance of visual tasks of high contrast or large size: e.g., reading printed material, typed originals, handwriting in ink and good xerography; rough bench and machine work; ordinary inspection; rough assembly
E	500–750–1000† (50–75–100)†	Performance of visual tasks of medium contrast or small size; e.g., reading medium-pencil handwriting, poorly printed or reproduced material; medium bench and machine work; difficult inspection; medium assembly
F	1000–1500–2000† (100–150–200)†	Performance of visual tasks of low contrast or very small size: e.g., reading handwriting in hard pencil on poor-quality paper and very poorly reproduced material; highly difficult inspection
G	2000–3000–5000‡ (200–300–500)‡	Performance of visual tasks of low contrast and very small size over a prolonged period: e.g., fine assembly; very difficult inspection; fine bench and machine work
H	5000–7500–10,000‡ (500–750–1000)‡	Performance of very prolonged and exacting visual tasks: e.g., the most difficult inspection; extra fine bench and machine work; extra fine assembly
I	10,000–15,000–20,000‡ (1000–1500–2000)‡	Performance of very special visual tasks of extremely low contrast and small size: e.g., surgical procedures

* General lighting throughout room.
† Illuminance on task.
‡ Illuminance on task, obtained by a combination of general and local (supplementary) lighting.
Source: RQQ, 1980, Table 1.

Although high levels of illumination might be warranted in some circumstances, there are definite indications that for certain tasks high levels of illumination can weaken information cues (Logan and Berger, 1961) by suppressing the "visual gradients" or the pattern density of the objects being viewed. An illustration of this effect is shown in Figure 16-10, which reveals the effects of general versus "surface-grazing" illumination. Thus, in some circumstances lower levels of illumination (albeit of the appropriate type) would be better than high levels. The illumination recommendations, for at least some tasks, may have to be tempered by experience.

TABLE 16-4
WEIGHTING FACTORS TO BE CONSIDERED IN SELECTING THE SPECIFIC ILLUMINATION LEVELS WITHIN EACH CATEGORY OF TABLE 16-3

Task and worker characteristics	**Weight**		
	−1	**0**	**+1**
Age	Under 40	40–55	Over 55
Speed or accuracy	Not important	Important	Critical
Reflectance of task background	Greater than 70%	30 to 70%	Less than 30%

Source: RQQ, 1980, Table 2.

TABLE 16-5
RULES FOR SELECTING THE SPECIFIC ILLUMINATION LEVEL WITHIN EACH CATEGORY OF TABLE 16-3 BASED ON THE TOTAL WEIGHTING FACTOR (TWF) CALCULATED USING TABLE 16-4

Illumination value from Table 16-3 to use	**Categories from Table 16-3**	
	A – C	**D – I**
Low value	−1 or −2	−2 or −3
Middle value	0	−1, 0, or +1
High value	+1 or +2	+2 or +3

TABLE 16-6
RECOMMENDED LEVELS OF ILLUMINATION FOR SAFETY PURPOSES (lx)

Activity level	**Level of hazard requiring visual detection**	
	Low hazard	**High hazard**
Low	5.4	22
High	11	54

Source: Adapted from IES, 1982, Table 7.

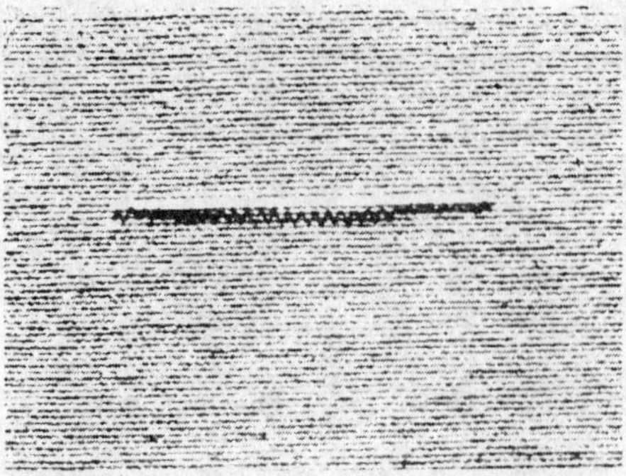

FIGURE 16-10
Illustration of the effects of general versus surface-grazing illumination on the visibility of a loose thread on cloth. (*Source: Faulkner and Murphy, 1973, Figs. 3 and 4.*)

DISTRIBUTION OF LIGHT

In addition to the overall level of illumination in the environment, one must also consider the distribution of light and its effects on visual comfort and task performance. The eye adapts to the ambient luminance it encounters (see Chapter 4). If there are great differences in luminance between areas in the work environment, the eye must adapt to these different levels as one shifts one's gaze from place to place. This *transient adaptation* reduces visibility momentarily until the eye fully adapts to the new luminance level. In addition, large differences in luminance can be a source of glare and cause visual discomfort.

Luminance Ratio

The *luminance ratio* is the ratio of the luminances of any two areas in the visual field. The IES (1982) recommendations for maximum luminance ratios in office environments (these also hold for other types of work environments) are as follows:

Task : adjacent surroundings	3:1
Task : Remote darker areas	10:1
Task : remote lighter areas	1:10

Brass (1982) contends, however, (1) that there are practically no real-life luminous environments providing overall luminance ratios of less than 20:1 and (2) that such ratios do not cause undue visual discomfort or transient adaptation problems. Kokoschka and Haubner (1985) provide supporting evidence that transient adaptation is probably not a problem with small luminance ratios. They had subjects enter numbers into a computer and varied the luminance ratio (paper source document to screen). Time taken to complete the task did not increase until the luminance ratio was 100:1, and even then, the increase was small. This would appear to be one area where intelligence and judgment should play a role in applying the IES recommendations.

Reflectance

The distribution of light within a room not only is a function of the amount of light and location of the luminaires, but also is influenced by the reflectance of the walls, ceilings, and other room surfaces. To maximize the illuminance on working surfaces, it is generally desirable to use rather light walls, ceilings, and other surfaces. However, areas of high reflectance in the visual field can become sources of reflected glare. For this and other reasons (including practical considerations), the reflectances of surfaces in a room (such as an office) generally increase from the floor to the ceiling. Figure 16-11 illustrates the IES recommendations, indicating for each type of surface the range of acceptable reflectance levels. Although Figure 16-11 applies specifically to offices, essentially the same reflectance values can be applied to other work situations, as in industry.

GLARE

Glare is produced by brightness within the field of vision that is sufficiently greater than the luminance to which the eyes are adapted so as to cause annoyance, discomfort, or loss in visual performance and visibility. *Direct glare* is caused by light sources in the field of view, and *reflected glare* is caused by light being reflected by a surface in the field of view. Reflected glare can be *specular* (as from smooth, polished, mirrorlike surfaces), *spread* (as from brushed, etched, or pebbled surfaces), *diffuse* (as from flat-painted or matte surfaces), or *compound* (a combination of the first three).

Glare is also classified according to its effects on the observer. Three types are recognized: *discomfort glare* produces discomfort, but does not necessarily

FIGURE 16-11
Reflectances recommended for room and furniture surfaces in an office.
(*Source: IES, 1982, Table 1.*)

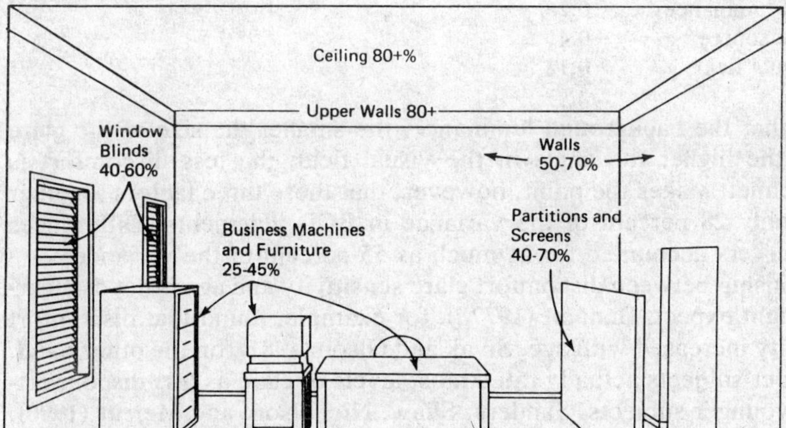

interfere with visual performance or visibility; *disability glare* reduces visual performance and visibility and often is accompanied by discomfort; and finally *blinding glare* is so intense that for an appreciable length of time after it has been removed no object can be seen.

Discomfort Glare

Discomfort glare is a sensation of annoyance or pain caused by high or nonuniform distributions of brightness in the field of view. Discomfort glare may be accompanied by disability glare, but it is a distinctly different phenomenon. The process by which glare causes discomfort is not well understood, but is believed to be related to the constriction of the pupil exposed to bright light sources. Visual discomfort from glare is, unfortunately, a common experience. Olson and Sivak (1984), for example, report on the discomfort caused by even moderate-intensity glare sources reflected in automobile rearview mirrors.

There has been considerable research related to glare and its effects on the subjective sensations of visual comfort and discomfort. Since discomfort glare is a subjective experience, it must be assessed by asking people to rate the degree of discomfort felt while in the presence of a particular glare source. A common metric used for comparing the degree of discomfort glare produced by different sources or for comparing the sensitivity of different people to glare is the *borderline between comfort and discomfort (BCD)*. The BCD is the luminance of a glare source that is judged by a person to be just necessary to cause that person feelings of discomfort. The higher the BCD, the less glaring was the source or the less sensitive the person was to glare. In one study, Bennett (1977*b*) varied background luminance, size of the glare source, and angle in the visual field and found the following correlations between the variables and BCD:

Background luminance	0.26
Size of glare source	−0.41
Angle in visual field	0.12

Thus, the higher the background luminance, the smaller the size of the glare source, and the higher the angle in the visual field, the less discomfort is produced. Bennett makes the point, however, that these three factors together account for only 28 percent of the variance in BCD judgments. Differences between observers accounted for as much as 55 percent of the variance.

The relationship between discomfort glare sensitivity and age is not as clear-cut as one might expect. Bennett (1977*a*), for example, found that discomfort glare sensitivity increased with age. Sivak and Olson (1987), on the other hand, found that older subjects actually rated most levels of glare as *less* discomforting than did younger subjects. Sanders, Shaw, Nicholson, and Merritt (1990), however, found no relationship at all between age and sensitivity to discomfort glare. These mixed results are probably a result of the large individual differences in sensitivity even among people of the same age. At best, we should consider the relationship between discomfort glare sensitivity and age as rather weak.

Other demographic variables have been investigated. Bennett (1977*a*), for example, found that blue-eyed people are more sensitive to glare than are brown-eyed people, but the difference was small. As another example, Sanders, Shaw, Nicholson, and Merritt (1990) found no difference in discomfort glare sensitivity between people who wore glasses and those that did not.

The IES has adopted a procedure for computing a *discomfort glare rating* (DGR) for luminaires in an interior space (Kaufman and Christensen, 1984). The DGR is a numeric assessment of the capacity of a number of luminaires in a given arrangment for producing discomfort. The major variables considered in computing the DGR are the average luminance of the visual field and the luminance, size, and position of each light source. DGRs are more or less arbitrary numbers, but they can be converted to a *visual comfort probability* (VCP). The VCP is the percentage of people who would rate the lighting as comfortable when looking in a specific direction from a specific position in the environment. Usually the VCP is computed by assuming that people are looking horizontally from a seated position at the center of the rear of the room (the least advantageous location). The IES considers a VCP of 70 to be satisfactory from a discomfort glare standpoint. Because 70 percent of the people would be expected to rate the environment comfortable in the worst position, it is assumed that most occupants would be in comfortable visual locations.

Disability Glare

Glare that directly interferes with visibility and ultimately with visual performance is considered to be *disability glare*. A good example of disability glare is experienced when viewing a person standing in front of a window on a bright day. The person is frequently seen as a silhouette because the glare from the window makes it impossible to see the features of the person. Trotter (1982) distinguishes two causes of disability glare: intraocular veiling luminances and transient adaptation. Light that enters the eye is scattered within the eyeball by "inhomogeneities" in the lens and the fluid that fills the eyeball. This scattered light creates a "veiling luminance" on the retina and reduces the contrast of the target being viewed; in effect, it "washes out" the image. Any source of light, direct or reflected, in the field of view causes some veiling luminances on the retina. The effect on foveal vision is a function of the intensity of the glare source and the angle of the source to the line of sight. The smaller the angle, the greater the effect on visibility.

Such effects are illustrated by the results of a study in which the subjects viewed test targets with a glare source of a 100-W inside-frosted tungsten-filament lamp in various positions in the field of vision (Luckiesh and Moss, 1927–1932). The test targets consisted of parallel bars of different sizes and contrasts with their background. The glare source was varied in position in relation to the direct line of vision, these positions being at 5, 10, 20, and 40° with the direct line of vision, as indicated in Figure 16-12. The effect of the glare on visual performance is shown as a percentage of the visual effectiveness that would be possible without the glare source. With the glare source at a 40° angle, the visual effectiveness is 58 percent, this being reduced to 16 percent at an angle of 5°.

Whereas the relationship between discomfort glare sensitivity and age is weak at best, there appears to be a strong, consistent, relationship of age to disability glare. Sanders, Shaw, Nicholson, and Merritt (1990), for example, found correlations from -0.42 to -0.53 between age and the luminance of a glare source just capable of obscuring a target. Higher levels of glare were required to obscure the target for younger subjects than for older subjects. Figure 16-13 presents the results of another study that investigated the effect of age on disability glare (Olson and Sivak, 1984). Figure 16-13 shows, for younger and older subjects, how much brighter a target disc had to be set to be just visible in the presence of a glare source (headlights reflected in a rearview mirror) compared to when no glare was present. In general, the difference between younger and older subjects is about 2:1. Therefore, with a given level of glare, if younger subjects had to make the target disc X times brighter than they did in the no-glare condition, then the older subjects had to make the disc 2X times brighter than their no-glare level. The greater disability glare sen-

FIGURE 16-12
Effects of direct glare on visual effectiveness. The disability effects of glare become greater as the glare source gets closer to the line of sight. (*Source: Luckiesh and Moss, 1927–1932.*)

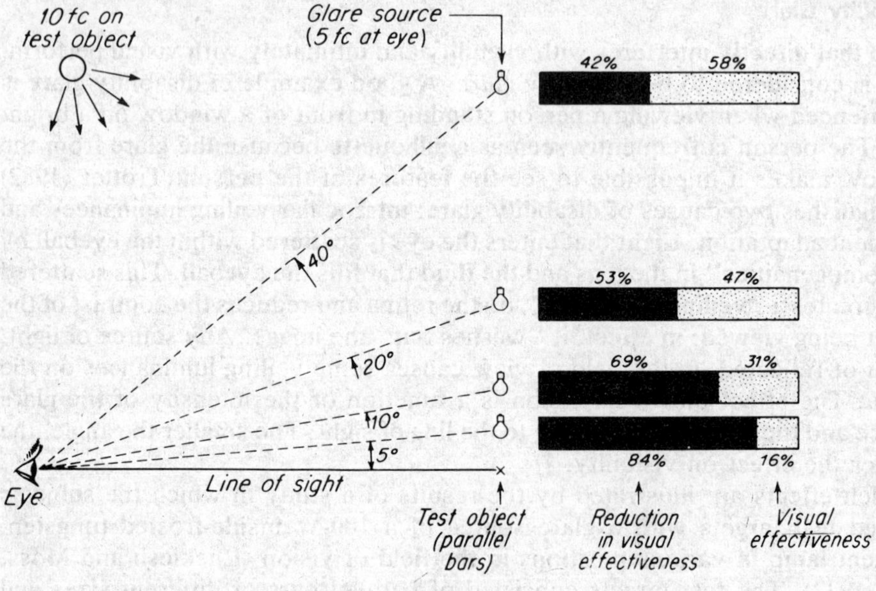

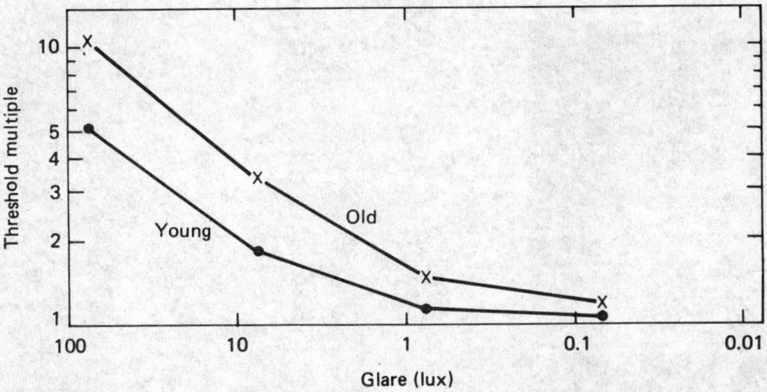

FIGURE 16-13
Relationship of age to disability glare. Shown for younger and older subjects is the number of times a target's luminance had to be increased above the no-glare threshold to be just visible again in the presence of glare (threshold multiple). The angle between the glare source and target disc is 35°. (*Source: Adapted from Olson, 1988, Fig. 6. Reprinted with permission, copyright 1988. Society of Automotive Engineers, Inc.*)

sitivity of older subjects is undoubtedly due to age-related changes in the eye that increase the scattering of light within the eyeball.

In addition to greater sensitivity to disability glare, Olson and Sivak also found that with a sudden onset of glare, older subjects took about 50 percent more time to adapt to the glare than did younger subjects.

For those of us who wear glasses, it appears that we are at a disadvantage when it comes to disability glare. Sanders, Shaw, Nicholson, and Merritt (1990) found that people who wore glasses experienced more disability glare from a source than did people who do not wear glasses. Freivalds, Harpster, and Heckman (1983) found that the order, from greatest to least disability glare was for wearers of hard contact lenses, wearers of soft contact lenses, wearers of frame glasses, and finally, those who did not wear corrective lenses. This is probably due, in large part, to the light-scattering properties of various types of corrective lenses.

There is a tendency for the eyes to turn toward a bright light (such as an oncoming automobile headlight); this phenomenon is called *phototropism*. Store owners and department store window designers take advantage of this human tendency when they direct bright lights toward a particular part of the store (perhaps where the high-profit items are) or at a particular item in the window. This is but one example of the artistic aspects of illuminating engineering we talked about at the beginning of this chapter.

Phototropism, however, can have negative effects on task performance and safety if it draws the eyes away from the area of greatest importance. Figure 16-14, for example, illustrates this phenomenon. In Figure 16-14*a*, the lights are arranged to produce a reflected glare source of high luminance, and the eyes

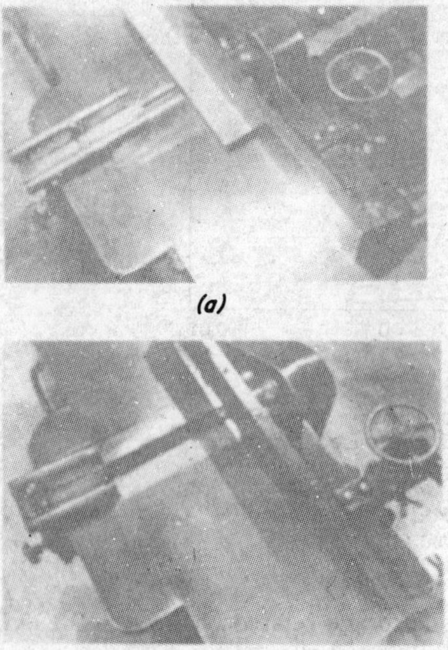

(a)

(b)

FIGURE 16-14
Effects of change of location of luminaires on a wood planer. *(a)* Original position caused reflection from polished surface, which distracted attention from the cutters *(b)* Relighting gives lower brightness reflection, which makes the cutter the center of attention. (*Source: Hopkinson and Longmore, 1959, by permission of the Controller of H. M. Stationery Office.*)

are drawn to that spot rather than to the middle of the machine where the cutting blade is located. By moving the luminaires, the glare source is removed (Figure 16-14*b*) and it is easier to concentrate on the cutting blade.

As a result of phototropism, the eyes are drawn to a bright glare source and become adapted to the higher luminance level. This is called *transient adaptation*. This in turn reduces the eyes' sensitivity to light. When the person looks back at the task, the eyes are less sensitive and visibility is reduced. It is like going into a dark movie theater on a bright sunny day, only the effect is not so pronounced. Windows are an especially good source of disability glare and transient adaptation because the illumination levels can be 1000 times more intense than normal indoor light levels.

Phototropism and transient adaptation played a role in a personal injury lawsuit in which your author served as an expert witness. The placement of lights in a stairwell created a transient adaptation problem that made it difficult to see a small object lying on one of the stairs. The plaintiff slipped on the object and sustained severe back injuries.

Reduction of Glare

• *To reduce direct glare from luminaires:* (1) Select luminaires with low DGR; (2) reduce the luminance of the light sources (for example, by using several low-intensity luminaires instead of a few very bright ones); (3) position luminaires as far from the line of sight as feasible; (4) increase the luminance of the area around any glare source so the luminance ratio is reduced; and (5) use light shields, hoods, visors, diffusing lenses, filters, or cross-polarizers.

• *To reduce direct glare from windows:* (1) Use windows that are set some distance above the floor; (2) construct an outdoor overhang above the window; (3) construct vertical fins by the window extending into the room (to restrict direct line of sight to the windows); (4) use light surrounds (to minimize contrast with light from the window); and (5) use shades, blinds, louvers, or tinting.

• *To reduce reflected glare:* (1) Keep the luminance level of luminaires as low as feasible; (2) provide a good level of general illumination (such as with many small light sources and use of indirect lights); (3) use diffuse light, indirect light, baffles, window shades, etc.; (4) position light source or work area so reflected light will not be directed toward the eyes; and (5) use surfaces that diffuse light, such as flat paint, nonglossy paper, and crinkled finish on office machines; avoid bright metal, glass, glossy paper, etc.

LIGHTING AND THE ELDERLY

It comes as no surprise to anyone over 40 that people's visual faculties decline with age. Most apparent is our tendency to become farsighted. (This tendency means not that one's far vision has improved, but rather that one's near vision has deteriorated.) However, certain other physiological changes in our visual system that occur with age have implications for lighting design (Boyce, 1981; Hughes and Neer, 1981; Wright and Rea, 1984).

For one, the lens of the eye continues its cellular growth throughout life. New cellular layers are continually being added to the outer parts of the lens. Cells at the center of the lens atrophy and harden; opacities develop, and the pliability of the lens decreases. In addition, the muscles controlling the diameter of the pupil begin to atrophy with age. This reduces the size of the pupil and decreases the range and speed with which the pupil can adjust to differing levels of illumination.

The thickening of the lens is the primary cause of farsightedness among the elderly. The lens can no longer conform to the shape necessary to focus a close object onto the retina; this defect can, however, be corrected with glasses. The increased opacity of the lens, coupled with the smaller pupil diameter, also reduces the amount of illumination reaching the retina. Weale (1961) determined that there is a 50 percent reduction in retinal illumination at age 50, compared to age 20, and this reduction increases to 66 percent at age 60. This reduced level of retinal illumination also plays a role in slowing the rate and reducing the level of dark adaptation among the elderly. Further, as the lens thickens, it yellows and thereby reduces the transmission of blue light through it. For this reason, elderly people have more trouble sorting or matching colors, making more errors in the blue-green and red regions than in the other regions of the hue circle (Verriest, Vandevyvere, and Vanderdonck, as cited in Boyce, 1981).

As the lens thickens, there is an increase in the scattering of light passing through the lens and in the intraocular fluids. There are widely differing estimates as to the degree of scatter that occurs with age. Wright and Rea (1984) cite studies showing a 2- to 10-fold increase between ages 20–30 and 60–70. This scattering, as with disability glare, produces a veiling luminance over the retinal image, effectively reducing its contrast. Further, scattering blurs the image on the retina (since the primary target is also scattered), thereby reducing visual acuity. This reduction in visual acuity cannot be corrected with glasses. The National Center for Health Statistics (1977) reported that the percentage of people with defective visual acuity (20/50 or poorer in the better eye with usual corrective lenses) increases from 0.7 percent at ages 35 to 44, to 14 percent at ages 65 to 74.

The reduced pupillary diameter response and the increased intraocular scattering of light combine to increase human sensitivity to disability glare with advancing age. The slower response of the pupil to changing levels of illumination also increases the magnitude and duration of transient adaptation effects.

What this all means is that lighting environments for the elderly must be designed with special care to ensure visual comfort and effective task performance. Specifically, the elderly need more light than the young to see, and they gain more from increased task illumination than do the young. The contrast of the task is also more critical to the elderly, who need higher contrasts in order to maximize their visual performance. Figure 16-15, for example, shows the dramatic increase in contrast necessary to allow people over 50 years old to see as well as people 20 to 30 years old.

Glare control is also more important in environments used by the elderly. Lower luminance ratios are needed to reduce transient adaptation problems and visual discomfort. Full-spectrum light sources with good color rendering indices are recommended for color perception as well as for psychological and biological reasons (Hughes and Neer, 1981). In the final analysis, designing a lighting environment for the elderly benefits everyone.

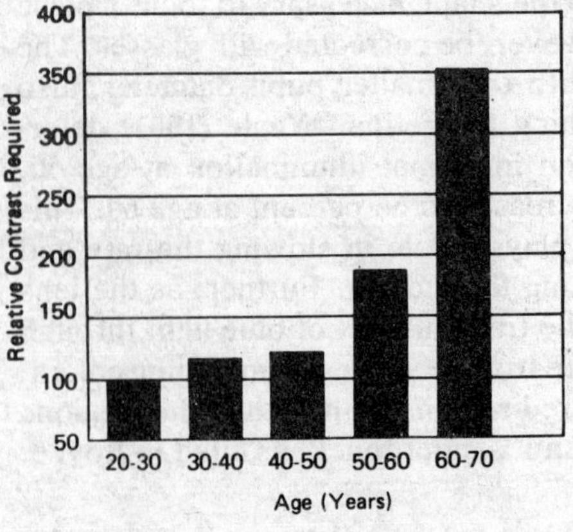

FIGURE 16-15
Effect of age on contrast needed to maintain visual performance. Age group from 20 to 30 years is considered as baseline, 100. (*Source: Based on data from Blackwell and Blackwell, 1971.*)

SPECIAL APPLICATION: LIGHTING FOR VIDEO DISPLAY TERMINALS

Video display terminals are everywhere. Traditionally we think of VDTs in offices and maybe homes, but increasingly we find them on the manufacturing floor, in warehouse storage areas, and even in automobiles. Police cars in Los Angeles, for example, are equipped with a computer terminal in the front seat. The advent of light-weight laptop computers has further expanded where we use computers. We cannot hope to cover the lighting implications of all these various situations, but the general principles discussed here should apply in most circumstances.

VDTs differ from other more traditional sources of visual information in a number of ways. Except for light-absorbing displays such as liquid crystal displays, VDTs emit light rather than depending on reflected light to be seen. Most VDTs have highly specular curved glass faces which present some interesting optical effects. And finally, the task surface is more vertical than with most other "desktop" tasks. All these factors must be considered in the setup of a lighting environment where VDTs will be used.

We focus our attention in this section on three main lighting considerations as they apply to VDTs: illumination level, luminance ratios, and screen reflections. In Chapter 4 we discussed other aspects of VDTs such as character font, color, contrast with the background, and flicker. Many of these factors, of course, interact with lighting to influence visual performance. In general, the less legible the characters on the screen, the more important is the quality of the lighting environment.

Illuminance Level

Since most VDT screens emit light, they do not require ambient illumination in order to be read. But alas, people do not work by VDT alone. General ambient illumination is required to carry on other tasks performed in conjunction with VDT work, including reading handwritten notes, looking at the keyboard, watching the clock, and moving around without tripping over the wastepaper basket.

With most office visual tasks, the higher the level of ambient illumination (up to a point) the easier it is to see the task. The opposite is true for VDT screens. Ambient illumination reflects off the screen and makes it more difficult to see the information on the screen. Recommended illumination levels, therefore, are a compromise between the higher levels demanded for such tasks as reading handwritten, pencil copy and the low levels required to read the VDT screen. Numerous standards specify illumination requirements for VDTs (Helander and Rupp, 1984). The majority specify levels between 300 and 500 lx, although some specify lower levels. The IES (1982), for example, specifies 50 to 100 lx for general lighting where VDTs are used. AT&T (1983) recommends 200 lx measured in the horizontal plane near the center of the VDT screen. The Human Factors Society (1988) recommends 200 to 500 lx. Where paper copy must be used in conjunction with VDT work, additional light directed on the paper copy (task lighting) is usually recommended. Depending on the difficulty of the visual task, 500 to 1500 lx may be required on the task itself (IES, 1982). One must be careful, however, not to create glare sources with the task lighting. Surveys of actual illumination levels found in VDT-equipped offices reveal that the majority are within the range of 300 to 500 lx (Stammerjohn, Smith, and Cohen, 1981; Laubli, Hunting, and Grandjean, 1982) with an occasional report of lower levels (Shahnavaz, 1982). By the way, for light-absorbing display terminals, the Human Factors Society (1988) recommends that incident illumination be at least 110/*R* lx (where *R* is the reflectivity of the display in its most reflective state).

Grandjean (1987) cited the work of Benz, Grob, and Haubner (1983), which showed that 40 percent of VDT users prefer illuminances between 200 and 400 lx, and 45 percent prefer illuminances between 400 and 600 lx. Shahnavaz and Hedman (1984) found that with a higher amount of light striking the screen, which would be indicative of higher levels of ambient illumination, there was less visual accommodation change after working for 6 h. It may well be that with the higher levels of ambient illumination, operators looked around the room more while working on the VDTs and thus relaxed the muscles of the eye that control and shape the lens. These results would give some support for keeping ambient illuminance levels in the range of 300 to 500 lx, rather than reducing them to lower levels.

Luminance Ratios

As we discussed previously, high luminance ratios between the task and adjacent and remote areas can cause transient adaptation problems and disability or discomfort glare. With VDTs, the problem is especially acute because of the low luminance levels of the VDT screen (assuming light characters on a dark background). The principal sources of high luminance in the office environment are task lighting, windows, and bare fluorescent bulbs used for general area lighting.

The general recommendations for luminance ratios are 1:3 between the screen and immediate surrounds and 1:10 for remote areas. As we discussed previously, there is some controversy about these recommendations. Ratios of 100:1 do not seem to have large effects on visual performance (Kokoschka and Haubner, 1985). Figure 16-16, for example, shows the percentage of VDT users with visual-accommodation changes after 6 h of work as a function of the luminance contrast (screen to surrounds). When the ratio exceeds 100:1 the percentage increases dramatically.

Actual field measurements confirm that luminance ratios in offices often exceed 1:10 (Stammerjohn, Smith, and Cohen, 1981; Fellmann et al., 1982). Laubli, Hunting, and Grandjean (1982), for example, reported screen-to-source document ratios of 1:10 to 1:87 with the median being 1:26. Screen-to-window ratios ranged from 1:87 to 1:1450 with a median of 1:300! It should be pointed

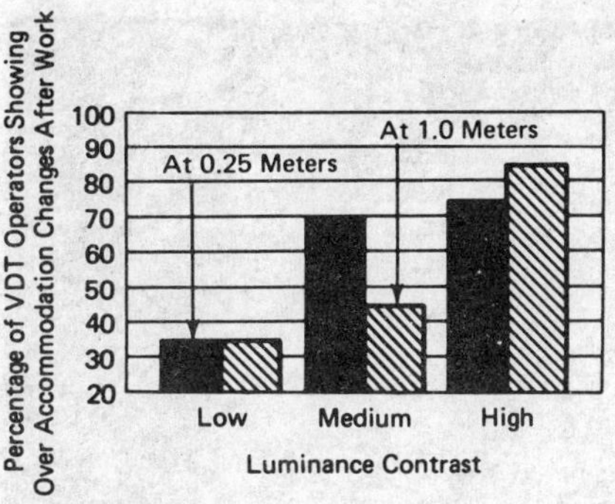

FIGURE 16-16
Percentage of VDT operators showing overaccommodation changes after 6 h of work in relation to luminance contrasts between the screen and surrounds. Contrast was computed as $(L_2 - L_1)/L_1$, where L_1 = luminance of screen and L_2 = luminance of surrounds. Low < 100; medium = 100 to 150; high > 150. (*Source: Shahnavaz and Hedman, 1984, Fig. 7.*)

out that these ratios are computed based on the integrated (average) luminance of the screen (background plus characters). Rupp (1981), however, believes that the visual system's level of adaptation is determined not by the average (background plus characters) luminance of the screen, but rather by the luminance of the bright characters on the screen. If true, the effective luminance ratios are lower than those calculated based on average screen luminance. The National Academy of Sciences (1983), however, did not accept Rupp's hypothesis and concluded that the matter needs further study.

If the 10:1 ratio recommendation is too stringent, what ratio is recommended? The Human Factors Society (1988) says only that high-luminance sources in the peripheral field of view should be avoided. The rationale is that luminance differences large enough to result in a noticeable effect on a VDT user's performance or comfort are unlikely if the overall illuminance level is between 200 and 500 lx. Grandjean (1987) is more specific. He recommends that the maximum luminance ratio within an office should not exceed 40:1. He also suggests that the reflectance of walls, ceilings, etc., be about 0.10 less in offices with VDTs than in offices without VDTs. This serves to reduce luminance ratios between the screen and the surrounds.

Reducing high luminance ratios can be accomplished by increasing screen luminance, using negative-polarity displays (dark characters on light background), decreasing the luminance of the offending area, or increasing the overall level of illumination. Increasing the overall level of illumination reduces the discomfort caused by "hot spots" in the environment, but adds reflected luminance to the screen which creates additional problems.

Screen Reflections

Screen reflections are one of the most prevalent problems associated with VDT use. Laubli, Hunting, and Grandjean (1982), for example, reported that 45 percent of the operators they interviewed complained of annoying reflections on their VDT screens. Typically reflections of windows, overhead lights, and even the operator are seen on the face of the screen. Figure 16-17 shows an example of typical screen reflections.

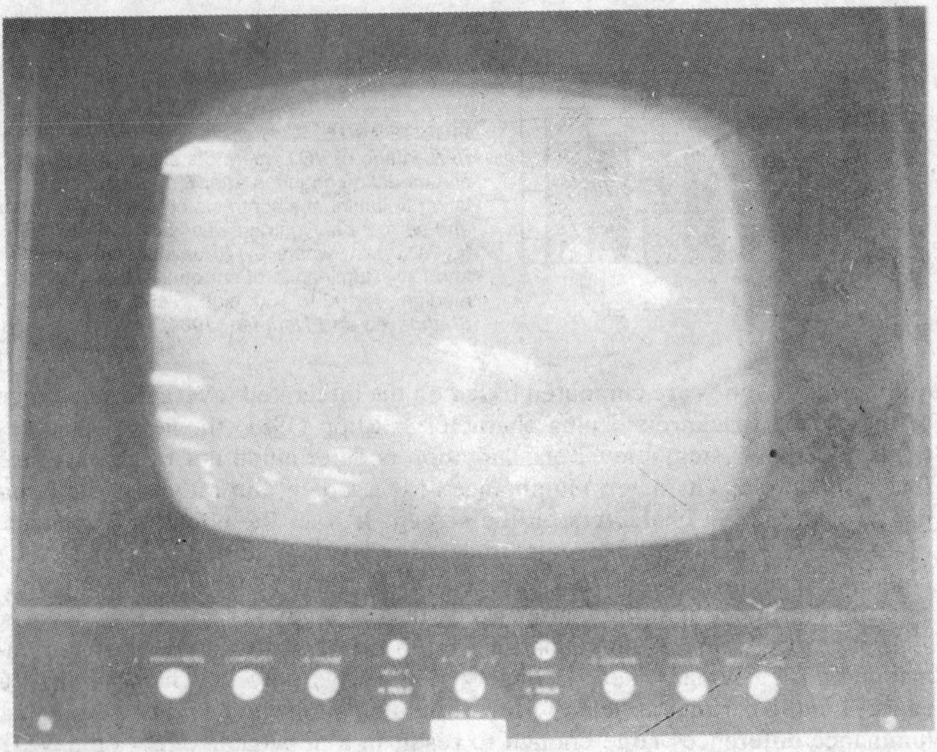

FIGURE 16-17
Example of typical reflections on a VDT screen.

Reflections are generally specular or diffuse. Specular reflections produce mirrorlike images. Such images are reflected off the smooth front surface of the screen. Since VDT screens are convex (with a radius of curvature of approximately 64 cm), distant objects reflected from the face of the screen are optically projected and appear to be about 32 cm behind the screen. Diffuse reflections produce a veiling luminance over parts of the screen and can wash out the information underneath. Such reflections are off the irregular phosphor surface behind the glass screen or, in some cases, off the front of the glass screen itself. To determine the source of screen reflections, either specular or diffuse, place a small mirror parallel to the screen, assume the operator's position(s), and slowly move the mirror over the surface of the screen. The source of the reflections will be clearly visible in the mirror.

Screen reflections affect VDT users in several ways. Reflections can be distracting (remember phototropism?) and annoying. Laubli, Hunting, and Grandjean (1982), for example, found a positive correlation between the luminance level of reflections and reported annoyance, but no relation between luminance and eye impairments. Reflections reduce the contrast between the characters and their background, making it more difficult to read information from the screen. Stammerjohn, Smith, and Cohen (1981) reported reflected luminance levels of up to 50 cd/m^2. Of the screens they evaluated, 17 percent had reflections that made it difficult to read characters on parts of the screen. Reflection can cause shifts in convergence between the characters on the screen and specular reflections optically located behind the screen.

Methods of reducing screen reflections are directed at the source of the reflections, between the source and the screen, and at the screen itself. Techniques that attempt to correct the problem at the source include (1) covering windows, (2) repositioning sources of reflections, (3) reducing intensity or luminance of sources, and (4) using diffuse, indirect lighting. Reflections can be reduced by erecting shields or partitions between the source and the screen to block the light directed at the screen. Hoods can also be attached to the VDT. However, these tend to restrict the posture of the operator, and they can cast annoying shadows on the screen and keyboard. Correcting the problem at the screen involves using such techniques as (1) moving or tilting the screen so reflections are not in the field of view, (2) positioning the workstation between rows of overhead lights rather than directly under them, (3) positioning the screen so it is perpendicular to windows rather than facing toward or away from them, (4) using reverse video (dark characters on a light background), and (5) applying antireflection treatment to the screen.

There are a number of antireflection techniques available for VDT screens (but they have their limitations):

1 *Etching or frosting* the front surface of the screen reduces specular reflections by making them diffuse, but does not reduce the diffuse reflections. In addition, the light from the characters on the screen is scattered, and the edges of the characters appear blurred, thus reducing legibility.

2 *Quarter-wave thin-film coatings* "ease" the transition of light from air to the glass, thereby reducing reflections (both specular and diffuse) from the front surface of the screen. These coatings do not, however, reduce diffuse reflections from the phosphor surface. An advantage is that the light emitted by the characters on the screen is not scattered, and character sharpness is not degraded. The disadvantage is that fingerprints and scratches defeat the system.

3 *Neutral-density filters* reduce the amount of light passing through them. This acts to reduce the luminance of the characters on the screen, but at the same time increases their contrast. The explanation is that the ambient light is reduced twice—once going through the filter to the phosphor surface and again when it is reflected from the phosphor surface back through the filter—whereas the light from the characters on the screen is reduced only once. Unless the surface of the filter is treated (etched, frosted, quarter-wave-coated), the filter itself can be a source of specular reflections.

4 *Circular polarizers* act as neutral-density filters but usually have a highly polished reflective front surface which makes them highly susceptible to specular reflections.

5 *Micromesh filters* look like a black nylon stocking stretched over the screen. Only light perpendicular to the mesh can pass through to the screen. This reduces both specular and diffuse reflections and enhances contrast. The problems with them are that the luminance of the characters on the screen drops off drastically when the screen is not viewed head on and that they tend to collect dirt.

One effect of neutral-density filters, circular polarizers, and micromesh filters is that they darken the screen. This can have the effect of increasing luminance ratios between the screen and surrounds above some recommended levels. Cakir, Hart, and Stewart (1980), for example, found that using such filters increased luminance ratios between the screen and source document from 1:6 to 1:100!

Habinek et al. (1982) compared reading performance with, and preference for, three types of antireflection treatments: micromesh filters, quarter-wave coating, and etching. Under the high-reflection condition, reading rates were higher (as much as 2.5 times higher) with the filters than without, as shown in Figure 16-18. Subjects preferred the treated screens to the untreated screens with little differences in preference among the three treatments.

VDTs present some interesting challenges in lighting design. Successfully balancing illumination levels, luminance ratios, and screen reflections is a difficult task that often requires delicate trade-offs to be made. Actually the problems are not unique to VDTs; they are only exaggerated by them. Undoubtedly, many of the visual problems and complaints associated with VDTs are due to improper lighting conditions. Judicious application of human factors and lighting principles can go a long way toward alleviating these problems.

DISCUSSION

We have come a long way in this chapter and have introduced many concepts which were probably unfamiliar to most readers. Yet we have just barely scratched the surface of the field of illuminating engineering. To design a lighting system to supply the proper amount of illumination, with the proper spectral composition, without creating glare—and to do it all in an energy-

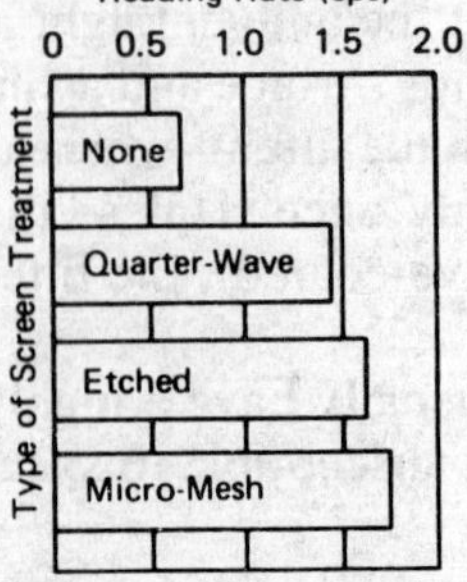

FIGURE 16-18
Effect of type of VDT antireflection treatment on reading rate. Reading rate, in characters per second (cps), was computed as follows: (number of characters read − number of errors and omissions)/time. (*Source: Adapted from Habinek et al., 1982.*)

efficient manner—is quite an achievement. Obviously, our goal in this chapter was not to develop this talent, but rather to acquaint you with the major concepts, to make you a more intelligent consumer of the expertise offered by illuminating engineers.

One last word of caution. It is easy to be swept away by the elegant formulas and sophisticated measuring techniques used by illuminating engineers. Remember, however, that often these techniques are based on simplifying assumptions and often the techniques ignore individual differences in visual capabilities. Their value lies more in making relative comparisons between lighting conditions than in specifying absolute requirements. Recommendations that are made must be tempered with good judgment and should be looked upon as first approximations that will undoubtedly require adjustment to fit the particular situation.

REFERENCES

AT&T (1983). *Video display terminals: Preliminary guidelines for selection, installation and use*. Short Hills, NJ: Bell Laboratories, Circulation Department.

Bennett, C. (1977*a*, January). The demographic variables of discomfort glare. *Lighting Design and Application*, pp. 22–24.

Bennett, C. (1977*b*). Discomfort glare: Concentrated sources parametric study of angularly small sources. *Journal of the Illuminating Engineering Society*, 7(1), 2–14.

Bennett, C., Chitlangia, A., and Pangrekar, A. (1977). Illumination levels and performance of practical visual tasks. *Proceedings of the Human Factors Society 21st Annual Meeting*. Santa Monica, CA: Human Factors Society, pp. 322–325.

Benz, C., Grob, R., and Haubner, P. (1983). *Designing VDU workplaces*. (German edition: *Gestaltung von Bildschirm-Arbeitsplätzen*). Köln: Verlag TÜV Rheinland.

Blackwell, H. R. (1959). Development and use of a quantitative method for specification of interior illumination levels on the basis of performance data. *Illuminating Engineering*, 54, 317–353.

Blackwell, H. R. (1961). Development of visual task evaluators for use in specifying recommended illumination levels. *Illuminating Engineering*, 56, 543–544.

Blackwell, H. R. (1964). Further validation studies of visual task evaluation. *Illuminating Engineering*, 59(9), 627–641.

Blackwell, H. (1967). The evaluation of interior lighting on the basis of visual criteria. *Applied Optics*, 6(9), 1443–1467.

Blackwell, H. R., and Blackwell, O. M. (1968). The effect of illumination quantity upon the performance of different visual tasks. *Illuminating Engineering*, 63(3), 143–152.

Blackwell, O., and Blackwell, H. (1971). IERI report: Visual performance data for 156 normal observers of various ages. *Journal of the Illuminating Engineering Society*, 1, 2–13.

Boyce, P. (1971). *Illumination and the sensitivity of performance measures* (Rept. R412). Capenhurst, UK: Electricity Council Research Centre.

Boyce, P. (1981). *Human factors in lighting*. New York: Macmillan.

Boyce, P. (1985). Movement under emergency lighting: The effect of illuminance. *Lighting Research and Technology*, 17, 51.

Boyce, P. (1988). Lighting and visual performance. In D. Oborne (ed.), *International reviews of ergonomics*, vol. 2. London: Taylor & Francis.

Brass, J. (1982, November). Discarding ESI in favor of brightness contrast engineering—A "wide-angle" view. *Lighting Design and Applications*, 12(11), 30–34.
Cakir, A., Hart, D., and Stewart, T. (1980). *Visual display terminals: A manual covering ergonomics, workplace design, health and safety, task organization*. New York: Wiley.
Collins, B., and Worthey, J. (1985). Lighting for meat and poultry inspection. *Journal of the Illuminating Engineering Society*, 15, 21.
Commission Internationale de l'Eclairage (1971). *Colorimetry*, CIE Publ. 15. Paris.
Commission Internationale de l'Eclairage (1972). *A unified framework of methods for evaluating visual performance aspects of lighting*, CIE Publ. 19. Paris.
Commission Internationale de l'Eclairage (1974). *Methods of measuring and specifying color rendering properties of light sources*, CIE Publ. 13.2. Paris.
Container Corporation of America. (1942). *Color harmony manual*. Chicago: Container Corporation of America.
DiLaura, D. (1978, February). Visibility, human performance, and lighting design—Equivalent contrast as a prelude to visibility level. *Lighting Design and Application*, p. 8.
Faulkner, T. W., and Murphy, T. J. (1973). Lighting for difficult visual tasks. *Human Factors*, 15(2), 149–162.
Fellmann, Th., Brauninger, U., Gierer, R., and Grandjean, E. (1982). An ergonomic evaluation of VDTs. *Behaviour and Information Technology*, 1, 69–80.
Freivalds, A., Harpster, J., and Heckman, L. (1983). Glare and night vision impairment in corrective lens wearers. *Proceedings of the Human Factors Society 27th Annual Meeting*. Santa Monica, CA: Human Factors Society, pp. 324–328.
General Electric Company (1965, March). *Light measurement and control* (TP-118). Nela Park, Cleveland, OH: Large Lamp Department, G.E.
General Electric Company (1968, August). *Light and color* (TP-119). Nela Park, Cleveland, OH: Large Lamp Department, G.E.
Grandjean, E. (1987). *Ergonomics in computerized offices*. London: Taylor & Francis.
Habinek, J., Jacobson, P., Miller, W., and Suther, T. (1982). A comparison of VDT antireflection treatments. *Proceedings of the Human Factors Society 26th Annual Meeting*. Santa Monica, CA: Human Factors Society. pp. 285–289.
Hard, A., and Sivik, L. (1981). NCS—Natural color system: A Swedish standard for color notation. *Color Research and Application*, 6, 129–138.
Hardesty, G., and Projector, T. (1973). *NAVSHIPS display illumination design guide*. Washington, DC: Government Printing Office.
Helander, M., and Rupp, B. (1984). An overview of standards and guidelines for visual display terminals. *Applied Ergonomics*, 15, 185–195.
Hesselgren, S. (1984). Why colour order systems? *Color Research and Application*, 9(4), 220–228.
Hopkinson, R. G., and Longmore, J. (1959). Attention and distraction in the lighting of workplaces. *Ergonomics*, 2, 321–334.
Hughes, P., and McNelis, J. (1978, May). Lighting, productivity, and work environment. *Lighting Design and Application*, pp. 37–42.
Hughes, P., and Neer, R. (1981). Lighting for the elderly: A psychobiological approach to lighting. *Human Factors*, 23, 65–85.
Human Factors Society (HFS) (1988). *American national standard for human factors engineering of visual display terminal workstations* (ANSI/HFS Standard No. 100–1988). Santa Monica, CA: Human Factors Society.

IES Nomenclature Committee. (1979). Proposed American national standard nomenclature and definitions for illuminating engineering (proposed revision of Z7.1R1973). *Journal Illuminating Engineering Society*, 9(1), 2–46.

Illuminating Engineering Society (1976). *IES lighting fundamentals course*. New York.

Illuminating Engineering Society (1972). *IES lighting handbook* (5th ed.). New York.

Illuminating Engineering Society (1981). *IES lighting handbook* (application vol.). New York.

Illuminating Engineering Society (1982). *Office lighting* (ANSI/IES RP-1-1982). New York.

Jaschinski, W. (1982). Conditions of emergency lighting. *Ergonomics*, 25, 363–372.

Kaufman, J., and Christensen, J. (eds.). (1984). *IES lighting handbook* (*ref. vol.*). New York: Illuminating Engineering Society of North America.

Knave, B. (1984). Ergonomics and lighting. *Applied Ergonomics*, 15(1), 15–20.

Kokoschka, S., and Haubner, P. (1985). Luminance ratios at visual display workstations and visual performance. *Lighting Research and Technology*, 17, 138.

Laubli, Th., Hunting, W., and Grandjean, E. (1982). Visual impairments in VDU operators related to environmental conditions. In E. Grandjean and E. Vigliani (eds.), *Ergonomic aspects of visual display terminals*. London: Taylor & Francis, pp. 85–94.

Logan, H. L., and Berger, E. (1961). Measurement of visual information cues. *Illuminating Engineering*, 56, 393–403.

Luckiesh, M., and Moss, F. K. (1927-1932). The new science of seeing. In *Interpreting the science of seeing into lighting practice*, vol. 1. Cleveland: General Electric Co.

Merritt, J., Perry, T., Crooks, W., and Uhlaner, J. (1983). *Recommendations for minimal luminance requirements for metal and nonmetal mines* (contract J0318022). Pittsburgh: U.S. Bureau of Mines.

Munsell Color Co. (1973). *Munsell book of colors*. Baltimore, MD: Munsell Color Co.

National Academy of Science (1983). *Video displays, work and vision*. Washington, DC: National Academy Press.

National Center for Health Statistics (1977). *Monocular visual acuity of persons 4–74 years* (DHEW Publ. HRA 77-1646). Washington, DC: U.S. Department of Health, Education, and Welfare.

Nimeroff, I. (1968). *Colorimetry*. National Bureau of Standards Monograph 104. Washington, DC: Government Printing Office.

Olson, P. (1988). Problems of nighttime visibility and glare for older drivers. In Society of Automotive Engineers, *Effects of aging on driver performance* (SP-762). Warrendale, PA, pp. 53–60.

Olson, P., and Sivak, M. (1984). Glare from automobile rear-vision mirrors, *Human Factors*, 26, 269–282.

Rea, M. (1986). Toward a model of visual performance: Foundations and data. *Journal of Illuminating Engineering Society*, 10, 41.

Richter, M. (1955). The official German standard color chart. *Journal of Optical Society of America*, 45, 223–226.

Ross, D. (1978, May). Task lighting—yet another view. *Lighting Design and Application*, pp. 37–42.

RQQ. (1980). Selection of illuminance values for interior lighting design (RQQ Rept. 6). *Journal Illuminating Engineering Society*, 9(3), 188–190.

Rupp, B. (1981). Visual display standards: a review of issues. *Proceedings of the Society for Information Display*, 22, 63–72.

Sanders, M., Gustanski, J., and Lawton, M. (1974). Effect of ambient illumination on noise level of groups. *Journal of Applied Psychology*, 59(4), 527–528.

Sanders, M., Shaw, B., Nicholson, B., and Merritt, J. (1990). *Evaluation of glare from center-high-mounted stop lights*. Washington, DC: National Highway Traffic Safety Administration.

Santamaria, J., and Bennett, C. (1981, March). Performance effects of daylight. *Lighting Design and Application*, 11(3), 31–34.

Shahnavaz, H. (1982). Lighting conditions and workplace dimensions of VDU-operators. *Ergonomics*, 25, 1165–1173.

Shahnavaz, H., and Hedman, L. (1984). Visual accommodation changes in VDU-operators related to environmental lighting and screen quality. *Ergonomics*, 27, 1071–1082.

Sivak, M., and Olson, P. (1987). *Toward the development of a field methodology for evaluating discomfort glare from automobile headlamps* (Rept. no. UMTRI-87-41). Ann Arbor, MI: University of Michigan Transportation Research Institute.

Stammerjohn, L., Smith, M., and Cohen, B. (1981). Evaluation of work station design factors in VDT operations. *Human Factors*, 23, 401–412.

Stenzel, A. (1962). Experience with 1000 lx in a leather factory. *Lichttechnik*, 14, 16.

Trotter, D. (1982). *The lighting of underground mines*. Clausthal-Zellerfeld, Germany: Trans Tech Publications.

Weale, R. (1961). Retinal illumination and age. *Transactions of the Illuminating Engineering Society*, 26, 95.

Wotton, E. (1986). Lighting the electronic office. In R. Lueder (ed.), *The ergonomics payoff, designing the electronic office*. Toronto, Ont., Canada: Holt, Rinehart & Winston. pp 196–214.

Wright, G., and Rea, M. (1984). Age, a human factor in lighting. In D. Attwood and C. McCann (eds.), *Proceedings of the 1984 International Conference on Occupational Ergonomics*. Rexdale, Ont., Canada: Human Factors Association of Canada. pp. 508–512.

Zahn, J., and Haines, R. (1971). The influence of central search task luminance upon peripheral visual detection time. *Psychonomic Science*, 24(6), 271–273.

CHAPTER **17**

CLIMATE

Humans are a product of millions of years of evolution. As such, we have adapted to a wide range of conditions. For example, the range of climates in which people live is impressive, ranging from the cold-dry arctic to hot-humid jungles and hot-dry deserts. The variation in climatic conditions between summer and winter at any given location can be 100°F (55°C) or more. Added to the natural variation in climatic conditions is the range made possible by technology: from cold-storage warehouses with low humidity and near-freezing temperatures to deep underground mines with temperatures near 90°F (32°C) and high humidity. We explore the effects of climate in this chapter, focusing on heat and cold stress. Climate, however, is but one aspect of our atmospheric environment that has implications for our health and level of performance. Other aspects include barometric pressure, air ionization, and air pollution. We have chosen to concentrate on climate because it has a powerful influence on our comfort, health, and performance; because there has been considerable research conducted in the area; and because it can be a problem in our work and everyday lives.

HEAT EXCHANGE PROCESS

Despite millions of years of evolution, humans have adapted to varying climatic conditions principally by altering the conditions under which they live rather than through physiological adaptation. We build shelters, wear appropriate clothing, and provide heating and air conditioning. Without these things, humans could only survive in a narrow zone along the equator where the ambient temperature is between 82 and 86°F (28–30°C).

The human body has a complex thermal regulation system that tries to maintain a relatively stable internal temperature of around 98.6°F ± 1.8°F (37°C ± 1°C). One can think of the body as composed of a core and a shell. The shell is the part of the body within about 1 in (2.54 cm) of the skin, while the core is essentially the vital organs of the body (heart, lungs, abdominal organs, and brain). It is the core temperature that the thermal regulation system maintains. Under normal conditions, the difference between core temperature and skin temperature is about 7°F (4°C), but the difference may be as great as 36°F (20°C) under severe climatic conditions.

The thermal regulation system is controlled principally by that part of the brain called the hypothalamus. Heat-sensitive nerves exist throughout the body including the skin, muscles, and stomach. The hypothalamus itself contains neurons that are sensitive to changes in arterial blood temperature. Through a process that is still not fully understood, the hypothalamus integrates the information from these various sensory nerves and regulates heat loss from the body. The mechanisms available to the hypothalamus are principally constriction or dilation of the blood vessels in the core and skin, sweating, and shivering.

Under most circumstances we generate more heat than we need to maintain our core temperature. The heat we generate comes from the metabolic process that converts chemical energy (from food) to heat and mechanical energy (or work). About 75 percent of the energy from metabolizing food is given off as heat, only about 25 percent goes into mechanical energy. Even under comfortable climatic conditions, this heat must be dissipated; otherwise it will continue to build up within our bodies and eventually we would die. Under cold conditions, of course, the body tries to limit the amount of heat that it loses to the environment.

Avenues of Heat Exchange

Keep in mind that cold is the absence of heat; it is heat that is exchanged between things. When you put a cold towel on your face to cool you off, technically, the towel is absorbing heat from you. That is, you are heating the towel.

There are four avenues of heat exchange between the body and the environment:

1 *Conduction:* heat transfer by direct contact with a solid or fluid, including air. The hotter material transfers heat to the cooler material. Conduction occurs when you touch a hot stove, put a cold towel on your face, or walk on a cold floor. Conductive heat exchanges with the air (cooler or hotter air contacting the skin) are usually included with convective heat exchange. Under most situations, other sources of conductive heat exchange are minor and are usually not even considered in most analyses of the body's heat exchange process.

2 *Convection:* the transfer of heat by currents of air. If the air temperature is cooler than the skin temperature, the air in contact with the skin is heated (by conduction). The heated air is then mixed and carried away from the body by convective air currents. If the air is hotter than the skin, the process is reversed and the skin absorbs heat from the air.

3 *Radiation:* the transmission of heat between objects by electromagnetic radiation. This is the way the sun heats the earth and our bodies. When you stand in sunlight even during winter, you can feel the warmth of the sun even though the temperature of the air is cold. This is due to radiation from the sun.

4 *Evaporation:* perspiration (sweat) is heated by the skin and is transformed into water vapor which is carried away from the skin by currents of air. It takes a certain amount of heat to change water into water vapor and that heat is carried away from the body with the evaporation of the sweat. For example, at normal skin temperature the evaporation of one liter (about a quart) of sweat requires 580 kcal (2428 kJ).

Heat Exchange Equation

The heat exchange process of the body can be characterized by the following formula:

$$\Delta S = (M - W) \pm R \pm C - E$$

where ΔS = change in body heat content (storage)
M = metabolism
W = work performed
R = radiative heat exchange
C = convective heat exchange
E = evaporative heat loss

Metabolism will always be positive because heat is generated by the process. Evaporation is always negative because heat is always lost through the process.

If the body is in a state of thermal equilibrium, ΔS will be zero. In conditions of imbalance, the body temperature can increase (ΔS positive) or decrease (ΔS negative), and if the imbalance is extreme enough, death can result.

Environmental Factors Influencing Heat Exchange

The heat exchange process is, of course, very much affected by four environmental conditions: air temperature; humidity; air flow; and the temperature of surrounding surfaces (walls, ceilings, windows, furnaces, etc; sometimes such temperature is called *wall temperature,* for short). The interaction of these conditions is quite complex and cannot be covered in detail here. But we can point out that under high air temperature and high wall temperature conditions, heat loss by convection and radiation is minimized, so that heat *gain* to the body may result. Under such circumstances the only remaining means of heat loss is by evaporation. But if the humidity is also high, evaporative heat loss will be minimized, with the result that body temperature rises.

Such effects are illustrated in Figure 17-1. This shows, for each of five combinations of air and wall temperature, the percentage of heat loss by evaporation, radiation, and convection. In particular, conditions *d* and *e* illustrate the fact that, with high air and wall temperatures, convection and radiation cannot dissipate much body heat, and the burden of heat dissipation is thrown on the evaporative process. But, as we know, evaporative heat loss is limited by the humidity; there is indeed truth to the old statement that "it isn't the heat—it's the humidity."

Effect of Clothing on Heat Exchange

The clothes we wear can have a profound effect on the heat exchange process. It is the insulating effects of clothing that reduce heat loss to the environment. When it is cold, of course, reduced heat loss is beneficial, but when it's hot, clothing interferes with heat loss and can be harmful.

The insulation value of most materials is a direct linear function of its thickness. The material itself (whether it's wool, cotton, or nylon) plays only a minor role. It is the amount of trapped air within the weave and fibers that provides the insulation. If the material is compacted or gets water-soaked, it loses much of its insulating properties because of the loss of trapped air.

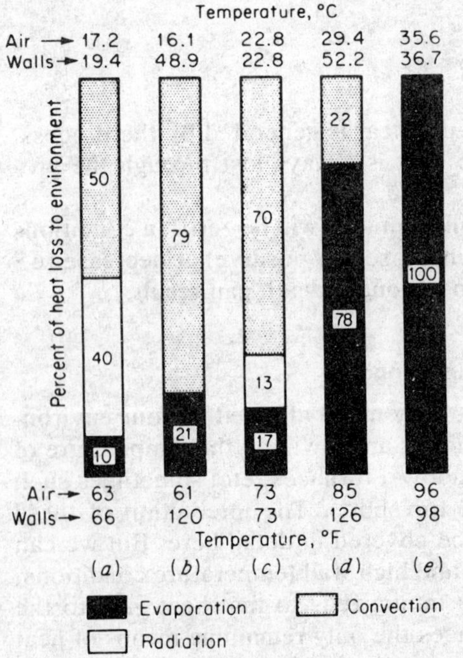

FIGURE 17-1
Percentage of heat loss to environment by evaporation, radiation, and convection under different conditions of air and wall temperatures. (*Source: Adapted from Winslow and Herrington, 1949.*)

The unit of measure for the insulating properties of clothing is the clo. The *clo unit* is a measure of the thermal insulation necessary to maintain in comfort a sitting, resting subject in a normally ventilated room at 70°F (21°C) and 50 percent relative humidity. Because the typical individual in the nude is comfortable at about 86°F (30°C), one clo unit has roughly the amount of insulation required to compensate for a drop of about 16°F (9°C). The typical value of clothing insulation is about 4 clo per inch of thickness (1.57 clo/cm). Clo values for some typical articles of men's and women's clothing are given in Table 17-1 along with a formula for computing overall clo values for a clothing ensemble. To lend support to the old adage that there is nothing new under the sun, the Chinese have for years described the weather in terms of the number of suits required to keep warm, such as a one-suit day (reasonably comfortable), a two-suit day (a bit chilly), up to a limit of a twelve-suit day (which would be really bitter weather). Perhaps coincidentally, the clo value of the traditional double-layer caribou clothing of the Eskimo is 12 clo.

Another feature of clothing that affects heat transfer is the permeability of the material to moisture. It is the permeability that permits evaporative heat transfer through the fabric. In general, the greater the clo value of the fabric, the lower is its permeability. The index of permeability (i_m) is a dimensionless unit which ranges from 0.0 for total impermeability to 1.0 if all moisture that could be evaporated into the air could pass through the fabric. Typical i_m values for most clothing materials in still air is less than 0.5. Weather-repellent treatments, very tight weaves, and chemical protective impregnation can reduce the i_m values significantly. In hot environments, evaporation of sweat is vital to maintain thermal equilibrium, and materials that interfere with this process can result in heat stress. In a cold environment, if evaporation of sweat is impeded, a garment can become soaked with perspiration, thus reducing its insulating capacity.

TABLE 17-1
CLO INSULATION VALUES FOR INDIVIDUAL ITEMS OF CLOTHING

Men's clothing	Clo	Women's clothing	Clo
T-shirt	0.09	Bra and panties	0.05
Underpants	0.05	Half-slip	0.13
Light-weight short-sleeve shirt	0.14	Light-weight blouse	0.20
Light-weight long-sleeve shirt	0.22	Light-weight dress	0.22
Light-weight trousers	0.26	Light-weight slacks	0.26
Light-weight sweater	0.27	Light-weight sweater	0.17
Heavy sweater	0.37	Heavy sweater	0.37
Light jacket	0.22	Light jacket	0.17
Heavy jacket	0.49	Heavy jacket	0.37
Socks	0.04	Stockings	0.01
Shoes (oxfords)	0.04	Shoes (pumps)	0.04

Total clo units = 0.8 × (sum of individual items) + 0.8

Source: Adapted from American Society of Heating, Refrigeration, and Air Conditioning Engineers, 1981. Reprinted with permission from 1989 ASHRAE Handbook-Fundamentals.

MEASUREMENT OF THERMAL CONDITIONS

There are five basic components that make up the thermal environment. The first is air temperature, or *dry-bulb temperature*. This is usually measured with a simple glass thermometer shielded from any major sources of radiant heat. The second component is *relative humidity* of the air. This is a measure of the amount of water vapor in the air relative to the maximum amount of water vapor the air could hold at that temperature. Relative humidity is measured with a *hydrometer*. The third component is *wet-bulb temperature*, which is measured by exposing a thermometer that has a wet wick placed over the sensory element to a forced convective air flow. This is done with a *sling psychrometer*, a wet wick-covered thermometer that is swung around on the end of a cord. A related measure, *natural wet-bulb temperature*, is like wet-bulb temperature except that the wet wick-covered thermometer is exposed only to the natural prevailing air movement: you don't swing it around. The thermometer must be shielded from radiant heat sources without blocking the natural air movement. The fourth component in the thermal environment is *mean radiant temperature*. This is commonly obtained as globe temperature. *Globe temperature* is measured using a thermometer placed in the center of a black copper sphere. The black sphere absorbs radiant heat, and the air temperature inside indicates the amount of heat absorbed. The last component of the thermal environment is *air velocity*, normally measured with an *anemometer*.

There are several heat-stress indices that have attempted to capture some of these components into a single composite index of the thermal environment. A few of these indices are discussed below.

Effective Temperature

There are two indices of effective temperature, both of which were developed under the sponsorship of the American Society of Heating, Refrigerating, and Air-Conditioning Engineers (ASHRAE, 1981). The original effective-temperature (ET) scale was intended to equate varying combinations of temperature, humidity, and air movements in terms of equal sensations of warmth or cold. For example, an ET of 70°F (21°C) characterized the thermal sensation of a 70°F (21°C) temperature in combination with 100 percent humidity. Equal thermal sensations [ET = 70° F (21°C)] can also be achieved with other (higher) temperatures in combination with other (lower) humidities [such as 81°F (27°C) with 10 percent humidity]. Although ET has been widely used and has been a useful index, it overemphasizes the effects of humidity in cool and neutral conditions and underemphasizes the effects in warm conditions; also, it does not fully account for air velocity in hot-humid conditions.

Because of these limitations of the original ET, a newer index, indentified as ET* was developed. It is derived by a complicated formulation based on the effects of environmental variables on the physiological regulation of the body. Figure 17-2 includes the new ET* values. For the moment consider the dry-

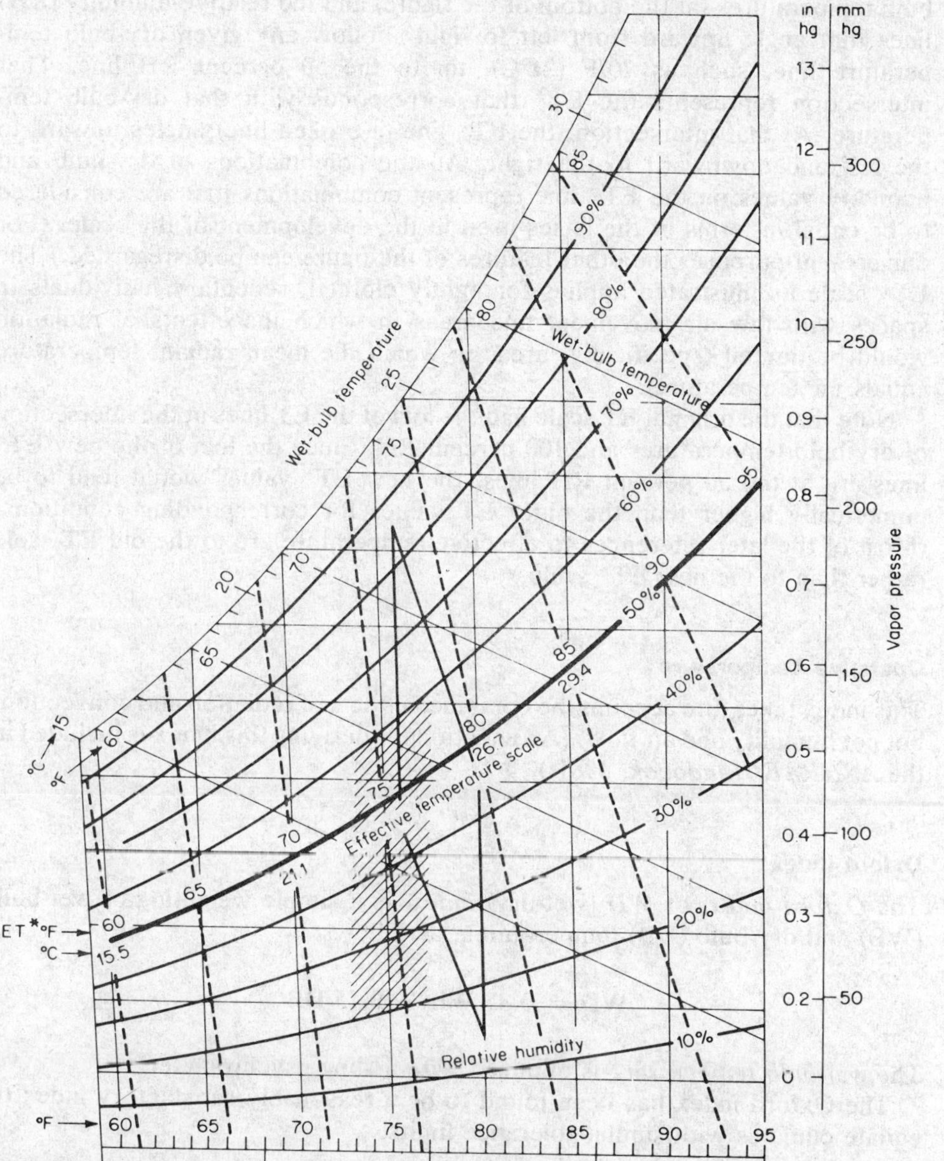

FIGURE 17-2
New effective temperature (ET*) scale. The broken lines represent the ET* values. See text for discussion. Vapor pressure is an index of humidity. Wet-bulb temperature is a measure of temperature under conditions of 100 percent RH, obtained with an aspirated wet-wick temperature or sling thermometer. (*Source: Adapted from ASHRAE, 1981, Fig. 16, p. 8.21.*) The shaded area represents the ASHRAE comfort standard. The diamond-shaped area represents the modular comfort envelope (MCE) proposed by Rohles (1974).

bulb temperatures (at the bottom of the figure) and the relative-humidity (RH) lines that angle upward from left to right. Follow any given dry-bulb temperature line, such as 70°F (21°C), up to the 50 percent RH line. That intersection represents the ET* that corresponds with that dry-bulb temperature. At that intersection, the ET* line (a broken line) angles upward to the left and downward to the right. All the combinations of dry-bulb and humidity values on the ET* line represent combinations that are considered to be equal in terms of the bases used in the development of the scale. (For our present purposes the other features of the figure can be disregarded.) The ET* scale as illustrated applies for lightly clothed, sedentary individuals in spaces with low air movement and areas in which the effects of radiation would be limited (specifically, areas in which the mean radiant temperature equals air temperature).

Note that the original ET scale had the loci of the ET lines at the intersection of dry-bulb temperatures and 100 percent RH. Since the loci of the new ET* lines are at the 50 percent RH lines, the new ET* values would tend to be numerically higher than the older ET values for corresponding conditions. (Most of the later references to effective temperature are to the old ET scale rather than to the new ET* scale.)

Operative Temperature

This index takes into account the combined effects of radiation and convection, but not humidity and air flow. (An equation for deriving this index is included in the *ASHRAE Handbook*, 1981.)

Oxford Index

The *Oxford index,* or *WD* (wet-dry) *index,* is a simple weighting of wet-bulb (WB) and dry-bulb (DB) temperatures, as follows:

$$WD = 0.85\ WB + 0.15\ DB$$

The *wet-bulb temperature* is obtained with a sling psychrometer.

The Oxford index has been found to be a reasonably satisfactory index to equate climates with similar tolerance limits.

Wet-Bulb Globe Temperature

This index (also called the WBGT index) is a weighted average of natural wet-bulb temperature (NWB), globe temperature (GT), and dry-bulb temperature (DB). For indoor, night, or sunless days, the formula is:

$$WBGT = 0.7\ NWB + 0.3\ GT$$

For outdoor measurements where the sun is a factor in heat load (radiant heat), the formula becomes:

$$WBGT = 0.7\ NWB + 0.2\ GT + 0.1\ DB$$

One feature of this measure that makes it attractive is the fact that the air velocity need not be measured directly, since its value is reflected in the measurement of the natural wet-bulb temperature. The WBGT index is coming into more common use because it does take into account the combination of variables mentioned above. However, WBGT may not adequately compensate for air velocity (ISO, 1982). In addition, Ramsey (1987) indicates that studies have shown differences in human physiological response obtained at "equal" WBGT levels but at different combinations of radiant heat, humidity, and air velocity. WBGT provided the poorest indicator when humidity was high and air velocity was low. Under such conditions, WBGT tends to underestimate the severity of the situation.

Botsball

The Botsball (BB) index is a special type of thermometer that produces a single index of the combination of air temperature, humidity, wind speed, and radiation variables (Botsford, 1971). On the basis of an extensive experimental use of the Botsball index in many industrial workstations, Beshir, Ramsey, and Burford (1982) report that Botsball readings were highly correlated with WGBT values (the correlation being .96). Thus these two measures are virtually interchangeable. They can be converted from one to the other with the following simplified (Ramsey, 1987) equation (in degrees Celsius):

$$WBGT = BB + 3^\circ \qquad BB = WBGT - 3^\circ$$

These equations are accurate for moderate levels of both humidity and radiant temperature. At the extremes, the formulas will be off by a few degrees.

Beshir, Ramsey, and Burford point out that the BB index is a practical simple device that can be used near workers, but without interfering with the workers. However, it does have some operational limitations.

Discussion of Composite Indices

The composite indices discussed above are expressed in terms of temperature (degrees Fahrenheit or Celsius). In using these, however, keep in mind that they cannot be interpreted in terms of conventional dry-bulb temperatures, because they also take into account other variables (such as humidity, air flow, etc.).

We discuss other composite indices in our discussion of heat and cold stress because they were developed specifically to assess such conditions.

THERMAL COMFORT AND SENSATIONS

The concept of thermal comfort is admittedly elusive. First, there are distinct individual differences, which are in part the consequence of differences in metabolism. Otherwise, comfort can be influenced by the work being performed, the clothing worn, and even the season of the year. ASHRAE (1981) established a comfort standard that is represented in Figure 17-2 by the shaded area. In turn, Rohles (1974) carried out an extensive study of 1600 subjects that resulted in the development of a modular comfort envelope (MCE). This envelope falls between the 75 and 80°F (24 and 27°C) ET* lines, forming a diamond, which is also shown in Figure 17-2. Although these two areas do overlap somewhat, the MCE is generally higher than the ASHRAE standard. Rohles suggests that this difference arises from differences in the experimental conditions used in their development. The MCE study was carried out with subjects who wore clothing of 0.6 clo unit, whereas those used in the ASHRAE standard study wore clothing of 0.8 to 1.0 clo unit. In addition, the MCE applies to "sedentary" activity whereas the ASHRAE standard applies to office work, which is above the sedentary level. Rohles makes the point that the ASHRAE standard needs to be revised. However, since the two areas do overlap, he recommends that that area be considered as the "design conditions for comfort," which is specified as an ET* of 76°F (24°C). Rohles points out that the MCE applies only to situations in which dry-bulb temperature and humidity are varied, and not to conditions of increased air movement or higher activity levels or when different clothing is used.

Some indication of the influence of level of activity and of clothing on sensations of comfort is shown in Figure 17-3. This figure shows "comfort lines" for each of three levels of activity (sedentary, medium, and high) for persons with light and medium clothing. The differences in these lines (they are actually ET* lines) reflect the very definite effects on comfort of work activity and clothing.

Although much has been found out about the sensations of thermal comfort under various conditions, there are still many gaps in the available knowledge. This is especially true with regard to the interactions among some of the variables that influence such sensations. Acclimatization, habits, and established traditions may also affect the comfort temperature. This may explain the fact that the preferred room temperature is higher in Singapore than in the United States, and higher in the United States than in England. It is also higher in the summer than in the winter in both the United States and England (Astrand and Rodahl, 1986).

Effect of Draught on Thermal Comfort

Draught is defined as an unwanted local cooling of the human body caused by air movement. Draught can be a serious problem in many ventilated or air-conditioned buildings. Draught has been identified as one of the two most

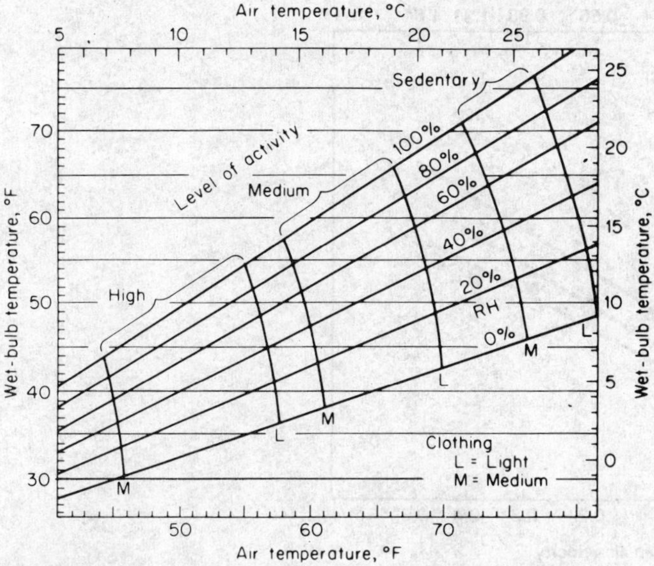

FIGURE 17-3
Comfort lines for persons engaged in three levels of work activity with light clothing (0.5 clo) and medium clothing (1.0 clo). These data are for a low level of air velocity. (*Source: Based on data from Fanger, 1970; reprinted, with adaptations, with permission from ASHRAE, 1981, Chap. 8, Fig. 18.*)

annoying environmental factors in workplaces and the most annoying factor in offices (Olesen, 1985).

Fanger and Christensen (1986) exposed 100 subjects to various air velocities at each of three air temperatures [20, 23, 26°C (68, 73, 79°F)]. The subjects were seated and dressed to keep their bodies thermally neutral at each temperature. The air flow simulated typical turbulent air flows occurring in real spaces. It was found that subjects were most sensitive to air movement at the head region (i.e., head, neck, shoulders, and upper back). Figure 17-4 shows the percentage of subjects that rated the various air velocities at the head as uncomfortable (i.e., they felt a draught) at each level of air temperature. Surprisingly, those with short hair (less hair to protect the neck from cooling) were *less* sensitive to draught than were subjects with long or medium-length hair. It should be noted that Fanger et al. (1988) have refined the study of draught by including a turbulence intensity factor to predict the percentage of people experiencing draught dissatisfaction.

Effect of Low Humidity on Thermal Comfort

Exposure to low humidity can occur outdoors in most desert regions during the summer and in air-conditioned buildings. Similar dry conditions can occur during winter in heated buildings. Low humidity can result in dryness in the

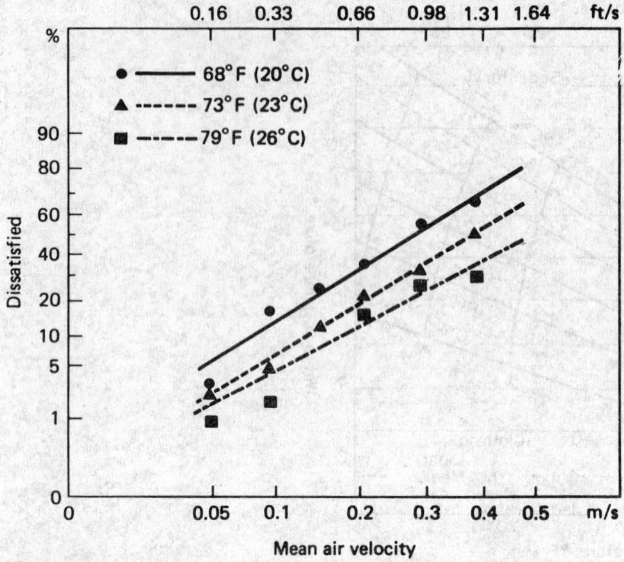

FIGURE 17-4
Percentage of subjects feeling a draught at the head region as a function of air velocity at three air temperatures. (*Source: Fanger and Christensen, 1986, Fig. 9.*)

nose and throat, dry skin, and chapped lips. We also seem to be more sensitive to odors when the humidity is low. Laviana, Rohles, and Hoffberg (1987) report that a perceivable level of eye irritation is experienced by both wearers and nonwearers of contact lenses when the relative humidity is at or below 30 percent and that this effect becomes most pronounced after 4 h.

Discussion

Thermal comfort is a somewhat elusive concept because it is basically a subjective evaluation of the effects of many interacting variables. Most thermal comfort studies have subjects performing sedentary or light work. Little data exists for establishing comfort zones for more intense levels of physical activity.

HEAT STRESS

Under conditions of heat stress, the body absorbs and/or produces more heat than it can dissipate. The result is that core temperature rises and illness and even death may result. In many parts of the world, outdoor temperatures during summer can produce heat stress in individuals engaged in moderate physical activity. Jensen (1983) indicates that the highest rates of reported worker's compensation cases for heat illness were in agriculture (9.16 cases per 100,000 workers), construction (6.36 per 100,000), and mining (5.01 per

100,000). Another industry associated with heat illness is the iron and steel industry (Dinman, Stephenson, Horvath, and Colwell, 1974). Heat stress can also occur in moderate or even cold climates where impermeable protective garments are worn and heavy physical activity is required. Examples include firemen and toxic waste disposal workers. The National Institute of Occupational Safety and Health (NIOSH, 1986) conservatively estimates that 5 to 10 million workers in the United States work in industries where heat stress is a potential safety and health hazard.

Physiological Effects of Heat Stress

One of the most direct effects of heat stress is a rise in core (rectal) temperature. The rise in core temperature results in an increase in metabolism. This is called the *Q10 effect;* for every 1°C increase in core temperature, metabolism increases about 10 percent (7 percent for every 1°F increase). The increased metabolism generates more heat that must be dissipated; otherwise, core temperature will increase further, resulting in a further increase in metabolism and heat production. A vicious cycle will result that can lead to death. Figure 17-5 shows the relationship between effective temperature (ET) and rectal temperature for one individual engaged in three levels of physical activity and for the average of three individuals at a single level of activity. Rectal temperature remains constant over a wide range of ETs, but eventually heat cannot be dissipated in sufficient quantities and rectal temperature begins to increase. The ET at which this occurs is lower the more strenuous the work activity. Figure 17-5 shows that at a work level of 420 kcal/h rectal temperature begins to increase at about 80°F ET, while at a lower level of work (180 kcal/h) the increase in rectal temperature does not occur until about 85°F ET.

Cardiovascular Response The first line of defense against heat stress is the cardiovascular system. Two basic responses are involved. For one, the blood vessels in the skin dilate. This brings the blood, heated in the core of the body, to the skin and increases skin temperature. The increased blood flow causes one to become flushed when in a hot environment. If the temperature of the air

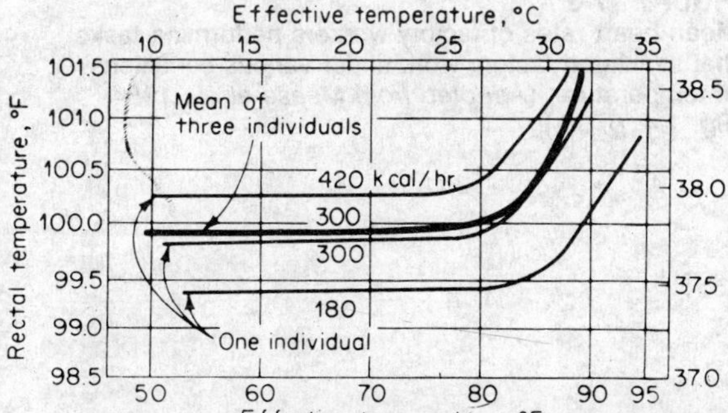

FIGURE 17-5
Rectal temperature as related to effective temperature for one individual working at three levels of work activity and for three individuals working at the same level (300 kcal/h). (*Source: Adapted from Lind, 1963a.*)

and surroundings is lower than that of the skin, heat is lost through convection and radiation. When in a comfortable environment, the skin blood flow amounts to about 5 percent of the cardiac output. In extreme heat, it can increase to 20 percent or more.

The second cardiovascular response to heat stress is an increase in heart rate. This can be seen in Figure 17-6, which shows heart rates of factory workers performing simulated factory tasks as a function of ambient temperature. The increased heart rate results in greater cardiac output (amount of blood pumped by the heart per minute). Under heat stress, cardiac output can increase 50 to 75 percent, with the excess output diverted to the skin for cooling purposes. The need to dissipate heat in hot environments creates competition for blood between the skin and muscles performing work. There may even be insufficient oxygen delivered to working muscles during heat stress thus causing lactic acid to build up and muscle pain and fatigue to occur. Recovery from physical work also takes longer in hot environments because blood is needed to both dissipate heat and remove the lactic acid.

Sweating Heat loss from convection and radiation is usually not sufficient to maintain thermal equilibrium in most situations, especially when engaging in physical activity. The body's second line of defense against heat stress is increased sweating. Heat is lost by evaporation of the sweat. Sweat that drips off the body rather than being evaporated results in virtually no heat loss from the body; it is just wasted water loss.

There are about 2 million sweat glands in the skin of the average person. The activation of these glands from different areas of the body shows wide variation among individuals. Some sweat more from the head, others from the torso. Workers exposed to intense heat stress can lose 6 to 7 L of sweat per day. Sweat loss up to 10 to 12 L in 24 hours has been reported (Leithead and Lind, 1964). Sweat glands can fatigue, and during prolonged exposure to a hot environment there is a gradual reduction in the sweat rate even if water loss is replaced at the same rate by drinking. This seems to be related to skin wetness because if the skin is dried with a towel at regular intervals, the sweat rate will be increased.

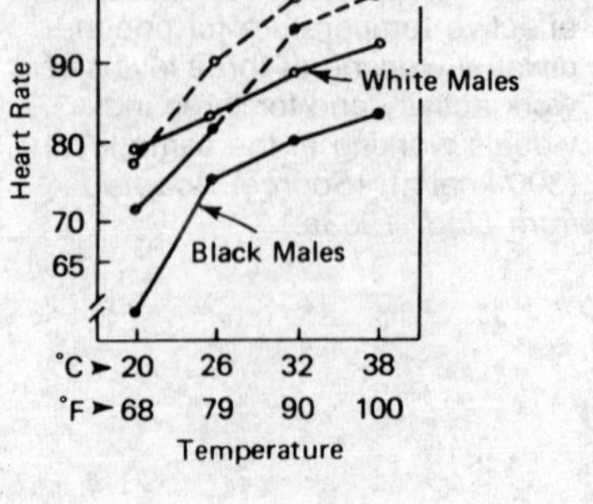

FIGURE 17-6
Mean heart rates of factory workers performing tasks that simulated factory work under various conditions of temperature. (*Adapted from Meese et al., 1984, Fig. 1.4, p. 36.*)

Excessive sweating can cause hypohydration or dehydration if fluids are not replaced by drinking. Hypohydration itself affects thermoregulation and results in a rise of core temperature. Hypohydration can also lead to muscle cramps and reduced endurance for physical work. People working in a hot environment must be encouraged to drink because people do not always feel thirsty during hypohydration. Water should be taken at frequent intervals. In hot environments, NIOSH (1986) recommends drinking 5 to 7 oz. (150 to 200 mL) of water [50°–60°F (10°–15°C)] every 15 to 20 min.

Sweat contains salt, and a fair amount can be lost from sweating in a hot environment. It is now generally accepted that given the relatively high salt content of the average worker's diet in industrialized countries, salt supplements are neither required or recommended. Salt tablets can irritate the stomach. Too much salt contributes to high blood pressure and may even retard acclimatization.

Heat Illness Under extreme or prolonged heat stress, several disorders can result. The most severe, heat stroke, can lead to death. The following are several forms of heat illness:

1 *Heat rash:* Areas of the skin break out in small red, blisterlike, raised bumps caused by plugging of sweat gland ducts, retention of sweat, and inflammation. This condition is also called "prickly heat" because of the pricking sensation felt during heat exposure.

2 *Heat cramps:* There may be painful spasms of muscles used during work (notably in the arms, legs, and abdomen). The cramps occur during or hours after working in a hot environment. Heat cramps are associated with restricted salt intake and profuse sweating causing excessive loss of salt.

3 *Heat exhaustion:* Heat exhaustion is characterized by muscular weakness, nausea, vomiting, dizziness, and fainting. This is caused principally by dehydration and is more likely to occur to individuals who are not heat-acclimatized and are in poor physical condition.

4 *Heat stroke:* Heat stroke is an acute medical emergency due to excessive rise in body temperature and failure of the temperature regulatory mechanisms. Symptoms include nausea, headache, cerebral dysfunction, and bizarre behavior followed by sudden and sustained loss of consciousness. The most typical cause is a fatiguing of the sweat glands and a cessation of sweat production.

Individual Differences and Heat Stress

There are considerable differences between people with respect to their tolerance to heat stress. Several factors that account for some of these differences include:

1 *Physical fitness:* The more physically fit a person is, the greater his or her tolerance to working in hot environments. Physically fit individuals can perform a given level of physical work with smaller increases in heat rate and less heat production than unfit individuals. This is due in part to the greater capacity of fit individuals to deliver oxygen to the working muscles. The fit individual can then divert more blood to the skin for cooling without sacrificing the oxygen needs of the muscles.

2 *Aging:* Aging results in more sluggish response of the sweat glands and less total body water. This results in less effective thermal regulation. Among South African gold miners, for example, men over 40 were 10 times more likely to experience heat stroke than were men under 25 years of age (Strydom, 1971).

3 *Gender:* Burse (1979) reviewed the literature on differences in response to thermal stress among men and women. Under heat stress, women generally show greater increases in heart rate, lower maximal sweat rates, higher skin temperatures, and in severe wet heat, higher core temperatures. They are affected more by dehydration, and fewer women than men can be successfully heat-acclimatized. A major factor accounting for these results are differences in physical fitness between men and women. Physically fit women respond to heat stress in a manner similar to men of equal fitness, but men tend to be more physically fit than women. The effects of a woman's menstrual cycle on heat tolerance has been found to be minimal.

4 *Body fat:* Carrying around excess body fat causes a hardship on a person working in a hot environment. Additional fat means additional weight must be carried, and the body surface-to-body weight ratio becomes less favorable for heat dissipation. The fat also creates an insulating layer between the skin and deep body tissues, thereby hindering heat transfer to the skin and dissipation.

5 *Alcohol:* Alcohol has been commonly associated with heat stroke. It interferes with central and peripheral nervous functioning and is associated with hypohydration. Ingestion of alcohol prior to or during work reduces heat tolerance and increases the risk of heat stroke.

Acclimatization to Heat Stress

With repeated exposure to work in a hot environment, people show marked adaptation in their response to heat stress. People become acclimatized to the heat. The physiological changes accompanying acclimatization include increased sweating efficiency (earlier onset, greater sweat production, and lower salt concentration), reduced heart rate, and lower core and skin temperatures. Some of these effects are illustrated in Figure 17-7, which shows rectal temperature, heart rate, and sweat loss over 9 days of working in a hot climate [120°F dry bulb, 80°F wet bulb (48.9°C dry bulb, 26.7°C wet bulb)].

Within 4 to 7 days of exposure to a hot environment, most of the acclimatization changes have taken place. At the end of 12 to 14 days, the process is complete. Relatively brief daily exposures are necessary for acclimatization. Continuous daily exposures of 100 min are sufficient (Leithead and Lind, 1964). Failure to replace water during exposure may retard acclimatization. Acclimatization also occurs more quickly if one is physically fit. Vitamin C may also enhance heat acclimatization (Strydom, Kotze, van der Walt, and Rogers,

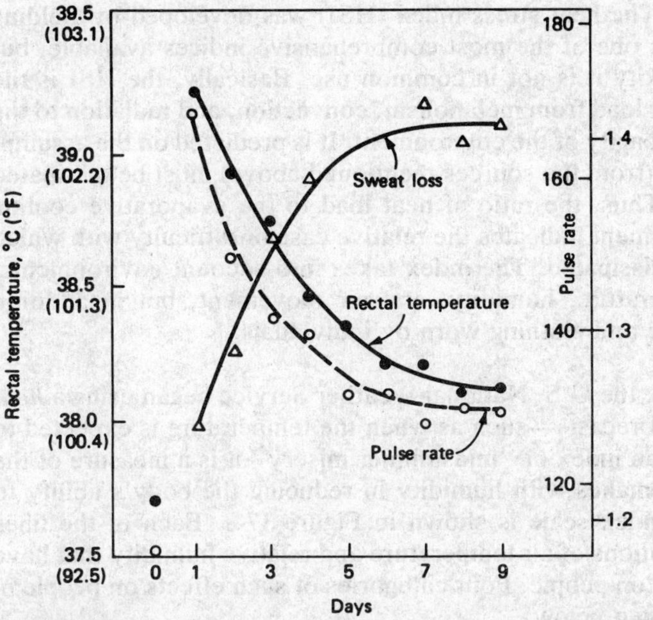

FIGURE 17-7
Effects of heat acclimatization on rectal temperature, heart rate, and sweat loss. A group of men exercised for 100 min each day for 9 days at a rate of energy expenditure of 300 kcal/h in a hot climate (120°F dry-bulb, 80°F wet-bulb [48.9°C dry-bulb, 26.7°C wet-bulb]). (*Source: Adapted from Lind and Bass, 1963.*)

1976). NIOSH (1986) recommends for acclimatization that new workers work at only 20 percent time the first day on the job and that the time be increased 20 percent each day so that on the fifth day the person is working full time.

The effects of acclimatization persist for several weeks following heat exposure, but some reduction in heat tolerance may be detected after only a few days without exposure. NIOSH recommends the following back-to-work schedule for individuals who were acclimatized but were off work for a period of time: 50 percent on day 1, 60 percent on day 2, 80 percent on day 3, and 100 percent on day 4.

There is wide variability among people in terms of their ability to acclimatize to heat. Therefore, even with a regimented program of heat acclimatization, individuals working in hot climates must be physiologically monitored during acclimatization to assess their level of physiological strain.

Indices of Heat Stress

We have already presented several measures of thermal conditions, including effective temperature. Those measures were developed to cover the full range of climatic conditions from cold to hot. Here we present a couple of indices developed specifically for assessing heat stress.

Heat Stress Index The heat stress index (HSI) was developed by Belding and Hatch (1955). It is one of the most comprehensive indices available, but because of its complexity it is not in common use. Basically, the HSI is the ratio of the body's heat load from metabolism, convection, and radiation to the evaporative cooling capacity of the environment. It is predicted on the assumption that the heat load (from the sources mentioned above) must be dissipated through evaporation. Thus, the ratio of heat load to the evaporative cooling capacity of the environment indicates the relative ease or difficulty with which the heat load can be dissipated. The index takes into account environmental factors such as temperature, humidity, and air movement, but in addition includes metabolic rate and clothing worn by individuals.

Heat Index In 1985, the U.S. National Weather Service began using a *heat index* in certain of its forecasts—such as when the temperature is expected to reach 95°F (35°C). As an index of "midsummer misery" it is a measure of the contribution that heat makes with humidity in reducing the body's ability to cool itself. The heat index scale is shown in Figure 17-8. Each of the lines represents the combinations of air temperature and relative humidity that have the same general effect on people. Four categories of such effects on people in high-risk groups are given below:

	Category	Heat index	General effects
I	Extremely hot	130°F + 55°C +	Heat/sunstroke highly likely with continued exposure
II	Very hot	105–130°F 41–55°C	Sunstroke, heat cramps, or heat exhaustion likely, and heatstroke possible with prolonged exposure and/or physical activity
III	Hot	90–105°F 32–41°C	Sunstroke, heat cramps, and heat exhaustion possible with prolonged exposure and/or physical activity
IV	Very warm	80–90°F 27–32°C	Fatigue possible with prolonged exposure and/or physical activity

The intent of the National Weather Service in using the heat index is to make people aware that the heat-humidity combination is dangerous and can lead to heat stroke or heat cramps.

Effect of Heat Stress on Performance

The results of studies investigating the effects of heat stress on performance have been somewhat contradictory. Some of this is undoubtedly due to differences among studies with regard to exposure conditions; the motivation, acclimatization, and skill level of the subjects used; and the nature of the tasks

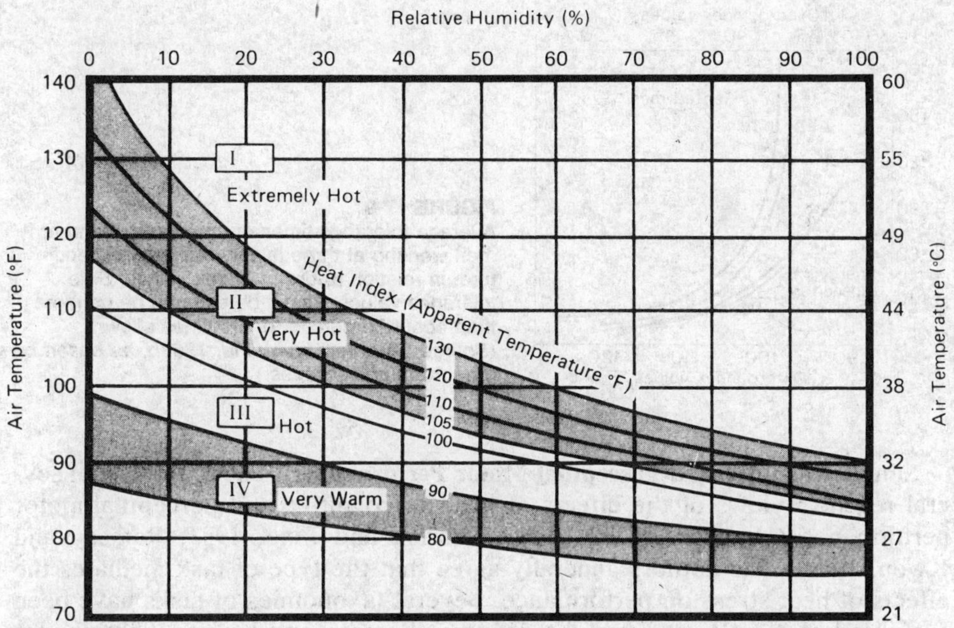

FIGURE 17-8
The heat index of the U.S. National Weather Service. (*Steadman, 1979.*)

performed. Despite this state of affairs, a few conclusions and reasonable hypotheses can still be extracted from the studies and reviews of the literature. We discuss briefly the effects of heat stress on performance of physical work, simple and complex cognitive and perceptual motor tasks, and safety.

Physical Work As discussed previously, heat stress creates competition for blood between the working muscles and the skin. Exhaustion from heavy physical work therefore occurs much sooner in a hot environment than in a temperate one. Figure 17-9 illustrates this. Shown are tolerance times for men sitting at rest and working at three levels of energy expenditure under various levels of heat stress (as measured by the Oxford index). For example, men sitting at rest can tolerate 100°F Oxford index for 3 or more hours. Working at 280 kcal/h (4.67 kcal/min), however, reduces tolerance time to about 30 minutes under such conditions.

Another example of the debilitating effects of heat stress on physical work is shown in Figure 17-10. Here the productivity of acclimatized men engaged in shoveling rock is shown as a function of effective temperature. Productivity is maintained up to about 82.4°F (28°C) ET and then begins to drop. At about 84°F (30°C) ET productivity is around 90 percent of prestress levels. The rate of decline in productivity continues to increase so that at about 93°F (34°C) ET, productivity is only 50 percent of prestress levels.

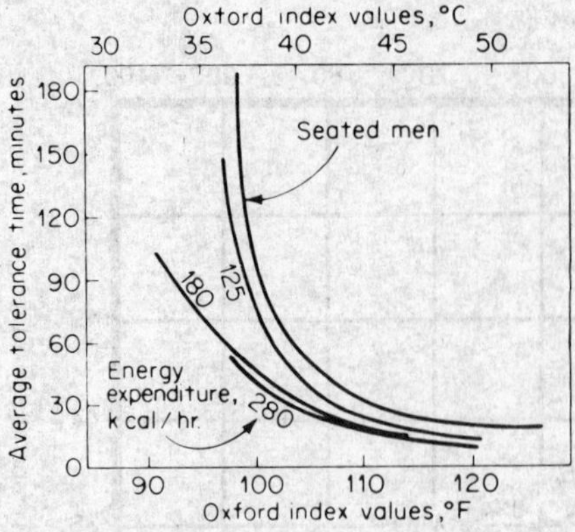

FIGURE 17-9
Average tolerance times of men seated, and of men working at three levels of energy expenditure, in relation to Oxford index values. Safe tolerance values should be taken to be no more than about 75 percent of the times shown. (*Source: Adapted from Lind, 1963b, as based on data from other studies.*)

Simple Cognitive and Perceptual-Motor Performance There have been several recent reviews of the effects of heat on cognitive and perceptual-motor performance (Hancock, 1981, 1982; Kobrick and Fine, 1983; Ramsey and Kwon, 1988). The authors generally agree that the type of task mediates the effects of heat stress on performance. Several taxonomies of tasks have been suggested. A simple, and useful taxonomy (Ramsey and Kwon, 1988) divides cognitive and perceptual-motor tasks into two categories: simple and complex. We discuss simple tasks here and complex tasks in the next section.

Simple tasks include visual and auditory reaction time, arithmetic problem solving, coding, and short-term memory tasks. Ramsey and Kwon reviewed 183 sets of data relating thermal stress and duration of exposure to performance on such tasks. (One study often provided more than one data set at different levels or durations of exposure.) Of the 183 sets of data, only 14 (7.3 percent) showed a statistically significant or even a partial, nonsignificant decrement in

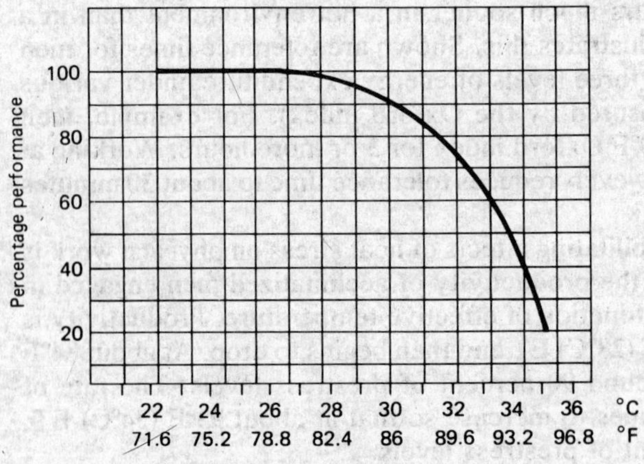

FIGURE 17-10
Productivity of acclimatized men engaged in shoveling rock as a function of effective temperature. The curve is the average performance across various conditions of air velocity [0.5–4.0 m/s (1.6–13 ft/s)]. (*Source: Adapted from Wyndham, 1974, Fig. 18.*)

performance. The conditions ranged from 59 to 111°F (15 to 44°C) WBGT with exposures from 5 min to 8 h. Ramsey and Kwon concluded that performance decrements on simple cognitive and perceptual-motor tasks are commonly not observed in the heat, and in fact, an enhancement in performance is frequently observed during brief exposures for these types of tasks. In support, Hancock (1981, 1982) notes that performance decrements on such tasks do not occur until conditions are close to the physiological heat-stress tolerance limits.

Complex Cognitive and Perceptual-Motor Performance Ramsey and Kwon put tracking, vigilance, and complex dual-tasks into this category. Their synthesis of the literature shows rather consistent performance decrements starting at about 86 to 91.4°F (30 to 33°C) WBGT. This is illustrated in Figure 17-11, which plots the percentage of data sets showing significant or partial performance decrements as a function of WBGT. An interesting finding is that there appears to be no effect of exposure time on performance decrement. Decrements occurred at the same WBGT temperature after 30 min as they did after several hours. This is in contrast to the conclusions reached by Hancock (1981, 1982) that for exposures under about 1 h, higher levels of heat stress were required before performance decrements were found.

Several theoretical explanations have been put forth to explain the decrements in performance due to heat stress: arousal, resource demands, perceived control, and body temperature. The arousal explanation considers heat stress to initially increase arousal and then over time to reduce arousal. This would

FIGURE 17-11
Percentage of data sets showing statistically significant or partial, but non-significant, performance decrements on complex cognitive and perceptual-motor tasks. One study often provided more than one data set. (*Source: Based on data from Ramsey and Kwon, 1988, Fig. 3.*)

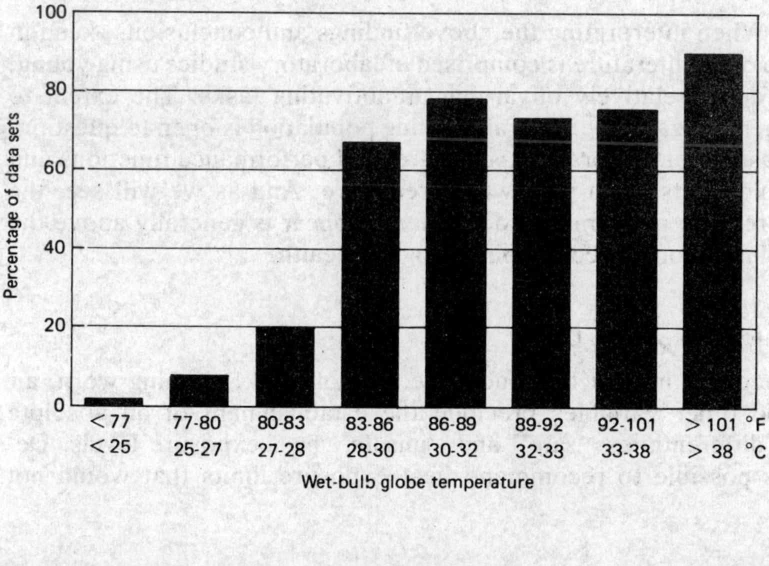

explain the initial facilitating effect of heat stress on performance of simple tasks. The heat stress initially raises the overall level of arousal closer to optimum levels.

The resource demand explanation suggests that cognitive resources (see Chapter 3) are required to adapt to a stressor and performance would therefore decline on complex tasks but would be little affected on simple tasks. This explanation, however, has trouble explaining the facilitating effects of heat stress. Bell and Greene (1982) invoke perceived control as an explanation and suggest that some of the performance decrement associated with heat stress may result from individuals reducing their level of effort, i.e., "giving up," to uncontrollable thermal stress.

The idea that performance decrement is related to deep body temperature (Hancock, 1981, 1982) has some merit. Deep body temperature, however, does not immediately rise upon exposure to a hot climate; rather it lags somewhat and then tends to elevate slightly with continued exposure. The research evidence, in contrast, does not show such a pattern between performance and exposure time. Ramsey and Kwon (1988) propose that rather than deep body temperature, performance may be related to cranial (brain) temperature, which responds almost immediately to large changes in external heat.

Safety Behavior Ramsey, Burford, Beshir, and Jensen (1983) observed workers' behaviors over a 14-month period under varying climatic conditions. Over 17,000 observations were made. The ratio of observed unsafe behaviors to the total number of observed behaviors formed a U-shaped curve when plotted over ambient temperature. The ratio of unsafe behaviors was at a minimum between 63 to 73°F (17 to 23°C) WBGT. When climatic conditions were below or above this generally preferred range, the incidence of unsafe behaviors increased. This is further support for the value of providing people with a comfortable climate in which to work.

Discussion When interpreting the above findings and conclusions, keep in mind that much of the literature is comprised of laboratory studies using young, healthy males doing relatively unvarying unmotivating tasks. The extent to which we can generalize to the general working population is open to question. It is for this reason that elaborate exposure-related performance functions and precise threshold limits seem somewhat premature. And as we will see, the level at which reliable performance decrements appear is generally above the recommended limits for protecting physiological health.

Recommended Heat Exposure Limits

Individual differences in heat tolerance, the type of work, clothing worn, air circulation, and other variables preclude the establishment of an absolute threshold that differentiates "safe" and "unsafe" heat exposure levels. Despite this, it is possible to recommend heat exposure limits that would not endanger most normal healthy workers, even though there might still be a risk for the few individuals with an abnormally low tolerance for heat. Several agencies have proposed such standards including the American Conference on Governmental Industrial Hygienist (ACGIH, 1985), American Industrial Hygiene Association (AIHA, 1975), International Organization for Standardization (ISO, 1982), and American Society of Heating, Refrigeration, and Air-Conditioning Engineers (ASHRAE, 1985). There are subtle differences between these recommendations, but overall they agree rather well (Ramsey and Chai, 1983).

We have chosen to present the recommendations of the National Institute of Occupational Safety and Health (NIOSH, 1986). The NIOSH recommendations take into account energy expenditure (called metabolic heat), acclimatization, and work-rest cycle. The recommendations assume the workers are wearing no more than the equivalent of a long-sleeved work shirt and trousers. Figure 17-12 presents the recommendations in terms of 1 h time-weighted average wet-bulb globe temperature (WBGT). Recommended alert limits (RAL) are for unacclimatized workers, although whenever the exposure levels

FIGURE 17-12
Recommended heat stress levels based on metabolic heat (1-h time-weighted average), acclimatization, and work-rest cycle. Temperature limits are 1-h time-weighted average WGBT. RAL = recommended alert limit for unacclimatized workers. REL = recommended exposure limit, for acclimatized workers. (*Source: NIOSH, 1986, Figs. 1 and 2.*)

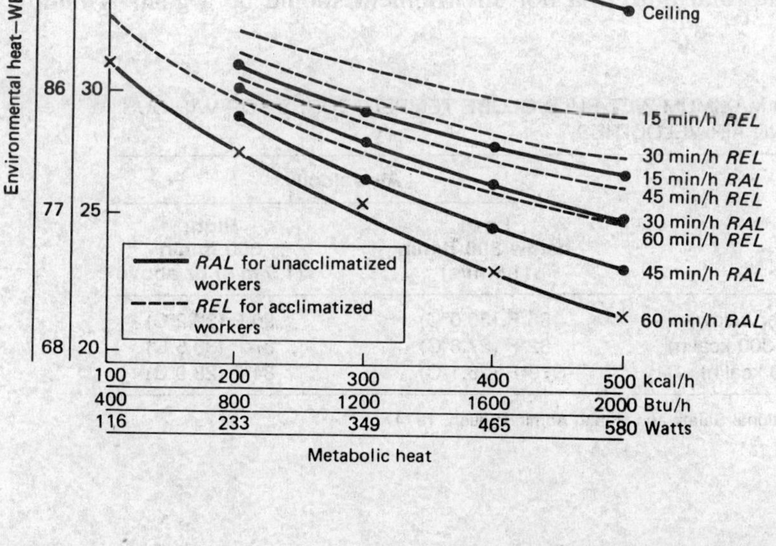

exceed the RAL, medical surveillance of all workers, acclimatized or not, is recommended. The recommended exposure limits (REL) are for acclimatized workers. The ceiling limit cannot be exceeded by anyone who is not wearing appropriate heat protection clothing and equipment.

For comparison, Table 17-2 presents the Occupational Safety and Health Administration recommended levels (OSHA, 1974). The newer NIOSH recommendations are somewhat lower than the older OSHA recommendations.

Comparing the NIOSH recommendations to the levels at which complex cognitive and perceptual-motor performance deteriorates, one can see that RALs exceed the lower estimate of performance decrements only for continuous very light work (100 kcal/h) or short duration (15 min/h) light work (200 kcal/h). Therefore, in most situations, protecting the health of workers will keep the level of heat stress below that associated with performance decrements.

Reducing Heat Stress

The reduction of heat stress can best be accomplished by a systematic and comprehensive approach. Changes can be made to the atmospheric conditions, the task, and the worker and through the use of appropriate protective equipment.

Atmospheric conditions can often be altered through the use of air conditioning, fans, or dehumidifiers. When air temperature is less than skin temperature, convective heat loss can be increased by increasing air velocity. Evaporative heat loss can be increased by increasing air movement and reducing relative humidity and moisture in the work area. Radiant heat sources can be shielded from workers with appropriate barriers.

Task variables can also be changed to reduce heat stress. Reductions in the level of energy expenditure required to perform tasks can greatly reduce the level of stress. Frequent rest breaks in a cool environment should be instituted and limiting the total time in a hot environment should be a goal. A buddy

TABLE 17-2
RECOMMENDED MAXIMUM WET-BULB GLOBE TEMPERATURES FOR VARIOUS WORKLOADS AND AIR VELOCITIES

	Air velocity	
Workload	**Low: below 300 ft/min (1.5 m/s)**	**High: 300 ft/min (1.5 m/s) or above**
Light (200 kcal/h or below)	86°F (30.0°C)	90°F (32.2°C)
Moderate (201 to 300 kcal/h)	82°F (27.8°C)	87°F (30.6°C)
Heavy (above 300 kcal/h)	79°F (26.1°C)	84°F (28.9°C)

Source: Occupational Safety and Health Administration, 1974.

system can be implemented in which workers are responsible for observing fellow workers for early signs and symptoms of heat strain. Providing adequate amounts of cool water near the work area and encouraging all workers to drink should be standard practice in all hot working environments.

Workers should be trained in the proper work, hygiene, and first-aid practices and procedures for working in hot environments. Systematic acclimatization and physical fitness programs are important for increasing tolerance to heat stress. Medical surveillance should also be instituted to identify individuals who are less tolerant of heat and to provide alternative work assignments for them.

Various types of protective equipment have been developed to reduce the stress of working in hot environments. These include water- or air-cooled vests or hoods, and ice-bag vests. Epstein, Shapiro, and Brill (1986) compared these types of protective equipment as well as using only a fan in a hot-dry environment. Figure 17-13 presents the mean composite physiological index of heat strain for each device. The smaller the number, the less strain experienced by the subjects. The ice-bag vest resulted in the least amount of heat stress. In general, vests were more efficient than hoods. Strydom (1982) reported that workers can actually become acclimatized to heat doing their normal work while wearing a dry-ice vest. Not only does such equipment reduce heat strain; it also appears not to significantly interfere with the acclimatization process.

FIGURE 17-13
Physiological strain experienced by acclimatized subjects wearing various types of protective equipment for 4 h in a hot-dry (122°F [50°C]; 30 percent RH) climate. Physiological strain is measured by a composite index of changes in heart rate, rectal temperature, and sweat rate. (*Source: Based on data from Epstein, Shapiro, and Brill, 1986, Table 1.*)

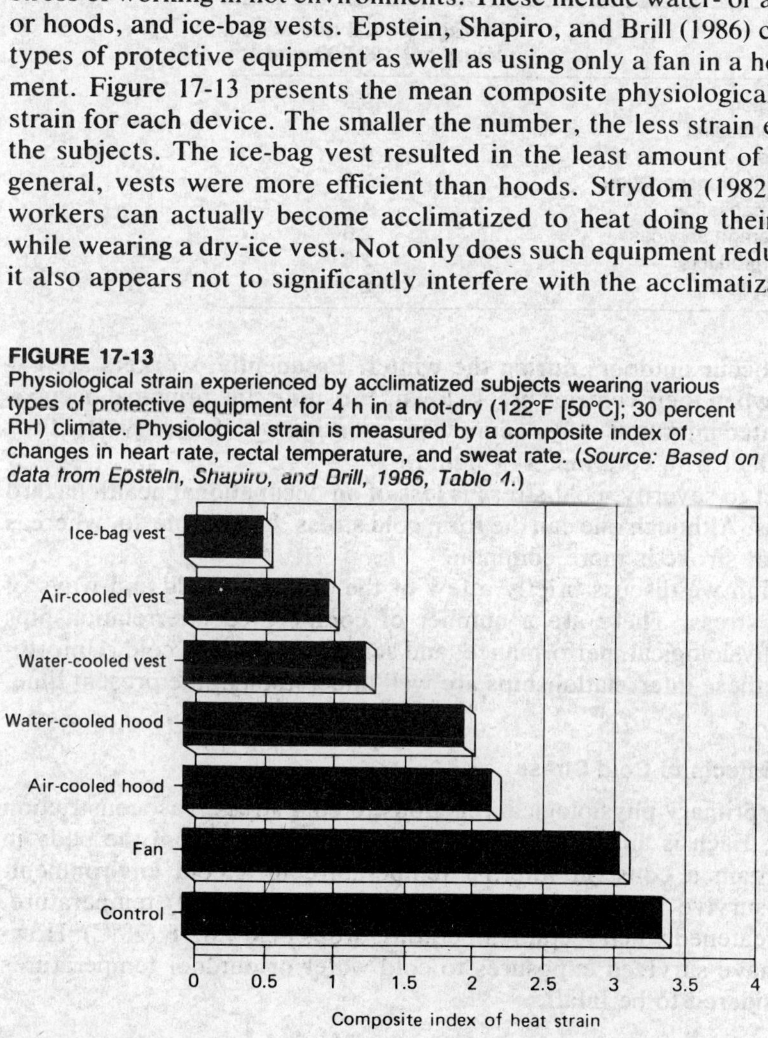

COLD STRESS

Although civilization is generally reducing the requirement for many people to work in cold environments, there are still some circumstances where people must work and live in such environments. The situations include outdoor work in winter, work in arctic locations (especially in the military and oil and gas exploration), and work in cold-storage warehouses. In most cases, people who must work in cold environments have the opportunity to dress appropriately and usually do. An analysis of workers' compensation claims (Sinks, Mathias, Halpern, Timbrook, and Newman, 1987) for cold-related injury revealed the following industries at most risk:

Industry	Average annual rate of cold injury per 100,000 workers
Oil and gas extraction	19.5
Trucking and warehousing	9.2
Protective services	8.1
Local and interurban transportation	7.6
Electric, gas, and sanitation	4.7
Auto repair and garage services	2.8
Food and kindred products	2.8
Heavy construction	2.1

Most injuries occur outdoors during the winter. Frequently, workers are exposed to cold when their vehicles break down. Frostbite, the principal occupational cold-related injury, usually occurs when working with frozen products or coming in contact with cold metal or liquid.

With respect to severity, cold stress is less of an occupational health hazard than heat stress. Although one can die from cold stress, few people do, whereas death from heat stroke is more common.

In this section we discuss briefly a few of the effects of cold and ways of reducing cold stress. There are a number of complicated interrelationships between the physiological, performance, and sensory aspects of cold. Unfortunately, not all these interrelationships are well understood at the present time.

Physiological Effects of Cold Stress

There are two primary physiological reactions to cold stress: vasoconstriction and shivering. Each is aimed at reducing the net heat loss from the body in order to maintain a constant internal temperature in a cold environment. Humans can survive rather large drops in deep body (rectal) temperature. Survival is threatened when rectal temperature drops below 82°F (28°C). However, people have survived exposures to cold water or outdoor temperatures normally considered to be fatal.

Vasoconstriction The body's first defense against cold stress is to constrict the blood vessels in the skin and extremities. The blood flow can be virtually shut off to these regions. This serves two purposes. First, the warm blood is kept away from the cold skin so that less heat is lost to the environment. Second, the insulating capacity of the skin is increased up to sixfold by cutting off the flow of blood. Because of this, the temperature of the fingers and toes can rapidly approach the temperature of the surrounding air, causing local cold injury (e.g., frostbite). Vasoconstriction also causes more blood to be distributed to the internal organs. As a consequence, more blood is passed through the kidneys, which causes an increased production of urine. This is why people urinate more frequently in cold environments than in warm environments.

Vasoconstriction also deprives the smooth muscles that constrict the blood vessels of oxygen. Eventually they fatigue and allow the blood vessels to dilate. The blood that flows in has not been circulating and has become deoxygenated. It is this deoxygenated blood that gives the skin that characteristic blue appearance we associate with cold stress.

Shivering If core temperature cannot be maintained through vasoconstriction, the body will attempt to increase production of metabolic heat by shivering. Shivering consists of synchronous activation of practically all muscle-groups: antagonistic muscle groups are made to contract against one another. The increased muscular activity increases metabolic heat from 2 to 4 times what is was at thermal equilibrium at rest. This can be enough to offset the loss of heat to the environment and bring the net heat storage to zero. However, it can never be enough to turn the heat storage positive. That is, shivering will never heat you up; it can only stop you from cooling down.

The better the physical condition of a person, the more efficiently the person can produce heat by shivering and the longer the person can maintain the shivering reflex without suffering exhaustion. Good physical conditioning can mean the difference for surviving severe cold stress.

Severe Cold Stress The most common cold injury is frostbite. *Frostbite* is the freezing of body tissue and the formation of ice crystals in the tissue cells. Frostbite can result from local cooling, most commonly to the hands and feet. Even if core temperature is maintained at normal levels, contact with metal or liquid (good conductors of heat) can cause the skin to freeze. The real danger comes when freezing occurs to the deeper tissue (second- and third-degree frostbite). The tissue cells can be ruptured by the ice crystals and blood cells can aggregate in the vessels, leading to gangrene.

Hypothermia is generalized cold stress and is usually characterized by a body temperature below 95°F (35°C). The initial stages are marked by disorientation, apathy, hallucinations, aggression, or euphoria. As body temperature drops further, the person may become unconscious; cardiac arrythmia may develop; the pupils of the eye become dilated; and finally the heart stops. In extreme cold stress, the body's rate of metabolism slows (the Q10 effect in reverse) and the body needs less oxygen. This, in part, explains the miraculous survivals of people (most notably children) who "drown" in frozen lakes and even after 10 minutes under water are revived with little or no brain damage. The frigid water lowers body temperature and reduces the brain's need for oxygen. At normal body temperatures, 3 min without oxygen would mean almost certain brain damage. In addition, there is also a reflex, called the *dive reflex,* that slows the heart rate when the face is exposed to cold and breathing is stopped. This can be demonstrated by submerging your head in a tub of ice water. Your heart rate will immediately slow down, but will immediately return to normal when you remove your head from the water and start to breathe.

Acclimatization to Cold Stress

It is difficult to demonstrate definite evidence of general physiological acclimatization to cold in humans. There is some scant evidence that humans repeatedly exposed to cold may increase their metabolic rates when exposed to cold without pronounced shivering (Astrand and Rodahl, 1986). There is also some evidence for local acclimatization to cold. Fish filleters, for example, who habitually expose their hands to cold water have an increased blood flow through their hands so that their hands remain warm and are not so apt to become numb when exposed to the cold water (Nelms and Soper, 1962). However, as Astrand and Rodahl (1986) so aptly point out, what physiological adaptation to cold there is, is of little practical value compared with the importance of know-how, experience, and state of physical fitness.

An Index of Cold Stress: Wind Chill Index

The most commonly used index for cold conditions is the *wind chill index* (WCI) or, more typically, a derivative of it, the *equivalent wind chill temperature*. The original basis for the WCI was developed by Dr. Paul Siple (who, as an Eagle Scout, was in one of Admiral Byrd's expeditions to Antarctica). He developed the WCI for the Army in 1945 (Siple and Passel, 1945). The WCI is an index of the combined effects of air temperature and wind speed on subjective discomfort, and thus it is useful as an idea of the relative severity of cold environments, especially with wind velocities below 50 mi/h (80 km/h).

Actually, the WCI as such is not often used directly. Rather it is common practice to convert such values to corresponding *equivalent wind chill temperatures*. The equivalent wind chill temperatures for selected air temperatures and wind speeds are given in Table 17-3. Any given value in the body of the table represents the air temperature (under calm conditions) that would produce the same "subjective discomfort" as the specific combination of air temperature and wind speed for that value. Thus, with an air temperature of 30°F (−1°C) and a wind speed of 40 mi/h (64 km/h), discomfort would be the

TABLE 17-3
EQUIVALENT WIND CHILL TEMPERATURES FOR CERTAIN AIR TEMPERATURES AND WIND SPEEDS

Wind speed, mi/h	Air temperature, °F 30	20	10	0	−10	−20
Calm	30	20	10	0	−10	−20
5	27	16	7	−6	−15	−26
10	16	2	−9	−22	−31	−45
20	3	−9	−24	−40	−52	−68
30	−2	−18	−33	−49	−63	−78
40	−4	−22	−36	−54	−69	−87
Wind speed, km/h	**Air temperature, °C −1**	**−7**	**−12**	**−18**	**−23**	**−29**
Calm	−1	−7	−12	−18	−23	−29
8	−3	−9	−14	−21	−26	−31
16	−9	−16	−23	−28	−36	−43
32	−15	−23	−31	−39	−47	−55
48	−19	−28	−36	−45	−54	−61
64	−22	−30	−38	−47	−56	−65

Source: Environmental Sciences Services Administration.

same as with an air temperature of −4°F (−22°C) with calm wind. Similarly, an air temperature of 0°F (−18°C) and a wind speed of 40 mi/h (64 km/h) would seem like a chilling −54°F (−47°C) with calm wind.

Subjective Sensations to Cold

We all have experienced sensations of discomfort in cold conditions. In part our sense of comfort or discomfort is related to skin temperature. This is shown in Table 17-4, which shows the mean skin temperature (the mean of skin temperatures of various body areas) and hand-skin temperature for various sensations.

Effects of Cold Stress on Performance

The effects of cold on performance of various types of human functions are not yet fully understood and, in any event, are rather intricate. Some of the interacting factors that complicate such understanding include the specific type of task or function; the interaction of air temperature, humidity, air flow, and radiation; duration of exposure; whether the body is being warmed or cooled; rate of cooling; differential exposure to different parts of the anatomy (as the body versus the hands); acclimatization; and individual differences.

TABLE 17-4
SUBJECTIVE SENSATIONS ASSOCIATED WITH SKIN TEMPERATURES

Sensation	Mean skin temperature	Hand-skin temperature
Comfortable	92°F (33.3°C)	
Uncomfortably cold	88°F (31°C)	68°F (20°C)
Shivering cold	86°F (30°C)	
Extremely cold	84°F (29°C)	59°F (15°C)
Painful		41°F (5°C)

Source: ASHRAE, 1981, pp. 8–15.

Given this assortment of interacting variables, however, we can distill at least a few generalizations from the available research. A few such generalizations are given below, most of these being extracted from Enander's (1984) comprehensive review.

Physical Work Cold has a detrimental effect on physical work. A reduction in core temperature and/or muscle temperature reduces physical functions resulting in a decrease in muscle strength and endurance (Enander, 1989). Bergh (1980), for example, found a 4 to 6 percent decrease in maximal exercise (less than 3 min duration) for every 1.8°F (1°C) drop in core temperature. The decrease was 8 percent per 1°C for exercise lasting 3 to 8 min.

A couple of factors account, at least in part, for this reduced physical capacity in the cold. For one, reduced core temperature reduces the rate of metabolism within the muscle. For another, cold reduces the speed of neural conduction in the peripheral motor nerves (Astrand and Rodahl, 1986).

Tactile Sensitivity Probably all of us, at one time or another, have experienced the loss in tactile sensitivity when our hands have become cold. Under severe conditions, a person may not even be able to determine what end of the screw they are touching without visual reference. Tactile sensitivity is related to finger skin temperature. Enander (1984) believes that it is actually the more slowly changing temperature of the deeper tissue rather than the actual surface temperature that affects sensitivity. In most cases, however, the two temperatures are closely related. Morton and Provins (1960) proposed that sensitivity is an L-shaped function of skin temperature and that each individual has a relatively sharp critical temperature at which performance deteriorates markedly.

This reduction in sensitivity makes performance of tasks that require manipulation of small objects difficult. Assembly and repair tasks, for example, are adversely affected by the cold. To demonstrate this to yourself, immerse your hands in ice water and then try to tie your shoelaces. (Perhaps you might want to just take our word for it; tying your shoes with cold hands is difficult to do.)

Manual Performance Related to tactile sensitivity, manual performance is without doubt adversely affected by the cold. Manual performance seems to be related to hand-skin temperature (HST). Figure 17-14, for example, illustrates the relationship between HST and performance on two manual dexterity tasks. A general finding, also illustrated in Figure 17-14, is that slow cooling results in a greater performance decrement than fast cooling at equivalent HSTs. This indicates that it is the temperature of the deeper tissue that is probably responsible for the decline in performance.

The lower bounds of unimpaired performance are still questionable. In terms of hand-skin temperatures, those from about 55 to 65°F (13 to 18°C) have been suggested as being such lower bounds, but there is still some question about such limits. In terms of ambient temperature, Riley and Cochran (1984*a*) reported a moderate drop on several dexterity tests when the ambient temperature dropped from 75 to 55°F (24 to 13°C) but a precipitous drop between that level and 35°F (1.7°C); where the precipitous drop starts, however, is not known.

Early work (Clark and Jones, 1962) suggested that practicing a manual task in a cold environment would improve performance in the cold more than

FIGURE 17-14
Deterioration in performance on two tasks as associated with reduction in hand-skin temperature (HST) and with fast cooling of the hands (within 5 min) or slow cooling (within 50 min). (*Source: Adapted from Lockhart, Kiess, and Clegg, 1975, Fig. 2, p. 111. Copyright 1975 by the American Psychological Association. Reprinted by permission.*)

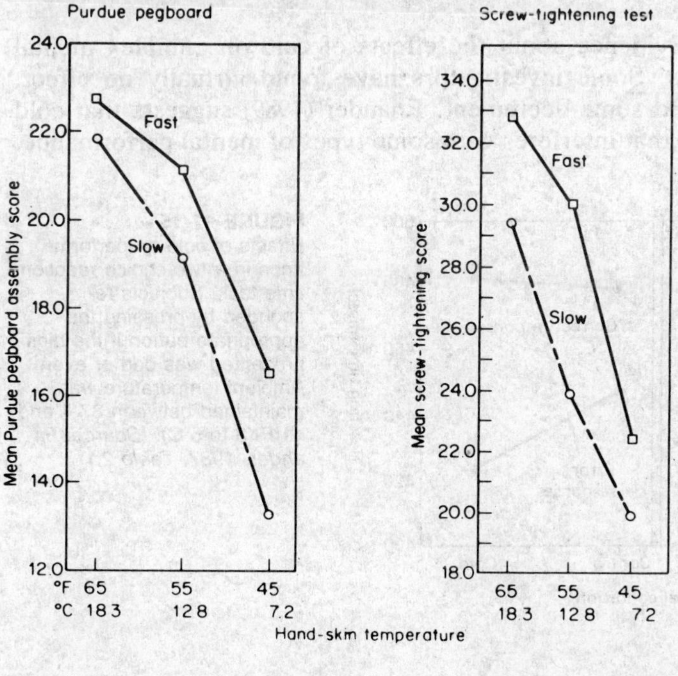

practicing in a neutral climate. Enander (1989), however, found opposite results when initially untrained subjects were used. Enander concludes that initial learning of a task in the cold may be less advantageous than initially learning the task in a neutral climate. Once the skill is acquired, however, practice in the cold may be more beneficial than practice in a neutral environment for subsequent performance in the cold.

Tracking Performance Tracking performance is adversely affected by the cold. Although the lower limit of unimpaired performance is still uncertain, it probably is between an ambient temperature of 39 to 55°F (4 to 13°C). At low temperatures there is often a marked loss of motivation and increase in apathy among subjects (Payne, 1959), which may account, in part, for the decrement in performance.

Reaction Time There is no systematic evidence that simple reaction time is affected by cold. There is, however, evidence that choice reaction time is adversely affected. What typically is found is that cold increases the number of errors made on the task but decreases the reaction time on trials in which errors are made. Reaction time on correct trials, however, is not affected. This is shown in Figure 17-15, which shows results on a digit classification task during exposure to cold (determine if digit presented is odd or even and press the appropriate button). Enander (1987) interprets such results to indicate a decreased ability to inhibit erroneous responses and is thus most evident in tasks requiring rapid accurate responses.

Mental Activities Evidence about the effects of cold on complex mental activities is ambiguous. Some investigators have found virtually no effect, while others have found some decrement. Enander (1989) suggests that cold may act as a distractor that interferes with some types of mental performance.

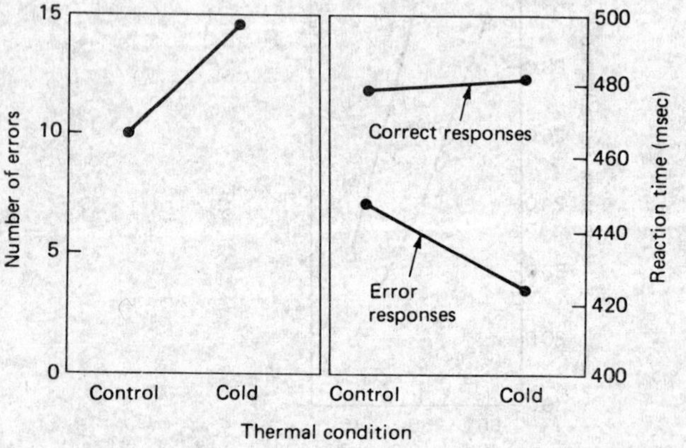

FIGURE 17-15
Effects of cold on performance in a two-choice reaction time task. Subjects responded by pressing the appropriate button if the digit presented was odd or even. Ambient temperature was maintained between 37.4 and 41°F (3 to 5°C). (*Source: Enander, 1987, Table 2.*)

The effect would, of course, depend on the type of task, the severity of the cold, the skill level of the subjects, and the prior experience of the subjects with performing in the cold.

Protection from Cold Stress

When people are in the cold environments, for whatever reason (work, recreation, or otherwise), efforts should be made to optimize their comfort, protect their health, and when relevant, maximize their performance.

Proper Clothing We discussed the insulating capacity of clothing earlier in this chapter. Needless to say, proper clothing is very important for reducing the adverse effects of cold on comfort, health, and performance. The use of warm clothing can extend the tolerance of people for cold conditions, as illustrated in Figure 17-16. This shows the tolerable combination of exposure time and temperature for each of four insulation levels; the difference in tolerance between 1 and 4 clo units is upward of 60°F (16°C). This demonstrates the trade-off effects, in terms of maintaining heat balance, of exposure time and insulation.

In connection with the use of apparel in cold conditions, however, McIntyre and Griffiths (1975) report that adding garments does not compensate fully for the discomfort of cool conditions. Although their subjects did report increased feelings of warmth with long-sleeved woolen sweaters, the sweaters did not fully alleviate the feelings of discomfort. The authors attributed this to the fact that the subjects still reported that their feet felt cold. Thus, for adequate protection in the cold, attention should be given to providing footwear that is as warm as possible.

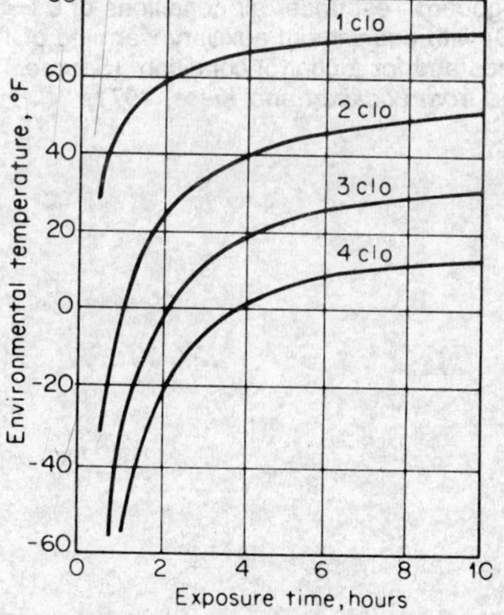

FIGURE 17-16
Exposure time and temperature that are tolerable for different levels of insulation (clo units). (*Source: Adapted from Burton and Edholm, 1955, as presented in Webb, 1964, p. 125.*)

Use of Gloves When the work or other activities involve primarily the hands (rather than the fingers, as such), gloves usually can be worn. When finger dexterity is required, however, regular gloves may not be feasible. In such instances partial gloves may be useful. These are gloves with the extreme and middle phalanges (sections) of the fingers exposed. The protection of the rest of the hand, however, seems to provide at least some carry-over protection to the exposed portions of the fingers under at least cool (if not cold) conditions. In an investigation with such gloves, Riley and Cochran (1984*b*) had subjects perform several manipulative tasks with and without such gloves at 60, 50, and 40°F (16, 10, and 4°C). They reported that partial gloves can protect the hands and fingers with minimal detrimental effects on performance of the tasks in question, at least down to about 40°F (4°C). Some indication of the protection is shown by the following data on mean finger-skin temperature:

	Without gloves	With gloves
Males	49°F (9.4°C)	66°F (18.9°C)
Females	53°F (11.7°C)	71°F (21.7°C)

Use of Auxiliary Heaters Most work in the cold involves some form of manual activity. In such cases efforts must be made to minimize reduction of hand-skin temperature. One such procedure is the use of infrared auxiliary heaters as proposed by Lockhart and Kiess (1971). In their investigation, one group used these heaters under 0°F (−18°C) conditions, another had no heaters under the same conditions, and a third control group performed the task indoors. The results for the Purdue Pegboard Test are shown in Figure 17-17

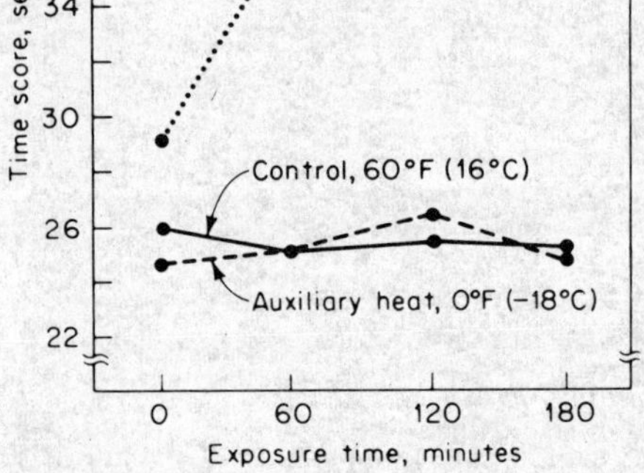

FIGURE 17-17
Time to perform the assembly part of the Purdue Pegboard Test under air conditions of 0°F (−18°C) with and without auxiliary warming of the hands and for a control condition. (*Source: Adapted from Lockhart and Kiess, 1971, Fig. 1.*)

and indicate that the auxiliary heating was very effective in maintaining task performance. The use of such auxiliary heaters would not be feasible, of course, in some situations (such as in some construction or logging operations).

Use of Rewarming Facilities There are times when exposure to cold exceeds reasonable tolerance levels of one or more relevant criteria (such as skin temperature, hand-skin temperature, core temperature, air temperature, work performance, etc.). When such tolerance levels are exceeded, individuals should go to a location (such as a warm room) where rewarming of the body or hands can take place. But there are still questions about how such rewarming should be carried out. In citing the results of various relevant investigations, for example, Enander (1984) reports that the effects of intermittent warming and cooling remain unclear. The ambiguities presumably revolve around the questions of optimum schedules for doing so, by considering such factors as the level of deep-body cooling that has taken place and the relative importance of hand cooling and body cooling in maintaining reasonable thermal balance.

DISCUSSION

Variations in climatic conditions will be with us forever and undoubtedly we will continue to work, live, and play in climates which are physiologically nonoptimum for us. Considerations of comfort, health, and human performance must be included in the design of facilities for hot and cold climates and the tasks that will be performed in such environments. An understanding of the effects of temperature, humidity, and air flow on human health and performance, as well as the effectiveness of various protective measures is essential for efficient and safe design.

REFERENCES

American Conference of Governmental Industrial Hygienists (ACGIH) (1985). *Threshold limit values for chemical substances and physical agents in the work environment with intended changes for 1985–1986*. Cincinnati.

American Industrial Hygiene Association (AIHA) (1975). Heat exchange and human tolerance limits. In *Heating and cooling for man in industry* (2d ed.). Akron, OH: AIHA, pp. 5–28.

American Society of Heating, Refrigeration, and Air-Conditioning Engineers (ASHRAE) (1981). *Handbook of fundamentals*. New York.

American Society of Heating, Refrigerating, and Air-Conditioning Engineers (ASHRAE) (1985). *ASHRAE handbook. 1985 fundamentals*. New York.

Astrand, P., and Rodahl, K. (1986). *Textbook of work physiology*. New York: McGraw-Hill.

Belding, H. S., and Hatch, T. F. (1955, August). Index for evaluating heat stress in terms of resulting physiological strains. *Heating, Piping and Air Conditioning*, pp. 129–136.

Bell, P., and Greene, T. (1982). Thermal stress: Physiological comfort, performance, and social effects of hot and cold environments. In G. Evans (ed.), *Environmental stress*. Cambridge: Cambridge University Press, pp. 75–104.

Bergh, U. (1980). Human power at subnormal body temperature. *Acta Physiologicia Scandinavian*, Suppl. 478.

Beshir, M. Y., Ramsey, J. D., and Burford, C. L. (1982). Threshold values for the Botsball: A field study of occupational heat. *Ergonomics*, 25(3), 247–254.

Botsford, J. H. (1971). A wet globe thermometer for environmental heat measurement. *American Industrial Hygiene Association Journal*, 32, 1–10.

Burse, R. (1979). Sex differences in human thermoregulatory response to heat and cold stress. *Human Factors*, 21(6), 687–699.

Burton, A. C., and Edholm, O. G. (1955). *Man in a cold environment: Physiological and pathological effects of exposure to low temperatures*. London: Edward Arnold.

Clark, R., and Jones, C. (1962). Manual performance during cold exposure as a function of practice level and the thermal conditions of training. *Journal of Applied Psychology*, 46, 276–280.

Dinman, B., Stephenson, R., Horvath, S., and Colwell, M. (1974). Work in hot environments: I. Field studies of work load, thermal stress and physiologic response. *Journal of Occupational Medicine*, 16, 785–791.

Enander, A. (1984). Performance and sensory aspects of work in cold conditions: A review. *Ergonomics*, 27(4), 365–378.

Enander, A. (1987). Effects of moderate cold on performance of psychomotor and cognitive tasks. *Ergonomics*, 30(10), 1431–1445.

Enander, A. (1989). Effects of thermal stress on human performance. *Scandinavian Journal of Work and Environmental Health*, 15 (suppl. 1), 27–33.

Epstein, Y., Shapiro, Y., and Brill, S. (1986). Comparison between different auxiliary cooling devices in severe hot/dry climate. *Ergonomics*, 29(1), 41–48.

Fanger, P. O. (1970). *Thermal comfort, analysis and applications in environmental engineering*. Copenhagen: Danish Technical Press.

Fanger, P., and Christensen, N. (1986). Perception of draught in ventilated spaces. *Ergonomics*, 29(2), 215–235.

Fanger, P., Melikov, A., Hanzawa, H., and Ring, J. (1988). Air turbulence and sensations of draught. *Energy and Building*, 12, 21–39.

Hancock, P. A. (1981). The limitation of human performance in extreme heat conditions. *Proceedings of the Human Factors Society 25th Annual Meeting*. Santa Monica, CA: Human Factors Society, pp. 74–78.

Hancock, P. (1982). Task categorization and limits of human performance in extreme heat. *Aviation, Space, and Environmental Medicine*, 53(8), 778–784.

Hohnsbein, J., Piekarski, C., Kampmann, B., and Noack, Th. (1984). Effects of heat on visual acuity. *Ergonomics*, 27(12), 1239–1246.

International Organization for Standardization (ISO) (1982). *Hot environments—Estimation of the heat stress on working man, based on the WBGT index (wet bulb globe temperature)* (ISO/TC 159, ISO-DIS 7243). Geneva.

Jensen, R. (1983, September). Workers' compensation claims attributed to heat and cold exposure. *Professional Safety*, 19–24.

Kobrick, J., and Fine, B. (1983). Climate and human performance. In D. Oborne, and M. Gruneberg, (eds.), *The physical environment at work*. Chichester, England: Wiley.

Laviana, J., Rohles F., Jr., and Hoffberg, L. (1987). Dry environments: The influence of low humidity on comfort and health. *Proceedings of the Human Factors Society 31st Annual Meeting*. Santa Monica, CA: Human Factors Society, pp. 1101–1104.

Leithead, C., and Lind, A. (1964). *Heat stress and heat disorders*. Philadelphia: FA Davis.

Lind, A. R. (1963*a*). A physiological criterion for setting thermal environmental limits for everyday work. *Journal of Applied Physiology,* 18, 51–56.

Lind, A. R. (1963*b*). Tolerable limits for prolonged and intermittent exposures to heat. In J. D. Hardy (ed.), *Temperature—Its measurement and control in science and industry,* vol. 3. New York: Reinhold.

Lind, A., and Bass, D. (1963). Optimal exposure time for development of acclimatization to heat. *Federal Proceedings,* 22, 704–708.

Lockhart, J. M., and Kiess, H. O. (1971). Auxiliary heating of the hands during cold exposure and manual performance. *Human Factors,* 13(6), 457–465.

Lockhart, J. M., Kiess, H. O., and Clegg, T. J. (1975). Effect of rate and level of lowered finger-surface temperature on manual performance. *Journal of Applied Psychology,* 60(1), 106–113.

McIntyre, D. A., and Griffiths, I. D. (1975). The effects of added clothing on warmth and comfort in cool conditions. *Ergonomics,* 18(2), 205–211.

Meese, G. B., Kok, R., Lewis, M. I., and Wyon, D. P. (1984). A laboratory study of the effects of moderate thermal stress on the performance of factory workers. *Ergonomics,* 27(1), 19–43.

Morton, R., and Provins, K. (1960). Finger numbness after acute local exposure to cold. *Journal of Applied Physiology,* 15, 149–154.

National Institute of Occupational Safety and Health, (1986). *Criteria for a recommended standard . . . Occupational exposure to hot environments, Revised Criteria 1986*. Washington, DC: Superintendent of Documents.

Nelms, J., and Soper, J. (1962). Cold vasodilation and cold acclimatization in the hands of British fish filleters. *Journal of Applied Physiology,* 49(4), 444.

Occupational Safety and Health Administration (OSHA) (1974, Jan. 9). *Recommendation for a standard for work in hot environments* (draft no. 5). Washington, DC: Department of Labor.

Olesen, B. (1985). *Local thermal discomfort* (Tech. Rev. no. 1-1985). Marlborough, MA: Bruel & Kjaer Instruments, Inc.

Payne, R. (1959). Tracking performance as a function of thermal balance. *Journal of Applied Physiology,* 14, 387–389.

Ramsey, J. (1987, February). Practical evaluation of hot working areas. *Professional Safety,* 42–48.

Ramsey, J., Burford, C., Beshir, M., and Jensen, R. (1983). Effects of workplace thermal conditions on safe work behavior. *Journal of Safety Research,* 14, 105–114.

Ramsey, J., and Chai, C. (1983). Inherent variability in heat-stress decision rules. *Ergonomics,* 26(5), 495–504.

Ramsey, J., and Kwon, Y. (1988). Simplified decision rules for predicting performance loss in the heat. *Proceedings on heat stress indices*. Luxembourg: Commission of the European Communities.

Riley, M. W., and Cochran, D. J. (1984*a*). Dexterity performance and reduced ambient temperature. *Human Factors*. 26(2), 207–214.

Riley, M. W., and Cochran, D. J. (1984*b*). Partial gloves and reduced temperature.

Proceedings of the Human Factors Society 28th Annual Meeting. Santa Monica, CA: Human Factors Society, pp. 179–182.

Rohles, F. H., Jr. (1974). The modal comfort envelope and its use in current standards. *Human Factors*, 64(3), 314–323.

Sinks, T., Mathias, C., Halpern, W., Timbrook, C., and Newman, S. (1987). Surveillance of work-related cold injuries using worker's compensation claims. *Journal of Occupational Health and Safety*, 29, 505–509.

Siple, P. A., and Passel, C. F. (1945). Movement of dry atmospheric cooling in subfreezing temperatures. *Proceedings of the American Philosophical Society*, 89, 177–199.

Steadman, R. C. (1979). The assessment of sultriness. *Journal of Applied Meteorology*, 18(7), 861–884.

Strydom, N. (1971). Age as a causal factor in heat stroke. *Journal of the South African Institute of Mining and Metallurgy*, 72, 112–114.

Strydom, N. (1982). Developments in heat-tolerance testing and acclimatization procedures since 1961. In H. Glen (ed.), *Proceedings 12th CMMI Congress*. Johannesburg, South Africa: South African Institute of Mining and Metallurgy.

Strydom, N., Kotze, H., van der Walt, W., and Rogers, G. (1976). Effect of ascorbic acid on rate of heat acclimatization. *Journal of Applied Physiology*, 41, 202–205.

Winslow, C. E. A., and Herrington, L. P. (1949). *Temperature and human life*. Princeton, NJ: Princeton University Press.

Wyndham, C. (1974). Research in the human sciences in the gold mining industry. *American Industrial Hygiene Association Journal*, 113–136.

CHAPTER **18**

NOISE

Before the days of machines and mechanical transportation equipment, our noise environment consisted of noises such as those of household activities, animals, horsedrawn vehicles, hand tools, and nature. But human ingenuity changed all that by creating machines, motor vehicles, radios, guns, bombs, sirens, jet aircraft, and rock concerts. Noise has become such a pervasive aspect of working situations and community life that we refer to it as *noise pollution*, and some consider it to be a health hazard.

Although noise has commonly been referred to as *unwanted sound*, a somewhat more exact definition is the one proposed by Burrows (1960), in which noise is considered in an information-theory context, as follows: Noise is "that auditory stimulus or stimuli bearing no informational relationship to the presence or completion of the immediate task." This concept applies equally well to attributes of task-related sounds that are informationally useless as well as to sounds that are not task-related.

We start this chapter with a discussion of various measures of the loudness of noise. The effects of noise on hearing loss, physiology, performance, and feelings of annoyance are then discussed. Noise exposure limits and methods of handling noise problems are also presented.

HOW LOUD IS IT?

We touched briefly in Chapter 6 on the fact that the human ear is not equally sensitive to all frequencies of sound. In general, we are less sensitive to low frequencies (below 1000 Hz) and more sensitive to higher frequencies. Thus a low-frequency tone will not sound as loud to us as a high-frequency tone of equal intensity (i.e., sound pressure). To put it another way, a low-frequency tone must have more intensity than a higher-frequency tone to be of equal loudness. This particular fact has led investigators to search for a metric to measure the subjective quality of sound. We discuss several basic measures here. When we discuss the annoyance quality of noise, we introduce several other measures that are derived from these basic measures.

Sound Level Meter Scales

As we indicated in Chapter 6, sound-pressure meters built to American National Standards Institute (ANSI) specifications contain frequency-response weighting networks (designated A, B, and C). Each network electronically attenuates sounds of certain frequencies and produces a weighted total sound-pressure level. Figure 18-1 shows the relative response curves of the A, B, and C scales and the response characteristics of the human ear at threshold. As can be seen, the C scale weights all frequencies almost equally. The B scale, originally intended to represent how people might respond to sounds of moderate intensity, is rarely (if ever) used. The most commonly used scale is the A scale. The Occupational Safety and Health Administration (OSHA) standards

FIGURE 18-1
Relative response characteristics of the A, B, and C sound-level meter scales and the human ear at threshold. (*Source: Jensen, Jokel, and Miller, 1978, Fig. 2.4.*)

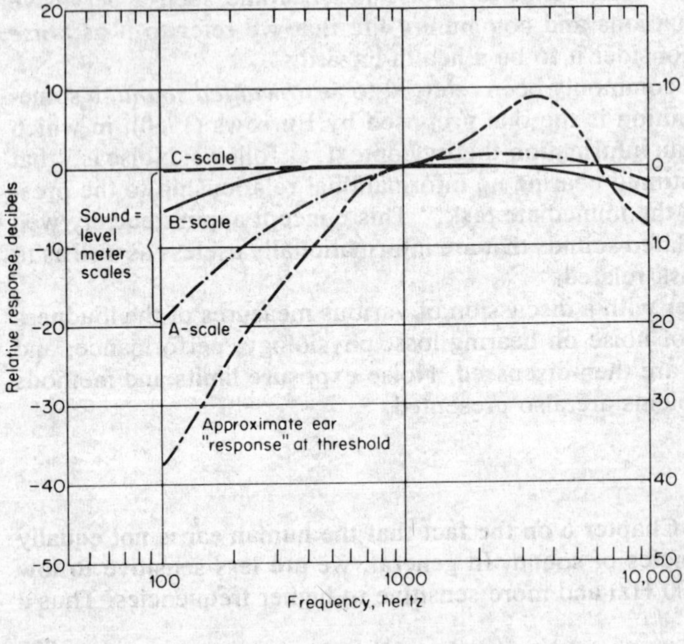

for daily occupational noise limits are specified in terms of this measure, and the Environmental Protection Agency (1974) has selected the A scale as the appropriate measure of environmental noise. As we will see, the many indices of loudness, noisiness, and annoyance are all based on the A scale (the unit is dB*A*). Of the three scales, the A scale comes closest to approximating the response characteristics of the human ear.

As if the A, B, and C scales were not enough, there exist on some meters D scales (there is more than one D scale). The scales were designed primarily to provide a measure of aircraft noise but have yet to gain complete universal acceptance, and currently they are used only rarely and for very specific measurement applications.

Psychophysical Indices

Loudness is a subjective or psychological experience related to both the intensity and the frequency of sound. Researchers have tried to develop scales or indices based on the *physical* properties of sound that will measure this *psychological* experience, hence the term *psychophysical*. Among the oldest and most widely recognized psychophysical indices of loudness are the *phon* and *sone*. The methodology used to define the phon and sone is common to other loudness indices and involves subjects matching comparison sounds to a reference sound in terms of subjective loudness (or, in one case, noisiness).

In the case of phons, Robinson and Dadson (1957) presented to subjects a 1000-Hz pure tone (the reference sound) at different sound-pressure levels and had the subjects adjust the intensity of various frequency pure tones (the comparison sounds) until the comparison sound was judged to be of equal loudness to the reference sound. The decibel level of the comparison sound was recorded. From these data, equal-loudness curves were generated, and they are shown in Figure 18-2. Each curve shows the decibel intensity of different frequencies that were judged to be equal in loudness to that of a 1000-Hz tone of a specified intensity level. To illustrate, one can see in Figure 18-2 that a 50-Hz tone of about 65 dB is judged equal in loudness to a 1000-Hz tone of only 40 dB. The unit *phon* was designated as the measure of loudness and was set equal to the decibel level of the 1000-Hz tone. All tones, for example, judged equal in loudness to a 60-dB, 1000-Hz tone are designated as having a loudness level of 60 phons. Our 65-dB, 50-Hz tone would, therefore, have a loudness level of 40 phons because it is equal in loudness to a 1000-Hz tone of 40 dB.

So far, so good. The phon tells us about the subjective *equality* of various sounds, but it tells us nothing about the *relative* loudness of different sounds. That is, we cannot say how many times louder a 40-phon sound is compared to a 20-phon sound. We know it is louder, but we do not know if it is, say, twice as loud or 4 times as loud. For such comparative judgments we need still another yardstick. Fletcher and Munson (1933) developed such a scale, and Stevens (1936) named it the *sone*. In developing this scale (as with the phon) a reference sound was used. One *sone* is defined as the loudness of a 1000-Hz tone of 40 dB

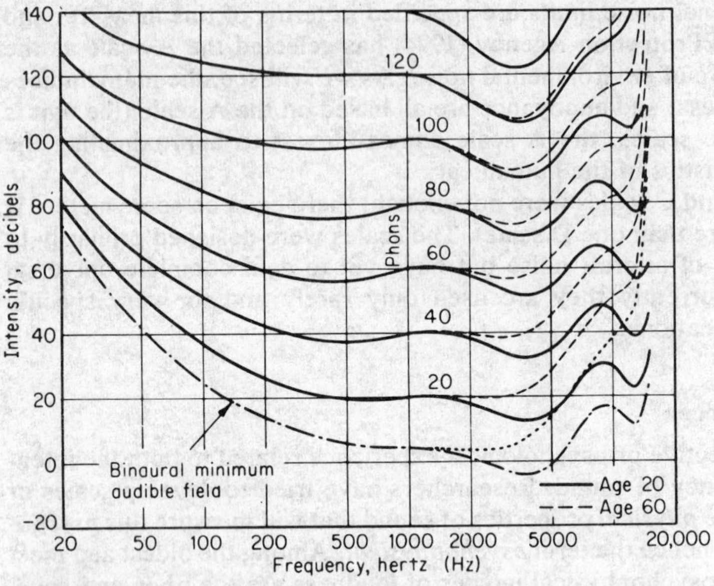

FIGURE 18-2
Equal-loudness curves of pure tones. Each curve represents intensity levels of various frequencies that are judged to be equally loud. The lowest curve shows the minimum intensities of various frequencies that typically can be heard. (*Source: Robinson and Dadson, 1957. Crown copyright reserved. Courtesy National Physical Laboratory, Teddington, Middlesex, England.*)

(that is, 40 phons). A sound that is judged to be twice as loud as the reference sound has a loudness of 2 sones, a sound that is judged to be 3 times as loud as the reference sound has a loudness of 3 sones, etc. In turn, a sound that is judged to be half as loud as the reference sound has a loudness of 0.5 sone.

Manufacturers of stove-top exhaust fans in the United States list the loudness of their product in sones; unfortunately, most consumers have no idea what it means! To provide some basis for relating sones to our own experiences, consider the examples given in Table 18-1.

There is a relationship between phons and sones: 40 phons = 1 sone, and every 10 additional phons doubles the number of sones. For example, 50 phons = 2 sones, 60 phons = 4 sones, and 70 phons = 8 sones. In like manner 30 phons = 0.5 sone, and 20 phons = 0.25 sone. Given this, we can determine that our 40-phon sound is 4 times as loud as a 20-phon sound.

There is another set of indices used to measure loudness which was developed as a refinement to the original phon and sone. The two measures, analogous to the phon and sone, are the *perceived level of noise* (PLdB) and the *Mark VII sone* (Stevens, 1972). There is also a set of indices used to measure noisiness which is not quite the same as loudness. The two measures are the *perceived noise level* (PNdB) and the *noy* (Kryter, 1970). These various indices

TABLE 18-1
EXAMPLES OF LOUDNESS LEVELS

	Loudness	
Noise source	**Decibels**	**Sones**
Residential inside, quiet	42	1
Household ventilating fan	56	7
Automobile, 50 ft (15 m)	68	14
"Quiet" factory area	76	54
18-in (46-cm) automatic lathe	89	127
Punch press, 3 ft (1 m)	103	350
Nail-making machine, 6 ft (2 m)	111	800
Pneumatic riveter, 4 ft (1.2 m)	128	3000

Source: Bonvallet, 1952, p. 43.

differ from the original phon and sone in terms of the reference sound used and the relationship between the equality and relative measures. Table 18-2 summarizes these differences and lists references where computational procedures can be found.

Equivalent Sound Level

Noise often varies in intensity over time. Fortunately, a baby's scream will change to a whimper and with luck to a yawn. Over the years many single-number measures have been proposed for time-varying sound levels. We discuss a few when we talk about the annoyance of noise. The Environmental Protection Agency (1974), however, concluded that, insofar as cumulative noise effects are concerned, the long-term average sound-level was the best measure for the magnitude of environmental noise. This long-term average is designated the *equivalent sound level* (L_{eq}) and is equal to the sound-pressure level (usually measured in dB*A*) of a *constant* noise that, over a given time, transmits to the receiver the same amount of acoustic energy as the actual time-varying sound.

L_{eq} depends on the time interval and acoustic events occurring during that period. For example, if a 100-dB*A* noise occurred for 1 h, the L_{eq} for that hour would be 100 dB*A*. Consider, however, a situation in which it was quiet during the next 4 h. The L_{eq} for the total 5-h period would now be less than 100 dB; in fact, it would be 94 dB*A*. What this says is that 5 h of 94-dB*A* noise is equivalent in acoustic energy to 1 h of 100-dB*A* noise and 4 h of quiet. There are hand-held sound-measuring instruments available that directly display L_{eq} values.

An alternative measure to L_{eq} is the *sound exposure level* (SEL or L_e), which is defined as the sound-pressure level (usually measured in dB*A*) of a constant noise that *over a period of 1 second* transmits to the receiver the same amount of acoustic energy as does the actual time-varying sound during the

TABLE 18-2
SOME PSYCHOPHYSICAL INDICES OF SOUND LOUDNESS AND NOISINESS

	Psychophysical indices		
Characteristics	**Phon and sone**	**Stevens, 1972**	**Kryter, 1970**
Measure of:	Loudness	Loudness	Noisiness
Equality scale measured in:	Phons	Perceived level of noise (PLdB)	Perceived noise level (PNdB)
Reference sound for equality scale:	100-Hz pure tone	⅓ octave band of noise centered at 3150 Hz	Octave band of random noise centered at 1000 Hz
Relative scale measured in:	Sones	Mark VII sones	Noys
Reference level for relative scale:	1 sone = 40 phon	1 Mark VII sone = 32 PLdB	1 noy = 40 PNdB
Relationship between equality and reference scales:	Each 10-phon increase doubles the number of sones	Each 9-PLdB increase doubles the number of Mark VII sones	Each 10-PNdB increase doubles the number of noys
Reference for computational procedures:	Peterson and Gross, 1978	Peterson and Gross, 1978	Williams, 1978

sampling period. For sampling periods longer than 1 s, SEL will be higher than L_{eq}. The SEL is often used for describing the noise energy of a single event such as a vehicle passing by or an aircraft flying overhead.

NOISE AND LOSS OF HEARING

Of the different possible effects of noise, one of the most important and clearly established is hearing loss. There are really two primary types of deafness: *nerve* deafness and *conduction* deafness. Nerve deafness usually results from damage or degeneration of the hair cells of the organ of Corti in the cochlea of the ear (see Chapter 6). The hearing loss in nerve deafness is typically uneven, being greater in the higher frequencies than in the lower ones. Normal deterioration of hearing through aging is usually of the nerve type, and continuous exposure to high noise levels also typically results in nerve deafness. Once nerve degeneration has occurred, it can rarely be remedied.

Conduction deafness is caused by some condition of the outer or middle ear that affects the transmission of sound waves to the inner ear. It may be caused by different conditions, such as adhesions in the middle ear that prevent the vibration of the ossicles, infection of the middle ear, wax or some other substance in the outer ear, or scars resulting from a perforated eardrum. Conduction deafness is more even across frequencies and does not result in complete hearing loss. It is only a partial loss because airborne sound waves strike the skull and are transmitted to the inner ear by conduction through the bones of the skull. People with conductive deafness sometimes are able to hear reasonably well, even in noisy places, if the sounds to which they are listening (for example, conversation) are at intensities above the background noise. This type of deafness can sometimes be arrested or even improved. Hearing aids are more useful in this type of deafness than they are when deafness is caused by nerve damage.

Measuring Hearing

Before reviewing the effects of noise on hearing, we shall first see how hearing (or more properly, hearing loss) is measured. The instrument used to measure a person's threshold of hearing (that is, the minimum sound-pressure level that is just audible) at selected frequencies is called an *audiometer*.

An audiometer presents tones of various frequencies to the subject through an earphone. By varying the sound-pressure level of the tone, the minimum audible sound-pressure level (threshold) is determined for each frequency. These are compared to the average threshold of hearing for young persons with no hearing impairment. The difference in decibels between the two thresholds, at each frequency tested, is reported on an *audiogram* as the *hearing level* or *hearing loss* at each frequency and for each ear.

There are both manual and automatic audiometers. Manual audiometers require a trained technician to administer the tones. With an automatic audiometer, the frequencies and sound-pressure level changes are controlled by the subject using a hand-held remote control. The subject is told simply, "Press the button when you hear the signal and release it when you can't hear it." Audiometers commonly measure hearing loss at frequencies of 500, 1000, 2000, 3000, 4000, 6000, and 8000 Hz.

Normal Hearing and Hearing Loss

Before we see what effect noise has on hearing, we should first see what normal hearing is like. Normal, nonoccupational, hearing loss is considered to be due to two sources: presbycusis and sociocusis. *Presbycusis* is hearing loss due to the normal process of aging. *Sociocusis* refers to hearing loss due to nonoccupational noise sources, such as household noises, television, radio, traffic, etc. Admittedly, what is considered nonoccupational to one person may be occupational to another. In essence, sociocusis excludes excess exposure to loud noises on a regular basis.

Kryter (1983*a*, 1983*b*) summarized various population surveys of hearing loss and derived idealized hearing loss curves due to presbycusis and sociocusis. Figure 18-3 shows the median hearing loss of males and females at various ages. [Figure 18-3 actually includes hearing loss due to pathological conditions (nosocusis), but Kryter indicates that this accounts for only a few decibels of the hearing loss shown.] It is clear that with age, hearing loss becomes increasingly severe at the higher frequencies. The difference between

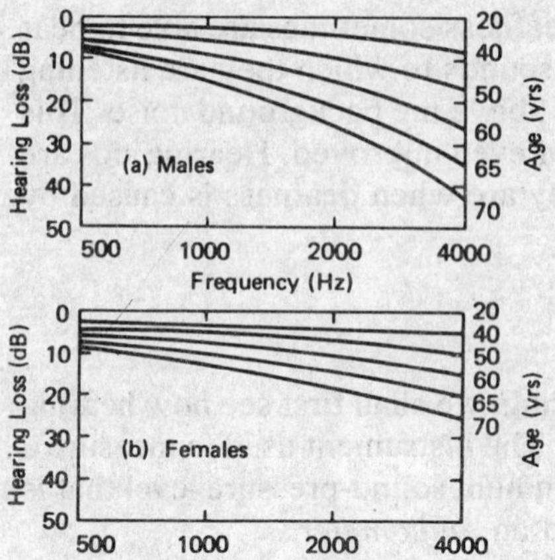

FIGURE 18-3
Idealized median (50th percentile) hearing loss due to presbycusis, sociocusis, and nosocusis for males and females as a function of age. (*Source: Kryter, 1983a, Fig. 20.*)

males and females is attributed to sociocusis; that is, males tend to be exposed to higher levels of nonoccupational noise than females. There are, of course, exceptions to this, as any mother who has taken care of two 3-year-olds can tell you. Based on Kryter's estimates, approximately 55 percent of normal hearing loss at higher frequencies is due to presbycusis, with sociocusis accounting for approximately 45 percent in industrialized countries.

Occupational Hearing Loss

Most hearing loss other than presbycusis and sociocusis can be considered occupationally related. Most such hearing loss is from continuous exposure over time, although exposure to noncontinuous noise (i.e., occasional or infrequent high levels of noise) also can take its toll.

After exposure to continuous noise of sufficient intensity there is some temporary hearing loss which is recovered a few hours or days after exposure. However, with additional exposure the amount of recovery gradually becomes less and less, and the individual is left with some permanent loss. The implications of permanent hearing loss are obvious. Temporary hearing loss, however, can also have serious consequences if a person depends on auditory information in the performance of a job or task.

Temporary Hearing Loss from Continuous Noise Since hearing generally recovers with time after exposure, the measurement of hearing loss must take place at a fixed time after exposure to be comparable. Traditionally, this has been done 2 min after the end of exposure. Any shift in threshold (from preexposure levels) is called the *temporary threshold shift at 2 min* (abbreviated TTS_2).

The relationship between TTS_2 and the sound level of the noise to which one is exposed is not a simple one. Some sound levels will not produce any measurable TTS_2 regardless of the duration of exposure. These sound-pressure levels define *effective quiet* where the hazardous effects of noise on hearing are concerned. This lower limit depends somewhat on frequency but is generally considered to be around 60 to 65 dB*A*. For exposures to noises of moderate intensity (80 to 105 dB*A*) TTS_2 increases in proportion to the logarithm of the sound-pressure level (SPL) of the exposure noise. Kryter (1985), after reviewing the literature, concluded that for exposures resulting in up to 10 db*A* of TTS_2, $TTS_2 = 10 \log SPL + k$. For exposures resulting in 10 to 40 dB*A* of TTS_2, $TTS_2 = 20 \log SPL + k$ (where k is a constant that depends on the specifics of the exposure, such as intermittency, duration, etc.).

The growth, or acquisition, of TTS_2 is proportional to the logarithm of exposure time, building up quickly at first and then more slowly as exposure time increases. Recovery from temporary threshold shifts, when the level of TTS_2 is less than 40 dB*A*, also follows a logarithmic function and is proportional to the logarithm of recovery time. Although both the growth and recovery of temporary threshold shifts are proportional to the logarithm of time, recovery takes longer than acquisition. For example, it may take less than an hour to acquire a TTS_2 of 25 dB*A*, but complete recovery may take up to 16 h (Davies and Jones, 1982).

There is an interesting relationship between the frequency of the exposure noise and the TTS_2 produced. The maximum threshold shift is produced not at the frequency of the exposure noise, but at frequencies well *above* the exposure noise; e.g., exposure to a pure tone of 700 Hz will produce a maximum TTS_2 at 1000 Hz or higher.

There are, of course, marked individual differences in TTS_2 that result from a given noise. Some people may experience considerable TTS_2 while others may experience hardly any. Further, some people are more sensitive to high-frequency sounds, others to low-frequency sounds. These individual differences must be considered when criteria are set to protect ears from damage.

Permanent Hearing Loss from Continuous Noise With repeated exposure to noise of sufficient intensity, a *permanent threshold shift* (PTS, or NIPTS for noise-induced permanent threshold shift) will gradually appear. Usually the PTS occurs first at 4000 Hz. As the number of years of noise exposure increases, the hearing loss around 4000 Hz becomes more pronounced, but is generally restricted to a frequency range of 3000 to 6000 Hz. With further noise exposure, the hearing loss at 4000 Hz continues and spreads over a wider frequency range (Melnick, 1979). (Note that 4000 Hz is in that region of frequencies to which the human ear is most sensitive.) This can be seen in Figure 18-4, which shows hearing loss curves for jute weavers exposed to wideband continuous noise with a noise spectrum that peaked in the octave bands centered at 1000 and 2000 Hz. The overall sound level varied from 99 to 102 dB*A*.

A committee under the aegis of the American Industrial Hygiene Association has teased out and condensed, from many sources, data that show the incidence of hearing impairment, in a consolidated figure, for several age groups of individuals who have been exposed (in their work) to noise intensities of different levels. This consolidation is shown in Figure 18-5. Actually this

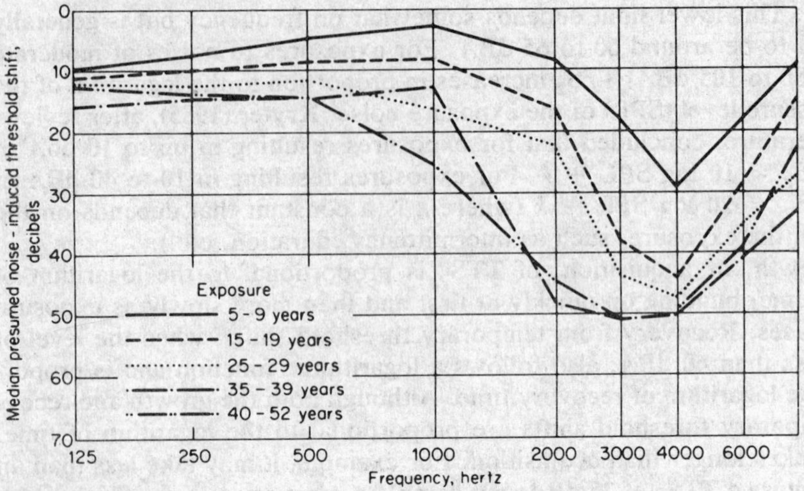

FIGURE 18-4
Median noise-induced permanent threshold shifts at various frequencies as a function of increasing exposure to noise among female jute weavers. The noise was wideband and continuous with a spectrum that peaked in the octave bands centered at 1000 and 2000 Hz, with overall sound level varying from 99 to 102 dBA. (*Source: Taylor, Pearson, Mair, and Burns, 1965, as presented by Melnick, 1979.*)

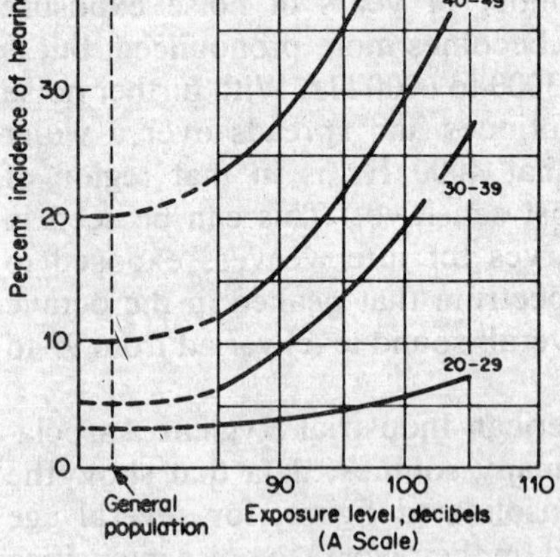

FIGURE 18-5
Incidence of hearing impairment in the general population and in selected populations by age group and occupational noise exposure; impairment is defined as a hearing threshold shift in excess of an average of 15 dB at 500, 1000, and 2000 Hz. (*Source: American Industrial Hygiene Association, 1966, Fig. 1, p. 420.*)

figure shows, for any group, the probabilities of individuals having hearing impairment (impairment being defined as an average hearing threshold shift in excess of 15 dB at 500, 1000, and 2000 Hz). Although the probabilities of impairment (so defined) are not much above those of the general population for individuals exposed to 85 dB, the curves shoot up sharply at higher levels, except for the youngest group (but their time will come!).

Must we wait 10 or 20 years to discover that a noisy environment is potentially harmful? Perhaps not. It is widely accepted that the average TTS_2 from an 8-h exposure to noise in young, normal ears is similar in magnitude to the average permanent threshold shift found in workers after 10 to 20 years of exposure to the same levels of noise (Kryter, 1985). Kryter indicates that there are no data available to seriously challenge this basic assumption.

Hearing Loss from Noncontinuous Noise The gamut of noncontinuous noise includes intermittent (but steady) noise (such as machines that operate for short, interrupted periods), impact noise (such as that from a drop forge), and impulsive noise (such as from gunfire). In heavy doses, such noise levies its toll in hearing loss, but the combinations and permutations of intensity, noise spectrum, frequency, duration of exposure, and other parameters preclude any simple, pat descriptions of the effects of such noise.* In the case of impact and impulsive noise, however, the toll sometimes is levied fairly promptly. For example, 35 drop-forge operators showed a noticeable increase in hearing threshold within as little as 2 years, and 45 gunnery instructors averaged 10 percent hearing loss over only 9 months, even though most had used hearing protection devices (Machle, 1945).

PHYSIOLOGICAL EFFECTS OF NOISE

Permanent hearing loss is, of course, the consequence of physiological damage to the mechanisms of the ear. Aside from possible damage to the ear itself, one would wonder whether continued exposure to noise might induce any other temporary or permanent physiological effects.

The onset of a loud noise will cause a *startle response*, characterized by muscle contractions, blink, and head-jerk movement. In addition, larger and slower breathing movements, small changes in heart rate, and dilation of the pupils occur. There is also a moderate reduction in the diameter of blood vessels in the peripheral regions, particularly the skin (Burns, 1979). All these responses are relatively transient and settle back to normal or near normal levels very quickly. With repeated exposure to this noise, the magnitude of the initial response is also diminished. Although transient in nature, the startle response can disrupt perceptual motor performance. Foss, Ison, Torre, and Wansack (1989), for example, found that a noise burst (peak SPL of 110 to 130 dB) disrupted aiming of a rifle for 1 to 2 s.

* A thorough treatment of exposure to intermittent noise is presented by Kryter et al. (1966).

Although there is considerable agreement on the startle-response reactions to the onset of noise, there is less agreement on the long-term physiological effects of repeated exposure to noise. There is considerable evidence that indicates that exposure to high noise levels (such as 95 dB*A* or more) is associated with generalized stress responses among those exposed to the noise (Jansen, 1969; Burns, 1979; Gulian, 1974). Kryter (1985), however, questions whether such effects are due to the direct overarousal of the autonomic nervous system by noise itself or to psychological responses indirectly associated with the noise. In residential communities, for example, high noise levels (e.g., aircraft noise) disrupt sleep, cause annoyance, and may cause concern for one's safety. In industrial settings, high noise is often associated with moving and dangerous equipment; noise may mask sounds relevant to job performance, thus increasing the difficulty of the job; and workers may be concerned about the damaging effects of noise on hearing. These indirect consequences of noise may be more important than the actual noise itself as the cause of stress. Whether stress associated with noise is the consequence of direct overstimulation of the autonomic nervous system or some other indirect cause, one cannot deny that stress and high noise often go hand in hand. It should be pointed out and stressed (excuse the pun) that adverse health effects associated with noise (such as hypertension, gastrointestinal problems, electrocardiogram irregularities, and complaints of headaches) are usually evidenced only with relatively high levels of noise, often over 95 dB*A*.

As an example of the kinds of effects reported in studies of residential community noise, Meecham (1983) reported on an 8-year study conducted on 200,000 residents living near Los Angeles International Airport. He found death rates from heart attacks among those over 75 years of age and suicides among those 45 to 54 years old to be higher in an area exposed to jet aircraft landing noise than in a nearby matched "quiet" area. In the high-risk area, jets passing overhead generated noise levels of 115 dB, and during peak daytime hours they passed overhead every 2.5 min, enough to rattle anyone.

EFFECTS OF NOISE ON PERFORMANCE

Aside from the startle-response effect on perceptual motor performance mentioned above, the effects of noise on performance are not clear-cut. By choosing the right studies, one can show that noise produces a decrement, no effect, or even an improvement in performance. Gawron (1982), for example, reviewed 58 noise experiments and found 29 showed noise hindered performance, 22 showed no effect, and 7 showed noise facilitated performance. This apparent state of confusion is due, in part, to the wide variability of conditions tested in noise experiments. The noises tested were intermittent in some cases and continuous in others. Electronically generated white noise and pure tones; tape-recorded machine noise, office noise, and street sounds; bells, buzzers, and air horn blasts; tank, helicopter, airplane, and rocket noise; music; and even gibberish have been used at one time or another. The definition of "quiet" has ranged from 0 to 90 dB and often is not even specified by the experimenter.

A variety of tasks have been studied, including tracking, reaction time, digit recall, arithmetic, letter cancellation, number checking, maze tracing, monitoring, distance judgment, hand steadiness, and accuracy of shooting. These tasks differ in difficulty and in the relative demands they place on the perceptual, cognitive, memory, and motor capabilities of the subjects. Even the experimental design used to test the effects of noise can influence the results. Adoption of a repeated-measures design (where the same subjects perform under both noise and quiet) may lead to different conclusions compared to a randomized-groups design (where different groups of subjects perform under noise and quiet) (Poulton, 1973; Poulton and Freeman, 1966).

General Conclusions Regarding Effects of Noise

Despite, or because of, this conglomeration of tasks and noise characteristics, we can make a few guarded conclusions about the effects of noise on performance: (1) With the possible exception of tasks that involve short-term memory, the level of noise required to obtain reliable performance effects is quite high, generally over 95 dB*A*. (2) Performance of simple, routine tasks may show no effect and often will even show an improvement as a result of noise. (3) If a person has to react at certain definite times, receives clear warning signals, and has an easily visible stimulus to respond to, then there will be little effect of continuous loud noise (over 95 dB*A*) on performance; (4) Sensory functions such as visual acuity, contrast discrimination, dark vision, accommodation, and speed of eye movements all show little effect from noise (Smith, 1989); (5) Motor performance is rarely impaired by noise unless balance is involved; (6) Simple reaction time is unimpaired provided the subjects have adequate warnings about when to respond; (7) The detrimental effects of noise are usually associated with tasks performed continuously without rest pauses between responses (Davies and Jones, 1982) and difficult tasks that place high demands on perceptual and/or information processing capacity (Eschenbrenner, 1971).

Specific Effects of Noise

In discussing the detrimental effects of noise, Broadbent (1976) identifies three effects which he has gleaned from the research available. First, he reports that certain decisions become more confident in the presence of noise. For example, if a person is presented a visual display and is asked to report signals that occur, that person will be more confident of the responses (whether correct or incorrect) when the task is performed in high levels of noise. Easily detected signals are usually not missed any more often in high-noise conditions than in quiet conditions. Doubtful and uncertain signals, however, may be missed more often in noisy conditions because the person sometimes is "confident" that they did not occur. In quiet conditions, the person may reconsider such signals and report their presence.

Broadbent also reports that there is a "funneling of attention" on the task. At high noise levels, a person typically focuses attention on the most important aspects of a task or on the most probable sources of information. If relevant task information is missed owing to this funneling phenomenon, then performance suffers. This phenomenon may explain why noise improves performance on simple, routine tasks. The person's attention is funneled onto the task rather than wandering because of possible boredom.

Broadbent also reports that continuous work in which there is no opportunity for relaxation will often show occasional moments of low performance and gaps in performance where no recorded response is made. The overall, average performance may not suffer, but the variability in performance increases.

These detrimental effects (increased confidence, funneling, and gaps), as we said, have typically been associated with high levels of noise. An apparent exception is performance on verbal tasks. Smith (1989), summarizing the literature, presents evidence that noise at levels below 90 dB*A* impairs semantic comprehension of verbal materials. An example of this can be seen in the results of Weinstein (1974, 1977). He reported that 68- to 70-dB*A* noise significantly impaired the detection of grammatical errors in a proofreading task (which requires semantic processing and short-term memory) but did not adversely affect the detection of spelling errors. This is consistent with a result reported by Smith and Stansfeld (1986) that people living in high aircraft noise areas more frequently reported that they would read something but fail to retain its meaning than did people living in low noise areas.

Basis for Specific Effects of Noise Although there is beginning to be some consensus on the effects of noise, there still remains considerable controversy over why, or how, noise has the effects it does. Poulton (1976, 1977, 1978) and Broadbent (1976, 1978) carried on a lively "discussion" of the mechanisms underlying the effects of noise on performance. Poulton (1978) contends that all the known effects of noise on performance can be explained by four determinants which combine to affect performance: (1) masking of acoustic task-related cues and inner speech; (2) distraction; (3) a beneficial increase in arousal when noise is first introduced, which gradually lessens and falls below normal when the noise is first switched off; and (4) positive and negative transfer from performance in noise to performance in quiet.*

Broadbent (1976, 1978, 1979), however, rejects the notion that detrimental effects of noise are caused by masking of acoustic task cues or inner speech. Rather, Broadbent attributes most of the detrimental effects of noise to "overarousal." It is a long-standing observation that there exists an inverted U relationship between arousal and performance. Too little or too much arousal results in lower performance than does a moderate level of arousal. Poulton rejects the overstimulation hypothesis when it is applied to noise.

Whether masking plays a major role in determining the effects of noise remains to be seen. Probably, as is so often the case, both sides are right—to a

* Positive transfer results from the better learning of the task in noise under the influence of the increased arousal. Negative transfer results from techniques of performance in noise used to counteract the masking or distraction being used in quiet where they are not appropriate.

degree. Undoubtedly if task-related acoustic cues are present and noise masks them, a decrement may well result. The question is whether noise can have a detrimental effect on a task in which no acoustic cue, inner speech, distraction, or transfer is present. Poulton would probably say no; Broadbent probably would contend that it would still be possible owing to overstimulation. Kryter (1985) seems to side with Poulton and, while acknowledging the role of arousal, believes much of the noise-induced performance decrements reported in the literature can be explained by masking inner speech or task-relevant auditory cues. In addition, Kryter also suggests that subjects in noise experiments construe various meanings to the noise which can affect motivation or cause stress. For example, subjects may believe that when a loud noise is present, it is a signal of importance of some kind, or they may believe that loud noise is supposed to affect them in some way. Some subjects may be concerned that the noise will be harmful to their hearing or that the experimental condition will become overbearing; such concerns may cause some physiological stress.

Discussion

Most of the evidence relating to either the degrading or the enhancing effects on performance is based on experimental studies, not on actual work situations. And extrapolation from such studies to actual work situations is probably a bit risky. Subjects used in most noise experiments are either students or military personnel under 30 years of age, and they are hardly representative of an industrial population. Subjects in experiments are usually highly motivated and will make an extra effort to maintain performance under adverse conditions. For simple or short-duration tasks, this may be adequate to mask any negative effects of noise. The long-term effects of noise on performance (i.e., those more representative of industrial exposure such as 5 to 7 h/d, 5 d/wk, for weeks on end) have been rarely studied in the laboratory. Further, tasks used in laboratory experiments often bear little resemblance to the types of tasks performed in noisy industrial environments, making generalization tenuous.

We reiterate one thing in closing: The level of noise required to exert measurable degrading effects on performance is, with the exception of verbal comprehension tasks, considerably higher than the highest levels that are acceptable by other criteria, such as hearing loss and effects on speech communications. Thus, if noise levels are kept within reasonable bounds in terms of, say, hearing loss considerations, the probabilities of serious effects on performance probably would be relatively nominal.

NOISE EXPOSURE LIMITS

In typical work situations, hearing loss is perhaps the prime criterion for acceptable noise levels. Standards that differentiate between continuous noise, impulse noise, infrasonic noise, and ultrasonic noise, have been set by various organizations. We do not attempt to review all the various standards; rather we concentrate on those promulgated by the Occupational Safety and Health

Administration (OSHA) of the U.S. Department of Labor. Actually, other national and international standards follow the OSHA recommendations rather closely.

Continuous and Intermittent Noise

OSHA has established permissible noise exposures for persons working on jobs in industry (OSHA, 1983). The permissible levels depend on the duration of exposure and are shown in Table 18-3. A key concept in the OSHA requirements is *noise dose*. Exposure to any sound level at or above 80 dB*A* causes the listener to incur a *partial dose* of noise. (Exposures to sound levels less than 80 dB*A* are ignored in calculating doses.) A partial dose is calculated for each specified sound-pressure level above 80 dB*A* as follows:

$$\frac{\text{Time actually spent at sound level}}{\text{Maximum permissible time at sound level (see Table 18-3)}}$$

The total or daily noise dose is equal to the sum of the partial doses. The noise dose can then be converted to an *8-h time-weighted average (TWA) sound level* by using Table 18-4. The TWA is the sound level that would produce a given noise dose if an employee were exposed to that sound level continuously over an 8-h workday.

A noise dose of 50 percent (TWA = 85 dB*A*) is designated as the *action level,* or the point at which the employer must implement a continuing, effective hearing conservation program. The program must include exposure monitoring, audiometric testing, hearing protection, employee training, and

TABLE 18-3
PERMISSIBLE NOISE EXPOSURES ACCORDING TO OSHA

Sound level, dB*A*	Permissible time, h
80	32
85	16
90	8
95	4
100	2
105	1
110	0.5
115	0.25
120*	0.125*
125*	0.063*
130*	0.031*

* Exposures above 115 dB*A* are not permitted regardless of duration; but should they exist, they are to be included in computations of the noise dose.
Source: OSHA, 1983.

TABLE 18-4
CONVERTING NOISE DOSE TO TWA

Noise dose	TWA,* dBA
10	73
25	80
50 (action level)	85
75	88
100 (permissible exposure level)	90
115	91
130	92
150	93
175	94
200	95
400	100

* Values are rounded to nearest decibel. The exact conversion from noise dose, *D*, to TWA is given by

$$\text{TWA} = 16.61 \log \frac{D}{100} + 90$$

Source: OSHA, 1983.

record keeping. A noise dose of 100 percent (TWA = 90 dB*A*) is designated as the *permissible exposure level,* or the point at which the employee must use feasible engineering and administrative controls to reduce noise exposure. OSHA (1981) estimated that there are 2.9 million production workers in the United States with TWAs in excess of 90 dB*A* and an additional 2.3 million workers with TWAs exceeding 85 dB*A*.

The concept of noise dose produces a curious situation. Consider a worker who, during a workday, is exposed to the following noise levels:

95 dB*A*	for 3.5 h
105 dB*A*	for 0.5 h
85 dB*A*	for 4.0 h

Such exposure is within the *individual* OSHA permissible time limits set forth in Table 18-3, but the total noise dose is equal to 163.5 [that is, 100(3.5/4.0 + 0.5/1.0 + 4.0/16.0)]. This represents a TWA of approximately 93.5 dB*A* and thus exceeds the permissible exposure level.

Impulse Noise

OSHA defines *impulse noise* as "a sound with a rise time of not more than 35 ms to peak intensity and a duration of not more than 500 ms to the time when the level is 20 dB below the peak" (OSHA, 1974). Table 18-5 lists the maximum number of permissible impulses for various peak intensity values. Note that OSHA (1983) is moving toward including impulse noise into the noise dose concept and doing away with regulating impulse noise by limiting the number of exposures.

TABLE 18-5
MAXIMUM NUMBER OF PERMISSIBLE IMPULSES IN AN 8-H DAY AS A FUNCTION OF PEAK SOUND-PRESSURE LEVEL

Peak sound-pressure level, dB	Maximum number of impulses per 8 h*
140	100
135	316
130	1,000
125	3,162
120	8,913
115	31,623
112.4	57,600†

* Based on following formula: number = $10^{16-P/10}$ (where P = peak decibels).

† This would be considered continuous noise.

Source: Leavitt, Thompson, and Hodgson, 1982, based on OSHA, 1981. Reprinted with permission by American Industrial Hygiene Association Journal.

Infrasonic Noise

Infrasonic noise is noise with frequencies below the audible range, typically less than 20 Hz. Currently, there are no national or international standards for permissible exposure limits to infrasonic noise. Von Gierke and Nixon (1976) present a review of the effects of infrasound and conclude that it is not subjectively perceived and has no effect on performance, comfort, or general well-being. To protect the auditory system, however, they recommend 8-h exposure limits ranging from 136 dB at 1 Hz to 123 dB at 20 Hz. If the level is increased 3 dB, then the permissible duration must be halved.

Ultrasonic Noise

Ultrasonic noise is noise with frequencies above the audible range, typically greater than 20,000 Hz. Action (1983) reviewed the topic and the various standards that do exist for ultrasonic exposure. The criteria are similar, typically limiting exposures to 110 dB for frequencies at and above 20,000 Hz. This translates to a one-third octave-band criterion of 75 dB at 20,000 Hz and 110 dB for bands at and above 25,000 Hz.

THE ANNOYANCE OF NOISE

No one need tell us that noise can be annoying; we have all had the experience at some time. Annoyance is not the same as loudness. Loud noises are usually more annoying than soft noises, but there are exceptions. Consider, for example, the slow, rhythmic drip of a water faucet versus the roar of the ocean surf—which would annoy you more? In general, more people are annoyed by aircraft and automobile noises than by loud neighbors, loud radios, or children (Kryter, 1985).

Annoyance is measured by having subjects rate noises on a verbal scale, such as noticeable–intrusive–annoying–very annoying–unbearable. There are a host of factors, both acoustic and nonacoustic, that influence the annoying quality of a noise. A list of some of these is presented in Table 18-6.

Measures of Noise Exposure

Considerable work has been done to develop a measure of noise exposure that would represent, in a single number, many of the important acoustic factors and some of the nonacoustic factors influencing the annoyance of noise. Sperry (1978) lists 13 different measures which are used around the world to assess community exposure to noise. Figure 18-6 lists these measures and shows the relationships among them. Most of these measures were developed in the context of community exposure to aircraft noise. From Figure 18-6 we see that all the measures are based on the A-weighted sound level (dB*A*). Equivalent sound level L_{eq} and perceived noise level (PNL), discussed previously, have formed the bases for a variety of other measures. The various measures make corrections for such factors as time of day, season of year, variability in the noise, and number of aircraft flyovers.

We do not take the time to describe each of the measures in Figure 18-6. However, one measure has been related to annoyance and community action and deserves further attention. The *day-night level* (L_{dn}) is used by the Environmental Protection Agency to rate community exposure to noise. The day-night level is the equivalent sound level (L_{eq}) for a 24-h period with a correction of 10 dB added to noise levels occurring in the nighttime (10 p.m. to 7 a.m.). Kryter (1985), however, recommends that 5dB be added to the noise levels occurring between 7 p.m. and 7 a.m. He proposes that the measure be called DENL (day-evening-night level).

It may be encouraging to note that many of the measures shown in Figure 18-6, when they are taken over a 24-h period, correlate almost perfectly with

TABLE 18-6
SOME FACTORS THAT INFLUENCE THE ANNOYANCE QUALITY OF NOISE

Acoustic factors	Nonacoustic factors
Sound level	Past experience with the noise
Frequency	Listener's activity
Duration	Predictability of noise occurrence
Spectral complexity	Necessity of the noise
Fluctuations in sound level	Listener's personality
Fluctuations in frequency	Attitudes toward the source of the noise
Risetime of the noise	Time of year
	Time of day
	Type of locale

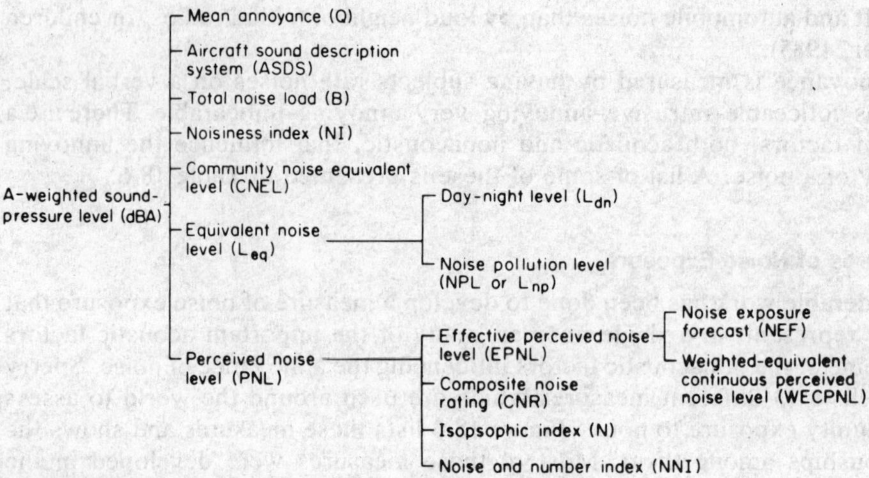

FIGURE 18-6
Various measures of exposure to noise. Many were developed in the context of community exposure to aircraft noise. Measures on the left of the figure are primary measures; those on the right are derived from, or based on, those to the left.

one another. For example, L_{eq}, L_{dn}, and CNEL rarely differ by more than ±1 dB (Fidell, 1979). This should not be too surprising since they are all measuring the same noise environment.

Annoyance and Community Response

From Table 18-6 we see that a host of nonacoustic factors influence the annoyance quality of noise. Most of these are not taken into consideration by the various noise exposure measures. Nevertheless, the Environmental Protection Agency, from a review of British and U.S. community noise surveys, found a linear relationship between L_{dn} and the percentage of people in the survey who were highly annoyed. This relationship is shown in Figure 18-7. Typically, such relationships show correlations on the order of .90.

Different noises having the same L_{dn}, as measured outdoors, however, are not necessarily equally annoying. For example, Kryter (1985) summarized several studies showing that people are more sensitive to aircraft noise than to ground-based vehicular noises, such as automobiles and trucks. Vehicular noises need to be about 10 dB higher than aircraft noises to be equally annoying. Kryter believes that this result is due to the difference in sound level, produced by these two types of noise, at the listeners' ears inside homes. More of the noise from aircraft is transmitted inside a house than is the case for vehicular noises. It is also possible that fears for one's safety are greater with respect to aircraft noise than with vehicular noise, which would add to the annoyance quality of the noise.

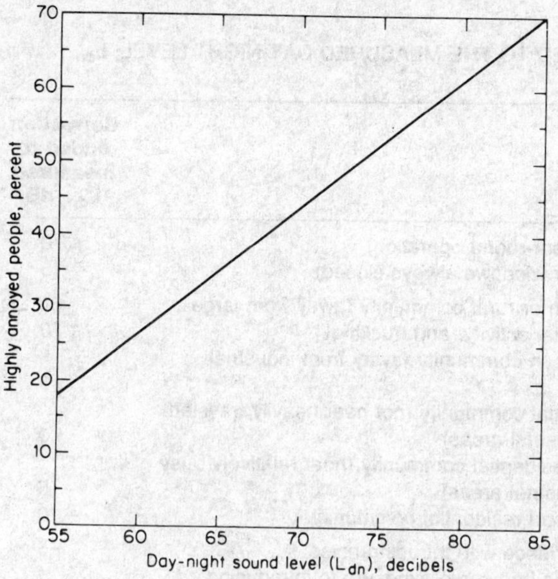

FIGURE 18-7
Relationship between noise exposure and percentage of community highly annoyed. (*Source: Environmental Protection Agency, 1974.*)

To take it a step farther, attempts have been made to relate noise exposure to community reactions such as complaints, threats, and legal action. There exists a giant step between being annoyed and taking action. Fidell (1979), for instance, found that indices of noise exposure were not highly correlated to community responses. A host of sociological factors, political factors, and psychological factors intervene in such a decision. Fidell indicates that the noise exposure itself usually does not account for even half the variance in community reactions.

The Environmental Protection Agency, using L_{dn}, however, has devised a procedure for predicting community reactions to noise. The procedure attempts to take into account some of the sociopolitical factors that influence community reactions. Table 18-7 lists corrections that are to be added to L_{dn} to obtain a *normalized* L_{dn}. The normalized L_{dn} is then used with Table 18-8 to predict community reactions.

Generally, a normalized L_{dn} of 55 dB*A* or lower will not result in complaints. It must be remembered, however, that these predictions are not precise, but rather are only rough indications of the probable community reaction.

HANDLING NOISE PROBLEMS

When a noise problem is suspected or exists, there is no substitute for having good, solid information to bring to bear on the problem and for attacking the problem in a systematic manner. We provide an overview of some noise control techniques, but our discussion cannot substitute for the expertise of a qualified acoustic engineer.

TABLE 18-7
CORRECTION FACTORS TO BE ADDED TO THE MEASURED DAY-NIGHT LEVEL, L_{dn}, TO OBTAIN NORMALIZED L_{dn}

Type of correction	Description	Correction added to measured L_{dn}, dB
Seasonal correction	Summer (or year-round operation)	0
	Winter only (or windows always closed)	−5
Correction for outdoor residual noise level	Quiet suburban or rural community (away from large cities, industrial activity, and trucking)	+10
	Normal suburban community (away from industrial activity)	+5
	Urban residential community (not near heavily traveled roads or industrial areas)	0
	Noisy urban residential community (near relatively busy roads or industrial areas)	−5
	Very noisy urban residential community	−10
Correction for previous exposure and community attitudes	No prior experience with intruding noise	+5
	Community has had some exposure to introducing noise; little effort is being made to control noise. This correction may also be applied to a community which has not been exposed previously to noise, but the people are aware that bona fide efforts are being made to control it.	0
	Community has had considerable exposure to introducing noise, noisemaker's relations with community are good.	−5
	Community is aware that operation causing noise is necessary but will not continue indefinitely. This correction may be applied on a limited basis and under emergency conditions.	−10
Pure tone or impulse	No pure tone or impulsive character	0
	Pure tone or impulsive character present	+5

Source: Environmental Protection Agency, 1974.

TABLE 18-8
EXPECTED COMMUNITY RESPONSE BASED ON NORMALIZED L_{dn}

Community response	Normalized L_{dn}, dBA
No reaction or sporadic complaints	50–60
Widespread complaints	60–70
Severe threats of legal action or strong appeals to local officials	70–75
Vigorous action	75–80

Source: Environmental Protection Agency data as presented by R. Taylor, 1978, Table 8-4.

Defining the Noise Problem

Usually, the first step in defining a noise problem is to measure the overall sound-pressure level (usually in dB*A*) of the situation with a sound-level meter. This gives a gross indication of the potential noise problem. There are a few principles to follow when making sound measurements. First, use the proper equipment and microphone, and be sure the equipment is calibrated and in good working order. Keep away from reflecting surfaces unless you intend to measure their influence. Hold the meter at arms length to avoid sound reflections from your body and the blocking of sound from particular directions. If wind noise is present, shield the meter.

When determining sound emissions from a single source such as a machine, do not take the reading too close to the machine because the readings will vary significantly with small changes in the position of the meter. This will occur within a distance of about twice the greatest dimension of the machine being measured (or the wavelength of the lowest frequency sound being emitted, whichever is greater). This area is called the *near field*. On the other hand, if you measure too far away from the machine, sound reflections from walls and other objects will make it impossible to determine the sound coming directly from the machine. This region is called the *reverberant field*. Between the reverberant field and the near field is the *free field*, where sound-pressure level readings should be made. It is possible that in some situations conditions are so reverberant or the room is so small that no free field exists. In such cases, corrections can be made to account for the effects of the reflected sound.

If a more detailed picture of the noise is desired, an octave-band analysis (or even a one-third octave-band analysis) of the noise can be made. This is done with electronic filters that reject sounds with frequencies outside the selected band. The frequency range from 20 Hz to 20,000 Hz is divided into the following octave bands:

Center frequency (Hz)	Band limits (Hz)
16	14–18
31.5	28–36
63	56–71
125	112–141
250	224–281
500	447–561
1000	895–1120
2000	1790–2240
4000	3550–4470
8000	7100–8950
16000	14100–17900

Notice that the center frequency of an octave band is double that of the prior band's center frequency.

Recent technological advances have made it possible to measure sound power by measuring sound intensity directly, rather than by measuring sound pressure as is done with a sound level meter. *Sound intensity* describes the rate of energy flow through a unit area and is measured in watts per square meter (W/m^2). A discussion of sound intensity is beyond the scope of this book, but suffice to say that the measurement of sound power using sound intensity is independent of the characteristics of the sound field or the presence of other sound sources. This is not the case when sound pressure is measured. Measuring sound intensity is especially valuable for determining the sound power of individual sources (such as one machine) under real-world conditions (such as in a machine shop with other machines). It is also useful for pinpointing the surfaces of a machine most responsible for the noise.

After the noise situation is measured, the next step in defining the noise problem is to determine what noise level would be acceptable, in terms of hearing loss, annoyance, communications, etc. Such limits normally would be adapted from relevant criteria such as discussed above. An example is shown in Figure 18-8. This figure shows the spectrum of the original noise of a foundry cleaning room and a tentative-design baseline that was derived from a set of relevant noise standards. Incidentally, part of the original spectrum was below this level, but the high frequencies were not; the difference represents the amount of reduction that should be achieved (in this case, the shaded area). The third line shows the noise level after abatement.

FIGURE 18-8
Noise spectrum of a foundry cleaning room before abatement, the baseline that represents a desired upper ceiling, and the spectrum after abatement. The shaded area represents the desired reduction. The overall noise level before and after abatement are represented by the (■) symbols at the left of the figure. The abatement consisted primarily in spraying a heavy coat of deadener on the tumbling barrels and surfaces of tote boxes. (*Source: American Foundryman's Society, 1966, p. 52.*)

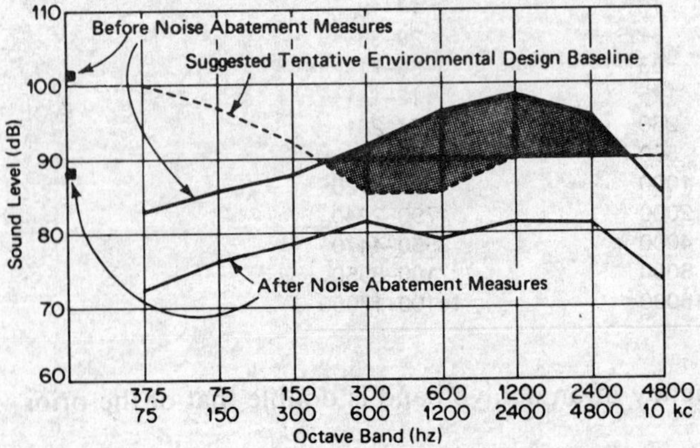

Noise Control

A noise problem can be controlled by attacking the noise at the source, along its path from the source to the receiver, and at the receiver. Often a combination of noise control techniques is required to achieve the desired level of abatement.

Control at the Source Noise is caused by vibration and can be reduced by decreasing either the amount of vibration or the surface area of the vibrating parts. Vibration can be reduced by proper design, maintenance, lubrication, and alignment of equipment. Isolating vibrating parts from other machine parts or structures by use of resilient materials such as rubber or elastomers reduces the number, and hence the surface area, of vibrating sources. This is illustrated in Figure 18-9 by the noise reduction accomplished by isolating the machine from the floor by resilient pads (treatment *a* in the figure). Adding damping materials to machine parts to increase their stiffness or mass can reduce the amplitude of vibrations as well.

Often, a potential noise problem can be averted by selecting quieter equipment initially. It is often more economical to pay extra for quieter equipment than to purchase noisier equipment that will require additional expenditures for noise control. Figure 18-10, for example, shows the noise spectra for two pneumatic screwdrivers and illustrates the considerable noise reduction that can be achieved simply by choosing the proper equipment. Low-frequency noise is less annoying and is tolerated better than high-frequency noise. Therefore, where possible, equipment that generates low-frequency noise should be selected over equipment that generates high-frequency noise. For example, use

FIGURE 18-9
Illustrations of the possible effects of some noise control measures. The lines on the graph show the possible reductions in noise (from the original level) that might be expected by vibration insulation, *a;* an enclosure of acoustic absorbing material, *b;* a rigid, sealed enclosure, *c;* a single combined enclosure plus vibration insulation, *a* + *b* + *c;* and a double combined enclosure plus vibration insulation, *a* + 2*b* + 2*c*. (*Source: Adapted from Peterson and Gross, 1978.*)

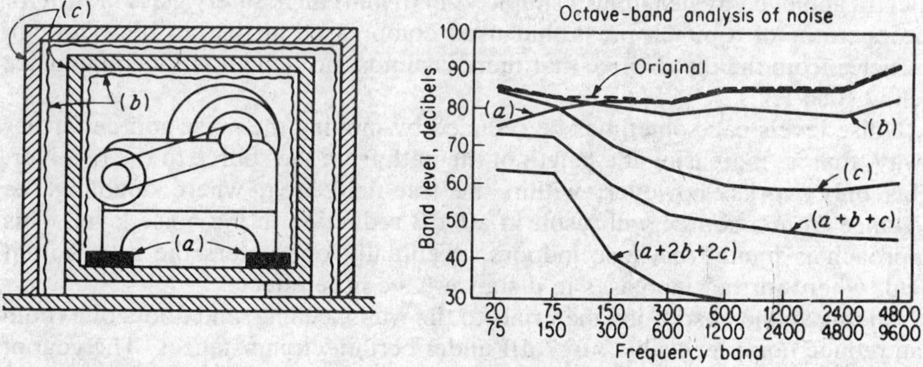

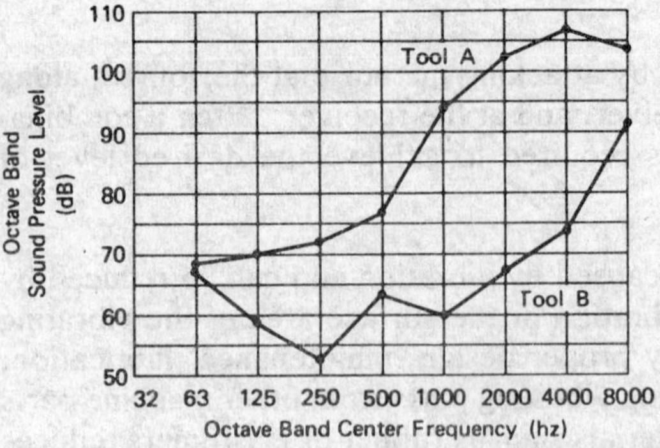

FIGURE 18-10
Octave-band analysis of two pneumatic screwdrivers illustrating the considerable noise reduction that can be achieved by choosing the proper tool. (*Source: American Industrial Hygiene Association, 1975, Fig. 11.3. Reprinted with permission by American Industrial Hygiene Association.*)

of a large, slow-speed blower would be preferred over a smaller, high-speed blower. Mufflers are effective in reducing exhaust noise, and reducing turbulences in pipes will decrease noise in fluid flows.

Control along the Path High-frequency noise is more directional than low-frequency noise and is more easily contained and deflected by barriers. Acoustic linings and materials can reduce noise, as illustrated in Figure 18-9. The noise reduction effect of acoustic material (treatment *b* in Figure 18-9) and rigid sealed enclosures (treatment *c*) is primarily in the high-frequency range. When full enclosures are designed, maintenance requirements of the enclosed equipment must be kept in mind. Although the noise from the machine in Figure 18-9 has been drastically reduced by enclosing it in an acoustically lined, sealed, double vault, pity the maintenance person who has to get inside to change the fan belt!

Full enclosures are not necessary to reduce high-frequency noise. A single wall, shield, or barrier placed between the source and the receiver will deflect much of this noise. Low-frequency noise, however, will not be reduced at all by such barriers, because such noise will easily go over or around the barrier. This is illustrated in Figure 18-11, which shows the before and after noise spectra achieved by installing a simple ¼-in (6 mm) thick safety glass in front of the operator of a punch press that used compressed air jets to blow foreign particles from the die. Notice that there is almost no reduction for frequencies below 1000 Hz.

Noise levels can sometimes be reduced by moving the noise source farther away, that is, increasing the length of the path from the source to the receiver. This only works, however, within the free-field area, where doubling the distance from a source will result in a 6-dB reduction in the noise level. This approach is limited because indoors, eventually one enters the reverberant field, where further increases in distance have little effect.

Adding sound absorption materials to the walls, ceiling, and floors of a room can reduce noise levels by 3 to 7 dB under certain circumstances. The goal of using such materials is to reduce the noise caused by reverberation rather than

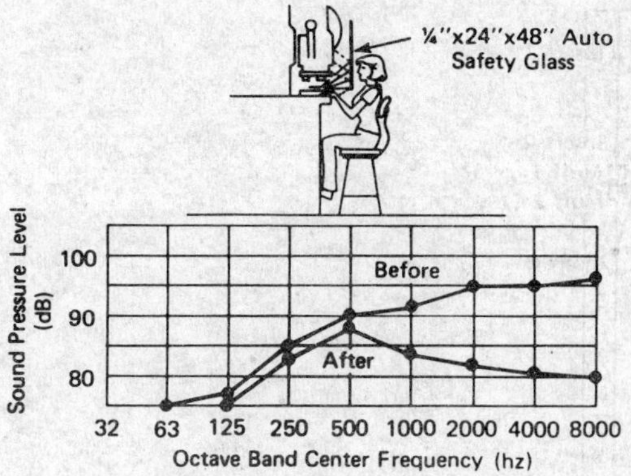

FIGURE 18-11
Use of a 1/4-in (6-mm) thick safey glass barrier to reduce high-frequency noise. The machine is a punch press which uses compressed air to blow foreign particles from the die. (*Source: American Industrial Hygiene Association, 1975, Fig. 11.73. Reprinted with permission by American Industrial Hygiene Association.*)

to reduce the noise of the equipment per se. For this reason the overall reduction is rather limited.

Control at the Receiver Controlling noise at the receiver involves primarily the use of hearing protection but can include, secondarily, audiometric testing of exposed workers with job reassignment or reduced exposure times for those showing signs of hearing loss. OSHA, for example, requires employers to make available hearing protection devices to all employees whose noise dose exceeds 50 percent (TWA = 85 dB*A*), and workers must wear hearing protection if their noise dose is above 100 percent (TWA = 90 dB*A*).

Hearing protection devices come in two general types: insert type and muff type. Insert types can be premolded or custom-molded, can be made of an expandable foam or plastic, or can be a simple fiber plug. Muff types can be liquid-filled or foam-filled and are mounted on a headband or helmet. Gasaway (1988) presented a compendium of noise attenuation data for virtually all hearing protection devices on the market at that time. The effectiveness of hearing protection devices varies widely from one type to another and even between brands within a specific type. Figure 18-12 illustrates the attenuation characteristics of several types of hearing protectors. In general, attenuation is greatest in the higher frequencies where the ear is most sensitive to sound. Figure 18-12 underscores the need to select hearing protection to match the characteristics of the intruding noise.

The Environmental Protection Agency (1979) requires manufacturers to compute and display the *noise reduction rating* (NRR) of their hearing protectors. The NRR is computed from an octave-band analysis of the attenuation effectiveness of the device and is based on the attenuation value 2 standard deviations below the mean at each octave band. The NRR of a hearing protection device can be used to estimate the noise exposure of the wearer. If the sound-pressure level of the noise is measured in decibels on the C-weighted scale, then the noise being experienced by the wearer is simply the sound level in dB*C* minus the NRR. If the sound-pressure level of the noise is measured in

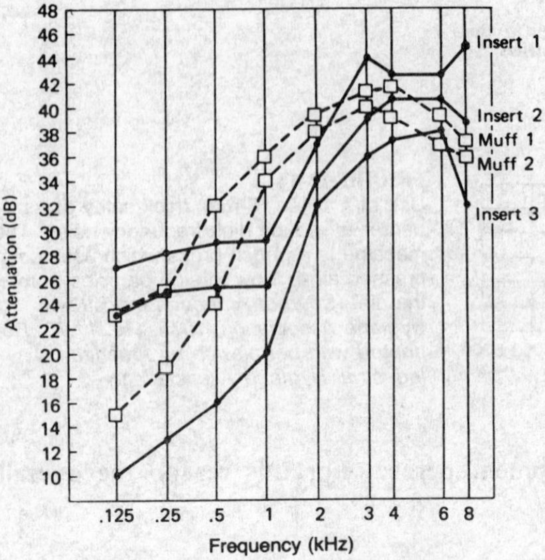

FIGURE 18-12
Noise attenuation of several insert- and muff-type hearing protection devices. (*Source: Bilsom International, Inc., 1984.*)

d*BA*, then the noise exposure is equal to the sound level (in d*BA*) plus 7 minus the NRR. The +7 is included to adjust for "spectral uncertainty" (OSHA, 1983).

Gasaway (1987) plotted the NRR values of 108 insert-type and 293 muff-type protectors, as shown in Figure 18-13. In general, insert-type devices provided better protection than muff-type devices, with 43 percent of the inserts having NRR values above 25 compared to 14 percent of the muff types. The best insert types were the expandable foam inserts. These are small, disposable pieces of foam that are rolled (compressed) between the fingers to form a thin cylinder which is inserted in the ear canal. In a few seconds the foam cylinder begins to expand and fills the ear canal, providing an effective barrier to noise. The four lowest NRR ratings in Figure 18-13 (insert-type) include nonlinear, level-dependent, premolded inserts that reportedly alter the degree of hearing protection as the level of the ambient noise increases beyond some critical point. For these types of devices, the NRR values shown in Figure 18-13 underestimate the actual degree of hearing protection afforded. Additional protection can be achieved in high-noise environments by combining an insert device with a muff. Berger (1984) reported that combining a low-NRR (17) insert device with a moderate-NRR (21) muff device yielded a relatively high level of protection (NRR = 29).

In using NRR data, one must be aware that the indicated attenuation, in the words of Gasaway (1984), may be "dangerously optimistic." NRR values are determined under ideal laboratory and fitting conditions. In real-world industrial settings, however, people often do not wear hearing protection devices properly; hair, beards, and eyeglass frames interfere with good fit, and insets are often not inserted properly. One reason for improper fit is that workers

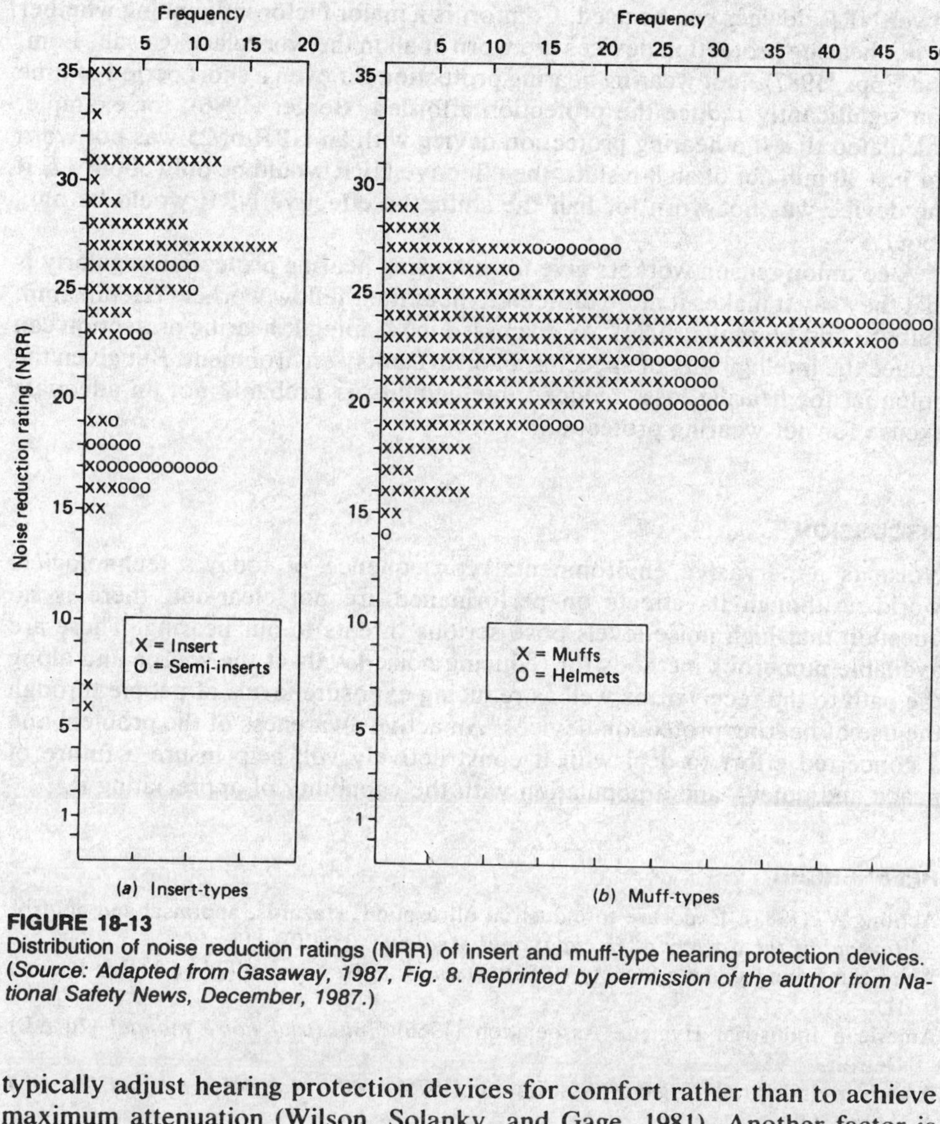

FIGURE 18-13
Distribution of noise reduction ratings (NRR) of insert and muff-type hearing protection devices. (*Source: Adapted from Gasaway, 1987, Fig. 8. Reprinted by permission of the author from National Safety News, December, 1987.*)

typically adjust hearing protection devices for comfort rather than to achieve maximum attenuation (Wilson, Solanky, and Gage, 1981). Another factor is often the lack of instruction given workers concerning how to properly use their hearing protection devices. Berger (1983) suggests subtracting 10 from the reported NRR values to account for these realities. One final word on selecting hearing protection devices based on NRR values: The device with the highest NRR is not always the best choice. NRR is only one factor, although it is an important one, to consider. Other factors that should be considered include comfort, wearability, ease of use, sizing and cleaning requirements, durability, and compatibility with other safety gear and clothing. In many noise environments the highest-NRR devices are not really necessary and more comfortable, lower NRR devices can be used. Comfort is a major factor influencing whether or not hearing protection devices are worn at all in the workplace (Casali, Lam, and Epps, 1987). Not wearing hearing protection for even a short period of time can significantly reduce the protection afforded. Berger (1986), for example, calculated that if a hearing protection device with an NRR of 25 was not worn for just 30 min out of an 8-h shift, the effective NRR would be only about 17. If the device was not worn for half the shift, the effective NRR would be only about 5.

A common reason workers give for not using hearing protection regularly is that they say it makes it more difficult to hear their fellow workers (Helmkamp, Talbott, and Margolis, 1984). As discussed in Chapter 7, hearing protection can reduce the intelligibility of speech spoken in a noisy environment. But given the potential for hearing loss, reduced intelligibility is probably not an adequate excuse for not wearing protection.

DISCUSSION

Noise is a pervasive environmental consequence of today's technological world. Although its effects on performance are not clear-cut, there is no question that high noise levels pose serious threats to our hearing. There are available numerous methods for reducing noise levels at the source and along the path to the receiver, as well as reducing exposure levels of people through the use of hearing protection devices. An active awareness of the problem and a concerted effort to deal with it constructively will help insure a future of peace and quiet—and a population with the capability of appreciating it.

REFERENCES

Action, W. (1983). Exposure to industrial ultrasound: Hazards, appraisal and control. *Journal of the Society of Occupational Medicine*, 33, 107–113.

American Foundryman's Society (1966). *Foundry noise manual* (2d ed.). Des Plaines, IL.

American Industrial Hygiene Association (1966). *Industrial noise manual* (2d ed.). Detroit.

American Industrial Hygiene Association (1975). *Industrial noise manual* (3d ed.). Akron, OH.

Berger, E. (1983). Using the NRR to estimate the real world performance of hearing protectors. *Sound and Vibration*, 18(5), 26–39.

Berger, E. (1984). *E.A.R.LOG 13: Attenuation of earplugs worn in combination with earmuffs*. Indianapolis, IN: E.A.R. Division, Cabot Corporation.

Berger, E. (1986). Hearing protection devices. In E. Berger, W. Ward, J. Morrill, and L. Royster (eds.), *Noise and hearing conservation manual*. Akron, OH: American Industrial Hygiene Association, pp. 319–381.

Bilsom International, Inc. (1984). Attenuation data for Bilsom products. *Bilsom hearing protection catalog*. Reston, VA.

Bonvallet, S. (1952, February). *Noise*. Lectures presented at the inservice training course on the acoustical spectrum. Ann Arbor: University of Michigan Press.

Broadbent, D. (1976). Noise and the details of experiments: a reply to Poulton. *Applied Ergonomics*, 7, 231–235.

Broadbent, D. (1978). The current state of noise research: Reply to Poulton. *Psychological Bulletin*, 85, 1052–1067.

Broadbent, D. (1979). Human performance and noise. In C. Harris (ed.), *Handbook of noise control*. New York: McGraw-Hill.

Burns, W. (1979). Physiological effects of noise. In C. Harris (ed.), *Handbook of noise control*. New York: McGraw-Hill.

Burrows, A. A. (1960). Acoustic noise, an informational definition. *Human Factors*, 2(3), 163–168.

Casali, J., Lam, S., and Epps, B. (1987). Rating and ranking methods for hearing protector wearability. *Sound and Vibration*, 21(12), 10–18.

Davies, D., and Jones, D. (1982). Hearing and noise. In W. Singleton (ed.), *The body at work*. New York: Cambridge University Press.

Environmental Protection Agency (1974). *Information on levels of environmental noise requisite to protect public health and welfare with an adequate margin of safety* (EPA 550/9-74-004). Washington, DC.

Environmental Protection Agency (1979). Noise labeling requirements for hearing protectors. *Federal Register*, 42, 56139–56147.

Eschenbrenner, A. J., Jr. (1971). Effects of intermittent noise on the performance of a complex psychomotor task. *Human Factors*, 13(1), 59–63.

Fidell, S. (1979). Community response to noise. In C. Harris (ed.), *Handbook of noise control*. New York: McGraw-Hill.

Fletcher, H., and Munson, W. A. (1933). Loudness, its definition, measurement, and calculation. *Journal of the Acoustical Society of America*, 5, 82–108.

Foss, J., Ison, J., Torre, J., and Wansack, S. (1989). The acoustic startle response and disruption of aiming: I. Effect of stimulus repetition, intensity, and intensity changes. *Human Factors*, 31(3), 307–318.

Gasaway, D. (1984, November). 1984 NIOSH compendium hearing protector attenuation. *National Safety News*, pp. 26–34.

Gasaway, D. (1987). Noise reduction ratings describe current hearing protection devices. *Occupational Health & Safety*, December, 42–50.

Gasaway, D. (1988). Compendium of data and information on hearing protection devices. *Occupational Health & Safety*, April.

Gawron, V. (1982). Performance effects of noise intensity, psychological set, and task type and complexity. *Human Factors*, 24, 225–243.

Gulian, E. (1974). *Noise as an occupational hazard: Effects on performance level and health. A Survey of the European literature*. Cincinnati, OH: National Institute for Occupational Safety and Health.

Helmkamp, J., Talbott, E., and Margolis, H. (1984). Occupational noise exposure and hearing loss characteristics of blue-collar population. *Journal of Occupational Medicine*, 26(12), 885–891.

Jansen, G. (1969, February). Effects of noise on physiological state. In W. D. Ward and J. E. Frick (eds.), *Noise as a public health hazard* (ASHA Rept. 4). Washington, DC: American Speech and Hearing Association.

Jensen, P., Jokel, C., and Miller, L. (1978). *Industrial noise control manual* (ref. ed.). Cincinnati, OH: National Institute for Occupational Safety and Health.

Kryter, K. (1970). *The effects of noise on man*. New York: Academic.

Kryter, K. (1983*a*). Presbycusis, sociocusis and nosocusis. *Journal of the Acoustical Society of America*, 73(6), 1897–1917.

Kryter, K. (1983*b*). Addendum and erratum: "Presbycusis, sociocusis, and nosocusis" [J. Acoust. Soc. Am. 73, 1897–1919 (1983)]. *Journal of the Acoustical Society of America*, 74(6), 1907–1909.

Kryter, K. (1985). *The effects of noise on man* (2d ed.). Orlando, FL: Academic.

Kryter, K., Ward, W., Miller, J., and Eldredge, D. (1966). Hazardous exposure to intermittent and steady-state noise. *Journal of the Acoustical Society of America*, 39, 451–463.

Leavitt, R., Thompson, R., and Hodgson, W. (1982). Computation of permissible noise exposure. *American Industrial Hygiene Association Journal*, 43, 371–373.

Machle, W. (1945). The effect of gun blast on hearing. *Archives of Otolaryngology*, 42, 164–168.

Meecham, W. (1983, May 10). Paper delivered at Acoustical Society of America Meeting, Cincinnati, Ohio, as reported in *Science News*, 123, p. 294.

Melnick, W. (1979). Hearing loss from noise exposure. In C. Harris (ed.), *Handbook of noise control*. New York: McGraw-Hill.

Occupational Safety and Health Administration (1974, Oct. 24). Occupational noise exposure: Proposed requirements and procedures. *Federal Register*, 39, 37773–37777.

Occupational Safety and Health Administration (1981, Jan. 16). Occupational noise exposure: Hearing conservation amendment. *Federal Register*, 46, 4078–4179.

Occupational Safety and Health Administration (1983). Occupational noise exposure: Hearing conservation amendment. *Federal Register*, 48, 9738–9783.

Peterson, A., and Gross, E., Jr. (1978). *Handbook of noise measurement* (8th ed.). New Concord, MA: General Radio Co.

Poulton, E. (1973). Unwanted range effects from using within-subject experimental designs. *Psychological Bulletin*, 80, 113–121.

Poulton, E. (1976). Continuous noise interferes with work by masking auditory feedback and inner speech. *Applied Ergonomics*, 7, 79–84.

Poulton, E., (1977). Continuous intense noise masks auditory feedback and inner speech. *Psychological Bulletin*, 84, 977–1001.

Poulton, E. (1978). A new look at the effects of noise: A rejoinder. *Psychological Bulletin*, 85, 1068–1079.

Poulton, E., and Freeman, P. (1966). Unwanted asymmetrical transfer effects with balanced experimental designs. *Psychological Bulletin*, 66, 1–8.

Robinson, D., and Dadson, R. (1957). Threshold of hearing and equal-loudness relations for pure tones, and the loudness function. *Journal of the Acoustical Society of America*, 29(12), 1284–1288.

Smith, A. (1989). A review of the effects of noise on human performance. *Scandinavian Journal of Psychology*, 30, 185–206.

Smith, A., and Stansfeld, S. (1986). Aircraft noise exposure, noise sensitivity, and everyday errors. *Environment and Behavior*, 18, 214–216.

Sperry, W. (1978). Aircraft and airport noise. In D. Lipscomb and A. Taylor (eds.), *Noise control: Handbook of principles and practices*. New York: Van Nostrand Reinhold.

Stevens, S. S. (1936). A scale for the measurement of a psychological magnitude: Loudness. *Psychological Review*, 43, 405–416.

Stevens, S. S. (1972). Perceived level of noise by Mark VII and decibels (E). *Journal of the Acoustical Society of America*, 51(2, pt. 2), 575–601.

Taylor, R. (1978). Exterior industrial and commercial noise. In D. May (ed.), *Handbook of noise assessment*. New York: Van Nostrand Reinhold.

Taylor, W., Pearson, J., Mair, A., and Burns, W. (1965). Study of noise and hearing in jute weavers. *Journal of the Acoustical Society of America*, 38, 113–120.

von Gierke, H., and Nixon, C. (1976). Effects of intense infrasound on man. In W. Tempest (ed.), *Infrasound and low frequency vibration*. New York: Academic, pp. 115–150.

Weinstein, N. (1974). Effects of noise on intellectual performance. *Journal of Applied Psychology*, 59(5), 548–554.

Weinstein, N. (1977). Noise and intellectual performance: A confirmation and extension. *Journal of Applied Psychology*, 62, 104–107.

Williams, K. (1978). An introduction to the assessment and measurement of sound. In D. Lipscomb and A. Taylor (eds.), *Noise control: Handbook of principles and practices*. New York: Van Nostrand Reinhold.

Wilson, C., Solanky, H., and Gage, G. (1981, October). Hearing protector effectiveness. *Professional Safety*, 19–24.

CHAPTER **19**

MOTION

Technological ingenuity in recent times has resulted in the creation of methods of travel that our ancestors probably never dreamed about. These include space shuttles, aircraft, zero-ground-pressure vehicles, rockets strapped to one's back, and, of course, various earthbound vehicles such as automobiles, buses, and trucks. Many of these make it possible for people to move at speeds and in environments never before experienced and to which they are not biologically adapted. The disparity between people's biological and physical nature, on the one hand, and the environmental factors imposed on them by their "exotic" modes of travel, on the other, defines a domain within which human factors can make a contribution.

The variables imposed by our increased mobility include vibration, acceleration and deceleration, weightlessness, and an assortment of more strictly psychological phenomena associated with these factors, such as disorientation and other illusions. All these involve, to one degree or another, the sense of motion and orientation. Therefore, before discussing these topics, we review briefly the sensory receptors associated with motion and body orientation.

MOTION AND ORIENTATION SENSES

The *five senses,* in the Aristotelian tradition (vision, audition, smell, taste, and touch), basically deal with stimuli external to the body. The sensory receptors involved (the eyes, ears, etc.) are referred to as *exteroceptors*. There are, however, a number of other sensory receptors related to motion and body orientation.

Proprioceptors

The *proprioceptors* are sensory receptors of various kinds that are embedded within the subcutaneous tissues, such as in the muscles and tendons, in the coverings of the bones, and in the musculature surrounding certain of the internal organs. These receptors are stimulated primarily by the actions of the body itself. A special class of these are the *kinesthetic receptors,* which are concentrated around the joints and which are primarily used to tell us where our limbs are at any given movement, to coordinate our movements, and to sense forces developed by our muscles. We discussed kinesthesis in Chapter 9.

Semicircular Canals

The three semicircular canals in each ear are interconnected, doughnut-shaped tubes that form, roughly, a three-coordinate system, as shown in Figure 19-1 (along with the vestibular sacs). With changes in acceleration or deceleration, the fluid shifts position in these tubes; this stimulates nerve endings that then transmit nerve impulses to the brain. Note that movement of the body at a constant rate does not cause any stimulation of these canals. Rather, they are sensitive only to a *change* in rate (acceleration or deceleration).

Vestibular Sacs: Utricle and Saccule

The *vestibular sacs* (also called the *otolith organs*) are two organs with interior hair cells that contain a gelatinous substance. The vestibular sacs together with the semicircular canals are only about the size of a pea. The *utricle* is generally positioned in a horizontal plane, and the *saccule* is more in a vertical plane. As the body changes position, the gelatinous substance is affected by gravity, triggering nerve impulses via the hair cells. The utricle apparently is the more important of the two organs. The primary function of these organs is the sensing of body posture in relation to the vertical, so they serve as something of

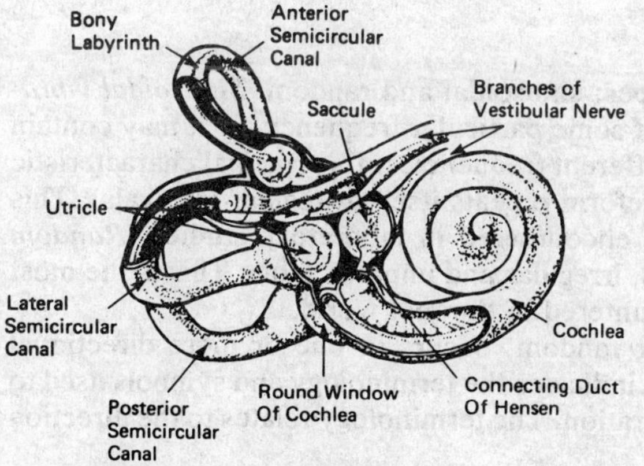

FIGURE 19-1
The body's orientation organs. The semicircular canals form roughly a three-coordinate system that provides information about body movement. The vestibular sacs (the utricle and saccule) respond to the forces of gravity and provide information about body position in relation to the vertical. (*Source: Geldard, 1972.*)

a gyroscope that helps to keep us on an even keel. Although their dominant role is that of aiding in sensing postural conditions of the body, they are also somewhat sensitive to acceleration and deceleration and presumably supplement the semicircular canals in sensing such changes.

Interdependence of Motion and Orientation Senses

In the maintenance of equilibrium and body orientation, or in reliably sensing motion and posture, all these senses play a role, along with the skin senses, vision, and sometimes audition. The importance of vision in orientation was illustrated in experiments with people in a "tilting room," in some cases with a chair that also could be tilted within the room (Witkin, 1959). Subjects seated in the room were tilted to various angles and asked to indicate what direction they considered vertical. When blindfolded, they were able to indicate the vertical more accurately than when they could see the inside of the room; when the subjects could see, they tended to indicate that the ceiling of the room was upright even if the room itself was tilted and the chair was actually vertical. The implication of such investigations is that misperceptions of the *true* upright direction may occur when there is a *conflict* between the sensations of gravity and visual perceptions; in such a case one's visual perceptions usually dominate, even when they are erroneous.

WHOLE-BODY VIBRATION

"If it moves, it probably vibrates." This saying is especially true in the field of transportation; be it car, truck, train, airplane, or boat, all subject occupants to vibration to some degree. Vibration is not a new topic for us. In Chapter 18 we discussed noise, which is a form of vibration that is audible, and in Chapter 12 we discussed vibration induced from hand tools. Our concern in this chapter, however, is low-frequency (generally less than 100 Hz), whole-body vibration typically encountered in trucks, tractors, airplanes, etc.

Vibration Terminology

Vibration is primarily of two types: sinusoidal and random. *Sinusoidal vibration* may be a single sine wave of some particular frequency, or it may contain combinations of sine waves of different frequencies. Its principal characteristic is its regularity (that is, the waveform repeats itself at regular intervals). This type of vibration is most often encountered in laboratory studies. *Random vibration* is, as the name implies, irregular and unpredictable. This is the most common type of vibration encountered in the real world.

Vibration, be it sinusoidal or random, occurs in one or more directional planes. Table 19-1 lists these and indicates the terminology and symbols used to designate the direction of the vibration. The terminology relates to the direction

TABLE 19-1
VIBRATION TERMINOLOGY WITH RESPECT TO DIRECTION OF VIBRATION

Direction of motion	Heart motion	Other description	Symbol
Forward-backward	Spine-sternum-spine	Fore-aft	$\pm g_x$
Left-right	Left-right-left	Side-to-side	$\pm g_y$
Headward-footward	Head-feet-head	Head-tail	$\pm g_z$

Source: Adapted from Hornick, 1973, Table 7-1, p. 229.

of the vibration relative to the human body. Thus, a person standing on a platform which is vibrating up and down experiences head-tail vibration ($\pm g_z$). If, however, the person lies down on his or her back, the vibration is then considered fore-aft ($\pm g_x$).

In addition to type and direction, vibration is described in terms of *frequency* and *intensity*. Frequency is measured in hertz (i.e., cycles per second), as described in Chapter 18. Intensity is measured in a variety of ways such as peak or maximum: (1) *amplitude* (in or m); (2) *displacement* (in or m); (3) *velocity*, the first derivative of displacement (in/s or m/s); (4) *acceleration*, the second derivative of displacement (in/s^2 or m/s^2) (sometimes acceleration is expressed in terms of numbers of gravities* and is labeled in terms of lowercase g units as $\pm g_x$, $\pm g_y$, or $\pm g_z$, depending on the direction of the oscillation as given in Table 19-1); or (5) *rate of change of acceleration*, also called *jerk*, the third derivative of displacement (in/s^3 or m/s^3).

In the case of random vibration, maximum amplitude or acceleration is somewhat inappropriate because the oscillations vary randomly in terms of frequency and intensity. To handle this variability, the frequency spectrum is usually indicated by *mean-square spectral density* and expressed as *power spectral density* (PSD) in g^2 per hertz. The PSD defines the power at discrete frequencies within the selected bandwidth. A plot of PSD (g^2/Hz) versus frequency illustrates the power distribution of the vibration environment. Figure 19-2, for example, shows PSD plots for an automobile and an aircraft.

Intensity is usually expressed as a root-mean-square (rms) value of acceleration (in/s^2, m/s^2, or g).† *Root-mean-square acceleration* defines the total energy across the entire range. Figure 19-3 shows rms acceleration values (actually ranges of rms g) for various air and surface vehicles during cruise. Measures of rms do not give a good estimate of subjective responses to vibration when the vibration contains occasional high-peak values or gross fluctuations in magnitude. One measure of fluctuation is the *crest factor*, which is the ratio of the peak amplitude to the rms amplitude.

* Note that $1g$ = 386 in/s^2 = 9.8 m/s^2.

† *Root mean square* (rms) is the square root of the arithmetic mean of instantaneous values (amplitude or acceleration) squared. In the case of a simple sine wave, rms g = 0.707 × peak g.

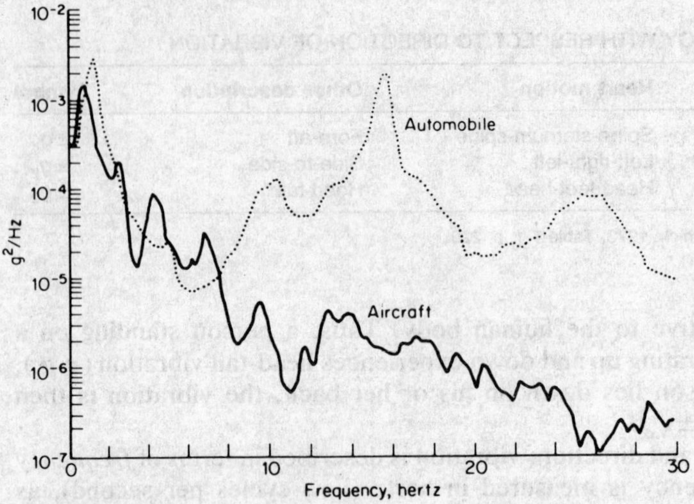

FIGURE 19-2
Power spectral density plots for an automobile and an aircraft in cruise. (*Source: Stephens, 1979, Fig. 2.*)

One last point should be made about vibration measurement: Acceleration, displacement, and frequency are all related. If, for example, acceleration is held constant and frequency is varied, displacement must also vary. It is impossible to hold two of the quantities constant and vary only the third. This makes it difficult to isolate the specific parameter that is affecting performance, subjective reactions, or physiological reactions.

FIGURE 19-3
Root-mean-square acceleration ranges of various air and surface vehicles during cruise. The variability within a vehicle type is due to such factors as road quality, air turbulence, and age and general condition of the test vehicles. (*Source: Stephens, 1979, Fig. 4b.*)

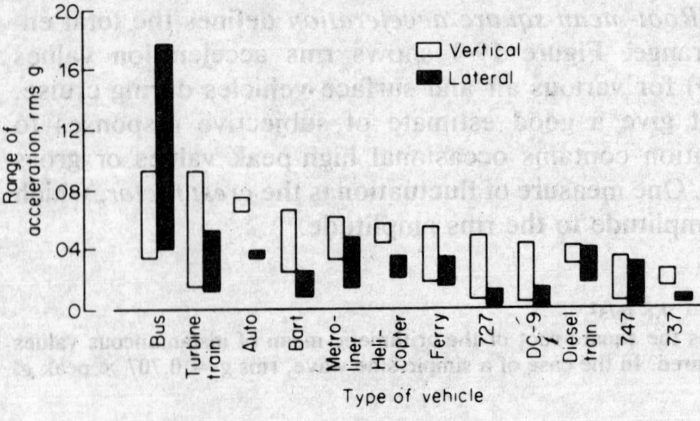

Attenuation, Amplification, and Resonance

As vibration is transmitted to the body, it can be amplified or attenuated as a consequence of body posture (e.g., standing or sitting), the type of seating, and the frequency of the vibration. Every object (or *mass*) has a resonant frequency. When an object is vibrated at its resonant frequency, the object will vibrate at maximum amplitude, which is larger than the amplitude of the original vibration. This phenomenon is called *resonance*. For example, soldiers break cadence when crossing a bridge because the low-frequency vibrations produced by marching in unison could set the bridge into resonance and it could collapse.

The human body and the individual body members and organs have their own resonant frequencies. Since the body members and organs have different resonant frequencies, and since they are not attached rigidly to the body structure, they tend to vibrate at different frequencies rather than in unison. In general, the larger the mass of the structure, the lower its resonant frequency. For a sitting person, the following are the resonant frequencies of various body structures (Grandjean, 1988):

3–4 Hz	Resonance in cervical (neck) vertebrae
4 Hz	Peak resonance in lumbar (upper torso) vertebrae
5 Hz	Resonance in shoulder girdle
20–30 Hz	Resonance between head and shoulders
60–90 Hz	Resonance in eyeballs

In general, for a seated person, vibrations in the range of 4 to 8 Hz cause the entire upper torso to resonate and, hence, should be avoided or reduced in intensity as much as possible.

The amplification or attenuation that would result from a particular vibratory frequency is also influenced by *damping*, either by intent or otherwise. For example, there is a general attenuation effect when a person is standing on a vibrating platform. The legs absorb (dampen) the vibration by bending and straightening in response to the movement within a range of 1 to 6 Hz. Another example of damping is the effect of muscle tension, as shown in Figure 19-4. Tensing the trunk muscles significantly reduces the degree of amplification and increases the resonance frequency by a few Hz.

Needless to say, the type and design of a seat (spring design, cushioning, etc.) can have an impact on the transmission of vibration to the person sitting in it. Figure 19-5 illustrates this effect for various seats. Notice the substantial attenuation of vibration afforded by the suspension seat, especially in the critical frequency range of 4 to 6 Hz.

Physiological Effects of Vibration

The evidence suggests that *short-term exposure* to vibration causes only small physiological effects which are of little practical significance (Hornick, 1973). A slight degree of hyperventilation has been reported and increases in heart rate

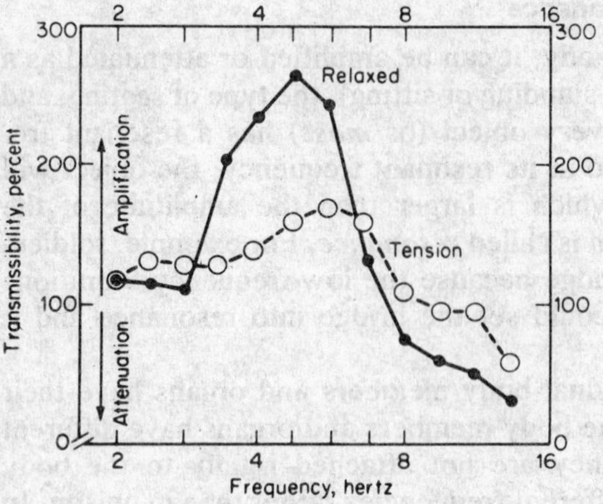

FIGURE 19-4
The effect of tensed and relaxed trunk muscles on the amplitude of vertical vibration of the shoulders of a seated man. (*Source: Adapted from Guignard, 1965, Fig. 380.*)

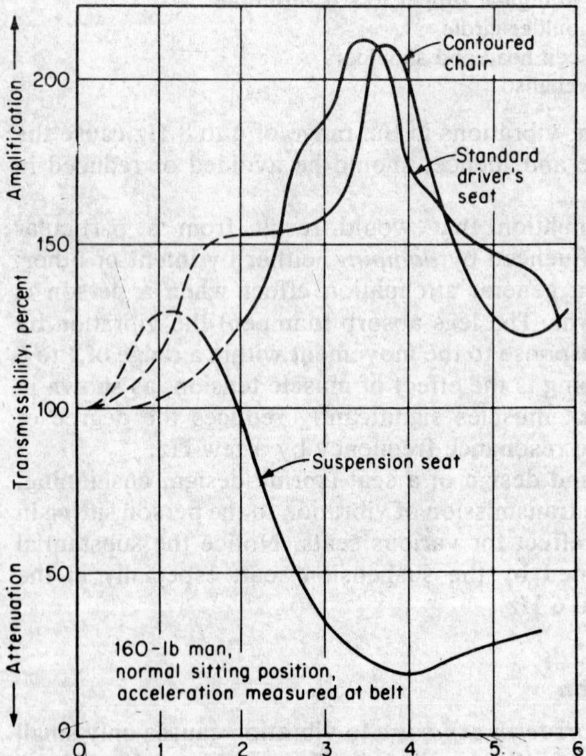

FIGURE 19-5
Mechanical response of a person's body to vibrations, when seated in different seats. (*Source: Simons, Radke, and Oswald, 1956.*)

are found during the early periods of exposure. The elevated heart rate appears to be an anticipatory general stress response. It is also well established that vibration increases muscle tension. This occurs because exposed persons tend to tense their muscles to dampen the vibration. There appears, however, to be no effect of vibration on blood chemistry or endocrine chemical composition (Gell and Moeller, 1972).

The causal link between *long-term exposure* to whole-body vibration and specific physiological damage is not clear (Carlsoo, 1982). Most of the evidence comes from epidemiological investigations of truck drivers and heavy-equipment operators (Seidel and Heide, 1986). Workers exposed to intense, long-term, whole-body vibration show an increased risk to the spine and the peripheral nervous system. The digestive system, peripheral veins, the female reproductive organs, and the vestibular system may also be affected. On average, the health risk seems to increase with higher intensity or duration of exposure (Seidel and Heide, 1986). The problem with these sorts of studies is that it is difficult to separate the effects due to vibration from the effects due to sitting all day or manually loading and unloading a vehicle, all of which are part of operating trucks and heavy equipment.

Burton and Sandover (1987) investigated the effect of a 1982 rule change that reduced the stiffness of Formula 1 *Grand Prix* car suspensions without changing the nature of the driving task. They found that, although Formula 1 race drivers were more likely to have lower back pain than the general driving public, both the incidence and severity of their back pain decreased significantly with the introduction of the softer riding suspensions.

Grandjean (1988) lists common physiological complaints associated with various frequencies of whole-body vibration. The frequencies generally correspond to the resonant frequencies of the various body structures being complained about. For example, people complain of pains in the chest and abdomen chiefly at 4 to 10 Hz. Backaches seem to occur particularly at 8 to 12 Hz. Headaches, eyestrain, and irritations in the intestines and bladder are usually associated with frequencies between 10 and 20 Hz.

Performance Effects of Vibration

Vibration appears to principally affect visual and motor performance. Random vibration is often found to produce less severe degradations in both types of performance than does sinusoidal vibration of equal intensity (Moseley, Lewis, and Griffin, 1982; Hornick, 1973).

Visual Performance Visual performance is generally impaired most by vibration frequencies in the range of 10 to 25 Hz. The amplitude of the vibration seems to be a key factor and the degradation in performance is probably due to the movement of the image on the retina, which causes the image to appear blurred. Moseley and Griffin (1986) found that, for frequencies below 3 Hz, reading time and errors are greater when only the display is vibrated than when only the reader is vibrated. Apparently, at these frequencies the person can engage in compensatory head and eye movements that stabilize the image on the retina. Interestingly, vibrating both the display and the reader at less than 3 Hz results in the best performance. Unfortunately, at higher frequencies it makes little difference who or what is vibrated; performance is equally poor under all conditions.

The characteristics of the material being viewed have significant effects on performance in a vibrating situation. Characters less than 10 min, and particularly those less than 6 min of visual angle are very difficult to read when the observer is being vibrated (Lewis and Griffin, 1979).

Motor Performance Considerable research has demonstrated the effects of vibration on the tracking performance of seated subjects. Hornick (1973) has reviewed much of this body of literature and has gleaned a few generalizations from it. The effects of vibration are somewhat dependent on the difficulty of the tracking task, type of display, and type of controller used. For example, the use of side-stick and arm support can reduce vibration-induced error by as much as 50 percent compared with conventional center-mounted joysticks. Detrimental effects of vertical sinusoidal vibration generally occur in the 4- to 20-Hz range with accelerations exceeding 0.20 *g*. These conditions can produce tracking errors up to 40 percent greater than under nonvibratory control conditions. Vertical vibration is generally more debilitating then lateral or fore-aft vibration. Somewhat unexpectedly, the duration of exposure to vibration does not seem to affect performance on manual control tasks. The effects of vibration on tracking performance, however, do not end with the cessation of the vibration; residual effects may last up to 30 min after exposure.

Neural Processes Tasks that involve primarily central-neural processes, such as reaction time, monitoring, and pattern recognition, appear to be highly resistant to degradation during vibration. In fact, Poulton (1978) points out that vibration between 3.5 and 6 Hz can have an alerting effect on subjects engaged in boring vigilance tasks. Figure 19-4 indicated that, within this frequency range, tensing the trunk muscles attenuates the amplitude of shoulder vibration. Tensing the muscles is a good method for maintaining alertness. Outside of the 3.5- to 6-Hz range the subject can attenuate shoulder vibration more by relaxing the trunk muscles—a good way to fall asleep.

Subjective Responses to Whole-Body Vibration

The subjective response most often assessed in vibration studies is comfort. *Comfort*, of course, is a state of feeling and so depends in part on the person experiencing the situation. We cannot know directly or by observation the level of comfort being experienced by another person; we must ask people to report to us how comfortable they are. This is usually done by using adjective phrases such as *mildly uncomfortable, annoying, very uncomfortable,* or *alarming.* Investigators have tried to link the physical characteristics of vibration, most notably frequency and acceleration, to subjective evaluations of comfort. Their studies usually result in equal-comfort contours for combinations of frequency and acceleration. Unfortunately, the resulting contours differ widely from study to study (e.g., see Oborne, 1976, for a review). This discrepancy is due to different methodologies, subject populations, vibration environments, and semantic comfort descriptors. It appears that individuals react in markedly different ways to whole-body vibration, and it is difficult to determine the causes of these differences (Oborne, Heath, and Boarer, 1981). In addition, an individual's response to vibration is quite stable over time (Oborne, 1978). The issue of individual differences in response to vibration provides a challenge for researchers and designers alike. Oborne, Heath, and Boarer (1981), for example, determined equal-sensation contours for 100 subjects. They found that 70 percent of the subjects did not respond in the same way as depicted by the average contour. Designing for the average would, in this case, "inconvenience" more than half the population!

Griffin and his colleagues at the University of Southampton's Institute of Sound and Vibration in England have carried out an extensive and continuing program of research on human response to whole-body vibration. [See Griffin, Parsons, and Whitham (1982) for a summary of many of the studies conducted.]

In one recent investigation, Corbridge and Griffin (1986) presented various magnitudes and frequencies of random and sinusoidal vibration to people seated on a hard wooden seat. From the data collected, Corbridge and Griffin constructed *equivalent-comfort contours* that show combinations of intensity (magnitude) and frequency judged by the study subjects to be of equal comfort to a reference vibration. They found that the shape of the contours obtained using sinusoidal and random vibration were similar. Carrying the comparison a bit further, Griffin (1976) reported that people are more sensitive to one-third octave-band random vibration than to sinusoidal vibration. Thus, random vibration may be less disruptive to performance (as we stated previously), but more uncomfortable than sinusoidal vibration.

Corbridge and Griffin (1986) combined their data with those from other studies and developed the simplified equivalent-comfort contours shown in Figure 19-6. Two vertical vibration contours are shown, one equivalent in comfort to a 2 Hz 0.25 m/s^2 rms vibration and the other equivalent in comfort to a 2 Hz 0.75 m/s^2 rms vibration. The contour for lateral vibration is equivalent in comfort to a 2 Hz 0.75 m/s^2 rms vibration. Within a contour, the lower the magnitude of the vibration, the greater is the sensitivity to the vibration at that particular frequency.

People are most sensitive to vertical vibration between 5 and 16 Hz, and to lateral vibration between about 1 and 2 Hz. Although not shown in Figure 19-6, Corbridge and Griffin found that women are more sensitive than men to vertical vibration above about 3 Hz.

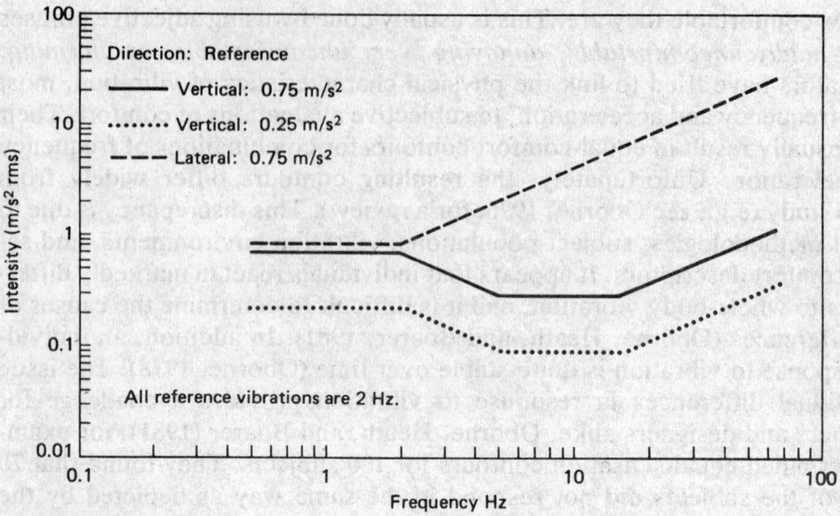

FIGURE 19-6
Simplified equivalent-comfort contours for vertical and lateral vibration. Subjects were seated on a wooden seat. All combinations of intensity and frequency connnected by a line would be judged equal in discomfort. The reference intensity and frequency for each contour is shown. (*Source: Adapted from Corbridge and Griffin, 1986, Figs. 9 and 11.*)

A more elaborate and complete method for estimating discomfort from vibration is presented by Griffin (1986). The method assesses the contribution to discomfort of vibration at the feet, seat, and seat back in each of three linear directions (x, y, z) and of rotational vibration at the seat around three axis (i.e., roll, pitch, and yaw). Most readers will be pleased that the method is beyond the scope of this book and won't be discussed further.

It is not easy to identify the bodily sensations that form the basis for comfort or discomfort judgments. Whitham and Griffin (1978) asked subjects to indicate which specific body locations were uncomfortable at various vibration frequencies (acceleration was held constant at 1.0 m/s^2 rms). In general, most responses of seated subjects implicated the lower abdomen at 2 Hz, moving up the body at 4 and 8 Hz, with most responses implicating the head at 16 Hz. At 32 Hz the responses were divided between the head and lower abdomen, while at 64 Hz they were mostly located near the principal vibration input site, i.e., the seat of the pants.

In actual vehicle environments, subjective evaluations of comfort depend on the expectations, anxiety, past experiences, and other psychological factors of the passengers. In addition to vibration, other physical environmental factors also affect comfort evaluations. Richards and Jacobson (1977), for example, found that among airline passengers the presence of smoke, lighting, and work space were important factors in their comfort ratings.

Limits for Exposure to Whole-Body Vibration

Specifying the limits for exposure to whole-body vibration depends on the criterion adopted. The criterion can be based on comfort, task performance, or physiological response. The equal-comfort contours discussed above are, in essence, limits based on the criterion of comfort.

The International Organization for Standardization (ISO), after wrestling with the problem for 10 years, developed a standard (ISO 2631)* for exposure to whole-body vibration (ISO 1978, 1982*a*, 1982*b*). The standard is most applicable for transportation and industrial type vibration exposures. The standard specifies the limits in terms of acceleration, frequency, and exposure duration. Limits are specified for comfort, task proficiency, and physiological safety. Figure 19-7 illustrates the fatigue-decreased proficiency (FDP) boundaries for vertical vibration. [An addendum to the standard (ISO, 1982*a*) tentatively suggests that the boundaries be extended to 0.63 at the same values as those at 1 Hz; however, there is still considerable controversy surrounding this.] The trough between 4 and 8 Hz corresponds to the resonant frequencies of the body (see Figures 19-4 and 19-6). Each line in Figure 19-7 represents an estimate of the upper limits that people generally can tolerate before the effects of fatigue catch up with them. Amplitudes about 10 dB less than those indicated (i.e., divide the acceleration values by 3.15) tend to characterize the upper boundary of comfort. Safe physiological exposure limits are about 6 dB higher than the values shown (i.e., multiply the acceleration values by 2.0). The standard allows one to raise or lower the FDP and comfort boundaries depending on the nature of the environment and task. The FDP boundaries can be altered by +3 to −12dB depending on the task; the comfort boundaries can be changed by + 3 to −30dB depending on environmental factors. The standard, however, does not specifically state what task or environmental factors are associated with these adjustments.

These standards, although heralded as a major contribution, have nonetheless been criticized (Morrissey and Bittner, 1975; Sandover, 1979; Oborne, 1983). Some of the criticisms are that (1) the comfort and FDP limits for short time exposures, less than 1 h, may be too high; (2) the limits appear to be based on mean results and do not take into account variability in the population (i.e., what percentage of the exposed population will experience fatigue? 50 percent? 100 percent?); (3) the proposals imply that the effects of combinations of single-axis vibration are additive, despite strong evidence to the contrary (e.g., Leatherwood, Dempsey, and Clevenson, 1980); (4) the use of the same shaped contours for the comfort, performance, and tolerance boundaries is probably an oversimplification; (5) comfort contours may underestimate the severity of high frequencies and overestimate the severity of low frequencies (Corbridge

* Also adopted by the American National Standards Institute as ANSI S3.18-1979 (Acoustical Society of America, 1980).

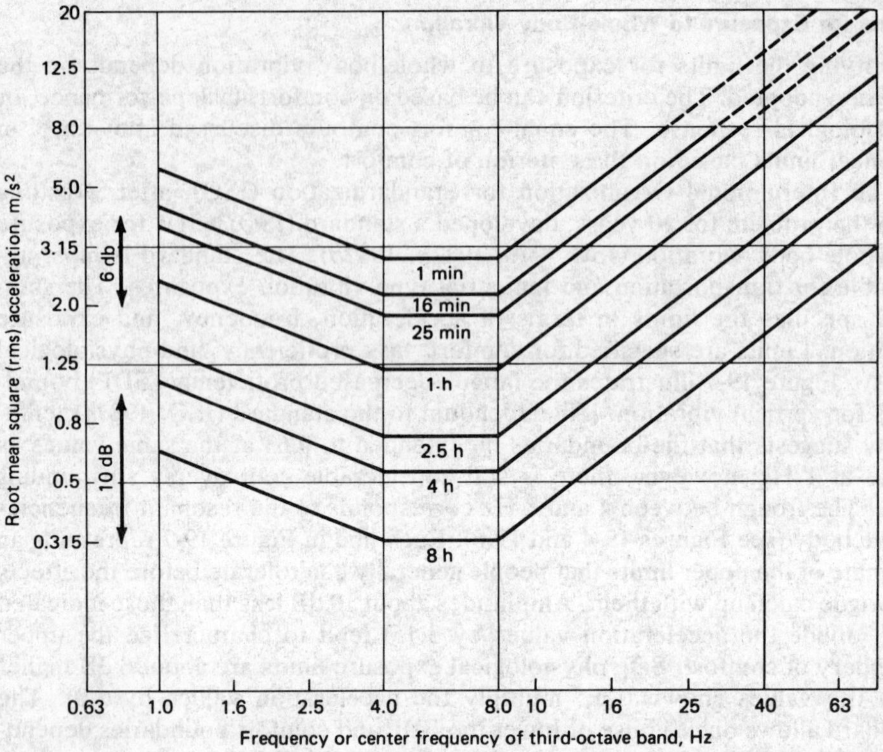

FIGURE 19-7
The fatigue-decreased proficiency boundary for vertical vibration contained in ISO 2631 and ANSI S3.18-1979. To obtain the boundary for reduced comfort, subtract 10 dB (i.e., divide each value by 3.15); to obtain the boundary for safe physiological exposure, add 6 dB (i.e., multiply each value by 2.0). (*Source: Acoustical Society of America, 1980.*)

and Griffin, 1986); (6) little, if any, evidence is available to support the assumption that there exists a time-intensity trade-off for comfort or performance or for the specific function depicted by the standard (Kjellberg and Wikström, 1985); and (6) the entire standard was based on insufficient and inadequate experimental evidence, although it was the best information available at the time.

Given these criticisms, one may wonder what purpose the standard serves. Clearly the standard was a major step in the right direction, and it has stimulated discussion and research in an area that had been poorly investigated prior to its release. The standard represents a good first approximation of human response to whole-body vibration, but it will need additional modifications as new research findings accumulate. In fact, an ISO study group has been formed to recommend what some people think might well be radical changes to the standard.

ACCELERATION

Acceleration of the human body is part and parcel of riding in any moving vehicle. In most vehicles, such as automobiles, trains, buses, or commercial aircraft, the levels of acceleration are moderate, and the effects are nominal. There are, however, vehicles that produce very high levels of acceleration for their occupants, such as high-performance jet aircraft and space rockets, with effects that can be of some consequence. Before talking about these, however, we should discuss certain basic concepts and terminology used to characterize acceleration.

Terminology

Acceleration is a rate of change of motion of an object having some mass. The rate of change of motion is expressed as feet per second per second (ft/s^2) or meters per second per second (m/s^2). The basic unit of acceleration, *G*, is derived from the force of gravity in our earthbound environment. The acceleration of a body in free-fall is 32.2 ft/s^2 (9.81 m/s^2), this being 1 *G*. If a person is accelerated at 2 *G*, the body effectively weighs twice its normal amount. Today's fighter aircraft can generate up to 9 *G* and more before risking structural damage; unfortunately, pilots are not so well constructed. Since 1982, the U.S. Air Force has attributed 14 of its worst-category aircraft accidents to *G-induced loss of consciousness* (*GLOC*). An F-16 or F-18 fighter plane can go from 0 to 9 *G* in less than 3 s. Under such conditions, pilots lapse into unconsciousness before experiencing any physiological symptoms (Kitfield, 1989).

Acceleration forces applied to an object, like a person, can be either linear (as when being propelled forward in a high-speed dragster racing car) or rotational (as when going around a tight turn in a Formula 1 racing car). Although rotational acceleration is important in some circumstances, such as in certain aircraft maneuvers, here we discuss only linear acceleration.

Linear acceleration can occur in any of three directions relative to the body, as shown in Figure 19-8. Rarely, if ever, is someone exposed to a simple unvarying linear acceleration in one direction. Instead, acceleration may vary in magnitude or direction and may be accompanied by complex oscillations and vibration. For purposes of summary, however, it is simpler to consider the effects of each direction of acceleration separately without adding to the complexity.

Effects of Headward ($+G_z$) Acceleration

In headward acceleration, there is an increase in apparent body weight, a tendency for soft tissue to droop, and a tendency for blood to pool in the lower parts of the body. Figure 19-9 summarizes these effects as a function of acceleration magnitude. The predominant effects of headward acceleration are on gross body movement and vision.

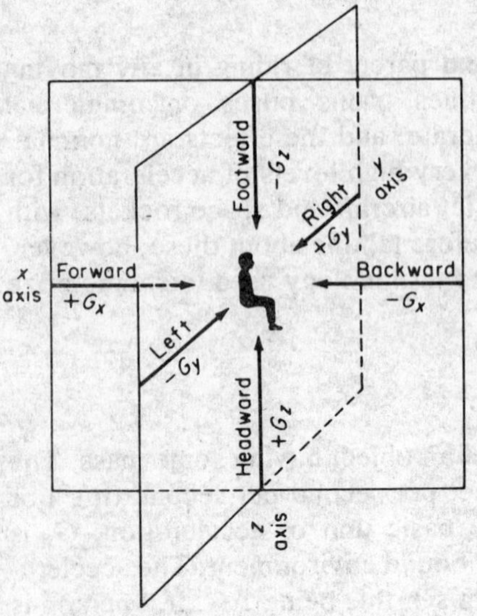

FIGURE 19-8
Illustration of three directions of linear acceleration. The direction of displacement of the heart and other organs is opposite to that of the motion of the body.

Headward acceleration results in decreased ability to detect targets, especially in the peripheral visual field. Complete loss of peripheral vision generally occurs at about $+4.1\ G_z$, and loss of central vision occurs at about $+5.3\ G_z$ (Zarriello, Norsworthy, and Bower, 1958).

These effects of $+G_z$ acceleration have consequences on the performance of tasks which involve movement, vision, or both, including such tasks as visual reaction time, reading, tracking, and certain higher mental processes (Fraser, 1973).

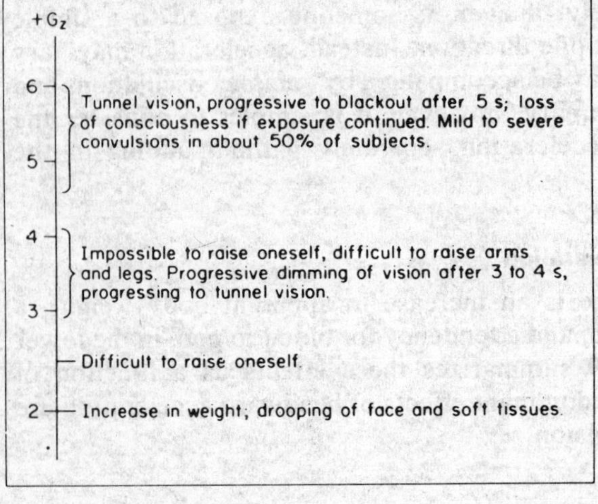

FIGURE 19-9
Summary of effects of headward ($+G_z$) acceleration. (*Source: Fraser, 1973, p. 150.*)

In addition to these effects, various cardiovascular and respiratory effects of headward acceleration have been noted (Fraser, 1973). Cardiac output and stroke volume decrease, while heart rate, aortic pressure, and systemic vascular resistance increase (Wood et al., 1961).

Effects of Footward ($-G_z$) Acceleration

Far less research has been conducted on the effects of footward acceleration, undoubtedly because it is far less common than headward acceleration. In footward acceleration, the force is acting up toward the head.

At $-1\ G_z$ there is unpleasant, but tolerable, facial congestion. At $-2\ G_z$ to $-3\ G_z$, the facial congestion becomes severe. At these levels one experiences throbbing headaches, progressive blurring, graying, or occasionally reddening of vision after about 5 s. The limit of tolerance at $-5\ G_z$ is about 5 s, reached only by the most exceptional subjects (Fraser, 1973).

Effects of Forward ($+G_x$) Acceleration

Seated subjects experiencing forward ($+G_x$) acceleration are pushed back in their chairs. Figure 19-10 summarizes some of the known effects of forward acceleration. A more detailed portrayal of the effects of vehicular (forward) acceleration $+G_x$ on gross body movements is shown in Figure 19-11. This figure shows the typical levels of G that are near the threshold of ability to perform the acts indicated. Some implications for design decisions are fairly obvious; it would seem unwise, for example, to place the ejection control lever of an airplane over the pilot's head.

In addition to the effects on gross body movement and vision, acceleration produces a marked effect on respiration. At approximately $+6\ G_x$, one begins to experience sensory impairment, with mental impairment due to lack of oxygen occurring around $+8\ G_x$.

Effects of Backward ($-G_x$) Acceleration

The effects of backward acceleration are similar to those experienced in forward acceleration. Chest pressure is reversed, making breathing easier. There still exists, however, pain and discomfort from outward pressure against the restraint harness being worn (and one would not want to be in such an environment without some type of upper-body restraint). The effects on vision are similar for both forward and backward acceleration (Fraser, 1973).

Effects of Lateral ($\pm G_y$) Acceleration

Little information is available on the effects of lateral acceleration (Fraser, 1973), but high levels of lateral acceleration are relatively rare in operational environments.

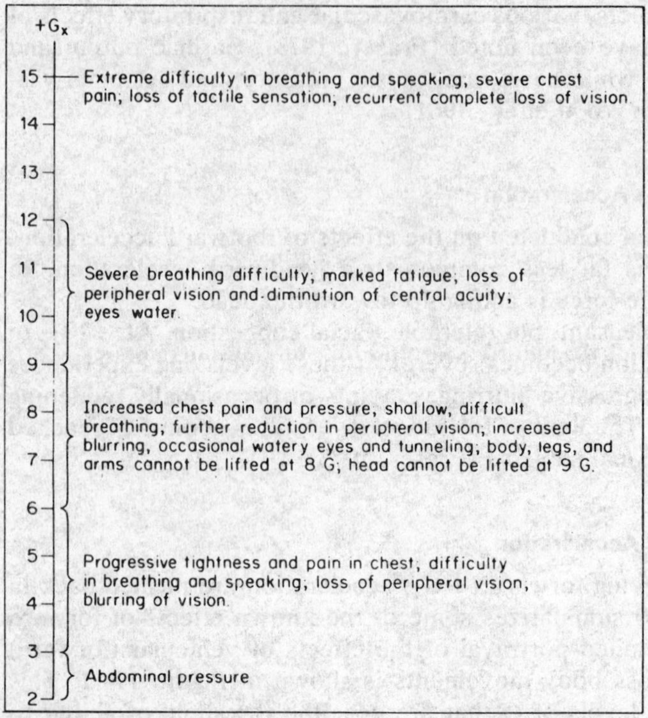

FIGURE 19-10
Summary of the effects of forward ($+G_x$) acceleration. (*Source: Fraser, 1973, p. 151.*)

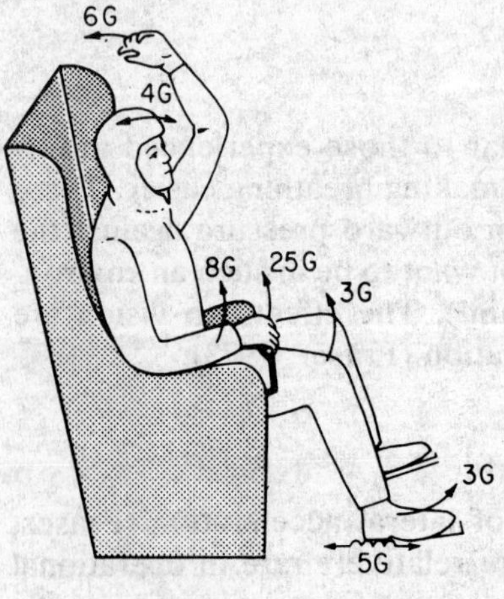

FIGURE 19-11
The forward ($+G_x$) forces that are near the threshold of various body movements. For any given motion indicated, the movement is just possible at the G_x forces indicated; greater G_x forces usually would make it impossible to perform the act. (*Source: Chambers and Brown, 1959.*)

Voluntary Tolerance to Acceleration

The physiological symptoms brought about by acceleration, of course, have a pretty direct relationship to the levels that people are willing to tolerate voluntarily. The physiological effects of linear acceleration in various directions, mentioned above, are paralleled by somewhat corresponding differences in average voluntary tolerance levels, such as shown in Figure 19-12. This figure shows the average levels for the various directions that can be tolerated for specific times. Tolerance to footward acceleration $-G_z$ is least, followed by headward acceleration, $+G_z$, with forward acceleration, $+G_z$ being the most tolerable. Individual differences are of considerable magnitude; trained and highly motivated personnel frequently can endure substantially higher levels than the average person.

Protection from Acceleration Effects

Although minor doses of acceleration (as in most land vehicles and most commercial airplanes) pose no serious problems for safety or performance, the effects of higher levels (especially with long exposures, as in space travel) usually necessitate some protective measures. One way to provide protection

FIGURE 19-12
Average levels of linear acceleration, in different directions, that can be tolerated on a voluntary basis for specified periods. Each curve shows the average G load that can be tolerated for the time indicated. The data points obtained were actually those on the axes; the lines as such are extrapolated from the data points to form the concentric figures. (*Source: Adapted from Chambers, 1963, Fig. 6., pp. 193–320.*)

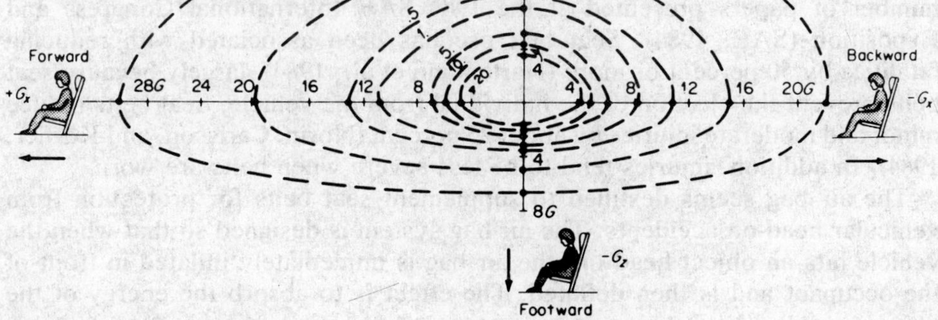

is to assume a posture which increases the tolerance to the direction of acceleration being experienced. For example, sitting and bending forward about 20 to 25° provides greater tolerance to forward acceleration then does sitting upright or even reclining 30° or more (Bondurant et al., 1958). Aside from the use of posture, presently available protection schemes include restraining devices, special contour or net couches, and anti-*G* suits.

Anti-*G* suits generally have their greatest utility in connection with headward ($+G_z$) acceleration, with some designs being more effective than others (Nicholson and Franks, 1966). Anti-*G* suits typically apply pressure to the lower abdominal area and legs, thus reducing the tendency of the blood to drain to those areas during headward ($+G_z$) acceleration. There is no particularly effective scheme for providing protection against footward acceleration ($-G_z$); so if you insist on being shot out of a cannon, have them shoot you out headward, and don't forget your anti-*G* suit!

Deceleration and Impact

The normal deceleration of a vehicle imposes forces upon people that are essentially the same as those of actual acceleration, but in reverse. Acceleration usually is a gradual affair (except in unusual instances, such as with astronauts during blast-off), but deceleration can be extremely abrupt, especially in the case of vehicle accidents. When a vehicle hits a solid object or another vehicle head on, an unrestrained occupant will continue forward at the initial velocity until the body strikes some part of the interior of the vehicle or is thrown out of the vehicle. This is sometimes called the *second collision*. The deceleration of the occupant is a function of the deformation, if any, of the part of the interior which is hit by the occupant's body. To be protected from this second collision, the occupant must be "tied" to the vehicle (as with restraining devices) and decelerate with the vehicle (which may occur through a distance of a couple of feet or more) or must be provided with some energy-absorbing object that results in more gradual deceleration by increasing the travel distance of the occupant during deceleration (air bags, collapsible steering wheels, etc.).

Some indication of the protection provided by seat belts comes from a number of papers presented at the 1984 SAE International Congress and Exposition (SAE, 1984). Seat belt use has been associated with reducing fatalities by 50 percent or more (Hartemann et al., 1984), largely because seat belts prevent the ejection of the individual from the vehicle. Seat belts reduce minor and moderate injuries by about 25 percent (Norin, Carlsson, and Korner, 1984). In addition, injuries tend to be less severe when belts are worn.

The air bag seems destined to supplement seat belts for protection from vehicular head-on accidents. The air bag system is designed so that when the vehicle hits an object head on, the air bag is immediately inflated in front of the occupant and is then deflated. The effect is to absorb the energy of the occupant being catapulted forward on impact. In other words, the deceleration takes place over a longer travel distance than if the occupant were to strike a rigid object such as the windshield or the steering wheel.

WEIGHTLESSNESS

Since human beings have evolved in an earthbound environment, the force of gravity has been our constant companion. Its constant presence has influenced our physiological makeup and is basic to all our activites. Except for the case of a few roller coasters, it is only in space flight and in certain aspects of aircraft flight that the "natural" phenomenon of weightlessness, or reduced gravity, is experienced. For those few who do venture into outer space, two aspects of the weightlessness, or reduced-weight, state are particularly important. The first is the absence of weight itself; the removal from the normal gravitational environment could be expected to have an impact on the human organism, such as on its physiological functioning and on perceptual-motor performance and sensory performance. Second, the weightless condition is accompanied by a tractionless condition when the person is moving and working.

Physiological Effects of Weightlessness

In summarizing the early space flight experience of both the United States and the Soviet Union, Berry (1973) indicates that although some physiological changes have been consistently noted, none have been permanently debilitating. Some of the temporary effects that have been observed include aberrations in cardiac electrical activity, changes in the number of red and white blood cells, loss of muscle tone, and loss of weight.

More recently, space sickness (also called *space adaptation syndrome*) has been reported by roughly half of all space travelers during the first 2 to 4 days in orbit (Oman, 1984; Oman, Lichtenberg, and Money, 1984). Symptoms include sensations of tumbling, overall discomfort, prolonged nausea, and vomiting. Although the symptoms often disappear after a few days in orbit, task performance can be adversely affected during that time.

Thornton (1978) reports various anthropometric changes that occur in a weightless environment. Space travelers grow in height by approximately 3 percent [approximately 2 in (5 cm)]. This has implications for, among other things, pressure-suit design and control stations with critical eye-level requirements. Even more significant is the natural relaxed posture assumed in a weightless environment, as shown in Figure 19-13. Notice the lower line of sight under zero-*g* conditions, the angled foot, and the height of the arms. Seats, workstations, etc., must be designed to accommodate this unique posture. Thornton reports that the body "rebels" with fatigue and discomfort against any attempt to force it into a more normal earthbound posture.

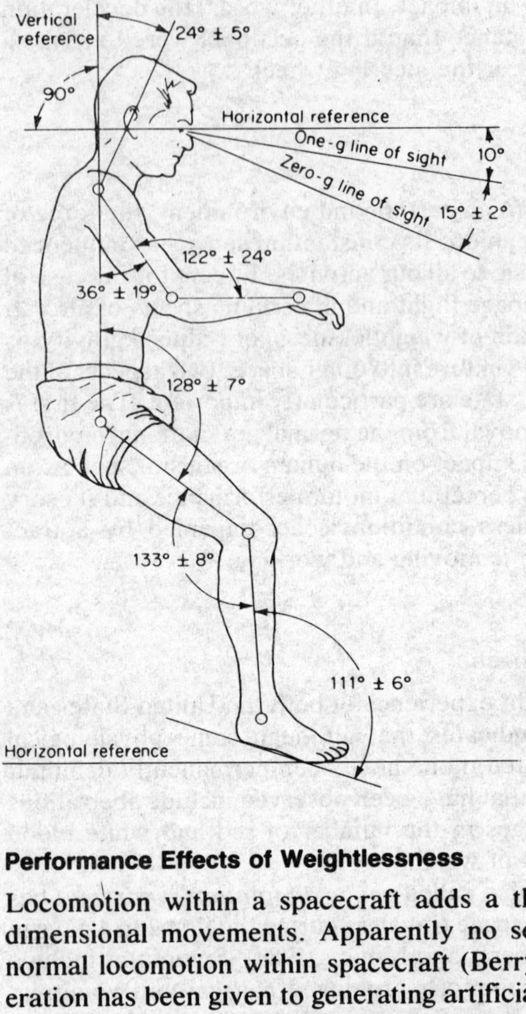

FIGURE 19-13
Typical relaxed posture assumed by people in weightless conditions. (*Source: Thornton, 1978, Fig. 16.*)

Performance Effects of Weightlessness

Locomotion within a spacecraft adds a third dimension to our usual two-dimensional movements. Apparently no serious problems have occurred in normal locomotion within spacecraft (Berry, 1973). Although serious consideration has been given to generating artificial gravity in spacecraft (by rotating the entire space vehicle or station or by having an on-board centrifuge), the experience of U.S. astronauts so far has suggested that this would be unnecessary.

Performance outside a spacecraft in an extravehicular activity (EVA) is a bit of a different story, since some astronauts have experienced considerable exhaustion in such activities. As Berry points out, this probably argues for careful consideration of workload in planning EVA missions.

ILLUSIONS DURING MOTION

When humans are in motion, they receive cues regarding their whereabouts and motion from sense organs, especially the semicircular canals, the vestibular sacs (the otolith organs), the eyes, the kinesthetic receptors, and the cutaneous senses. These sense organs and the intricate interactions among them were specifically designed for a self-propelled terrestrial animal. They were not intended to cope with the unusual and prolonged forces encountered in three-dimensional flight. Such forces can push our sense organs beyond their capabilities to function accurately and cause them to signal erroneous information concerning the position and motion of the body.

Pilots in airplanes sometimes experience *disorientation* regarding their position or motion with respect to the earth. Actually, numerous phenomena cause disorientation; some give rise to sensations that the person is spinning (true vertigo), and others cause dizziness (a feeling of movement within the head). Indeed, the three terms *disorientation, vertigo,* and *dizziness* are usually, though inaccurately, used interchangeably to describe a variety of symptoms such as false sensations of turning, linear velocity, or tilt (Kirkham et al., 1978).

Disorientation can be the basis for fatalities in an aircraft environment. Kirkham et al. (1978), for example, report that disorientation was the cause of 16 percent of all fatal general aviation aircraft accidents in the United States. Reason (1974) reports that virtually all pilots experience some form of disorientation at least once in their flying careers—for some, it is their last.

There are a multitude of disorienting effects, and we cannot hope to present or discuss them all. To understand some of them requires a knowledge of physics and physiology that is more than we wish to tackle here. Reason (1974) distinguishes two types of disorientation effects: those arising from false sensations and those arising from misperception.

Disorientation from False Sensations

The mental confusion or misjudgment resulting from false sensations is due to inaccurate or inappropriate sensory information coming from the semicircular canals, otolith organs, or both. That which results from false sensations from the semicircular canals involves angular acceleration cues. One such situation occurs when a pilot executes a roll at an angular acceleration that is below the threshold of perception of the semicircular canals. The pilot is not aware of the full extent of the roll and so overcorrects when attempting to reestablish straight, level flight. The aircraft ends up in a bank turned in the opposite direction, but the pilot thinks it is flying straight and level.

Normally, when you look at a stationary object and move your head, the *vestibulo-ocular reflex* moves your eyes in the opposite direction of your head movement. This serves to maintain clear vision of the object by fixing the object on the retina. That's fine for everyday circumstances, but when the vestibular system is subjected to extraordinary stimulation, such as would occur when an aircraft goes into a spin, the beneficial coordination of the visual and vestibular systems can be disrupted and visual acuity degraded. During a spin, the visual angle of a target necessary to see it clearly may be twice what was needed under normal conditions. In addition, if the pilot was spinning for longer than about 20 s, *postrotary nystagmus* occurs in which the eyes move repeatedly in the opposite direction of the spin and the visual angle required to see a target clearly may be five times as large as under normal conditions (Guedry, Lentz, and Jell, 1979). The postrotary nystagmus also gives the pilot the impression that he or she is actually spinning in the opposite direction. Attempting to recover from this false spin will often send the aircraft into another real spin. Pilots call this the *graveyard spin,* and for good reason.

Another form of disorientation arising from inaccurate semicircular canal information is called the *Coriolis illusion,* or *cross-coupling effect.* This can occur when the head is tilted during a long-established turning or circling maneuver. The head must be tilted in a different plane from the plane of the aircraft's turn. The experience is an illusion of roll, often accompanied by dizziness. This can be demonstrated by closing your eyes, spinning around, and tipping your head forward. You should have the sensation of falling sideways, and may experience nausea.

False sensations from the otolith organs often give rise to false sensations of tilt. The *oculogravic illusion* is one example. Whenever an aircraft's forward speed is suddenly increased (i.e., accelerated), the acceleration force vector combines with the gravity force vector to fool the otoliths into signaling that the body (and hence the aircraft) is tilted in a "nose-up" attitude. The eyes also roll back, thus adding to the impression of tilt. Attempting to level the aircraft (when, in fact, it is already level) can send it into the ground rather abruptly. Incidentally, decelerating often gives rise to the sensation of tilting nose-down. Another example of the oculogravic illusion occurs when a person is exposed to centrifugal forces, as when executing a turn in an aircraft. The centrifugal forces, acting outward, interact with the downward forces of gravity, causing the person to feel pressed downward and inclined in the direction of the resultant force. Vertical objects also appear to be inclined in the same direction.

These false-sensation forms of disorientation tend to occur under conditions of poor visibility, e.g., at night, in clouds, or in fog. It is for this reason that pilots are trained to trust their instruments rather than their senses. Usually a visual reference can reduce or eliminate disorientation, but not always.

Disorientation Resulting from Misperception

These forms of disorientation arise because the brain misinterprets or misclassifies perfectly accurate sensory information, usually provided by the visual sense. One form is *autokinesis,* in which a fixed light appears to move against its dark background. It has been reported that pilots have attempted to "join up" in formations with stars, buoys, and streetlights which appear to be moving (Clark and Graybiel, 1955). Another common visual misperception occurs under conditions of limited visibility wherein a pilot will accept a sloping cloud bank as an indication of the horizontal and, as a result, align the wings with the cloud bank and fly along at a "rakish" tilt.

Discussion

The reduction of disorientation and illusions generally depends more on procedural practices and training than on the engineering design of aircraft. Among such practices are the following (Clark and Graybiel, 1955): understanding the nature of various illusions and the circumstances under which they tend to occur, maintaining either instrument *or* contact flight, avoiding night aerobatics, shifting attention to different features of the environment, learning to depend on the correct cue to orientation (such as visual cues or instruments), avoiding sudden accelerations and decelerations at night, and avoiding prolonged constant-speed turns at night.

MOTION SICKNESS

We would be remiss if we failed to include motion sickness in a chapter on humans in motion.* Although motion sickness may never actually kill us, there are times when we wish it would! Motion sickness, of course, is not really a sickness in the pathological sense at all; it only makes us "feel sick." Motion sickness is associated with most forms of travel—cars, boats, trains, and even camels, but not horses. (We explain the camel-horse paradox shortly.) People differ in their susceptibility to motion sickness, but that susceptibility is a relatively stable and enduring characteristic of the individual (Reason, 1974). Further, a person susceptible to one type of motion sickness is probably susceptible to all types except, for some unexplained reason, space sickness.

The symptoms of motion sickness are familiar to us all. Reason (1974) puts them into two classes; head and gut. *Head symptoms* include drowsiness and a general apathy, together called the *sopite syndrome* by Graybiel and Krepton (1976). Headaches are also experienced. *Gut symptoms* range from a disconcerting awareness of the stomach region to acute nausea and vomiting (or, more politely, *emesis*).

The most widely accepted theory of motion sickness, called *sensory rearrangement theory* (Reason and Brand, 1975), considers motion sickness as a consequence of incongruities among the spatial senses, i.e., the organs of balance, the eyes, and the nonvestibular position senses (joints, muscles, and tendons). The incongruity among the senses is incompatible with what we have come to expect on the basis of our past experience. Reason (1974, 1978) points out that the vestibular system, i.e., the semicircular canals and otolith organs,

* See Kennedy and Frank (1984) and Money (1970) for reviews of the motion sickness literature.

must be implicated for motion sickness to be an outcome. And, since the vestibular system responds only to accelerations (linear and angular), acceleration (or apparent acceleration) must always be involved in motion sickness.

Reason (1974, 1978) identifies two classes of sensory rearrangement: *visual-inertia rearrangements* and *canal-otolith rearrangements*. Further, these rearrangements, or incongruities, are of two types: (1) both systems (e.g., canal and otolith) simultaneously signal contradictory information, and (2) one system signals in the absence of an expected signal from the other. For example, some people can get motion sickness from watching a movie filmed inside a roller coaster. In this case, the visual system senses accelerations, but the semicircular canals and otoliths sense no motion.

Low-frequency oscillations of less than 1 Hz, as in the up-and-down motion of a ship, usually cause motion sickness. Reason and Brand (1975) believe this is because the otolith signals are out of phase with the semicircular canal signals; however, the exact mechanism is unclear. McCauley and Kennedy (1976) summarized the results of exposing over 500 subjects to 2 h of vertical sinusoidal vibration of various combinations of frequency (<1 Hz) and acceleration. The dependent variable was the percentage of subjects who vomited—not the kind of research you dream about doing after graduate school! Figure 19-14 shows the percentage of subjects vomiting at various frequency-acceleration combinations. Vibration from 0.15 to 0.25 Hz (or 9 to 15 cycles per min) is, for some reason, especially effective for inducing gut symptoms of motion sickness.

Lawther and Griffin (1988) surveyed over 20,000 passengers on more than 100 voyages made by various seagoing vessels and found results consistent with those depicted in Figure 19-14. The vertical motions of the vessels, over widely varying sea conditions, were concentrated between 0.1 and 0.3 Hz, which, according to Figure 19-14, are the most upsetting (at least to the stomach). Under such conditions, vomiting and illness ratings were linearly related to the rms acceleration in the vertical axis. The more intense the motion, the more sick the people became. Of the total number of passengers surveyed, 7 percent reported vomiting at some time during their journey and about one-quarter reported feeling "slightly" or "quite" ill. Incidentally, McCauley and Kennedy indicate that those who are going to get sick generally do so within the first hour when they are exposed to very regular sinusoidal motion. Lawther and Griffin (1988), on the other hand, found a cumulative effect of time at least for exposure up to 6 h; the longer the exposure, the greater the incidence of vomiting and the more sick everyone felt.

Were you wondering about the camel-horse paradox? It has nothing to do with whether horses smell better than camels; actually, the answer lies in Figure 19-14. The characteristic frequency of a camel's gait happens to be in the 0.2 Hz range, while a horse tends to oscillate at higher frequencies.

You will recall that frequency, amplitude, and acceleration are all related in sinusoidal vibration. For very low frequency vibration, relatively large ampli-

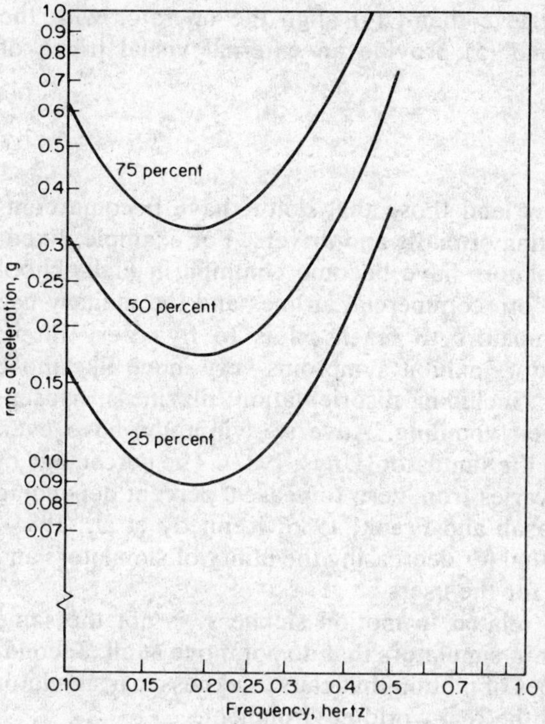

FIGURE 19-14
Equal motion sickness contours, based on percentage of emesis within 2 h for 500 male subjects exposed to vertical sinusoidal motion. (*Source: McCauley and Kennedy, 1976, Fig. 4.*)

tudes are required to achieve relatively low levels of acceleration. For cxample, a 0.2-Hz vibration, having a 0.09 rms *g*, would have a peak-to-peak displacement of 5.2 ft (1.58 m).*

Now that we know what causes motion sickness, what can we do to reduce its effects? Reason (1974) lists several practical suggestions, including getting off the boat. In the case of seasickness, lying down on your back will often help; holding your head still will also reduce the symptoms. Several drugs are available which seem to be effective against motion sickness. Most people will eventually adapt to the motion environment and "get their sea legs" in a few days. We can all add a few home-grown preventive measures to the list.

In addition to the remedies listed above, Bittner and Guignard (1985) discuss several work-space design approaches for minimizing seasickness based on human factors principles. Included are these suggestions: (1) Locate the workstation near the ship's effective center of rotation to reduce rotational heave; (2) group displays and controls to minimize the need for large-angle head turning; (3) place tools and equipment within close reach so that bending and

* Peak-to-peak displacement (in) for a sine wave = (27.675 × rms *g*)/[frequency (Hz)^2].

twisting are not required to retrieve them; (4) align the operator with the longitudinal axis of the ship; and (5) provide an external visual frame of reference.

Simulator Sickness

Simulators, both those that move and those that don't, have become commonplace in the training and testing of pilots and drivers. For example, fixed-based (nonmoving) driving simulators have become common in high school driver's education classes, and both commercial airlines and the military use sophisticated moving-based simulators to teach pilots to fly. Very often, though, the people using simulators exhibit symptoms very much like those associated with motion sickness, including disorientation, dizziness, nausea, postural disequilibrium, and even vomiting. Adverse symptoms have even been reported days after being in the simulator (Ungs, 1987). The percentage of people experiencing symptoms varies from zero to near 90 percent depending on the simulator being used (Casali and Frank, 1986; Kennedy et al., 1984). Simulator sickness has the potential for decreasing the utility of simulators and may even create safety hazards for the users.

Simulator sickness, although related to motion sickness, is not the same thing. First of all, people get sick in simulators that do not move at all. Second, when motion is involved, the types of motion that cause sickness in a simulator do not always cause sickness in the real-world environment.

We do not know exactly what causes simulator sickness, but it does appear related to such things as a wide visual field of view, realistic out-of-window scenes (Hettinger et al., 1987), and visual and motion simulator-system delays (Frank, Casali, and Wierwille, 1988). The consensus appears to be that simulator sickness is caused by some type of conflict or incongruity between either (1) the visual and vestibular senses or (2) what is experienced by the senses and what is expected by the senses based on prior learning or innate biological connections (Kennedy and Frank, 1984; Casali and Frank, 1986). Experienced pilots and drivers, for example, are often more likely to experience simulator sickness then are novices. Presumably this is because experienced people have stronger expectations about how the aircraft or automobile should respond and those expectations are not always fulfilled in true-to-life fashion in the simulator, hence, stronger incongruities.

DISCUSSION

The common denominator of the exotic experiences of astronauts in outer space, of the more mundane processes of moving around in automobiles or other vehicles, and of the operation of tractors and other related items of equipment is the fact that the human body is being moved somehow, e.g., accelerated, decelerated, transported, jostled, or shaken. Although some forms and degree of such shuffling are inconsequential, other forms or degrees obviously have potentially adverse effects in terms of physiological, performance, or subjective criteria. The challenge to the human factors disciplines is to design the physical systems or protective devices in question so as to minimize such effects and, where they cannot be eliminated, to design the system to match the reduced capabilities of the operator and to make whatever motion is involved more tolerable for passengers.

REFERENCES

Acoustical Society of America (1980). *American national standard: Guide for the evaluation of human exposure to whole-body vibration (ANSI S3.18-1979)*. New York: Acoustical Society of America.

Berry, C. A. (1973). Weightlessness. In *Bioastronautics data book* (2d ed.) (NASA SP-3006). Washington, DC: National Aeronautics and Space Administration.

Bittner, A., and Guignard, J. (1985). Human factors engineering principles for minimizing adverse ship motion effects: Theory and practice. *Naval Engineers Journal*, 97(4), 205–213.

Bondurant, S., Clark, N. P., Blanchard, W. G., et al. (1958). *Human tolerance to some of the accelerations anticipated in space flight* (TR 58-156). Wright-Patterson Air Force Base, OH: WADC.

Burton, A., and Sandover, J. (1987). Back pain in Grand Prix drivers: A "found" experiment. *Applied Ergonomics*, 18, 3–8.

Carlsoo, S. (1982). The effect of vibration on the skeleton, joints and muscles: A review of the literature. *Applied Ergonomics*, 13, 251–258.

Casali, J., and Frank, L. (1986). Perceptual distortion and its consequences in vehicular simulation: Basic theory and incidence of simulator sickness. *Transportation Research Record*, 1059, 57–65.

Chambers, R. M. (1963). Operator performance in acceleration environments. In N. M. Burns, R. M. Chambers, and E. Hendler (eds.), *Unusual environments and human behavior*. New York: Free Press.

Chambers, R. M., and Brown, J. L. (1959). *Acceleration*. Paper presented at Symposium on Environmental Stress and Human Performance. American Psychological Association.

Clark, B., and Graybiel, A. (1955). *Disorientation: A cause of pilot error* (Res. Project NM 001 110 100.39). U. S. Navy School of Aviation Medicine.

Corbridge, C., and Griffin, M. (1986). Vibration and comfort: Vertical and lateral motion in the range 0.5 to 5.0 Hz. *Ergonomics*, 29, 249–272.

Frank, L., Casali, J., and Wierwille, W. (1988). Effects of visual display and motion system delays on operator performance and uneasiness in a driving simulator. *Human Factors*, 30, 201–217.

Fraser, T. (1973). Sustained linear acceleration. In *Bioastronautics data book* (2d ed.) (NASA-SP-3006). Washington, DC: National Aeronautics and Space Administration.

Geldard, F. (1972). *The human senses* (2d ed.). New York: Wiley.

Gell, C., and Moeller, G. (1972). The biodynamics aspects of low altitude high speed flight. *Ergonomics*, 15, 655–670.

Grandjean, E. (1988). *Fitting the task to the man, (4th ed.). London: Taylor & Francis.*

Graybiel, A., and Krepton, J. (1976). Sopite syndrome: A sometimes sole manifestation of motion sickness. *Aviation, Space, and Environmental Medicine*, 47, 873–882.

Griffin, M. (1976). Subjective equivalence of sinusoidal and random whole-body vibration. *Journal of Acoustical Society of America*, 60, 1140–1145.

Griffin, M. (1986). Evaluation of vibration with respect to human response. *SAE Tech. Paper Series* No. 860047. Warrendale, PA: Society of Automotive Engineers.

Griffin, M., Parsons, K., and Whitham, E. (1982). Vibration and comfort. IV. Application of experimental results. *Ergonomics*, 25, 721–739.

Guedry, F., Lentz, J., and Jell, R. (1979). Visual-vestibular interactions: I. Influence of peripheral vision on suppression of the vestibulo-ocular reflex and visual acuity. *Aviation, Space and Environmental Medicine*, 50, 205–211.

Guignard, J. (1965). Vibration. In J. Gillies (ed.), *A textbook of aviation physiology*. Oxford, England: Pergamon.

Hartemann, F., Henry, C., Faverjon, G., Tarriere, C., Got, C., and Patel, A. (1984). Ten years of safety due to the three-point seat belt. In SAE (ed.), *Advances in belt restraint systems: Design, performance and usage* (SAE/P-84/141). Warrendale, PA: Society of Automotive Engineers, pp. 7–14.

Hettinger, L., Nolan, M., Kennedy, R., Berbaum, K., Schnitzius, K., and Edinger, K. (1987). Visual display factors contributing to simulator sickness. *Proceedings of the Human Factors Society 31st Annual Meeting*. Santa Monica, CA: Human Factors Society, pp. 497–501.

Hornick, R. (1973). Vibration. In *Bioastronautics data book* (2d ed.) (NASA SP-3006). Washington, DC: National Aeronautics and Space Administration.

ISO (1978). *Guide to the evaluation of human exposure to whole-body vibration* (International Standard 2631). Geneva: International Standards Organization.

ISO (1982*a*). *Guide for the evaluation of exposure to whole-body vibration. Addendum 2. Evaluation of exposure to whole-body z-axis vertical vibration in the frequency range 0.1 to 0.63 Hz* (International Standard 2631/AD2). Geneva: International Standards Organization.

ISO (1982*b*). *Guide for the evaluation of exposure to whole-body vibration. Amendment 1* (International Standard 2631/A1). Geneva: International Standards Organization.

Kennedy, R., and Frank, L. (1984, Mar. 9). *A review of motion sickness with special reference to simulator sickness* (rev.). Prepared for distribution at the National Academy of Sciences/National Research Council Committee on Human Factors, Workshop on Simulator Sickness, September 26–28, 1983, held at Naval Postgraduate School, Monterey, CA.

Kennedy, R., Frank, L., McCauley, M., Bittner, A., Jr., Root, R., and Binks, T. (1984, April). Simulator sickness: Reaction to a transformed perceptual world. VI. Preliminary site surveys. Paper presented at the AGARD Aerospace Medical Panel Symposium, Williamsburg, VA, and in conference *Proceedings* No. 392.

Kirkham, W., Collins, W., Grape, P., Simpson, J., and Wallace, T. (1978). Spatial disorientation in general aviation accidents. *Aviation, Space, and Environmental Medicine*, 49, 1080–1086.

Kitfield, J. (1989, May/June). Danger inside the cockpit. *Military Forum*.

Kjellberg, A., and Wikström, B. (1985). Whole-body vibration: Exposure time and acute effects—a review. *Ergonomics*, 28(3), 535–544.

Lawther, A., and Griffin, M. (1988). Motion sickness and motion characteristics of vessels at sea. *Ergonomics,* 31, 1373–1394.

Leatherwood, J., Dempsey, T., and Clevenson, S. (1980). A design tool for estimating passenger ride discomfort within complex ride environments. *Human Factors,* 22, 291–312.

Lewis, C., and Griffin, M. (1979). The effect of character size on the legibility of a numeric display during vertical whole-body vibration. *Journal of Sound and Vibration,* 67, 562–565.

McCauley, M., and Kennedy, R. (1976). *Recommended human exposure limits for very low frequency vibration* [TP-76-36(U)]. Point Mugu, CA: Pacific Missile Test Center.

Money, K. (1970). Motion sickness. *Physiological Reviews,* 50(1), 1–38.

Morrissey, S., and Bittner, A. (1975). *Effects of vibration on humans: Performance decrements and limits* (TP-75-37U). Point Mugu, CA: Pacific Missile Test Center.

Moseley, M., and Griffin, M. (1986). Effects of display vibration and whole-body vibration on visual performance. *Ergonomics,* 29, 977–983.

Moseley, M., Lewis, C., and Griffin, M. (1982). Sinusoidal and random whole-body vibration: Comparative effects of visual performance. *Aviation, Space, and Environmental Medicine,* 53, 1000–1005.

Nicholson, A. N., and Franks, W. R. (1966). Devices for protection against positive (long axis) acceleration. In P. I. Altman and D. S. Dittmer (eds.), *Environmental biology* (Tech. Rept. 66–194). Aerospace Medical Research Laboratory, pp. 259–260.

Norin, H., Carlsson, G., and Korner, J. (1984). Seat belt usage in Sweden and its injury reducing effect. In SAE (ed.), *Advances in belt restraint systems: Design, performance and usage* (SAE/P-84/141). Warrendale, PA: Society of Automotive Engineers, pp. 15–28.

Oborne, D. (1976). A critical assessment of studies relating whole-body vibration to passenger comfort. *Ergonomics,* 19, 751–774.

Oborne, D. (1978). The stability of equal sensation contours for whole-body vibration. *Ergonomics,* 21, 651–658.

Oborne, D. (1983). Whole-body vibration and international standard ISO 2631: A critique. *Human Factors,* 25, 55–69.

Oborne, D., Heath, T., and Boarer, P. (1981). Variation in human response to whole-body vibration. *Ergonomics,* 24, 301–313.

Oman, C. (1984, August). Why do astronauts suffer space sickness? *New Scientist,* 23, 10–13.

Oman, C., Lichtenberg, B., and Money, K. (1984, May 3). Space motion sickness monitoring experiment: Spacelab 1, Paper 35. NATO-AGARD Aerospace Medical Panel Symposium: *Motion sickness; Mechanisms, prediction, prevention and treatment.* Williamsburg, VA.

Poulton, E. (1978). Increased vigilance with vertical vibration at 5 Hz: An alerting mechanism. *Applied Ergonomics,* 9, 73–76.

Reason, J. (1974). *Man in motion: The psychology of travel.* London: Weidenfeld and Nicolson.

Reason, J. (1978). Motion sickness: Some theoretical and practical considerations. *Applied Ergonomics,* 9, 163–167.

Reason, J., and Brand, J. (1975). *Motion sickness.* London: Academic.

Richards, L., and Jacobson, I. (1977). Ride quality assessment III: Questionnaire results of a second flight programme. *Ergonomics*, 20, 499–519.

SAE (1984). *Advances in belt restraint systems: Design, performance and usage* (SAE/P-84/141). Warrendale, PA.

Sandover, J. (1979). A standard on human response to vibration—One of a new breed? *Applied Ergonomics*, 10, 33–37.

Seidel, H., and Heide, R. (1986). Long-term effects of whole-body vibration: A critical survey of the literature. *International Archives of Occupational and Environmental Health*, 58, 1–26.

Simons, A. K., Radke, A. O., and Oswald, W. C. (1956, Mar. 11). *A study of truck ride characteristics in military vehicles* (Rep. 118). Milwaukee: Bostrom Research Laboratories.

Stephens, D. (1979). Developments in ride quality criteria. *Noise Control Engineering*, 12, 6–14.

Thornton, W. (1978). Anthropometric changes in weightlessness. In Anthropology research staff (eds.), *Anthropometric source book*, vol. I: *Anthropometry for designers* (NASA RP-1024). Houston: National Aeronautics and Space Administration.

Ungs, T. (1987). Simulator induced syndrome: Evidence for long term simulator aftereffects. *Proceedings of the Human Factors Society 31st Annual Meeting*. Santa Monica, CA: Human Factors Society, pp. 505–509.

Whitham, E., and Griffin, M. (1978). The effects of vibration frequency and direction on the location of areas of discomfort caused by whole-body vibration. *Applied Ergonomics*, 9, 231–239.

Witkin, H. A. (1959, February). The perception of the upright. *Scientific American*.

Wood, E., Sutterer, W., Marshall, H., Lindberg, E., and Headley, R. (1961). *Effects of headward and forward accelerations on the cardiovascular system* (Tech. Rept. 60–634). Wright-Patterson Air Force Base, OH: WADC.

Zarriello, J., Norsworthy, M., and Bower, H. (1958). *A study of early grayout thresholds as an indicator of human tolerance to positive radial acceleratory force* (Project NM-11-02-11). Pensacola, FL: Naval School of Aviation Medicine.

PART SIX

HUMAN FACTORS APPLICATIONS

20 Human Error, Accidents, and Safety

21 Human Factors and the Automobile

22 Human Factors in Systems Design

CHAPTER **20**

HUMAN ERROR, ACCIDENTS, AND SAFETY

The major thrust of this text has been to emphasize the importance of human factors in designing equipment, facilities, procedures, and environments for people involved in the production of goods and services, as in manufacturing industries, service industries, military services, government, and other types of organizations. In large part, human factors efforts are directed toward designing things people use in order to enhance performance and minimize errors. (This takes us back to the matter of *human reliability* discussed in Chapter 2; such reliability, of course, being the complement of error.)

This chapter deals with topics that are central to this endeavor: human error, accidents, and safety. Reducing human error and accidents and improving safety remain top priorities in any human factors effort. For example, the National Research Council of the National Academy of Sciences conducted meetings to outline research questions in the area of human error at the individual, team, and system level. We conclude this chapter with a discussion of the effectiveness of warnings and the topic of product liability litigation (the lawyers always get the last word). Both of these topics are getting increased attention in the United States and elsewhere, and an understanding of the role played by human factors may be enlightening.

HUMAN ERROR

To some people the term *human error* has a connotation of blame or cause. A much more productive approach, however, is to consider human error simply as an event whose cause can be investigated. Numerous definitions have been proposed for human error, but the following embodies the essence of most of them: *Human error* is an inappropriate or undesirable human decision or behavior that reduces, or has the potential for reducing, effectiveness, safety, or system performance. Two things should be noted about this definition. First, an error is defined in terms of its undesirable effect or potential effect on system criteria or on people. Forgetting to pack cookies in a lunch would not be considered a human error in the context of a construction crew building a bridge, but forgetting to take safety shoes and glasses to the work site would be. Second, an action does not have to result in degraded system performance or an undesirable effect on people to be considered an error. An error that is corrected before it can cause damage is an error nonetheless. The important point is that an action must have the potential for adversely affecting system or human criteria.

Although there is a tendency among some to view errors as those of "operators," other people involved in the design and operation of systems also can make errors, such as equipment designers, managers, supervisors, and maintenance personnel. Therefore, in talking about human error, we should consider the entire system and not focus only on the operator.

If human error involves inappropriate or undesirable behavior, then it is important to understand how one determines what behavior is appropriate or desirable. Rasmussen (1979) points out that such determinations are often set by someone conducting a rational, careful evaluation of the behavior after the fact. In essence, what is considered to be a human error is somewhat arbitrary because the determination of what is appropriate may not have been established until the error was identified. In addition, Rasmussen (1987) feels that the identification of an event as a human error depends entirely upon the stop-rule applied during the investigation. If system performance is judged to be lower than some standard, someone will typically try to backtrack to find the causes. How far back to go is an open question. One could stop at the operator's actions and call the event a human error, or one could investigate what caused the human to act as he or she did. The cause may then be traced to other factors, such as faulty equipment, poor management practices, inaccurate or incomplete procedures, etc. Rasmussen (1982) makes a provocative point that an action might become an error only because the action is performed in an unkind environment that does not permit detection and reversal of the behavior before an unacceptable consequence occurs.

Human Error Classification Schemes

Various error classification schemes have been developed over the years. An effective classification scheme can be of value in organizing data on human errors and for giving useful insights into the ways in which errors are caused and how they might be prevented. Over the years, there have been numerous attempts at developing a practical taxonomy of human errors. We will briefly discuss a few such schemes to illustrate the thinking in the area.

Discrete-Action Classifications One of the simplest classification schemes for individual, discrete actions is that used by Swain and Guttman (1983):

- Errors of omission
- Errors of commission
- Sequence errors
- Timing errors

Errors of omission involve failure to do something. For example, an electrician was electrocuted while attempting to position himself on the steel framework of an electrical substation. There were several points to disconnect in order to shut off power completely to the substation, and he apparently forgot to disconnect one of them.

Errors of commission involve performing an act incorrectly. For example, a mechanic sitting on a conveyer belt called for his partner to lightly hit the start button to jog the belt forward a few inches. The helper lost his balance momentarily and hit the button hard enough to actually start the belt moving at full speed, rather than just jogging it forward. The mechanic was pulled between the belt and a steel support 9 in (23 cm) above it.

A *sequence error* (really a subclass of errors of commission) occurs when a person performs some task, or step in a task, out of sequence. An example occurred in the case of a crane operator who was lifting a 24-ton block of stone. Rather than lifting the boom and then rotating it 90 degrees, he rotated the near-flat extended boom first, and before he could lift it, the crane overturned.

A *timing error* (also a subclass of errors of commission) occurs when a person fails to perform an action within the allotted time, either performing too fast or too slowly. Taking too long to remove one's hand from a workpiece in a drill press, for example, is a timing error that can result in a nasty injury.

Information Processing Classifications Several authors have used an information processing model to classify human errors. One example is the scheme proposed by Rouse and Rouse (1983) and shown in Table 20-1. The scheme follows the information processing assumed to occur when a human operates and controls systems such as aircraft, ships, or power plants. An operator would observe the state or condition of the system, formulate a hypothesis, test the hypothesis, choose a goal, select procedures to attain the goal, and finally execute the procedures. For example, an operator may notice that the pressure in a boiler is high. One hypothesis would be that a high temperature is causing the increased pressure. This could be checked by observing the temperature gauge. The goal might then be to reduce the temperature, and a set of procedures would be selected and executed to do that. Table 20-1 lists some specific categories of errors that can occur at each of these stages of information processing. Notice that the specific categories of errors under the general category *execution of procedure* are very similar to those proposed by Swain and Guttman and discussed previously.

TABLE 20-1
HUMAN ERROR CLASSIFICATION SCHEME PROPOSED BY ROUSE AND ROUSE (1983)

1. Observation of system state
   - Improper rechecking of correct readings
   - Erroneous interpretation of correct readings
   - Incorrect readings of appropriate state variables
   - Failure to observe sufficient number of variables
   - Observation of inappropriate state variables
   - Failure to observe any state variables
2. Choice of hypothesis
   - Hypothesis could not cause the values of the state variables observed
   - Much more likely causes should be considered first
   - Very costly place to start
   - Hypothesis does not functionally relate to the variables observed
3. Testing of hypothesis
   - Stopped before reaching a conclusion
   - Reached wrong conclusion
   - Considered and discarded correct conclusion
   - Hypothesis not tested
4. Choice of goal
   - Insufficient specification of goal
   - Choice of counterproductive or nonproductive goal
   - Goal not chosen
5. Choice of procedure
   - Choice would not fully achieve goal
   - Choice would achieve incorrect goal
   - Choice unnecessary for achieving goal
   - Procedure not chosen
6. Execution of procedure
   - Required stop omitted
   - Unnecessary repetition of required step
   - Unnecessary step added
   - Steps executed in wrong order
   - Step executed too early or too late
   - Control in wrong position or range
   - Stopped before procedure complete
   - Unrelated inappropriate step executed

Source: Rouse and Rouse, 1983. © 1983 IEEE.

A somewhat different approach to classifying human errors is taken by Rasmussen (1982). He identifies 13 types of errors as depicted in the decision flow diagram of Figure 20-1. The errors depend on the type or level of behavior involved: skill-based, rule-based, or knowledge-based behavior. *Skill-based behavior* is controlled by subconscious routines and stored patterns of behavior and is appropriate for skilled operators in routine situations. Errors involving skill-based behavior are primarily errors of execution. *Rule-based behavior* applies to familiar situations where stored rules for coordinating behavioral subroutines can be applied. Errors involving rule-based behavior usually involve recognizing the salient features of the situation and remembering and applying the correct rules. *Knowledge-based behavior* occurs in unique, unfamiliar situations for which actions must be planned in relation to goals. Errors result from inadequate analysis or decision making.

Discussion Although error classification schemes abound in the literature, no scheme has really been particularly useful. Part of the problem is that human error is complex, and simple classification schemes such as those described above do not capture that complexity. Often information is not available to

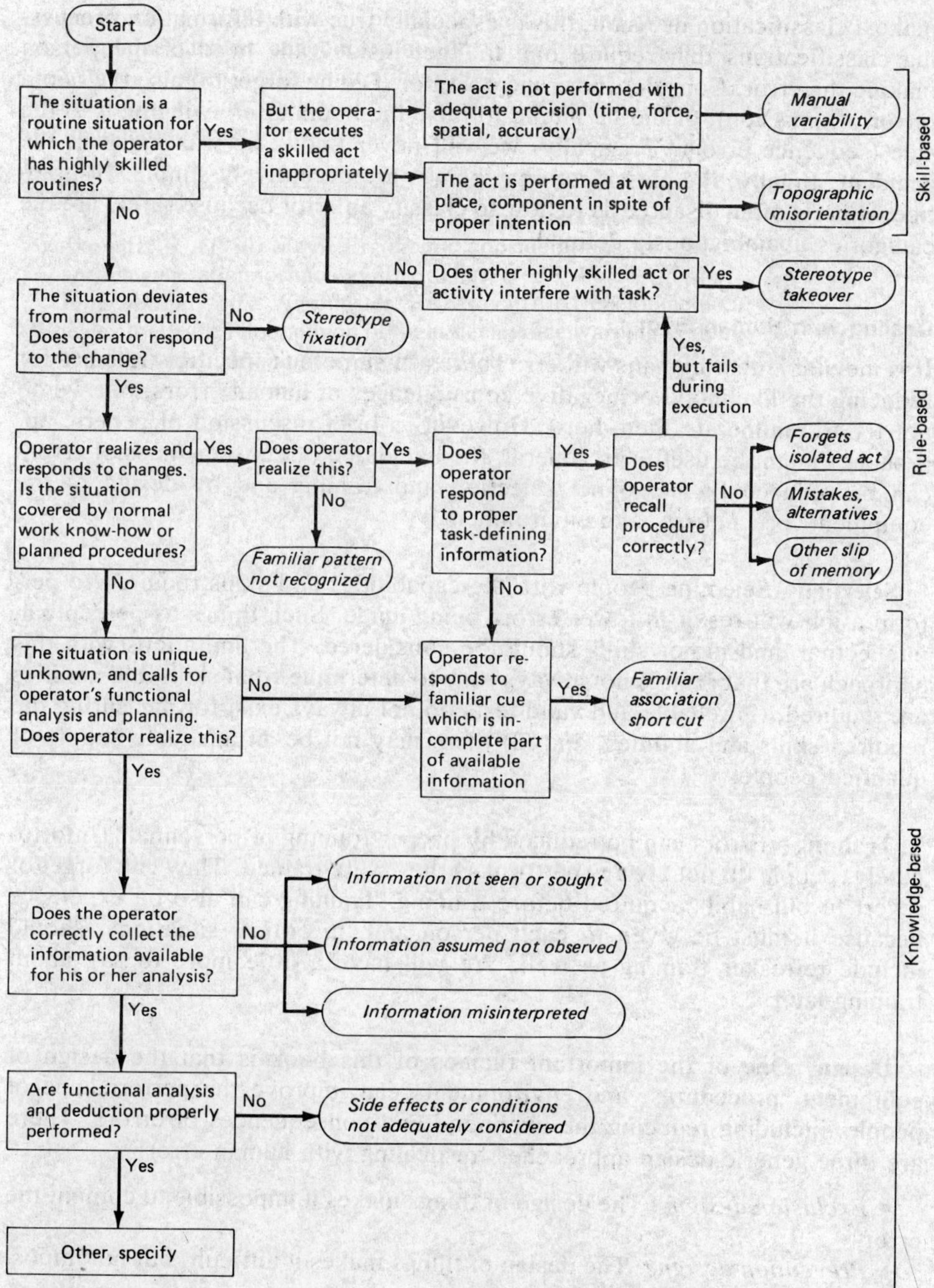

FIGURE 20-1
Decision flow diagram for analyzing an event into one of 13 types of human error. (*Source: Rasmussen, 1982.*)

make a classification decision; this is especially true with information processing classifications that require one to "get inside" the head of the person making the error. Consider the crane operator. Did he forget to raise the boom (error of omission), or did he intend to raise the boom after swinging it to the side (sequence error)? Tragically, we will never know; he was killed in the accident. Finally, the classification schemes are deceptively simple. In practice, experts often disagree as to how to classify an error because rarely are the categories unambiguously defined.

Dealing with Human Error

It is inevitable that humans will err. There are numerous specific strategies for reducing the likelihood or negative consequences of human errors, but we do not try to enumerate them here. However, a brief discussion of generic approaches might be useful. In general, the likelihood or consequences of errors can be reduced by personnel selection and training and by design of the equipment, procedures, and environment.

Selection Selecting people with the capabilities and skills required to perform a job will result in fewer errors being made. Such things as perceptual, intellectual, and motor skills should be considered. The limitations with this approach are that (1) it is not always easy to determine what skills and abilities are required, (2) reliable and valid tests do not always exist for measuring the required skills and abilities, and (3) there may not be an adequate supply of qualified people.

Training Errors can be reduced by proper training of personnel. Unfortunately, people do not always perform as they were trained. They can forget or revert to old habits acquired before training. Training can also be expensive because it must be given to each person and, in critical situations, should include refresher training as well. We will have a little more to say about training later.

Design One of the important themes of this book is that the design of equipment, procedures, and environments can improve the performance of people, including reducing the likelihood and consequences of errors. There are three generic design approaches for dealing with human error:

- *Exclusion designs:* The design of things makes it impossible to commit the error.
- *Prevention designs:* The design of things makes it difficult, but not impossible, to commit the error.
- *Fail-safe designs:* The design of things reduces the consequences of errors without necessarily reducing the likelihood of errors.

One of the authors was an expert witness in a product liability case involving a roller press in which exclusion, prevention, and fail-safe designs were issues. In this case a worker was seriously injured because the safety guards were rendered inoperative because of incorrect wiring of a three-phase motor used to operate the press. The very nature of three-phase motors makes it a 50-50 chance that the wiring will be done incorrectly. An exclusion design would have made it impossible to start the machine if the motor were wired incorrectly. Prevention designs would have included such things as instructions on how to test if the motor was wired correctly or a warning system that signaled when the motor was incorrectly wired. A fail-safe design would dictate a system in which the safety guards would remain operable no matter which way the motor was wired. The particular roller press involved in this case, however, provided none of these designs.

Designing to reduce errors or their consequences can often be the most cost-effective approach to the problem of human error. A system need be designed only once, while selection and training must be repeated as new people become part of a system. In essence, it is easier to bend metal than to twist arms.

Senders (1983) and Rasmussen (1987) make an interesting point with respect to eliminating human error. They point out that humans are servomechanisms and must experiment with their environment to learn and acquire skill. They maintain that errors are a natural consequence of this experimentation and trial-and-error learning. Errors, they contend, are necessary for the development of skilled performance. Their emphasis would be on how to provide "safe" opportunities for making errors (e.g., in training) or how to structure the environment to improve error detection and correction.

ACCIDENTS

One objective of human factors is to reduce accidents and improve the safety and physical well-being of people. In this regard, previous chapters have made reference to accidents and have cited accident statistics. To a very considerable extent, many of the design features discussed in previous chapters are directed toward the creation of safer conditions for people in their work and everyday lives.

In this section we discuss the nature of accidents in a general frame of reference. Our emphasis in on the reporting and analysis of accidents and the factors contributing to accidents, rather than on a recitation of boring accident statistics.

Definition of the Term *Accident*

Dictionaries use such phrases as "without apparent cause," "unexpected," "unintentional act," "mishap," and "chance" to define an accident. In some cultures, accidents are attributed to "acts of God" with no further attempts to determine the causation. The idea that an accident is totally without cause, or is due solely to chance or an act of God, however, is a serious impediment to further scientific inquiry. Probably no single definition of accident will satisfy all the people interested in the causes and prevention of accidents. Suchman (1961) produced a list of indicators of the accidental nature of an event. The more indicators that are present, the more likely the event is to be called an accident. The indicators are (1) low degree of expectedness; (2) low degree of avoidability; and (3) low degree of intention.

Most definitions of the term *accident* also include reference to the consequence of the event. Meister (1987), for example, defines an accident as "an unanticipated event which damages the system and/or the individual or affects the accomplishment of the system mission or the individual's task." This covers a broad spectrum of possible outcomes. In many practical settings, the term *accident* is synonymous with injury. Accidents, however, often occur without injury. Rather, they result in only property damage (the freeway fender-bender accident is a perfect example). Often, whether injury results or not is just a matter of chance and luck.

Human Error and Accidents

What percentage of accidents is caused by human error? This is a question that has vexed researchers for years. One often hears that approximately 85 percent of accidents are due to human error. This figure came from an analysis of insurance company records carried out by Heinrich (1959) and probably should not be taken too seriously. The percentage of accidents one attributes to human error depends on several factors. For example, which humans are to be considered when searching for errors? Traditionally, human error has been used to describe operator error, or errors of the injured employee. This is a very narrow view of human error. Other humans whose errors can contribute to an accident are managers, system designers, maintainers, and coworkers. When this broader perspective is considered, it is no wonder that Petersen (1984) concluded that "human error is the basic cause behind *all* accidents" (emphasis added). Rarely is the broader perspective of human error adopted when assessing its role in accident causation.

A second factor that influences the proportion of accidents attributed to human error is what other factors, besides human error, are considered in making the determination. In the simplest model, accidents are categorized as caused by either unsafe acts of persons (read operator error) or by unsafe conditions (Heinrich, 1959). The consequence of using this dichotomy is often to blame the individual who was injured or who was in charge of the machine that was involved in the accident. This occurs because of the tendency to direct our attention to the fault of the person and the unsafe act of that person. Shealy (1979) suggests four reasons why this tends to happen: (1) it is just human nature to blame what appears to be the active operator when something goes wrong; (2) our legal system is geared toward the determination of responsibility, fault, and blame; (3) it is easier for management to blame the worker than to accept the fact that the workplace, procedure, or environment might need improving; and (4) the forms that we fill out when we investigate accidents are usually modeled after the "unsafe act–unsafe condition" dichotomy. The major emphasis tends to be on describing the person who was injured and the injury-producing events rather than on finding the aspects of the situation that contributed to the accident or that disposed the individual to have the accident.

Given the differences in philosophy regarding the role of human error in accidents and the differences in the scope of accident investigations, it is somewhat meaningless to synthesize the studies that have attempted to determine the proportion of accidents due to human error. In fact, it is probably meaningless even to ask what proportion of accidents were due to human error. A more meaningful question to ask would be how much does human error contribute to accidents relative to other contributing factors? Despite the difficulties inherent in specifying the percentage of accidents attributable to human error, numerous investigators have done so. Sanders and Shaw (1988), for example, reviewed 15 studies and found the percentages of accidents or "incidents" attributed to human error ranged from 4 to 90 percent, with a median of 35 percent. In their analysis of 338 underground mining injury accidents, they found that human error of the injured employee was involved to a degree in about 80 percent of the accidents and was judged to be a primary contributing factor in about 50 percent of the cases. In no case, however, was human error the only factor involved in the accident.

Collection and Analysis of Accident and Injury Data

Data on accidents and injuries are compiled as a matter of routine by various organizations, including insurance companies, police departments, the Occupational Safety and Health Administration (OSHA), the Mine Safety and Health Administration (MSHA), various state agencies, the National Safety Council (NSC), and many trade associations. Often accident data are collected to fulfill a legal requirement or to assess blame for the damages resulting from an accident. Unfortunately, data collected to satisfy such aims are rarely useful for directing accident prevention efforts. It is rare to see an accident reporting form that requires specification of countermeasures for reducing similar accidents in the future. Often reporting forms contain predefined classification schemes for recording the basic elements of an accident in a form suitable for computer analysis. Almost all forms include the following categories, with some examples:

- Nature of injury (amputation, contusion, strain)
- Part of body (head, back, finger)
- Type of accident (struck by, caught between, fell)
- Source of injury (equipment, hand tools, body movement)

Many sets of injury data involve tabulating frequencies and rates of injuries as related to the above categories and their specific subcategories.

Accident data bases such as those maintained by OSHA or NSC serve a useful purpose in identifying trends in accident data. They do have limitations; for example, many accidents result from complex chains of events that cannot be adequately described by existing classification systems. Further, these data bases rely for their data on individuals who are often not trained in accident investigation, and the coding of the basic data is sometimes inaccurate and incomplete. In addition, the data tend to focus on the injured person rather than on the person(s) causing the accident. It is also recognized that not all accidents are reported. In some manufacturing industries, for example, accidents may be underreported by 15 to 20 percent (Beaumont, 1980). Even with these limitations, large-scale accident data bases are valuable sources of information. Without them, we would effectively be at a loss in our accident reduction efforts.

Accidents, fortunately, are rather rare events. Relying on accident reports to discover accident trends often requires more data than is available to a company. Relying on accident reports to uncover unsafe situations is analogous to closing the barn door after the horses have left. One technique that gets around these problems is the use of the *critical-incident technique*. The technique involves the description of observed unsafe acts or near-miss accidents. These are described in detail. Because there are many more critical incidents than there are accidents, enough data can be collected to reflect patterns of behavior and events that can be useful in developing preventive measures. Further, the barn door can be closed before the horses leave. There is evidence from various sources that observed unsafe acts and conditions are definitely related to accidents and injuries. Such confirmation, for example, comes from a survey by Edwards and Hahn (1980) of over 4000 workers in 19 plants. They used a variation of the critical-incident technique for obtaining data on observations by workers of unsafe acts and conditions. The number of such acts correlated, on an across-the-board basis, .61 with accidents and .55 with disabling injuries. The critical-incident technique is not without its problems. First, there may be selective recall, that is, workers may tend to recall critical incidents caused by forces over which they had no control and may tend to forget those for which they felt responsible. Second, the definition of "critical" or "near-miss" is vague. How critical or near does an incident have to be to get reported? This leads to an underreporting bias with respect to critical-incident data.

There are various other methods for collecting and analyzing data relevant to accidents and injuries, including various forms of task analysis, fault-free analysis, and the use of accident review teams (as in industry and aviation). Some of these and other techniques are discussed by Ramsey (1973) and Christensen (1980) and are not described here. In general terms, however, the various techniques provide for obtaining and analyzing data relating to the behaviors, to the physical conditions, or to both that presumably contribute to accidents or injuries. Such data, when appropriately developed, can serve as the basis for taking remedial action to reduce the incidence of such events.

Theories of Accident Causation

A wide array of accident-causation theories have been proposed. Each theory emphasizes the orientation of its author, be it psychological, sociological, or statistical. For convenience, the various theories can be grouped into three broad categories: accident-proneness theories, job demand versus worker capability theories, and psychosocial theories.

Accident-Proneness Theories The oldest and probably the most influential accident causation theory is that of *accident proneness*. In its pure form it hypothesizes that some people are more prone to have accidents than others because of a peculiar set of constitutional characteristics. Further, accident proneness is considered to be a permanent characteristic of the individual. The support for this theory has come from statistical comparisons between the distribution of accidents in a population of workers and the distribution expected by pure chance. What was often found was that more people than expected had multiple accidents. More recent authors (e.g., McKenna, 1983) have challenged these early statistical studies, pointing out that to accept accident proneness one must accept the underlying assumption that all people in a population of workers are exposed to the same job and environmental hazards. The fact that more people have multiple accidents than expected by chance may only indicate that some people are exposed to more hazards on the job than others.

A restrictive, and more realistic, view of accident proneness is that people are more or less prone to accidents in given specific situations and that this proneness is not permanent but changes over time. This has been called *accident-liability theory*. The relationship between age and accidents is often cited as evidence of this theory. The general finding is that younger workers have higher accident rates than older workers (e.g., National Academy of Sciences, 1982; Shahani, 1987). Besides the fact that younger workers generally have less experience than older workers, Lampert (1974) cites the following factors to account for the higher rates among younger workers: inattention, lack of discipline, impulsiveness, recklessness, misjudgment, overestimation of capacity, and pride. Of interest is that some investigators have found an increase in accident rates among workers over 50 or 60 years of age (although it is still lower than the rates for the youngest groups) (e.g., Shahani, 1987; Broberg, 1984). This may be due to deterioration in motor skills, sensory functions, and mental agility (DeGreen, 1972).

It appears that indeed, accident liability fluctuates with time, especially as related to age (and although not discussed, specific job experience). The notion that some people are naturally more accident-prone than others across all situations and at all times is probably a less tenable position.

Job Demand versus Worker Capability Theories This class of theories shares some common ground with the accident-liability theories. Simply put, accident liability increases when job demands exceed worker capabilities. For example, if a job requires greater psychomotor skills or strength than workers have, accidents are expected to increase.

Another theory within this category is the *adjustment-to-stress theory*. This theory postulates that accident rates will be higher in situations where the level of stress (physiological or psychological) exceeds the level of capacity of the people to meet it. Stressors include such things as noise, poor illumination, anxiety, lack of sleep, anger, etc. The research data, however, are mixed with respect to the relationship between accidents and many of these stressors. A related theory is the *arousal-alertness theory*. This theory predicts that accidents are more likely to occur both when arousal is too low (e.g., person is underloaded or bored) or too high (e.g., overloaded or excessively motivated). Brown (1990) cautions that stress and arousal must not be confused. Stress is by definition harmful; arousal may or may not be harmful depending on its level. Increasing stress, therefore, should not be considered as a way to reduce accidents attributed to underarousal.

Psychosocial Theories One theory, *goals-freedom-alertness theory,* holds that greater freedom among workers to set reasonably attainable goals is accompanied by high-quality work performance, and accidents are viewed as examples of low-quality work performance (Kerr, 1957). Sanders, Patterson, and Peay (1976) found some evidence supporting this theory among underground coal miners. They found that when decisions were decentralized, when management was flexible and innovative in trying new procedures and programs, and when morale was high, disabling injuries decreased.

Another class of psychosocial theories is *psychoanalytic theories* that view accidents as self-punitive acts caused by guilt and aggression. Such theories may account for isolated accidents but are of no real value in explaining the typical accident event.

Discussion There is no one really good theory of accident causation that adequately explains the complexity of the accident situation. Each has a ring of truth to it, but each by itself is incomplete. Perhaps a more fruitful approach to accident causation is to assess the factors that contribute to accidents. To the degree possible, these factors can be controlled and accident frequency reduced.

Factors Contributing to Accidents

There are several models that attempt to portray the various factors that contribute to unsafe behavior or accidents. Sanders and Shaw (1988) review several different models and the factors that each includes. Each, by itself, was deemed incomplete. Based on their review, Sanders and Shaw propose the model of *contributing factors in accident causation* (CFAC) shown in Figure 20-2. The CFAC categories are sufficiently broad to encompass virtually all the factors included in the other models reviewed. The unique features of the CFAC model are the emphasis given to management and social-psychological

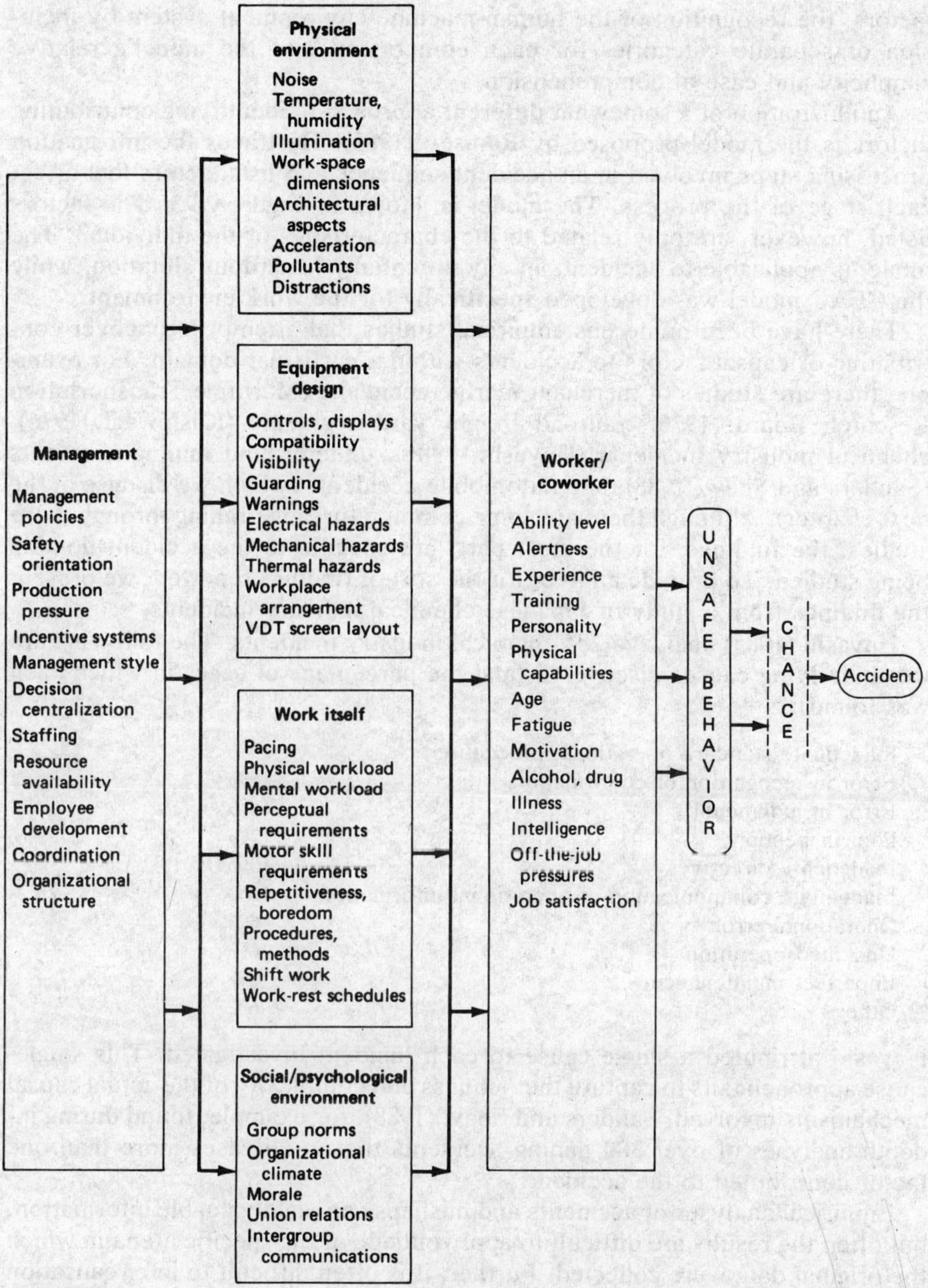

FIGURE 20-2
A model of contributing factors in accident causation (CFAC). Management's role is stressed along with the contribution of the social-psychological environment. Classical human factors variables are recognized in the physical environment and equipment design categories and in the work itself. Recognition is also given to the influence of coworkers. (*Source: Adapted from Sanders and Shaw, 1988, Fig. 7.*)

factors, the recognition of the human-machine-environment system by inclusion of separate categories for each component, and the model's relative simplicity and ease of comprehension.

An illustration of a somewhat different approach to identifying contributing factors is the model proposed by Ramsey (1985). He traces the information processing steps involved in an accident sequence and lists factors that affect each stage of the process. The model is shown in Figure 20-3. The factors listed, however, are only related to the characteristics of the individual. The model is applicable to accidents in any potentially hazardous situation, while the CFAC model was developed specifically for the work environment.

There have been numerous empirical studies that attempt to uncover contributing or causal factors to accidents within a particular domain. For example, there are studies of merchant marine accidents (Maritime Transportation Research Board, 1976), railroad freight yard accidents (Kashiwagi, 1976), chemical industry incidents (Hayashi, 1985), underground mining accidents (Sanders and Shaw, 1988), and automobile accidents (which we discuss in the next chapter). Although there are some common threads running through these studies, the findings, for the most part, are specific to the accident domain being studied. To provide a flavor for the sort of findings reported, we present the findings from a study of Japanese chemical industry incidents.

Hayashi (1985) analyzed 284 chemical industry incidents. The following are the underlying causes discovered (and the percentage of cases in which each was found):

Inadequate standard operational procedure	19%
Error in recognition or confirmation	15%
Error in judgment	14%
Poor inspection	12%
Inadequate directives	10%
Inadequate communication of operational information	10%
Operational error	6%
Unskilled operation	6%
Imperfect maintenance	2%
Other	6%

Hayashi attributed a single cause to each incident investigated. This single-cause approach fails to capture the richness and complexity of the actual causal mechanisms involved. Sanders and Shaw (1988), for example, found during in-depth analyses of over 300 mining accidents that in all cases more than one factor contributed to the accident.

Empirical analyses of accidents and mishaps can yield valuable information, but often the results are difficult to apply outside of the specific arena in which the original data were collected. Further, it is often difficult to infer causation from the analysis of accidents because the necessary information is usually not available. The people involved in the incident may be unable or unwilling to reveal information or to explain the motivations and rationale behind their behavior. In some cases, after-the-fact explanations are created by the in-

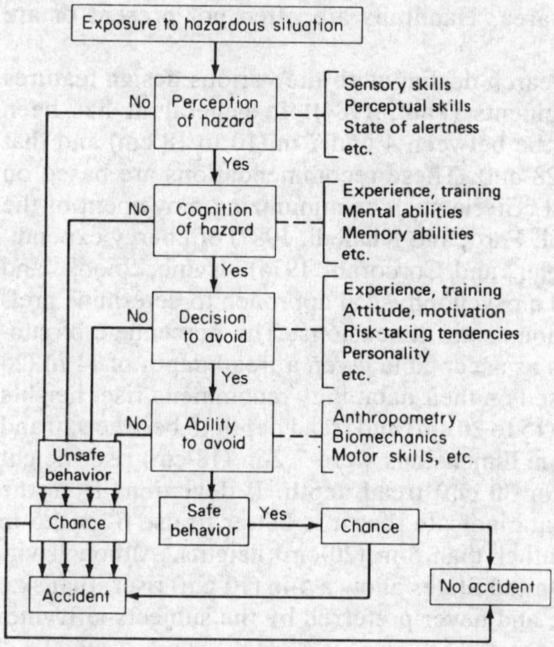

FIGURE 20-3
Accident sequence model representing various stages in the occurrence or avoidance of accidents in a potentially hazardous situation. (*Source: Adapted from Ramsey, 1985.*)

volved people to rationalize their behavior, or they will create information because it seems as though it must have been that way to make sense out of the situation. These sorts of biases are inherent in any eyewitness report and become especially acute when the witness may also share in the blame.

Special Accident Situations

We have selected a couple of examples of accident situations to illustrate the kinds of human factors research that are carried out and the types of recommendations that can be gleaned from such research.

Accidents Involving Stairs The Consumer Product Safety Commission (CPSC) has reported that stair-related falls are associated with an estimated 2 million injuries and 1000 deaths each year in the United States (Product Safety and Liability Reporter, 1983). Most stair accidents occur in the home, where they often rank as the number one type of accident.

Your author has been involved as an expert witness in numerous stair-related accident litigations. A few common elements run through many of them. Often there are only one or two risers (steps) comprising the stairway. The stairs are usually not conspicuous, but rather are the same color as the top and bottom levels to which they connect, and there is no distinctive marking on the nose edge of the steps. In most cases, there are few indications that the steps are present. In many cases there is a low level of illumination and visual distractions are present in the area. Handrails are often not present or are improperly placed.

There has been extensive research dealing with the various design features of stairs that contribute to accidents (Paul, 1984). In general, it has been recommended that riser heights be between 4 and 7 in (10 to 18 cm) and that tread depths be at least 11 in (28 cm). These recommendations are based on physiological and biomechanical criteria such as minimizing movement of the ankle, knee, and hip joints (Mital, Fard, and Khaledi, 1987) or energy expenditure and missteps (Fitch, Templer, and Corcoran, 1974). Irvine, Snook, and Sparshatt (1990), however, used a psychophysical approach to determine preference and acceptability of various stair dimensions. The percentage of subjects rating various riser heights as acceptable given a tread depth of 11 in (28 cm) is shown in Figure 20-4. Based on their data, they recommend riser heights be limited to between 6 and 8 in (15 to 20 cm) and tread depth to between 10 and 13 in (25 to 33 cm). The optimum dimensions were 7.2-in (18-cm) riser height with either an 11- or 12-in (28 or 30 cm) tread depth. If deviations from the optimum are necessary, their data indicate that it is better to use 6- or 6.5-in (15- to 16.5-cm) riser heights rather than 8-in (20-cm) heights. Although virtually all building codes in the United States allow a 4-in (10-cm) riser, that size was almost totally unacceptable and never preferred by the subjects in Irvine, Snook, and Sparshatt's study.

Other recommendations from the literature include: dimensional uniformity between successive risers and treads [a dimensional nonuniformity of as little

FIGURE 20-4
Percentage of subjects rating various stairways "acceptable" as a function of riser height. Data are shown for only those cases in which tread depth was 11 in (279 mm). The percentages are the average for ascending and descending acceptability. (*Source: Adapted from Irvine, Snook, and Sparshatt, 1990, Fig. 3. By permission of Butterworth-Heinemann Ltd. ©.*)

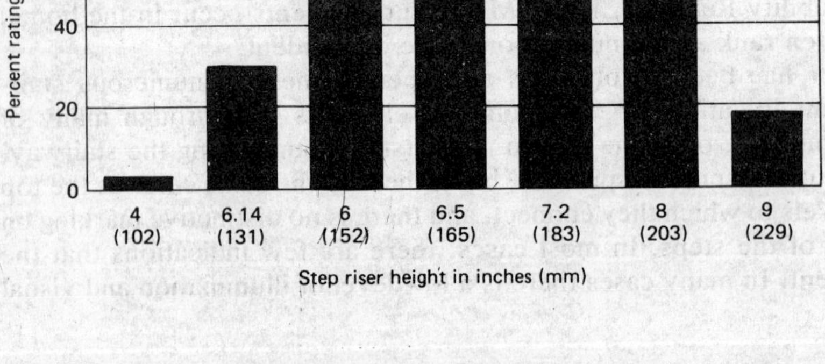

as 0.25 in (6 mm) between adjacent riser heights can cause an accident]; proper placement of handrails; use of a nonslip surface on tread surfaces; and enhanced visibility of the tread nose edges. Finally, attention should be drawn to the existence of the stairs, especially when there are only one or two risers.

Accidental Poisoning of Children Kessler and Reiff (1980) report that during just 3 months in 1980, approximately 28,000 children under the age of 5 received emergency room treatment for the ingestion of hazardous substances. Godfrey et al. (1986) conducted scenario analyses of 86 accidental-poisoning cases in which the Texas State Poison Center was called for help. *Scenario analysis* is a time-line analysis of the sequence of subevents that culminate in an accident. Described in such an analysis are the victim, product, environment, and task. The data for the Godfrey et al. study came from telephone interviews and relied on the memory of the respondent as to the events leading up to and surrounding the accident. All cases involved children under 5 years of age. Here are some of the findings from the study:

- Fully 93 percent of cases happened in the child's home.
- In 79 percent of cases parent and child were in different rooms.
- In 89 percent of cases children were engaged in an activity that is not usually closely monitored, such as watching TV; in 36 percent of cases, the child changed rooms without the parent's knowledge.
- In 87 percent of cases parents were engaged in some routine activity.
- In 59 percent of cases the substance was left out or was in use; in 28 percent of cases the substance was in normal storage.
- In 31 percent of cases the child encountered no barrier in accessing the substance; in 36 percent, only the height of a table or counter served as the barrier.
- In 57 percent of cases the substance was in a container equipped with a safety (child-resistant) cap; children were reported to have removed safety caps as often as nonsafety caps to ingest the substance.

Godfrey et al. postulate that safety caps are not always replaced properly by adults because sometimes it is difficult to tell when such a cap is properly closed. They also suggest that the standard for selecting a child-resistant cap be made more stringent. (Current Consumer Product Safety Commission standard is that 85 percent of the children tested under the age of 5 must not be able to open the container within 5 min and 80 percent must be unable to open it within 10 min following a demonstration of the proper means of opening the package.)

Reducing Accidents by Altering Behavior

We have stressed the importance of applying relevant human factors principles and data to the design of things people use in their jobs and everyday lives. Doing this can reduce both the likelihood and severity of accidents and injuries. Aside from strictly design considerations, however, other actions can be taken to reduce accidents and injuries by altering the behavior of people in potentially hazardous situations. Altering behavior is a tricky business. We have no intention of providing a how-to manual on modifying human behavior, but a few examples will illustrate some approaches that have been successful.

Procedural Checklists One technique that has become standard practice in aircraft operation, certain military operations, and certain other situations is the use of checklists that are to be followed in executing some specified routine. The individual follows a list of steps or functions that are to be executed, checking off each one as it is completed. This is, in effect, a substitute for memory in carrying out the stated activities.

Training Training is, of course, one of the standard procedures used to aid people in acquiring safe behavior practices. Cohen, Smith, and Anger (1979), on the basis of a review of approaches for fostering self-protective measures against workplace hazards, concluded that training remains the fundamental method for effecting such self-protection. However, they point out that the success of such training depends on (1) positive approaches that stress the learning of safe behaviors (not the avoidance of unsafe acts); (2) suitable conditions for practice that ensure the transferability of these learned behaviors to the real settings and their resistance to stress or other interferences; and (3) the inclusion of means for evaluating their effectiveness in reaching specified protection goals with frequent feedback to mark progress.

Feedback Closely related to training is the matter of feedback, specifically feedback following desirable work behaviors on the part of workers. Cohen, Smith, and Anger (1979), for example, report the effects of feedback following training in practices that would reduce exposure to styrene (a suspected carcinogen) in a work situation. After initial training, the instructor visited each worker once or twice each day to provide encouragement or feedback in observing the work procedures in question. Observers, from remote vantage points, logged the number of times that the behaviors in question occurred both before and after training and found reductions of 36 to 57 percent in exposure levels to styrene.

In another situation, in two departments of a large bakery, a brief safety training program was instituted. This was followed by an "intervention" period in which there was an updated posting of the percentage of behavioral incidents that were performed safely (based on the observations of an observer). As another form of feedback the supervisors took occasion to recognize workers when they performed certain selected incidents safely.

The results are summarized in Figure 20-5, which shows the percentage of incidents performed safely for "baseline" observation sessions, for "intervention" sessions, and a "reversal" period at the end of the 25-week study period,

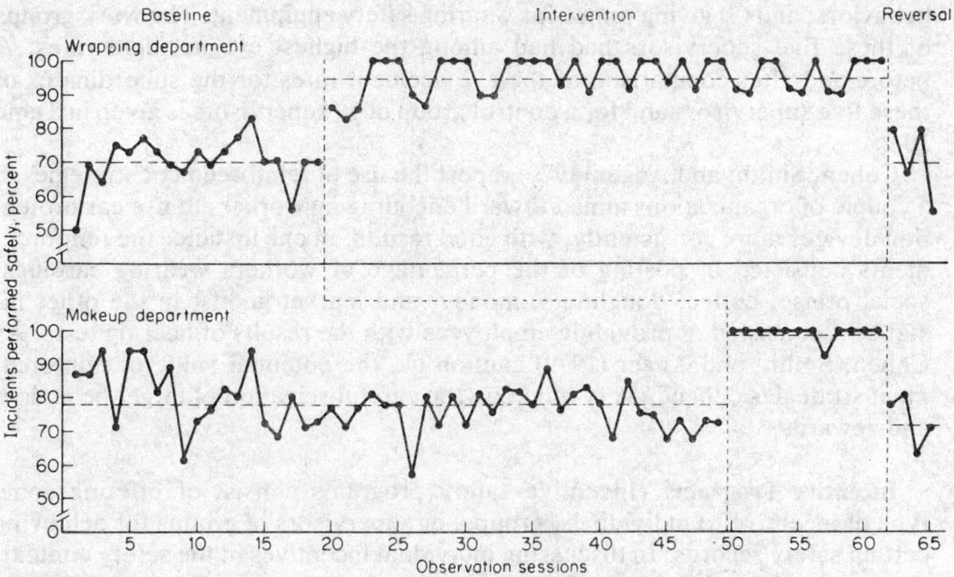

FIGURE 20-5
Percentage of behavioral incidents performed safely by bakery employees during a baseline period, an intervention period (during which feedback was provided), and a reversal period (after removal of feedback). (*Source: Komaki, Barwick, and Scott, 1978, Fig. 1, p. 439. Copyright 1978 by the American Psychological Association. Reprinted by permission.*)

the four reversal observation sessions being those after the feedback of the intervention sessions was terminated. The immediate drop following the intervention sessions was so pronounced that within 2 weeks the management assigned and trained an employee to reinstitute the posting of data on a weekly basis. Within a year the injury frequency rate had stabilized at less than 10 lost-time accidents per million hours worked, as contrasted with the previous rates of 35 and above. In connection with feedback regarding safety practices, clearly this must be a continuing process and not a one-shot proposition.

Contingency Reinforcement Strategies Contingency reinforcement has its roots in the early work of Skinner, whose extensive research demonstrated that the probability of repetition of specific behaviors may be increased by providing rewards (i.e., reinforcement) when the specific behaviors occur.

As an example of the effectiveness of such a strategy, Smith, Anger, and Uslan (1978) report the results of their study with a shipyard employing 20,000 workers. They concentrated on the use of reinforcement (what they referred to as *behavioral modification*) in the reduction of eye injuries, which accounted for over 60 percent of all injuries. The program consisted of training five supervisors in the fundamentals of behavior modification: (1) observing worker behaviors (specifically the use of safety equipment), (2) recording worker behaviors; and (3) giving praise for wearing safety equipment. The work groups of these five supervisors had had among the highest eye accident rates. A before-and-after comparison of the eye accident rates for the subordinates of these five supervisors and for a control group of 39 supervisors is given in Table 20-2.

Cohen, Smith, and Anger (1979) report the use of reinforcement strategies in a couple of organizations aimed toward encouraging workers to use ear protection devices more consistently, with good results. In one instance the reinforcements consisted of posting of the percentage of workers wearing earplugs, social praise, coffee, doughnuts, money, and market goods; in the other instance it consisted of providing employees with the results of hearing tests. As Cohen, Smith, and Anger (1979) caution us, the potential value of reinforcement strategies depends very much on the careful selection of target behaviors and rewards.

Incentive Programs Incentive safety programs consist of offering some type of incentive to individuals, groups, or supervisors of groups for achieving certain safety records. In discussing individual incentives in the safety context, Cohen, Smith, and Anger (1979) suggest that the incentives for safe behaviors might include bonuses, promotions, and special privileges (time off, preferred parking locations, etc.) Among other incentives cited by Cohen, Smith, and Anger (1979) are the use of group safety records as a part of supervisors' performance evaluations; tokens redeemable for catalog merchandise; trading stamps given to individual employees with bonus stamps to work groups with no accidents; and direct payment of money to individuals and groups with good safety records. Cohen, Smith, and Anger (1979) cite the experiences of certain organizations that have been very successful in reducing accident rates. However, they caution against excessive preoccupation with or continued use of such programs, on the grounds that a succession of such efforts, each requiring a bigger prize than the one before, would appear unwise. They therefore recommend limited use of this approach.

Discussion

There is no such thing as a completely accident-free circumstance in human life. However, the human race should not assume a completely fatalistic attitude toward the inevitability of accidents. The first step is to understand the nature of the accidents one wants to reduce and to identify the contributing factors involved. One should not wait for accidents to happen before directing attention to reducing the likelihood of them. A comprehensive analysis of the situation, task, equipment, or product can uncover potential accident scenarios and contributing factors that can be reduced through application of human factors principles. Any given accident has multiple contributing factors. The key to reducing accidents is to attack the problem from multiple perspectives.

TABLE 20-2
EYE ACCIDENT RATES OF SHIPYARD WORKERS BEFORE AND AFTER BEHAVIOR MODIFICATION TRAINING

	Eye accident rates		
Work group	**Before**	**After**	**Change**
Experimental (5)	11.8	4.3	−7.5%
Control (39)	5.8	4.7	−1.2%

Source: Smith, Anger, and Uslan, 1978.

Designing for safety and designing to promote safe behavior is a must. In conjunction, modifying behavior through training, reinforcement, and feedback can further reduce the likelihood of accidents. Neither design nor behavior modification can do the job alone. A combination of approaches is usually necessary. Good design can reduce hazards and the extent of training required. A good system of training, reinforcement, and feedback can make up, in part, for a hazardous situation. Our bias, of course, is to eliminate the hazard through design rather than depending on behavior modification techniques to solve the problem.

PERCEPTION OF RISK

In this section we briefly discuss definitions of risk, hazard, and danger and how people evaluate these concepts in everyday life. The perceived risk or hazardousness of a situation or product influences our behavior and the degree to which we take precautionary actions to prevent injuries and accidents.

Definitions

Words such as hazard, danger, and risk are bandied around by safety professionals as well as by lay people when discussing unsafe situations and behaviors. Although there is some diversity of opinion on the specific definitions of these terms, most safety professionals would probably accept the following (Christensen, 1987):

Hazard: a condition or set of circumstances that has the potential of causing or contributing to injury or death. An electric circular saw contains several hazards such as a sharp spinning blade, electricity, and flying particles of the wood being cut. One might think that the more hazards a product or situation contains, the greater its perceived hazardousness would be, but this does not appear to be how laypeople conceive of the word "hazardousness."

Risk: the probability or likelihood of injury or death.

Danger: the product of hazard and risk.

Confusion occurs when dealing with these terms, especially when talking about reducing accidents and injuries. Does putting a guard on a circular saw reduce the hazard, the risk, or both? Would it be more correct to say that it reduces the risk by eliminating a hazard?

To add to the confusion, there is considerable evidence that laypeople do not define or conceptualize these terms the same way as safety professionals do. For example, Wogalter, Desaulniers, and Brelsford (1987) found that people's perception of the hazardousness of various products was related to their perceptions of the severity of injuries that could result from the product. The perceived likelihood of injury added little to the subjects' perceptions of hazardousness. Desaulniers (1989), however, found that the perceived likelihood of a *major* injury, but not a minor injury, was related to the perception of a product's hazardousness.

There are other factors that are related to people's perception of the hazardousness of a product (Wogalter, Desaulniers, and Brelsford, 1987). The more familiar the person is with the product and the more contacts the person has with the product, the less hazardous the product is perceived to be. Also, the more technologically complex the product, the more hazardous it is perceived to be.

Slovic, Fischhoff, and Lichtenstein (1979, 1980) suggest that people's perception of risk is determined by a combination of severity of injury and likelihood of injury. Finally, Young, Brelsford, and Wogalter (1990) found that perceptions of hazardousness, risk, and danger were all highly correlated when college students evaluated common products. One consequence of this state of affairs is that safety professionals may not be communicating risk information to the public in a way that is meaningful and apt to influence perceptions.

Evaluation of Risk

People's ability to estimate the risk (likelihood of injury) or the frequency of injury associated with technological risks (such as, pollution or nuclear power) has been extensively studied by Slovic, Fischhoff, and Lichtenstein (1980). Risk perception for common accidents and injuries has been less thoroughly studied.

There is evidence that people are pretty good at estimating the *relative* risk involved in using various products. That is, for example, they recognize that television sets are associated with fewer injuries than lawn mowers. How good they are, of course, depends on the products they have to rank. However, when asked to estimate the actual frequency of injury associated with various products, people tend to overestimate relatively low frequencies and underestimate relatively high frequencies. This effect is illustrated in Table 20-3. College students estimated the annual frequency of emergency room injuries associated with each of the categories shown in the table. As a base of reference, the subjects were told that annually, approximately 115,000 emergency room injuries were associated with swimming pools and accessories. After a series of tasks which included generating accident scenarios for each of the product categories and being read a list of all reasonable scenarios, the subjects estimated the frequency of emergency room injuries for each category. The results are shown in Table 20-3. The low frequencies tended to be overestimated and the high frequencies tended to be underestimated.

There are several sources of bias that appear to influence perceptions of risk. People appear to overestimate the value of their own experience. If they have not been injured or have not known people who were injured, they tend to underestimate the risk involved. Tversky and Kahneman (1973) refer to an *availability heuristic* that people use to estimate risks. People tend to give higher probabilities to events they can easily remember (that is, are available in memory). News stories also bias perceptions of risk. People grossly overestimate risks associated with situations or products that have received considerable attention in the media, although the media tends to report on unusual risks rather than normal risks. (The risk of being injured in a car accident is much higher than being injured in a commercial plane crash, yet look at the media coverage given to plane crashes versus automobile accidents.)

People tend to adopt an "It can't happen to me" bias when assessing risks and are often overconfident of their ability to avert injury. For example, various surveys have revealed that from 75 to 90 percent of automobile drivers believe that they are above average in their driving skill (Christensen, 1987).

A number of factors influence the acceptability of risk (Slovic, 1978; Starr, 1969). For example, people will accept a higher level of risk if the activity is voluntary, the level of risk is controllable, the hazards are known and understood, and the consequences are immediate. This may explain, in part, why

TABLE 20-3
ACTUAL AND ESTIMATED ANNUAL EMERGENCY ROOM INJURY FREQUENCIES ASSOCIATED WITH VARIOUS HOUSEHOLD PRODUCTS

Category	Actual frequency	Estimated frequency	Over or under estimate
Fans	14,078	38,478	+ 24,400
Television sets	24,962	26,361	+ 1,399
Razors and shavers	37,859	59,694	+ 21,835
Cooking ranges	48,026	165,366	+117,340
Lawn mowers	72,543	72,593	+ 50
Bathtubs and showers	100,618	136,628	+ 36,010
Glass bottles and jars	101,678	132,663	+ 30,985
Glass doors and windows	205,537	67,958	−137,579
Cutlery and knives	350,075	145,971	−204,104
Bicycles	556,682	127,403	−429,279

Source: Adapted from Breıns, 1986, Table 1. Reprinted with permission of The Human Factors Society, Inc. All rights reserved.

people become so concerned about risks associated with nuclear or electromagnetic radiation. The risk is perceived to be outside their control, it is not well understood, it cannot be seen or felt, and the consequences are slow to materialize.

Risk Perception and Accidents

The model of accident causation presented in Figure 20-3 stresses the importance of hazard perception in the occurrence of accidents (remember, for most people, the terms *hazard* and *risk* have similar connotations). An example of a study that focused on hazard perception and accidents was carried out by Lawrence (1974) in an investigation of 405 South African gold-mining accidents. The percentage of accidents attributed to factors related to hazard and risk perception were as follows:

Failed to perceive hazard	36%
Underestimated hazard	25%
Failed to respond to a recognized hazard	17%
Responded to hazard, but ineffectively	14%

Altering Hazard and Risk Perception

It appears that many accidents occur because people do not recognize or underestimate the hazards and risks involved in a situation or course of action. One way to alter a person's perceptions of hazards and risks is through training. Another way is through the use of safety communications. Two general classes of safety communication can be distinguished, although the distinction can become blurred. The first is safety information usually presented with posters, information sheets, or news articles. The purpose is to promote safe behavior or heighten awareness of general classes of hazards. The second type of communication is specific warnings such as found on countless products. The purpose is also to promote safe behavior, but more so to alert people of the existence of a specific hazard and to provide information regarding the consequences of not heeding the warning. So much has been written about the design and effectiveness of warnings that we have devoted the next section of this chapter to the topic. Here we deal with the effectiveness of safety communications (called propaganda by some).

On the basis of an analysis of various studies dealing with safety communications directed toward workers, Sell (1977) emphasizes (1) the need to try to modify the attitudes of workers that, in turn, influence their behavior and (2) for any communication to have an effect, it must be perceived and understood by the individuals in question. Drawing further from his survey, he believes that such communications, to be really effective, must:

1 Be specific to a particular task and situation.
2 Back up a training program.

3 Give a positive instruction.
4 Be placed close to where the desired action is to take place.
5 Build on existing attitudes and knowledge.
6 Emphasize nonsafety aspects.

The communication should not:

1 Involve horror, because in the present state of our knowledge this appears to activate defense mechanisms in the people at whom the communication is most directed.
2 Be negative, because this can show the wrong way of acting when what is required is the correct way.
3 Be general, because almost all people think they act safely. This type of communication is thus seen as relevant only to other people.

Studies that have assessed the effectiveness of safety communications (or campaigns) have produced mixed results (Lehto and Miller, 1986). Often the introduction of safety communications is coupled with reinforcement of safe behavior, training or counseling sessions, or free safety products. The range of behaviors targeted include use of seat belts, reduction of infant falls, reduction of burn injuries, and safe operation of cranes. Given these variations between studies, it is no wonder that results have been mixed.

Saarela, Saari, and Aaltonen (1988), for example, found no effect on accidents in a shipbuilding company of posting safety slogan signs such as "Keep gangway clear," "Wear your safety helmet," or "Safety is our common interest." Saarela (1989), however, found that a safety campaign dealing with hazards associated with the use of scaffolds that included a poster, information circular, and newsletter article increased the level of awareness among workers of scaffold-related hazards. There was also some evidence of improved scaffold safety practices among the workers as well. Before the safety campaign, only 48 percent of scaffolds and related items observed were considered safe. After the campaign, 76 percent of the observed items were considered safe.

In another study carried out in a steel mill (Sell, 1977), safety posters were used to encourage crane slingers to hook back the chain slings onto the crane hook when they were not in use, as a safety precaution. Three types of large posters with instructions and illustrations to "hook that sling" were posted in relevant areas where gantry cranes were used. The percentage of slings hooked back before and after posting of the signs, as shown in Figure 20-6, shows a systematic difference, with the improvement actually carrying over to a follow-up period. Such data suggest that posters can be useful in changing safety behavior if they are used where workers have the opportunity to control the occurrence of accidents by their own actions.

It appears that safety information campaigns can alter awareness and behavior. Oftentimes the effect is small and short-lived. To be effective, however, information campaigns should be reinforced with training, feedback, and encouragement from management.

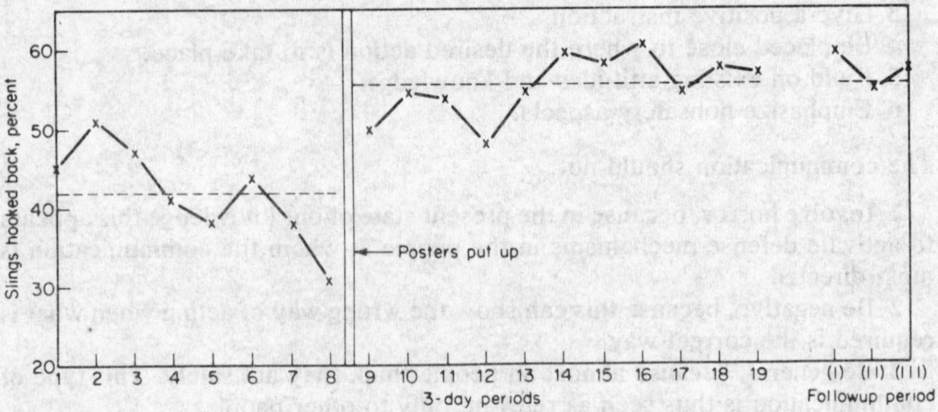

FIGURE 20-6
Percentage of crane slings hooked back as a safety precaution by crane operators in a British steel mill before and after the posting of safety posters. (*Source: Sell, 1977, Fig. 3, p. 212. This figure appeared as Fig. 3 of vol. 8, no. 4, p. 212, of* Applied Ergonomics, *published by IPC Science and Technology Press, Ltd., Guilford, Surrey, U.K.*)

WARNINGS

Warnings are becoming more and more prevalent in our environment. Even as your author sits quietly in his office writing this chapter, he is confronted with warnings of danger. A bottle of correction fluid warns that if the contents are deliberately concentrated and inhaled, death could result. A bottle of rubber cement warns that the contents are EXTREMELY FLAMMABLE, and further, prolonged exposure may result in permanent damage to the nervous system. There are many reasons why warnings are appearing on more and more products: scientists are more aware of the hazards contained in products we use; there is a greater concern for public health and safety among consumers, government officials, and, yes, even manufacturers; and there is a desire among manufacturers to reduce their legal liability when sued for failure to warn.

Some manufacturers are reluctant to place warnings on products because they think warnings about product hazards will deter consumers from purchasing the product. Evidence, however, suggests otherwise. Ursic (1984), for example, found that products with a warning were perceived as safer than the same product without the warning. Ursic postulates that this may be because people take the warning as an indication that the manufacturer is careful in producing and marketing the product. Laughery and Stanush (1989) found results supporting Ursic's findings; further, they found that the more explicit the warning, the safer the product is perceived to be.

Previously, we made a distinction between warnings and safety information. We should also make a distinction between warnings and instructions, although it is rather fuzzy. *Warnings* inform the user of the dangers of improper use and, if possible, tell how to guard against those dangers. *Instructions* tell the user how to use the product effectively. In addition, instructions themselves may, and often do, contain warnings. We concentrate on warnings and indicate some of the human factors and legal issues involved in their use.

There are three basic approaches for making a product safer:

- Design the dangerous feature out of the product.
- Protect against the hazard by guarding or shielding.
- Provide adequate warnings and instructions for proper use and reasonable foreseeable misuse.

In general, designing the dangerous feature out of a product is the most effective means of making the product safer. Often, however, it is not possible or economically feasible to do that. In such cases guarding and shielding should be implemented where possible. Warnings should be considered after the first two design methods have been applied and where unreasonable dangers still exist.

Purposes of Warnings

We can identify four principal purposes for warnings: (1) Inform the users or potential users of a hazard or danger, of which they may not be aware, that is inherent in the use or reasonably foreseeable misuse of the product. (2) Provide users or potential users with information regarding the likelihood and/or severity of injury from the use or reasonably foreseeable misuse of the product. (3) Inform users or potential users how to reduce the likelihood and/or severity of injury. (4) Remind users of a danger at the time and place where the danger is most likely to be encountered.

Designing Warnings

Designing appropriate warnings is a complex task, and where the risks of serious injury or death are involved, warnings should be tested for effectiveness by using people representative of the foreseeable user population. We cannot hope to discuss in detail all the various aspects of warning design; however, we outline the major sorts of considerations involved.

The ultimate aim of a warning is to alter behavior by encouraging the user either to not engage in a particular act or to change the manner in which the act is performed. For a warning to change behavior, the warning must be sensed (seen or heard), received (read or listened to), understood, and finally heeded. Each of these steps involve human factors design considerations.

Sensing a Warning

The warning must catch the attention of the consumer under the circumstances in which the product will be used. The following considerations are important:

size; shape; color; graphical design; contrast; placement; use of "active" attention getters such as bells, waving flags, or blinking lights; and the physical durability of the warning itself to such things as weather and physical abuse. In addition, in Chapter 6 we discuss some design considerations for auditory warnings.

Receiving a Warning

Sensing the presence of a warning does not ensure that it will be read or listened to. For example, Otsubo (1988) found that 74 percent of her subjects noticed a warning on a circular saw, but only 52 percent of the subjects read it (and it only contained seven words!). With visual warnings, the length of the warning may influence the willingness of a user to read it. Probably, the longer and more complicated the warning, the less likely it is that people will invest the time and effort required to read it. Desaulniers (1987), for example, found that subjects were more likely to read multiple warnings if they were arranged in an outline format rather than in a paragraph format. The perceived level of hazard inferred by the warning may also influence whether it is read. The warning should, of course, be commensurate with the dangers inherent in using the product. Traditionally, three levels of hazard have been differentiated:

- *Danger* is used where there is an immediate hazard which, if encountered, will result in severe personal injury or death.
- *Warning* is the signal word for hazards or unsafe practices which *could* result in severe personal injury or death.
- *Caution* is for hazards or unsafe practices which could usually result in minor personal injury, product damage, or property damage.

Presumably a person would be more likely to read a warning that started out "DANGER! THIS PRODUCT CAN KILL YOU!" than one that said "NOTE: IF THIS PRODUCT IS USED IN AN IMPROPER MANNER, IT IS POSSIBLE THAT INJURY COULD OCCUR."

The degree of familiarity with a product and its perceived hazardousness are probably among the most potent variables affecting whether people will read warnings (Donner and Brelsford, 1988; Friedmann, 1988; Godfrey and Laughery, 1984; LaRue and Cohen, 1987; Otsubo, 1988; Wogalter, Desaulniers, and Brelsford, 1986). The more familiar people are with a product, the more confident they are of their ability to use the product safely; and the less dangerous the product appears, the less likely it is that they will read permanently affixed warnings. As just one example of this principle, Otsubo (1988) found that while 52 percent of her subjects read a warning attached to a circular saw, only 25 percent of another group of subjects read the same warning when attached to a less hazardous jigsaw. There is some evidence, however, that people will read a *temporarily* posted warning that informs them of the development of a hazard in a familiar product or situation (Godfrey, Rothstein, and Laughery, 1985).

Understanding the Warning The words and/or graphics of a warning must be understood by the user population. If words are used, they must be chosen carefully and tested for comprehension. Do not use vague, ambiguous, or ill-defined terms; highly technical words or phrases; double negatives; or long, grammatically complex phrasing. Consideration must also be given to the primary languages of the user population. Symbols must be carefully chosen to convey the intended message, and this can be more difficult than one may think. Collins (1983), for example, tested comprehension of standard hazard symbols among mining industry personnel. In several cases respondents interpreted the symbol to mean the opposite of its intended meaning. The symbol for corrosive hazard, for example, was thought by 21 percent of the people to denote "emergency hand-wash location!" The symbol for emergency eye-wash location was interpreted by 24 percent of the people to mean "eye irritant located here"!

At a minimum, a warning should contain the following fundamental elements:

- *Signal word:* to convey the gravity of the risk, for example, "danger," "warning," "caution"
- *Hazard:* the nature of the hazard
- *Consequences:* what is likely to happen if the warning is not heeded
- *Instructions:* appropriate behavior to reduce or eliminate the hazard

An example of a minimum warning would be (Wogalter, Desaulniers, and Godfrey, 1985)

> DANGER
> HIGH VOLTAGE WIRES
> CAN KILL
> STAY AWAY

How specific a warning must be in spelling out hazards is a difficult question. Courts have often opted for a great deal of specificity in warnings. For example, the following warning was found to be inadequate in a court case:*

> Caution: Flammable mixture. Do not use near fire or flame. Caution! Warning! Extremely Flammable! Toxic! Contains naphtha, acetone, and methyl ethyl ketone. Although this adhesive is no more hazardous than drycleaning fluids or gasoline, precautions must be taken. Use with adequate ventilation. Keep away from heat, sparks, and open flame. Avoid prolonged contact with skin and breathing of vapors. Keep container tightly closed when not in use.

* *Florentino v. A.E. Dtaley Mfg. Co.* Massachusetts 416 N.E. 2d, p. 998.

It was found to be inadequate because it did not mention the danger of using the product in the vicinity of closed and concealed pilot lights (Kreifeldt and Alpert, 1985).

There is a problem, however, in warning about all possible hazards inherent in the use and reasonable foreseeable misuse of a product. The problem is *warning overload*. That is, if there is a warning against every conceivable hazard, the effect of all the warnings may be so diluted as to make any one of them ineffective. Take, for example, a large industrial machine with electrical hazards, flammable liquids, high-pressure lines, sharp edges, pinch points, slippery surfaces, etc. Warning against all such hazards and, in addition, reasonable foreseeable misuses could result in virtually the entire machine being plastered with warning labels. It is doubtful that people working in and around such a machine would pay much attention to so many warnings.

Heeding the Warning Just because a warning is sensed, received, and understood does not ensure that a person will heed or comply with the warning. These are some factors that may influence whether people heed a warning: whether the user is capable of performing the action required; whether the warning is remembered at the time and place where the action is required; whether heeding the warning will result in additional time, cost, inconvenience, or discomfort to users; and the riskiness and general carelessness of the individual.

Effectiveness of Warnings

The ultimate measure of effectiveness is whether the warning was heeded by the user. Unfortunately, there has not been a great deal of research assessing the effectiveness of warnings by using behavioral measures. Most research on warning effectiveness assesses the perceived effectiveness by asking subjects which of several warnings they *think* would be most effective. There are, however, some behavioral studies which may shed some light on the issue of effectiveness.

Warnings in Instructions Warnings properly placed in instruction manuals that are used by people to assemble, operate, or repair a product can be quite effective. This is due, in part, to the fact that people usually read instructions when they are not confident about their ability to do the task and are unfamiliar with the procedures. Under such conditions people are more likely to notice and read the warnings. Wogalter et al. (1987), for example found that placing a warning ("Wear rubber gloves and mask") at the beginning of an instruction sheet used to perform a chemistry experiment resulted in 80 percent of the subjects complying with the warning. When the warning was placed at the end of the instructions, however, only 50 percent complied. With no warning present, only 10 percent used the gloves and mask. Using a survey technique, the Consumer Product Safety Commission (1980) reported that 69 percent of people who felt they had been exposed to warning labels or hazard-avoidance instructions related to installing CB antennas near power lines said that the warnings had caused them to consider the proximity of the power lines when selecting an installation site. There was also evidence that CB antennas installed *before* warning labels and hazard-avoidance instructions were made mandatory were mounted *closer* to power lines than were CB antennas installed *after* the warnings and instructions became mandatory.

Warnings within instruction, however, may be effective only when the instructions are being used. Unfortunately, people, for a variety of reasons, do not always read instructions (see for example, Figure 20-7). Further, even if instructions are read, if a product is used again without rereading of the instructions, there is a good probability that the warning may be forgotten. For this reason, a warning is often placed on the product itself.

On-Product Warnings The literature addressing the effectiveness of on-product warnings indicates that people often do not heed such warnings. Dorris and Purswell (1977) reported that of 100 subjects, not one even noticed a warning placed on the handle of a hammer they were asked to use to perform a task. The widely recognized lack of effectiveness of the warning on cigarette packages* ("Warning: The Surgeon General has determined that cigarette smoking is dangerous to your health.") has often been cited as evidence for the general lack of effectiveness of product warnings (McCarthy et al., 1984). To be fair, however, there are features of cigarette smoking that are not at all comparable to other on-product applications of warnings. Cigarette smoking, for example, may be addictive, it is pleasurable (at least to those who do the smoking), and considerable advertising is aimed at promoting smoking. In how many other situations is noncompliance with a warning addictive, pleasurable,

* New warnings have been mandated, but their effectiveness has not, as yet, been assessed.

FIGURE 20-7
People, for a variety of reasons, do not always read instructions. (*Reprinted with the permission of Universal Press Syndicate, 1988.*)

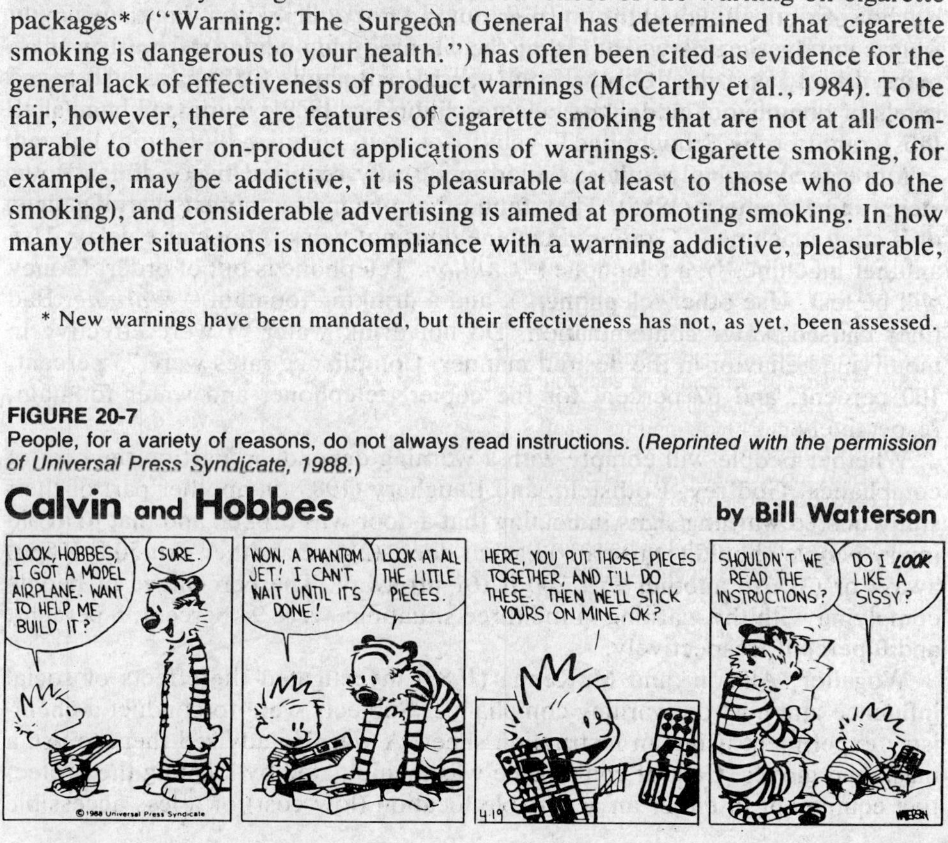

and encouraged in the media? The lack of effectiveness of cigarette warnings, therefore, may be a special case rather than a general indictment of on-product warnings.

In a revealing study of on-product warnings Strawbridge (1986) had subjects scan a label on the back of a bottle of liquid adhesive to determine whether the adhesive could be used to glue two specific materials and if so, to do so. The label contained the following warning:

DANGER: Contains acid. To avoid severe burns, shake well before opening.

Various conditions were tested including placement of the warning at the top, middle, or bottom of the label; highlighting with inverse type (white letters on a black background); and embedding the warning in a paragraph of text or starting the paragraph with the warning. Surprisingly, the addition of highlighting and the position of the warning (top-middle-bottom) had little effect on compliance. Embedding the warning, however, did reduce compliance. For the nonembedded conditions, 94 percent of the subjects noticed the warning, but only 81 percent read it. Even more surprising, although 81 percent of the subjects read the warning, only 47 percent complied with it. The reason given by the subjects who read the warning but did not comply was that they simply forgot; yet the time between reading the warning and opening the bottle was no more than 10 seconds! Otsubo (1988) and Friedmann (1988) have also found levels of compliance under the best conditions in their experiments similar to those reported by Strawbridge.

More encouraging results came from a study reported by Godfrey, Rothstein, and Laughery (1985). They found that placing temporary warning signs on a copy machine (''*Caution:* Machine does not work. May cause delay. Use another machine.''), a telephone (''*Caution:* Telephone is out of order. Money will be lost. Use other telephone.''), and a drinking fountain (''*Warning:* Bad filter caused water contamination. Do not drink water.'') were effective in modifying behavior in the desired manner. Compliance rates were 73 percent, 100 percent, and 67 percent for the copier, telephone, and water fountain, respectively.

Whether people will comply with a warning depends in part on the cost of compliance. Godfrey, Rothstein, and Laughery (1985) in another part of their study posted warning signs indicating that a door was broken and that to avoid injury people should (1) use an adjacent door, (2) use another exit 50 ft (15 m) away, or (3) use another exit 200 ft (61 m) away. The percentage of people complying with the warning in the three situations were 94 percent, 6 percent, and 0 percent, respectively.

Wogalter, Allison, and McKenna (1989) investigated the effects of social influence and cost on warning compliance. Subjects were to conduct a chemistry experiment using an instruction sheet. A warning advised them to use a mask and gloves. Cost of compliance was manipulated by locating the protective equipment in either an accessible location (low cost) or a less accessible location (high cost) Social influence was manipulated by the presence of a confederate who either did or did not comply with the warning. In support of previous research, compliance was lower in the high-cost situations (overall 28 percent complied) than in the low-cost situations (68 percent). Further, when the confederate complied (combining high- and low-cost situations) 83 percent complied compared to 16 percent when the confederate did not comply. The combined effect of the variables was dramatic. When cost was low and the confederate complied, all the subjects complied, but when the cost was high and the confederate did not comply, none of the subjects complied.

PRODUCT LIABILITY

The production and the sale of products (be they hair dryers or nuclear power plants) are not necessarily the end of the line for the producer or seller. For certain items the producer or seller maintains relationships with the buyers for service, maintenance, or repair, for example. In recent years in the United States, the matter of product liability has taken on increasing importance, to the point that producers and sellers must give advance attention to such liability in designing, producing, and selling products. The liability issue applies equally to products that are used in industry (machines, tools, etc.) and to those purchased by consumers (household appliances, utensils, etc.).

Product liability is the legal term used to describe an action in which an injured party (the plaintiff) seeks to recover damages for personal injury or loss of property from a manufacturer or seller (the defendant) because the plaintiff believes that the injuries or damages resulted from a defective product. Today, more than ever, people look to the courts for redress when they suffer injury or damage. This attitude has been acutely felt in the area of product liability. Each year more and more cases are brought to court for adjudication. This growth in product liability cases has created a greater demand for human factors experts, both in the initial design to make products safer and in the courtroom as expert witnesses.

Human factors specialists have served as expert witnesses in cases involving such products as automobiles, forklift trucks, medical equipment, boats, conveyers, industrial machinery, railroad crossings, CB antennas, and even stairs. The issues that human factors experts have addressed in such cases include hearing discrimination, visual acuity, visual illusions, eye-hand coordination, reaction time, temperature sensitivity, biomechanics of lifting, reading speed, control and display design, and the adequacy and effectiveness of warnings and instructions. Several cases have been decided principally on the testimony of human factors experts (Wallace and Key, 1984).

In this section we present some basic concepts in product liability that are relevant to human factors people. We do not dwell on the innumerable nuances of the law involved in product liability cases; our intention here is not to make the reader a lawyer. (God knows we have enough lawyers already!) Product liability is case law; that is, each new court decision adds, changes, clarifies, or sometimes clouds the accumulated legal precedents. Each state has different legal precedents; hence, cases may be tried and decided differently in different states. Even as you read this, a case may be decided that drastically alters the nature and course of product liability cases. In addition, Congress has been considering, for a number of years, a comprehensive product liability bill that would alter the scope and nature of product liability litigations. The fate of the bill, however, is uncertain.

Product liability cases are usually tried under one of the following bodies of law: (1) *negligence,* which tests the conduct of the defendant; (2) *strict liability,* which tests the quality of the product; (3) *implied warranty,* which also tests the quality of the product; and (4) *express warranty* and *misrepresentation,* which tests the performance of a product against the explicit representations made about it by the manufacturer or sellers.

Product liability cases typically involve three types of defects: manufacturing defects, design defects, and warning defects. The Interagency Task Force on Products Liability (1977) reported that 35 percent of all product liability cases involve manufacturing defects, 37 percent involve design defects, and 18 percent involve warning defects. The absence of a proper warning can easily be seen as a kind of design defect. Weinstein et al. (1978), for example, repeatedly point out the trade-off between providing warning to decrease the danger of a product and designing to reduce the danger. They rightly stress the desirability of designing dangers out of the product rather than warning of their presence.

Making a Case

Certain assertions must be established to make a product liability case, regardless of the body of law under which the case is tried (Weinstein et al., 1978). It must be established that the product was defective in manufacture or design. This requires a definition of when a product is defective. We discuss this further in the next section. Suffice it to say, however, that a product can be dangerous without being defective; a knife is a good example. It must be established that the product was defective at the time the product left the defendant's hands. Products are often abused by users, and through abuse the product becomes defective. One might think that in such a case the manufacturer would not be held liable; actually, the manufacturer could still be held liable if the abuse was foreseeable. The mere presence of a defect in a product at the time of injury is not enough to make a case. It must be established that the defect was involved in the injury. The injury, after all, may have had no relationship whatever to the defect. A closely related, and more difficult, question that must also be addressed is whether the defect actually caused the harm. Here the *but-for test* is applied: "But for the presence of the defect, product failure, or malfunction, would the injury have occurred?" Imagine, for example, that a car is sold with defective brakes. The plaintiff was driving down a street, encountered a patch of ice, slammed on the brakes, which failed, and hit the car in front. Is this a product liability case? Not necessarily; the court could find that even with no defect, slamming on the brakes would not have stopped the car under the icy conditions.

Where a product is found to lack a warning, a case may turn on whether the person under the circumstances would have heeded it. That is, even if there was a warning, would it have prevented the injury?

Let us now discuss how to determine when a product is defective. The rules here have changed over the years and will undoubtedly continue to do so. However, a few common threads emerging from recent decisions probably will set the trend for the near future at least.

When Is a Product Defective?

We said that a product can be dangerous without being defective in design. In fact, for many years the legal precedent, established in *Campo v. Scofield** held that a manufacturer was not responsible for dangers in a product that were open and obvious. This came to be called the *patent-danger rule,* i.e., a patently obvious danger. It was not until 26 years later that the patent-danger rule was rejected in the landmark case of *Micallef v. Miehle Company.*† The plaintiff was employed as an operator on a huge photo-offset printing press. One day he discovered a foreign object on the printing plate, called a *hickie* in the trade, which causes a blemish on the printed page. To correct the situation, the plaintiff informed his supervisor that he intended to ''chase the hickie,'' a common practice wherein a piece of plastic is inserted against the plate which is wrapped around a cylinder that spins at high speed. The plastic caught and drew the plaintiff's hand into the unit between the cylinder and an ink roller. The machine had no safety guards to prevent such an occurrence, and the plaintiff was unable to stop the machine quickly because the shut-off button was distant from his position at the machine.

The plaintiff was fully aware of the obvious danger, but it was the custom to ''chase hickies on the run'' because once the machine was stopped, it required 3 h to start it up again. The court, if it maintained the patent-danger rule, would have been forced to deny payment to the plaintiff. The court, however, rejected the rule and stated that it would judge a product for *reasonableness*. Therefore, a product is defective if it presents an *unreasonable danger* to the user.

The question, then, is, What is unreasonable? or What is reasonable? First, most products present some risks to the user. These risks, however, must be balanced against the functions the product performs and the cost of providing for greater safety. The California supreme court, in *Barker v. Lull Engineering Company,*‡ went even further in specifying the conditions under which a product would be found defective in design. The court set out a two-pronged

* *Campo v. Scofield,* New York, vol. 301, p. 468, Northeastern 2d, vol. 95, p. 802 (1950).

† *Micallef v. Miehle Company,* New York 2d, vol. 39, p. 376, Northeastern 2d, vol. 348, p. 571, New York Supplement 2d, vol. 384, p. 115 (1976).

‡ *Barker v. Lull Engineering Company,* California 3d, vol. 20, p. 413, California P. 2d, vol. 573, p. 431, California Reporter vol. 143, p. 225 (1978).

test. A product would be found defective (1) if it failed to perform safely as an ordinary user would expect when it was used in an intended or *reasonably foreseeable manner* (emphasis added) or (2) if the risks inherent in the design outweighed the benefits of that design. In determining whether the benefits of the design outweigh such risks the jury may consider, among other things, the gravity of the danger posed by the design, the likelihood that such danger would cause damage, the feasibility of a safer alternate design at the time of manufacture, the financial cost of an improved design, and the adverse consequences to the product and the user that would result from an alternate design.

The idea that a manufacturer can be held liable even if a user abuses or misuses a product, as long as the abuse or misuse was reasonably foreseeable, has considerable implications for human factors. Human factors deals with human behavior and how people respond in certain situations. The courts have now extended their concern to include consideration of what unusual, perhaps even harebrained, things people might do with a product for which it was not intended. Weinstein et al. (1978) present the example of a hair dryer. Is it reasonable to expect that a hair dryer might be used to defrost a freezer, or thaw frozen water pipes, as well as dry things other than hair? It will be used in humid environments and near water. Even in drying hair, it may be reasonable to foresee the user sitting in a bathtub when he or she turns on the unit. Surely it is foreseeable that users will be male and female young and old, and have a broad range of manual dexterity and levels of understanding and awareness.

*Ritter v. Narragansett Electric Company** is a case that hinged on the issue of foreseeable use. The defendant manufactured a small freestanding 30-in (76-cm) gas range. The plaintiff, a 4-year-old girl, opened the oven door and used it as a step stool to look into a pot on top of the stove. The stove tipped forward, seriously injuring the plaintiff. Expert testimony concluded that the oven door could not hold a weight of 30 lb (14 kg) without tipping. The issue was whether the use of the open door as a step stool was so unforeseeable that the manufacturer should not be held liable. Note, however, that had the stove tipped because a homemaker used the open door as a shelf for a heavy turkey, there probably would have been no question of the manufacturer's liability since it could be argued that one of the intended uses of the door was a shelf for checking food during preparation. But, as a step stool, was that foreseeable? The jury said yes and attached liability to the manufacturer.

One last point before we leave the question of when a product is defective. Designing a product to meet government or industry standards does not guarantee that it will not be found unreasonably dangerous. Weinstein et al. (1978) point out that traditionally courts have taken the position that all standards provide, at best, lower limits for product acceptability. Consider a case in point, *Berkebile v. Brantly Helicopter Corporation.*† The plaintiff took off in a

* *Ritter v. Narragansett Electric Company,* Rhode Island, vol. 109, p. 176, Atlantic 2d, vol. 285, p. 255 (1971).

† *Berkebile v. Brantly Helicopter Corporation,* Pennsylvania Superior, vol. 219, p. 479, Atlantic 2d, vol. 281, p. 707 (1971).

helicopter with a nearly empty gas tank. Shortly after takeoff the helicopter crashed. The charge of defect was that the helicopter system of autorotation (wherein the aircraft "glides" to the ground) required that the pilot be able to throw the helicopter into autorotation within 1 s. The plaintiff contended that this was too short a time and that this was the cause of the crash. The defendant argued that the 1-s time met Federal Aviation Administration regulations. The court agreed with the defendant, but the appeals court stated that "such compliance (with regulations) does not prevent a finding of negligence where a reasonable man [and we assume woman] would take additional precautions."

Designing a Reasonably Safe Product

Products must be designed for reasonably foreseeable use, not solely intended use. This requires that an analysis be made to determine the types of use and misuse a product could be subjected to; this may require a survey of users of similar products or potential users of a new product. Laboratory simulation tests might also be conducted on product prototypes to gain insight into user behaviors and product performance. Weinstein et al. (1978) list seven steps that should be included in the design process of a product to ensure that a reasonably safe product evolves:

1 Delineate the scope of product uses.
2 Identify the environments within which the product will be used.
3 Describe the user population.
4 Postulate all possible hazards, including estimates of probability of occurrence and seriousness of resulting harm.
5 Delineate alternative design features or production techniques, including warnings and instructions, that can be expected to effectively mitigate or eliminate the hazards.
6 Evaluate such alternatives relative to the expected performance standards of the product, including the following:
  a Other hazards that may be introduced by the alternatives
  b Their effect on the subsequent usefulness of the product
  c Their effect on the ultimate cost of the product
  d A comparison to similar products
7 Decide which features to include in the final design.

DISCUSSION

We covered a lot of ground in this chapter under the topics of human error, accidents, perception of risk, warnings, and product liability. Improving safety and reducing human errors are central criteria used by human factors practitioners in designing and evaluating equipment, facilities, procedures, etc. We discussed methods for improving safety including training, feedback, and warnings. It is important, however, not to lose sight of the fact that it is far better to design out hazards than to rely on techniques of behavioral modification and persuasion to improve safety. This is not to say that such techniques should not be employed; they should and they can be effective, but they are not as effective as eliminating the hazard through design. Remember, it is easier to bend metal than to twist arms (Sanders' Law).

REFERENCES

Beaumont, P. (1980). Analysis of the problem of industrial accidents in Britain. *International Journal of Manpower,* 1(1), 28–32.

Brems, D. (1986). Risk estimation for common consumer products. *Proceedings of the Human Factors Society 30th Annual Meeting*. Santa Monica, CA: Human Factors Society, pp. 556–560.

Broberg E. (1984). Use of census data combined with occupation and accident data. *Journal of Occupational Accidents,* 6, 147–153.

Brown, I. (1990). Accident reporting and analysis. In J. Wilson and E. Corlett (eds.), *Evaluation of human work*. London: Taylor & Francis, pp. 755–778.

Christensen, J. M. (1980). Human factors in hazard/risk evaluation. In *Proceedings of the Symposium—Human Factors and Industrial Design in Consumer Products*. Medford, MA: Tufts University, pp. 442–447.

Christensen, J. (1987). Comments on products safety. *Proceedings of the Human Factors Society 31st Annual Meeting*. Santa Monica, CA: Human Factors Society, pp. 1–14.

Cohen, A., Smith, M. J., and Anger, W. K. (1979). Self-protective measures against workplace hazards. *Journal of Safety Research,* 11(3), 121–131.

Collins, B. (1983). *Use of hazard pictorials/symbols in the minerals industry* (NBSIR 83–2732). Washington, DC: Department of Commerce, National Bureau of Standards.

Consumer Product Safety Commission (1980, December). *Evaluation of the impact of the communications antenna labeling rule*. Washington, DC.

DeGreen, K. (1972). *Systems psychology*. New York: McGraw-Hill.

Desaulniers, D. (1987). Layout, organization and the effectiveness of consumer product warnings. *Proceedings of the Human Factors Society 31st Annual Meeting*. Santa Monica, CA: Human Factors Society, pp. 56–60.

Desaulniers, D. (1989). Consumer product hazards: What will we think of next? *Proceedings of the Interface 89 Annual Meeting*. Santa Monica, CA: Human Factors Society, pp. 115–120.

Donner, K., and Brelsford, J. Jr. (1988). Cuing hazard information for consumer products. *Proceedings of the Human Factors Society 32d Annual Meeting*. Santa Monica, CA: Human Factors Society, pp. 532–535.

Dorris, A., and Purswell, J. (1977). Warnings and human behavior: Implications for the design of product warnings. *Journal of Products Liability*. 1, 255–264.

Edwards, D., and Hahn, C. (1980). A chance to happen. *Journal of Safety Research,* 12(2), 59–67.

Fitch, J., Templer, J., and Corcoran, P. (1974). The dimensions of stairs. *Scientific American,* 231, 82–90.

Friedmann, K. (1988). The effect of adding symbols to written warning labels on user behavior and recall. *Human Factors,* 30, 507–515.

Godfrey, S., Fontenelle, G., Brems, D., Brelsford, J. Jr., and Laughery, K. (1986).

Scenario analysis of children's ingestion accidents. *Proceedings of the Human Factors Society 30th Annual Meeting*. Santa Monica, CA: Human Factors Society, pp. 566–569.

Godfrey, S., and Laughery, K. (1984). The biasing effect on familiarity on consumers' awareness of hazard. *Proceedings of the Human Factors Society 28th Annual Meeting*. Santa Monica, CA: Human Factors Society, pp. 483–486.

Godfrey, S., Rothstein, P., and Laughery, K. (1985). Warnings: Do they make a difference? *Proceedings of the Human Factors Society 29th Annual Meeting*. Santa Monica, CA: Human Factors Society, pp. 669–673.

Hayashi, Y. (1985). Hazard analysis in chemical complexes in Japan—Especially those caused by human error. *Ergonomics*, 28(6), 835–841.

Heinrich, H. (1959). *Industrial accident prevention* (4th ed). New York: McGraw-Hill.

Interagency Task Force on Products Liability (1977). Final Rept. II-54. Washington, DC: Department of Commerce.

Irvine, C., Snook, S., and Sparshatt, J. (1990). Stairway risers and treads: Acceptable and preferred dimensions. *Applied Ergonomics*, 21(3), 215–225.

Kashiwagi, S. (1976). Pattern-analytic approach to analysis of accidents due to human error: An application of the ortho-oblique-type binary data decomposition. *Journal of Human Ergology*, 5(1), 17–30.

Kerr, W. (1957). Complementing theories of safety psychology. *Journal of Social Psychology*, 45, 3–9

Kessler, E., and Reiff, L. (1980, October). *A special study of ingestions to children under 5 years of age treated in NEISS hospital emergency departments*. Washington, DC: U.S. Consumer Product Safety Commission.

Komaki, J., Barwick, K. D., and Scott, L. R. (1978). A behavioral approach to occupational safety: Pinpointing and reinforcing safe performance in a food manufacturing plant. *Journal of Applied Psychology*, 63(4), 434–445.

Kreifeldt, J., and Alpert, M. (1985). Use, misuse, warnings: A guide for design and the law. In T. Kvalseth (ed.), *Interface 85: Fourth Symposium on Human Factors and Industrial Design in Consumer Products*. Santa Monica, CA: Consumer Products Technical Group of the Human Factors Society. pp. 77–82.

Lampert, U. (1974). Age and the predisposition to accidents. *Archives des Maladies Professionelles*, 62, 173.

LaRue, C., and Cohen, H. (1987). Factors affecting consumers' perceptions of product warnings: An examination of the differences between male and female consumers. *Proceedings of the Human Factors Society 31st Annual Meeting*. Santa Moncia, CA: Human Factors Society, pp. 610–614.

Laughery, K., and Stanush, J. (1989). Effects of warning explicitness on product perceptions. *Proceedings of the Human Factors Society 31st Annual Meeting*. Santa Monica, CA: Human Factors Society.

Lawrence, A. (1974). Human error as a cause of accidents in gold mining. *Journal of Safety Research*, 6, 78–88.

Lehto, M., and Miller, J. (1986). *Warnings*, vol. 1: *Fundamentals, design, and evaluation methodologies*., Ann Arbor, MI: Fuller Technical Publications.

Maritime Transportation Research Board (1976). *Human error in merchant marine safety*. Washington, DC: National Academy of Sciences.

McCarthy, R., Finnegan, J., Krumm-Scott, S., and McCarthy, G. (1984). Product information presentation, user behavior, and safety. *Proceedings of the Human Factors Society 28th Annual Meeting*. Santa Monica, CA: Human Factors Society, pp. 81–85.

McKenna, F. (1983). Accident proneness: A conceptual analysis. *Accident Analysis and Prevention,* 15, 65–71.

Meister, D. (1987). *Behavioral analysis and measurement methods*. New York: Wiley.

Mital, A., Fard, H., and Khaledi, H. (1987). A biomechanical evaluation of staircase riser heights and tread depths during stair-climbing. *Clinical Biomechanics,* 2, 162–164.

National Academy of Sciences (1982). *Toward safer underground coal mines*. Washington, DC: National Academy of Sciences.

Otsubo, S. (1988). A behavioral study of warning labels for consumer products: Perceived danger and use of pictographs. *Proceedings of the Human Factors Society 32d Annual Meeting*. Santa Monica, CA: Human Factors Society, pp. 536–540.

Paul, J. (1984). Stair safety: Review of research. *Proceedings of the 1984 International Conference on Occupational Ergonomics,* vol. 1: *Reviews*. Rexdale, Ont., Canada: Human Factors Association of Canada.

Petersen, D. (1984). *Human-error reduction and safety management*. New York: Aloray.

Product Safety and Liability Reporter (1983). Cited in Irvine, C., Snook, S., and Sparshatt, J. (1990), op. cit.

Ramsey, J. D. (1973). Identification of contributory factors in occupational injury. *Journal of Safety Research,* 5(4), 260–267.

Ramsey, J. (1985). Ergonomic factors in task analysis for consumer product safety. *Journal of Occupational Accidents,* 7, 113–123.

Rasmussen, J. (1979). Notes on human error analysis and prediction. In G. Apostalakis and G. Volta (eds.), *Synthesis and analysis methods for safety and reliability studies*. New York: Plenum.

Rasmussen, J. (1982). A taxonomy for describing human malfunction in industrial installations. *Journal of Occupational Accidents,* 4, 311–333.

Rasmussen, J. (1987). The definition of human error and a taxonomy for technical system design. In J. Rasmussen, K. Duncan, and J. Leplat (eds.), *New technologies and human error*. New York: Wiley.

Rouse, W., and Rouse, S. (1983). Analysis and classification of human error. *IEEE Transactions on Systems, Man, and Cybernetics,* SMC-13(4), 539–549.

Saarela, K. (1989). A poster campaign for improving safety on shipyard scaffolds. *Journal of Safety Research,* 20(40), 177–185.

Saarela, K., Saari, J., and Aaltonen, M. (1988). The effects of an informational safety campaign in the shipbuilding industry. *Journal of Occupational Accidents,* 10, 255–266.

Sanders, M., Patterson, T., and Peay, J. (1976). *The effect of organizational climate and policy on coal mine safety* (BuMines OFR 108–77). Pittsburgh, PA: Bureau of Mines.

Sanders, M., and Shaw, B. (1988). *Research to determine the contribution of system factors in the occurrence of underground injury accidents*. Pittsburgh, PA: Bureau of Mines.

Sell, R. G. (1977). What does safety propaganda do for safety? A review. *Applied Ergonomics,* 8(4), 203–214.

Senders, J. (1983). On the nature and source of human error. In R. Jensen (ed.), *Proceedings of the Second Symposium on Aviation Psychology*. Columbus, OH: Ohio State University.

Shahani, C. (1987). Industrial accidents: Does age matter? *Proceedings of the Human Factors Society 31st Annual Meeting*. Santa Monica, CA: Human Factors Society, pp. 553–557.

Shealy, J. (1979). Impact of theory of accident causation on intervention strategies. *Proceedings of the Human Factors Society 23d Annual Meeting*. Santa Monica, CA: Human Factors Society, pp. 225–229.

Slovic, P. (1978). The psychology of protective behavior. *Journal of Safety Research*, 10(2), 58–68.

Slovic, P., Fischhoff, B., and Lichtenstein, S. (1979). Rating of risks. *Environment*, 21, 14–39.

Slovic, P., Fischhoff, B., and Lichtenstein, S. (1980). Facts and fears: Understanding perceived risk. In R. Schwing and W. Albers, Jr. (eds.), *Societal risk assessment*. New York: Plenum.

Smith, M. J., Anger, W. K., and Uslan, S. S. (1978). Behavioral modification applied to occupational safety. *Journal of Safety Research*, 10(2), 87–88.

Starr, C. (1969, September). Social benefit versus technological risk. *Science*, 165, 1232–1238.

Strawbridge, J. (1986). The influence of position, highlighting, and imbedding on warning effectiveness. *Proceedings of the Human Factors Society 30th Annual Meeting*. Santa Monica, CA: Human Factors Society, pp. 716–720.

Suchman, E. (1961). On accident behavior. In *Behavioural Approaches to Accident Research*. Washington, DC: Association for the Aid to Crippled Children.

Swain, A., and Guttmann, H. (1983). *Handbook of human reliability analysis with emphasis on nuclear power plant applications* (NUREG/CR-1278). Washington, DC: Nuclear Regulatory Commission.

Tversky, A., and Kahneman, D. (1973). Availability: A heuristic for judging frequency and probability. *Cognitive Psychology*, 5, 207–232.

Ursic, M. (1984). The impact of safety warnings on perception and memory. *Human Factors*, 26, 677–682.

Wallace, W., and Key, J. (1984, December). Human factors experts: Their use and admissibility in modern litigation. *For the Defense*, 16–24.

Weinstein, A., Twerski, A., Piehler, H., and Donaher, W. (1978). *Products liability and the reasonably safe product*. New York: Wiley.

Wogalter, M., Allison, S., and McKenna, N. (1989). Effects of cost and social influence on warning compliance. *Human Factors*, 31(2), 133–140.

Wogalter, M., Desaulniers, D., and Brelsford, J. Jr. (1986). Perceptions of consumer products: Hazardousness and warning expectations. *Proceedings of the Human Factors Society 30th Annual Meeting*. Santa Monica, CA: Human Factors Society, pp. 1197–1201.

Wogalter, M., Desaulniers, D., and Brelsford, J. Jr. (1987). Consumer products: How are hazards perceived? *Proceedings of the Human Factors Society 31st Annual Meeting*. Santa Monica, CA: Human Factors Society, pp. 615–619.

Wogalter, M., Desaulniers, D., and Godfrey, S. (1985). Perceived effectiveness of environmental warnings. *Proceedings of the Human Factors Society 29th Annual Meeting*. Santa Monica, CA: Human Factors Society, pp. 664–668.

Wogalter, M., Godfrey, S., Fontenelle, G., Desaulniers, D., Rothstein, P., and Laughery, K. (1987). Effectiveness of warnings. *Human Factors*, 29, 599–612.

Young, S., Brelsford, J., and Wogalter, M. (1990). Judgments of hazard, risk, and danger: Do they differ? *Proceedings of the Human Factors Society 34th Annual Meeting*. Santa Monica, CA: Human Factors Society, pp. 503–507.

CHAPTER 21

HUMAN FACTORS AND THE AUTOMOBILE

When people are asked to list technological developments that have profoundly altered the way we live, work, and die, inevitably the automobile is among the items listed. The automobile has expanded our personal sphere of influence. We routinely travel in one hour what would have taken a day or more to travel by horse or stagecoach—and we do it a lot more comfortably too. Unfortunately, as we shall see in the next section, the automobile has also become a major cause of death and injury in the world.

In this chapter we discuss a potpourri of human factors topics related to the automobile. We set the stage with a brief discussion of vehicle accidents and their causes. We then discuss human factors issues related to the driver, the automobile, and the road environment. An emphasis will be placed on driver performance and safety.

VEHICULAR ACCIDENTS

Hutchinson (1987) estimates that worldwide, half a million people are killed *each year* in traffic accidents. In addition, about 15 million people are injured each year (Trinca et al., 1988). *Each year* in the United States, almost 50,000 people lose their lives in traffic accidents—about equal to the total number of Americans who died during the entire Vietnam war.

Evans (1991) summarizes fatality rates from 21 countries. In terms of deaths per 1000 registered vehicles, the United States had the lowest rate (0.24). The greater the degree of motorization (number of vehicles) of a country, the lower the fatality rate per 1000 vehicles. In general, fatalities per registered vehicle decline by about 5 percent per year in a country. This decline may be due to a number of factors, including safer cars, better road systems, stricter enforcement of traffic laws, and a more experienced driving population.

Terminology dealing with accidents is used rather loosely, with such terms as *accidents, collisions,* and *crashes* sometimes being used interchangeably. For our purpose we consider accidents to be those traffic mishaps that involve some property damage, personal injury or death, or both.

Causes of Accidents

The causes of accidents can be characterized in various ways and at different levels of specificity. At a very general level, accidents can be attributed to human behavior, the environment, and the vehicle, usually with some interaction among these. In discussing accident causation, Older and Spicer (1976) suggest that accidents can be the consequence of conflict situations involving the driver and the environment (and presumably the vehicle) that lead to evasive actions on the part of the driver. Such evasive actions may or may not result in what they call a collision (or, in our terms, an accident).

In connection with accident causation, Shinar (1978) presents data on the percentage of accidents from two samples that were attributed to human, vehicle, and environmental causes and their combinations. One sample of 2258 accidents was investigated on site, and the other sample of 420 was investigated more thoroughly, in depth. The results, given in Figure 21-1, show that human behavior was clearly the dominant cause, being involved, to some extent, in over 90 percent of the accidents. The combination of the human and environment is also a major contributor to accidents.

Evans (1991) points out some limitations in interpreting results such as those in Figure 21-1. Identifying the factors contributing to an accident is not the same as identifying how to reduce accidents. Both accidents and nonaccidents must be analyzed to do that. For example, head-on collisions while passing a vehicle on a two-lane road might be attributed to the human, but such accidents do not happen on divided highways. When one looks at the bigger picture, other factors—the environment (roadway configuration) in this example—may present themselves and suggest ways to prevent accidents.

THE DRIVER: PERFORMANCE AND BEHAVIOR

Driving an automobile requires the full range of human capabilities, including perception, decision-making, and motor skills. These capabilities must be performed in a highly coordinated fashion often under stressful conditions. As Evans (1991) points out, it is quite remarkable that such a large proportion of people can drive a car and can learn to do it in a matter of weeks or months.

There has been a considerable amount of research dealing with driver performance and behavior. The research has been directed toward describing the performance level of drivers and relating driving behavior to accidents.

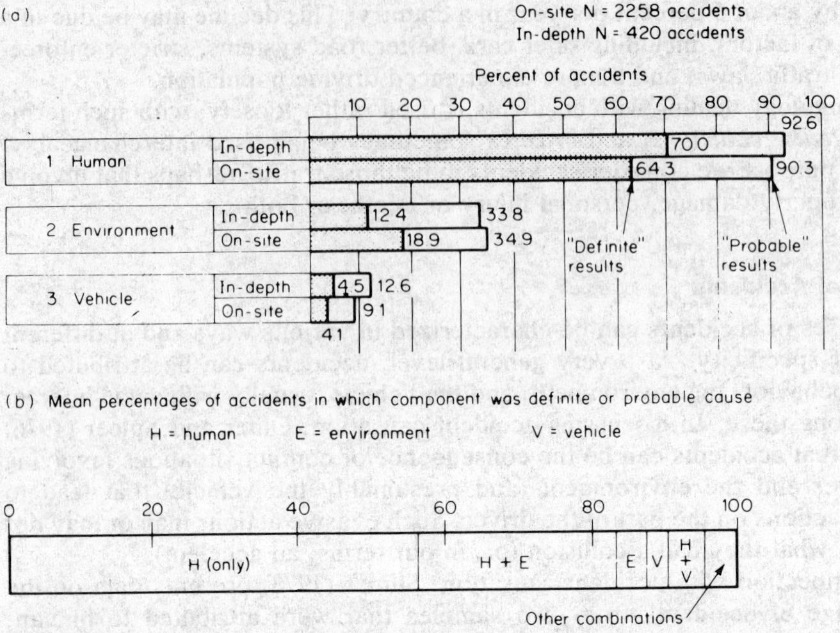

FIGURE 21-1
(*a*) The percentage of accidents caused by human, environmental, and vehicular factors and (*b*) the relative proportion of combinations of these causes. Because many accidents have multiple causes, the sum of the percentages in (*a*) exceeds 100 percent, while the sum in (*b*) is 100 percent. (*Source: Shinar, 1978, Fig. 5.2, p. 111.*)

Behaviors investigated have been characterized in various ways and at various levels of specificity such as inattention, risk taking, judgments of speeds and distances, etc. We cannot deal extensively with all such behaviors, but we discuss a few for illustrative purposes after briefly discussing the role of behavioral errors in accidents.

Behavioral Errors and Accidents

As something of an overview of the driving behaviors associated with accident occurrences, we present data from two studies. The results of a study by Fell (1976) are summarized in Figure 21-2. Shown are the percentages of accidents investigated in which each of four direct-cause categories of behavioral errors was considered to have "definite involvement" or "definite or probable involvement."

Figure 21-3 presents the results of another study (Malaterre, 1990) in which 72 accidents involving 115 drivers and pedestrians were analyzed. The behavioral error categories are more specific than those shown in Figure 21-2. The results highlight the importance of information perception and interpretation in accidents. In addition, performance errors occurred more often as a factor than

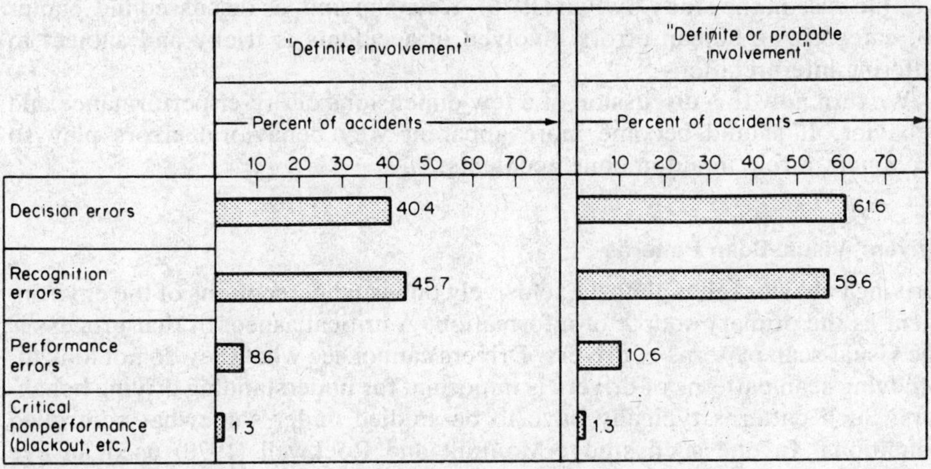

FIGURE 21-2
Summary of the causal accident involvement of four human direct-cause categories, expressed as percentages of accidents in which they were implicated. (*Source: Fell, 1976, Fig. 2, p. 92.*)

FIGURE 21-3
Summary of the involvement of various driver failures and errors in accidents. The percentages of failures attributed to category are shown. A total of 159 failures and errors played a part in the 72 accidents investigated. (*Source: Adapted from Malaterre, 1990, Fig. 5.*)

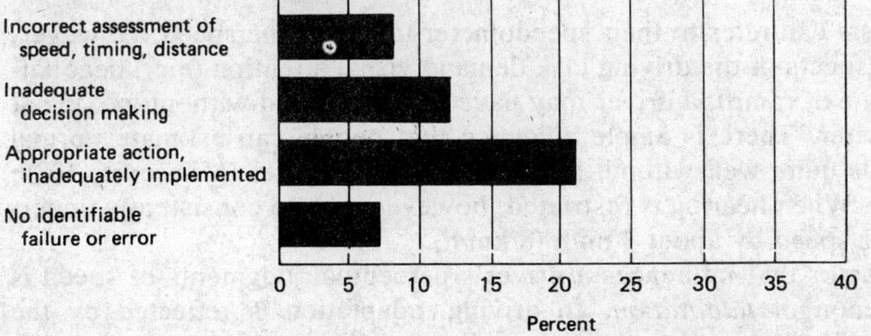

was the case in the study by Fell (1976). Keep in mind, as discussed in Chapter 20, categorizing human errors involved in accidents is tricky and subject to differing interpretations.

We turn now to a discussion of a few dimensions of driver performance and behavior. It should become more apparent why behavioral errors play so prominent a role in automobile accidents.

Drivers' Visual-Scan Patterns

Driving behavior relies almost exclusively on visual perceptions of the environment as the primary source of information. A critical aspect of that process is the visual-scan patterns of drivers. Drivers cannot see what they do not look at. Studying scan patterns of drivers is important for understanding driving behaviors. Such patterns typically have to be studied under somewhat controlled conditions. In one such study, Mourant and Rockwell (1970) used an eye camera to record the eye movements of subjects driving on an expressway at 50 mi/h (80 km/h). One implication of the study was that route familiarity plays a role in such patterns. Over unfamiliar routes the drivers typically "sampled" a wide area in front of them, but with increasing familiarity their eye movement tended to be confined to a smaller area. Other studies of driver eye-scanning behavior have shown that novice drivers sample the roadway environment more narrowly than experienced drivers; thus novices tend to receive less information from the periphery of the visual field (Mourant and Rockwell, 1972). While following a car, drivers tend to focus on the center of the road, closer to the car ahead of them, and spend less time looking at traffic control devices. Alcohol and fatigue also reduce visual scanning of the environment. In addition, increasing speed is accompanied by a narrowing of visual attention. These sorts of results have implications for placement and design of road signs and road-edge markers.

The perceptions of drivers form the basis for making judgments about aspects of the driving task, including perceptual judgments of speed and spacing. These in turn play into the subjective evaluation of risk involved in various driving situations and behaviors. We discuss each of these in turn.

Perceptual Judgments of Speed

Usually drivers can refer to their speedometer to determine speed. However, when other aspects of the driving task demand visual attention (e.g., negotiating a freeway exit ramp), a driver may have to judge speed without looking at the speedometer. There is ample evidence that people can estimate normal driving speeds quite well without the aid of a speedometer (Milosevic, 1986; Evans, 1970). When hearing is restricted, however, people consistently underestimate their speed by about 3 mi/h (8 km/h).

Another factor that influences a driver's perceptual judgments of speed is the phenomenon of *adaptation*. In driving, adaptation is reflected by the tendency to perceive a given speed to be *less* when a person has previously adapted to a *higher* speed and to be *higher* when a person has previously adapted to a *lower* speed. Various investigators have confirmed this idea (Casey and Lund, 1987; Mathews, 1978). This effect is especially noticeable in changing from a high speed to a lower speed—people tend to think they are going slower than they really are (Schmidt and Tiffin, 1969). Further, the longer the exposure to the high speed, the greater the underestimation of the lower speed. Thus people who slow down (as on an exit ramp of a superhighway) would tend to drive at a higher speed than they think they are driving. In this and other driving circumstances, a person might be driving at a higher speed than is warranted, thereby increasing the likelihood of accidents.

Judging the relative speed of a car ahead (i.e., closure rate) involves the perceptions of changes in the gap between the two vehicles and changes in the visual size (angle) of the car ahead. People are very good at determining the direction of closure, i.e., whether the car ahead is going slower or faster than their vehicle. However, people are not very good at determining the actual closure rate. It is harder to tell whether one is closing fast or slow than just whether one is closing or not.

Many rear-end collisions involving slow-moving or stopped vehicles on the highway are probably caused by inaccurate perceptions of closure rates. Usually, the slow or stopped vehicle is struck as the following vehicle tries to go around it, but misjudges the clearance. In such situations, the driver probably determined that he or she was closing on the vehicle ahead but realized too late that the closing speed was higher than initially perceived.

As Evans (1991) points out, an important factor in such accidents is the expectation that the vehicle ahead is moving at a reasonable (although slow) speed. This expectation dominates the perceptual process and leads to erroneous interpretations of the situation. Your author has been involved, as an expert witness, in several such accidents. Usually they involve a truck stopped on the highway with no indications (flares, flashers, triangles, etc.) that it is stopped. Trucks are expected to be traveling slow, but not to be stopped in the roadway without indicators.

People's inability to judge relative speeds of vehicles probably accounts for many accidents in which a car trying to pass another car has a head-on collision with an oncoming car. The research evidence is somewhat mixed with respect to people's ability to judge the speed of an oncoming vehicle (Triggs, 1988). Some studies find overestimations, others find underestimations. Expectation seems to play a major role here as well. How fast a person expects the oncoming vehicle to be traveling can bias judgments one way or the other.

Perceptual Judgments of Spacing

Have you ever observed that drivers of small cars tend to follow closer to the car ahead than do drivers of large cars? They do (Herman, Lam, and Rothery, 1973). Evans and Rothery (1976) found a major cue to perceptions of spacing that may help explain this phenomenon. The amount of road surface visible beyond the hood of one's car to the car ahead is the main cue to judging distance. The visual angle (size) of the car ahead did not appear to be a factor. Now, in a small car, the hood is lower and smaller and hence, more road is exposed ahead of the car than is the case when driving a large car. Therefore if drivers of large and small cars are trying to maintain the same subjective distance from the car ahead, the driver of the small car will drive closer than will the driver of the large car. This may also explain why small drivers (those who can barely see over the steering wheel) tend to follow further behind the car ahead and wait longer to start out after stopping behind another car.

Risk Taking

Undoubtedly people's subjective perception of the risk of injury and death influences their behavior. We discussed risk taking in Chapter 20. The vast majority of drivers are not thrill seekers or daredevils. The typical driver engages in behavior that he or she perceives will not result in an accident (Summala, 1988). The problem is that sometimes drivers are wrong. There is some evidence that drivers adapt to perceived risks on the road. Behaviors that have not resulted in an accident (or near accident) are perceived as less risky, and so more risky behavior may be engaged in in the future. In addition, people will accept more risk if there is a payoff for doing it or a loss for not doing it. For example, people will take more risks when they are in a hurry. However, even when in a hurry, people believe that they will not get into an accident. Although risky, the risk is not perceived as high enough to result in an accident. As discussed in Chapter 20, the "it can't happen to me" bias and the overestimation of one's ability play a major role in such perceptions.

A theory that has been bandied around for a decade or so is *risk homeostasis theory* (Wilde, 1982; 1986). This theory claims that drivers have a target level of risk per unit time: changes in the design of vehicles or highways that increase safety are offset by drivers engaging in more risky behavior so that the target level of risk is maintained. The consensus [see Evans (1991) for a summary] is that there is no validity to the theory and that even its theoretical formulation is without merit.

Reaction Time

One factor that separates the quick from the dead in traffic is reaction time in emergency conditions. Reaction time is generally slower when one is surprised than when one is prepared to make a response. Surprisal reaction time is often twice as long or longer than prepared reaction time. Reaction time also increases as the stimulus event and required response become more complex. For these reasons, reaction time must be measured under realistic conditions by using representative drivers who are not anticipating the need to make a response. Summala (1981) summarized a series of studies in which unobtrusive measurements were made of the steering response of drivers to roadside stimuli. In one daytime situation, the door of a car parked on the side of the road was opened as cars approached, and the evasive steering response (turning toward the center of the road) was measured as a function of time from onset of the stimulus (i.e., the opening of the door). In another study performed at night, a light was turned on by the side of a dark road as cars approached, and again steering response was measured. In these studies drivers did not know they were participating in an experiment, so most likely they were responding in a natural manner.

Some of the results from these studies are shown in Figure 21-4. The evasive response started in no case until more than 1 s had passed from the onset of the stimulus. The halfway point of the steering response was reached at about 2.5 s, and maximum steering deflection occurred between 3 and 4 s. Also, the response latencies were very similar in both daytime and nighttime situations.

The U.S. standard for perceptual reaction time used by traffic engineers to design highways is 2.5 s (Oglesby, 1975), which corresponds well with the halfway point for evasive steering responses reported by Summala. Hooper and McGee (1983), however, argue, from a review of the literature, that the standard should be 3.2 s to be appropriate for the 85th percentile driver.

Discussion

People generally do not realize their limitations in judging speed and spacing, responding to unexpected events, or assessing the risks involved in their behavior. This can lead to overconfidence and increased accident potential. We must consciously adopt a more cautious attitude toward driving to compensate for the natural tendency to overestimate our capabilities and lack of understanding of the factors that create inaccurate perceptions of the driving environment.

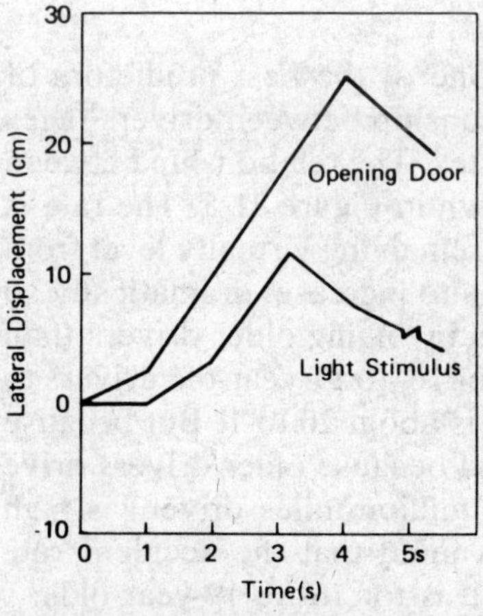

FIGURE 21-4
Average lateral displacement of cars in response to the opening of a car door at the side of the road during the day and to a light appearing at the side of the road at night as a function of the time available from the onset of the stimulus. (*Source: Adapted from Summala, 1981, Fig. 1. Copyright by the Human Factors Society, Inc. and reproduced by permission.*)

THE DRIVER: PERSONAL CHARACTERISTICS

Although driving an automobile is a skill acquired by many, all do not acquire the skill equally. There are many personal characteristics that can affect driving performance, and we discuss a few of the major ones in this section.

Experience and Skill

One would expect that the greater the experience and skill of drivers, the lower would be their accident rate. This has been found among bus drivers (Hakkinen, 1979; Lim and Dewar, 1989), although not universally (McKenna, Duncan, and Brown, 1986). Bus drivers, however, are constrained to a great degree in the range of driving behaviors they can exhibit. Schedules and fixed routes, for example, reduce the self-paced nature of the driving task. In contrast, for the rest of us daily drivers, as we gain in experience and skill we can adjust our behavior if we so desire. We can, for example, drive faster, follow more closely, change lanes more often, and pass more aggressively. To the extent that we engage in such behaviors, the relationship between experience and skill and accidents will be muddied.

The ultimate in driving skill and experience is the race car driver. One often hears claims that licensed race car drivers are safer on the road than other drivers. In fact, however, this may be one of the great myths of the automotive age. Williams and O'Neill (1974) compared national-competition race car drivers from the Sports Car Club of America to comparison groups of regular drivers. They found that the race car drivers had substantially *more* accidents and violations (especially for speeding) than did the comparison groups. Evans (1991) points out, however, that we cannot say whether race car drivers drive more aggressively because of their additional skill or whether risk-taking drivers are attracted to racing. Nonetheless, the study does provide evidence that higher skill levels are not necessarily associated with lower accident rates.

Age

Age has been found, quite consistently, to be one of the best predictors of accident rates. Cerrelli (1989) analyzed the relationship between drivers' ages and accident involvement using data from five states. The relationship between age and crashes per million miles traveled is shown in Figure 21-5. The rate is high for 16- to 19-year-olds and declines sharply, remaining virtually level from 30 to 69 years of age. After age 70, the rate begins to increase, dramatically so after age 80. Actually there are far fewer crashes involving older drivers than involving young drivers. The ratio of the number of 16- to 19-year-old drivers to drivers 80 years old or older involved in crashes is about 20 to 1. But because there are more young drivers than old drivers and because older drivers drive fewer miles than younger drivers, the rate (per million miles driven) is high among older drivers. It is important to keep in mind that the accident rate among 80- to 84-year-olds is still only half what it is for 16- to 19-year-olds.

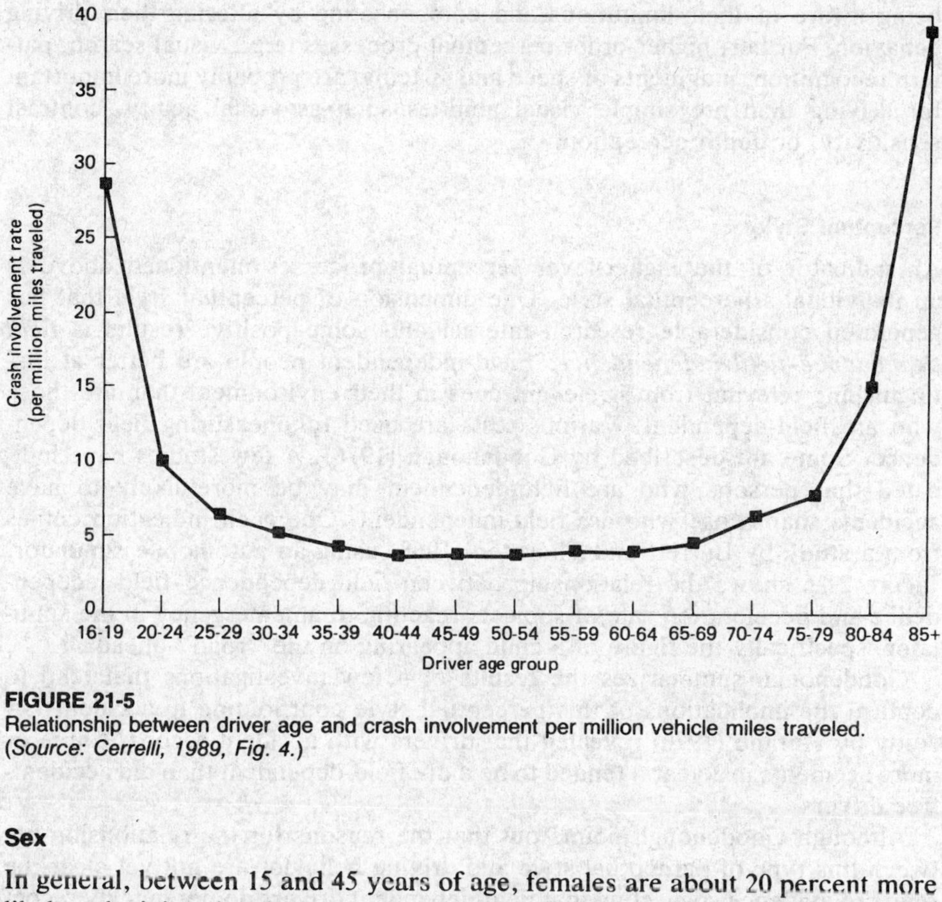

FIGURE 21-5
Relationship between driver age and crash involvement per million vehicle miles traveled. (*Source: Cerrelli, 1989, Fig. 4.*)

Sex

In general, between 15 and 45 years of age, females are about 20 percent more likely to be injured and 25 percent more likely to be killed than males if both are involved in a similar accident (Evans, 1991). Comparisons of accident involvement, on the other hand, consistently show that males have higher rates (per million miles driven) than do females. The largest differences are among the young, while only slight differences are found after age 40.

Vision

Given the importance of vision in the driving task, one would expect that drivers' visual abilities would be related to accident involvement. This does not seem to be the case. Accident rates are highest among the young and lowest among those 40 to 60 years old. Yet visual abilities are keenest among the young and deteriorate between age 40 and 60. In fact, studies have found that monocular drivers (those with vision in only one eye) are no more likely to be involved in accidents than are normal-sighted drivers (Evans, 1991).

Boyce (1981) believes that the lack of relationship between visual abilities and accident rates may be due, in part at least, to drivers with poor abilities being aware of their limitations and compensating by altering their driving behavior. Further, higher-order perceptual processes (e.g., visual search, pattern recognition, judgments of speed and spacing) are probably more important for driving than are simple visual abilities such as visual acuity, contrast sensitivity, or depth perception.

Perceptual Style

An indicator of the higher-level perceptual processes mentioned above is an individual's perceptual style. One dimension of perceptual style that has generated considerable research interest and some positive results is *field dependence–field independence*. Field-independent people are better at distinguishing relevant from irrelevant cues in their environment than are those who are field-dependent. Various tests are used for measuring field dependence. Some are described by Goodenough (1976). A few studies have indicated that persons who are field-dependent may be more likely to have accidents than those who are field-independent. One such indication comes from a study by Barrett and Thornton (1968) using an automobile simulator. Figure 21-6 shows the relationship between field dependence–field independence and deceleration rate of subjects reacting to an emergency in the simulator, specifically the figure of a child appearing on the "road" ahead.

Goodenough summarizes the results of a few investigations that tend to confirm the implications of this perceptual style contributing to accidents. A study by Harano (1970) revealed that drivers with accident records (three or more accidents in 3 years) tended to be more field-dependent than did accident-free drivers.

Although Goodenough points out that the reasons for the relationship between this type of perceptual style and driving behavior are not yet clear, he refers to evidence suggesting that field-dependent drivers do not quickly recognize developing hazards, are slower in responding to embedded road signs (those surrounded by many other stimuli), have difficulty in learning to control

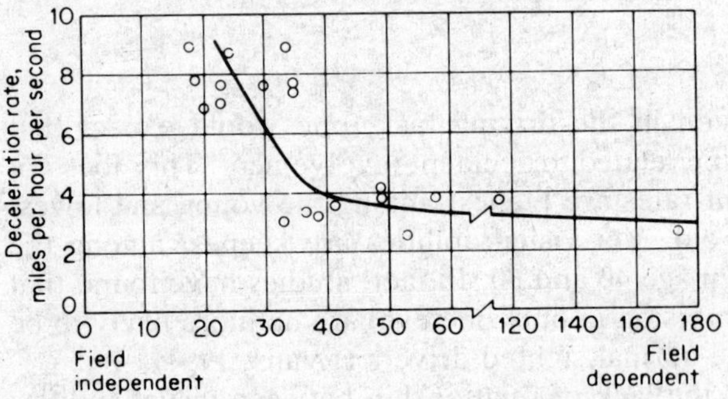

FIGURE 21-6
Relationship between field dependence–field independence style and deceleration rate of subjects reacting to an emergency situation in a vehicle simulator. (*Source: Barrett and Thornton, 1968, Fig. 3.*)

a skidding vehicle, and fail to drive defensively in high-speed traffic. In further efforts to "explain" the relationship, Shinar et al. (1978) report data that show a relationship between field dependence and eye-movement behavior. The more field-dependent the driver, the longer his or her eye-fixation duration, thus the longer it takes to pick up relevant information. Dewar (1984) indicates that field-dependent individuals tend to concentrate their visual attention within a narrow field of view, thus reducing peripheral vision.

Although it has been suggested that training might help field-dependent individuals to become more field-independent, there is as yet no strong evidence to indicate that this would be possible.

Personality

There is an increasing body of evidence that manifestations of certain personality characteristics are associated with accident behavior. In summarizing the results of various studies, for example, McGuire (1976) concludes that some highway accidents are just another correlate of being emotionally unstable, unhappy, asocial, antisocial, impulsive, or under stress—or a host of similar conditions referred to by other labels. Such attributes presumably cause some people to be less cautious, less attentive, less responsible, less caring, less knowledgeable, or less capable of driving—and thus increase the risks of accident.

Evans (1991) reviews additional studies that show a link between personality and accident involvement. Tsuang, Boor, and Fleming (1985) concluded that such personality characteristics as low tolerance for tension, immaturity, and paranoid conditions were associated with increased risk of accident involvement.

Discussion

Except for sex and age, personal characteristics of drivers do not seem to be highly predictive of accident involvement. Personality variables may hold some promise but do not explain the accident involvement of so many "normal" people in traffic accidents. One also has to be cautious not to infer causation from correlation. It is possible that some other variable is causing the relationship being observed. For example, DiFranza et al. (1986) report that smokers have 50 percent more traffic accidents and 46 percent more traffic violations than do nonsmokers. This does not necessarily mean that if one quits smoking, the probability of being in an accident will decrease.

THE DRIVER: TEMPORARY IMPAIRMENTS

The primary factors that contribute to temporary impairment of driving performance are fatigue, alcohol, and drug use.

Fatigue

The primary culprit in generating fatigue is travel time and associated loss of sleep. The degradation in driving performance because of fatigue is reported by Shinar (1978) to account for a small, but significant, percent of highway accidents. However, among long-haul commercial drivers, who often must drive long hours, fatigue can be a significant cause of accidents. The Bureau of Motor Carrier Safety, for example, investigated 286 commercial vehicle accidents and found that 38 percent of them were categorized as attributable to the driver's being either asleep at the wheel or inattentive (Harris and Mackie, 1972).

Some drivers are often aware of the early stages of fatigue and can take steps to minimize it, as by stopping for a snooze or a cup of coffee. Hulbert (1972) lists other behaviors as preceding the final stage of actually falling asleep.

- Longer and delayed deceleration reactions in response to changing road demands
- Fewer steering corrections
- Reduced galvanic skin response (GSR) to emerging traffic events
- More body movement such as rubbing the face, closing the eyes, and stretching

Presently the primary basis for reducing the risk from fatigue lies with the driver, who must be sensitive to the signs of the onset of fatigue and must stop driving. On the freeways in certain locations *Botts dots* are used to demarcate the lanes. These are small, raised bumps, and they are arranged so that if a driver changes lanes, the tires roll over the lanes with a "roar." In parts of Canada, as one approaches stop signs there is a change in the road surface that produces a loud noise. Both these techniques are quite effective in alerting drowsy or inattentive drivers.

Alcohol and Drugs

After being consumed, alcohol is absorbed rapidly into the blood and distributed throughout the body, including the brain. The best indicator of the effects of alcohol on performance and mood is the amount of alcohol in the blood, or *blood alcohol content* (BAC). BAC is often expressed as a percentage and is equal to 100 times the mass of alcohol per mass of blood. Thus, 1 part alcohol per 1000 parts blood is equal to a BAC level of 0.10 percent. BAC can also be estimated by measuring the alcohol content of a person's breath. In most American states the legal limit for driving under the influence of alcohol is 0.10 percent BAC. In Britain and several other American states, the limit is 0.08 percent BAC.

Alcohol and Performance Moskowitz and Robinson (1987) conducted a very thorough review of the literature dealing with the effects of alcohol on driving related task performance. Nine behavioral categories were reviewed:

reaction time, tracking, concentrated attention, divided attention, information processing, visual functions, perception, psychomotor skills, and driving performance. The BAC level at which statistically significant impairment was demonstrated varied among studies and by type of behavior investigated. Figure 21-7 attempts to summarize the results by showing the BAC levels at which 25, 50, 75, and 95 percent of the studies within a behavioral category found significant results. For example, 50 percent of the studies within various categories showed significant impairment at BAC levels between 0.05 and 0.08 percent. It is obvious from Figure 21-7 that there is considerable evidence for performance decrements at BAC levels far below 0.10 percent.

Alcohol and Traffic Accidents First of all, let us explode another common myth. Alcohol-impaired drivers are 3.85 times *more* likely to die than the alcohol-free driver in crashes of comparable severity (Waller et al., 1986). Alcohol-impaired drivers do not escape injury by being relaxed and "rolling with the punches" of a crash.

The involvement of alcohol in traffic accidents is legendary. During the period 1982–1987, for example, approximately 142,500 people in the United States lost their lives in alcohol-related traffic accidents (1 every 22 minutes). Each year about 595,000 police reported motor vehicle accidents (10 percent of the total) are alcohol-related (NHTSA, 1988). One bit of good news: the proportion of all people killed in crashes in which at least one person had a BAC of 0.10 percent or higher has declined over the past few years as shown in Figure 21-8.

FIGURE 21-7
Range of BAC levels at which various percentages of studies show statistically significant impairment due to alcohol. See text for types of behaviors studied. (*Source: Adapted from Moskowitz and Robinson, 1987.*)

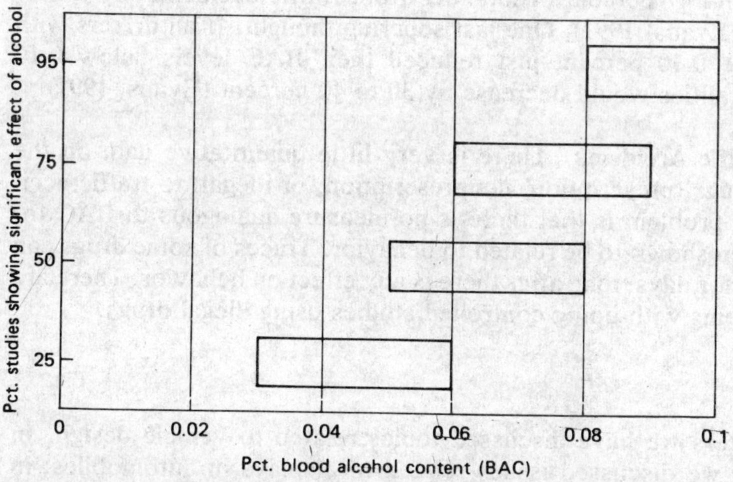

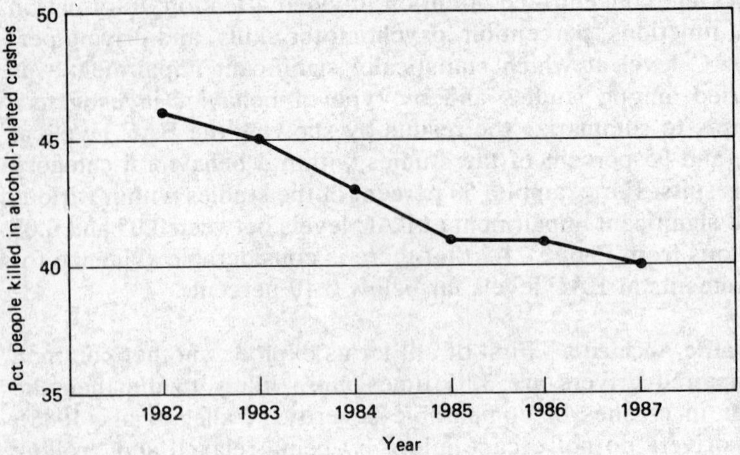

FIGURE 21-8
Percentage of people killed in traffic accidents from 1982 through 1987 in which at least one person had a BAC level of 0.10 percent or higher. (*Source: National Center for Statistics and Analysis, 1989, p. iii.*)

The percentage of crashes that are alcohol-related does not tell us whether or not alcohol caused the crash. To make such a determination requires a control group to determine the incidence of alcohol impairment in noncrash drivers, and for each crash we need to determine who was at fault. Such a study (the largest and probably the last) was carried out by Borkenstein et al. (1964). Using their data Evans (1991) estimated the risk that drivers with various BAC levels would *cause* a crash. The results are shown in Figure 21-9. The risk factor increases from 2 times more likely to 10 times more likely as the BAC level goes above 0.10 percent. About 10 percent of property damage, 20 percent of injuries, and 47 percent of fatalities from traffic accidents are attributable to alcohol (Evans, 1991). One last sobering thought. If all drivers with BAC levels above 0.10 percent just reduced their BAC levels below 0.10 percent, traffic fatalities would decrease by 30 to 40 percent (Evans, 1991)!

Drugs and Traffic Accidents There is very little quantitative data on the contribution of drugs (prescription, nonprescription, or illegal) to traffic accidents. Part of the problem is that there is no measure analogous to BAC for drugs that has been shown to be related to behavior. Traces of some drugs can linger in the body for days, long after there is any effect on behavior. There are also ethical problems with doing controlled studies using illegal drugs.

VEHICLE DESIGN

In previous chapters we have discussed topics related to vehicle design. In Chapters 4 and 5 we discussed issues related to displays in automobiles; in

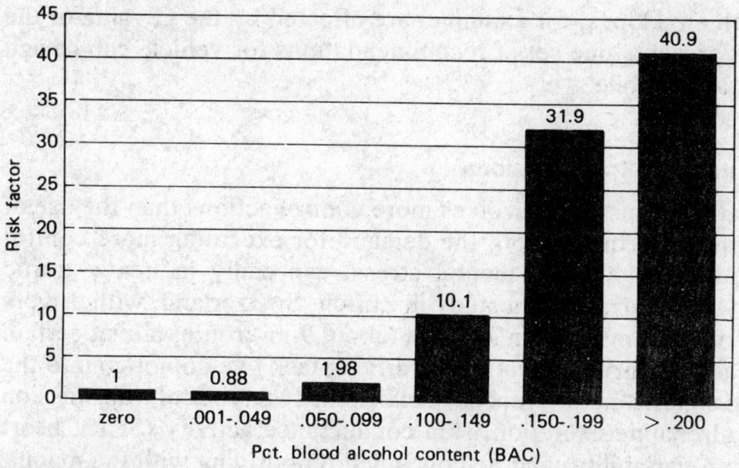

FIGURE 21-9
Relative risk of causing an accident as a function of BAC level of the driver (*Source: Adapted from Evans, 1991, Table 7-2.*)

Chapter 11 we discussed some automobile controls; and in Chapter 19 we discussed whole-body vibration and ride comfort. In this chapter, we deal with a few vehicle design issues that affect driver performance and accidents.

Cab Design

Vehicle cabs represent a type of work space that presents some interesting human factors considerations. The layout must provide clearance for large drivers yet present controls (hand and foot) close enough for small drivers to reach. Consideration must be given to the visual field of the driver as well as the driver's comfort. Vehicle cab design involves applications of human factors information dealing with anthropometry and seating (see Chapter 13), placement of controls (see Chapter 14), and compatibility (see Chapter 10), among other things.

Vehicles come in a variety of cab configurations; for example, compare the configuration of a sports car with that of a sedan or even a small truck. Sports cars tend to place the driver in a more reclined posture with the steering wheel in a more vertical plane. Sedans usually sit the driver more erect, placing the steering wheel in a more horizontal position. Also larger steering wheels are usually found in sedans. Truck cabs are at the extreme: upright posture with a large steering wheel in a horizontal orientation. Automotive engineers have devised a way to numerically describe the cab configuration of a vehicle in a single number called the *general package factor* (G) (SAE, 1980). G is computed from a linear equation which includes such variables as seat position, seat back angle, steering wheel position, and steering wheel size. Large positive G values are typical of trucks. Negative G values are associated with sports cars. Reach envelopes, for example, are affected by the G value of the cab. Figure 21-10 presents one set of recommendations for vehicle cab design for a sedan-style automobile.

Manual versus Automatic Transmissions

The use of manual transmissions involves more control actions than the use of automatic transmissions. In addition, the demand for executing more control actions might add at least some mental stress, especially in heavy traffic situations. Zeier (1979) carried out a study in Zurich, Switzerland, with drivers using both types of transmissions in a 14-km (about 9-mi) route, taking certain physiological measures during and after the driving task (we do not go into the details). He found significant differences between the two types of transmission users in rate of adrenaline excretion, skin conductance activity (SCR), heart rate, and heart rate variability, and he concluded that driving with the manual

FIGURE 21-10
A set of recommended design features for vehicle cabs. These features are designed to be suitable for persons between the 5th and 95th percentiles of operators. Dimensions are given in inches, with centimeters in parentheses. (*Adapted from Van Cott and Kinkade, 1972, Fig. 9-11.*)

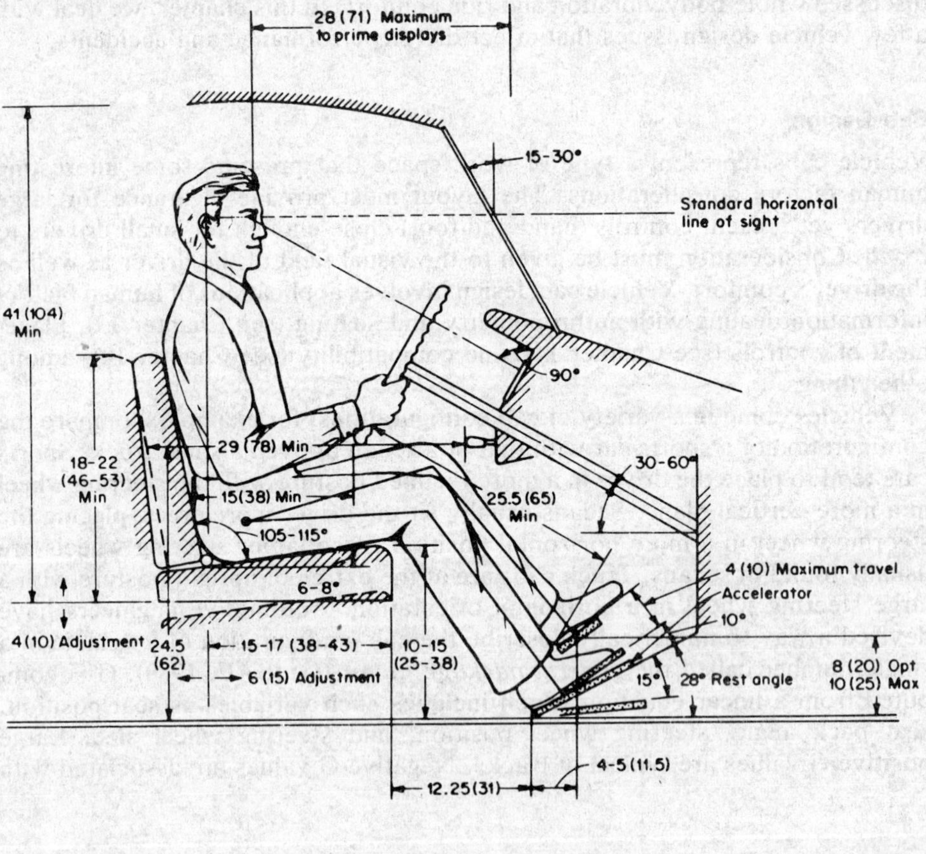

transmission produces greater activation of the sympathetic nervous system, in effect reflecting a higher level of stress. He suggests that the stress reduction due to the use of an automatic transmission may constitute a valuable measure for improved health and safety since it enables a driver to concentrate more on traffic events.

Auxiliary Information Systems

In the very near future we will see a new class of information being presented to drivers in their vehicles. Most likely a visual display terminal will be used to present the information. If one is willing to pay for it, moving-map displays showing the driver's position on the map are currently available for automobiles. Work is progressing on in-vehicle systems to provide routing instructions, information on traffic conditions and roadside services, and even collision-avoidance warnings. Such systems are also referred to as *intelligent vehicle highway systems* (IVHS).

IVHS raise important human factors issues such as information overload and competition for visual resources. Dingus, Antin, Hulse, and Wierwille (1988), for example, compared the visual-attention demands for navigation system tasks using a commercially available moving-map display and conventional driving-related tasks. Figure 21-11 presents the average total time spent

FIGURE 21-11
Total time looking at display and/or controls to perform navigation tasks using a moving-map display and conventional driving-related tasks. (*Source: Adapted from Dingus, Antin, Hulse, and Wierwille, 1988, Table 1. Reprinted with permission of the Human Factors Society, Inc. All rights reserved.*)

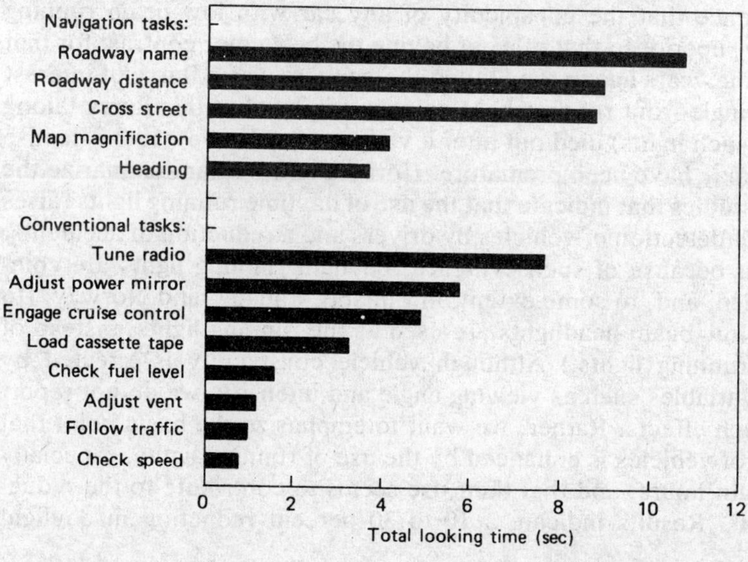

looking at a display or control to perform various tasks. Several of the navigation tasks required 8 to 11 s of looking time to perform—time when the driver's eyes were not on the road. The most visually demanding conventional driving-related tasks were tuning the radio, adjusting power mirror, and engaging cruise control. These sorts of tasks are more apt to be performed infrequently, at the discretion of the driver, or under open-road driving. On the other hand, the demanding navigation tasks are likely to be performed in all types of traffic and when driving in an unfamiliar environment.

Auxiliary information systems will require extensive research and testing to determine the information needs of drivers, how best to display the information, and how best to access the information and operate the device. This promises to be an exciting area for the application of human factors expertise.

Vehicle Lighting

Vehicle lighting serves two main purposes: it provides information to other drivers regarding the presence and movements of one's vehicle; and it provides illumination of the environment to aid drivers to see ahead. We briefly discuss three types of vehicle lighting systems: running lights, rear lights, and headlights.

Running Lights It stands to reason that the more adequately drivers are able to identify the presence and movements of vehicles in their traffic area, the more likely they are to avoid accidents. One of the most important factors influencing the detectability of an object in the environment is the contrast between the object and the background. Although the colors of vehicles as contrasted with their backgrounds can be useful in this regard, the use of lights on vehicles can also provide desired contrast. In fact, Hörberg and Rumar (1979) cite evidence that the conspicuity of any car with low-beam running lights is equal or superior to that of a car having the best color contrast for that background. Some years ago in the United States there was a flurry of interest in the use of a single front running light on automobiles, but this flurry (along with the sale of such lights) died out after a year or two. However, the demise of that practice may have been premature. Hörberg and Rumar summarize the results of a few studies that indicate that the use of daytime running lights raises the probability of detection of vehicles by drivers and a reduction of accidents. They report that because of such evidence, daylight running lights are compulsory in Sweden, and, to some extent, in Finland, Canada, and Norway. (In these countries low-beam headlights are used as the running lights, instead of having special running lights.) Although vehicle conspicuity is affected by certain related variables such as viewing angle and intensity, we do not report the details of such effects. Rather, we want to emphasize the basic point that the conspicuity of vehicles is enhanced by the use of running lights, especially during the twilight hours, and that their use seems to contribute to the reduction of accidents. Results indicate a 10 to 30 percent reduction in daylight collisions.

Rear Lights Another aspect of vehicle conspicuity is the brake light system used on automobiles. Some years ago Malone et al. (1978) found that rear-end accident rates of taxicabs were significantly reduced by adding a center, high-mounted stop light (CHMSL) to the conventional rear light system. Sivak et al. (1980) measured reaction times of unsuspecting drivers to conventional stop light systems with and without a single high-mounted (on the top of the trunk) stop light. Drivers following the experimental car applied their brakes in response to the onset of the experimental car's brake lights 31 percent of the time when the experimental car was equipped with only conventional rear stop lights. When the CHMSL was added to the experimental car, 55 percent of the time the following car responded. On those trials where the following car responded, however, the mean reaction times (1.39 s) were the same for both light configurations. In part as a result of these studies and others, all new cars sold in the United States are now required to have, in addition to the conventional stop lights, a single CHMSL on the rear of the car. Kahane (1989) in a study of the effectiveness of CHMSLs found that CHMSL-equipped cars were 17 percent less likely to be struck in the rear while braking than cars without a CHMSL. The CHMSL is especially effective in preventing chain collisions involving three or more vehicles, and more daylight crashes are prevented than nighttime crashes. It is estimated that when all cars on the road are equipped with a CHMSL, 126,000 accidents, 80,000 nonfatal injuries, and almost $1 billion in property damage will be prevented *each year*. At a lifetime cost of just over $10, the CHMSL is truly a human factors success story.

Headlights Headlights serve both purposes of vehicle lighting presented above. People tend to turn on headlights even before the sun sets with over half of all cars having their headlights on 10 min before sunset (Johnson, 1990). The primary purpose for headlights under such conditions is to increase the conspicuity of one's vehicle rather than to provide any discernable increase in the detectability of roadway hazards.

Designing the beam pattern of headlights is tricky. To maximize detection of roadway hazards, more light should be directed to the sides of the road and at or above the horizon. However, headlights are a major source of glare to oncoming cars. To reduce the glare potential, most of the light from low-beam headlights is directed down, below the horizon, and to the right. The result is that low beams are not very good for illuminating obstacles or roadway hazards.

Olson and Sivak (1984) present data on the expected distances at which a driver using low or high beams would respond to a pedestrian in dark or light clothes standing on the right or left side of the road. The distances were corrected to approximate an unexpected situation; however, the subjects were probably more alert and careful (because they knew they were in an experiment) than usual. The results are shown in Table 21-1. Because low beams direct more light to the right, response distances are longer. That is, the driver detects the target sooner when it is on the right than when it is on the left and low beams are being used. There is no right-left difference with high beams. High beams, as one would expect, increase the response distance.

TABLE 21-1
EXPECTED RESPONSE DISTANCE TO PEDESTRIAN TARGETS AS A FUNCTION OF HEADLIGHT BEAM, TARGET LOCATION, AND CLOTHING

		Target position relative to car			
		Left		Right	
Clothing	Beam	ft	(m)	ft	(m)
Dark	Low	60	(18.3)	80	(24.4)
	High	120	(36.6)	120	(36.6)
Light	Low	120	(36.6)	160	(48.8)
	High	240	(73.2)	240	(73.2)

Source: Olson and Sivak, 1984, Table 15-1.

Olson and Sivak's (1984) data also reinforce the admonition to pedestrians to always wear light-colored (high-reflectance) clothing when walking on the side of the road at night. The average response distance for drivers doubled when the pedestrian wore light-colored clothing. If a driver is going more than about 40 mi/h and low beams are being used, the best chance a pedestrian has to avoid being hit is to wear light-colored clothes. There is simply insufficient distance to bring a car, traveling at that speed, to a stop within the response distance if low beams are used and the pedestrian is clad in dark clothing.

Discussion

We have presented a few topics dealing with the design of vehicles that impact safety and comfort. Advancements in the design of vehicles continue to improve the safety of automobiles. The wider use of air bags, antilock brakes, CHMSL, and improved headlights should reduce the frequency and severity of automobile accidents. The big question is the impact of auxiliary information systems on safety. Undoubtedly, human factors will play a significant role in the design and testing of such systems.

DRIVING ENVIRONMENT

The driving environment includes the roads and highways, street and highway lighting, road markings, road signs, and traffic as well as the natural features of the ambient environment such as rain and snow.

Road Characteristics

There are various indications regarding the effects of road characteristics on driving performance. One of the most clear-cut factors involves the use of superhighways versus conventional two-way roads. The accident rate on superhighways is clearly lower, this lower incidence being attributed to the reduced number of circumstances in which any given vehicle encounters other vehicles moving in other directions.

An example of a different kind of "effect" of road design characteristics is reported by Rutley and Mace (1972), who rigged electrodes to record the heart rates of a few drivers approaching the London airport. An example of these recordings for one driver is shown in Figure 21-12. This figure shows the heart rate in relation to the road being traversed and the "events" that occurred in driving over the section of the road in question. One can consider such physiological responses on the part of the driver (in this instance, changes in heart rate) as indices of strain that reflect varying levels of stress generated by the driving situation.

Traffic Events

Although it is not feasible to discuss many implications of the traffic environment in driving behavior, we mention one interesting procedure used by Helander (1978). He placed electrodes on 60 subjects driving a rural test route

FIGURE 21-12
Heart rate of one driver on a section of a road near the London airport. The horizontal scale represents the time during which 160 heartbeats were recorded. The features of the road are sketched at the top, and the written entries indicate the "events" occurring during the time period. (*Source: Rutley and Mace, 1972, Fig. 3.*)

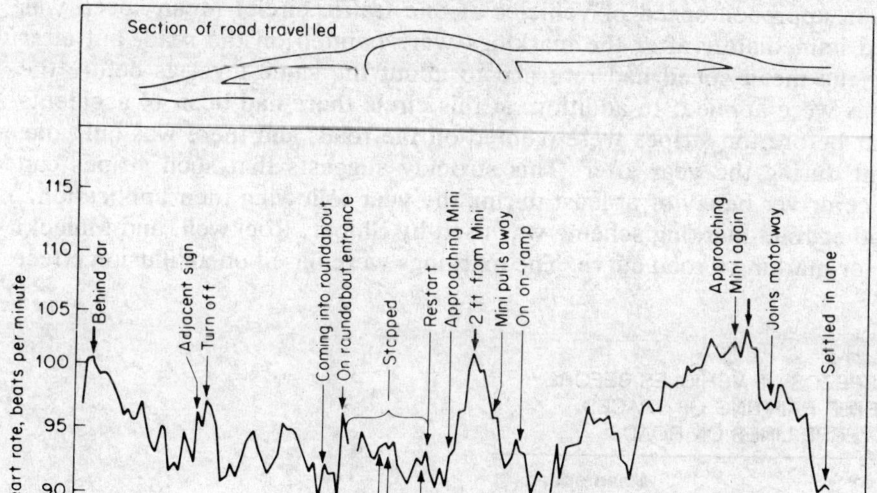

and obtained measures of brake pressure. The electrodes recorded heart rate; electrodermal response (EDR), which is the same as galvanic skin response (GSR); and two electromyograms. The various measures were then related to 15 types of "traffic events" such as "other car passes own car" and "meeting other car." Brake pressure was considered as a criterion of the *mental difficulty* of the various traffic events. The event that was most difficult was "cycle or pedestrian meeting other car," and the next one was "other car merges in front of own car."

Road Markings

Everyday experience in driving clearly demonstrates the value of road markings, such as edge markings and center lines, especially in night driving. Aside from such markings, however, there are some interesting developments in the value of special markings in specific situations. In Great Britain, for example, Rutley (1975) experimented with the use of yellow transverse lines on the road starting 0.4 km. (about ¼ mi) ahead of traffic circles (called *roundabouts* in Great Britain). The spacing of these lines was reduced as they got closer to the traffic circle. The driver, first becomes "adapted" to lines being crossed at a given (slow) rate, then becomes aware of lines being crossed at ever-increasing rates. This creates an illusion of speed (like telephone poles being passed at a high rate of speed) that typically causes the driver to slow down as he or she approaches the traffic circle. Rutley presents the data shown in Table 21-2 on the mean approach speed of vehicles at one traffic circle. Mean speed was reduced immediately after the markings were painted on the road, but after 1 year, the mean speed had returned to about the same level as before the markings were applied. In addition, at this circle there had been 14 accidents the year before the stripes were painted on the road, and there was only one accident during the year after. This strongly suggests that such stripes can influence driver behavior at least during the year following their application.

Another road-marking scheme was used by Shinar, Rockwell, and Malecki (1980) for marking a road curve. The markings were based on an illusion effect

TABLE 21-2
MEAN SPEEDS OF VEHICLES BEFORE AND AFTER PAINTING OF SPACED TRANSVERSE LINES ON ROADS

	Mean speed	
	km/h	**mi/h**
Before	57.0	35.4
Immediately after	44.1	27.3
1 year after	52.4	32.5

Source: Rutley, 1975.

discovered many years ago by a German psychologist, Wundt, who found that V-shaped lines between parallel lines make the parallel lines appear closer together at the center than they really are, as illustrated in Figure 21-13. Shinar, Rockwell, and Malecki (1980) painted V-shaped lines in a herringbone pattern on a road starting 97.5 m (about 320 ft) ahead of an obscured curve and on the curve itself. They hypothesized that the herringbone pattern of lines would create the illusion of narrowing of the road and cause drivers to slow down.

The herringbone pattern presumably did have the intended effect. Before the painting of the pattern, the mean reduction in speed of vehicles approaching the curve was 4.0 mi/h (6.4 km/h). After the pattern was painted, the mean speed reduction was 7.3 mi/h (11.7 km/h).

Road Signs

Considerable research has been directed toward improving the effectiveness of road signs. In Chapter 4 we discussed several factors that influence the effectiveness of signs and made some recommendations for their design. We cannot hope to cover, even briefly, all the aspects of roadway sign design. We do, however, briefly discuss driver's response to signs and legibility of signs at night.

Response to Road Signs Several studies over the years have attempted to measure drivers' recall of road signs previously encountered on a stretch of road. Typically, a sign is placed on the road, then around a curve cars are stopped at random (often by a police officer), and the drivers are asked to recall the sign that was on the road. Generally, the percentage of drivers who can recall a sign is small, usually less than 50 percent and often less than 20 percent. Recall, however, is directly related to the degree of urgency of the information contained in the sign; the greater the urgency, the higher the probability of recall (Johansson and Backlund, 1970). The generally poor recall performance may, in part, be due to the emotional upset caused by being stopped on the road.

Drory and Shinar (1982) suggest that in daylight driving, on an open road, drivers can obtain most of the information contained in warning signs from their view of the road ahead. Therefore, only the information they consider

FIGURE 21-13
The Wundt illusion causes parallel lines to appear closer together than they are at the center. This illusion was used by Shinar, Rockwell, and Malecki (1980) as the basis for drawing a herringbone pattern of lines approaching, and on, a curve in a road to cause drivers to slow down on the curve.

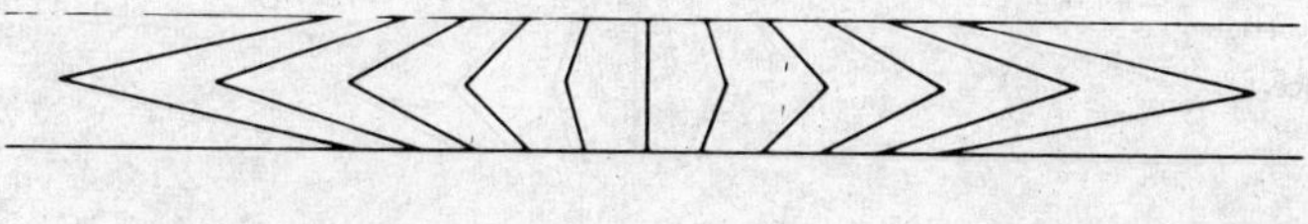

relevant and that cannot be obtained directly from the roadway is registered and recalled. This would explain why signs indicating police patrol and change in speed limits are recalled better than general warning and pedestrian-crossing signs (Johansson and Rumar, 1966). Shinar and Drory (1983) tested this hypothesis by comparing sign recall during the day and at night. Mean percentage recall was higher during the night (20 percent) than during the day (6 percent). This finding, then, tends to support the notion that signs are recalled better if they present information that is relevant and not otherwise easily obtainable.

Recall of signs is important; however, failure to recall does not necessarily mean that a sign was not perceived. Summala and Hietamaki (1984), for example, unobtrusively measured speed changes of 2185 drivers at a place in the road where an experimental sign ("*Danger*: Children" or "Speed Limit 30 km/h") suddenly came into view. The change (decrease) in speed was dependent on the significance of the sign; however, all the drivers responded to each sign by releasing the accelerator pedal, indicating that the sign was perceived.

Legibility of Signs at Night The legibility of road signs at night depends on several factors, such as the size of symbols, the colors of sign legends and of their backgrounds, the luminance of the legends and backgrounds, the type of material used, any ambient illumination, and the lights from approaching cars.

An illustration of the effect of sign brightness on legibility of numerals on signs of high and low reflectance is based on a controlled experiment by Hicks (1976). In this study, the drivers used both low-beam and high-beam headlights on different nights. Legibility was measured in terms of the distance at which the numerals could be read by subjects when sober and when alcohol-impaired. For one group of subjects, the mean legibility distances under the different experimental conditions are shown in Figure 21-14. Both reflectance of signs and the position of the headlights influenced legibility. Alcohol impaired the performance of the drivers, and high sign reflectance and use of high-beam headlights made the signs more legible than low sign reflectance and use of low-beam headlights.

FIGURE 21-14
Legibility of numerals by sober and alcohol-impaired drivers under specified reflectance and headlight conditions. (*Source: Adapted from Hicks, 1976.*)

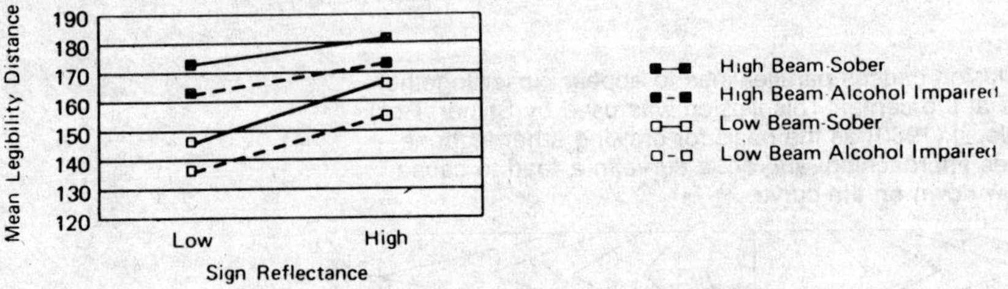

Sivak, Olson, and Pastalan (1981) measured the nighttime legibility distance for groups of drivers under 25 and over 61 years of age with equal visual acuity as measured under high luminance. The legibility distances for the older subjects were 65 to 77 percent of those for the younger subjects. This implies that older drivers are likely to have less distance, and hence less time, in which to act on information contained in road signs.

Roadway Illumination

Only about one-quarter of all road miles are driven during darkness, yet the number of road fatalities at night often is equal to or greater than the number occurring during the day (Boyce, 1981). Roadway illumination, or lack of it, undoubtedly contributes to the higher accident rates at night. In general, fixed overhead roadway luminaires are superior to vehicle headlights for target detection at night; under some circumstances overhead lights can be 300 times more effective (Staplin, 1985). The before-and-after accident records of streets and highways which have been illuminated provide extremely persuasive evidence that this pays off. Table 21-3 summarizes the data compiled by the Street and Highway Lighting Bureau of Cleveland for 31 thoroughfare locations throughout the country, showing traffic deaths for the year before they were illuminated and for the year after. Such evidence is fairly persuasive and argues for the expansion of highway illumination programs.

Factors of Roadway Lighting The CIE (1977) has published a set of recommendations for roadway lighting. Criteria are given for good roadway lighting in terms of average road surface luminance, uniformity of luminance, extent of discomfort and disability glare, and directional guidance provided to the driver. We do not discuss these criteria, but we briefly discuss a few aspects of roadway lighting that affect visibility. One is transition lighting. Where roadway luminaires are to be used (such as at an intersection), it is desirable to provide some "transition" on the approaches and exits by graduating the size of lamps used. This helps facilitate visual adaptation to and from the highly illuminated area. Another factor is the *system geometry*, which refers in particular to the

TABLE 21-3
TRAFFIC DEATHS BEFORE AND AFTER ILLUMINATION OF 31 THOROUGHFARE LOCATIONS

		Reduction	
Year before illumination	Year after	Number	Percentage
556	202	354	64

Source: Public Lighting Needs, 1966.

mounting height of luminaires; in some installations the mounting height is 35 ft (10.7 m) or more, and in some European systems installations at heights of 40 ft (12 m) or more are used. Such heights (while requiring larger lamps) help to increase the *cutoff* distance, that is, the distance from the light source at which the top of the windshield cuts off the view of the luminaire from the driver's eyes. This has an impact on visual comfort and visibility because overhead lamps can be a source of glare. Under average conditions this cutoff distance is about 3.5 times the mounting height of the luminaires.

In addition, the reflectivity of roadway surfaces varies substantially, a factor which affects visibility. For example, asphalt pavement surface (about 6 years old) reflects approximately 8 percent of the light, whereas concrete pavement has a surface reflectance of about 20 percent. It has been estimated that the amount of illumination required for equal brightness of asphalt pavements, compared with concrete, is roughly 2:1 (Blackwell, Prichard, and Schwab, 1960). Boyce (1981) also points out that more work is needed to assess the influence of wet roads on the effectiveness and design of roadway and vehicle lighting. We know that night visibility is reduced when the roads are wet; perhaps, this is due in part to the glare of lights reflecting off the surface water on the roadway.

DISCUSSION

Human factors has played, and will continue to play, an important role in defining the capabilities and limitations of drivers and in designing automobiles and road environments to match them. We have touched on only a few examples of human factors contributions in this chapter. Many accidents occur because a driver overestimates his or her own abilities or the abilities of other drivers (e.g., in terms of reaction time, detection of objects and signs, or estimation of closure rates). Human factors professionals are often called upon to serve as expert witnesses in personal injury litigations resulting from automobile accidents. Their role usually involves explaining the perceptual, cognitive, and motor skills of drivers and explaining how various aspects of automobile and roadway design affect drivers' behavior.

The advances being made in intelligent vehicle highway systems will provide stimulating challenges for the human factors professional for many years. Especially important will be the cognitive demands such systems place on drivers.

REFERENCES

Barrett, G. V., and Thornton, C. L. (1968). Relationship between perceptual style and driver reaction to an emergency situation. *Journal of Applied Psychology*, 52(2), 169–176.

Blackwell, H., Prichard, B., and Schwab, R. (1960). *Illumination requirements for roadway visual tasks*. Highway Research Board Bull. 255, Publ. 764.

Borkenstein, R., Crowther, R., Shumate, R., Ziel, W., and Zylman, R. (1964). *The role of the drinking driver in traffic accidents*. Bloomington, IN: Department of Police Administration, Indiana University.

Boyce, P. (1981). *Human factors in lighting*. New York: Macmillan.

Casey, S., and Lund, A. (1987). Three field studies of driver speed adaptation. *Human Factors,* 29(5), 541–550.

Cerrelli, E. (1989, February). *Older drivers, the age factor in traffic safety* (DOT-HS-807-402). Washington, DC: National Highway Traffic and Safety Administration.

Commission Internationale de l'Eclairage (CIE) (1977). *International recommendations for the lighting of roads for motorized traffic*, CIE Publ. 12/2, Paris.

Dewar, R. (1984). Ergonomics in highway transportation. In M. Matthews and R. Webb (eds.), *Proceedings of the 1984 International Conference on Occupational Ergonomics*, vol. 1: *Reviews*. Rexdale, Ont., Canada: Human Factors Association of Canada.

DiFranza, J., Winters, T., Goldberg, R., Cirillo, L., and Biliouris, T. (1986). The relationship of smoking to motor vehicle accidents and traffic violations. *New York Journal of Medicine,* 86, 464–466.

Dingus, T., Antin, J., Hulse, M., and Wierwille, W. (1988). Human factors issues associated with in-car navigation system usage. *Proceedings of the Human Factors Society 32d Annual Meeting*. Santa Monica, CA: Human Factors Society, pp. 1448–1452.

Drory A., and Shinar, D. (1982). The effect of roadway environment and fatigue on sign perception. *Journal of Safety Reseach,* 13, 25–32.

Evans, L. (1970). Speed estimation from a moving automobile. *Ergonomics,* 13, 219–230.

Evans, L. (1991). *Traffic safety and the driver*. New York: Van Nostrand Reinhold.

Evans, L., and Rothery, R. (1976). The influence of forward vision and target size on apparent intervehicular spacing. *Transportation Science,* 11, 60–72.

Fell, J. C. (1976). A motor vehicle accident causal system: the human element. *Human Factors,* 18(1), 85–94.

Goodenough, D. (1976). A review of individual differences in field dependence as a factor in auto safety. *Human Factors,* 18(1), 53–62.

Hakkinen, S. (1979). Traffic accidents and professional driver characteristics: A follow-up study. *Accident Analysis and Prevention,* 11, 7–18.

Harano, R. M. (1970). Relationship of field dependence and motor-vehicle accident involvement. *Perceptual and Motor Skills,* 31, 372–374.

Harris, W., and Mackie, R. (1972, November). A study of the relationships among fatigue, hours of service, and safety of operations of truck and bus drivers (Rept. BMCS-RD-71-2). Washington, DC: Bureau of Motor Carrier Safety.

Helander, M. (1978). Applicability of drivers' electrodermal response to the design of the traffic environment. *Journal of Applied Psychology,* 63(4), 481–488.

Herman, R., Lam, T., and Rothery, R. (1973, June). An experiment on car size effects in traffic. *Traffic Engineering and Control,* 15, 90–93.

Hicks, J., III. (1976). The evaluation of the effect of sign brightness on the sign-reading behavior of alcohol-impaired drivers. *Human Factors,* 18, 45–52.

Hooper, K., and McGee, H. (1983). Driver perception-reaction time: Are revisions to current specification values in order? *Transportation Research Board,* 904, 21–30.

Hörberg, V., and Rumar, K. (1979). The effect of running lights in vehicle conspicuity in daylight and twilight. *Ergonomics,* 22, 165–173.

Hulbert, S. (1972). Effect of driver fatigue. In J. Forbes (ed.), *Human factors in highway traffic research.* New York: Wiley.

Hutchinson, T. (1987). *Road accident statistics.* Adelaide, Australia: Rumsby Scientific Publishing.

Johansson, G., and Backlund, F. (1970). Drivers and road signs. *Ergonomics.* 13, 749–759.

Johansson, G., and Rumar, K. (1966). Drivers and road signs: A preliminary investigation of the capacity of car drivers to get information from road signs. *Ergonomics,* 9, 57–62.

Johnson, D. (1990). Headlight use. *Proceedings of the Human Factors Society 34th Annual Meeting.* Santa Monica, CA: Human Factors Society, pp. 1086–1090.

Kahane, C. (1989). *An evaluation of center high mounted stop lamps based on 1987 data* (DOT-HS-807-446). Washington, DC: National Highway Traffic Safety Administration.

Lim, C., and Dewar, R. (1989). *Driver cognitive ability and traffic accidents.* Alberta, Canada: University of Calgary.

Malaterre, G. (1990). Error analysis and in-depth accident studies. *Ergonomics,* 33, 1403–1421.

Malone, T., Kirkpatrick, M., Kohl, J., and Baker C. (1978). *Field test evaluation of rear lighting systems.* Alexandria, VA: Essex Corp.

Mathews, M. (1978). A field study of the effects of drivers' adaptation to automobile velocity. *Human Factors,* 20, 709–716.

McGuire, F. L. (1976). Personality factors in highway accidents. *Human Factors,* 18(5), 433–442.

McKenna, F., Duncan, J., and Brown, I. (1986). Cognitive abilities and safety on the road: A re-examination of individual differences in dichotic listening and search for embedded figures. *Ergonomics,* 29, 649–663.

Milosevic, S. (1986). Perception of vehicle speed. *Revija za Psihologijy,* (Yugoslavia) 16, 11–19. Cited in Evans (1991).

Moskowitz, H., and Robinson, C. (1987). Driving-related skills impairment at low blood alcohol levels. In P. Noordzij and R. Roszbach (eds.), *Alcohol, drugs and traffic safety—T86.* Amsterdam: Elsevier.

Mourant, R., and Rockwell, T. (1970). Mapping eye-movement patterns to the visual scene in driving: An exploratory study. *Human Factors,* 12(1), 81–87.

Mourant, R., and Rockwell, T. (1972). Strategies of visual search by novice and experienced drivers. *Human Factors,* 14, 325–335.

National Center for Statistics and Analysis (NCSA). (1989, March). *Alcohol involvement in fatal traffic crashes 1987* (DOT-HS-807-401). Washington, DC: National Highway Traffic and Safety Administration.

National Highway Traffic and Safety Administration (NHTSA) (1988, August). *Drunk driving facts.* Washington, DC.

Oglesby, C. (1975). *Highway engineering.* New York: Wiley.

Older, S. J., and Spicer, B. R. (1976). Traffic conflicts: A development in accident research. *Human Factors,* 18(4), 335–350.

Olson, P., and Sivak, M. (1984). Visibility problems in nighttime driving. In G. Peters and B. Peters (eds.), *Automotive engineering and litigation,* vol. 1. New York: Garland.

Public Lighting Needs. (1966). *Illuminating Engineering,* 61(9), 585–602.

Rutley, K. S. (1975). Control of drivers' speed by means other than enforcement. *Ergonomics,* 18(1), 89–100.

Rutley, K. S., and Mace, D. G. W. (1972). Heart rate as a measure in road layout design. *Ergonomics,* 15(2), 165–173.

Schmidt, F., and Tiffin, J. (1969). Distortion of drivers' estimates of automobile speed as a function of speed adaptation. *Journal of Applied Psychology,* 53, 536–539.

Shinar, D. (1978). *Psychology on the road: The human factor in traffic safety.* New York: Wiley.

Shinar, D., and Drory, A. (1983). Sign registration in daytime and nighttime driving. *Human Factors,* 25, 117–122.

Shinar, D., McDowell, E. D., Rackoff, N. J., and Rockwell, T. H. (1978). Field dependence and driver visual search behavior. *Human Factors,* 20(5), 553–559.

Shinar, D., Rockwell, T. H., and Malecki, J. A. (1980). The effects of changes in driver perception on road curve negotiation. *Ergonomics,* 23(3), 263–275.

Sivak, M., Olson, P., and Pastalan, L. (1981). Effect of driver's age on nighttime legibility of highway signs. *Human Factors,* 23, 59–64.

Sivak, M., Post, D., Olson, P., and Donohue, R. (1980). Brake responses of unsuspecting drivers to high-mounted brake lights. *Proceedings of the Human Factors Society 24th Annual Meeting.* Santa Monica, CA: Human Factors Society, pp. 139–142.

Society of Automotive Engineers (SAE) (1980). *Driver hand control reach* (SAE J287). Warrensdale, PA.

Staplin, L. (1985). Nighttime hazard detection on freeways under alternative reduced lighting conditions. *Proceedings of the Human Factors Society 29th Annual Meeting.* Santa Monica, CA: Human Factors Society, pp. 725–729.

Summala, H. (1981). Latencies in vehicle steering: It is possible to measure drivers' response latencies and attention unobtrusively on the road. *Proceedings of the Human Factors Society 25th Annual Meeting.* Santa Monica, CA: Human Factors Society. pp. 711–715.

Summala, H. (1988). Zero-risk theory of driver behaviour. *Ergonomics,* 31, 491–506.

Summala, H., and Hietamaki, J. (1984). Drivers' immediate responses to traffic signs. *Ergonomics,* 27, 205–216.

Triggs, T. (1988). Speed estimation. In G. Peters and B. Peters (eds.), *Automotive engineering and litigation,* vol. 2. New York: Garland.

Trinca, G., Johnston, I., Campbell, B., Haight, F., Knight, P., Mackay, G., McLean, A., and Petrucelli, E. (1988). *Reducing traffic injury—A global challenge.* Melbourne, Australia: A. H. Massina.

Tsuang, M., Boor, M., and Fleming, A. (1985). Psychiatric aspects of traffic accidents. *American Journal of Psychiatry,* 142, 538–546.

Van Cott, H., and Kincade, R. (eds.) (1972). *Human engineering guide to equipment design* (rev. ed.). Washington, DC: U.S. Government Printing Office.

Waller, P., Stewart, J., Hansen, A., Stutts, J., Popkin, C., and Rodgman, E. (1986). The potentiating effects of alcohol on driver injury. *Journal of the American Medical Association,* 256, 1461–1466.

Wilde, G. (1982). The theory of risk-homeostasis: Implications for safety and health. *Risk Analysis,* 2, 209–255.

Wilde, G. (1986). Notes on the interpretation of traffic accident data and of risk homeostasis theory: A reply to L. Evans. *Risk Analysis,* 6, 95–101.

Williams, A., and O'Neill, B. (1974). On-the-road driving records of licensed race drivers. *Accident Analysis and Prevention,* 6, 263–270.

Zeier, H. (1979). Concurrent physiological activity of driver and passenger when driving with and without automatic transmission in heavy city traffic. *Ergonomics,* 20, 799–810.

CHAPTER 22

HUMAN FACTORS IN SYSTEM DESIGN

Previous chapters have dealt, in part, with human factors aspects of the design of certain types of things people use (computer workstations, displays, controls, hand tools, etc.) and of the environments in which people work and live. Previous chapters have presented and referenced a wealth of principles and information on human capabilities and limitations. The process by which that information is brought to bear on the design of things and systems is the focus of this chapter. We review the system design process and discuss the role played by human factors in the various stages.

There is no shortage of horror stories resulting from the failure of designers to include human factors in the design process. The problem is not so much one of a lack of data or human factors principles, but rather the lack of attention paid to it by designers. Beevis and Hill (1983) cite numerous problems resulting from a failure to incorporate elementary human factors data into designs of military hardware. Examples include insufficient room in a truck to operate the pedals, inability to see traffic signals from the cab of a truck, inability to operate controls with gloved hands, insufficient room to carry required crew and equipment in an armored vehicle, inaccessible placement of an emergency fuel cutoff switch in a hovercraft, and color coding that made it near impossible to distinguish between caution and danger conditions.

SYSTEM DESIGN PROCESS

In Chapter 1 we discussed briefly some characteristics of systems. In the design of such systems (including equipment, facilities, and other physical items that people use) typically certain basic stages or processes have to be carried out.

There are various ways to depict the major stages in the system design process. One rather straightforward representation is shown in Figure 22-1. This representation (and, in fact, much of the following discussion of the system design process) applies to the design of very complex systems that often include major hardware components. The comprehensive, systematic approach to the design process discussed here has been developed and used primarily in the design of major military and space systems. Although the "systematic" features of such an approach have not been widely used in the design of other types of systems and products, often the basic stages and considerations involved need to be addressed somehow, even if in a nonsystematic fashion. In the development of some simple systems or products some aspects of the design process may simply be irrelevant. Elaborate design processes may not be warranted in many circumstances, but the basic point is this: Each aspect of the design process must be examined to determine if it is relevant, and, if so, appropriate attention should be paid to it.

Characteristics of the System Design Process

Although representations of the system design process such as the one shown in Figure 22-1 imply an orderly, structured process, in practice the stages often overlap and are performed in iterative fashion. In the design of a system, decisions in later stages often necessitate the modification and refinement of information and decisions from earlier stages. Meister (1985) lists the following characteristics of the process:

1 *Molecularization.* The process works from broad molar functions to more molecular tasks and subtasks.

2 *Requirements are forcing functions.* Design options are developed to satisfy system requirements. (For this reason formal behavioral requirements must be included in the initial design specifications.)

3 *System development is discovery.* Initially there are many unknowns about the system, and during the design process these unknowns are clarified and addressed.

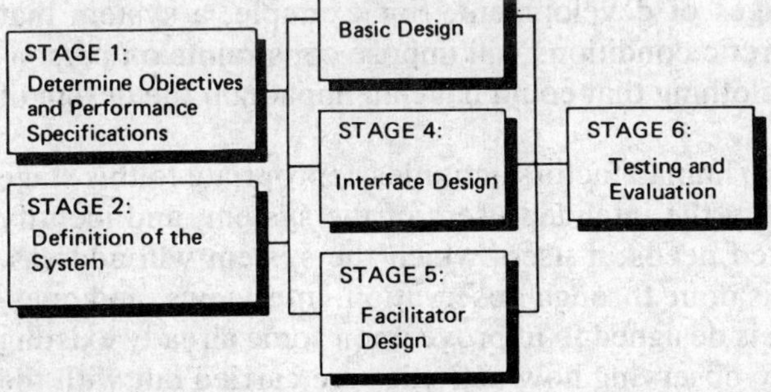

FIGURE 22-1
Major stages in the system design process. These stages are carried out in iterative fashion as the system develops. Later stages may modify decisions made in earlier stages. For some simple systems, these stages are carried out informally and, in some cases, not at all. (*Source: Adapted from Bailey, 1982.*)

4 *System development involves transformation.* There is a transformation from physical requirements to behavioral implications of those requirements, and from behavioral implications to the physical mechanisms (e.g., controls, displays) required for implementing the behavioral implications.

5 *Time.* There is never enough time for practitioners to perform the analyses, studies, and tests that they would like to do.

6 *Cost.* There is usually not enough money to support the design effort, and behavioral recommendations, if too costly, will be rejected automatically.

7 *Iteration.* Activities are repeated as more detailed information about the system becomes available.

8 *Design competition.* Large systems are designed by groups or teams of specialists, each with their own concerns and criteria. Less influential groups (often including human factors people) must function under constraints established by the dominant groups.

9 *Relevance.* Relevance to design, as perceived by engineers, is critical for the acceptance and judged value of behavioral inputs.

We now turn our attention to the design stages depicted in Figure 22-1, and we discuss some of the human factors activities, concerns, and tools associated with each.

STAGE 1: DETERMINE OBJECTIVES AND PERFORMANCE SPECIFICATIONS

Before a system can be designed, it must have a purpose or reason for being. The purposes of a system are stated as *objectives* and usually in rather general terms. For example, the objective of a system might be to provide communication links between offices in different cities. System *performance specifications* detail what the system must do to meet its objectives. Examples might include transmitting both voice and visual information across the country, linking up to 100 offices, providing data security for transmitting sensitive information, etc. System performance specifications should reflect the context in which the system will operate, including consideration of the skills available in the existing or projected user population, organizational constraints, etc. System performance specifications often have behavioral implications that will have an impact on later stages of development. For example, a system that must be able to operate in arctic conditions will impose constraints on personnel and will require special clothing that could have an impact on the design of controls and work spaces.

Bailey (1982) identified two human factors activities appropriate to this stage of the design process: Identify the intended users of the system, and identify and define the activity-related needs of users which the system will address. Identification of user needs is done through observation, interviews, and questionnaires. Usually a system is designed to improve upon some already existing system or set of systems. By observing how activities are carried out with the existing system and by interviewing and surveying users of the existing system, information can be gathered to ensure that the new system's specifications meet the needs of existing and potential users. This information will also be available in later stages of the system's development.

STAGE 2: DEFINITION OF THE SYSTEM

The principal activity during stage 2 is the definition of the functions which the system has to perform to meet its objectives and performance specifications. In theory, no attempt is made during this stage to assign functions to hardware, software, or humans. Functions can be instantaneous (e.g., to power up equipment) or prolonged (e.g., to store incoming messages) and simple (e.g., to determine message destination) or complex (e.g., to prioritize incoming messages). In general, functions are relatively molar behaviors; the individual tasks and behaviors needed to implement or carry out the functions are more detailed. There are, however, no rules for deciding whether a set of behaviors should be called a function or a task (Meister, 1985). To aid in the conceptualization of system functions, *functional flow diagrams* are often developed; an example is shown in Figure 22-2. Such diagrams depict the interrelationships among the system functions, with each box representing one function. Often subsidiary flow diagrams are developed to analyze the subfunctions necessary to carry out each function identified in the higher-order diagrams. In the case of an urban transportation system, for example, some functions that would need to be performed include the sale of tickets, collection of tickets, movement of passengers to loading platforms, opening and closing of vehicle doors to permit entrance and exit of passengers (possibly to take into account handicapped

FIGURE 22-2
An example of a functional flow diagram used to graphically depict the functions of a system. (LRU = line replaceable unit) (*Source: Geer, 1981.*)

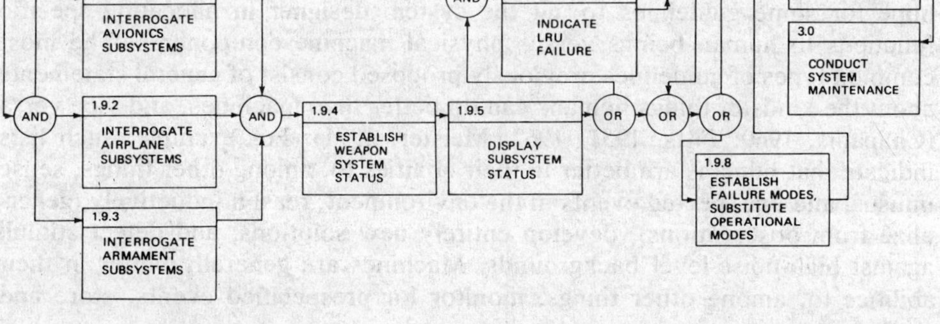

passengers), vehicle control, movement of passengers to exits of stations, providing relevant information and directions for passengers, and a communications system for use in the spacing of vehicles. This function analysis initially should be concerned with *what* functions need to be performed to fulfill the objectives, and not with the *way* in which the functions are to be performed (such as whether they are to be performed by individuals or by machine components).

During this stage, human factors specialists help insure that the functions identified match the needs of the intended users. Further, these specialists may collect more detailed information regarding the characteristics, capabilities, and limitations of the intended user population which will be used in the next stage of development.

STAGE 3: BASIC DESIGN

It is in this stage of system development that the system begins to take shape. The principal human factors activities are (1) allocation of functions to humans, hardware, and software; (2) specifications of human performance requirements; (3) task analysis; and (4) job design. During this stage usually numerous changes are made to the system which necessitate revisions and updating of the human factors activities. Rarely are the four activities listed above done only once; rather, they are performed in iterative fashion as the design matures.

Allocation of Functions

Given functions that have to be performed, in some instances there may be an option as to whether any particular function should be allocated to a human being or to some physical (machine) component(s). In this process, the allocation of certain functions to human beings, and of others to physical components, is virtually predetermined by certain manifest considerations, such as obvious superiority of one over the other or economic considerations. Between these two extremes, however, may lie a range of functions that are within the reasonable repertoire of both human beings and physical components. (The control of vehicles in an urban transportation system, for example, could be achieved by individual operators or by a central computer system.)

Because of the importance of these decisions for some systems, one would hope for some guidelines to aid the system designer in allocating specific functions to human beings versus physical machine components. The most common types of guidelines previously proposed consist of general statements about the kinds of things humans can do better than machines, and vice versa (Chapanis, 1960; Fitts, 1951, 1962; Meister, 1971). For example, such lists indicate that humans are better in their abilities to, among other things, sense unusual and unexpected events in the environment, reason inductively (generalize from observations), develop entirely new solutions, and detect stimuli against high-noise-level backgrounds. Machines are generally better in their abilities to, among other things, monitor for prespecified events, store and retrieve coded information, exert considerable physical force, and perform repetitive activities reliably.

As Chapanis (1965) points out, however, such comparisons serve a useful function in only the most elementary way. The practical utility of general comparisons of human and machine capabilities is rather limited. First, machines are evolving much faster than are humans so that the relative advantages of one over the other are constantly changing. In addition, such lists do not take into account the costs involved with employing a human or machine to do the function. Machines are gaining on humans, but often at great cost. It is not always important to provide for the "best" performance. Just because a machine can do something better than a human is no reason to assign the function to a machine every time. Do you need a computer whenever computations are required?

Probably the most important limitation of using human-machine comparison lists is that they do not recognize that function allocation directly predetermines the role of human beings in systems and thereby raises important issues of a social, cultural, economic, and even political nature. The basic roles of human beings in the production of the goods and services of the economy have a direct bearing on such factors as job satisfaction, human motivation, and the values of individuals and of the culture. Because our culture places a premium on certain human values, the system should not require human work activities that are incompatible with them. In this vein, Jordan (1963) sets forth the premise that humans and machines should be considered not comparable, but rather complementary. Whether one would agree with this or not, the fact does remain that decisions (if not allocations) need to be made concerning the relative roles of humans and machines. In this context and given that the objective of most systems is not the entertainment of the operators (pinball machines and gambling devices excepted), it would seem that, within reason, the human work activities generated by a system should provide the opportunity for reasonable intrinsic satisfaction to those who perform them. That is, we must design jobs that people can do and want to do.

Although general human-machine comparison lists are of limited value, specific lists tailored to specific applications may be more useful. For example, Barfield et al. (1987) developed a proposed allocation list for computer-assisted design (CAD) systems, and Nof (1985) developed a human-versus-robot comparison list.

When in Doubt, Automate? One approach, understandably advocated by many engineers, is to allocate functions to the machine whenever possible. That is, rely on automation to solve the allocation-of-function problem. This simplistic approach to the problem has an intuitive appeal given the achievements of automation, but in practice it can be dangerous. First, not all functions can be automated. The functions that are left for the human must comprise a meaningful set of tasks. Leaving the human with bits and snippets of unrelated functions can lead to cognitive underload, resulting in boredom, inattention, lack of motivation, and overall poor performance. Second, no machine is perfectly reliable, and the more complex the machine, generally the less reliable it is. Machines break, and when they do, a human often has to take over the operation of the system, as would be the case in modern jet aircraft. Automation often results in increased training costs. Humans have to know how to operate the automated system and, in addition, be given practice to maintain their skills for manual operation. Humans also have to be kept in the loop; they need to understand what the machine is doing so that if it fails they can take over its operation.

When automated equipment fails, as it often does in the early stages, operators become mistrustful of it and, if given an option, they will not rely upon the equipment, preferring to do it themselves. Sometimes, the opposite is true; operators rely too heavily on the automated system and do not adequately monitor system performance. When the system fails, they may not realize it. The crash of Eastern Airlines flight 401 could be attributed, in part, to just such a reliance on automation (Danaher, 1980). The crew set the autopilot to cruise at an altitude of 2000 ft while they checked on an apparent landing gear malfunction. Somehow the autopilot was switched off and the plane slowly descended into the ground. The crew, preoccupied with the landing gear problem, failed to notice the disengaged autopilot, probably assuming that the autopilot would perform its function.

A Strategy for Allocating Functions The discussion above suggests that functions cannot be allocated by formula (Price and Pulliam, 1983). The allocation process must rely on expert judgment as the final means for making meaningful decisions; allocation of function is as much an art as a science. Although there are no clear-cut guidelines for making specific allocation decisions, Price (1985) suggests four rules for developing an allocation strategy:

1 *Mandatory allocation*. Some functions or portions of functions may have to be allocated to a human or machine because of system requirements (e.g., a human must maintain abort control), hostile environments, safety considerations, or legal or labor constraints. Mandatory allocations should be identified and made first.

2 *Balance of value*. The relative goodness of a human and that of the to-be-available machine (including software) technology for performing the function are estimated and represented as a point in the decision space of Figure 22-3. If the point is in the unacceptable (*Umh*) region (i.e., it cannot be satisfactorily performed by either human or machine), then either the function must be redefined or the system requirements and constraints must be modified. If the point is in region *Uh* (unacceptable for humans) or region *Um* (unacceptable for machines), then the function should be treated as a mandatory allocation. If the function is in the *Ph* (better performed by humans) or *Pm* (better performed by machines) region, it can be tentatively allocated to the preferred alternative, but that allocation may be changed by the next two rules.

3 *Utilitarian and cost-based allocation*. In utilitarian allocation a function may be allocated to humans simply because human beings are present and

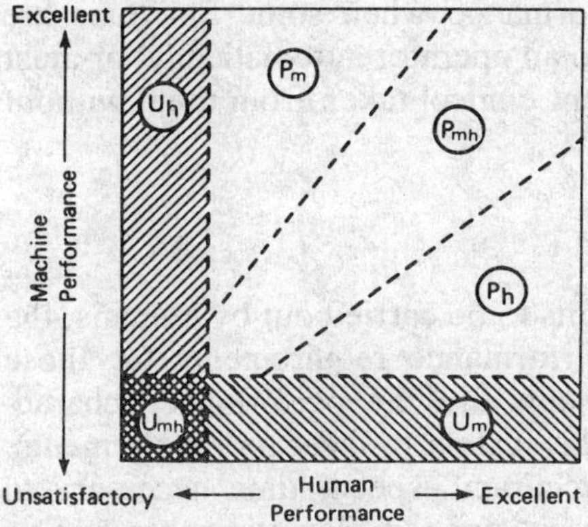

FIGURE 22-3
"Decision space" for representing relative machine and human performance as an aid in making allocation of function decisions. (*Source: Price, Maisano, and Van Cott, 1982.*)

there is no compelling reason why they should not perform the work. Notice, this is the antithesis of "when in doubt-automate." The relative costs of human and machine performance must be considered, and allocations can be made on the basis of least cost.

4 *Affective and cognitive support allocations.* The final rule recognizes the unique needs of humans. Allocation decisions may have to be revised to provide affective and cognitive support for the humans in the system. *Affective support* refers to the emotional requirements of humans, such as their need to do challenging work, to know their work has value, to feel personally secure, and to be in control. This has a bearing on issues of job satisfaction and motivation. *Cognitive support* refers to the human need for information so as to be ready for action or to make decisions that may be required. The human must maintain an adequate "mental model" of the system and its condition in order to take control in an emergency. Another consideration in cognitive support is that the human be given sufficient activity to ensure alertness.

The allocation of functions can best be accomplished by a team of individuals, including engineers and human factors specialists. The rules outlined above should be viewed as a reasonable starting point with the understanding that the detailed decisions still depend on the judgments of experts.

Dynamic Allocation Rather than make final allocation decisions during design, Kantowitz and Sorkin (1987) suggest that the system can be designed to allow the operator to make allocation decisions during the operation of the system. They call this *dynamic allocation*. When the operator feels a need for the machine to perform certain functions he or she was doing, the automatic system can be engaged. Autopilots on aircraft and cruise controls in automobiles are examples of dynamic allocation. A futuristic application would be a system that senses when an operator is being overloaded (or underloaded) and takes over (or gives back) functions from (to) the operator—automatically. Such a system would raise many interesting human factors issues. For example, how do you tell an overloaded operator that the system is taking control?

What happens to integrated human performance when some functions are taken over? Can you give functions back to an operator automatically, or must it be operator-initiated? Will operators want control taken from them without recourse?

Human Performance Requirements

After the design team identifies the functions to be carried out by humans, the next step is to determine the human performance requirements for those functions. Human performance requirements are the performance characteristics which must be met in order for the system to meet its requirements. Addressed are such things as required accuracy, speed, time necessary to develop performance proficiency, and user satisfaction. Sometimes the human performance requirements are beyond the anticipated capacity of the user population. In such cases it is necessary to reallocate functions or even to redefine the system requirements developed in stage 1.

Task Description and Analysis

Meister (1985) makes a distinction between task description and task analysis, although other authors combine the two processes and speak only of task analysis. Even Meister, however, admits that the dividing line between the two is unclear. Yet both must be carried out during the system design process. It is not appropriate here to include an extensive discussion of various task description and analysis procedures. [The interested reader can consult Meister (1985) or Drury et al. (1987) for more information.] Our intent here is to present the general nature of the process, the types of information collected, and the purposes to which it can be applied.

Nature of the Process The general procedure for developing a task description and analysis (hereafter referred to as simply task analysis) is to list in sequence all the tasks that must be performed to accomplish the functions in which a human plays a part. Each task is then further broken down into the steps required to carry out the task. Each step, then, is usually analyzed to determine such things as the stimuli that initiate the step; decisions the human must make in performing the step; actions required during the step; information necessary to carry out the step; feedback information resulting from performing the step; potential sources of error or stress; and the criterion for successful performance. In addition, determinations of task criticality and difficulty can be made, the number and skill level of people needed to operate the system can be estimated, and projections regarding training requirements can be outlined.

Often one product of a task analysis is an *operational-sequence diagram* (OSD). An OSD is a graphical depiction of the interactions between people and equipment in the performance of a task over time. An example of an OSD is shown in Figure 22-4. Figure 22-5 summarizes the standard symbols used to

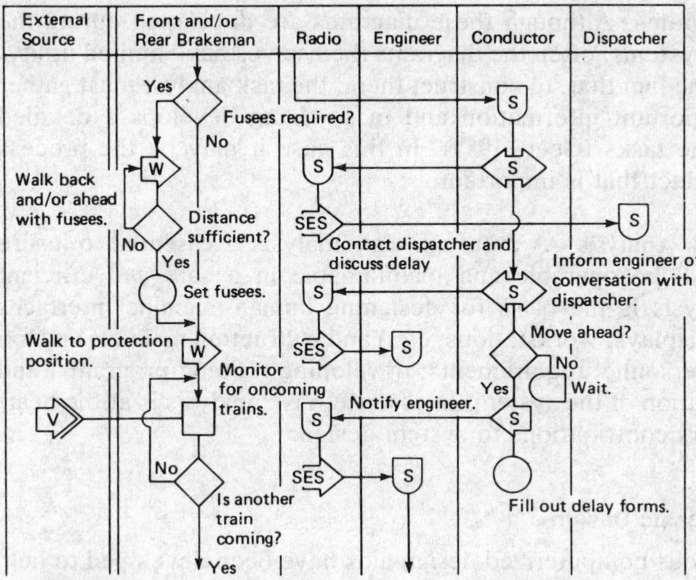

FIGURE 22-4
An example of an operational-sequence diagram for a task involved in railroad operations. The symbols used to construct the diagram are explained in Figure 22-5. (*Source: Sanders, Jankovich, and Goodpaster, 1974.*)

FIGURE 22-5
Standard symbols and notation used to construct operational-sequence diagrams.

OPERATE – An action function to accomplish or continue a process.

INSPECT – To monitor or verify quantity or quality.

TRANSMIT[1] – To pass information without changing its form. Also used to denote travel between two locations.

RECEIPT[1] – To receive information or objects.

DECISION – To evaluate and select a course of action or inaction based on receipt of information.

STORAGE – To retain information or objects.

DELAY – Period of inactivity.

---

1. Mode of transmission and receipt is indicated by a code letter within the TRANSMIT and RECEIPT symbols:

V – Visual	T – Touch
E – Electrical/Electronic	M – Mechanically
S – Sound (Verbal)	W – Walking
	H – Hand deliver

construct such diagrams. Although these diagrams are developed during the design of complex systems, often the diagrams themselves have limited utility. Their value lies in the fact that, to construct them, the task analyst must gather a great deal of important information and in so doing develops a detailed understanding of the tasks (Geer, 1981). In this case it may be the process rather than the product that is important.

Purposes for Task Analysis A thorough task analysis is essential to insure that the system will be operable and maintainable in a safe and efficient manner. Task analysis is the basis for designing human-machine interfaces (i.e., the controls, displays, workstations, etc.) and instruction manuals and job aids; determining personnel requirements; developing training programs; and designing the evaluation of the system. In essence, task analysis is at the heart of the human factors contributions to system design.

Computer Aids for Basic Design

Over the years various computerized design aids have been developed to help the designer make some of the basic design decisions discussed above. Although several of these design aids have been developed [see, for example, Geer (1981) or Meister (1985)], we present a brief overview of only one.

CAFES (computer-aided function allocation evaluation system), probably the most comprehensive and sophisticated of the computerized design aids, contains several modules that are useful throughout the system design process (Parks and Springer, 1975). The most complicated of the CAFES modules is the function allocation model (FAM). The purpose of FAM is to guide the designer in allocating functions and tasks. To use it requires a great deal of input data, such as task start and stop times, task attributes (e.g., criticality), functional flow diagrams, mission time lines, candidate allocations, and various weighting criteria. The designer inputs the data, and the computer calculates such useful information as the probabilities of successful task and mission performance, gross workload for each person in the system, a rank ordering of the various allocation strategies, a time line showing who in the system is doing what and when (useful for developing operational sequence diagrams), and the percentage of tasks completed, interrupted, or being performed simultaneously. With such information the designer can assess potential problems, alter the allocation of functions, and rerun the program until a satisfactory allocation strategy has been developed.

Another CAFES module is the workload assessment model (WAM). The designer inputs information regarding the task sequences and durations as well as the sensory and motor channels used to perform each task. WAM compares the time required to perform a task sequence with the time available for performance. This is done for eyes, hands, feet, and auditory, verbal, and cognitive channels. The program identifies time segments and tasks in which some criterion level of workload is exceeded.

The results from programs such as FAM and WAM should not be considered the final word on function allocation. The data they generate can be valuable, but such programs, as yet, do not take into account aspects such as the affective and cognitive support requirements of function allocation. In the end, experts will allocate functions based on their judgment and, perhaps, computer analyses similar to those of FAM and WAM. In this regard, Meister (1985) points out that despite their potential, programs such as FAM and WAM are used much less frequently than the traditional manual methods.

Job Design

To a very considerable degree the design of the equipment and other facilities that people use in their jobs predetermines the nature of the jobs they perform. Thus, the designers of some types of equipment are, in effect, designing the jobs of people who use the equipment. They may, or may not, be aware of this, but they certainly should be.

Human Values in Job Design As discussed above in the context of function allocation, there are certain philosophical considerations in the design of jobs. In this regard there has been considerable interest in the notion of job enlargement and job enrichment, this concern being based on the assumption that enlarged or enriched jobs generally bring about higher levels of motivation and job satisfaction. There are a number of variations on this theme, such as increasing the number of activities to be performed, giving the worker responsibility for inspection of her or his work, delegating responsibility for a complete unit (rather than for a specific part), providing opportunity to the worker to select the work methods to be used, rotating jobs, and placing greater responsibility on work groups for production processes.

The human factors approach has placed primary emphasis on efficiency or productivity and has tended toward the creation of jobs that are more specialized and that require less skill than jobs that are enlarged. In discussing this, McCormick and Ilgen (1985) emphasize that individuals differ in their system of values and that some people do not like "enlarged" or "enriched" jobs.

The degree of possible conflict between these two approaches is not yet clear, and there are as yet no clear-cut guidelines for designing jobs that would be compatible with the dual objectives of work efficiency and opportunity for motivation and job satisfaction on the part of workers. Although we cannot offer any pat solutions to this problem, it is to the credit of the human factors clan that increased attention is being given to this issue.

STAGE 4: INTERFACE DESIGN

After the basic design of the system is defined and the functions and tasks allocated to humans are delineated, attention can be given to the characteristics of the human-machine and human-software interfaces. These include work spaces, displays, controls, consoles, computer dialogs, etc. Many of the earlier chapters have dealt with the design and human factors considerations involved in such interfaces. Although the actual design of the physical components is predominantly an engineering chore, this stage represents a time when human factors inputs are of considerable importance. If inappropriate design decisions, in terms of human considerations, are made during this phase they will forever plague the user and cause decrements in system performance. If, on the other hand, appropriate decisions are made, the use of the system will be facilitated and improved performance will result. Human factors specialists usually work with engineers and designers in a team during the interface design stage of system development. The human factors specialist is primarily responsible for defending and promoting the human considerations of the design, while engineers are primarily responsible for defending and promoting the technological considerations of the design. Although this representation of the division of responsibility is somewhat simplistic, it does capture the essence of the roles played by human factors specialists and engineers during the design effort.

Human factors supports the design process by providing information to the engineers about the human performance implications of various design alternatives. Human factors specialists suggest design alternatives that match the capabilities and limitations of people and evaluate the human performance implications of designs suggested by engineers. When a human factors specialist suggests a design, often it is presented more as a set of specifications than as a detailed engineering proposal. A human factors specialist may suggest that a handle be placed within a specific area to facilitate access and that no more than *x* amount of force be required to operate it. It is left to the engineer to design the handle and to decide how it will be mounted and where within the recommended area it will be placed.

During the interface design, human factors members of the design team will perform three primary activities to support the design effort: (1) gather and interpret human factors and human performance data; (2) conduct attribute evaluations of suggested designs; and (3) conduct human performance studies. Before discussing these activities, we review some aspects of how engineers design to provide some context for the role of human factors.

How Engineers Design

Because the design process is mainly a cognitive process and thus somewhat covert, we really do not know a lot about how engineers make design decisions. A few studies of the process have been conducted (e.g., Meister, 1971; Rogers and Armstrong, 1977; Rouse and Boff, 1987). Meister (1989) summarizes some of the important findings that impact how engineers use human factors inputs.

1 *Engineers are experience-oriented.* They tend to repeat design approaches and solutions they have previously found effective. Although there is nothing inherently wrong with this approach, it does make it difficult to suggest alternatives that might improve human performance.

2 *Engineers are often intuitive in their thinking.* Often they do not systematically apply multiple criteria to evaluate design alternatives, but rather rely on their own intuition. Often they do not include behavioral criteria other than whether it makes sense to them.

3 *Engineers get down to the nitty-gritty aspects of hardware and software design as quickly as possible.* Often there is no time to conduct the human factors tests that one would like before the design is set in concrete, or at least plaster of Paris.

4 *Engineers often do not know where to find information they need.* This is especially true for information of a behavioral nature.

Klein and Brezovic (1986) surveyed training-device designers about sources of information they use during the design process. Figure 22-6 presents the percentage of design decision points where the various sources were rated as helpful by the designers. The high regard given to personal experiences and similar or familiar projects attests to the experience orientation of engineers discussed above. The low regard given to human factors experts, colleagues, and technical literature speaks to the difficulty in finding appropriate experts or locating relevant literature.

Meister (1989, p. 71) summarizes the situation by saying "design is at bottom a highly judgmental process in which all the engineer's biases and

FIGURE 22-6
Percent of design decision points where training-device designers rated various information sources as helpful. (*Source: Klein and Brezovic, 1986, Fig. 1. Reprinted with permission of the Human Factor Society, Inc. All rights reserved.*)

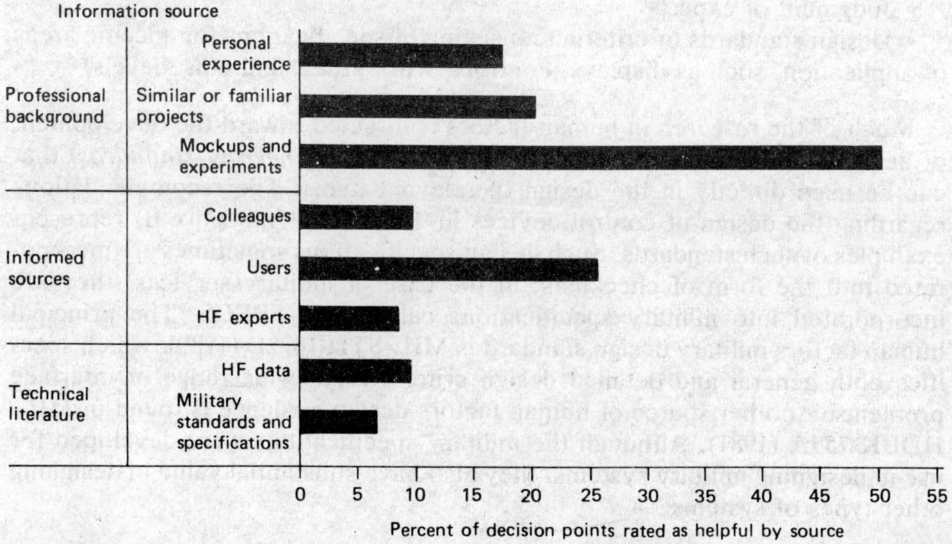

predilections have the opportunity to influence the final configuration." The implication of this is that human factors specialists need to have good people skills: they need to know how to negotiate, compromise, and persuade. Human factors specialists must also provide information that is relevant to the needs of the engineer and that will be useful for making design decisions.

Gathering and Interpreting Human Factors Data

A major and continuing effort of the human factors members of a design team during this stage of system development lies in tracking down and interpreting existing human factors data potentially relevant to the design problem. The data relevant to various human factors design problems cover a wide range (with many gaps yet to be filled) and exist in many forms, such as the following:

- Common sense and experience
- Comparative quantitative data (such as relative accuracy in reading two types of visual instruments, as illustrated in Chapter 5)
- Sets of quantitative data (such as anthropometric measures of samples of people, as given in Chapter 13, and error rates in performing various tasks)
- Principles (based on substantial experience and research, which provide guidelines for design, such as the principle of avoiding or minimizing glare when possible)
- Mathematical functions and equations (which describe certain basic relationships with human performance, such as in certain types of simulation models)
- Graphical representations (nomographs or other representations, such as tolerance to acceleration of various intensities and durations)
- Judgment of experts
- Design standards or criteria (consisting of specifications for specific areas of application, such as displays, controls, work areas, and noise levels)

Much of the research in human factors is directed toward the development of design standards or criteria (sometimes called *engineering standards*) that can be used directly in the design of relevant items. The recommendations regarding the design of control devices in Table B-2 (Appendix B) represent examples of such standards. Such design specifications sometimes are incorporated into the form of checklists; in the case of military services, they are incorporated into military specifications called MIL-SPECS. The principal human factors military design standard is MIL-STD-1472D (1989), which specifies both general and detailed design criteria for a wide range of interface problems. Another source of human factors design guidance is found in MIL-HDBK-759A (1981). Although the military specifications were developed for use in designing military systems, they also have substantial value in designing other types of systems.

Use of Human Factors Data It would be very satisfying to say that available human factors data are being extensively applied to the design of the things people use in their work and daily living. But such is not the case. Meister (1971), for example, reported that (unfortunately) many design engineers did *not* consider human factors in their design procedures. And Rogers and Armstrong (1977), on the basis of a study dealing with the use of human factors design standards in engineering design, came to the dismal conclusion that such standards have little, if any, effect on product design. They suggested that the apparent reasons for this are varied and complex, involving resistance on the part of designers and managers, education of human factors specialists and designers, the standards themselves, and interdisciplinary communication. Klein and Brezovic (1986) found that designers had difficulty in extrapolating from published studies because they thought the information was too general or they considered the methods used as irrelevant. The latter occurred when the experimental conditions differed from the operational design conditions, and the designers did not know how to take such differences into account, so they made only limited use of the data or rejected it altogether.

The human factors research community probably has contributed to this state of affairs in at least a couple of ways, in particular in failing to communicate adequately with people in other disciplines (such as engineering, architecture, and industrial design) and in failing to present human factors data in a reasonably useful form.

Presentation of Human Factors Data Blanchard (1975) proposes a set of guidelines for the development of human factors data banks that would better contribute to the use of such data. In connection with the development of human factors design standards, Rogers and Armstrong (1977) state that such standards would be enhanced if the following suggestions were incorporated:

1 Eliminate general or ambiguous terms such as "proper feel" or "high torque."

2 Present quantitative data in a manner consistent with designer preference, i.e., graphical or pictorial first, followed by tabulations.

3 Eliminate the use of narrative statements when data can be presented quantitatively.

4 Eliminate inconsistencies in data within and among standards.

5 Provide revisions and updating of standards on a more timely basis.

Rogers and Pegden (1977) elaborate on some of these recommendations and add certain others, such as providing a quick-access system, including relevant definitions, and providing adequate cross-references. An excellent example of this approach is the *Engineering Data Compendium* edited by Boff and Lincoln (1988); this three-volume set contains over 2000 figures and tables and an index of over 10,000 entries.

Considerations in Applying Human Factors Data The application of relevant human factors data to some design problems is a fairly straightforward proposition. More generally, however, it is necessary to use judgment in this process, in particular, in evaluating the applicability of potentially relevant data to specific design problems. In making such evaluations at least four considerations should be taken into account.

Practical Significance One consideration is the practical significance of the application of some relevant human factors information. For example, although the time required to use control device A might be significantly less (statistically) than that for device B, the difference might be so slight, and of such nominal utility, that it would not be worthwhile to use device A if other factors (such as cost) argued against it.

Extrapolation to Different Situations Much of the available human factors data has been based on research findings or experience in certain specific settings. The second consideration in the application of human factors data deals with possible extrapolation of such data to other settings. For example, could one assume that the reaction time of astronauts in a space capsule would be the same as in an earthbound environment? Referring specifically to laboratory studies, Chapanis (1967) urges extreme caution in applying the results of such studies to real-world problems.

Perhaps three points might be made about these sobering reflections. First, despite Chapanis's words of caution, there are some design problems for which available research data or experience is fully adequate to resolve the design problem. Second, in certain areas of investigation it may be possible to simulate the real world with sufficient fidelity to derive research findings that can be used with reasonable confidence. Third, there are undoubtedly circumstances in which one might extrapolate from data that are not uniquely appropriate to the design problem at hand, on the expectation that this would increase the odds of achieving a better design than if such data were not used. Often published studies provide a ball-park estimate that can be refined by human factors expertise and judgment. Clearly this strategy involves risks and should be followed only when the risks can be justified.

Consideration of Risks A third consideration in applying human factors data is associated with the seriousness of the risks incurred in making bad guesses. Bad guesses in designing a shuttle for shipping an astronaut into space obviously would be a matter of greater concern than those in designing a hat rack. The more serious the risks, the greater the need for relevant, high-quality data.

Consideration of Trade-Off Function A fourth consideration is trade-off values. Since frequently it is not possible to achieve an optimum in all possible criteria for a given design of a human-machine system, some give and take must be accepted. The possible payoff of one feature (as suggested by the results of research) may have to be sacrificed, at least in part, for some other more desirable payoff.

Conducting Attribute Evaluations

In addition to the application of relevant human factors data, human factors specialists use other methods to assist in the interface design. *Attribute evaluation* consists of assessing whether or not a design meets appropriate human factors standards and guidelines. Each attribute of the design is evaluated. Attributes might include such things as the readability of displays, accessability of controls, amount of resistance in controls, adjustability of seats, space available for performing maintenance, etc. Attribute evaluations are performed on 2-D engineering drawings, mock-ups of the equipment, and prototype equipment. Usually the human factors specialist uses some sort of checklist to ensure that all relevant attributes are considered.

Use of Checklists More often than not, attribute evaluations are conducted rather informally and unsystematically, with a checklist acting only as a guide to the evaluator. Formal and strict item-by-item use of checklists can be very time-consuming and has its limitations. Checklists are usually verbal, stated as a standard (''Minimum character height is 16 min of visual angle'') or as a question (''Is character height adequate at anticipated viewing distances?''), yet what is being evaluated is a physical entity or a pictorial representation of the entity. Thus, easy-to-use checklists are often difficult to construct and difficult to apply. Checklists foster a pass-fail approach to evaluation, yet there are degrees of acceptability. For example, if the checklist item indicates a minimum character height of 16 min of visual angle, does the system fail that attribute if the character height is 15.5 min of visual angle? Checklists do not handle interactions between attribute items very well. If the illumination is barely adequate, how does that affect the evaluation of the 15.5 min character height? Checklists also have difficulty dealing with cumulative effects of several just-adequate attributes. If temperature, noise, and illumination are each just within acceptable ranges, it may still be possible that the environment as a whole is unacceptable. There is no question that checklists are valuable instruments, but their limitations must be recognized and the results obtained must be tempered by good expert judgment.

Walk-Through Evaluations In addition to paper-and-pencil attribute evaluations using checklists, people representative of the user population can be placed in mock-ups of the proposed equipment and walked through the planned task sequences. Clearances, reach envelopes, visibility envelopes, etc. can be systematically assessed by using people of different sizes. Sometimes rudimentary performance evaluations can be conducted measuring such things as time required to exit or movement time in reaching for a control.

Computer-Aided Design and Evaluation Aids Construction and modification of mock-ups to reflect alternative designs and the evaluation of each version by using human subjects are both expensive and time-consuming. In practice, only a few alternative designs may be considered. An alternative to constructing mock-ups is to use computer simulation with its graphics capabilities to compare alternatives early in the design process. Several computer-based workstation design and evaluation aids are available, including such acronym-based programs as HECAD (Topmiller and Aume, 1978), COMBIMAN (Evans, 1978), CHESS (Jones, Jonsen, and Van, 1982) and SAMMIE (Bonney and Case, 1976). To illustrate the nature of such design aids, we briefly discuss the SAMMIE system (SAMMIE CAD Ltd., 1987). SAMMIE is a 3-D representation of the human body. The dimensions and body shape of the model can be varied to reflect the range of sizes and shapes in a particular population. SAMMIE can be placed in a 3-D representation of the system being evaluated and moved into various postures and positions. Figure 22-7 presents a few examples from SAMMIE. Figure 22-7*a* shows SAMMIE assuming a driving position, looking rearward while reversing and maintaining control of the vehicle speed and direction. This was taken from an evaluation of an English car with the driver in the right-hand seat. The driver's left shoulder is shown to intersect the top of the backrest, indicating that a soft foam would be desirable in this location. SAMMIE can display what a person would see if positioned within the system, an example of which is shown in Figure 22-7*b*. In addition, as shown in Figure 22-7*c*, reach volumes for various sizes of people can also be displayed to assess control accessability. SAMMIE can also be used to design multiperson facilities such as an office. What would have taken months to evaluate using mock-ups can be done in hours or days using computer-based workstation design and evaluation aids such as SAMMIE.

FIGURE 22-7
Examples of applications of SAMMIE, a computer-based design and evaluation tool. (*a*) SAMMIE in an automobile preparing to move in reverse (in England, drivers sit on the right-hand side); (*b*) visual field as seen from the operator's position; (*c*) reach volume for an operator in a helicopter. (*Source: SAMMIE CAD Ltd., 1990.*)

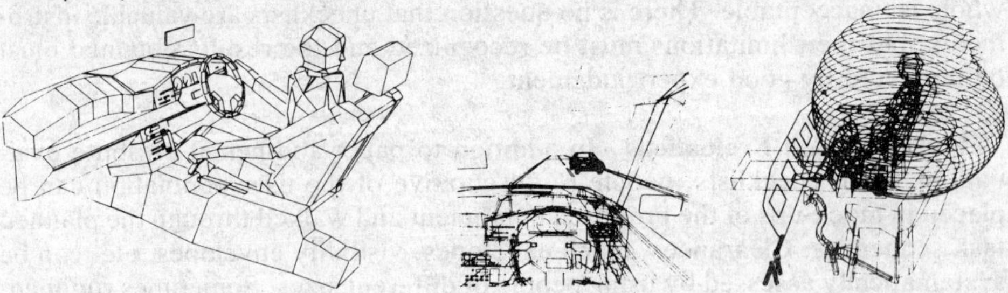

(*a*) (*b*) (*c*)

Conducting Human Performance Studies

When the available literature cannot supply the solution to a design problem and the risks involved in making an incorrect decision are high, research should be undertaken to provide an answer. These studies are usually small experiments involving several subjects representative of the user population. The studies are more elaborate than simple walk-throughs and usually involve measurement of human performance such as time and errors in completing a task. For example, a comparison might be made between several screen layouts in terms of the time required to find target information. Various control-response ratios for a trackball might be tested in terms of target acquisition time or tracking performance. Unfortunately, often there is not enough time or money available to perform multifactor research designs with large sample sizes and rigorous experimental controls. However, important information can be obtained from even the simplest small-scale studies that, along with the expertise of a human factors specialist, can provide valuable input to the design process.

STAGE 5: FACILITATOR DESIGN

The main focus of this stage of the system design process is to plan for materials that will promote acceptable human performance. Included are instruction manuals, performance aids, and training devices and programs. Although not substitutes for good design, these support materials can have an impact on the decisions made during the interface design stage. Most, if not all, complex systems require some training and properly designed documentation to aid system users. It is important that these aspects of the system's design be given the same consideration and time as the design of the human-hardware and human-software interfaces.

In some systems a concept referred to as *embedded training* is utilized. Embedded training means that training programs are "built into" the system so that the operational equipment can be switched over to a training mode during periods when it is not needed for operational use.

One form of system documentation that specifies how to operate and, perhaps, maintain the system is instruction manuals. Instruction manuals can be elaborate books or a simple piece of paper packed with a product. We take a moment to discusss a few issues involved in the design of instruction manuals.

Instruction Manuals

The design of effective instruction manuals is more an art than a science. Although we cannot discuss the many human factors issues involved in the design of documentation, listing a few will illustrate the complexity of the task. One needs to consider such things as whether the documentation will be hard copy or online (i.e., part of the software) or some combination; how to utilize illustrations; how to combine text and illustrations; how much detail to present; and how to index, tab, and organize the material. In regard to some of these issues Coskuntuna and Mauro (1980) offer these overly simplified rules of thumb:

- Less is more: avoid informational overload.
- Avoid abstract information; use only concrete information.
- Forget why's; concentrate on how's.
- Remember that learning will come from doing.
- Forget the *hype*. The users already have the product, and what they want to do now is set it up or use it with minimal hassle.

As an aside, they urge putting important material at the front or in a prominent position.

Use of Illustrations Illustrations have their place either in connection with manuals or instructions or simply by themselves to give instructions or to serve as warnings. In this regard Johnson (1980) differentiates between pictures and symbols: a *picture* is a realistic photograph or drawing of an object about which information is to be conveyed; a *symbol* is a photograph or drawing that represents something *else,* such as the Red Cross symbol. In turn he describes a *pictogram* as a series of associated pictures intended to give information about performing a series of actions.

In discussing the use of illustrations to convey information to users, Johnson makes the point that concepts about which information is to be conveyed range from simple to complex, as indicated by the following ends of the continuum:

- Simple concept: instructions are given to perform a single action (why and when are obvious).
- Complex concept: multiple actions are required, and multiple results can ensue (when and why an action is to be performed may not be obvious).

He also makes the point that the concepts in question can range from concrete (such as an emergency door) to abstract (with no physical referent, such as the general notion of danger). The various combinations of these distinctions, along with proposed methods of conveying information about them, are given in Table 22-1.

STAGE 6: TESTING AND EVALUATION*

Evaluation in the context of system development has been defined as the measurement of system development products (hardware, procedures, and personnel) to verify that they will do what they are supposed to do (Meister and

* For a thorough treatment of human factors evaluation in system development, see Meister and Rabideau (1965) and Meister (1986).

TABLE 22-1
PROPOSED METHODS FOR CONVEYING INSTRUCTIONAL INFORMATION

	Concrete data	Abstract data
Simple	Use pictures/words	Use symbols/words
Complex	Use pictograms/words	Use words

Source: Johnson, 1980

Rabideau, 1965, p. 13). In turn, *human factors evaluation* is the examination of these products to ensure the adequacy of attributes that have implications for human performance. Actually, almost every decision during the design of a system includes some evaluation, such as deciding whether to use a visual signal or an auditory signal. Although many such evaluative decisions need to be made as part of the ongoing development cycle, for most items developed for human use there should be some systematic evaluation upon the completion of the development stage and before the item goes into production. Although such evaluation should be carried out, unfortunately many things people use in their work and everyday lives do not benefit from such evaluation. Thus many things do not serve their purposes effectively or safely and in some instances have to be recalled for expensive retrofitting or replacement.

The Nature of Human Factors Evaluation

The evaluation of some items in terms of human factors considerations should be carried out systematically, in much the same manner as an experiment. In many instances, however, probably the evaluation procedures leave much to be desired in terms of acceptable procedures.

In connection with experimental procedures, Meister (1978) differentiated between what he called controlled experimentation (CE) and personnel subsystem measurement (PSM). The CE experimenter deals largely with the *explanation* of relationships and phenomena, such as explaining why reaction time is a function of the numbers of stimuli and the possible responses to them. The PSM investigator, however, must be concerned with the measurement of human performance (and related criteria) *in operational terms,* that is, in terms relevant to the system, subsystem, or item in question, specifically to the functions of the personnel involved in the operation and maintenance of the system.

Special Problems in Human Factors Evaluation

The above comments may give some inklings about the problems of human factors evaluation. We can touch here on only a few problems that haunt those responsible for human factors evaluation in actual operational situations or in circumstances that approximate operational conditions. Although much of the research on which human factors data and principles are developed is carried out in laboratory settings, DeGreene (1977) makes the point that the real laboratory for human factors research should be the real world. This is indeed so in the final evaluation and testing of things to be used by people.

The overview of research methods discussed in Chapter 2 is as applicable to the final evaluation of the things people are to use as it is to the research aimed toward developing human factors data and principles. However, the problems involved in evaluation in the real world usually are much more complex than those in the laboratory. In the design of evaluation procedures, consideration must be given to four factors: subjects, criteria, experimental procedures, and research setting.

Subjects The evaluation should be carried out with the same types of individuals who are expected to use the item in question, such as homemakers, trained pilots, persons with driver's licenses, factory assemblers, handicapped persons, or the public at large.

Criteria The criteria (the dependent variables) should be in some manner relevant to the operational use of the item, and, depending on the item, they can include work performance, physiological effects, accidents, effects on health, learning time, job satisfaction, attitudes and opinions, and economic considerations, to mention a few. The problems of measuring some of these can drive an experimenter crazy. The use of multiple criteria is characteristic of most evaluation research. Rarely, if ever, is a single criterion sufficient basis for an evaluation.

Experimental Procedures and Controls The problem of carrying out appropriate experimental procedures, including adequate controls (involving in some instances control groups), is much greater in the real world than in a laboratory. We cannot discuss this problem in detail here, but it should be recognized by the investigator.

Johnson and Baker (1974) point out that in many field studies it is not possible to replicate the study, control groups cannot always be used, and the physical environment cannot be completely controlled. Recognizing these and other limitations of evaluation in real-world circumstances, the investigator should make every reasonable effort to follow appropriate experimental procedures.

Research Setting To properly evaluate a system it should be tested under conditions as close as possible to those under which it ultimately will be used. To not do so could result in release of a system or product that neither satisfies its users' needs nor meets its intended purpose (see Figure 22-8). The term *operational fidelity* refers to the relationship between the operational environ-

FIGURE 22-8
To properly evaluate a system it should be tested under conditions as close as possible to those under which it will ultimately be used. (*Reprinted with special permission of King Features Syndicate.*)

ment and the testing environment (or in the case of training systems, the training environment). The goal is to maximize the closeness of the two environments. The two do not have to be identical in order to produce valuable test data. What is needed is that the essential features of the operational environment be replicated in the testing environment.

Meister (1989) identifies several dimensions of operational fidelity: *physical fidelity,* which relates to the physical elements, equipment, technical data, and environmental parameters (e.g., temperature, lighting, vibration); *task fidelity,* which refers to the similarity of the tasks to be performed; *organizational fidelity,* which includes the similarity of team and group organization; and *personnel fidelity,* which relates to the personnel background, training, skills, and aptitudes of the people. Fidelity is not an all-or-none affair. There are degrees of operational fidelity, and the question is how much is necessary to provide valuable data on system performance. Unfortunately, there is no way to quantify fidelity and there is no easy answer to how much is enough.

Although it may be possible to duplicate the operational environment, it is more difficult to duplicate the motivation and attitudes of the ultimate users as they would be in the operational setting. Subjects in an evaluation, even if selected from the actual user population, know they are evaluating a new system and probably act and react somewhat differently than they would in a nonevaluative setting. One would expect, for example, that subjects in an evaluation would act more carefully and safely than they would in an operational setting in which they were not being observed by a researcher. This, however, should not diminish the value of a well-constructed evaluation. Often people are more critical of a system in their role as subjects than they would normally be and can thus offer important insights into potential system problems. In the final analysis, nothing substitutes for a good system evaluation using people from the intended user population. It never ceases to amaze how different people are in their perceptions and interpretations of information. Something that seems so obvious to the design team may be totally confusing to users.

DISCUSSION OF HUMAN FACTORS AND SYSTEM DESIGN

Clearly the application of human factors data to design processes does not (at least yet) lend itself to the formulation of a completely routine, objective set of procedures and solutions. However, a systematic consideration of the human factors aspects of a system at least focuses attention on features that should be designed with human beings in mind. In this connection, it might be useful to list at least some reminders that are appropriate when one is approaching a design problem. These reminders are presented in the form of a series of questions (with occasional supplementary comments). Two points should be made about these questions. First, some of these would not be pertinent in the design of particular items; in turn, this is not intended as an all-inclusive list of questions. Second (and as indicated frequently before), the fulfillment of one objective may of necessity be at the cost of another. Nevertheless, this list of questions should serve as a good start in the design process.

1 What functions need to be carried out to fulfill the system objective?

2 If there are any reasonable options available, which should be performed by human beings?

3 For a given function, what information external to the individual is required? Of such information, what information can be adequately received directly from the environment, and what information should be presented through the use of displays?

4 For information to be presented by displays, what sensory modality should be used? Consideration should be given to the relative advantages and disadvantages of the various sensory modalities for receiving the type of information in question.

5 For a given type of information, what type of display should be used? The display generally should provide the information when and where it is needed. These considerations may reflect the general type of display, the stimulus dimension and codes to be used, and the specific features of the display. The display should provide for adequate sensory discrimination of the minimum differences that are required.

6 Are the various visual displays arranged for optimum use?

7 Are the information inputs collectively within reasonable bounds of human information-receiving capacities?

8 Do the various information sources avoid excessive time-sharing?

9 Are the decision-making and adaptive abilities of human beings appropriately utilized?

10 Are the decisions to be made at any given time within the reasonable capability limits of human beings?

11 In the case of automated systems or components, do the individuals have basic *control* so that they do not feel that their behavior is being controlled by the system?

12 When physical control is to be exercised by an individual, what type of control device should be used?

13 Is each control device easily identifiable?

14 Are the controls properly designed in terms of shape, size, and other relevant considerations?

15 Are the operational requirements of any given control (as well as of the controls generally) within reasonable bounds? The requirements for force, speed, precision, etc., should be within limits of virtually all persons who are to use the system. The human-machine dynamics should capitalize on human abilities so that, in operation, the devices meet the specified system requirements.

16 Is the operation of each control device compatible with any corresponding display and with common human response tendencies?

17 Are the control devices arranged conveniently and for reasonably optimum use?

18 Is the work space suitable for the range of individuals who will use the facility?

19 Are the various components and other features of the facility arranged in a satisfactory manner for ease of use and safety?

20 When relevant, is the visibility from the work station satisfactory?

21 If there is a communication network, will the communication flow avoid overburdening the individuals involved?

22 Are the various tasks to be done grouped appropriately into jobs?

23 Do the tasks which require time-sharing avoid overburdening any individual or the system? Particular attention needs to be given to the possibility of overburdening in emergencies.

24 Is there provision for adequate redundancy in the system, especially of critical functions? Redundancy can be provided in the form of backup or parallel components (either persons or machines).

25 Are the jobs such that personnel can be trained to do them?

26 If so, is the training period expected to be within reasonable time limits?

27 Do the work aids and training complement each other?

28 If training simulators are used, do they achieve a reasonable balance between transfer of training and costs?

29 Is the system or item adequately designed for convenient maintenance and repair, including any individual components? For example, is there adequate clearance space for reaching parts that need to be repaired or replaced? Can individual parts be repaired or replaced easily? Are proper tools and adequate troubleshooting aids available? Are there adequate instructions for maintenance and repair?

30 Do the environmental conditions (temperature, illumination, noise, etc.) permit satisfactory levels of human performance and provide for the physical well-being of individuals?

31 In any evaluation or test of the system (or components), does the system performance meet the desired performance requirements?

32 Does the system in its entirety provide reasonable opportunity for the individuals involved to experience some form and degree of self-fulfillment and to fulfill some of the human values that we should all like to have the opportunity to fulfill in our daily lives?

33 Does the system in its entirety contribute generally to the fulfillment of reasonable human values? In systems with identifiable outputs of goods and services, this consideration would apply to those goods and services. In the case of systems that relate to our life space and everyday living, this consideration would apply to the potential fulfillment of those human values that are within the reasonable bounds of our civilization.

In the resolution of these and other kinds of human factors considerations, one should draw upon whatever relevant information is available. This information can be of different types, including principles that have been developed through experience or research, sets of normative data (such as frequency distributions of, say, body size), sets of factual data of a probability nature (such as percentage of signals detected under specified conditions), mathematical formulas, tentative theories of behavior, hypotheses suggested by research investigations, and even the general knowledge acquired through everyday experience.

With respect to information that would have to be generated through research (as opposed to experience), while very comprehensive information is available in certain areas of knowledge, in others it is pretty skimpy, and there are some areas in which one draws a complete blank. Where adequate information is not available, perhaps considered judgments based on partial information will, in the long run, result in better design decisions than those that are arbitrary. But let us reinforce the point that such judgments usually should be made by those whose professional training and experience put them in the class of experts, whether in the field of night vision, physical anthropometry, hearing disorders, perception, heat stress, acceleration, learning, or decision making.

Here, again, we mention the almost inevitability of having to trade certain advantages for others. The balancing of advantages and disadvantages generally needs to take into account various types of considerations—engineering feasibility, human considerations, economic considerations, and others. Given that there probably are few guidelines to follow, the general objective of this horse trading is fairly clear.

SOME CLOSING REFLECTIONS

Several earlier chapters dealt with certain features of human beings relevant to the design of the things people use in their work and everyday lives, including people's sensory and perceptual skills, psychomotor abilities, and anthropometric characteristics. In addition, various chapters have dealt with some of the implications of such human characteristics to the design of some of the things people use, as displays, controls, work-space arrangement and layout, and the physical environment.

In reflecting about this coverage, it should be kept in mind that the material included represents only a small sample of the relevant material that could have been included. This text offers an overview of the human factors field and is not a complete compendium of relevant information. (Such a compendium would occupy much more than the proverbial 5-ft bookshelf.) Given that this text represents a sample of the human factors domain, it has been an objective to help the reader become aware of, or become more sensitive to, the human factors implications of the things people use.

In connection with human factors research and applications to date, we could develop quite a list of areas that it has not been feasible to include in this text or that are touched on very briefly. Such a list could include such areas as consumer products, health services, recreation, the expanding use of computers, law enforcement, exploration of outer space, communication systems, mining, agriculture, (in the words of the king of Siam in *The King and I*) etcetera, etcetera, etcetera. (It should be added that the human factors inroads into some of these, and other, areas have to date still been fairly limited.)

But now, what of the future? There are many items of unfinished business relating to the many facets of present-day work and living that have not yet benefited from the possible application of human factors data and principles. Beyond such concerns, however, there are at least a couple of broad areas of attention that should serve as challenges to the human factors professional.

One of these deals with what (in present-day expressions) is called the *quality of working life* or, more broadly, the *quality of life*. There is some truth to the charges that human factors efforts have tended toward making work easier for people to do and perhaps toward greater work specialization. In recent years, however, we have heard much about job enrichment programs that are directed toward opposite objectives of broadening work responsibilities, increasing the decision-making aspects of jobs, etc. These efforts have as their aim the enhancement of job satisfaction and the improvement of the quality of working life. Although job enrichment probably is not for everybody, its general goals are on the positive side of the ledger. In recent years several people prominent in human factors affairs have expressed concern about the quality of working life and have urged the human factors community to work toward such a goal. However, there seems to be a basic need for efforts to blend the human factors and the job enrichment approaches into an integrated effort which would offer reasonable possibilities for people to engage in work activities that are efficient and safe and that also would enhance the quality of working life.

The second area of challenge to the human factors discipline deals with the shape of things to come and actually is related to the previous discussion of the quality of working life. The future is always somewhat shrouded, but we certainly can expect increased automation in the production of goods and services with increased use of computer-controlled processes, significant changes in technology and in the products and services that such changes may bring, and the development of new products and services that we simply cannot envision. The "brave new world" of the future should indeed be developed with people—all of us—in mind. Thus, the human factors discipline must be at the cutting edge of future developments to ensure that such developments will, in reality, contribute to the improvement of the quality of working life and of life in general.

REFERENCES

Bailey, R. (1982). *Human performance engineering*. Englewood Cliffs, NJ: Prentice-Hall.

Barfield, W., Chang, T., Majchrzak, A., Eberts, R., and Salvendy, G. (1987). Technical and human aspects of computer-aided design (CAD). In G. Salvendy (ed.), *Handbook of human factors*. New York: Wiley.

Beevis, D., and Hill, M. (1983). The designer as the limiting human factor in military systems. *Proceedings of the 24th Defense Research Group Seminar on the Human as a Limiting Element in Military Systems,* vol. 1. Toronto, Ont., Canada: NATO.

Blanchard, R. E. (1975). Human performance and personnel resource data store design guidelines. *Human Factors*. 17(1), 25–34.

Boff, K., and Lincoln, J. (1988). *Engineering data compendium: Human perception and performance*. Wright-Patterson Air Force Base, OH: Harry G. Armstrong Aerospace Medical Research Laboratory.

Bonney, M., and Case, K. (1976). The development of SAMMIE for computer aided work place and work task design. *Proceedings, 6th Congress of the International Ergonomics Association and Technical Program for the 20th Annual Meeting of the Human Factors Society*. Santa Monica, CA: Human Factors Society, pp. 340–348.

Chapanis, A. (1960). Human engineering. In C. D. Flagle, W. H. Huggins, and R. H. Roy (eds.), *Operations research and systems engineering*. Baltimore, MD: John Hopkins.

Chapanis, A. (1965). On the allocation of functions between men and machines. *Occupational Psychology,* 39, 1–11.

Chapanis, A. (1967). The relevance of laboratory studies to practical situations, *Ergonomics,* 10(5), 557–577.

Coskuntuna, S., and Mauro, C. L. (1980). Human factors and industrial design in consumer products. *Proceedings of the Symposium: Human Factors and Industrial Design in Consumer Products*. Medford, MA: Tufts University, pp. 300–313.

Danaher, J. W. (1980). Human error in ATC system operations. *Human Factors,* 22, 535–545.

DeGreene, K. B. (1977). Has human factors come of age? *Proceedings of the Human Factors Society 21st Annual Meeting*. Santa Monica, CA: Human Factors Society, pp. 457–461.

Drury, C., Paramore, B., Van Cott, H., Grey, S., and Corlett, E. (1987). Task analysis. In G. Salvendy (ed.), *Handbook of human factors*. New York: Wiley.

Evans, S. (1978). *Updated user's guide for the COMBIMAN* (Rept. AMRL-TR-78-31). Wright-Patterson Air Force Base, OH: Aerospace Medical Research Laboratory.

Fitts, P. M. (ed.) (1951). *Human engineering for an effective air-navigation and traffic-control system*. Washington, DC: NRC.

Fitts, P. M. (1962). Functions of men in complex systems. *Aerospace Engineering*, 21(1), 34–39.

Geer, C. (1981). *Human engineering procedures guide* (AFAMRL-TR-81-35). Wright-Patterson Air Force Base, OH: Aerospace Medical Research Laboratory.

Johnson, D. A. (1980). The design of effective safety information diplays. *Proceedings of the Symposium: Human Factors and Industrial Design in Consumer Products*. Medford, MA: Tufts University, pp. 314–328.

Johnson, E. M., and Baker, J. D. (1974). Field testing: The delicate compromise. *Human Factors*, 16(3), 203–214.

Jones, R., Jonsen, G., and Van, C. (1982). Evaluation of control station design: The crew human engineering software system. *Proceedings of the Human Factors Society 26th Annual Meeting*. Santa Monica, CA: Human Factors Society, pp. 40–43.

Jordan, N. (1963). Allocation of functions between man and machines in automated systems. *Journal of Applied Psychology*, 47(3), 161–165.

Kantowitz, B., and Sorkin, R. (1987). Allocation of functions. In G. Salvendy (ed.), *Handbook of human factors*. New York: Wiley.

Klein, G., and Brezovic, C. (1986). Design engineers and the design process: Decision strategies and human factors literature. *Proceedings of the Human Factors Society 30th Annual Meeting*. Santa Monica, CA: Human Factors Society, pp. 771–775

McCormick, E. J., and Ilgen, D. R. (1985). *Industrial psychology* (8th ed.). Englewood Cliffs, NJ: Prentice-Hall.

Meister, D. (1971). *Human factors: Theory and practice*. New York: Wiley.

Meister, D. (1978). A theoretical structure for personnel subsystem management. *Proceedings of the Human Factors Society 22d Annual Meeting*. Santa Monica, CA: Human Factors Society, pp. 474–478.

Meister, D. (1985). *Behavioral analysis and measurement methods*. New York: Wiley.

Meister, D. (1986). *Advances in human factors/ergonomics*, vol. 5: *Human factors testing and evaluation*. Amsterdam: Elsevier.

Meister, D. (1989). *Conceptual aspects of human factors*. Baltimore: Johns Hopkins University Press.

Meister, D., and Rabideau, G. F. (1965). *Human factors evaluation in system development*. New York: Wiley.

MIL-HDBK-759A (1981, June). *Military handbook: Human factors engineering design for army material*. Washington, DC: Department of Defense.

MIL-STD-1472D (1989, March). *Human engineering design criteria for military systems, equipment, and facilities*. Washington, DC: Department of Defense.

Nof, S. (1985). Robot ergonomics: Optimizing robot work. In S. Nof (ed.), *Handbook of industrial robotics*. New York: Wiley.

Parks, D., and Springer, W. (1975). *Human factors engineering analytic process definition and criterion development for CAFES* (Rept. D180-18750-1). Seattle: Boeing Aerospace Company.

Price, H. (1985). The allocation of functions in systems. *Human Factors*, 27(1), 33–45.

Price, H., Maisano, R., and Van Cott, H. (1982). *The allocation of functions in man-machine systems: A perspective and literature review* (NUREG-CR-2623). Oak Ridge, TN: Oak Ridge National Laboratories.

Price, H., and Pulliam, R. (1983). Control room function allocation—A need for man-computer symbiosis. *Proceedings of the 1982 IEEE Computer Forum*. Denver, CO: Institute of Electrical and Electronics Engineers.

Rogers, J. G., and Armstrong, R. (1977). Use of human engineering standards in design. *Human Factors,* 19(1), 15–23.

Rogers, J. G., and Pegden, C. D. (1977). Formatting and organization of a human engineering standard. *Human Factors,* 19(1), 55–61.

Rouse, W., and Boff, K. (1987). Designer tools and environments: State of knowledge, unresolved issues, and potential directions. In W. Rouse and K. Boff (eds.), *System design behavioral perspectives on designers, tools, and organizations.* Amsterdam: North-Holland.

SAMMIE CAD Ltd. (1990). *The SAMMIE system.* Loughborough, Leicestershire, England.

Sanders, M. S., Jankovich, J., and Goodpaster, P. (1974). *Task analysis for the jobs of freight train conductor and brakeman* (RDTR 263). Crane, IN: Applied Sciences Department, Naval Ammunition Depot.

Topmiller, D., and Aume, N. (1978, January). Computer graphic design for human performance. *Proceedings of the 1978 Annual Reliability and Maintainability Symposium.* Los Angeles, pp. 385–388.

APPENDIXES

A List of Abbreviations

B Control Devices

C NIOSH Recommended Action Limit Formula for Lifting Tasks

APPENDIX **A**

LIST OF ABBREVIATIONS

A	ampere
ACGIH	American Conference of Governmental Industrial Hygienists
ADP	adenosine diphosphate
AFB	Air Force Base
AFHRL	Air Force Human Resources Laboratory
AFSC	Air Force Systems Command
AI	articulation index
AI	artificial intelligence
AL	action limit
AMD	Aerospace Medical Division, Air Force Systems Command
AMRL	Aerospace Medical Research Laboratory
ANOVA	analysis of variance
ANSI	American National Standards Institute
APA	American Psychological Association
ASD	Aeronautical Systems Division, AFSC
ASHA	American Speech and Hearing Association
ASHRAE	American Society of Heating, Refrigerating, and Air-Conditioning Engineers.
ASHVE	American Society of Heating and Ventilating Engineers (now ASHRAE)
ATM	automated teller machine
ATP	adenosine triphosphate
B	Bel (unit of sound measurement)
BAC	blood alcohol content
BB	Botsball index
BCD	borderline between comfort and discomfort
BMR	basal metabolic rate

C	degrees Celsius
CAFES	computer-aided function-allocation evaluation system
CAD	computer aided design
Cal	calorie
CAM	computer assisted manufacturing
cd	candela
C/D	control-display ratio
CE	controlled experimentation
CFAC	contributing factors in accident causation (model)
CG	center of gravity
CHMSL	center high-mounted stop light
CIE	*Commission International de l'Eclairage*
cm	centimeter
CNEL	community noise equivalent level
CP	creatinine phosphate
cpd	cycles per degree
CPSC	Consumer Product Safety Commission
CR	contrast ratio
C/R	control-response ratio
CRI	color rendering index
CRT	cathode ray tube
CSCW	computer supported cooperative work
d	day
d′	measure of sensitivity in SDT
D	diopter
dB	decibel
DENL	day-evening-night (sound) level
DGR	discomfort glare rating
DHHS	Department of Health and Human Services
DV	dependent variable
ed.	edition
ed(s).	editor(s)
EDR	electrodermal response
EMG	electromyography
EPA	Environmental Protection Agency
EPRI	Electric Power Research Institute
ET	effective temperature (original)
ET*	effective temperature (new)
EVA	extravehicular activity
F	degrees Fahrenheit
FAA	Federal Aviation Administration
FAM	function allocation model
fc	footcandle
FDP	fatigue-decreasing proficiency
fL	foot-lambert
ft	foot
g	gram
G	general package factor (automobile design)
G	gravity (referring to acceleration)

g	gravity (referring to vibration)
gal	gallon
GLOC	G-induced loss of consciousness
GSR	galvanic skin response
h	hour
H	amount of information in bits
HAVS	hand-arm vibration syndrome
HECAD	human engineering computer-aided design
HFS	Human Factors Society
HID	high-intensity discharge (lamps)
hp	horsepower
HSI	heat stress index
HST	hand-skin temperature
HUD	head-up display
Hz	hertz (cycles per second)
ID	index of difficulty
IEEE	Institute of Electrical and Electronics Engineers
IES	Illuminating Engineering Society
im	index of permeability
in	inch
IRE	Institute of Radio Engineers
ISO	International Organization for Standardization
IV	independent variable
IVHS	intelligent vehicle highway system
J	joule
JND	just-noticeable difference
K	degree Kelvin
kcal	kilocalorie
kg	kilogram
km	kilometer
L_{dn}	day-night (sound) level
L_e	(sound) exposure level
L_{eq}	equivalent (sound) level
lb	pound
LCD	liquid crystal display
lm	lumen
lx	lux
m	meter
MAMA	minimum audible movement angle
MANOVA	multivariate analysis of variance
MAP	maximum aerobic power
MAWL	maximum acceptable weight of lift
MCE	modular comfort envelope
mi	mile
MIC	methylisocyanate
MIL-HDBK	military handbook
MIL-SPEC	military specification
MIL-STD	military standard
min	minute

mL	millilambert
mm	millimeter
MMH	manual materials handling
MPL	maximum permissible limit (lifting)
MRT	modified rhyme test
ms	millisecond
MT	movement time
MTF	mean time to failure
N	newton
NADC	Naval Air Development Center
NAS	National Academy of Sciences
NASA	National Aeronautics and Space Administration
NASA-TLX	NASA task load index
NC	noise criteria
NHTSA	National Highway Traffic and Safety Administration
NIOSH	National Institute for Occupational Safety and Health
NIPTS	noise-induced permanent threshold shift
nm	nanometer
NRC	National Research Council
NRR	noise reduction rating
NSC	National Safety Council
OSD	operational sequence diagram
OSHA	Occupational Safety and Health Administration
P300	positive polarity event-related brain potential
PLdB	perceived level of noise, decibel
PNC	preferred noise criteria
PNdB	perceived noise, decibel
PNL	perceived noise level
PSD	power spectral density
psi	pounds per square inch
PSIL	preferred-octave speech interference level
PSM	personnel subsystem measurement
PTS	permanent threshold shift (hearing)
PVC	polyvinylchloride
r	Pearson product-moment correlation
RAL	recommended alert limit (heat stress)
REL	recommended exposure limit (heat stress)
RH	relative humidity
rms	root mean square
RPE	rating of perceived effort
s	second
S	standard deviation
SAE	Society of Automotive Engineers
SCR	skin conductance activity
SDT	signal detection theory
SEL	sound exposure level
SI	International System of units
SIL	speech interference level
S/N	signal to noise ratio

SOS	spatial operational sequence
SPL	sound-pressure level
sr	steradian
SRP	seat reference point
SWAM	statistical workload assessment model
SWAT	subjective workload assessment technique
THERP	technique for human error rate prediction
TOT	time on target
TR	technical report (term used by various organizations)
TSD	Theory of Signal Detection
TTS	temporary threshold shift
TV	television
TVSS	tactile vision substitution system
TWA	time-weighted average
UMTA	Urban Mass Transportation Administration
USA	United States Army
USAF	United States Air Force
USASI	United States of America Standards Institute (now ANSI)
USN	United States Navy
USPHS	United States Public Health Service
UV	ultraviolet
VA	visual acuity
VCP	visual comfort probability
VDT	visual display terminal
VDU	visual display unit (same as VDT)
VWF	vibration-induced white finger
W	watt
WADC	Wright-Patterson Air Development Center, USAF (see AMRL and AFHRL)
WADD	Wright-Patterson Air Development Division, USAF (see AMRL and AFHRL)
WAM	workload assessment model
WBGT	wet-bulb globe temperature
WCI	wind chill index
WD	wet-dry index (Oxford)
W/INDEX	workload index
wpm	words per minute
μ	micrometer (10^{-6})
π	pi

APPENDIX **B**

CONTROL DEVICES

Table B-1 presents a brief evaluation of the operational characteristics of certain types of control devices.* Table B-2 summarizes recommendations regarding certain features of these types of control devices.† In the use of these and other recommendations, it should be kept in mind that the unique situation in which a control device is to be used and the purposes for which it is to be used can affect materially the appropriateness of a given type of control and can justify (or virtually require) variations from a set of general recommendations or from general practice based on research or experience. For further information regarding these, refer to the original sources given in the reports from which these are drawn.

COMMENTS REGARDING CONTROLS‡

- *Hand pushbutton:* Surface should be concave or provide friction. Preferably there should be an audible click when activated. Use elastic resistance plus slight sliding friction, starting low, building up rapidly, to a sudden drop. Minimize viscous damping and inertial resistance.
- *Foot pushbutton:* Use elastic resistance, aided by static friction, to support foot. Resistance to start low, build up rapidly, drop suddenly. Minimize viscous damping and inertial resistance.

* Adapted largely from A. Chapanis. "Design of Controls," Chap. 8 in H. P. Van Cott and R. G. Kinkade, *Human engineering guide to equipment design* (rev. ed.). Washington, DC: Government Printing Office, 1972.

† Adpated from ibid.

‡ Adapted from ibid.

TABLE B-1
COMPARISON OF COMMON CONTROL CHARACTERISTICS

Characteristic	Hand push-button	Foot push-button	Toggle switch	Rotary switch	Knob	Crank	Lever	Hand wheel	Pedal
Space required	Small	Large	Small	Medium	Small-medium	Medium-large	Medium-large	Large	Large
Effectiveness of coding	Fair-good	Poor	Fair	Good	Good	Fair	Good	Fair	Poor
Ease of visual identification of control position	Poor*	Poor	Fair-good	Fair-good	Fair-good†	Poor‡	Fair-good	Poor-fair	Poor
Ease of nonvisual identification of control position	Fair	Poor	Good	Fair-good	Poor-good	Poor‡	Poor-fair	Poor-fair	Poor-fair
Ease of check reading in array of like controls	Poor*	Poor	Good	Good	Good†	Poor‡	Good	Poor	Poor
Ease of operation in array of like controls	Good	Poor	Good	Poor	Poor	Poor	Good	Poor	Poor
Effectiveness in combined control	Good	Poor	Good	Fair	Good§	Poor	Good	Good	Poor

* Except when control is backlighted and light comes on when control is activated.
† Applicable only when control makes less than one rotation and when round knobs have pointer attached.
‡ Assumes control makes more than one rotation.
§ Effective primarily when mounted concentrically on one axis with other knobs.

TABLE B-2
SELECTED DATA REGARDING DESIGN RECOMMENDATIONS FOR CONTROL DEVICES

Device	Size, in		Displacement		Resistance	
	Minimum	Maximum	Minimum	Maximum	Minimum	Maximum
Hand pushbutton						
Fingertip operation	½	None	⅛ in	15 in	10 oz	40 oz
Foot pushbutton	½	None				
Normal operation			½ in			
Wearing boots			1 in			
Ankle flexion only				2½ in		
Leg movement				4 in		
Will *not* rest on control					4 lb	20 lb
May rest on control					10 lb	20 lb
Toggle switch			30°	120°	10 oz	40 oz
Control tip diameter	⅛	1				
Lever arm length	½	2				
Rotary selector switch					10 oz	40 oz
Length	1	3				
Width	½	1				
Depth	½					
Visual positioning			15°	40°*		
Nonvisual positioning			30°	40°*		
Knob, continuous adjustment†						
Finger-thumb						4½–6 in/oz
Depth	½	1				
Diameter	⅜	4				
Hand/palm diameter	1½	3				
Crank†						
For light loads, radius	½	4½				
For heavy loads, radius	½	20				
Rapid, steady turning						
3–5 in radius					2 lb	5 lb
5–8 in radius					5 lb	10 lb
For precise settings					2½ lb	8 lb

TABLE B-2 (continued)

Device	Size, in		Displacement		Resistance	
	Minimum	Maximum	Minimum	Maximum	Minimum	Maximum
Lever§						
Fore-aft (one hand)				14 in		
Lateral (one hand)				38 in		
Finger grasp, diameter	½	3			12 oz	32 oz
Hand grasp, diameter	1½	3			2 lb	20–100 lb
Handwheel†				90°–120°	5 lb	30 lb‡
Diameter	7	21				
Rim thickness	¾	2				
Pedal						
Length	3½					
Width	1					
Normal use			½ in			
Heavy boots			1 in			
Ankle flexion				2½ in		10 lb
Leg movement				7 in		180 lb
Will *not* rest on control					4 lb	
May rest on control					10 lb	

* When special requirements demand large separations, maximum should be 90°.

† Displacement of knobs, cranks, and handwheels should be determined by desired control-display ratio.

‡ For two-handed operation, maximum resistance of handwheel can be up to 50 lb.

§ Length depends on situation, including mechanical advantage required. For long movements, longer levers are desirable (so movement is more linear).

- *Toggle switch:* Use elastic resistance which builds up and then decreases as position is approached. Minimize frictional and inertial resistance.
- *Rotary selector switch:* Provide detent for each control position (setting). Use elastic resistance which builds up and then decreases as detent is approached. Minimize friction and inertial resistance. Separation of detents should be at least ¼ in (6 mm).
- *Knob:* Preferably code by shape if knob is used without vision. Type of desirable resistance depends on performance requirements.
- *Crank:* Use when task involves two rotations or more. Friction [2 to 5 lb (0.9 to 2.3 kg)] reduces effects of jolting but degrades constant-speed rotation at slow or moderate speeds. Inertial resistance aids performance for small cranks and low rates. Grip handle should rotate.
- *Lever:* Provide elbow support for large adjustment, forearm support for small hand movements, wrist support for finger movements. Limit movement to 90°.
- *Handwheel:* For small movements, minimize inertia. Indentations in grip rim aid holding. Displacement usually should not exceed ±60° from normal. For displacements less than 120°, only two sections need to be provided, each of which is at least 6 in (15 cm) long. Rim should have frictional resistance.
- *Pedal:* Pedal should return to null position when force is removed; hence, elastic resistance should be provided. Pedals operated by entire leg should have 2- to 4-in (5- to 10-cm) displacement, except for automobile-brake type, for which 2 to 3 in (5 to 7 cm) of travel may be added. Displacement of 3 to 4 in (7 to 10 cm) or more should have resistance of 10 lb (4.5 kg) or more. Pedals operated by ankle action should have maximum travel of 2.5 in (5.1 cm).

APPENDIX C

NIOSH RECOMMENDED ACTION LIMIT FORMULA FOR LIFTING TASKS

Factor	*U.S. Customary System units*	*Metric*
Horizontal location	$6/H$	$15/H$
Vertical location	$1 - (0.01 \times \lvert V - 30 \rvert)$	$1 - (0.004 \times \lvert V - 75 \rvert)$
Distance traveled (vertical)	$0.7 + 3/D$	$0.7 + 7.5/D$
Frequency of lift	$1 - F/F_{max}$	$1 - F/F_{max}$
Constant	90	40

Basic equations:

$$\text{AL} = (\text{constant}) \times (\text{horizontal location}) \times (\text{vertical location}) \times (\text{distance traveled}) \times (\text{frequency of lift})$$

$$\text{MPL} = 3 \times \text{AL}$$

where AL = action limit (lb or kg) = absolute value of quantity

H = horizontal location (in or cm) forward of midpoint between ankles at origin of lift [6 to 32 in (15 to 81 cm)]

V = vertical location (in or cm) at origin of lift [0 to 70 in (0 to 178 cm)]

D = vertical travel distance (in or cm) between origin and destination of lift [minimum 10 in (25 cm); if less than minimum, set at minimum]

F = average frequency of lift (lifts/min); 0.2, or 1 lift per 5 min, to F_{max}); if less than 0.2, set $F = 0$

MPL = maximum permissible limit

F_{max} = maximum frequency which can be sustained, taken from following chart:

		Average vertical location, in (cm) >30 (75)	Average vertical location, in (cm) ≤30 (75)
Period of performance	1 h or less	18	15
	More or less continuous during shift	15	12

Source: Adapted from NIOSH, 1981.

INDEXES

NAME INDEX

Aaltonen, M., 679, 694
Aatola, A., 398, 411
Abernethy, C., 449, 452
ACGIH, 270, 397, 409, 573, 585
Ackroff, J., 58, 89
Acoustical Society of America, 633, 634, 649
Action, W., 606, 618
Acton, M., 304, 333
Adams, J., 41, 42, 279, 298
Adams, S., 177, 179, 193
Aeronautical Systems Division, 471, 472, 474, 482
Aghazadeh, F., 383, 409
AIHA, 241, 269, 573, 585, 598, 614, 615, 618
Air Force Systems Command, 339, 378
Albert, A., 371, 378
Aldrich, F., 189, 193
Alfaro, L., 111, 129
Allison, S., 686, 695
Alluisi, E., 75, 87, 126, 130
Alpert, M., 684, 693
Altman, I., 503, 506, 507
Altman, P., 651
American Foundryman's Society, 612, 618
Anderson, C., 259, 269
Anderson, J., 47, 85
Anderson, W., 280, 299
Andersson, G., 249, 250, 252, 255, 270, 272, 273, 298, 438, 440, 444, 452
Andre, A., 143, 144, 159
Angalet, B., 487, 488, 507
Anger, W., 672–675, 692, 695
ANSI, 163, 193, 206, 219, 397, 409
Antin, J., 713, 723
Apostalakis, G., 694
Applied ergonomics handbook, 140, 158
Arbak, C., 377–379
Armstrong, J., 188, 193, 194
Armstrong, R., 738, 741, 756
Armstrong, T., 384, 387, 409, 411
Arnaut, L., 346, 368–370, 378, 380
Arndt, R., 442, 455
Arons, I., 402, 409
Asfour, S., 256, 257, 259, 270
Ashley, L., 402, 409
ASHRAE, 555–558, 560, 561, 573, 580, 585
Astrand, I., 233, 234, 235, 269
Astrand, P., 231, 233–235, 237, 238, 247, 254, 269, 271, 560, 578, 580, 585
Asuter, S., 442, 455
AT&T, 541, 547
AT&T Bell Labs, 449, 453
Atkinson, R., 89, 193, 194
Attwood, D., 196, 379, 550
Auflick, J., 369, 380
Aume, N., 744, 756
Ayoub, M., 31, 42, 245, 246, 255, 257, 259–261, 267–269, 356, 357, 378, 393, 409, 416, 424–426, 435, 453, 455, 468, 484
Bachtold, R., 498, 499, 506
Backlund, F., 719, 724
Baddeley, A., 65, 85
Badre, A., 220
Bailey, G., 291, 298
Bailey, R., 14, 21, 727, 728, 754
Baines, A., 135, 159
Baird, J., 48, 85, 500, 501, 506
Baker, C., 126, 138, 158, 715, 724
Baker, J., 748, 755
Bakken, G., 24, 42, 426, 453
Bangerter, B., 226, 271, 274, 276, 299
Banks, W., 98, 130, 462, 482
Barash, D., 458, 483
Barfield, W., 118, 128, 731, 754
Barnard, P., 58, 86
Barnes, R., 431, 432, 453
Barrett, G., 706, 722
Barrett, N., 292, 299
Barsley, M., 394, 409
Barth, J., 189, 193
Barwick, K., 673, 693
Bass, D., 567, 587

Baum, A., 503, 506
Bayerl, J., 303, 304, 331
Beare, A., 100, 130
Beaton, R., 368, 369, 379
Beatty, J., 80–81, 85
Beauchamp, G., 191, 193
Beaumont, P., 664, 692
Becker, C., 85, 86
Becker, F., 498, 500, 505, 506
Becker, J., 370, 379
Beevis, D., 726, 754
Beggs, W., 292, 298
Behrens, V., 397, 412
Belding, H., 568, 585
Bell, P., 502, 507, 572, 586
Bell, T., 206, 219
Bendix, T., 439, 453
Bennett, C., 486, 502, 506, 511, 526, 527, 534, 535, 547, 550
Bennett, J., 388, 389, 410
Bennett, R., 170, 193
Benz, C., 542, 547
Beranek, L., 208–212, 219
Berbaum, K., 648, 650
Berger, E., 530, 549, 616–618
Bergh, U., 580, 586
Bergman, M., 213, 219
Bergum, B., 302, 331
Bergum, J., 302, 331
Beringer, D., 312, 313, 333, 368, 379
Berkhout, J., 150, 158
Berrian, R., 280, 299
Berry, C., 641, 642, 649
Beshir, M., 559, 572, 586, 587
Bethea, N., 31, 42, 416, 453
Bex, F., 434, 453
Biddulph, R., 172, 173, 196
Bied, B., 9, 11, 22
Biers, D., 155, 159
Bilger, R., 201, 219
Biliouris, T., 707, 723
Bilson International, Inc., 616, 618
Binks, T., 648, 650
Bishu, R., 50, 85, 402, 412
Bisseret, A., 65, 85
Bittner, A., Jr., 422, 453, 633, 647–651
Black, J., 498, 506
Blackwell, H., 525, 540, 547, 722
Blackwell, O., 525, 540, 547
Blake, T., 365, 382
Blakemore, C., 98, 128
Blanchard, R., 741, 754
Blanchard, W., 640, 649
Blanche, K., 474, 483
Blazier, W., 209, 211, 212, 219
Blink, B., 245, 269
Bliss, J., 186, 187, 193
Blumenfeld, V., 213, 219
Boarer, P., 631, 651
Bobo, M., 31, 42, 416, 453
Boff, K., 88, 93, 128, 156, 158, 219, 299, 332, 333, 377, 379, 738, 741, 754, 756
Bondurant, S., 640, 649
Boni, A., 447, 454
Bonney, M., 467, 468, 482, 744, 754
Bonvallet, S., 593, 618
Booher, H., 118, 128
Boone, M., 462, 482
Boor, M., 707, 725
Borg, G., 237, 269
Borkenstein, R., 710, 723
BOSTI, 490, 506
Botsford, J., 559, 586
Boulette, M., 143, 158
Bowen, H., 361, 379
Bower, G., 88
Bower, H., 636, 652
Box, G., 33, 42
Boyce, P., 494, 506, 520, 522, 524, 526, 528, 529, 539, 547, 704, 721, 722, 723
Boyd, S., 82, 85
Brabyn, J., 181, 194, 216, 219
Brabyn, L., 216, 219
Bradley, J., 307, 308, 331, 339–341, 343, 379, 480, 482
Brand, J., 645, 646, 651
Brass, J., 532, 548
Braune, R., 76, 89
Brauninger, U., 542, 548
Breaux, R., 379
Brebner, J., 308, 309, 331, 332
Brelsford, J., Jr., 671, 676, 682, 692, 695
Brems, D., 671, 677, 692
Brezovic, C., 739, 741, 755
Bridgeman, B., 279, 298
Bridger, R., 433, 439, 440, 453
Briggs, G., 321, 331
Brill, S., 575, 586
Broadbent, D., 61, 72, 85, 110, 128, 601, 602, 619
Broberg, E., 665, 692
Brookes, M., 489, 493, 506
Brooks, V., 299
Brouha, L., 235, 245, 269
Brown, C., 114, 128
Brown, F., 472, 484
Brown, I., 666, 692, 704, 724
Brown, J., 294, 295, 298, 638, 649
Brown, O., Jr., 508
Browning, H., 504, 506
Brunner, H., 363, 364, 379
Buck, L., 346, 379
Buckhout, R., 65, 86
Buckley, E., 362, 382
Buggie, S., 288, 289, 299
Bui, S., 181, 194
Bullinger, H., 325, 353, 379
Bullock, M., 426, 453
Bureau of Labor Statistics, 225, 269
Burford, C., 416, 453, 559, 572, 586, 587
Burg, A., 99, 128
Burke, F., 475, 483
Burke, M., 189, 193
Burns, R., 649
Burns, W., 598, 599, 600, 619, 621
Burr, B., 372, 379
Burrows, A., 589, 619
Burse, R., 556, 586
Burton, A., 583, 586, 629, 649
Buskirk, E., 233, 272

Caddel, D., 416, 453
Caelli, T., 174, 193
Caillet, R., 254, 269
Cain, W., 191, 193
Cairney, P., 119, 121, 128
Cakir, A., 546, 548
Calhoun, G., 175, 193, 377–379
Cameron, N., 417, 453
Campbell, B., 696, 725
Campbell, D., 494, 495, 506
Campbell, F., 98, 128
Campbell, J., 189, 195
Cannon, M., 97, 129
Canter, D., 496, 508
Caplan, S., 408, 409
Card, S., 65, 85, 372, 379
Carlsoo, S., 629, 649
Carlsson, G., 640, 651
Carswell, C., 143, 158
Carter, R., 344, 379
Casali, J., 80, 82, 85, 89, 213, 219, 618, 619, 648, 649
Cascardo, D., 213, 219
Case, K., 744, 754
Casey, S., 358, 359, 379, 701, 723
Cassidy, B., 500, 501, 506
Castellan, N., 33, 43
Cerella, J., 76, 86
Cerrelli, E., 704, 705, 723
Chaffin, D., 248–250, 252, 255, 256, 259, 269, 270, 273, 291, 298, 299, 391, 393, 395, 410, 454
Chai, C., 573, 587

Chambers, J., 494, 506
Chambers, R., 638, 639, 649
Chan, W., 309, 333
Chang, T., 731, 754
Chapanis, A., 4, 5, 7, 15, 21, 304, 305, 311, 332, 347, 377, 379, 481, 483, 730, 731, 742, 754, 764
Chitlangia, A., 526, 527, 547
Chodorow, M., 216, 218, 220
Christ, R., 127, 128
Christensen, J., 517, 529, 535, 549, 561, 562, 586, 664, 675, 677, 692
CIE, 515, 520, 528, 548, 721, 723
Ciriello, V., 266, 272
Cirillo, L., 707, 723
Clark-Carter, D., 181, 193
Clark, B., 644, 645, 649
Clark, N., 640, 649
Clark, R., 581, 586
Clegg, T., 581, 587
Clevenson, S., 633, 651
Cochran, D., 581, 584, 587
Cohen, A., 672, 674, 692
Cohen, B., 542, 545, 550
Cohen, E., 139, 158
Cohen, H., 682, 693
Cohen, J., 31, 42
Cole, R., 197, 219
Coler, C., 376, 379
Coletta, G., 402, 409
Collins, B., 121, 122, 128, 494, 506, 520, 548, 683, 692
Collins, W., 643, 650
Colquhoun, W., 65, 86
Coluccio, T., 377, 380
Colwell, M., 563, 586
Congleton, J., 268, 270
Connor, M., 346, 347, 381
Conover, D., 149, 159
Conrad, R., 66, 70, 86, 360, 362, 380
Consumer Product Safety Commission, 684, 692
Container Corporation of America, 515, 548
Coombes, L., 87
Cooper, G., 82, 86
Cooper, J., 431, 455
Cooper, M., 362, 382
Corbridge, C., 631, 632, 634, 649
Corcoran, P., 670, 692
Corl, L., 155, 159, 327, 333
Corlett, E., 23, 43, 271, 692, 734, 754
Corliss, W., 373, 374, 380
Cornell, P., 450, 453, 495, 506
Cornog, D., 102, 128
Coskuntuna, S., 746, 754
Couchman, D., 99, 130
Coury, B., 143, 144, 158
Craig, A., 74, 86
Crawford, A., 151, 158
Creitz, W., 217, 219
Crooks, W., 525, 549
Crossman, E., 292, 298
Crowther, R., 710, 723
Cullinane, T., 463, 483
Curran, M., 379
Curran, P., 9, 11, 22
Cushman, W., 113, 128, 450, 453
Czaja, S., 76, 86
Dadson, R., 591, 592, 620
Dainoff, M., 441, 445, 449, 453
Damon, F., 388, 409
Danaher, J., 732, 754
Darnell, M., 24, 42
Das, B., 432, 438, 453
Dash, B., 213, 219
Dashevsky, S., 147, 158
Datta, S., 242, 243, 270
Davies, B., 258, 266, 270, 358, 380
Davies, D., 74, 75, 86, 166, 193, 597, 601, 619
Davis, H., 170, 196
Deatherage, B., 53, 86, 170, 172, 174, 180, 193
Decker, J., 113, 131
Deering, L., 99, 130
DeGreene, K., 665, 692, 748, 754
DeGroot, J., 442, 453
Deininger, R., 466, 467, 483
Dember, W., 74, 88
Dempsey, T., 633, 651
Dempster, W., 426, 427, 453
Derrick, W., 78, 83, 87
Desaulniers, D., 676, 682–684, 692, 695
Desor, J., 191, 193
Detweiler, M., 65, 88
Deutsch, S., 255, 271
Dewar, R., 119, 120, 128, 704, 707, 723, 724
Dey, D., 326, 332
Dhillon, B., 40, 42
DiFranza, J., 707, 723
DiLaura, D., 523, 548
Dingus, T., 713, 723
Dinman, B., 563, 586
Dionne, E., 386, 409
Dirks, D., 206, 219
Dismukes, K., 130
Dittmer, D., 651
Dobelle, W., 185, 193
Dodds, A., 181, 193
Dodfrey, S., 671, 692
Doll, T., 180, 193
Donaher, W., 688, 690, 691, 695
Donchin, E., 81, 82, 87
Donner, K., 682, 692
Donohue, R., 715, 725
Dorfman, P., 70, 86
Dorris, A., 685, 692
Downing, J., 478, 479, 483
Drennan, A., 402, 409
Dressel, D., 448, 453
Drory, A., 719, 720, 723, 725
Drury, C., 50, 65, 85, 86, 292, 298, 357, 380, 734, 754
Ducharme, R., 394, 409
Duncan, J., 704, 724
Duncan, K., 694
Dunnette, M., 21
Dupuis, H., 470, 471, 483
Durnin, J., 241, 271
Dutta, S., 262, 263, 270
Dvorak, A., 360, 380
Dye, C., 113, 131
Easterby, R., 122, 123, 128, 454
Eastman, M., 432, 453
Eberts, R., 84, 86, 731, 754
Eckel, J., 40, 43
Edholm, O., 6, 21, 240, 270, 583, 586
Edinger, K., 648, 650
Edman, T., 217, 219
Edwards, D., 664, 692
Edwards, R., 228, 271
Edwards, S., 493, 507
Eggemeier, F., 78, 82, 86, 88
Ehrenreich, S., 58, 85, 86
Eisner, E., 80, 88
Eklund, L., 399, 410
Eldredge, D., 599, 620
Electric Power Research Institute, 452, 453, 476, 483
Elfstrom, G., 439, 454
Elkerton, J., 468, 483
Elkin, E., 145, 158
Elkind, J., 85, 319, 332
Ellingstad, V., 32, 42, 305, 332, 369, 380
Ellis, J., 119, 120, 128
Ellis, S., 154, 158
Ellison, K., 65, 86
Elvers, G., 175, 196
Emanuel, J., 388, 389, 410
Enander, A., 580, 582, 585, 586
Endsley, M., 133, 152, 158
Engen, T., 191, 193

English, W., 372, 379
Environmental Protection Agency, 591, 593, 609, 615, 619
Epps, B., 372, 380, 618, 619
Epstein, Y., 575, 586
Era, P., 251, 272
Eschenbrenner, A., Jr., 601, 619
Evans, D., 97, 98, 108, 129
Evans, G., 586
Evans, L., 696, 697, 700–702, 704, 705, 707, 710, 711, 723
Evans, S., 744, 754
Evans, T., 404, 405, 410

Fanger, P., 561, 562, 586
Fard, H., 670, 694
Farkkila, M., 398, 411
Farley, R., 431, 453
Faulkner, T., 408, 410, 532, 548
Faverjon, G., 640, 650
Feallock, J., 126, 129
Feldman, M., 374, 380
Fell, J., 698–700, 723
Fellmann, Th., 542, 548
Feustel, T., 216, 220
Ficks, L., 173, 195
Fidell, L., 33, 43
Fidell, S., 170, 193, 609, 619
Figwer, J., 209, 211, 212, 219
Finck, A., 309, 310, 333
Fine, B., 570, 586
Fine, L., 387, 411
Finn, R., 111, 129
Finnegan, J., 685, 693
Fischhoff, B., 68, 89, 676, 695
Fisher, D., 116, 117, 129
Fisher, W., 375, 382
Fisk, A., 118, 129
Fitch, J., 670, 692
Fitts, P., 58, 86, 126, 130, 194, 290, 294, 298, 302, 303, 332, 333, 357, 380, 465, 467, 483, 730, 754, 755
Flach, J., 189, 194
Flagle, C., 754
Fleishman, E., 87, 237, 270
Fleming, A., 707, 725
Flesch, R., 110, 129
Fletcher, H., 212, 219, 591, 619
Fogel, L., 153, 158
Folds, D., 180, 193
Folkard, S., 86
Follert, R., 139, 158
Fontenelle, G., 671, 684, 692, 695
Forbes, J., 724
Foss, J., 599, 619
Foulke, J., 384, 409
Foulkes, J., 291, 299
Fowler, M., 459, 483
Fowler, R., 459, 483
Fracker, M., 152, 158
Francis, J., 448, 453
Francis, R., 468, 483
Frank, J., 295, 296, 300
Frank, L., 645, 648–650
Frankel, V., 273, 298
Franks, P., 350, 382
Franks, W., 640, 651
Fraser, T., 386, 410, 636–638, 649
Frederick, W., 266, 270
Freeman, P., 601, 620
Freivalds, A., 111, 129, 244, 270, 386, 410, 537, 548
French, N., 204, 208, 219
Frick, J., 619
Friedmann, K., 682, 686, 692
Frost, B., 185, 193, 195
Frymoyer, J., 272
Fugelli, P., 235, 269
Furness, T., 175, 193

Gage, G., 617, 621
Gage, H., 386, 411
Galer, M., 135, 159
Gales, R., 172, 194
Galitz, W., 114, 129, 487, 488, 506
Gamst, F., 336, 380
Garfield, E., 170, 194
Garg-Janardan, C., 84, 86
Garg, A., 24, 42, 259–261, 270, 426, 453
Garner, W., 7, 21
Garvey, P., 99, 130
Gasaway, D., 615–617, 619
Gawron, V., 600, 619
Gebhard, J., 54, 88
Gebhardt, D., 237, 270
Geer, C., 464, 465, 483, 729, 736, 755
Gehlen, G., 369, 380
Geldard, F., 167, 169, 182, 184, 194, 623, 649
Gell, C., 629, 649
Genaidy, A., 256, 257, 259, 270
General Electric Company, 512, 517, 548
Giambra, L., 74, 75, 86
Gian, K-C., 292, 298
Gierer, R., 366, 381, 542, 548
Gilbert, J., 389, 410
Gill, J., 188, 194
Gill, R., 24, 42
Gillies, J., 650
Gilliland, K., 469, 470, 483
Gilson, R., 189, 190, 193, 194
Ginsburg, A., 97, 98, 108, 129
Gish, K., 99, 130
Glaser, R., 236, 272
Glass, S., 358, 380
Glencross, D., 281, 292, 298, 299
Godfrey, S., 682–684, 686, 693, 695
Goebel, C., 99, 130
Goldberg, R., 707, 723
Goldstein, E., 197, 219
Goldstein, I., 32, 43, 70, 86
Goldstein, S., 384, 409
Gonzales, W., 466, 478, 484
Gonzalez, W., 342, 382
Goodenough, D., 706, 723
Goodeve, P., 292, 298
Goodman, L., 171, 195
Goodpaster, P., 735, 756
Gordon, E., 241, 270
Gordon, G., 196
Got, C., 640, 650
Gould, J., 85, 86, 111, 129
Grady, R., 432, 438, 453
Grandjean, E., 112, 129, 229, 238–240, 242, 244, 270, 309, 310, 332, 366, 381, 435, 436, 439, 443, 446, 449, 450, 453, 454, 492, 496, 498, 499, 507, 542–544, 548, 549, 627, 629, 650
Grantham, C., 175, 194
Grape, P., 643, 650
Graybiel, A., 644, 645, 649, 650
Green, D., 62, 84, 167, 194
Green, P., 141, 158
Green, T., 376, 377, 380, 381
Greenberg, L., 391, 393, 395, 410
Greene, B., 216, 220
Greene, T., 502, 507, 572, 586
Greenstein, J., 346, 368, 369, 370, 378–380
Grether, W., 126, 138, 158
Grey, S., 734, 754
Grieco, A., 440, 454
Griffin, M., 629–632, 634, 646, 649–652
Griffiths, I., 583, 587
Grillner, S., 280, 299
Grischkowsky, N., 111, 129
Grob, R., 542, 547
Gross, E., Jr., 164, 195, 206, 209, 210, 220, 594, 613, 620
Grudin, J., 58, 86
Gruneberg, M., 586
Guedry, F., 644, 650
Guignard, J., 628, 647, 649, 650
Guilfoyle, J., 403, 410

Guinnes, G., 361, 379
Gulian, E., 600, 619
Gustanski, J., 528, 550
Guttmann, H., 40, 43, 657, 695
Guyton, A., 166, 194

Haber, R., 61, 86
Habinek, J., 546, 548
Hahn, C., 664, 692
Haigh, R., 375, 382
Haight, F., 696, 725
Haines, R., 469, 470, 483, 528, 550
Hakkinen, S., 704, 723
Halcomb, C., 358, 381, 424–426, 455
Hale, G., 503, 507
Hallback, I., 233, 234, 269
Halpern, W., 576, 588
Hameluck, D., 98, 131
Hamill, B., 84, 87
Hancock, P., 86, 87, 570–572, 586
Hanley, T., 201, 219
Hansen, A., 709, 725
Hansson, J., 399, 410
Hanzawa, H., 561, 586
Hapeshi, K., 376, 380
Harano, R., 706, 723
Hard, A., 515, 548
Hardesty, G., 515, 516, 548
Harper, R., 82, 86
Harpster, J., 111, 129, 537, 548
Harris, C., 194, 619, 620
Harris, S., 221
Harris, W., 708, 723
Hart, D., 546, 548
Hart, S., 83, 87
Hartemann, F., 640, 650
Hartley, C., 373, 374, 382
Hartman, B., 353, 380
Harwood, K., 79, 89
Hatch, T., 568, 585
Haubner, P., 532, 542, 547, 549
Haupt, B., 111, 129
Hauske, G., 286, 299
Hawkes, G., 184, 194
Hawkins, B., 295, 296, 300
Hawkins, J., Jr., 205, 219
Hayashi, Y., 668, 693
Hays, W., 33, 42
Hazell, J., 404, 405, 410
Headley, R., 637, 652
Heath, T., 631, 651
Heckman, L., 537, 548
Hedge, A., 492, 507
Hedman, L., 542, 543, 550
Heglin, H., 103, 104, 107, 127, 129, 135, 151, 156, 157, 159
Heide, R., 629, 652
Heikkinen, E., 251, 272
Heinrich, H., 662, 693
Heise, G., 202, 220
Helander, M., 86, 88, 380, 449, 454, 495, 497, 507, 541, 548, 717, 723
Helmkamp, J., 618, 619
Hendler, E., 649
Hendrick, H., 508
Henry, C., 640, 650
Henschel, A., 233, 272
Herman, R., 701, 723
Heron, W., 408, 410
Herrin, G., 250, 270
Herrington, L., 554, 588
Herrnstein, G., 193
Herrnstein, R., 89, 194
Hershenson, M., 61, 86
Hertzberg, H., 475, 483
Hesselgren, S., 514, 548
Hettinger, L., 648, 650
Heyes, A., 181, 194
Hick, W., 50, 87
Hicks, J., III, 720, 723
Hietamaki, J., 720, 725
Hilgendorf, R., 149, 159
Hill, M., 726, 754
Hirsh-Pasek, K., 58, 87
Hitchock, M., 376, 379
Hitchock, R., 154, 158
Hochberg, J., 85
Hodges, J., 487, 488, 507
Hodgson, W., 606, 620
Hoffberg, L., 562, 587
Hoffmann, E., 292, 298
Hogan, J., 237, 270
Hohnsbein, J., 586
Holding, D., 309, 332
Holding, H., 299, 300
Hooper, K., 703, 723
Hopkinson, R., 538, 548
Horberg, V., 714, 723
Hormann, H., 213, 219
Hornick, R., 625, 627, 629, 630, 650
Hornseth, J., 473, 484
Horvath, S., 245, 270, 563, 586
Houy, D., 275, 299
Howarth, C., 181, 193, 292, 298
Howell, S., 503, 507
Howell, W., 87, 126, 129
Howett, G., 107, 129
Howland, D., 349, 380
Huebner, W., Jr., 467, 483
Huey, B., 9–11, 22, 85
Huff, E., 376, 379
Huggins, W., 754
Hughes, P., 527, 539, 540, 548
Hulbert, S., 708, 724
Hull, A., 201, 219, 362, 380
Hull, J., 24, 42
Hulse, M., 713, 723
Human Factors Society, 112, 129, 434, 435, 442, 449, 451, 454, 541–543
Hunsicker, P., 251, 252, 270
Hunt, D., 324, 332, 338, 341, 380
Hunter, J., 33, 42
Hunter, W., 33, 42
Hunting, W., 112, 129, 366, 381, 443, 446, 450, 454, 542–544, 549
Huston, R., 88
Hutchinson, T., 696, 724
Hutton, K., 245, 270
Hyman, R., 50, 87
Hyvarinen, J., 398, 411

Iavecchia, H., 156, 159
Iavecchia, J., 156, 159
Ilg, R., 361, 366, 380
Ilgen, D., 737, 755
Illuminating Engineering Society, 512, 519, 521, 523, 528, 529, 531–533, 541, 542, 549
Imamoglu, V., 494, 507
Inaranont, K., 31, 42
Industrial Noise Manual, 165, 194
Intaranont, K., 416, 453
International Business Machines, 449, 454
Irvine, C., 245, 272, 670, 693, 694
ISO, 397, 410, 559, 573, 586, 633, 650
Ison, J., 599, 619
Isreal, J., 73, 87
Ittelson, W., 505, 508

Jackson, A., 253, 270
Jacobson, I., 632, 652
Jacobson, P., 546, 548
Jagacinski, R., 189, 190, 193, 194
James, G., 188, 194
Jankovich, J., 735, 756
Jansen, G., 600, 619
Janson, W., 175, 193, 378, 379
Jantti, V., 398, 411
Jaschinski, W., 529, 549
Jell, R., 644, 650
Jenkins, J., 43
Jenkins, W., 186, 194, 337, 346, 347, 353, 380, 381
Jensen, C., 226, 271, 274, 276, 299
Jensen, P., 590, 619

Jensen, R., 155, 159, 225, 255, 270, 271, 327, 332, 333, 379, 562, 572, 586, 587, 694
Jex, H., 319, 332
Jiang, B., 260, 272
Johansson, G., 719, 720, 724
Johansson, R., 183, 184, 196
Johnson, D., 715, 724, 746, 747, 755
Johnson, E., 373, 374, 380, 748, 755
Johnston, I., 696, 725
Jokel, C., 590, 619
Jones, C., 581, 586
Jones, D., 166, 193, 268, 271, 376, 380, 597, 601, 619
Jones, M., 125, 126, 129
Jones, N., 228, 271
Jones, R., 465, 483, 744, 755
Jonsen, G., 744, 755
Jordan, N., 731, 755
Jorgensen, K., 263, 264, 272
Jorve, W., 145, 146, 159
Joseph, B., 384, 409
Jusczyk, P., 197, 219

Kahane, C., 715, 724
Kahn, M., 160, 194
Kahneman, D., 72, 87, 677, 695
Kamm, C., 206, 219
Kamon, E., 249, 272, 432, 453
Kampmann, B., 586
Kantowitz, B., 332, 733, 755
Kao, H., 323, 332, 406, 407, 410
Kaplan, A., 493, 506
Karlin, L., 289, 299
Karlsson, C., 235, 269
Karr, A., 353, 381
Kashiwagi, S., 668, 693
Katcher, M., 186, 187, 193
Katchmer, L., 351, 381
Kaufman, J., 517, 529, 535, 549
Kaufman, L., 88, 93, 128, 219, 299, 332, 333
Kay, L., 181, 182, 194
Kaye, S., 501, 507
Keegan, J., 439, 454
Keele, S., 281, 286, 287, 291, 292, 295, 296, 299
Kelley, C., 314, 325, 326, 332
Kelso, J., 276, 278, 299, 300
Kemper, H., 234, 271
Kennedy, R., 645–648, 650, 651
Keppel, G., 23, 42
Kerr, R., 292, 299
Kerr, W., 666, 693
Kershner, R., 180, 181, 196
Kessel, C., 73, 89
Kessler, E., 671, 693
Kestenbaum, R., 289, 299
Keuss, P., 178, 179, 196
Key, J., 687, 695
Keyserling, W., 250, 270
Khaledi, H., 670, 694
Khalil, T., 236, 271
Kiang, N., 167, 194, 199, 219
Kibler, A., 286, 300
Kidd, A., 84, 87
Kiess, H., 581, 584, 587
Kihlberg, S., 399, 410
Kilbom, A., 233–235, 237, 269, 271
Kinkade, R., 129, 158, 193, 196, 219, 311, 332, 347, 377, 379, 382, 481, 483, 484, 712, 725, 764
Kinney, J., 100, 130
Kirch, M., 279, 298
Kirk, N., 436, 437, 455
Kirkhan, W., 643, 650
Kirkpatrick, M., 715, 724
Kirman, J., 185, 194
Kistler, D., 175, 196
Kitfield, J., 635, 650
Kjellberg, A. 634, 650
Klapp, S., 280, 299
Klare, G., 110, 130
Kleeman, W., Jr., 441, 454, 487, 507
Klein, B., 255, 271
Klein, G., 739, 741, 755
Klein, R., 288, 289, 299
Kline, D., 130
Klockenberg, E., 366, 381
Knauft, E., 294, 295, 298
Knave, B., 525, 549
Knight, P., 696, 725
Knopf, C., 488, 489, 507
Knowlton, R., 389, 410
Kobayashi, M., 126, 129
Kobrick, J., 570, 586
Kobylak, R., 118, 129
Kochhar, D., 458, 483
Kodak, Eastman Company, 246, 249, 271, 386, 410
Kohl, G., 352, 381
Kohl, J., 715, 724
Kok, M., 234, 271
Kok, R., 564, 587
Kokoschka, S., 532, 542, 549
Kokot, D., 450, 453
Komaki, J., 673, 693
Koncelik, J., 504, 507
Konz, S., 389, 394, 410
Koradecka, D., 396, 397, 410
Korhonen, O., 398, 411
Korner, J., 640, 651
Kothari, N., 113, 131
Kotze, H., 566, 588
Kozinsky, E., 40, 43
Kramer, A., 81, 82, 87
Krebs, M., 148, 149, 159, 286, 288, 300
Kreifeldt, J., 684, 693
Krepton, J., 645, 650
Krestzschmer, H., 447, 454
Kroemer-Elbert, K., 240, 271
Kroemer, H., 240, 271
Kroemer, K., 27, 30, 42, 43, 240, 248, 249, 253, 255, 271, 272, 355, 357, 366, 381, 416, 418, 420, 454
Krohn, G., 23, 43
Krohn, R., 389, 410
Krumm-Scott, S., 685, 693
Kryter, K., 166, 194, 199, 200, 201, 206–208, 219, 592, 594–598, 600, 603, 607, 608, 619, 620
Kuller, R., 507
Kunishima, M., 502, 507
Kupec, J., 150, 159
Kurr, J., 500, 501, 506
Kvalseth, T., 320, 322, 332, 693
Kwon, Y., 570–572, 587

Lam, S., 618, 619
Lam, T., 701, 723
Lampert, U., 665, 693
Lane, C., 168, 196
Lane, D., 72, 87
Lang, C., 504, 507
Langolf, G., 291, 299
LaRue, C., 682, 693
Lauber, J., 32, 43
Laubli, Th., 542–544, 549
Laughery, K., 253, 270, 680, 682, 686, 693
Lavender, S., 401, 411
Laveson, J., 395, 410
Laviana, J., 562, 587
Lawrence, A., 678, 693
Lawther, A., 646, 651
Lawton, M., 503, 506, 507, 528, 550
Lazarus, H., 213, 219
Lazarus-Mainka, G., 213, 219
Leahy, W., 40, 43
Leatherwood, J., 633, 651
Leavitt, R., 606, 620
Lederman, S., 183, 189, 195
Lehmann, G., 242, 271
Lehto, M., 679, 693
Leibowitz, H., 99, 111, 129, 130

Leithead, C., 564, 566, 587
Lentz, J., 644, 650
Leplat, J., 694
Lerner, N., 121, 122, 128
Leskinen, A., 251, 272
Levine, J., 80, 88
Levinson, S., 376, 381
Levis, J., 221
Levitt, H., 213, 219
Lewis, C., 85, 86, 629, 630, 651
Lewis, J., 312, 332
Lewis, M., 564, 587
Lewis, S., 303, 304, 331
Liberman, M., 376, 381
Lichten, W., 202, 220
Lichtenberg, B., 641, 651
Lichtenstein, S., 68, 69, 676, 695
Lickhart, J., 581, 584, 587
Licklider, J., 176, 195, 205, 220
Lim, C., 704, 724
Lin, M., 292, 299
Lincoln, J., 156, 158, 741, 754
Lind, A., 236, 271, 563, 564, 566, 567, 570, 587
Lindberg, E., 637, 652
Lindenbaum, L., 304, 305, 332
Lindzey, G., 89, 193, 194
Linton, P., 79, 87
Linvill, J., 186, 195
Lippert, T., 114, 130
Lipscomb, D., 195, 621
Lloyd, C., 113, 131
Lo Presti, P., 393, 409
Loeb, K., 364, 365, 381
Loeb, M., 75, 87
Logan, H., 530, 549
Logan, J., 216, 220
Long, G., 65, 87, 99, 130
Longman, D., 360, 380
Longmore, J., 538, 548
Loomis, J., 185, 195
Louis Harris & Assoc., 491, 507
Lovasik, J., 114, 130
Lucaccini, L., 404, 405, 410
Lucas, R., 404, 405, 410
Luce, P., 216, 220
Luce, R., 89, 193, 194
Luchetii, R., 495, 506
Luckiesh, M., 535, 536, 549
Lueder, R., 441–444, 449, 454, 550
Lund, A., 701, 723
Lunn, R., 98, 130
Lupp, U., 286, 299

Maas, S., 234, 271
Mace, D., 717, 725
Machle, W., 599, 620
Mack, L., 495, 506, 507
Mackay, G., 696, 725
Mackie, R., 75, 87, 90, 708, 723
MacMillan, G., 88
Madni, A., 84, 87
Mahach, K., 366, 381
Mair, A., 598, 621
Maisano, R., 733, 755
Majchrzak, A., 731, 754
Malaterre, G., 698, 699, 724
Malecki, J., 718, 719, 725
Malone, T., 715, 724
Malven, F., 441, 454
Mandal, A., 439, 442, 454
Mandel, D., 441, 454
Manligas, C., 175, 195
Manning, P., 494, 507
Mansur, D., 182, 195
Margolis, H., 618, 619
Margulies, M., 213, 219
Maritime Transport. Res. Board, 668, 693
Mark, L., 88
Markowitz, J., 149, 159
Marras, W., 27, 42, 389, 401, 411
Marshall, C., 450, 455
Marshall, H., 637, 652
Mason, K., 377, 380
Mathews, M., 701, 724
Mathias, C., 576, 588
Matthews, M., 114, 130, 723
Mauro, C., 746, 754
May, D., 620
McBride, D., 171, 195
McCann, C., 196, 379, 550
McCarrey, M., 493, 507
McCarthy, G., 685, 693
McCarthy, R., 685, 693
McCauley, M., 214–217, 220, 375, 376, 381, 646–648, 650, 651
McCloy, T., 78, 83, 87
McConville, J., 422, 455
McCormick, E., 737, 755
McCracken, H., 283, 284, 299
McDowell, E., 707, 725
McFarland, R., 336, 381
McFarling, L., 32, 42
McGee, H., 703, 723
McGovern, D., 292, 299
McGreevy, M., 154, 158
McGuire, F., 707, 724
McIntyre, D., 583, 587
McKenna, F., 665, 694, 704, 724
McKenna, N., 686, 695
McLean, A., 696, 725
McLeod, W., 441, 454
McNelis, J., 527, 540, 548
McRurer, D., 319, 332
Mead, P., 296, 297, 299
Meadows, S., 255, 271
Meecham, W., 600, 620
Meese, G., 564, 587
Mehr, M., 353, 354, 381
Meister, D., 23, 26, 35, 37, 39–42, 662, 694, 727, 729, 730, 734, 736–739, 741, 746, 747, 749, 755
Melikov, A., 561, 586
Melnick, W., 597, 598, 620
Mercer, A., 491, 492, 507
Merritt, J., 525, 534–537, 549, 550
Mertins, K., 114, 130
Meshkati, N., 8, 21, 86, 87
Metz, S., 217, 219
Meyer, D., 292, 299
Meyer, R., 395, 410, 423, 454
Michael, E., 245, 270
Michaelis, P., 214, 220
Micro Switch, 335, 381
Millen D., 303, 304, 331
Miller, D., 40, 42, 189, 190, 194
Miller, G., 55, 67, 87, 172, 195, 202, 204, 205, 220
Miller, J., 599, 620, 679, 693
Miller, L., 590, 619
Miller, W., 546, 548
Mills, S., 388, 389, 410
Milosevic, S., 700, 724
Milton, J., 465, 483
Miner, F., 359, 381
Minnix, R., 165, 195
Minuto, A., 111, 129
Mital, A., 245, 246, 255, 257–261, 267–271, 383, 409, 670, 694
Mitchell, C., 21
Mitta, D., 84, 87
Modern Materials Handling, 376, 381
Moe, K., 21
Moeller, G., 629, 649
Molbech, S., 254, 271
Money, K., 641, 645, 651
Monk, T., 86
Moore, B., 167, 195
Moray, N., 71, 72, 78, 83, 87, 88, 329, 332
Morgan-Campbell, E., 228, 271
Morgan, C., 7, 21
Moriarty, S., 109, 130
Morrison, D., 376, 377, 380, 381
Morrison, R., 358, 381
Morrissey, S., 416, 453, 633, 651
Mortimer, R., 150, 159
Morton, R., 580, 587
Moseley, M., 629, 651
Mosier, J., 114, 130
Moskowitz, H., 708, 709, 724
Moss, F., 535, 536, 549

Mourant, R., 336, 342, 343, 381, 700, 724
Moussa-Hamouda, E., 336, 342, 343, 381
Mowbray, G., 54, 88
Moxley, S., 282, 299
Muckler, F., 9, 21, 42, 87, 220, 381
Mudd, S., 57, 88, 176, 180, 195
Muller, P., 126, 130
Mulligan, B., 171, 195
Mulvanny, P., 97, 129
Munsell Color Co., 515, 549
Munson, W., 170, 195, 591, 619
Muntzinger, W., 325, 353, 379
Murphy, R., 438, 452
Murphy, T., 532, 548
Murray, M., 501, 507
Murrell, K., 6, 21, 245, 271
Muto, W., 365, 382

Nachemson, A., 438, 439, 452, 454
Nakaseko, M., 366, 381
NASA, 446, 419, 454
National Academy of Sciences, 543, 549, 665, 694
National Center for Health Statistics, 540, 549
National Research Council, 94, 113, 130, 182, 195
Navon, D., 74, 88
NCSA, 710, 724
Neal, A., 24, 42
Neer, R., 539, 548
Nelms, J., 578, 587
Nemecek, J., 492, 507
Nemeth, S., 474, 483
Newman, R., 209, 210, 219
Newman, S., 503, 507, 576, 588
NHTSA, 359, 381, 709, 724
Nicholson, A., 640, 651
Nicholson, B., 534–537, 550
Nimeroff, I., 515, 549
Nimmo-Smith, M., 109, 131
NIOSH, 254, 255, 258, 264, 265, 268, 271, 396, 397, 411, 563, 565, 567, 573, 587, 770
Nixon, C., 606, 621
Noack, Th., 586
Noble, M., 349, 380
Nof, S., 84, 86, 731, 755
Nolan, M., 648, 650
Noordzij, P., 724
Nordin, M., 273, 298
Norin, H., 640, 651
Norman, D., 3, 22, 361, 381
Norsworthy, M., 636, 652
North, K., 420, 454
North, R., 79, 88
Noyes, J., 360, 375, 381, 382
Nuclear Regulatory Commission, 24, 43
Nudelman, S., 58, 87
Nusbaum, H., 216, 220

Obermayer, R., 32, 43
Oborne, D., 221, 410, 586, 631, 633, 651
O'Donnell, R., 78, 88
Ogden, G., 80, 88
Oglesby, C., 703, 724
Okolie, S., 260, 261, 271
Older, S., 697, 724
Olesen, B., 561, 587
Olson, G., 495, 506
Olson, P., 534, 536, 537, 549, 550, 715, 716, 721, 724, 725
Oman, C., 641, 651
O'Neill, B., 704, 725
Ortengren, R., 438, 440, 452
Osborne, D., 305, 332
Osburn, H., 253, 270
OSHA, 574, 587, 604, 605, 616, 620
Ostergren, C., 399, 410
Oswald, W., 628, 652
Otsubo, S., 682, 686, 694
Owsley, C., 97, 99, 129, 130

Palmiter, S., 468, 483
Pangrekar, A., 526, 527, 547
Paramore, B., 734, 754
Parasuraman, R., 74, 75, 86, 88
Parkin, A., 189, 193
Parks, D., 736, 755
Parng, A., 369, 380
Parsons, H., 504, 507
Parsons, K., 631, 650
Parsons, S., 342, 382, 466, 478, 484
Passel, C., 578, 588
Passmore, R., 241, 271
Pastalan, L., 721, 725
Patel, A., 640, 650
Patterson, R., 170, 171, 195
Patterson, T., 666, 694
Pattie, C., 292, 299
Paul, J., 670, 694
Paul, L., 362, 382
Paulson, H., 100, 130
Pawlak, U., 113, 130
Payne, R., 582, 587
Payne, S., 376, 377, 380, 381
Payykko, I., 398, 411
Peake, W., 167, 194
Pearson, J., 598, 621
Pearsons, K., 170, 193
Peay, J., 136, 159, 666, 694
Pegden, C., 741, 756
Perry, T., 525, 549
Peters, B., 724, 725
Peters, G., 724, 725
Petersen, D., 662, 694
Peterson, A., 164, 195, 206, 209, 210, 220, 594, 613, 620
Peterson, I., 493, 507
Peterson, J., 290, 298, 368, 379
Petrofsky, J., 236, 271, 272
Petropoulos, H., 308, 309, 332
Petrucelli, E., 696, 725
Pew, R., 84, 88, 320, 332
Phauth, M., 375, 382
Pheasant, S., 427, 428, 454
Phillips, C., 236, 272
Phillips, R., 108, 130
Pickett, J., 200, 220
Pickett, R., 65, 89
Piderman, M., 112, 129, 450, 454
Piehler, H., 688, 690, 691, 695
Piekarski, C., 586
Pisoni, D., 216, 220
Pitts, D., 99, 130
Plummer, R., 376, 379
Pollack, I., 57, 88, 170, 173, 195, 200, 220
Pollard, D., 362, 382
Pomerantz, J., 143, 159
Poock, G., 375, 382
Pope, M., 272
Popkin, C., 709, 725
Porter, D., 174, 193
Posner, M., 288, 289, 299
Post, D., 715, 725
Poulsen, E., 263, 264, 272
Poulton, E., 108, 130, 314, 318, 320–324, 327, 333, 347, 350, 382, 601, 602, 620, 630, 651
Poydar, H., 454
Presgrave, R., 291, 298
Preuschen, R., 470, 471, 483
Price, H., 732, 733, 755
Prichard, B., 722
Projector, T., 515, 516, 548
Proshansky, H., 505, 508
Provins, K., 580, 587
Prunier, T., 441, 454
Pulat, B., 468, 484
Pulliam, R., 732, 755
Purcell, J., 144, 158
Purswell, J., 390, 411, 685, 692
Pytel, J., 249, 272

Quilter, R., 74, 75, 86
Quinn, J., Jr., 295, 296, 300

Rabideau, G., 23, 42, 746, 747, 755
Rackoff, N., 707, 725
Radke, A., 628, 652
Radwin, R., 292, 299
Raibert, M., 282, 299
Ramanathan, N., 242, 243, 270
Ramsey, J., 559, 570–573, 586, 587, 664, 668, 669, 694
Rasmussen, J., 40, 43, 88, 329, 333, 656–658, 661, 694
Rasnow, R., 23, 43
Ray, R., 305, 333
Ray, W., 305, 333
Rayner, K., 101, 130
Rea, M., 526, 539, 540, 549, 550
Reason, J., 643, 645–647, 651
Rebiffe, P., 443, 455
Reddy, R., 375, 382
Regulinski, T., 41, 43
Reid, G., 82, 88
Reiff, L., 671, 693
Reising, J., 155, 159
Revensman, S., 359, 381
Reynolds, D., 397, 412
Reynolds, R., 149, 159
Rice, J., 373, 374, 382
Rice, T., 408, 410
Richards, L., 632, 652
Richardson, B., 185, 193, 195
Richardson, J., 191, 195
Richardson, R., 363, 364, 379
Richter, M., 515, 549
Riesz, R., 172, 195
Rigby, L., 431, 455
Riley, M., 581, 584, 587
Riley, V., 79, 88
Ring, J., 561, 586
Rivlin, L., 505, 508
Robertson, D., 228, 271
Robinette, K., 422, 455
Robinshaw, H., 111, 131
Robinson, C., 708, 709, 724
Robinson, D., 591, 592, 620
Robless, R., 118, 128
Roche, A., 417, 455
Rockway, M., 321, 323, 331, 333, 350, 382
Rockwell, T., 700, 707, 718, 719, 724, 725
Rodahl, K., 231, 233–235, 237, 238, 247, 254, 269, 271, 560, 578, 580, 585
Rodgers, S., 402, 412
Rodgman, E., 709, 725
Roe, C., 365, 382
Roebuck, J., 30, 43, 249, 272
Roelofs, J., 218, 220
Roger, M., 255, 271
Rogers, C., 186, 187, 193
Rogers, G., 566, 588
Rogers, J., 350, 382, 738, 741, 756
Rogers, S., 358, 359, 379
Rohles, F., Jr., 504, 508, 560, 562, 587, 588
Rohmert, W., 246, 272
Roland, E., 214, 216, 217, 220
Root, R., 648, 650
Roscoe, S., 24, 42, 99, 130, 141, 142, 153, 155, 156, 159, 327, 333
Rose, F., 102, 128
Rosenbaum, G., 294, 295, 298
Rosenthal, R., 23, 43
Ross, D., 527, 549
Rosson, M., 216, 218, 220
Roszbach, R., 724
Roth, J., 424–426, 455
Rothery, R., 701, 723
Rothstein, P., 682, 684, 686, 693, 695
Rourke, D., 99, 130
Rouse, S., 657, 658, 694
Rouse, W., 88, 657, 658, 694, 738, 756
Roy, R., 754
Rozier, C., 429, 455
RQQ, 530, 531, 549
Rubin, T., 450, 455
Rumar, K., 714, 720, 723, 724
Rupp, B., 449, 454, 541, 543, 548, 549
Russell, L., 181, 196
Ruth, J., 214, 216, 217, 220
Rutley, K., 717, 718, 724, 725
Ruys, T., 494, 508
Ryack, B., 467, 483

Saarela, K., 679, 694
Saari, J., 679, 694
Sabeh, R., 145, 146, 159
SAE, 640, 651, 652, 711, 725
Sairanen, E., 398, 411
Salan, J., 416, 453
Salaun, J., 111, 129
Salthouse, T., 76, 88
Salvendy, G., 42, 84, 86, 220, 333, 455, 731, 754, 755
Sampson, P., 296, 297, 299
Samueloff, S., 397, 412
Sanders, A., 61, 88
Sanders, M. S., 9, 11, 22, 23, 31, 43, 136, 159, 416, 426, 429, 455, 478, 479, 483, 528, 534–537, 550, 663, 666, 668, 694, 735, 756
Sanderson, L., 255, 271
Sandover, J., 629, 633, 649, 652
Sandow, B., 308, 309, 331
Sandry, D., 59, 89
Sanger, D., 366, 382
Santamaria, J., 511, 550
Saran, C., 504, 506
Sarlanis, K., 362, 382
Saunders, F., 185, 195
Saxena, U., 24, 42, 259–261, 270, 426, 453
Scerbo, M., 118, 129
Scha, R., 85, 88
Scheiner, E., 109, 130
Schmidt, F., 701, 725
Schmidt, R., 277–283, 295, 296, 300, 359, 382
Schmidtke, H., 290, 300
Schneider, M., 58, 87
Schneider, W., 65, 88
Schneiderman, B., 67, 88
Schnitzius, K., 648, 650
Schoenmarklin, R., 389, 401, 411
Schubeius, M., 213, 219
Schulte, B., 470, 471, 483
Schultz, G., 226, 271, 274, 276, 299
Schwab, E., 216, 220
Schwab, R., 722
Scialfa, C., 99, 130
Scott, L., 673, 693
Seeger, C., 58, 86, 302, 303, 332
Segel, L., 79, 89
Seibel, R., 359, 382
Seidel, H., 629, 652
Sekuler, R., 99, 130
Sell, R., 678–680, 694
Seminara, J., 342, 382, 466, 478, 484
Semple, C., 215, 220
Senders, J., 661, 694
Seven, S., 9, 21
Shaffer, L., 282, 300
Shahani, C., 665, 694
Shahnavaz, H., 542, 543, 550
Shapiro, Y., 575, 586
Sharer, R., 443, 446, 454
Sharp, E., 473, 484
Shaw, A., 376, 377, 380, 381
Shaw, B., 534–537, 550, 663, 666, 668, 694
Shealy, J., 662, 695
Shepard, R., 186, 187, 193
Sheridan, T., 82, 88, 328–331, 333
Sherman, W., 79, 89
Shinar, D., 304, 333, 697, 698, 708, 707, 718–720, 723, 725
Shingledecker, C., 82, 88
Shneiderman, B., 220
Shower, E., 172, 173, 196
Shulman, G., 111, 129
Shumat, R., 710, 723
Shurtleff, D., 112, 130

Shute, S., 441, 455
Sidorsky, R., 126, 130
Siegel, A., 40, 43, 472, 484
Siegel, S., 33, 43
Siemsen, D., 99, 130
Siess, D., 119, 121, 128
Silverstein, B., 387, 411
Simmonds, G., 135, 159
Simon, C., 29, 31, 43
Simons, A., 628, 652
Simonson, E., 246, 272
Simpson, C., 214, 216–218, 220
Simpson, G., 309, 333
Simpson, J., 643, 650
Simpson, R., 82, 88
Sinaiko, W., 88
Singleton, W., 6, 22, 193
Sinks, T., 576, 588
Siple, P., 578, 588
Sivak, M., 534, 536, 537, 549, 550, 715, 716, 721, 724, 725
Sivik, L., 515, 548
Slivinske, A., 126, 130
Slovic, P., 68, 89, 676, 677, 695
Small, D., 85, 89
Smith, A., 601, 602, 620
Smith, J., 260, 272, 292, 299
Smith, K., 323, 332
Smith, L., 9, 22
Smith, M., 542, 545, 550, 672–675, 692, 695
Smith, R., 143, 158
Smith, S., 114, 124, 125, 130, 309, 310, 333
Snook, S., 245, 255, 262, 263, 266, 268, 271, 272, 670, 693, 694
Snyder, H., 112, 113, 131, 358, 382
Solanky, H., 617, 621
Sommer, R., 490, 508
Soper, J., 578, 587
Sorkin, R., 175, 196, 733, 755
Southard, J., 126, 129
Sparrow, J., 118, 131
Sparshatt, J., 670, 693, 694
Sperling, A., 279, 298
Sperry, W., 607, 620
Spicer, B., 697, 724
Spickard, W., 431, 455
Splinter, W., 245, 272
Spraff, S., 309, 310, 333
Sprague, L., 319, 332
Springer, T., 441, 455
Springer, W., 736, 755
Squires, P., 432, 455
Stager, P., 98, 131
Stammerjohn, L., 542, 545, 550
Stansfeld, S., 602, 620
Stanush, J., 680, 693
Staplin, L., 721, 725
Starch, J., 398, 411
Starr, A., 375, 382
Starr, C., 677, 695
Starr, S., 441, 455
Staveland, L., 83, 87
Steadman, R., 569, 588
Steelcase, 489, 492, 508
Steer, M., 201, 219
Steinberg, J., 204, 208, 219
Steinfeld, E., 504, 508
Stelmach, G., 283, 284, 298, 299
Stenzel, A., 524, 550
Stephens, D., 626, 652
Stephenson, R., 563, 586
Stern, K., 217, 220
Sternberg, S., 67, 89
Stevens, J., 184, 196
Stevens, S., 170, 196, 205, 219, 591, 594, 620
Stewart, J., 709, 725
Stewart, T., 546, 548
Stier, F., 290, 300
Strawbridge, J., 686, 695
Strayer, D., 76, 89
Streeter, L., 58, 89
Strelow, E., 181, 194
Strobbe, T., 252, 272
Struyk, R., 503, 507
Strybel, T., 175, 196
Strydom, N., 566, 575, 588
Stutts, J., 709, 725
Suchman, E., 662, 695
Sudhakai, L., 401, 411
Suggs, C., 245, 272, 358, 380
Summala, H., 702, 703, 720, 725
Summers, J., 281, 288, 289, 299, 300
Sundstrom, E., 490, 508
Suther, T., 546, 548
Sutterer, W., 637, 652
Swain, A., 40, 42, 43, 657, 695
Sweere, J., 454
Swets, J., 62, 65, 86, 89
Swierenga, S., 369, 380
Swope, J., 358, 381

Tabachnick, B., 33, 43
Taboun, S., 262, 263, 270
Talamo, J., 171, 196
Talbott, E., 618, 619
Tan, K., 116, 117, 129
Tanner, J., 416, 455
Tarriere, C., 640, 650
Taylor, A., 621
Taylor, A., Jr., 195
Taylor, G., 58, 89, 112, 131
Taylor, H., 233, 272
Taylor, R., 610, 620
Taylor, W., 396, 397, 410–412, 598, 621
Teakevic, I., 79, 89
Teichner, W., 148, 149, 159, 286, 288, 300
Templer, J., 670, 692
Terrell, R., 390, 411
Thelen, E., 280, 300
Thomas, C., 216, 221
Thomas, D., 124, 125, 130
Thomas, J., 85, 88, 89, 93, 128, 216, 218–220, 299, 332, 333
Thompson, D., 429, 430, 455
Thompson, R., 606, 620
Thomson, W., 30, 43, 249, 272
Thornton, C., 706, 722
Thornton, W., 641, 642, 652
Tichauer, E., 385-388, 391, 392, 411, 432, 455
Tiffin, J., 701, 725
Tikkanen, K., 494, 508
Timbrook, C., 576, 588
Tolin, P., 150, 159
Topmiller, D., 40, 43, 286, 300, 744, 756
Torre, J., 599, 619
Triggs, T., 701, 725
Trinca, G., 696, 725
Trombley, D., 356, 357, 378
Trotter, D., 535, 550
Trucks, L., 177, 179, 193
Tsang, P., 83, 89
Tsuang, M., 707, 725
Tufte, E., 119, 124, 131
Tullis, T., 114–116, 131
Tversky, A., 677, 695
Twerski, A., 688, 690, 691, 695

U.S. Public Health Service, 503, 508
Uhlaner, J., 525, 549
Ungs, T., 648, 652
Ursic, M., 680, 695
Uslan, S., 673, 675, 695

Valencia, G., 175, 193
Valins, S., 503, 506
Vallbo, A., 183, 184, 196
Van, C., 744, 755
Van Balen, P., 21
Van Cott, H., 9–11, 22, 129, 158, 174, 193, 196, 219, 332, 379, 382, 481, 483, 484, 712, 725, 733, 734, 754, 755, 768
Van der Molen, M., 178, 179, 196

van der Walt, W., 566, 588
Van Kulmiz, P., 493, 507
Vanderheiden, G., 292, 299
Vanderplas, J., 145, 146, 159
Varnes, V., 111, 129
Vellinga, R., 442, 453
Vidulich, M., 59, 83, 89
Vigliani, E., 549
Viitasalo, J., 251, 272
Vokac, Z., 235, 269
Volta, G., 694
von Gierke, H., 606, 621
Vos, H., 242, 243, 272
Vreuls, D., 32, 43

Waag, W., 65, 87
Walchli, S., 155, 159
Wallace, T., 643, 650
Wallace, W., 687, 695
Waller, P., 709, 725
Walton, K., 397, 411
Waly, S., 257, 270
Wang, M-J., 402, 412
Wansack, S., 599, 619
Ward, J., 436, 437, 455
Ward, W., 599, 619, 620
Warm, J., 74, 88, 89
Warrick, M., 174, 196, 286, 300, 307, 333
Washburn, S., 383, 412
Wasserman, D., 397, 410, 412
Waterworth, J., 216, 217, 221
Watts, J., Jr., 358, 380
Weale, R., 539, 550
Webb, R., 723
Webster, J., 209, 221
Wegel, R., 168, 196
Weidman, B., 400, 401, 412
Weiman, N., 368, 369, 379
Weinstein, A., 688, 690, 691, 695
Weinstein, C., 277–279, 300
Weinstein, N., 602, 621
Weiss, C., 180, 181, 196
Weldon, L., 85, 89
Welford, A., 6, 22, 61, 89, 357, 382
Welford, H., 76, 89
Wells, B., 489, 491, 508
Werkowitz, E., 215, 221
Westra, H., 234, 271
Wheale, J., 218, 221
White, A., 268, 272
White, J., 468, 483
White, R., 149, 159
Whitehurst, H., 137, 139, 159
Whitham, E., 631, 632, 650, 652
Wichman, H., 492, 508
Wickens, C., 59, 61, 62, 65, 68, 71, 76–79, 81–83, 87, 89, 90, 142, 143, 144, 158, 159, 314, 318, 319, 321, 333
Wierwille, W., 80, 82, 85, 89, 355, 382, 648, 649, 713, 723
Wiesen, J., 40, 43
Wiggins, R., 214, 220
Wikstrom, B., 634, 650
Wilde, G., 702, 725
Wilkins, A., 109, 131
Wilkinson, R., 111, 131
Williams, A., 704, 725
Williams, J., 41, 43
Williams, K., 594, 621
Williams, R., 467, 468, 482
Williams, W., 459, 483
Williges, B., 214, 216, 217, 220
Williges, R., 153, 159
Williams, D., 218, 220
Wilson, C., 617, 621
Wilson, J., 23, 43, 271, 692
Winslow, C., 554, 588
Winters, T., 707, 723
Wise, B., 494, 508
Wise, J., 494, 508
Witkin, H., 624, 652
Wogalter, M., 676, 682–684, 686, 695
Wohlwill, J., 503, 506, 507
Wolf, C., 116, 131, 367, 382
Wolf, J., 40, 43
Wolf, W., 286, 299
Wood, E., 637, 652
Woodson, W., 107, 131, 149, 159
Woodworth, R., 286, 287, 300
Wools, R., 496, 508
Worringham, C., 312, 313, 333
Worthey, J., 520, 548
Wotton, E., 522, 550
Wotzka, G., 443, 446, 454
Wright, C., 292, 299
Wright, G., 539, 540, 550
Wrightman, F., 175, 196
Wuillemin, D., 185, 195
Wylie, C., 75, 90
Wyndham, C., 570, 588
Wyon, D., 564, 587

Yanase, T., 502, 507
Yeh, Y., 83, 90
Yorchak, J., 373, 374, 382
Young, D., 459, 483
Young, S., 676, 695

Zahn, J., 528, 550
Zais, J., 503, 507
Zarriello, J., 636, 652
Zeier, H., 712, 725
Zelaznik, H., 295, 296, 300
Zenyuh, J., 155, 159
Ziel, W., 710, 723
Zimolong, B., 84, 86,
Zucco, G., 191, 195
Zue, V., 375, 382
Zwahlen, H., 113, 131
Zweigenhaft, R., 495, 508
Zylman, R., 710, 723

SUBJECT INDEX

Abbreviations, 759–763
Abduction, 274–275
Absolute judgments
  (*see* Judgments, absolute)
Absolute positioning, 369–370
Acceleration, 635–641
  effects of, 635–638
  protection from, 639–640
  terminology, 635–636
  tolerance to, 639
  types of, 635
Accident proneness, 665
Accident-liability theory, 665
Accidents, 661–675
  alcohol and, 708–710
  automobile, 293, 696–700, 704–711
  causation theories, 665–666
  contributing factors, 666–667
  data collection, 663–664
  drugs and, 710
  experience and, 704
  eye, 675
  fatigue and, 708
  human error and, 662–663
  perceptual style and, 706–707
  personality and, 707
  poisoning, 671
  reducing income of, 671–675
  stairs, 669
Acclimatization, 575
  cold stress, 578
  heat stress, 566–567
Accommodation, 93–94
Acoustic reflex, 166–167
Actin, 226–227
Action level, 604

Action limit, 264–265, 769
Action office, 491
Activation time, 473
Acuity
(*see* Visual acuity)
Acuity gratings, 94, 96–97
Adaptation, visual, 99–100
Adaptation, speed, 700–701
Adduction, 274–275
Adjacency layout diagram, 464, 465
Adjustment-to-stress theory, 666
ADP, 227–229
Aerobic glycolysis, 228
Affective support, 733
Age, 503–504
anthropometry, 416
automobile accidents and, 704–705
basal metabolic rate, 229
contrast sensitivity and, 99
discomfort glare and, 534
glare sensitivity, 536–537
heat stress and, 566
information processing and, 76–77
lighting and, 539–540
maximum aerobic power, 233–234
speech intelligibility and, 213
strength, 250–251
vision and, 539–540
visual acuity and, 99
Aiding, 325
Aiming, 599
Air bags, 640
Air-cooled vests, 575
Aircraft controls, 339, 354–355
Aircraft displays 154
altimeters, 141–142
auditory, 180
bank angle, 152–153
head-up, 156
helmet-mounted, 156
principles of, 155
3-D, 154
Aircraft accidents, disorientation and, 643
Alarms, features of, 178–181
Alcohol
accidents and, 708–711
heat stress and, 566
performance and, 708–709
Alpha level, 34
Alphanumeric information, 359
Altimeters, 141–142
Anaerobic glycolysis, 228–229
Analysis of variance, 33
Anemometer, 556
Anthropometry, 415–420
application principles of, 420–423
weightlessness and, 641–642
Anti-G suits, 640
Anticipation, 320
Aqueous humor, 92
Arousal-alertness theory, 666
Arrangement of components, 456–469
Articulation index, 206–208
Artificial intelligence, 83
ATP, 227–229
Attention, 70–76
(*see also specific types*)
Attenuation, 627–628
Attribute evaluation, 743–744
Audiogram, 595
Audiometer, 595
Auditory displays, 169
for the blind, 181–182
principles of, 176–177
recommendations, 180
Auditory signals
coding of, 173–174
discrimination of, 171–172
identification of, 173
Aural reflex, 166–167
Autokinesis, 644
Automation, 495–497, 731–732
Automobile,
accidents, 696–700, 704–711
control coding, 342–343
lighting, 714–716
(*see also* Driver behavior)
Automobile controls, 358–359
Availability heuristic, 677

Back pain, vibration and, 629
Backlash, 350
Backrest angle, 438, 447
Ballistic movements, 292
Bandwidth, 50, 319
Basal metabolic rate, 229
Behavioral modification, 673–674
Beta, 64
and age, 77
Beyond-reach envelope, 429–430
Biased sample, 30
Biomechanics, 273
Bit, 49–50, 55
Blind reach, 341
Blind-positioning movements, 293–295, 467
Blind, aids for, 181–182
Blood alcohol content, 708–711
Blood flow, distribution of, 230–231
Body measurements, 416–418
Borderline between comfort and discomfort, 534
Borg-RPE scale, 237
Botts dots, 708
Bottsball index, 559
Braille, 186–187
Brake pedal, 475
Brightness, 513
Built environment, 485–505
Bürolandschaft, 489
But-for test, 688

C/R ratio
(*see* Control/response ratio)
Cab design, 711–712
CAD, 731
CAFES, 736
Camera, 407–408
Car phones, 73
Cardiac output, 230
acceleration effects, 637
Carpal tunnel, 384–385
Carpal tunnel syndrome, 387, 388
Carrying, 243
Carrying tasks, 261–263
Catecholamines, 230, 235
Ceiling height, 500
Center clipping, 204–205
Center high-mounted stop light, 715
CFAC Model, 666–667
Chairs, 444–447
Character
(*see* Text)
Check reading, 132, 147
Checklists, 6, 672, 743
CHESS, 744
Children, and accidents, 671
Chord keyboard, 360
Chromaticity, 515
Chromaticity diagrams, 516, 521
Chunking, 67
CIE colorimetric system, 515–516
Circumduction, 274
Clearance, 429, 431
Clipping of speech, 203–205
Clo unit, 555
Clo unit, 561
Clockwise-for-increase principle, 308–309
Closed loop, 277
Closed-loop system, 18
Clothing
cold stress and, 583
heat exchange and, 554–555
Coal mines, 311

Coarticulation, 215, 375
Codes, 124–127
auditory, 184
detectability of, 56
discriminability of, 57
meaningfulness of, 57
standardization of, 57
types of, 126–127
(*see also* Symbols)
Coding, 54–57
auditory signals, 173–174
characteristics of, 56–57
color, 125–126
controls, 185–186
of controls, 336–344
of qualitative displays, 145–146
(*see also* Symbols; Codes)
Cognitive engineering, 47
Cognitive performance
cold stress and, 582–583
heat stress and, 570–572
Cognitive psychology, 47
Cognitive support, 733
Cold stress, 576–585
acclimatization to, 578
index of, 578–579
performance and, 579–583
physiological effects of, 576–578
protection from, 583–585
sensations to, 579–580
Color, 502, 512–516
Color, 520, 522
of lamps, 520, 522
systems of, 514–516
warning light and, 149
Color coding, 125–126, 342
Color coordinates, 515
Color rendering index, 520, 522
Color temperature, 513
Color vision, 100
COMBIMAN, 744
Comfort, vibration and, 630–632
Compatibility, 176, 302–314
conceptual, 58
definition of, 58
identification of, 60
modality, 59
movement, 58, 306–313
origins of, 60
proximity principle, 142
reaction time and, 287–289
spatial, 59, 302–306
types of, 58–59
Compensatory display, 315–316, 320–321
Components, arrangement of, 456–469
Computer chairs, 444–446
Computer design tools, 744
Computer workstations
(*see* VDT workstations)
Computer-supported cooperative work, 494
Computerized design, 736
Conduction, 552
Cones, 92
Configural displays, 143
Construct deficiency, 38
Construct validity, 38
Contamination, in measurement, 38–39
Content validity, 38
Contrast, 523
types of 98–99
Contrast sensitivity, 96–98
age and, 99
factors affecting, 98–99
spatial frequency and, 97
training and, 99
Control, supervisory, 327–331
Control accessibility index, 462
Control order, 317–319
Control panel, 458, 472
Control/display ratio
(*see* Control/response ratio)
Control/response ratio, 344–347, 350, 353, 370–371
Controls
automobile, 358–359
coding of, 185–186, 336–344
design of, 764–768
displacement of, 766, 767
eye-activated, 377–378
foot, 474–475
function of, 334–335
identification of, 336–344
placement of, 469–475
resistance in, 347–349, 766, 767
size of, 766, 767
spacing of, 479–481
speech recognition, 374–377
speed of operation, 344
types of, 335–336
vehicles, 311
(*see also specific types*)
Convection, 552–554
Cooper-Harper Scale, 82
Coriolis illusion, 644
Cornea, 91–92
Correlation, 32
Counter displays
(*see* Display, digital)
Cranks, 351–352
Creatinine phosphate, 227–229
Crest factor, 625
Criterion measures, 28, 34–36
classes of, 35–36
Criterion measures (*Cont.*):
evaluation and, 748
intermediate, 36
requirement for, 37–39
terminal, 36
Critical band, 168–169
Critical-incident technique, 664
Cross-coupling effect, 644
Cumulative trauma injuries, 383–384
Cursor positioning devices, 335–337, 346, 368–372
Cutaneous sense, 182–184

d′, 64–65
and age, 77
Damping, 348
Danger 675, 689
Dark focus, 93
Data entry devices, 359–372
Data Store, 40
Day-night sound level, 607–610
Deadspace, 349–350
Deafness
(*see* Hearing loss)
Decibel, 162
Decision making, 68–69
computer aided, 69
Dental syringe, 404–406
Descriptive studies
choosing a research setting, 26
choosing subjects, 29–30
collection of data, 31
definition of, 24
variables in, 28
Desk, slanted, 432–433
Deskilling, 330
Desocialization, 330
Detection, with headlights, 715–716
Difference threshold, 57
Digitizing tablet, 369
DIN color system, 515
Diopter, 93
Disability glare, 535–538
Disc camera, 407–408
Discomfort glare, 534
Discomfort glare rating, 535
Discrimination
color, 100
of frequency differences, 172–173
of intensity differences, 172
of pitch, 57
Disorientation, 643
Displays
3-D, 154–155

aircraft, 141–142, 152–154, 156, 465
altimeters, 141–142
auditory versus visual, 53
check reading, 147
definition of, 51
digital, 134
electronic, 134
general guidelines, 157
head-coupled auditory, 175
head-up, 156
helmet-mounted, 156
inside-out, 153
moving pointer, 134–142
moving scale, 134–136
numeric progression, 137
object, 142–144
olfactory, 191–192
outside-in, 153
pointer design, 139
principles of auditory, 176–177
qualitative, 144–148
quantitative, 133–144
quantitative, features of, 136–142
representational, 152–156
scale markers, 138–139
scale size, 139
scale unit, 136–138
status indicators, 148
tactual, 184–190
visual, 469
Dive reflex, 578
Divided attention, 70, 72–74
Dizziness, 643
Dormitories, 503
Draught, 560–562
Driver behavior
reaction time, 702–703
risk taking, 702
spacing judgments, 701–702
speed judgments, 700–701
visual-scan patterns, 700
(*see also* Automobile)
Drugs,
accidents and, 701
Dvorak keyboard, 361
Dwelling units, 496, 498–505
Dynamic allocation, 733–734
Dynamic information, 51
Dynamic seat designs, 444–445
Dynamic strength, 247–248
Dynamic work, 234

Ear
anatomy of, 164, 166–167
bandwidth of, 50
Ear (*Cont.*):
inner, 166–167
middle, 166–167
outer, 164, 166
Earplugs
(*see* Hearing protection)
Echoic storage, 66
Effective quiet, 596
Effective temperature, 556–558
rectal temperature and, 563
shoveling performance, 569–570
Elderly, 503–504
(*see also* Age)
Electromyography, 226, 236, 440
Embedded training, 745
Emergency lights, 150
Emergent features, 143
Endurance, 253–254, 261
Energy, units of measure, 231–232
Energy consumption, 239–244, 266
factors affecting, 242–244
(*see also* Energy expenditure)
Energy costs, 259
Energy expenditure,
limits, 242, 245
(*see also* Energy consumption)
Engineering anthropometry, 415
Engineering psychology, 47
(*see also* Human factors)
Equilibrium, 624
Equivalent noise, 608
Equivalent sound level, 593–594
Equivalent wind chill temperature, 578–579
Equivalent-comfort contours, 631–632
Ergonomics
(*see* Human factors)
Ergonomics Research Society, 7
Evaluation, 743–744, 746–749
Evaluation research
choosing a research setting, 27
choosing subjects, 31
collection of data, 32
criteria in, 748
definition of, 25
setting, 748
subjects in, 748
variables in, 28
Evaporation, 553–554
Event-related brain potential, 81–82
Exit signs, 121–122
Experimental research
choosing subjects, 30–31
collection of data, 32
Experimental research (*Cont.*):
definition of, 24
variables in, 28
Expert systems, 84
Expert witness, 669, 687
Extension, 274–275
Exteroceptors, 622
Eye, 91–93
bandwidth of, 50
Eye irritation, 562
Eye-activated controls, 377–378

Face validity, 38
Factor analysis, 501
Fail-safe, 660–661
FAM, 736
Fans, 575
Far point accommodation, 93
Farsightedness, 539, 93–94
Fatigue, accidents and, 708
Fatigue-decreased proficiency boundary, 633–634
Feedback, 17, 18, 277–278, 365, 672–673
Feedforward, 278
Feel, 363
Fidelity, 748–749
Field dependence-independence, 706
Field of view, 24
Field research, versus laboratory research, 26
Fitts' law, 291–292, 357
Fixations, 101
Flares, design of, 27
Flash rate, of warning lights, 149, 151
Flesch reading ease score, 110
Flexion, 274–275
Focused attention, 70–72
Foot controls, 355–359
Force, 231
Fovea, 92
Free field, 611
Frequency-band analysis, 163
Frequency-of-use principle, 457, 459
Friction, 348
Frostbite, 576–577
Function allocation, 730–734
Functional flow diagrams, 729
Functional principle, 457–459, 476–477
Furnishings, office, 494–495

G-induced loss of consciousness, 635

Gain, 345
Ganged control knobs, 340–341
General package factor, 711
Gestalt, 147
Gestural data entry, 366–367
Gilbreth, Frank, 6–7
Glare, 533–539
 reduction of, 538–539
Glare screens, 545–546
Globe temperature, 556
Gloves, 399–402
 cold stress and, 584
Glycolysis, 228–229
Goals-freedom-alertness theory, 666
Golgi tendon organs, 279
Graduation interval, 136
Graphics tablet, 369–370
Graphs, 117–118
 auditory, 182
 tactual, 188–189
Graveyard spin, 644
Guidelines, 6

Hammer, 389
Hand, anatomy of, 384–385
Hand tools
 principles of design, 386–395
 vibration and, 395–399
Hand-arm vibration syndrome, 396–398
Handicapped, 503–504
Handles
 bent, 387–389
 design of, 389–393
Handwheels, 351
Hardcopy
 (*see* Text)
Hazard, 675
Hazardousness, 676, 682
Head-coupled auditory displays, 175
Head-up displays, 156
Headlights, 715–716
Hearing, 167–168
Hearing loss, 594–599
Hearing protection, 615–618, 674
 speech intelligibility and, 213–214
Heart rate, 235, 240–241, 245, 247
 acceleration effects, 637
 driving and, 717
 heat stress and, 564
 oxygen consumption and, 234–235
Heart rate recovery curve, 235
Heat cramps, 565
Heat exchange, 551–555
 avenues of, 552–553
 clothing and, 554–555
 equation, 553
 factors influencing, 553–554
Heat exhaustion, 565
Heat illness, 562–563, 565
Heat index, 568–569
Heat rash, 565
Heat stress, 562–575
 acclimatization to, 566–567
 incidence, 562–563
 indices of, 567–568
 individual differences and, 565–566
 performance effects of, 568–572
 physiological effects of, 563–565
 recommended limits, 572–574
 reduction of, 574–575
Heat stress index, 568
Heat stroke, 565
Heaters, 584–585
HECAD, 744
Helmet-mounted displays, 156
Hick-Hyman law, 50
Highlighting, 116–117
Highway illumination, 721–722
Horizontal-vertical illusion, 102
Hue, 513
Human engineering
 (*see* Human factors)
Human error, 655–663
 accidents, 662, 663, 698–700
 classification of, 656–660
 types of, 656–660
Human factors
 approach of, 5, 23
 case for, 13
 definition, 4, 5
 doctrines of, 5
 focus of, 4
 goal of, 3
 graduate education in, 9
 history of, 6–9
 in other countries, 12
 job satisfaction, 11–12
 objectives of, 4
 professional organizations, 13
 what it is not, 6
 work activities, 11–12
 work settings, 10–11
Human factors data, 741–743
Human Factors Society, 7–12
 composition of, 10
Human reliability, 36, 39–42, 655
 criticisms of, 41–42
 data bases, 40
Human reliability (*Cont.*):
 historical pespective, 40
Human-computer interaction, 328
 (*see also* Screen design; VDT; cursor controls)
Human-machine system
 (*see* Systems)
Humidity, low, 561–562
Hydrometer, 556
Hypothermia, 577–578
Hysteresis, 363

Ice vests, 575
Iconic storage, 66
Illness, heat stress and, 562–563, 565
Illuminance, 517
 VDT and, 542
Illuminating engineering, 511
Illumination, roadway, 721–722
Illusions, 642–645
 (*see also specific types*)
Illustrations, 746–747
Impact, 640–641
Impact noise, 599, 605–606
Importance principle, 457, 459
Impulse noise, 599, 605–606
Incentive programs, 674
Index of permeability, 555
Inertia, 349
Information
 definition of, 48–49
 dynamic, 132–158
 processing, 17
 receiving, 17
 storage, 17
 types of, 51–53
 unit of measure, 49
Information criteria, 524
Information processing
 accident sequence, 668–669
 age and, 76–77
 human error classification, 657–658
 model of, 61–62
Information revolution, 47, 83–84
Information sources, 739
Information theory, 48
Infrasonic noise, 606
Injuries
 back, 225
 cold stress, 576
 hand tools, 383–384
 manual materials handling, 254–255
Injury, 676
Instructions, 681, 684–685

Intelligent vehicle highway systems, 27, 713–714
Interface, natural language, 84–85
Interquartile range, 33
Inverse-square law, 518
Iris, 92
Irradiation, 103
Ischemia, 391
Ischial tuberosities, 443
Isokinetic strength, 247
Isometric controls, 347
Isometric strength, 248, 251, 254
Isotonic controls, 347
Isotonic strength, 248

Jitter, 113
Joints, 274
Joysticks, 353–354
Judgments, absolute
along multiple dimensions, 55–56
along single dimensions, 54–55
brightness, 55
loudness, 55
of pitch, 54–55
pure tones, 55
size, 55
versus relative, 54
Just noticeable difference, 5, 150, 171–173

Key pads, touch screen, 368–369
Keyboards, 359–369
membrane, 364–365
tilted, 365–367
Kinesiology, 273
Kinesthesis, 278
Kinesthetic receptors, 623
Kitchen, 437, 499
Knobs, 352–353
Knowledge-based behavior, 329–330, 658–659

Label coding, 342–343
Laboratory research, versus field research, 26
Lactic acid, 228–229, 247
Lamp, 519–522
Lamp efficacy, 520, 523
Landholt ring, 94–95
Lead-lag compensation, 370
Legibility, of signs, 720–721
Lifting, 257–261
biomechanical models, 255–256
limits, 263–267
recommended limits, 769–770
Light pen, 368–369
Lighting
effect on performance, 524–528
measurement of, 516–519
nature of, 512
recommended levels, 528–531
VDT and, 541–546
vehicle, 714–716
Lights, emergency, 150
Line of sight, 469
Linear programming, 466
Link analysis, 462–466
Link table, 463
Load stress, 70
Location coding, 340–342
Long-term memory, 68
Lordosis, 438–439
Loudness, 589–594
Luminaire, 519–523
Luminance, 98, 513, 518–519
warning lights and, 148–149
Luminance factor, 519
Luminance ratio, 532. 542–543
Luminous flux, 516
Luminous intensity, 517

Magical number 7 +/− 2, 55, 67
Manual material handling, 254–269, 769–770
Manual performance, cold stress and, 581, 582
Manuals, 745–747
Maps, tactual, 187–188
Mark VII sone, 592, 594
Masking, 168–169
MAWL, 257, 259–260, 262, 263, 266–267
Maximal oxygen uptake, 232–234
Maximum acceptable weight of load
(*see* MAWL)
Maximum aerobic power, 232–234, 265
age and, 233–234
measurement of, 233–234
sex and, 233–234
Maximum permissible lift, 769
Maximum permissible limit, 264–265
Mean time to failure, 19
Membrane keypads, 364–365
Memory, 65–68
heat stress and, 570
long-term, 68
Mental workload, 77–83
automobile driving and, 712
concept of, 78
driving and, 717–718
measurement of, 78–83
physiological measures, 80–82
Mental workload (*Cont.*):
primary task measures, 79
secondary task measures, 79–80
subjective measures, 82–83
(*see also specific measures*)
Menus, number of options, 67
Method of adjustment, 56
Method of constant stimuli, 57
Method of limits, 56
Method of serial exploration, 56
Michelson contrast
(*see* Modulation contrast)
Minimum audible movement angle, 175
Mining, 668
Mining equipment, 24
Mirror-imaged displays, 476, 478–479
Mnemonics, 68
Models, 61
of memory, 65
Modified rhyme test, 216
Modulation contrast, 96
Moon illusion, 24
Motion sickness, 645–649
Motor programs, 280–284
Motor responses
closed loop theories, 279–280
control of, 276–284
open loop theories, 280
types of, 276–277
Motor unit, 227
Mouse, 371
Movement compatibility, 306–313
Movement time, 285, 290–293
Movements
accuracy of, 293–297
blind-positioning, 293–295
range of, 274–275
speed of, 284–293
types of, 274
Multipurpose chairs, 446
Multivariate analysis of variance, 33
Munsell system, 515
Muscle fibers, 226–227
Muscle metabolism, 227–229
Muscle spindles, 278
Muscles
control of, 227
physiology of, 226–229
Myofibrils, 226–227
Myopia, 94
Myosin, 226–227

NASA Task Load Index (NASA-TLX), 83
Natural language interfaces, 84–85

Near field, 611
Near miss, 664
Near point accommodation, 93
Nearsightedness, 93–94
NIOSH lifting recommendations, 264–265
Noise
annoyance and, 606–609
control of, 613–618
effect on speech, 205–214
exposure limits, 603–606
exposure measures, 608
measurement of, 611–612
peformance effects, 600–603
stress and, 600
Noise attenuation
(*see* Noise, control of)
Noise dose, 604–605
Noise reduction rating, 615–618
Noy, 592, 594
Nuclear power, 24
human reliability in, 40
Nuclear power plant, 452, 460, 476–477, 482
Numbered interval, 136
Numeric keyboard, 362
Numeric keypad, membrane, 364–365

Object displays, 142–144
recommendations for, 144
Octave band analysis, 165
of speech, 200
Octave bands, 163–164, 611
Oculogravic illusion, 644
Odors, identification of, 191
Office
activities, 487–489
automation, 495–497
environment, 486–496
noise, 209–210
open plan, 489–493
tasks, 488–489
types, 489–493
workers, 486–487
Olfactory displays, 191–192
Olfactory sense, 190–191
Open-loop system, 18
Open-plan office, 489–493
Operational coding, 342
Operational sequence diagram, 464–466, 734–735
Operative temperature, 558
Optacon, 186–187
Optic nerve, 92
Orthogonal dimensions, 55–56
Ossicles, 166
Ostwald, 515

Otolith organs, 623, 645–646
Out-of-reach, 429–430
Oxford index, 558, 569–570
Oxygen consumption, 232, 240–241
heart rate and, 234–235
Oxygen debt, 229–230

P300 brain potentials, 81–82
Patent-danger rule, 689
Pathsounder, 181
Peak clipping, 203–205
Pedals, 355–359
Pens, 406–407
Perceived noise level, 592, 594, 608
Percentiles, 33, 422
Perception, 61–62, 101–102
Perceptual style, accidents and, 706–707
Perceptual-motor performance, heat stress and, 570–572
Performance measures, 35
Performance specifications, 728
Permanent threshold shift, 597–598
Permissible exposure level, 605
Personality, accidents and, 707
Personalization, 490
Phon, 591–592, 594
Phoneme, 198
Phosphor persistence, 113
Photometry, 516
Photopic vision, 512, 513
Phototropism, 537–538, 544
Physical fitness, 575
heat stress and, 565–566
Physical work
cold stress and, 580
heat stress and, 569–570
Physiological indices, classes of, 36
Pisiform bone, 385
Pitch, 161
Pixel, 111
Pliers, design of, 388–389
Pointer design, 139
Poisoning, 671
Polarity, 113
Population stereotypes, 306
Posters, 679–680
Postrotary nystagmus, 644
Postural fixity, 440–441, 446
Posture, 242–243
weightlessness and, 641–642
Power, units of measure, 232
Power spectral density, 625–626
Power switches, 312

Predictor display, 325–327
Preferred noise criteria curves, 209, 211–212
Preferred-octave speech interference level, 206–210
Presbycusis, 595–596
Preview, 320
Primary task workload margin, 79
Product liability, 661, 687–691
Pronation, 274–275
Proprioception, 278
Proprioceptors, 623
Prosody, 215
Protective equipment
cold stress and, 583–584
heat stress and, 575
Proximity principle, compatibility of, 142–143
Psychoanalytic theories, 666
Psychophysics, 56–57, 256–257, 591, 670
Psychrometer, 556
Pupil, 92
Pupillary response, 80–81
Purkinje effect, 513
Pursuit display, 315–316, 320–321
Push buttons, 473
Pushing tasks, 262
Pyruvic acid, 228–229

Q10 effect, 563
Qualitative displays, coding of, 145–146
Qualitative reading, type of display and, 132, 145
Quality of life, 753
Quantitative reading, type of scale and, 132, 145
Quickening, 189, 326–327
QWERTY keyboard, 360–362

Race, anthropometry, 416, 419
Radiation, 553–554, 678
Railroad, 668
Random sampling, 30
Range effect, 294
Reach envelopes, 424–430
Reaction time, 59, 285–290
cold stress and, 582
control activation, 473
driving and, 702–703
field of view, 469–470
heat stress and, 570
signal intensity, 178–180
to signs, 120
vibration effects, 630
(*see also* Response time)

Reaction time, choice
number of alternatives, 50
stimulus information and, 50
Readability, 109
indices of, 110
order of words and, 110
type of sentence and, 110
Reading, 100–101
Rear lights, 715
Recommended alert limits, 573
Recommended exposure limits, 574
Redundancy, 50
Redundant dimensions, 55–56
Reflectance, 518, 533
Reflections, 543–546
Refresh rate, 113
Reinforcement, 673–674
Relative positioning, 369–370
Reliability
psychometric, 37
system, 19–20
(*see also* Human reliability)
Representational displays, 152–156
Representative sample, 29
Research methods, 486
Resistance, 347–349, 766, 767
Resonance, 627–628
Resource models of attention, 72–74
Respiration, acceleration effects, 637–638
Response time, 292–293, 479
(*see also* Reaction time)
Retina, 92
Reverberant field, 611
Reverberation, 211–213
Risk, 675–679
acceptability of, 677
evaluation of, 676–677
Risk homeostasis theory, 702
Risk taking, driving and, 702
Road characteristics, 716–717
Road markings, 718–719
Roadway, illumination, 721–722
Robots, 15, 731
Rods, 92
Room arrangement, 498–499
Root-mean square, 625
Rule-based behavior, 329, 658–659
Running lights, 714

Saccades, 101
Saccule, 623
Safety
heat stress and, 572
lighting requirements, 529, 531
Safety communications, 678–680
SAMMIE, 744
Sample size
descriptive studies, 30
experimental research, 31
Saturation, 513
Scale markers, 138
Scale range, 136
Scale size, 139
Scale unit, 136–138
Scale-side principle, 308–309
Scan line, 111
Scenario analysis, 671
Schema, 282–284
Schmidt's law, 295
Scotopic vision, 512, 513
Screen design, 114–117
Screen polarity, 543
Seat
automobile, 711–712
seat pan angle, 439–440
vibration damping, 627–628
Seat back, 443–444
Seat belts, 640
Seat cushioning, 443
Seat depth, 443
Seat design, dynamic, 437–445
(*see also specific parameter*)
Seat height, 442, 447
Seat pan angle, 447
Seat pan slope, 439–440, 442
Seat width, 443
Seatbelts, 424–426
Selection, 268, 660
Selective attention, 70–71
Semicircular canals, 623, 643–646
Sensitivity, 64–65
Sensitivity of measurement, 39
Sensor lines, 305
Sensory rearrangement theory, 645–646
Sensory storage, 65–66
Sequence-of-use principle, 457–459
Setting, evaluation and, 748
Sex
anthropometry, 416, 418–419
automobile accidents and, 705
basal metabolic rate, 229
energy expenditure limits, 245
heat stress and, 566
maximum aerobic power, 233–234
response to vibration, 631
strength, 249–250
tool use and, 394
Shape coding, 337–340
Shipyard, 673–674, 679–680
Shivering, 577
Short-term memory
(*see* Working memory)
Shoveling, heat stress and, 569–570
Signal detection theory, 62–65
applications of, 65
Signal lights
(*see* Warning lights)
Signal-to-noise ratio, 163, 205
Signals, 329
detection of, 170–171
discrimination of, 171–172
identification of, 173
Signs, 329
discrimination of, 98
legibility of, 720–721
roadway, 719–721
Simulation, 27
stochastic models, 41
Simulator sickness, 648
Single-object sensor, 181
Sinus arhythmia, 80
Sirens, localization of, 174–175
Situation awareness, 132–133, 152, 154
Size coding, 340–341
Skill, 276
Skill learning, 283
Skill-based behavior, 329, 658–659
Sleeper berth, 429–430
Smell
(*see* Olfactory sense)
Snellen acuity, 95, 107
Sociocusis, 595–596
Sociometry, 491
Somatography, 419–420
Sonar, human reliability, 41
Sone, 591–593
Sonic pathfinder, 181
Sonicguide, 182
Sopite syndrome, 645
Sound
frequency of, 161–162
intensity of, 162–164, 612
measure of, 161–164
Sound exposure level, 593–594
Sound intensity, 612
Sound localization, 174–175, 185
Sound pressure level, 162–163
Sound spectrogram, 198–199
Sound spectrum, 163
Sound-level meters, 163, 590–591
Space adaptation syndrome, 641
Space sickness, 641
Spatial adequacy, 498
Spatial compatibility, 302–306
Spatial frequency, 96–97
Spatial operational-sequence diagram, 464–466
Spectrum locus, 515

Speech
clipping of, 203–205
depicting, 198–199
filtering of, 203
frequency composition of, 199–200
intensity of, 199
nature of, 197–200
organs of, 198
perception of, 185
sounds of, 198
spectrum, 198–199
waveform, 198–199
Speech intelligibility, 200
age and, 213
effect of message, 201–202
effect of speaker, 201
effect of transmission system, 203–205
hearing protection and, 213–214
noise and, 205–214
reverberation and, 211–212
synthesized speech, 216–217
Speech interference level, 207–209
Speech quality, 200–201
Speech recognition, 374–377
Speed stress, 70
Stairs, 669–671
Standard deviation, 32
Standards, 740
Startle response, 599
Static information, 51
Static strength, 248, 250
Static work, 234
Statistical significance, 33–34
Statistical workload assessment model (SWAM), 79
Status indicators, 148
Status markers, 491
Steel mill, 679–680
Stereophony, 174–175
Stimulus dimension, 54
Stoves, 304–306
Strain, 225
Stratificaton variables, 28
Strength
age and, 250–251
definition of, 247–248
examples of, 251–252
grip, 390, 393, 401–402
leg, 475
measurement of, 248–249
of pull, 471
sex and, 249–250
Stress, 225
Stroke-width-to-height ratio, 103–104
Subjective measures, 36
Subjective Workload Assessment Technique (SWAT), 82–83
Supervisory control, 327–331
Supination, 274–275
Survey questionnaires, return rate from, 31
Sustained attention, 70, 74–76
SWAT, 82–83
Sweating, 564–565
Switch, toggle, 473, 768
Symbols, 119–124, 330
exit signs, 121–122
mandatory action, 121
principles of design, 122–123
versus text, 119
Synthesized speech
guidelines for, 218
intelligibility of, 216
intonation of, 217
memory for, 216
preference for, 217
types of, 214–215
uses of, 215–216
System design, 726–754
characteristics of, 727–728
checklist for, 750–752
stages in, 727
Systems
boundary of, 16
characteristics of, 16–19
classes of, 15
definition of, 13–14
environment of, 16
functions of components, 17–18
limit of resolution for, 16
parallel, 19–20

Tactile sensitivity, cold stress and, 580
Tactual displays
substitute for hearing, 184–185
substitute for seeing, 185–190
Task analysis, 460, 734–736
Teleoperators, 372–374
Telephone, 362
Temperature
indices, 556–559
measurement of, 556
Temporary threshold shift, 596–597
Tennis elbow, 388–389
Tenosynovitis, 387–388
Text
criteria for evaluation of, 102
interletter spacing, 109
interline spacing, 109
size of, 105–108, 112–113
stroke width to height, 103–104
styles of, 104–106
typography, 102–103
upper/lower case, 108
Text (*Cont.*):
versus symbols, 119
width to height ratio, 104–105
(*see also* VDT)
Texture coding, 338–340
Theory of signal detection (*see* Signal detection theory)
Thermal comfort, 560–562
THERP, 40
Three Mile Island, 40
Threshold, 56
difference, 57
Threshold contrast, 96
Time lags, 322–323
Time-sharing, 72–74
Toggle switch, 473
Tool design, 239, 244
Toothbrush, 403–404
Torque, 352–353
Touch screen, 368
Tracking, 314–327
cold stress and, 582
heat stress and, 571
tactual displays, 189–190
vibration effects, 630
Tractor, 292–293
Traffic events, 717–718
Training, 660, 672, 745
exercise, 247
lifting, 268–269
motor skills, 283
Transient adaptation, 532, 538
Transmissioin, 712–713
Travel aids, for the blind, 181–182
Tremor, 296–297
Trigger finger, 392
Two-point threshold, 183
Typing, 364–365
Typography, 102–103

Ultrasonic noise, 606
Unemployment, 330
Utility cost, 467
Utricle, 623

Validity, types of, 38
Variables
criterion, 28
dependent, 28–29
extraneous, 28
independent, 28
stratification, 28
VDT, 51, 511
color, 113–114
controls, 346
cursor positioning devices, 368–372

VDT (*Cont.*):
lighting and, 541–546
polarity, 113
reading distance with, 112
reading from, 111
screen design, 114–117
size of characters, 112–113
size of display, 24
spatial frequency and, 98
text, typography, 111–112
(*see also* Text)
VDT workstations, 447–451
productivity and, 448–449
recommendations, 451
standards, 449
user preferences, 449–450
VDU
(*see* VDT)
Vehicle, vibration, 626
Vertigo, 643
Vestibular sacs, 623, 643
Vestibulo-ocular reflex, 643
Vibration
hand, 395–399
motion sickness and, 646–647
types of, 624–625
units of measure, 625–626
Vibration induced white finger, 396
Vibration isolation, 613
Vibration, whole-body, 624–634
exposure limits, 633–634
performance effects of, 629–630
physiological effects of, 627, 629
subjective response to, 630–632
terminology, 624–626
Vigilance, 70, 74–76
heat stress and, 571
theories of, 75
vibration effects, 630
Vigilance decrement, 74–75
Visibility, 523–524
Visibility level model, 526
Vision
acceleration effects, 635–636
automobile accidents and, 705–706
(*see also* specific abilities)
Visual acuity, 94–95
age and, 99
factors affecting, 98–99
training and, 99
types of, 94–95, 99
Visual angle, 95
Visual comfort probability, 535
Visual demands, automobile controls, 713
Visual display terminal
(*see* VDT)
Visual performance, 526
vibration effects, 629–630
Vitreous humor, 92

Walk-through, 743
Walking, 238–239
WAM, 736–737
Warning lights, 148–152
background lights and, 151
color of, 149, 152
context of, 150–151
detectability of, 148–151
exposure time for, 148–149
flash rate, 149, 151
intensity of, 152
luminance of, 148–149
recommendations for, 151–152
size of, 148–149, 152
Warning overload, 684
Warning signals, 177–181
detection of, 170–171
recommendations, 180
synthesized speech, 218
Warnings, written, 678, 680–687
design of, 681–684
effectiveness of, 684–687
elements of, 683
purposes of, 681
Warrick's principle, 307–309
Washing machine, 327–328
Water-cooled vests, 575
Weightlessness, 641–642
Wet-bulb globe temperature, 558–559
Wet-bulb temperature, definition, 556
White light, 513
Wind chill index, 578–579
Windows, 494, 501
Word-spelling alphabet, 202
Work
grades of, 240–241
methods of, 242
units of measure, 231–232
Work efficiency, 237–239, 266
Work physiology, 229–236
Work rate, 243–244
Work space clearance, 429, 431
Work surface
height, seated, 433–435
height, standing, 435–437
horizontal, 431–433
Work-rest cycles, 245–247
Work-space envelope
seated, 424–427
standing, 427–428
Working memory, 66
capacity of, 67
time to search, 67
Workload, factors affecting, 237–238
Workload Index (W/INDEX), 79
Workload, mental
(*see* Mental workload)
Workstations
(*see* VDT workstations)
Wrist movements, 385–386
Writing instruments, 406–407

Zones of interest, 152